Coalbed Methane Resources of the United States

Edited by
Craig T. Rightmire
Greg E. Eddy
James N. Kirr

Published by
The American Association of Petroleum Geologists
Tulsa, Oklahoma 74101, U.S.A.

Published December 1984
Printed in the United States of America

Coalbed methane resources of the United States

(AAPG studies in geology series: 17)
Bibliography: p.
Includes index.
1. Coalbed methane — United States. 2. Coal mines and mining — United States. I. Rightmire, Craig T. II. Eddy, Greg E. III. Kirr, James N. IV. Series.
TN844.6.C63 1984 553.2'85'0973 84-45747
ISBN 0-89181-023-4

Association Editor: Richard Steinmetz
Science Director: Edward A. Beaumont
Project Editor: Douglas A. White
Typography: Eula Matheny and Tricia Kinion
Design and Production: Custom Editorial Productions, Cincinnati, Ohio

Contents

Publisher's Note:

The idea of producing methane from fractured coals is not new, and has been done on repeated occasions with great success. This book addresses that practice.

But we hope this book will also stimulate the reader to think of deep coal beds as the organic-rich source rocks that they are. The chapters herein contain statistics of volumetric methane generation and offer BTU ratings to methane resources found as such, around the United States. One can only imagine the volume of gas released through geologic history, only to migrate and be trapped elsewhere with gases from other organic decomposition.

To stimulate acceptance and development of an exploration alternative, we publish this book.

AAPG Publications
Tulsa, Oklahoma

All the work on the condensation of reports contained in this volume was conducted by TRW staff under the auspices of TRW. The following are the present addresses of the principal investigators involved in the effort:

Craig T. Rightmire
TRW Exploration and Production
P.O. Box 441807
Houston, Texas 77244-1807

Greg E. Eddy
BDM
7915 Jones Branch Drive
West Branch Building, 7B
McLean, Virginia 22102

James N. Kirr
BDM
7915 Jones Branch Drive
West Branch Building, 7B
McLean, Virginia 22102

Paul Archer
9505A Parkfield Drive
Austin, Texas 78758

M. A. Adams
TRW Exploration and Production
P.O. Box 77244
Houston, Texas 77244-1807

George Saulnier
Untere Halde 2
CH - 5400 Baden
Switzerland

J. L. Hewitt
N.Y. State Department of Environmental Conservation
Division of Mineral Resources
50 Wolf Road
Albany, New York 12233-0001

H. H. Rieke
Geofax, Inc.
P.O. Box 31103
Independence, Ohio 44131

R. Choate
TRW, Inc.
200 Union Boulevard
Suite 201
Lakewood, Colorado 80228

D. Jurich
Research Engineer
Earth Mechanics Institute
Colorado School of Mines
Golden, Colorado 80401

J. P. McCord
TRW, Inc.
200 Union Boulevard
Suite 201
Lakewood, Colorado 80228

Publication Supported
by

FOREWORD

The presence of methane in coalbeds has been known since miners first went underground to mine coal. The study of methane in coalbeds was initiated with the intent of reducing the hazard to miners and improving mine safety. As such, the original effort was expended in areas where coalbeds were minable—thick, shallow, and of sufficient quality to warrant extraction.

Early studies carried out by the U.S. Bureau of Mines, Pittsburgh Mining and Safety Research Center, focused on the mechanisms of gas retention and release, gas flow through the coalbeds, mine ventilation, and degasification both in advance of mining and during mining. The emphasis in these studies was on shallow (less than 2,500 ft deep) coalbeds economically minable with existing technology.

Following the oil and gas crisis of the early 1970s, this hazard to mining was recognized by some as a potential resource. The U.S. Department of Energy (DOE) established the Unconventional Gas Recovery Program to address the evaluation and potential production of natural resources from Devonian shales, tight gas sandstones, and coalbed reservoirs. The Methane Recovery from Coalbeds Project (MRCP) was established as part of the effort to determine the magnitude and distribution of the resource, prove its producibility, and generate interest in private industry in its production and utilization.

A part of this effort was the analysis of numerous sedimentary basins for their coalbed methane resource. Thirteen basins were analyzed by TRW geologists between 1979 and 1982 in an attempt to determine the magnitude, distribution, and most favorable areas for potential production of methane from coalbeds. This analysis included, in many instances, the first estimate in individual basins of deep coal resources (greater than 6,000 ft), the sampling of numerous coalbeds to measure their gas content, and the compilation of geologic, hydrologic, and resource data to define areas of potentially high coalbed methane production.

The MRCP has achieved one of its major objectives of generating interest in the oil and gas industry in the potential of coalbed methane. Commercial production of natural gas from coalbeds is occurring in the Warrior, San Juan, and Piceance basins. Much of the impetus for commercial development was provided by a set of 13 "basin" reports that presented an assessment of the coalbed methane resource in selected basins. A listing of the original MRCP reports compiled by TRW geologists concludes this foreword.

In an effort to provide for a greater appreciation of the resource and a wider distribution of the reports, the Gas Research Institute (GRI) provided support for the printing and distribution of a limited number of additional copies. With requests for the reports exceeding the supply, the idea of compiling—in a single volume—a discussion of the coalbed methane resource in the conterminous United States to provide background information to interested geologists took form. The approach taken and presented here was to provide condensations of the individual basin analysis reports, with an emphasis on the coal and coalbed methane resources. The effort to condense the reports and prepare art and tables for publication was supported by the GRI with the major costs borne by TRW. When possible, the authors of the original report provided the condensation. When not possible, a second TRW geologist provided the condensation.

The editors wish to thank TRW, GRI, and the AAPG for continued support and assistance in undertaking this effort and particularly Christine M. Fuller, James I. McComas, and Carolyn A. Starr for supporting the development of the final product.

It is our hope that this volume will provide the background and continued impetus for development of this underutilized resource within the United States.

Craig T. Rightmire
Greg E. Eddy
James N. Kirr

McLean, Virginia
November 1983

ORIGINAL MRCP REPORTS

Adams, M. A., 1982, Central Appalachian Basin report; a study of Carboniferous geology, coal, and the potential coalbed methane resource of the Central Appalachian Basin in Maryland, West Virginia, Kentucky, and Tennessee: 145 p.

——— and J. N. Kirr, 1980, Uinta Basin report; a study of Cretaceous geology, coal and the potential coalbed methane resource of the Uinta Basin in Utah and Colorado: 127 p.

———, G. E. Eddy, J. L. Hewitt, J. N. Kirr, and C. T. Rightmire, 1982, Northern Appalachian coal basin report; a study of Carboniferous geology, coal, and the potential coalbed methane resources of the northern Appalachian coal basin in Pennsylvania, Ohio, Maryland, West Virginia, and Kentucky: 179 p.

Archer, P. L., 1979, Illinois Basin report; a study of Pennsylvanian geology, coal, and coalbed methane resources in Illinois, Indiana, and western Kentucky: 70 p.

Choate, R., and C. A. Johnson, 1980, Powder River Basin report; a study of Early Tertiary geology, coal and the potential for methane recovery from coalbeds in Montana and Wyoming: 174 p.

——— and ———, 1980, Western Washington coal region report; a study of Early Tertiary geology, coal and coalbed methane resources in western Washington: 227 p.

———, D. Jurich, and G. J. Saulnier, Jr., 1981, Piceance Basin report; a study of Cretaceous and Tertiary geology, coal and the potential coalbed methane resource of the Piceance Basin, Colorado: 184 p.

———, J. Lent, and C. T. Rightmire, 1982, San Juan Basin report; a study of Upper Cretaceous geology, coal and coalbed methane resources of the San Juan Basin, in Colorado and New Mexico: 149 p.

Hewitt, J. L., 1980, Warrior Basin report; a study of Carboniferous geology, coal and the potential coalbed methane resource in Alabama and Mississippi: 120 p.

Jurich, J. M., 1980, Raton Mesa coal region report; a study of Upper Cretaceous and Early Tertiary geology, coal and the potential coalbed methane resource of the Raton Basin in Colorado and New Mexico: 137 p.

McCord, J. P., 1980, Greater Green River coal region report; a study of Upper Cretaceous and Early Tertiary geology, coal and the potential coalbed methane resource of the Greater Green River coal region in Colorado and Wyoming: 170 p.

Rieke, H. H., 1980, Arkoma Basin report; a study of Pennsylvanian geology, coal and coalbed methane resources in Arkansas and Oklahoma: 121 p.

———, 1981, Wind River Basin report; a study of Late Cretaceous and Early Tertiary geology, coal and the potential coalbed methane resource of the Wind River Basin in Wyoming: 160 p.

Coalbed Methane Resource

C. T. Rightmire

Methane has been observed in coalbeds since underground mining of that resource began. This gas, however, has only recently been recognized as an economically producible resource. Coal underlies approximately 360,000 sq mi of the conterminous United States but because of the paucity of data on coals deeper than 3,000 ft the size of the deep coal and therefore deep gas resource is not well known.

Coal responds to increased temperature over time by increasing in rank or thermal maturity. During this maturation process increased volumes of methane are generated. Coal is identified as a humic, Type III, kerogen and as such yields methane as its primary hydrocarbon product and water, carbon dioxide, and nitrogen as nonhydrocarbon products. It is estimated that more than 7,000 cubic feet of methane is generated for each ton of coal during coalification from lignite to anthracite rank.

Methane is found in coals either adsorbed on the coal surfaces, as free gas in fractures and large pores or dissolved in ground water in coalbeds. The amount of gas stored in the coals is influenced by depth of burial and its related pressure, rank of coal and its related porosity distribution, and a time-maturity relationship. Adsorption isotherm determinations show the maximum amount of methane that can be adsorbed on coals of various ranks. Desorption analyses of coal core samples indicate that for high rank coals adsorbed gas may exceed that predicted by isotherm analyses.

Methane has been produced from coals since the early 1900s. Producibility is influenced by depth, rank, permeability, water saturation, and other hydrogeologic characteristics. Current commercial gas production from coalbeds has been documented in the Warrior Basin, Northern Appalachian Basin, and the San Juan Basin. One well in the San Juan Basin is producing at a rate in excess of 1.5 MMcfd from a 17-ft coalbed.

Possible constraints to coalbed gas production involve completion technology and legal and institutional inconsistencies. Research is currently underway to enhance gas production from deep coals. Legal and regulatory constraints are being addressed through appropriate state and federal channels to resolve questions of gas ownership and price.

Data indicate that the coalbed methane resource is producible at currently economic rates.

INTRODUCTION

The presence of methane in coalbeds has been recognized for hundreds of years as a hazard to coal mining. Only recently has its potential as an energy resource begun to be recognized.

Early estimates of the coalbed methane resource, conducted as part of the Unconventional Gas Studies by the U.S. Department of Energy (DOE) and the Gas Research Institute (GRI), show the coalbed methane resource (gas-in-place) as high as 800 trillion cubic feet (Tcf) (Table 1). If only half of this resource is recoverable, it equates to a 20-year supply of gas, at current consumption of about 20 Tcf per year. With the exception of the Methane Recovery from Coalbeds Project (MRCP) basin analyses, these preliminary estimates were based on rough approximations of the coal resources in the conterminous United States—and an even rougher approximation of the gas-in-place in cubic feet per ton (cf/ton) for that coal. Since coal resource estimates have been based primarily on minable or potentially minable coals, little effort has been expended on the collection of data on coalbeds or coal-bearing strata deeper than 3,000 ft. When considering coal as a source rock and a potential reservoir, all coal must be considered in the establishment of a coalbed methane resource base.

Coal is known to underlie approximately 360,000 sq mi of the conterminous United States (Fig. 1). Where this coal is close to outcrop or in the shallow subsurface (less than 500 ft below water table), the potential for retaining the methane generated is significantly diminished. Where coal is found at greater depths, the amount of gas held in the coal is a function of the depth (i.e., hydrostatic pressure), the rank of coal, and the hydrogeologic environment of the area of interest. The amount of methane gas generated during the coalification process will depend primarily on the thermal maturity (rank) of coal and the amount of coal present. These points will be discussed in greater detail in subsequent parts of this chapter.

This Memoir will present the general background information on the coalbed methane resource and condensations of reports on the coalbed methane potential of 13 major coal-bearing basins in the conterminous United States. These reports were originally compiled by the geologists of the TRW Coalbed Methane Program for the U.S. DOE MRCP and were distributed to organizations interested in the development of the potential resource under the auspices of the Methane from Coal Seams Program of the GRI. Preparation of these condensations was supported by the TRW Coalbed Methane Program and GRI.

The 13 basins analyzed (Fig. 1) are underlain by a total of 235,420 sq mi of coal-bearing strata. Of that area, prior to the initiation of the basin analysis program, approximately 108,670 sq mi were initially considered to have high coalbed methane potential. Following a coal and coalbed methane resource assessment of these basins, the high-potential areas were

Table 1—Comparison of estimated total U.S. in-place coalbed methane.

Study	Total Resource In-Place (Tcf)	Recoverable Resource (Tcf)
Kuuskraa and Meyer (1980)	550	40-60
National Petroleum Council (1980)	398	45*
Sharer and Rasmussen (GRI) (1980)	500	10-60
Rosenberg and Sharer (GRI) (1979)	72-860	16-487
National Gas Survey Task Force on Nonconventional Natural Gas Resources		
Federal Energy Regulatory Commission (1978)	300-850	(not reported)
Deul and Kim (1978)	318-766	(not reported)
MRCP Basin Analysis (13 Basins)	72-400	(not estimated)

*Assumed a price of up to $9/Mcf and rate of return of 10% (1979 dollars).
Modified from Potential Gas Committee, 1981.

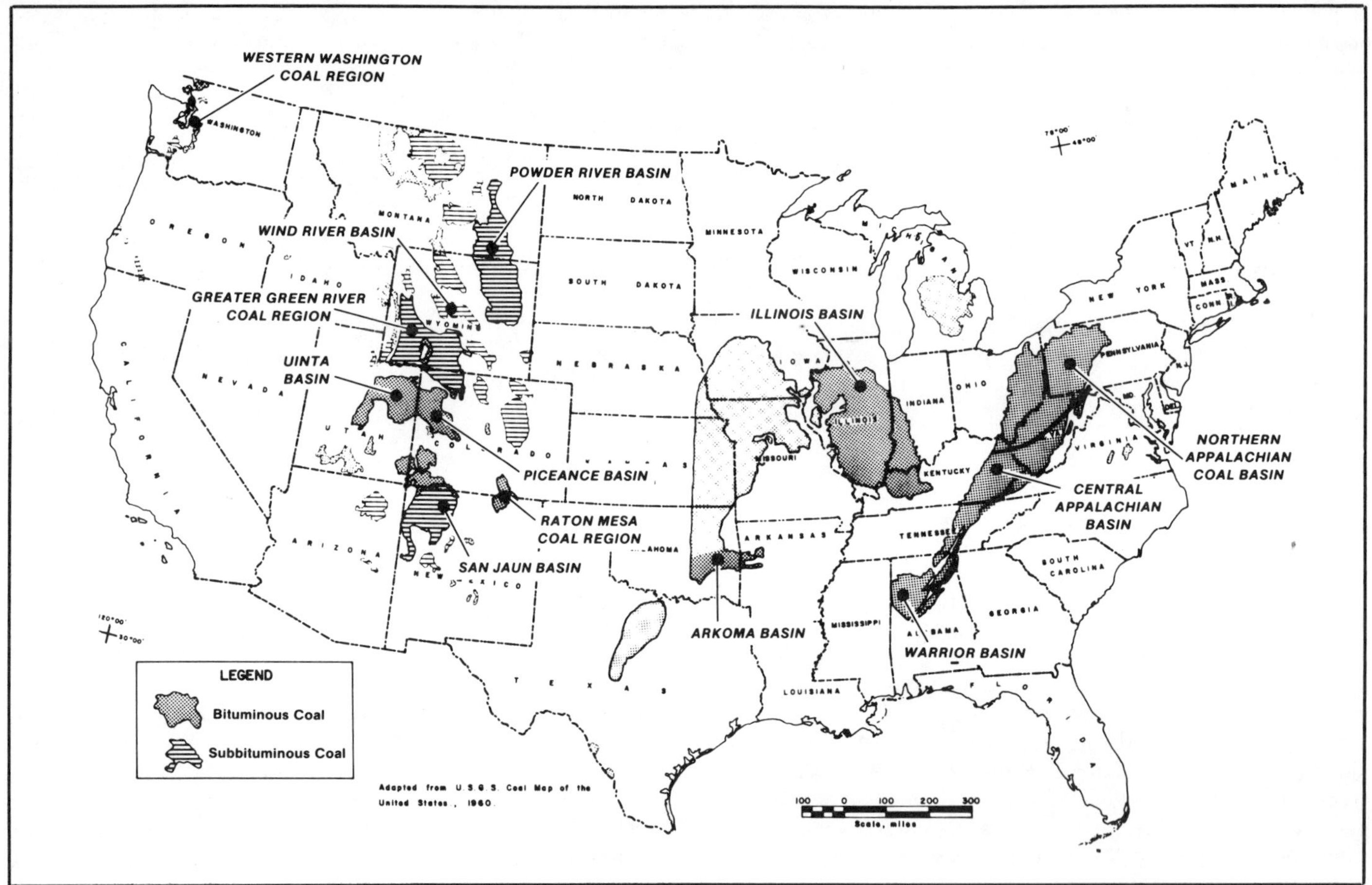

Figure 1—Map of major coal regions of conterminous United States.

narrowed to 40,415 sq mi, with an in-place-gas estimate ranging from 72 to 400 Tcf (Table 2). Because of the way methane is retained in a coalbed (primarily by adsorption), the gas-in-place is often referred to volumetrically as cubic centimeters of gas per gram of coal (cc/gm) or cubic feet of gas per ton of coal (cf/ton).

This chapter will address those characteristics of coal that influence a coalbed's ability to generate, trap, and produce natural gas. Of principal interest is the unique characteristic of coalbeds to act as both source rock and reservoirs for methane gas.

Background

Numerous studies have been conducted on the presence of methane in coalbeds, but the bulk of this work has concerned minable coals. The U.S. Bureau of Mines (USBM) has been conducting research in the mine safety aspects of coalbed methane since the early 1970s. These studies have dealt with the presence and distribution of gas in numerous, predominantly northeastern coalbeds in the United States (Kissell, 1972; Kim, 1974, 1975; Diamond and Levine, 1981), with the mechanisms by which the gas is generated, retained,

Table 2—MRCP coalbed methane basin analysis results. Modified from Rightmire and Byrer, 1981.

Basin	Date	Area Underlain By Coal (Approx. sq mi)	Initial Target Area (sq mi)	Current Target Area (sq mi)	Estimated Total Gas-In-Place (Tcf) Min.	Max.
Illinois	Mar. 1980	53,000	9,100	4,300	5.2	21.1
San Juan	July 1982	19,000	4,900	1,900	1.8	31.0
Powder River	Oct. 1979	25,800	12,800	6,750	5.9	39.4
Arkoma	Nov. 1979	5,300	5,300	3,600	1.6	3.6
Greater Green River	Nov. 1979	21,200	21,200	4,900	0.2	30.9
Western Washington	Jan. 1980	6,500	400	1,820	3.6	24.0
Raton Mesa Region	Sept. 1980	2,200	1,600	925	8.0	18.4
Uinta/Wasatch Plateau	Dec. 1980	11,100	11,100	620	0.2	0.8
Warrior	Dec. 1980	14,400	6,800	2,500	5.0	10.0
Wind River	Mar. 1981	3,800	3,800	1,500	0.5	2.2
Piceance	Mar. 1981	6,570	6,570	3,100	30.0	110.0
Central Appalachian	June 1982	22,850	5,500	4,000	10.0	48.0
Northern Appalachian	Aug. 1982	43,700	19,600	4,500	—	61.0
					72.0	400.4

and released by the coal (Kissell and Bielicki, 1972; Kim, 1973, 1974, 1978a; Kissell and Edwards, 1975), and potential methods of degasification of coalbeds relative to proposed or existing mining operations (Moore and Zabetakis, 1972; Fields et al, 1973; Cervik, Fields, and Aul, 1975; Fields, Cervik, and Goodman, 1976; Perry, Aul, and Cervik, 1978). Most research has not been concerned with coalbed methane as a potential natural gas resource, with its presence and potential producibility from unminable coalbeds or with techniques and procedures for the development of the most cost-effective methods for overall production of the gas resource.

COALIFICATION

The process by which plant material is progressively altered through peat, lignite, subbituminous, and bituminous, to anthracite coal, is termed "coalification." It describes the alteration of vegetal matter in a manner similar to diagenetic and metamorphic alteration of inorganic rock material.

As organic material is altered to coal through the rank sequence shown in Table 3, both physical and chemical changes are observed. These changes are analogous to increased levels of maturity, observed in marine kerogen-bearing source rocks, and can be utilized in an identical fashion to evaluate the coalbed methane potential of a coal-bearing region. Those changes most commonly taken as indicators of the maturity of the organic material are calorific value, moisture content or moisture-holding capacity, percent volatile matter, vitrinite reflectance, and fixed carbon content. Some chemical changes indicating level of maturity are more appropriate at some rank stages than at others (see Fig. 2). For instance, bed moisture (ash-free) and calorific value (moist; ash-free) are most meaningful in the peat to medium-volatile bituminous range. All of the above properties are altered measurably and predictably by increased temperature commonly brought about by increased depth of burial.

The formation of peat from vegetal remains has been termed "biochemical coalification" because of microbial activity. This

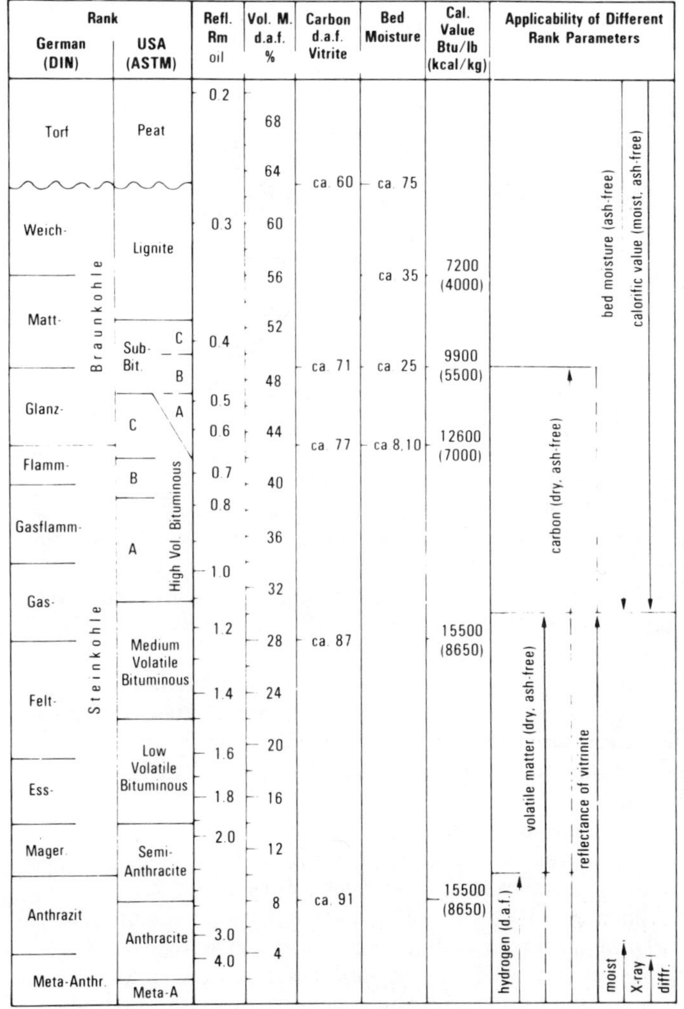

Figure 2—Stages of coalification according to German and American classifications. From Stach et al. Copyright © 1982 Gebruder Borntraeger Stuttgart. Used with permission.

Table 3—Classification of coals by ASTM ranking.

Class	Group	Fixed Carbon Limits, % (Dry, Mineral-Matter-Free Basis)		Volatile Matter Limits, % (Dry, Mineral-Matter-Free Basis)		Calorific Value Limits, Btu/lb. (Moist, Mineral-Matter-Free Basis)		Agglomerating Character
		Equal or Greater Than	Less Than	Greater Than	Equal or Less Than	Equal or Greater Than	Less Than	
I Anthracite	1. Meta-anthracite	98	—	—	2	—	—	
	2. Anthracite	92	98	2	8	—	—	Nonagglomerating
	3. Semianthracite	86	92	8	14	—	—	
II Bituminous	1. Low Volatile Bituminous Coal	78	86	14	22	—	—	
	2. Medium Volatile Bituminous Coal	69	78	22	31	—	—	
	3. High Volatile A Bituminous Coal	—	69	31	—	14,000	—	Commonly Agglomerating
	4. High Volatile B Bituminous Coal	—	—	—	—	13,000	14,000	
	5. High Volatile C Bituminous Coal	—	—	—	—	11,500	13,000	
		—	—	—	—	10,500	11,500	Agglomerating
III Subbituminous	1. Subbituminous A Coal	—	—	—	—	10,500	11,500	
	2. Subbituminous B Coal	—	—	—	—	9,500	10,500	
	3. Subbituminous C Coal	—	—	—	—	8,300	9,500	Nonagglomerating
IV Lignite	1. Lignite A	—	—	—	—	6,300	8,300	
	2. Lignite B	—	—	—	—	—	6,300	

Source: ASTM, Standard specification D388.

activity occurs in reducing environments where anaerobic bacteria consume the oxygen in the organic matter, leaving a hydrogen-enriched material. During peat formation, the carbon content increases from 45 to 50% [dry, ash-free, (daf)] to 55 to 60% (daf); the moisture content remains in excess of 75%; and free cellulose (i.e., not intimately mixed with lignin) is present (Stach et al, 1982).

Early decreases in porosity and particle alignment, induced by compaction with increased overburden thickness, are readily apparent physical changes.

Diagenetic changes occur up to the lignite-subbituminous coal boundary, which is observed at a formation temperature of about 50°C, depending on time-temperature relationships. Above subbituminous rank, changes can be equated to metamorphic alteration.

The major products of the coalification process are methane (CH_4), carbon dioxide (CO_2), nitrogen (N_2) and water (H_2O). Methane is generated by two mechanisms during the coalification process: biogenic and thermogenic. During the early stages, at temperatures below 50°C, biogenic methane is formed by microbial decomposition of the organic material. This gas is generally found as "swamp" or "marsh" gas and, where subsidence and burial are sufficiently rapid, may subsequently be trapped in shallow gas reservoirs. Studies conducted by the U.S. Geological Survey (USGS) and others (Claypool, Threlkeld, and Magoon, 1980; Fuex, 1977) have shown the possibility of differentiating between gas generated by the two mechanisms. Biogenic methane will form where reducing conditions induced by biological activity are sufficiently strong to remove dissolved oxygen and sulfate from the environment.

Figure 3 is a schematic diagram showing the relative volumes of methane generated by the biogenic and thermogenic mechanisms from humic and sapropelic organic material. Biogenic methane may comprise up to 10% of the methane generated at less than 200°C.

THERMOGENIC GAS GENERATION

As temperature increases above 50°C through increased depth of burial or increasing geothermal gradient, the coal rank also increases. Rank increase is not an instantaneous phenomenon but is related to the time that the material is maintained in a given thermal regime. This time-temperature relationship, often referred to as the "time-depth of burial," dictates the level of maturity of the coal that—among other things—controls the volume of methane generated. The principal product gases generated under this regime are methane, carbon dioxide, and nitrogen (Fig. 4).

Catagenesis is the process by which organic material is altered

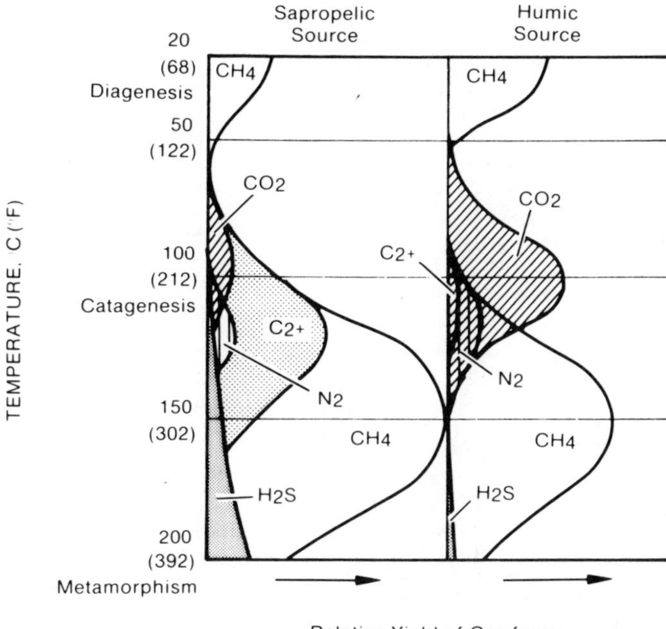

Figure 3—Generation of gases with depth, C_{2+} represents hydrocarbons heavier than CH_4 in gas phase. N_2 is generated initially as NH_3. From PETROLEUM GEOCHEMISTRY AND GEOLOGY by J. M. Hunt. Copyright © 1979 by W. H. Freeman and Company. Used with permission.

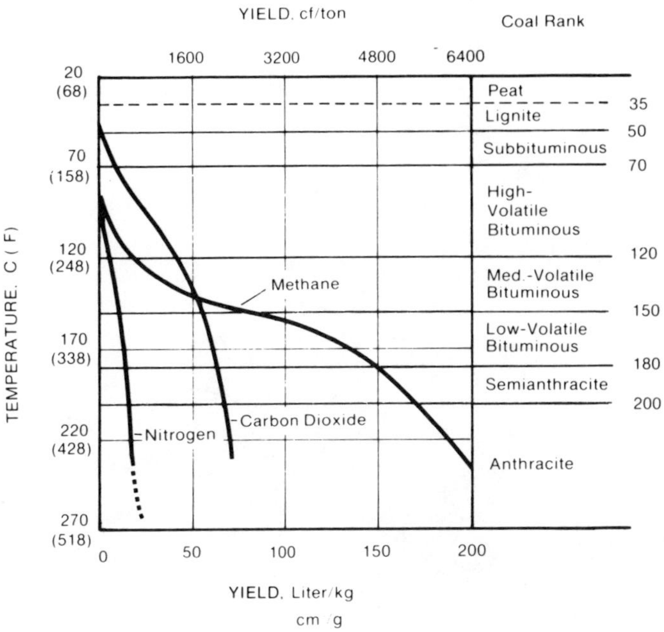

Figure 4—Calculated curves of gases generated from coal during coalification. Data from Karweil, 1969. Modified from PETROLEUM GEOCHEMISTRY AND GEOLOGY by J. M. Hunt. Copyright © 1979 by W. H. Freeman and Company. Used with permission.

as a result of the effect of increasing temperature. Methane generated at temperatures in excess of 50°C will be due to the catagenic processes; i.e., thermogenic methane.

During the coalification process, large volumes of methane and other gases are generated by the thermal alteration of the organic material. Chemical changes that occur in the organic material lead to a decrease in both atomic oxygen/carbon (O/C) and atomic hydrogen/carbon (H/C) ratios with increasing alteration. Humic material from which most coals are generated is composed largely of oxygen-rich lignin and cellulose. A common means of approximating the level of maturity of an organic material is to plot the material on a van Krevelen diagram (Fig. 5)—the branches of which represent different types of kerogen or source material. Coal is identified as a Type III kerogen. As coal matures, it continues to lose oxygen and moves to the left along the Type III branch of the diagram. As noted, the oxygen loss occurs more readily than hydrogen loss during the lower temperature, diagenetic portion of the maturation process. For a humic coal material, because of its chemistry, hydrogen loss will be less than that for sapropelic material. Figure 6 shows the maturation pathways, the principal products generated, and the related vitrinite reflectance (VR) and thermal alteration index (TAI) values. It must be noted that the "heavy hydrocarbon" product portion of Type III kerogen pathway is very small, corresponding to a very small volume of that product generated from this source material.

During coalification (as noted on Fig. 4), more than twice as much CO_2 as CH_4 is generated, up to a high-volatile/medium-volatile bituminous coal boundary. Methane volumes generated increase rapidly above that point. From the data of Karweil (1969), Table 4 presents the volumes of the various thermogenically generated gases within any particular coal rank in both cc/gm and cf/ton for comparative purposes. Other data (Tissot and Welte, 1978) suggest that the predicted volume of methane generated as shown in Table 4 may be high, above a coal rank equivalent to R_O approximately 2%.

As seen on Figure 3, the methane generation peak occurs at about 150°C, or at the medium-volatile bituminous/low-volatile bituminous boundary.

The two principal contaminants in methane from coalbeds are CO_2 and N_2. Both form from decomposition of organic material and should be expected at some levels in all coalbed methane gas. Nitrogen emissions begin as ammonia (NH_3) near the end of the high-volatile A bituminous stage at about 120°C (Hunt, 1979). It is often found as only a minor constituent of the produced gas because its smaller molecular size of 3.0 Å permits it to migrate from the system more readily than the gases with which it is generated.

Although a principal constituent of early thermogenic gases, CO_2 is commonly a relatively minor and extremely variable constituent in the produced gas. Although its molecular size inhibits its rapid migration as a gas, its high solubility in water significantly facilitates its mobility. Approximately one volume of CO_2 will dissolve in one volume of fresh water at earth's surface conditions of one atmosphere at 20°C. At 300 atmospheres and 100°C, conditions equivalent to a coalbed methane reservoir at 10,000 ft, about 30 volumes of CO_2 (STP)

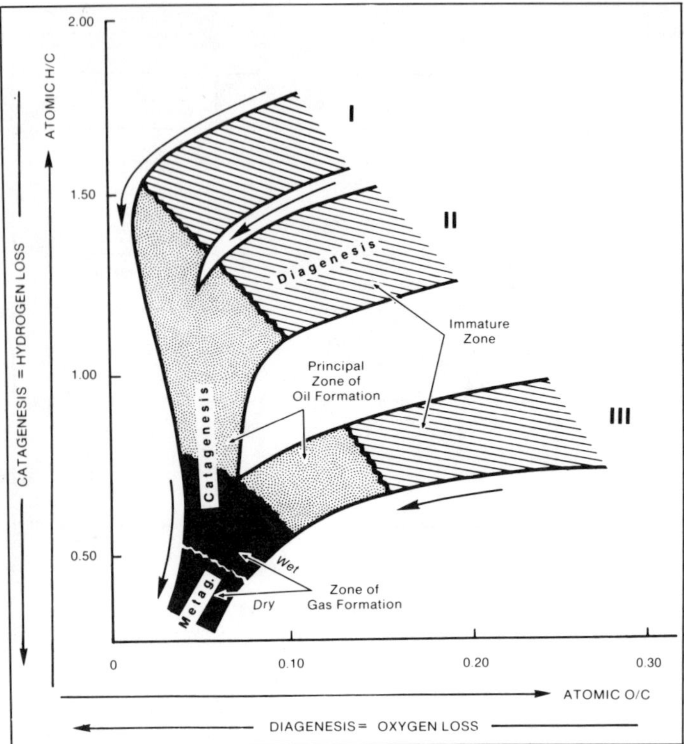

Figure 5—Schematic van Krevelan diagram showing relative chemical changes in oxygen, hydrogen, and carbon along designated maturation pathways. Coal is identified as a Type III kerogen.

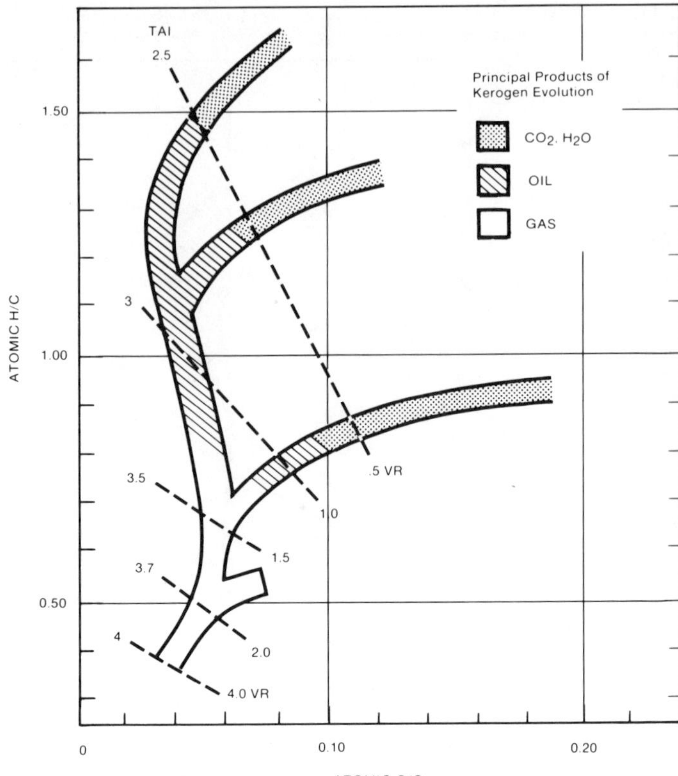

Figure 6—Kerogen maturation and type relative to vitrinite reflectance. Note long CO_2 and H_2O and CH_4 windows relative to oil window for Type III kerogen. Tissot et al, 1974.

will dissolve in that same one volume of water. As shown in Table 4, in excess of 2,300 cf/ton CO_2 are generated during the maturation of organic material from lignite through anthracite.

Hydrogen sulfide (H_2S) has been identified as being generated in trace amounts from humic source material. Because of its very high solubility in water (2.6 times that of CO_2 at 20°C and one atmosphere) and the fact that it is the last constituent of natural gas to form (starting at about 100°C), only trace amounts of H_2S have ever been identified in coalbed methane reservoirs.

Analysis of gas produced from coalbeds either in wells or during desorption testing shows that, with few exceptions, these gases contain in excess of 95% methane, trace to very minor amounts of higher hydrocarbons (ethane, propane, etc.), and less than 3% each of N_2 and CO_2. These gases have heating values of approximately 1,000 Btu/scf. Table 5 presents representative analyses of gases produced from coalbeds.

IDENTIFYING COALBED-GENERATED GAS

Stable isotope analysis of the methane is a principal means of identifying the natural gas in question as being thermally generated from humic material. When used in conjunction with gas constituent chemistry, the results of isotopic fractionation during the maturation process are quite apparent. Carbon isotope analysis (^{13}C) plotted vs. wetness $[C_1/(C_2 + C_3)]$ readily differentiates biogenic from other gases, as well as humic from sapropelic deposits.

Biogenic methane normally ranges for $\delta^{13}C = -55$ to -75 parts per thousand (‰). Methane generated by thermal cracking of parent material would range from -5 to -3‰, relative to that source material. Methane generated from coalbeds has a higher $\delta^{13}C$ content than marine kerogen methane derived at equivalent rank or thermal maturity. Coalbed methane can be expected to yield $\delta^{13}C$ values of -25 to -40 ‰, depending on source material, and to have a high $C_1/(C_2 + C_3)$ ratio.

METHANE RETENTION IN COALBEDS

Methane is retained in coals in one of three states: as adsorbed molecules on the organic surfaces, as free gas within the pores or fractures, and/or dissolved in solution (ground water) within the coalbed. These will be examined individually to determine their relative influence on the total gas volume in the coal.

Porosity exists in coals as fracture porosity and matrix porosity. The latter is most significant when considering the methane retention potential of coals. The primary mechanism of methane retention in coalbeds is adsorption on the coal surface within the matrix pore structure.

Pores in coals range in size from >500Å diameter (macropores); 20 to 500Å (mesopores); 8 to 20Å (micropores); and <Å (submicropores) (IUPAC, 1972). Following the suggestion by Mahajan (1982), no distinction is made in these studies between submicropores and micropores to permit utilization of pre-1972 research. In an attempt to determine porosity distribution and to verify sorption properties, much

Table 4—Approximate gas generation by coal rank—cc/gm (cf/ton).

Rank	Temp. (°C)	CH_4	CO_2	N_2	Vitrinite Reflect. R_o
Lignite		—	4 (120)		
	50				
Subbituminous		—	6 (200)		
	70				— .50±
Hi-Vol Bit.		20 (640)	34 (1,080)	5 (160)	C .60± B .75±
	120				1.20
MV Bit.		84 (2,680)	11 (360)	5 (160)	
	150				1.63
LV Bit.		68 (2,160)	8 (240)	3 (80)	
	180				2.10
Semi-Anth.		24 (760)	4 (120)	3 (80)	
	200				2.40
Anth.		28+ (880+)	7+ (216+)	10± (320±)	
Totals		223 (7,120+)	73 (2,336+)	25± (800±)	

Table 5—Composition and heating value of coalbed gas and natural gas (vol. pct.) (Kim, 1978b).

Source	Basin	CH_4	C_2+	H_2	Inerts[1]	O_2	Btu/Scf
Pocahontas No. 3	Central Appalachian	96.87	1.40	0.01	2.09	0.17	1,059
Pittsburgh	Northern Appalachian	90.75	.29	—	8.84	.20	973
Kittanning	Northern Appalachian	97.32	.01	—	2.44	.24	1,039
Lower Hartshorne	Arkoma	99.22	.01	—	.66	.10	1,058
Mary Lee	Warrior	96.05	.01	—	3.45	.15	1,024
Natural Gas[2]	—	94.40	4.90	—	.40	—	1,068

Notes:
1. N_2, CO_2, and He.
2. Moore, Miller, and Shrewsbury, 1966.

effort has gone into attempts to determine the porosity and internal surface areas of coal samples. The technique employed involves the measurement of volume of a fluid that can be sorbed on a coal surface at given pressures. Various sorbates, such as N_2, CO_2, and methanol, have been used with surface area measurement differing dramatically depending on the sorbate used. Estimates of pore volume and surface area depend on the techniques used and the properties of the adsorbate. Mahajan (1982) states that "the concept of a fixed definable internal surface area of microporous coals determined by physical adsorption measurements, has little significance." When addressing surface area measurements, it is important to be aware of coal particle size, adsorbate, outgassing time and temperature, adsorption time, and cross-sectional area of the adsorbate.

Some background information on coal porosity and adsorption properties is desirable to permit utilization of the available surface area data.

- Coals have an aperature-cavity type of porosity, with the size of the aperature controlling access to the pore surfaces. The volume adsorbed is controlled by the relationship of adsorbate size and pore size.
- Small changes in moisture content and/or degree of coalification may dramatically alter the measured surface area of a coal sample.
- An unambiguous value for the cross-sectional area of an adsorbed molecule cannot be assigned. A typical coal pore-adsorbed molecule will occupy two to three times the surface area that it would on a flat surface (Mahajan, 1982).
- Surface area appears generally related to carbon content (Table 6). In studies by Gan, Nandi, and Walker (1972) and Nandi and Walker (1971) using N_2, adsorption coals with carbon content less than 76% and greater than 83% generally have surface areas less than 1 sq m/g. Those within that range have areas greater than 10 sq m/g. Anthracite coals with carbon content >92% are the exception to this observation, with areas ranging from 5 to 8 sq m/g. Carbon dioxide surface areas for 18 different coals range from 110 to 310 sq m/g (Gan, Nandi, and Walker, 1972). The difference for coals with the same carbon content is as large as 90 sq m/g.
- Of additional interest is the conclusion by Van der Sommen et al (1955) that methane sorbed on a bituminous coal at

Table 6—Coal surface areas (N_2 adsorption). Data from Nandi and Walker, 1971.

Rank	Fixed Carbon Content (%)	Surface Area (m^2/g)
	76	1
Med Vol Bituminous	76-78	10
Low Vol Bituminous	78-83	
	83-92	1
Anthracite	92	5-8

Table 7—Gross open-pore distribution in coals. Modified from Gan et al, 1972; FUEL, v. 51, p. 272-277, by permission of the publishers, Butterworth & Co. (Publishers) Ltd. © 1972.

Rank	C (% daf)	Porosity Distribution (%)		
		<12Å	12-300Å	>300Å
Anthracite	90.8	75.0	13.1	11.9
Low Vol	89.5	73.0	nil	27.0
Med Vol	88.3	61.9	nil	38.1
Hi-Vol A	83.8	48.5	nil	52
Hi-Vol B	81.3	29.9	45.1	25.0
Hi-Vol C	79.9	47.0	32.5	20.5
Hi-Vol C	77.2	41.8	38.6	19.6
Hi-Vol B	76.5	66.7	12.4	20.9
Hi-Vol C	75.5	30.2	52.6	17.2
Lignite	71.7	19.3	3.5	77.2
Lignite	71.2	40.9	nil	59.1
Lignite	63.3	12.3	nil	87.7

Table 8—Effective molecular diameters. Data from Hunt, 1979.

Molecule	Approximate Molecular Diameter (Å)
He	2.6
N_2	3
H_2O	4.1
H_2S	4.1
CH_4	4.1
CO_2	4.7
C_2H_6	5.5
C_3H_8	6.5

25°C has a density well above its liquid density of 0.480.

- Porosity depends at least partially on maceral content in highly volatile bituminous coals, with vitrinite having the finest porosity in the range of 20 to 200Å diameter. The bulk of the porosity is in the smaller size range. Inertinite was the most porous, with pores ranging from 50 to 500Å (Harris and Yost, 1976; Harris, 1979).

Pore size distribution will effectively control the volume and rate at which sorbed gas is released. Table 7, modified from Gan et al (1972), shows pore size distribution with rank for 12 coal samples. This distribution confirms that the bulk of the surface area of moderate- to high-rank coals is in the micropores. Additional studies narrow the range of maximum pore volume to those with a radius of 7–8 Å (Debelak and Schrodt, 1979). Toda et al (1971) showed that the micropore volume decreases to about 85% carbon content, then increases with increasing carbon content.

Table 8 presents the effective molecular diameters of gas molecules of interest in the study of the retention, release, and migration of gases in coals. Examination of the data presented shows the effective diameter of a methane molecule to be between that of N_2 and that of CO_2.

Sorption of methane on coals as a function of pressure has been studied by several groups and has been observed to follow a Freundlich isotherm. As previously discussed in dealing with rank-porosity relationships, the adsorptive capacity of coal appears to increase with increasing rank. While the adsorption isotherms are determined on sized coal particles and, as such, do not represent in situ conditions, they do exhibit the relative ability of a coal to retain methane under given pressure conditions. Figure 7 presents a series of generalized adsorption isotherms for moderate- to high-rank coals (Kim, 1977). These represent the maximum amount of methane that can be sorbed on the internal surfaces, including micropores, of sized coal particles. Adsorption isotherms determined for several coals of different rank and from different geographic areas are presented in Figure 8. The two determinations for the Pocahontas No. 3 coal, at different moisture contents, show the effect of moisture on the adsorptive capacity of coals.

Gas generated in excess of that which can be adsorbed on the coal surfaces will initially be present as "free gas" within the porosity of the coal, most noticeably within the fracture porosity. This gas is available for dissolution in ground water moving through the coalbed, migration up-gradient, or under proper hydrogeologic conditions retention in the coalbed as trapped free gas. Relative volumes of adsorbed vs. free gas have been reported by Eddy, Rightmire, and Byrer (1982) for more than 300 coal samples tested for gas content. The "lost gas" column presented in Table 9 represents the free gas being discussed here. While the amount of lost gas reported is influenced by the length of time it takes to prepare a sample for desorption analysis, its calculation is a means of approximating the "free gas" component in the coalbed reservoir. Methane is soluble to a limited degree in ground water at the pressures (normally close to hydrostatic) and temperatures encountered in most coalbed methane reservoirs (Table 10). Increasing salinity will decrease this solubility, as shown in Table 11. To put these volumes in perspective, at 1,000 psi and 100°F, a maximum of 8.5 cu ft of methane could dissolve in one barrel of fresh water. While this is a small volume of gas in a dynamic ground-water system, the methane saturation of large volumes of water can remove large total volumes of gas from a potential reservoir.

The fracture porosity in coal is due primarily to the fractures called cleat in the coal mining industry. Cleat is a joint or set of joints perpendicular to the coalbed. There are usually two cleat sets developed nearly perpendicular to each other.

Face is the major cleat and may extend great distances. Butt cleat (also called end cleat) usually extends only from one face cleat fracture to the next. Because face cleat yields longer fractures, as well as usually being more prominent, horizontal holes drilled perpendicular to the face cleat drain a larger area and, therefore, produce more methane than do holes drilled

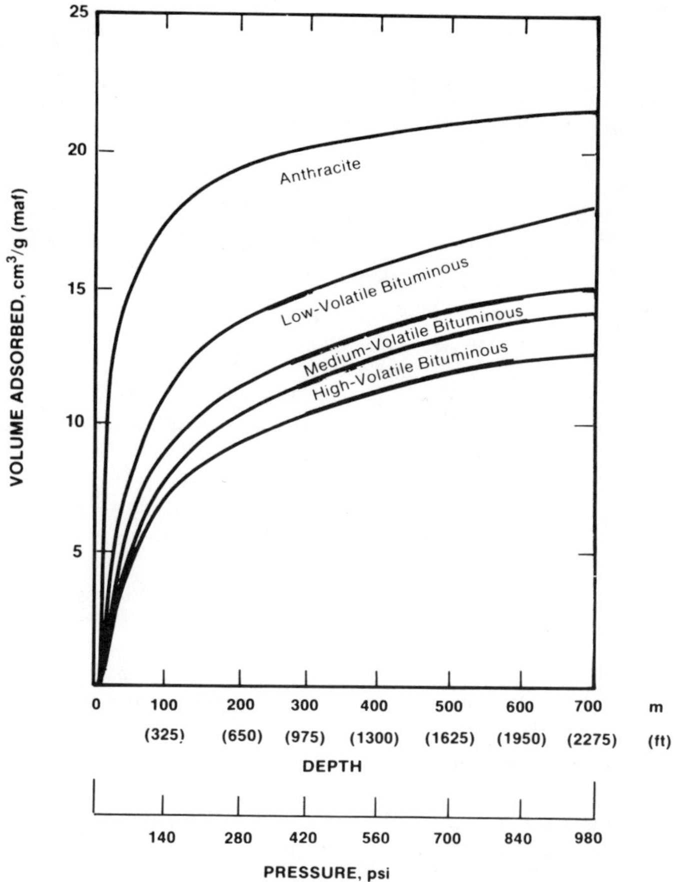

Figure 7—Adsorptive capacity of coal as a function of rank and depth. Kim, 1977.

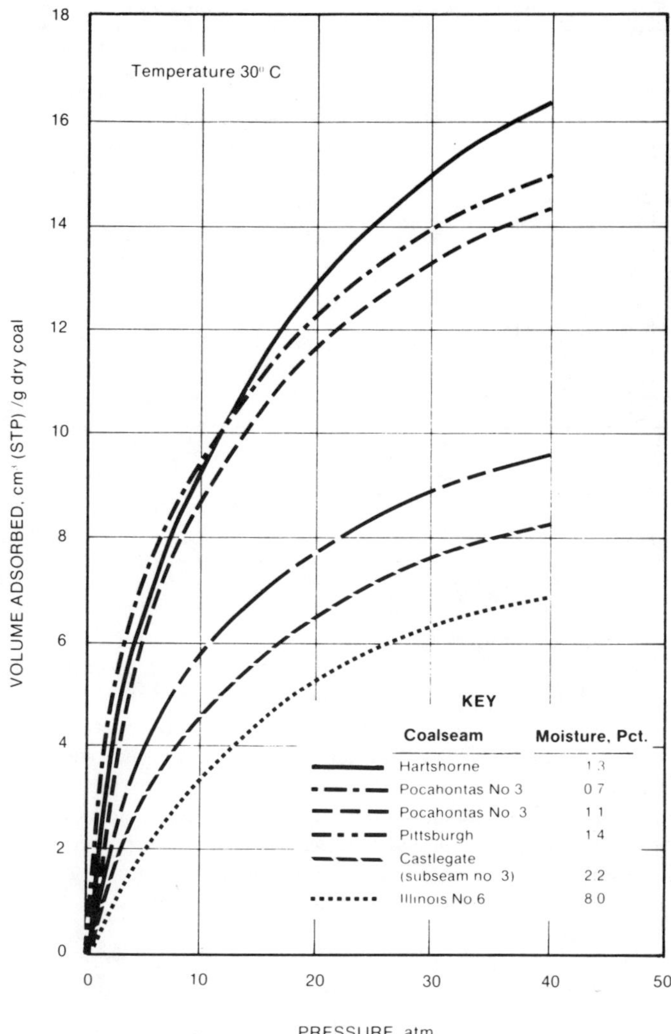

Figure 8—Equilibrium adsorption isotherms for different coal type and moisture content. Kissell et al, 1973.

parallel to the face cleat. Tests by the Bureau of Mines (McCulloch, Deul, and Jeran, 1974), have shown that the difference in production is 2.5 to 10 times as great in holes drilled perpendicular to face cleat. This is why vertical boreholes drain an elliptical area in a coalbed. The major axis of the ellipse is parallel to the face cleat, and the short axis is parallel to the butt cleat. This demonstrates that cleat is the major control on the directional permeability of coalbeds.

IN-PLACE GAS CONTENT DETERMINATIONS

Measurements of gas content in coals have been conducted for a number of years by the Direct Method adopted by the USBM (Kissell, McCulloch, and Elder, 1973; Diamond and Levine, 1981). This method reports total gas based on three component parts:

- Lost Gas: That gas lost between the time that the coalbed was penetrated by the bit and the sample was sealed in the canister. Determined by extrapolation of the first 2 hours of desorption data.
- Desorbed Gas: That gas given up by the coal against atmospheric pressure while it is sealed in the canister.
- Residual Gas: That gas remaining trapped in the coal matrix following completion of desorption. Determined by crushing the coal sample in a gas-tight container and measuring volume emitted.

The Direct Method provides analytical data quite acceptable in a relative sense (Kissell, McCulloch, and Elder, 1973) but with uncertainties as to its absolute accuracy. Direct method determinations probably have a 25 to 30% uncertainty based on unknowns related to "lost gas" extrapolations (Kim, 1977).

Figure 9 shows the maximum lost-plus-desorbed gas for samples of various rank analyzed as part of the MRCP (Eddy, Rightmire, and Byrer, 1982). It should be noted that these curves represent the upper limit of envelopes and that not all coal samples of the respective ranks will approach those limits. The shape of curves and their approximate positions do resemble those of the adsorption isotherms previously discussed. This figure provides an estimate of the maximum producible methane content for a specific rank of coal at any given depth, based on the most complete data base available.

Table 9—Coalbed methane desorption data averages by rank. Eddy et al, 1982.

Coal Rank	Lost Gas (cu m/g)	Desorbed Gas (cu m/g)	Residual Gas (cu m/g)	Total Gas (cu m/g)	% Residual	Number of Samples
Anthracite	.98	8.10	.61	9.69	6.31	9
Low-Volatile Bituminous	1.21	11.97	.25	13.43	1.86	21
Medium-Volatile Bituminous	1.33	6.31	.32	7.96	4.02	22
High-Volatile A Bituminous	.21	2.77	1.38	4.36	31.65	217
High-Volatile B Bituminous	.31	2.01	.47	2.79	16.85	86
High-Volatile C Bituminous	.12	1.09	.07	1.28	5.47	42

Individual gas content analyses have been reported in excess of 1,000 cu ft of methane per ton of coal (Choate, Jurich, and Saulnier, 1981). Samples of such high gas content are extremely rare and are considered the exception rather than the rule.

Data in this volume are derived from numerous sources, including the USBM, U.S. DOE, numerous state geological surveys, and many different institutional and industrial concerns. These data are reported in cu ft of gas per ton of coal, not standard cf/ton, unless otherwise noted. The reported gas contents may vary significantly, based on the elevation of sample collection and the elevation of the desorption laboratory.

Predicting in-place gas content based on rank-pressure relationships has been alluded to and is a good reconnaissance method of estimating gas content in a given seam, at a given depth. With specific knowledge of coal rank and depth below potentiometric surface, this method improves significantly. Knowledge of the rank permits determination of the gas volume generated; depth below the potentiometric surface gives the hydrostatic pressure; and these, in combination with the adsorption isotherm or the "Eddy" curve for the given rank, permit an in-place gas estimation. Rank can be determined from vitrinite reflectance measurements on coal cuttings, minimizing the requirement for expensive coring operations.

A third method for estimation of in-place coalbed methane was developed by Kim (1977) and is an equation based on a combination of adsorption isotherm and coal chemical data.

Mine emission data provide an additional in-place gas content estimator based on the volume of methane exhausted from coal mines in their ventilation returns. Figure 10 shows the relationship of gas emission from mines to gas content measured by the direct method for selected mines. The comparison provides only a generalized predictive technique in older, well-developed mines. Preliminary indications are that fairly large, deep mines with sustained and constant coal production of at least several thousand tons a day and that have been in operation for many years with extensive old workings and gas areas have a relatively fixed ratio defined as:

$$\frac{\text{Mine Emission (cu ft/ton mined)}}{\text{Direct Measurement (cu ft/ton)}}$$

The value of this ratio varies between 6 and 9 (Kissell, McCulloch, and Elder, 1973). Because gas emissions must be addressed relative to method of mining, age of the mine, overlying strata, means of abandoning mined-out areas, and other operational procedures, any attempt to equate emissions accurately to gas-in-place must be made on a case-by-case basis. In general, if 20% of the emitted gas is from mined coal, the estimated in-place gas more closely matches that determined by the direct method analysis for the greatest number of mines (Adams et al, 1982).

METHANE FROM COALBEDS PRODUCTION

Methane from coalbeds has been produced and utilized on at least a minor scale since the early 1900s, when a ranch in the Powder River Basin capped its water well, drilled into a coalbed, and started heating the buildings with the produced gas.

Coalbed methane has not been a target of major exploration efforts until recently because, generally, coalbed methane wells are considered low-pressure, low-flow producers. Most of the coalbed reservoirs are at about hydrostatic pressure. Once the free gas in the fractures is depleted, production is controlled by methane desorption and diffusion through the matrix to the fractures. Desorption and diffusion—and, therefore, long-term methane production—are controlled more directly by differential pressure than is a conventional reservoir.

Most of the permeability in a coalbed is fracture controlled, with the fracture spacing and orientation controlled by the local and regional stress patterns of the geologic formation. Permeability in bituminous and higher rank coals has been determined on laboratory samples to be in the low millidarcy range (Patching, 1965). Laboratory samples are generally of a size insufficient to adequately characterize in situ reservoir conditions, so care must be taken when attempting to utilize laboratory permeabilities in reservoir predictions.

In coalbed reservoirs, the relative permeabilities to gas and water are critical to the initiation of production. Where the coals are water saturated, water saturation must be reduced to permit desorption and flow of gas to the wellbore. Figure 11 shows the relationship of relative gas and water permeabilities

Table 10—Solubility of natural gas in water (Scf/bbl). Data from Craft and Hawkins, 1959.

Pressure (psia)	Temperature 100°F	160°F	220°F
15	0.2	—	—
500	5	4.2	4.0
1,000	8.5	7.5	7.2
1,500	11	10	10
2,000	14	12	12
3,000	17	15	16
4,000	20	17	18
5,000	23	19	21

Table 11—Natural gas solubility in water correction factor for total dissolved solids (TDS) and temperature. Data from Craft and Hawkins, 1959.

TDS (10^3 ppm)	Temperature 100°F	160°F	220°F
0	1.0	1.0	1.0
10	0.94	0.95	0.96
20	0.88	0.90	0.92
30	0.82	0.85	0.88

with water saturation for a Pittsburgh coal sample.

Pore size distribution creates a second porosity/permeability relationship in coals that must be addressed. Because the bulk of the matrix porosity is in pores with internal diameters of < 12 Å, diffusion through the matrix to the fracture surfaces is a rate controlling step. The pore throats of these micropores can be effectively blocked by water molecules, retarding gas diffusion out of the pores until the coals are dried.

Because of the dual matrix and fracture permeability systems exhibited by coalbed methane reservoirs, they tend to display an anomalous decline curve. Although the "free" gas present in the fractures and macropores may be produced in a typical reservoir response, the continual introduction of gas into the fractures by diffusion—as the zone of influence around the well expands—results in more gas being produced with time. This phenomenon,—a negative decline curve—was first observed for a well in the San Juan Basin, which has produced in excess of a billion cu ft of gas over a 30-year period, the first 20 years with a negative decline curve. This well exhibited very little pressure reduction over its life (Choate, Lent, and Rightmire, 1982).

While the size of the coalbed methane resource in the basins studied has been fairly well established, the producibility of the resource continues as a major unknown. The upper limit of producibility can be represented by the percentage of gas recovered as "lost" and "desorbed" gas from a desorption test, leaving only the residual gas-in-place. For high-volatile A bituminous coal, the residual gas accounts for about 32% of the total (Eddy, Rightmire, and Byrer, 1982), leaving the maximum producible gas at about 68%. The percent residual gas varies with rank and is shown on Table 9. For purposes of reserve estimation, where no additional data are available, it is normally considered that 50% of the gas is recoverable over the economic life of a coalbed methane well.

Where the geologic environment may have changed, causing a decrease in hydrostatic pressure since initial coalbed burial and gas sorption (i.e., uplift and erosion), or where more gas has been generated than is sorbable under reservoir conditions, overpressuring may occur. This has been observed in several basins, particularly parts of the San Juan, Greater Green River, and Wind River Basins, in the western United States. Where overpressuring occurs in the coalbeds, they will charge any adjacent overlying or underlying reservoir rock. This introduces the potential of producing more gas from a porous and permeable sandstone reservoir than it could hold if it were an infinite reservoir. Under these circumstances, the coalbed continues to recharge the sandstone reservoir as that reservoir is depleted by the well. Because of the potentially large area of

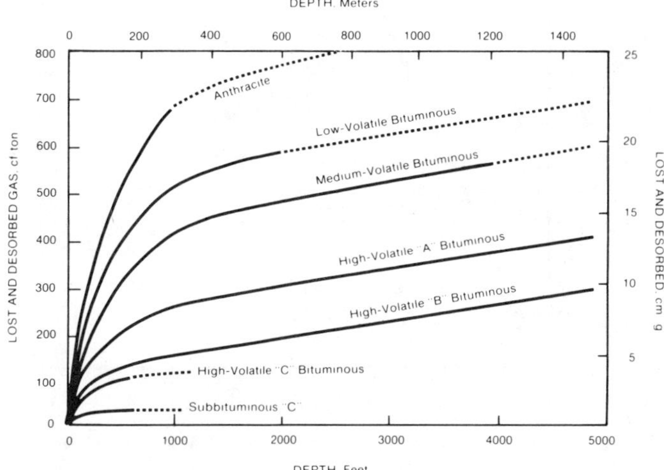

Figure 9—Estimated maximum producible methane content with depth and rank. Eddy et al, 1982.

contact between the coalbed and the sandstone, it is easier to flow gas out of the coal into the sandstone and to the well then it is to flow through the coal to the well.

CONSTRAINTS ON COALBED METHANE PRODUCTION

Presently, the primary constraints on coalbed methane production involve completion technology and legal and institutional inconsistencies.

Coalbeds should not be treated as conventional oil and gas reservoirs, with the same expectation of success. Coals are thin, compressible, chemically reactive reservoirs that require some amount of finesse to complete successfully. Production from coalbeds has begun, and techniques will be addressed under the individual basins in which those techniques are successful. Completion in deep coalbeds continues to be a potential problem. GRI is initiating research in the Piceance Basin in Colorado to attempt to solve some of the deep coal completion and production problems.

Completion from multiple coalbeds is necessary in some areas, particularly in parts of the eastern United States, to optimize the economic viability of resource recovery. A second GRI

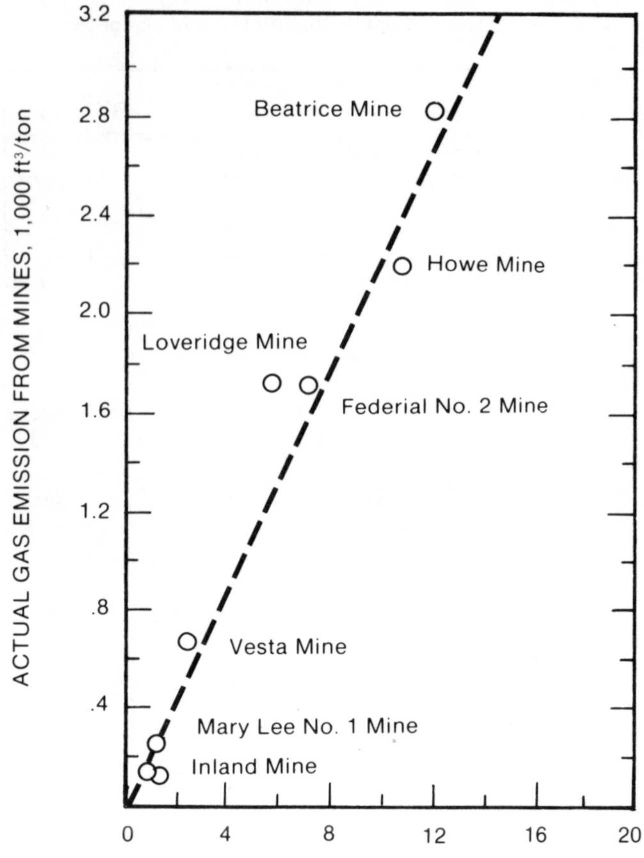

Figure 10—Gas content of coal vs. actual mine emissions. Modified from Kissell et al, 1973.

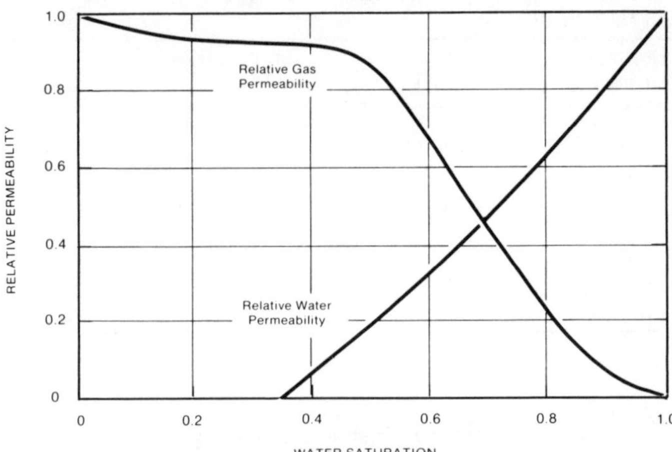

Figure 11—Relative gas and water permeabilities vs. water saturation for Pittsburgh coal samples under 200 psi overburden pressure. Samples were draining (decreasing water saturation) during these tests. From Taber et al, 1974.

research project in the Warrior Basin of Alabama will address means of maximizing the recovery of methane from multiple seams present in the area.

Legal constraints introduce difficulties with the recovery and utilization of the coalbed methane resource in some parts of the country. This is particularly true in the Northern Appalachian Coal Basin, where mining is occurring or is anticipated in some coalbeds from which methane can be produced. The confrontation between the coal miner (who owns the coal and is concerned about the economic viability of mining coal penetrated by numerous wells and, possibly, artificially stimulated by hydraulic fracturing) and the oil and gas lease holder (who owns the gas and is concerned about the mine operator venting that gas to the atmosphere to facilitate mining) is real and must be addressed in a rational manner. These two groups can work together to determine the best means of producing usable gas while minimizing the potential for interfering with mine operation and development.

REFERENCES CITED

Adams, M.A., et al, 1982, Geologic overview, coal resources, and potential methane recovery from coalbeds, Northern Appalachian Coal Basin, Pennsylvania, Ohio, Maryland, West Virginia and Kentucky: McLean, Virginia, TRW Coalbed Methane Program, for U.S. Department of Energy, Morgantown Energy Technology Center.

Cervik, J., H. H. Fields, and G. N. Aul, 1975, Rotary drilling holes in coalbeds for degasification: U.S. Bureau of Mines Report of Investigations 8097, 21 p.

Choate, R., D. Jurich, and G. J. Saulnier, Jr., 1981, Geologic overview, coal deposits and potential for methane recovery from coalbeds, Piceance Basin, Colorado: McLean, Virginia, TRW Coalbed Methane Program, for U.S. Department of Energy, Morgantown Energy Technology Center.

——, J. Lent, and C. T. Rightmire, 1982, San Juan Basin report—Upper Cretaceous geology, coal and the potential for methane recovery from coalbeds, San Juan Basin, Colorado and New Mexico, revision 1: McLean, Virginia, TRW Coalbed Methane Program, for U.S. Department of Energy, Morgantown Energy Technology Center.

Claypool, G. E., C. N. Threlkeld, and L. B. Magoon, 1980, Biogenic and thermogennic orgins of natural gases in the Cook Inlet Basin, Alaska: Bulletin of the American Association of Petroleum Geologists, v. 64, p. 1131-1139.

Craft, B., and M. Hawkins, 1959, Applied petroleum reservoir engineering: Englewood Cliffs, New Jersey, Prentice-Hall.

Debelak, D. A., and J. T. Schrodt, 1979, Comparison of pore structure in Kentucky coals by mercury penetration and carbon dioxide absorption: Fuel, v. 58, p. 732-736.

Deul, M., and A. G. Kim, 1978, Methane drainage—an update: Mining Congress Journal, July, 1978, p. 38-42.

Diamond, W. P., and J. R. Levine, 1981, Direct method determination of the gas content of coal, procedures and results: U.S. Bureau of Mines Report of Investigations 8515, 36 p.

Eddy, G. E., C. T. Rightmire, and C. Byrer, 1982, Relationship of methane content of coal, rank and depth: Proceedings of the SPE/DOE Unconventional Gas Recovery Symposium, Pittsburgh, Pennsylvania, SPE/DOE 10800, p. 117-122.

Federal Energy Regulartory Commission, U.S. Department of Energy, 1978, Natural Gas Survey Task Force on nonconventional natural gas resources: DOE/FERC-0010, June, 1978.

Fields, H. H., J. Cervik, and T. W. Goodman, 1976, Degasification and production of natural gas from an air shaft in the Pittsburgh coalbed: U.S. Bureau of Mines Report of Investigations 8173, 23 p.

——, et al, 1973, Degasification of virgin Pittsburgh coalbed through a large borehole: U.S. Bureau of Mines Report of Investigations 7800, 27 p.

Fuex, A. N., 1977, The use of stable carbon isotopes in hydrocarbon exploration; Application of geochemistry to search for crude oil and natural gas: Journal of Geochemical Exploration, v. 7, p. 155-88.

Gan, H., S. P. Nandi, and P. L. Walker, Jr., 1972, Nature of porosity in American coals: Fuel, v. 51, p. 272-277.

Harris, L. A., 1979, Ultrafine structure of coal determined by electron microscopy: Prepr. Paper, American Chemical Society, Division of Fuel Chemicals, v. 24, n. 1, p. 210-217.

——, and C. S. Yost, 1978, Transmission electron microscope observations of porosity in coals: Fuel, v. 55, p. 233-236.

Hunt, J. M., 1979, Petroleum geochemistry and geology: San Francisco, W. H. Freeman and Co., 617 p.

International Union of Pure and Applied Chemistry, 1972, Manual of symbols and terminology for physics chemical quantities and units: London, Butterworth.

Karweil, J., 1969, Aktuelle Probleme der Geochemie der Kohle, in P.A. Schenk and I. Havenaar, eds., Advances in organic chemistry, 1968: Pergamon Press, Oxford, England, p. 59-84.

Kim, A.G., 1973, The composition of coalbed gas: U.S. Bureau of Mines Report of Investigations 7762, 9 p.

——, 1974, Methane in the Pittsburgh coalbed, Washington County, Pennsylvania: U.S. Bureau of Mines Report of Investigations 7969, 16 p.

——, 1975, Methane in the Pittsburgh coalbed, Greene County, Pennsylvania: U.S. Bureau of Mines Report of Investigations 8026, 10 p.

——, 1977, Estimating methane content of bituminous coalbeds from adsorption data: U.S. Bureau of Mines Report of Investigations 8245, 22 p.

——, 1978a, Experimental studies on the origin and accumulation of coalbed gas: U.S. Bureau of Mines Report of Investigations 8317, 18 p.

——, 1978b, Methane drainage from coalbeds: research and utilization, *in* Methane gas from coalbeds—development, production and utilization: MERC/SP-78/1, p. 13-17.

Kissel, F.N., 1972, The methane migration and storage characteristics of the Pittsburgh, Pocahontas No. 3, and Oklahoma Hartshorne coalbeds: U.S. Bureau of Mines Report of Investigations 7677, 22 p.

——, and R. J. Bielicki, 1972, An in-situ diffusion parameter for the Pittsburgh and Pocahontas No. 3 coalbeds: U.S. Bureau of Mines Report of Investigations 7668, 13 p.

——, and J. C. Edwards, 1975, Two-phase flow in coalbeds, U.S. Bureau of Mines Report of Investigations 8066, 16 p.

——, C. M. McCulloch, and C. H. Elder, 1973, The direct method of determining methane content of coalbeds for ventilation design: U.S. Bureau of Mines Report of Investigations 7767, 17 p.

Kuuskraa, V. A., and R. F. Meyer, 1980, Review of world resources of unconventional gas: IIASA Conference on Conventional and Unconventional World Natural Gas Resources, Laxenburg, Austria, June 30-July 4, 1980, p. 27-43.

Mahajan, O. P., 1982, Coal porosity, in R. A. Meyers, ed., Coal structure: New York, Academic Press, p. 51-86.

McCulloch, C. M., M. Deul, and P. W. Jeran, 1974, Cleat in bituminous coalbeds: U.S. Bureau of Mines Report of Investigations 7910, 15 p.

Moore, B. J., R. D. Miller, and R. D. Shrewsbury, 1966, Analyses of natural gases of the United States: U.S. Bureau of Mines Information Circular 8302, 144 p.

Moore, Jr., T. D., and M. G. Zabetakis, 1972, Effect of a surface borehole on longwall gob degasification (Pocahontas No. 3 coalbed): U.S. Bureau of Mines Report of Investigations 7657, 9 p.

Nandi, S. P., and P. L. Walker, 1971, Adsorption of dyes from aqueous solution by coals, chars, and active carbons: Fuel, v. 50, p. 345-366.

National Petroleum Council (W. N. Poundstone, chairman), 1980, Committee on unconventional gas sources, in National Petroleum Council, v. 2, Coal seams.

Patching, T. H., 1965, Variations in permeability of coal: Proceedings of Rock Mechanics Symposium, Toronto, p. 185-199.

Perry, J. H., G. N. Aul, and J. Cervik, 1978, Methane drainage study in the Sunnyside coalbed, Utah: U.S. Bureau of Mines Report of Investigations 8323, 11 p.

Rightmire, C. T., and C. W. Byrer, 1981, Coalbed methane exploration and development: U.N. Conference on Small Energy Resources, U.N. Institute for Training and Research, September, 1981, Los Angeles, California UNITAR/CF9/II/14.

Rosenberg, R. B., and J. C. Sharer, 1979, Natural gas from geopressured zones: Gas Research Institute Digest, v. 2, n. 2, p. 5.

Sharer, J. C., and J. J. Rasmussen, 1980, Position paper: unconventional natural gas: Gas Research Institute.

Stach, E., et al, 1982, Coal Petrology, 3rd edition: Berlin, Gebruder Borntraeger, 535 p.

Taber, J., et al, 1974, Development of techniques and the measurement of relative permeability and capillary pressure relationships in coal: U.S. Bureau of Mines Contract G0122006 Final Report.

Tissot, B., and D. H. Welte, 1978, Petroleum formation and occurrence: Berlin, Springer-Verlag, 538 p.

——, et al, 1974, Influence of nature and diagenesis of organic matter in formation of petroleum: Bulletin of the American Association of Petroleum Geologists, v. 58, n. 3, p. 498-506.

Toda, Y., et al, 1971, Micropore structure of coal: Fuel, v. 50, n. 2, p. 187-200.

Van der Sommen, J., et al, 1955, Chemical structure and properties of coal, XII; Sorptian capacity for methane: Fuel, v. 34, p. 444-448.

Geologic Overview, Coal Resources, and Potential Methane Recovery from Coalbeds of the Northern Appalachian Coal Basin — Pennsylvania, Ohio, Maryland, West Virginia, and Kentucky

M. A. Adams
G. E. Eddy
J. L. Hewitt
J. N. Kirr
C. T. Rightmire

The Northern Appalachian Coal Basin, as discussed in this report, covers approximately 43,700 sq mi in parts of Ohio, Pennsylvania, Maryland, West Virginia, and Kentucky. Calculations on the major coals in the basin showed an estimated total resource for this basin of 578 billion tons of coal. Data collected from gassy mines, DOE/MRCP test projects, and U.S. Bureau of Mines studies indicate that these coals contain an estimated 61 Tcf of coalbed methane. An elliptical area covering about 4,500 sq mi around the southwestern corner of Pennsylvania and in northern West Virginia has been designated as having the highest coalbed methane potential.

INTRODUCTION

The Northern Appalachian Coal Basin covers approximately 43,700 sq mi of the Kanawha, Allegheny Mountain, and New York sections of the Appalachian Plateau physiographic province of eastern Ohio, western Pennsylvania, northwestern West Yirginia, western Maryland, and northeastern Kentucky. This region is a dissected plateau that slopes toward the west from elevations of 2,100 ft near the New York–Pennsylvania border, 1,600 ft at the boundary with the Central Appalachian Coal Basin in central West Virginia, to approximately 1,200 ft at the western border in Ohio, and 700 ft in northeast Ohio. Local relief throughout this area is generally 200 to 400 ft in the Kanawha and New York sections. The eastern edge of the basin in Pennsylvania, western Maryland, and northeastern West Virginia (Preston County)—the Allegheny Mountain Section of the Appalachian Plateau—is considerably higher. Elevations of 3,000 ft, with local relief of 1,000 ft, are common in that area (Lobeck, 1948).

Regional Geologic Setting

The Northern Appalachian Coal Basin is a northeast-southwest–trending basin centered on and including the Dunkard Basin and lying entirely within the Appalachian Plateau physiographic province. The Kanawha section of the Appalachian Plateau is gently folded, with dips measured up to 30 ft per mi. The only exception is the sharp, asymmetric, north-south–trending Burning Springs Anticline of Pleasant, Wirt, and Ritchie counties, West Virginia, and Washington County, Ohio (Fig. 1). In the Allegheny Mountain section, the folding is more intense, with dips of 15° to 20° common. The folds are eroded and dissected, imparting a very rugged mountainous appearance to the region.

Some structures were present by the beginning of the Pennsylvanian period and exerted some control on the deposition of strata during Pennsylvanian and Dunkard time (Fig. 1) (Arkle, 1974; Williams and Bragonier, 1974). According to recent tectonic theory, at least part of the gently folded plateau may be a thin cover over more intensely folded and faulted strata or less intensely folded strata. Decollement zones are present in the less competent strata in the subsurface such as shales, evaporites, bentonites, and unconformities over which the younger formations slid to the west (Dennison, 1978). The western limits of the decollements and their effects are not clearly defined but appear to be limited within part of this study area by the Burning Springs Anticline of west-central West Virginia (Rogers, 1963; Gwinn, 1964).

The basin is bounded on the west, north, and northeast by the outcrop of Pennsylvanian-age strata (Fig. 2). In the southeast and south, the boundary is structural and follows the Rome Trough hinge line. The trough, a graben structure formed during the Lower Cambrian, remained active throughout the Paleozoic and affected sedimentation patterns in the Devonian, Mississippian, and Pennsylvanian. The Northern Appalachian Coal Basin was a gently subsiding basin trending

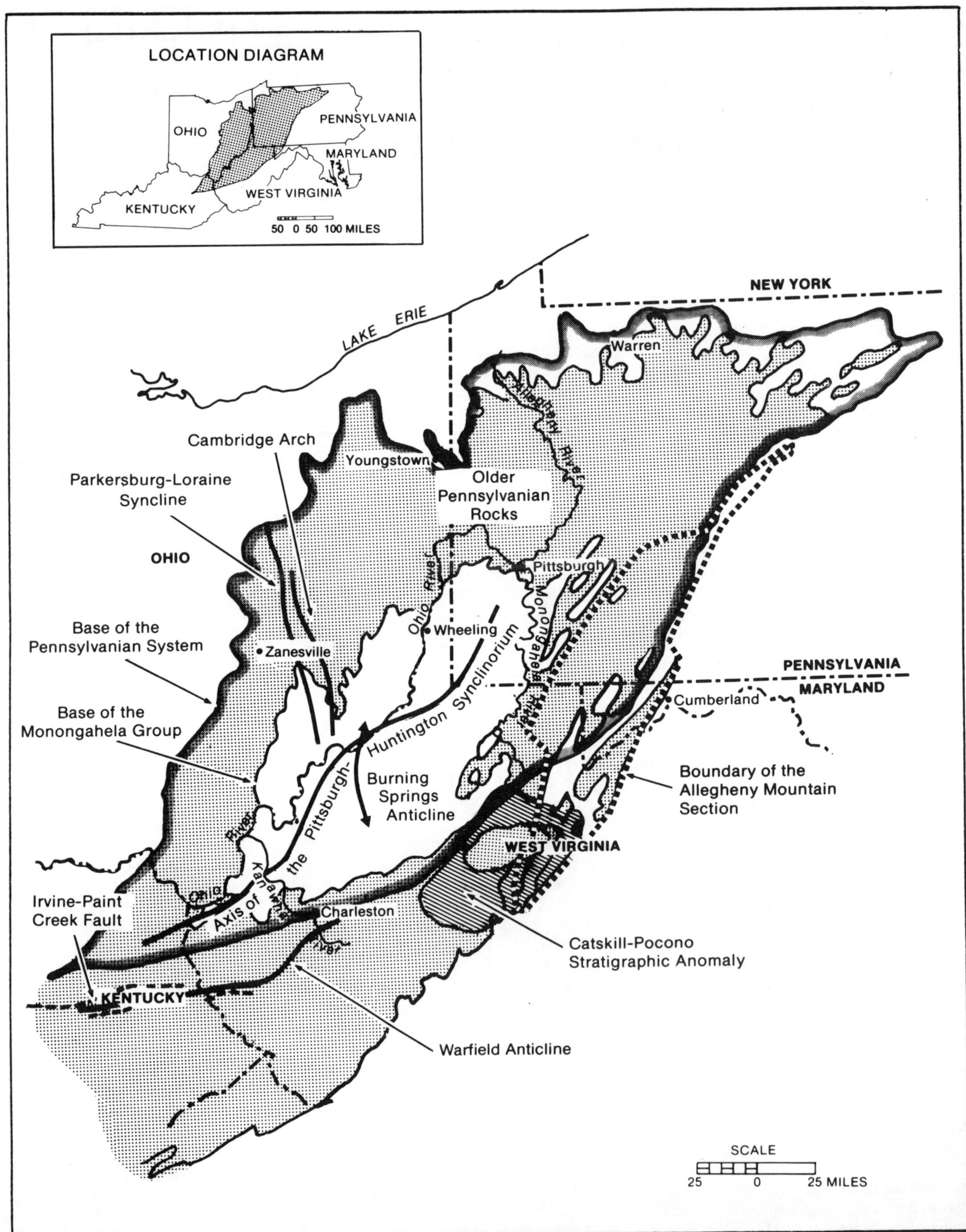

Figure 1—Geologic setting of late Paleozoic landscape. Arkle, 1974.

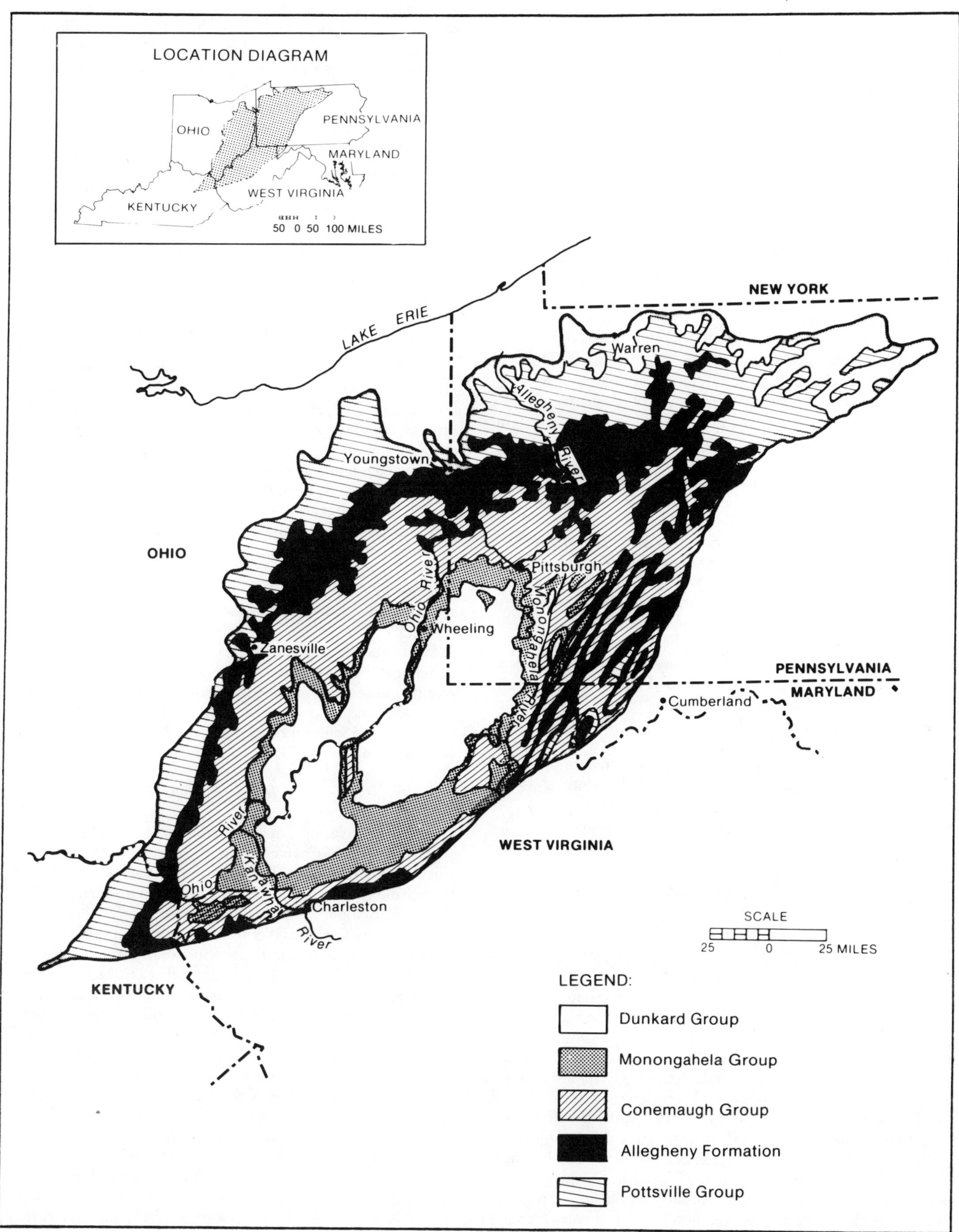

Figure 2—Geologic base map of the Northern Appalachian Coal Basin. From Barlow. Copyright © 1975 West Virginia Geological and Economic Survey. Used with permission.

north-northeast throughout Pennsylvanian and Permian time. To the southeast, the Central Appalachian Coal Basin was a rapidly but intermittently subsiding basin during the Pennsylvanian, and its character is unknown during the Permian (Arkle et al, 1979). The Northern Appalachian Coal Basin was a shelf-like area in comparison with the Central Appalachian Coal Basin during the time of major coal deposition.

Stratigraphy of Pennsylvanian Strata and the Dunkard Group

Pennsylvanian strata and the Dunkard Group in the Appalachian Basin have been the objects of much detailed study because all commercial coals in the basin are produced from these rocks.

In the eastern coal regions, the Pennsylvanian has been divided into four groups that are, in ascending order: the Pottsville, Allegheny, Conemaugh, and Monongahela. The Sharon conglomerate of northern Ohio and northwestern Pennsylvania and the Olean conglomerate of northern Pennsylvania and New York are now considered time-equivalent or older than the Pottsville Group. The Pocahontas and New River Formations of southern West Virginia and the time-equivalent Lee Formation of Kentucky and Virginia are also thought to be older than the Pottsville Group of central West Virginia. The Kanawha Formation of southern West Virginia may also be older than the Pottsville Group and is equivalent to the lower Breathitt of Kentucky. The Charleston Sandstone Group is time-equivalent to the Pottsville and part of the Breathitt Groups. The upper Breathitt is time-equivalent to the Allegheny Group.

The Dunkard Group—divided into the Waynesburg, Washington, and Greene Formations—was once thought to be Permian and is now partly assigned to the Pennsylvanian. The Waynesburg Formation is accepted by the present authors as Late Pennsylvanian and the Washington and Greene Formations as Early Permian. Thus, the Pennsylvanian in this study area is divided into the Pottsville, Allegheny, Conemaugh, and Monongahela Groups: the sandstone and conglomerate units older than the Pottsville; and the Waynesburg Formation is younger than the Monongahela Group (Fig. 3).

Pottsville Group

The Pottsville Group, older formations, and equivalents are principally nonmarine sandstone, conglomerate, and siltstone with some marine limestone and shale to the west. In the northern edge of this study area, these sediments were derived from earlier Paleozoic strata in the highlands to the north, but the bulk of the sediment was derived from the orogenic belts to the south and east (Meckel, 1967). The northwest and southeast alluvial fans merged to bury all earlier deposits except in a small area in west-central Pennsylvania. There, a cuesta-like outlier of the Burgoon Sandstone and Mauch Chunk Formation formed sandstone ridges along which the high-alumina Mercer flint clays formed (Edmunds and Berg, 1971). Further to the west and north at this time, alluvium from the Canadian Shield deposited coarse quartz conglomerate and sandstone, seen in the Sharon conglomerate of Ohio and northwestern Pennsylvania and the Olean conglomerate of Pennsylvania. Both are economically important sources of high-purity silica that range in thickness from 0 to 199 ft, but these tend to occur erratically owing to their deposition on a deeply dissected Mississippian surface. A sandy back-barrier system dominated the region at this time, preventing all but a few incursions of marine or terrestrial sediments.

Ferm (1974) and Donaldson (1974) have proposed that the Lee Formation of Kentucky and Virginia and the New River and Pocahontas Formations of West Virginia—both massive, pebbly, quartz conglomerates and sandstones—are lateral extensions of the regional beach and barrier system that developed before a southwest-prograding delta in Pottsville time. Although their combined thickness is over 1,800 ft at the southeastern Kentucky–southwestern Virginia border, in the Northern Appalachian Coal basin neither is over 500 ft thick nor contains minable coals. The Sewell coalbed of the New River Formation is believed to be time-equivalent to the Sharon coal at the northern edge of the basin. The overlying Breathitt Formation of Kentucky is a lithologically diverse unit that includes siltstone, clay shale, subgraywacke, ironstone, and limestone and interfingers with the Pottsville. It represents bay- and channel-fill sequences of the advancing delta and contains most of the minable coals in eastern Kentucky.

Within the study area, the Kanawha Formation is composed of shale with subgraywacke sandstones and thin, irregular coals and thins from a maximum thickness of 500 ft at the southeastern edge of the study area.

The Charleston Group of the southern part of the study area is time-equivalent to the Pottsville (Arkle, 1974). It is a subgraywacke with conglomeratic sandstone and some shale but no thick or traceable coals within the study area.

The Pottsville sediments that cover most of the study area are conglomeratic quartzose sandstones in the northeast, some marine limestone and shale to the northwest, and shale, marine limestone, and subgraywacke in the central area. The only Pottsville coal found throughout most of the study area is the Mercer seam.

Allegheny Formation

By Allegheny time, a regional transgression had created a shallow marine embayment trending northeast-southwest across the study area. A delta-plain environment was established on the northern delta and an alluvial plain environment lay on the delta advancing from the south and east; both are known to contain abundant coals (Ferm, 1974; Ferm and Cavaroc, 1969). On the eastern margin in the anthracite region in central Pennsylvania, the Llewellyn Formation is a thicker, coarser grained equivalent of the Allegheny, Conemaugh, and possibly Monongahela Groups owing to its proximity to the sediment source area. Most of the commercial anthracite coal is probably Alleghenian in age (Dennison, 1978). Throughout the rest of the study area, the Allegheny is traced from the base of the Brookville coal to the top of the Upper Freeport. It contains no red beds and few marine or brackish water deposits in the lower half. The upper half is entirely nonmarine and contains a fresh-water limestone that exists as the Hamden Limestone in Ohio and can be traced into West Virginia and Maryland. In the Allegheny Mountain region of West Virginia and Maryland, the Allegheny crops out and is somewhat thicker. In Kentucky,

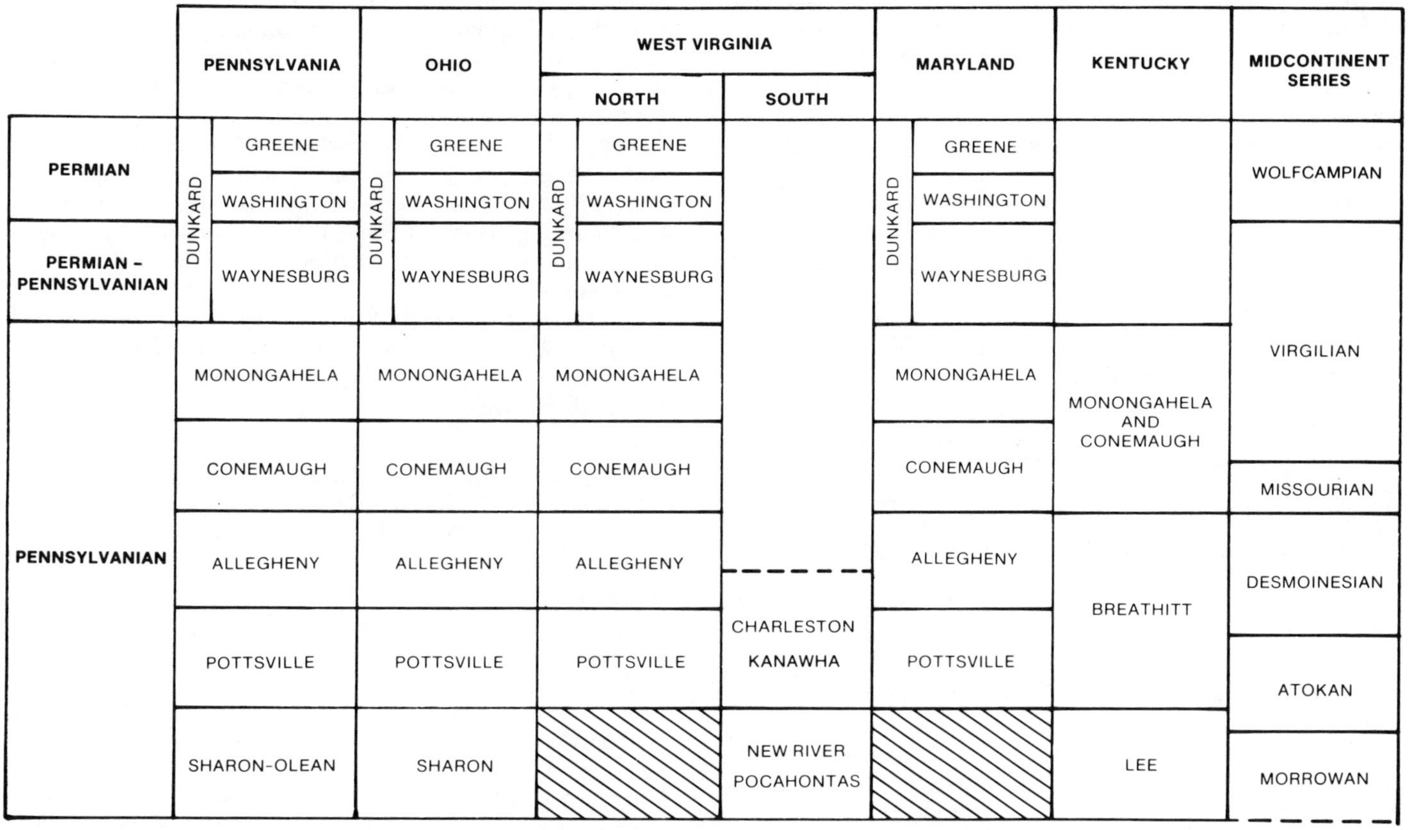

	PENNSYLVANIA	OHIO	WEST VIRGINIA NORTH	WEST VIRGINIA SOUTH	MARYLAND	KENTUCKY	MIDCONTINENT SERIES
PERMIAN	DUNKARD: GREENE	DUNKARD: GREENE	DUNKARD: GREENE		DUNKARD: GREENE		WOLFCAMPIAN
	DUNKARD: WASHINGTON	DUNKARD: WASHINGTON	DUNKARD: WASHINGTON		DUNKARD: WASHINGTON		
PERMIAN - PENNSYLVANIAN	DUNKARD: WAYNESBURG	DUNKARD: WAYNESBURG	DUNKARD: WAYNESBURG		DUNKARD: WAYNESBURG		VIRGILIAN
PENNSYLVANIAN	MONONGAHELA	MONONGAHELA	MONONGAHELA		MONONGAHELA	MONONGAHELA AND CONEMAUGH	
	CONEMAUGH	CONEMAUGH	CONEMAUGH		CONEMAUGH		MISSOURIAN
	ALLEGHENY	ALLEGHENY	ALLEGHENY	CHARLESTON	ALLEGHENY	BREATHITT	DESMOINESIAN
	POTTSVILLE	POTTSVILLE	POTTSVILLE	KANAWHA	POTTSVILLE		ATOKAN
	SHARON-OLEAN	SHARON		NEW RIVER POCAHONTAS		LEE	MORROWAN

Figure 3—Stratigraphic correlation chart for the Northern Appalachian Coal Basin.

the Upper Breathitt Formation is equivalent to the Allegheny Formation. The top of the Pottsville has not been positively identified there but is thought to occur at about the stratigraphic position of the Princess No. 5 coalbed (Rice et al, 1979). The most important coals in the Allegheny are the Brookville-Clarion; the Lower, Middle, and Upper Kittanning; and the Lower and Upper Freeport.

Conemaugh Group

In the Middle to Late Pennsylvanian, a series of rapid, eustatic marine transgressions encroached from the northwest across the basin, depositing a series of marine units within the lower delta plain sediments. As a result, the Conemaugh has few important coals but contains nodular iron ores deposited as siderite in the embayed margins of the sea. Although no longer mined, these ores are responsible for the early development of the iron industry in the Pittsburgh region. Within Pennsylvania, the Conemaugh includes all beds from the top of Upper Freeport coal to the bottom of the Pittsburgh. The Glenshaw Formation, Lower Conemaugh, contains four regional marine sequences. The topmost, the Ames marine event, is the most prominent and widespread and deposited the Mill Creek Limestone as far east as Wilkes-Barre, Pennsylvania (Chow, 1951). The Casselman Formation, Upper Conemaugh, is entirely nonmarine and contains a few scattered red beds and fresh-water limestones. It represents deposition of fluvial and deltaic sediments along an estuary margin and foreshadows the complete isolation of the Northern Appalachian Coal Basin that was to occur in Monongahela time. Further west in Ohio, the Conemaugh is the thickest of the four groups but contains almost no major coals. The lower units contain identifiable marine and brackish water fossils in thick sandstones and shales, and red mudstone and shale beds become increasingly abundant upward. To the south in West Virginia and Maryland, the Conemaugh units are transitional between marine and terrestrial but generally contain more coal, particularly in the Allegheny Mountain region. Conemaugh outcrops in northeastern Kentucky have been measured at 360 ft thick (Rice et al, 1979). Dominantly siltstone and shale, the Conemaugh deposits are various shades of red, green, and grey and may contain marine limestones, especially near the base. Because of their lithologic similarity, the Conemaugh and Monongahela are not differentiated in Kentucky, and the Conemaugh is often projected from other stratigraphic marker horizons or arbitrarily placed at the base of persistent red shales.

Monongahela Group

To most geologists, the Monongahela Group includes all strata from the base of the Pittsburgh coalbed to the top of the Waynesburg coalbed. Berryhill and Swanson (1962) divided it into two formations: the Pittsburgh, from the base of the Pittsburgh coalbed to the base of the Uniontown coalbed, and the Uniontown, from the base of the Uniontown coalbed to the base of the Waynesburg coalbed. Berryhill and Swanson's nomenclature will be used in this report.

During the Monongahela, the entire northeastern

Appalachian Basin is thought to have been a series of lakes and deltaic lake-margin environments with no marine sedimentation. These conditions persisted during deposition of the Monongahela Group and Dunkard Groups (Berryhill, Schweinfurth, and Kent, 1971; Donaldson, 1969, 1974), and some of the most economically important coals in this region were formed in the vast lakes existing at that time. To the west in Ohio, the Monongahela is thinner and the major coals are concentrated in the lower half of the group (Collins, 1979). To the south, this group records the transition from nonmarine red beds to swampy lake deposits in West Virginia and Maryland. A thick accumulation of fresh-water limestones near Wheeling, West Virginia, suggests that a brackish bay or lake persisted there for some time; this lacustrine limestone and associated thin mudstones tend to disappear to the southwest (Arkle et al, 1979). In northeastern Kentucky, the Monongahela and Conemaugh have a combined thickness of just over 500 ft and are not usually differentiated. Shallow lacustrine environments in Monongahela time were probably fed by streams flowing to the northwest (Wanless, 1975; Arkle, 1974).

Dunkard Group

The age of the Dunkard Group is an enigma; it is considered Pennsylvanian or Permian, or both, based on megafloral, spore, invertebrate, and vertebrate evidence (Barlow, 1975). Throughout all of the basin, the Dunkard lies conformably upon the Monongahela. Gillespie and Clendening (1969) concluded that their "current study of fossil spores and leafy remains substantiates a Late Pennsylvanian, rather than a Permian, age for even the youngest Dunkard strata" (Clendening, 1969). This report will use the classification of Berryhill and Swanson (1962), which defines the Waynesburg Formation that includes strata from the base of the Waynesburg coalbed to the base of the Washington coalbed (Berryhill, Schweinfurth, and Kent, 1971) as Late Pennsylvanian and Early Permian. The Washington and Greene Formations are here considered Early Permian and include strata from the base of the Washington coalbed to the base of the Greene Formation. The Greene Formation includes strata above the upper limestone member of the Washington Formation.

All three formations in the Dunkard Group are lithologically variable and contain few persistent mappable rock units. Shale is the most abundant rock type, grading from a red facies in the south through a transitional facies to become grey in the northern basin. Gypsum found within the shales suggests arid conditions at the time of deposition (Stauffer and Schroyer, 1920) and cyclic sedimentation has been observed. Most of the Dunkard cycles were probably caused by changing loci of deltaic deposition, but certain marker zones are quite widespread and may have been caused by sea-level changes that altered the base level.

Dunkard coals are thicker and more abundant in the northern part of the basin; in West Virginia and Maryland the Dunkard is over 1,100 ft thick (Arkle et al, 1979).

COAL RESOURCE

The coal-bearing strata of the Northern Appalachian Coal Basin occur in rocks of Pennsylvanian and Lower Permian age. In West Virginia, Ohio, Maryland, and Pennsylvania, the groups in which the principal coalbeds are located are the Pottsville, Allegheny, Conemaugh, Monongahela, and Dunkard. In the Princess reserve district of northeastern Kentucky, the main coal-bearing formations are the Middle Pennsylvanian Breathitt and the Upper Pennsylvanian Conemaugh. The Lower Pennsylvanian Lee Formation, a major coal-bearing series in the adjoining districts of Kentucky, is present in the Princess reserve, but the coalbeds are either too thin or too impersistent to have any commercial value. Original reserves of the Lee coals in this district are estimated at 9.7 million tons, with more than half occurring in beds less than 18 inches thick (Huddle et al, 1963). Therefore, the Lee coals will not be considered further in this report. Stratigraphic columns for the coal-bearing series of each state in the Northern Appalachian Coal Basin are shown in Figures 4 to 8.

Chemical Characteristics of the Coals

Coal rank changes in this basin from east to west. Along the Allegheny Front of Pennsylvania in Cambria and Somerset counties, the coal is low-volatile bituminous (Fig. 9). West of this area, an elongated band of medium-volatile bituminous coal is adjacent to and parallels the low-volatile coals and the Allegheny Front. This elongated narrow band of medium-volatile coals extends from the northeastern corner of the basin into the tricorner area of Pennsylvania, West Virginia, and Maryland. West of this band, coal rank is in the high-volatile bituminous range. The easternmost part is primarily high-volatile A bituminous with some high-volatile B bituminous occurring along the Ohio–West Virginia state line. Most Ohio coals fall in the high-volatile C or B rank (Brant and DeLong, 1960).

The sulfur and ash contents of the coals also change across the basin. The lowest sulfur contents (less than 1.5%) occur along the eastern side of the basin and increase to over 3% in Ohio. The ash content of the coals is also the lowest (under 6%) along the eastern side of the basin and increases to over 12% near its center.

Coal-Bearing Groups

Pottsville Group

In southeastern West Virginia, the Pottsville Group and older formations comprise a major coal-bearing sequence. The group thins to the northwest, and at points along the Ohio River near the Burning Springs Anticline, it is only several hundred feet thick (Cross and Schemel, 1956a). Further north in Pennsylvania and northwest in Ohio, the group is absent for the most part. Depositional conditions during Pottsville time resulted in a sporadic occurrence of coals with highly variable thicknesses and limited lateral extents. The main coals of this group are, in ascending order: the Sharon, Quakertown, and Lower and Upper Mercer.

Sharon Coal—The Sharon coal occurs sporadically and is known mainly from outcrops in northeastern and southeastern Ohio and northwestern Pennsylvania. The coal is related to a northern sediment source that controlled its rather localized occurrence along the northern and western part of the basin. The coal varies in thickness from a few inches to about 4 ft, but has limited areal extent and is cut out by channeling in many areas (Brant and DeLong, 1960). The Sharon was extensively mined in the past but is now considered to be depleted (Stout, 1946). Further to the southeast, in West Virginia, the Sewell coal—the lateral equivalent of the Sharon—is extensively developed. A discussion of the Sewell coal can be found in the Central Appalachian Basin Report (Adams, 1982).

Quakertown, and Lower and Upper Mercer Coals—The Quakertown and the Lower and Upper Mercer coals have characteristics similar to the Sharon. They have been mined locally but do not have adequate thickness, areal distribution, or quality to be of economic importance. Two or three thin coals occur in Pennsylvania, along the Beaver River, and in the vicinity of the Mercer horizon. In other areas, one seam may exist at that same horizon in Ohio and Kentucky (Ashley, 1928). These data suggest that the Pottsville Group will be insignificant in contributing to the coalbed methane resource of this basin.

Allegheny Group

In general, the group's thickness and number of coal seams increase from west to east in the basin (Ashley, 1928). The average thickness of this group in Maryland exceeds 250 ft; in Ohio, it is about 200 ft, and along the West Virginia side of the Ohio River Valley, the section is 200 to 250 ft in thickness (Cross and Schemel, 1956a). In Pennsylvania, this group contains more minable coalbeds than any other, and 50% of the bituminous coal production in this state is from Allegheny coals (Sisler, 1928). In ascending order, the most important Allegheny coals are: Brookville; Clarion; Lower, Middle, and Upper Kittanning; and Lower and Upper Freeport.

Brookville and Clarion—The Brookville coal in Pennsylvania ranges from a few inches to 8 ft thick locally (Sisler, 1928). Known in some areas as the "Dirty A," it is not mined on a large scale because of its high ash content. The coal lies along the southwest corner of the state except along the axes of the Laurel Ridge and Chestnut Ridge anticlines where it has been eroded. The most important and thickest coal occurs in Clarion and Somerset counties, where the seam may reach 5 ft in thickness (Ashley, 1928). Of Ohio's ten most valuable coalbeds, the Brookville ranks ninth (Stout, 1946). The coal crops out along the western part of the basin in Ohio, where it averages 1 to 2.5 ft in thickness. Estimated original reserves of this coal in Ohio are 450 million tons (Brant and DeLong, 1960).

In West Virginia, the Brookville coal is not recognized, although the overlying Clarion coal is present in minable quantities along the Pennsylvania–West Virginia state line. In the West Virginia counties adjacent to Pennsylvania, the Clarion is a multiple-bedded seam ranging from 2 to 10 ft in thickness (Headlee and Nolting, 1940). The minable extent of this coal lies mainly along the basin's southeastern boundary in Preston, Barbour, Upshur, and Clay counties, West Virginia. The interval between the Clarion and Brookville coals is about 25 ft along the basin's north boundary and thins to about 15 ft at the Pennsylvania–West Virginia state line (Ashley, 1928). The Clarion commonly is split into several seams so the Brookville in West Virginia may be a part of the multiple-bedded Clarion coal. In Pennsylvania and Ohio, the

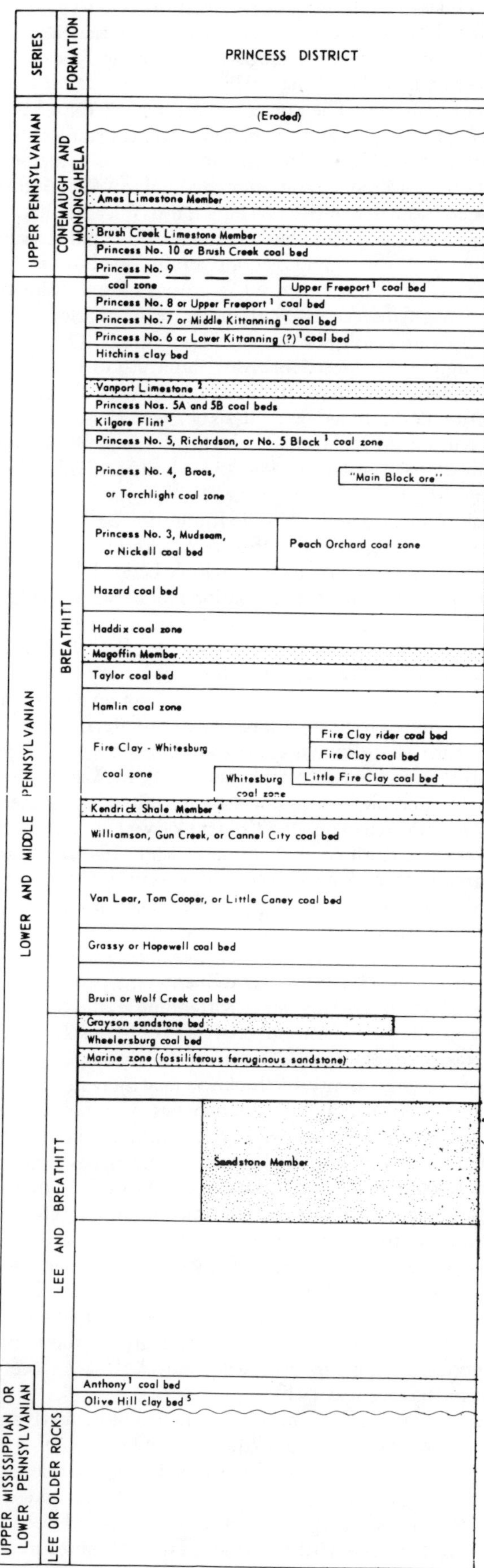

Figure 4—Stratigraphic column of the coal seams in the Princess Reserve District of eastern Kentucky. Rice and Smith, 1980.

Clarion is less well developed, ranging from zero to 4 ft in thickness. In Clarion County, Pennsylvania, it reaches its maximum thickness and averages 4 ft, attaining a local thickness of up to 7 ft (Sisler, 1928).

Lower Kittanning—The Lower Kittanning coal (No. 5 block, Princess 6 in Kentucky) is a persistent and important coal seam in the Allegheny Group. In some areas, the coal is double- or triple-bedded. Minable reserves of this coal in West Virginia have been estimated at over 10 billion tons (Headlee and Nolting, 1940), and Ohio has an estimated original reserve of 9.9 billion tons (Brant and DeLong, 1960). The Lower Kittanning is thickest in central West Virginia and thins very gradually in all directions. With minor local variations, thicknesses are as follows: central West Virginia, 12 ft; northern West Virginia, 4 ft; western Pennsylvania, 2.5 to 4 ft; Ohio, 2 to 4 ft; Maryland, less than 3 ft; and southern West Virginia, 3 to 7 ft (Averitt, 1975). Based on analysis of 292 samples, 84% of the Lower Kittanning is classified as high-volatile A bituminous; 12%, medium-volatile; and 4%, low-volatile coal (Headlee and Nolting, 1940).

Middle Kittanning—The Middle Kittanning coal (No. 6, Princess 7 in Kentucky), like the Lower Kittanning, is widespread across the basin and multiple bedded. In Clearfield County, Pennsylvania, there is an interval between the Upper and Lower Kittanning coals in which coal occurs at not less than three, and probably five, horizons (Ashley, 1928). In Pennsylvania and West Virginia, the coal is not generally thick enough to be of any commercial value except on a local basis. The coal appears to be best developed along the northern part of the basin, where an average thickness of 4.5 ft occurs in several counties (Sholes and Skema, 1974). In West Virginia, the coal is multiple bedded and attains a minable thickness and purity over an area of about 320 sq mi located in Clay, Nicholas, and Webster counties (Headlee and Nolting, 1940). Sections of coal up to 10 ft thick have been measured in Fayette County, West Virginia. In contrast, the Middle Kittanning in Ohio is persistent, with an average thickness of about 4 ft. From 1956 to 1979, 205 million tons of this coal were mined, which comprised 21% of Ohio's total production (Keystone, 1981).

Upper Kittanning—The Upper Kittanning coal is of minor importance in the bituminous coal fields of Pennsylvania. Best developed along the basin's eastern boundary and Chestnut Ridge in Fayette County, its thickness may increase to 12 to 20 ft by the presence of one or more benches of cannel coal. Most of the state's cannel coal appears to be at this horizon (Ashley, 1928). In West Virginia, the Upper Kittanning has a minable area of 1,400 sq mi and an estimated reserve of 4 billion tons (Headlee and Nolting, 1940). This minable area extends along the basin's southeast boundary from Preston to Clay counties, West Virginia. Along the northern West Virginia panhandle and in Ohio, scattered occurrences are generally less than 1 ft in thickness (Brant and DeLong, 1960). Based on analyses of 44 samples, 80% of this coal is classified as high-volatile A bituminous; 16%, medium-volatile; and 4%, low-volatile (Headlee and Nolting, 1940).

Lower Freeport—The Lower Freeport coal in Pennsylvania is variable and may reach a thickness of 16 ft but thins to a feather edge in a short distance. Data on this seam suggest that instead of a single coal, there may be several closely overlapping coal horizons in the stratigraphic interval 30 to 65 ft below the Upper Freeport Coal (Ashley, 1928). The coal mined in

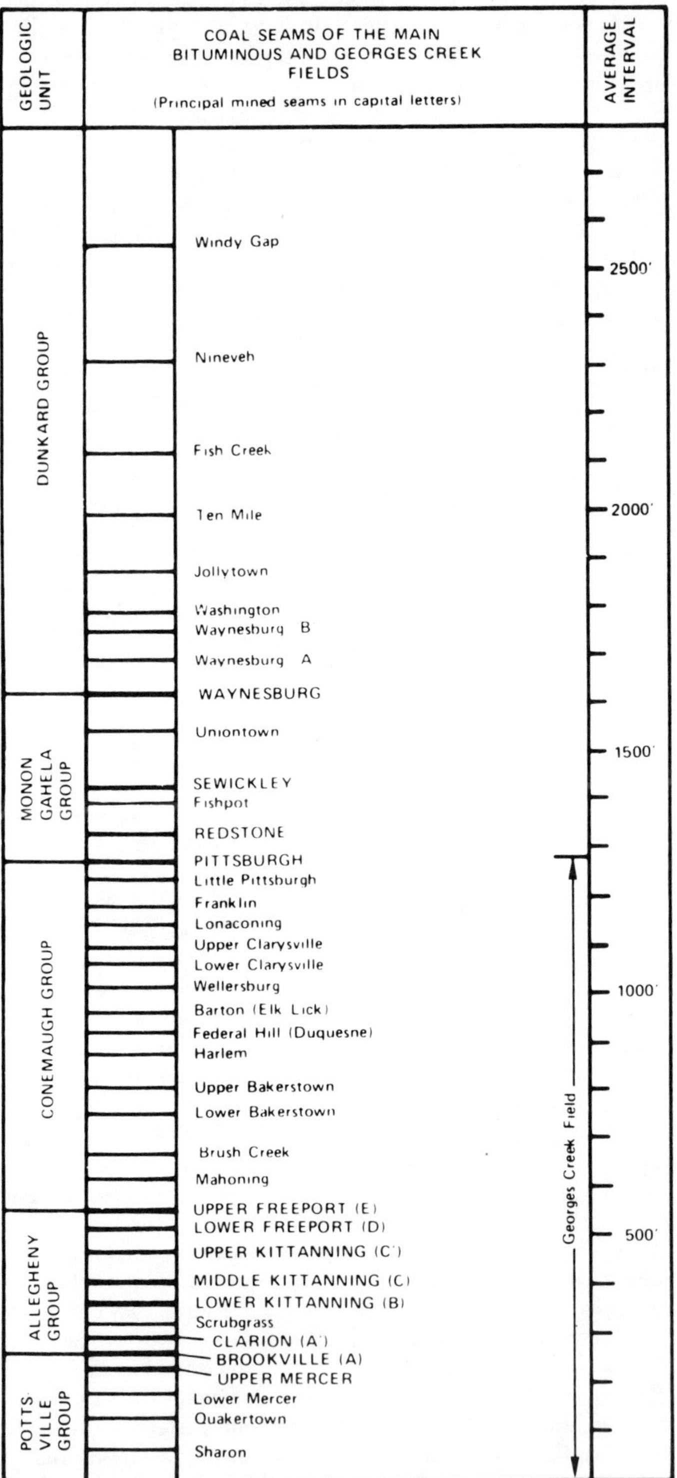

Figure 5—Stratigraphic column of coal seams in the main bituminous field of Pennsylvania. From Keystone Coal Industry Manual. Copyright © 1981 McGraw-Hill Mining Publications, New York. Used with permission.

BITUMINOUS COAL SEAMS IN WEST VIRGINIA

Geologic Units	Coal Map Index	Geologic Names: Minable Coal Seams (Thin seams, generally unminable)	Thickness Feet	Interval Feet	Total from Top Feet
Permian or Pennsylvanian System 1000' to 1180'; Dunkard Group 1000' to 1180'	Column 1	Upper Proctor Sandstone (Crest of Dunkard Group)			
		Windy Gap			
		Gilmore			
		Nineveh "A"			
		Nineveh		825 to 1025	
		Hostetter			
		Fish Creek			
		Dunkard			
		Jollytown			
		Hundred			
		Washington "A"			
		WASHINGTON	2 to 5		1030
		Little Washington			
		Waynesburg "A"		105 to 150	
Monongahela Group 250' to 410'		WAYNESBURG	3 to 5		1185
		Little Waynesburg		71 to 97	
		UNIONTOWN	1 to 3		1285
		Lower Uniontown		160 to 195	
		SEWICKLEY	3 to 5		1485
				40 to 55	
		REDSTONE	3 to 5		1545
				20 to 37	
		PITTSBURGH	5 to 8		1590
Conemaugh Group 500' to 800'	Column 2	Morantown			
				41 to 69	
		LITTLE PITTSBURGH	1 to 6		1665
		Second Little Pittsburgh		0 to 68	
		Franklin Rider			
		LITTLE CLARKSBURG	2 to 7		1740
		Normantown			
		Lower Hoffman			
		Upper Clarysville			
		Lower Clarysville		40 to 196	
		Wellersburg Rider			
		Wellersburg			
		Barton Rider			
		ELK LICK	2 to 4		1940
		West Milford			
		Federal Hill		10 to 98	
		Duquesne			
		HARLEM	1/2 to 2		2040
				56 to 95	
		UPPER BAKERSTOWN	1 to 5		2140
				45	
		BAKERSTOWN	2 to 5		2190
				89 to 74	
		BRUSH CREEK	1/2 to 1		2265
				50 to 74	
		MAHONING	0 to 6		2345
				35 to 45	
		UPPER FREEPORT	2 to 5		2395
Allegheny Formation 175' to 330'	Column 3			40 to 67	
		LOWER FREEPORT	1 to 3		2465
				40 to 75	
		Upper Kittanning Rider			
		UPPER KITTANNING	3 to 5		2545
				40 to 62	
		MIDDLE KITTANNING	0 to 8		2615
				15 to 47	
		LOWER KITTANNING No. 5 BLOCK	3 to 8		2670
				30 to 46	
		CLARION	1 to 4		2720
Pennsylvania System 1175 to 5300; Pottsville Group 250 to 3850; Kanawha Formation 150 to 2100'	Column 4	Tionesta			
				46 to 88	
		UPPER MERCER	0 to 2		2810
				20 to 35	
		STOCK-LEWISTON	3 to 10		2855
				5 to 125	
		COALBURG	3 to 10		2990
		Little Coalburg			
				45 to 95	
		BUFFALO CREEK	0 to 5		3090
				40 to 90	
		WINIFREDE	1 to 10		3190
		Lower Winifrede		35 to 47	
		CHILTON "A"	0 to 3		3240
				40 to 82	
		Chilton Rider			
		CHILTON	3 to 8		3330
		Little Chilton			
				30 to 86	
		HERNSHAW	2 to 4		3420
				0 to 26	
		DINGESS	0 to 4		3450
	Column 5			20 to 62	
		WILLIAMSON	1 to 8		3520
				40 to 94	
		CEDAR GROVE	3 to 6		3620
				0 to 70	
		LOWER CEDAR GROVE	3 to 5		3695
		Alma "A"		20 to 50	
		ALMA	2 to 5		3750
				49 to 76	
		Little Alma			
		PEERLESS (CAMPBELL CREEK)	2 to 4		3830
				0 to 25	
		CAMPBELL CREEK (No. 2 GAS)	2 to 10		3865
		Lower Campbell Creek			
		Powellton "A"		2 to 71	
		POWELLTON	0 to 4		3940
				30 to 58	
		MATEWAN	0 to 5		4003
	Column 6			20 to 86	
		Eagle "A"			
		EAGLE	1 to 6		4095
				0 to 32	
		BENS CREEK	0 to 3		4130
				30 to 77	
		LITTLE EAGLE	1 to 3		4210
				0 to 21	
		CEDAR	0 to 4		4235
				80 to 117	
		Little Cedar			
		LOWER WAR EAGLE	1 to 3		4355
				75 to 95	
		GLENALUM TUNNEL	0 to 5		4455
				60 to 91	
		Gilbert "A"			
		GILBERT	1 to 4		4550
				70 to 136	
		Douglas "A"			
		DOUGLAS	0 to 4		4690
				80 to 121	
		LOWER DOUGLAS	2 to 4		4815
New River Formation 100 to 1030	Column 7	Iaeger "B"		110 to 175	
		Iaeger "A"			
		IAEGER	2 to 5		4995
		Lower Iaeger			
				120 to 178	
		CASTLE	0 to 2		5175
				60 to 80	
		SEWELL "B"	0 to 5		5260
		Sewell "A"		0 to 80	
		SEWELL	3 to 10		5350
				40 to 60	
		WELCH	0 to 5		5415
				60 to 108	
		Little Raleigh "A"			
		LITTLE RALEIGH	0 to 2		5525
				40 to 75	
		Beckley Rider			
		BECKLEY	3 to 10		5610
				0 to 100	
		FIRE CREEK	0 to 5		5715
		Little Fire Creek			
				50 to 100	
Pocahontas Formation 0 to 720'	Column 8	No. 9 POCAHONTAS	2 to 5		5820
				0 to 28	
		No. 8 POCAHONTAS	0 to 2		5850
				30 to 67	
		No. 7 POCAHONTAS	0 to 3		5920
				60 to 105	
		No. 6 POCAHONTAS	0 to 5		6030
				50 to 72	
		No. 5 POCAHONTAS	0 to 5		6107
				0 to 20	
		No. 4 POCAHONTAS	0 to 8		6135
				40 to 70	
		No. 3 Pocahontas Rider			
		No. 3 POCAHONTAS	5 to 15		6220
				40 to 68	
		No. 2 POCAHONTAS	0 to 2		6290
		No. 1 Pocahontas			
		Landgraff			
		Keystone			
		Simmons			
				120 to 280	
		Squire Jim			
		Base of Pennsylvania System			6570

Figure 6—Stratigraphic column of the bituminous coal seams in West Virginia. From Headlee and Nolting. Copyright © 1940 West Virginia Geological and Economic Survey. Used with permission.

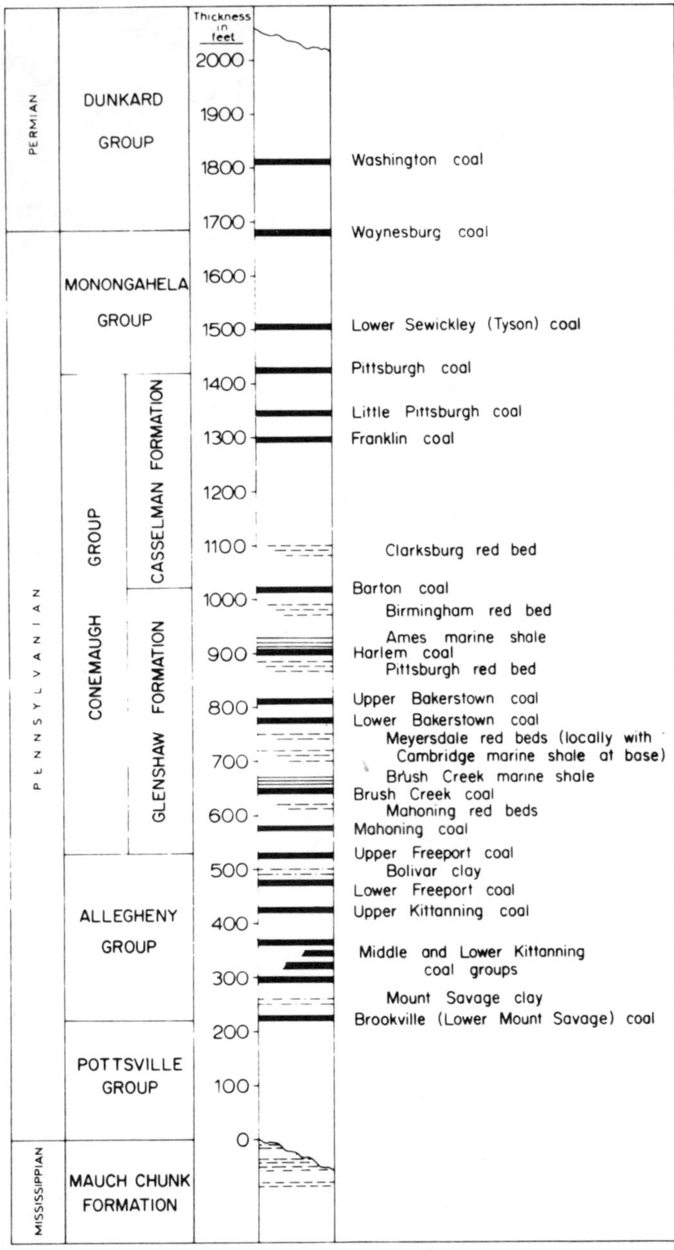

Figure 7—Stratigraphic column of coal seams in western Maryland. From Weaver, Coffroth, and Edwards, 1976. Used with the permission of the Maryland Geological Survey.

Clearfield and Jefferson counties is best developed in these counties and along the basin's northeastern boundary. Toward western Pennsylvania, the coal becomes thin and irregular. In West Virginia, the Lower Freeport is widely scattered along the basin's southeastern boundary and northern panhandle where it is also multiple bedded and variable. In Ohio as well, the Lower Freeport is patchy and averages 1 ft in thickness with the thickest coal in Harrison and Jefferson counties on the Ohio River adjacent to the northern panhandle of West Virginia.

Upper Freeport—The Upper Freeport coal lies at the top of the Allegheny Group and is one of its most persistent and valuable coals. Originally deposited extensively across the basin, it has been removed in some areas by uplift and erosion (Ashley, 1928). In Pennsylvania, it varies from a few inches up to 7 ft thick. In a small area of northeastern Allegheny County, the Upper Freeport is 2 to 10 ft thick and known locally as the "Thick Freeport." In general, the bed averages 6 to 8 ft thick across small areas, 3 to 6 ft in larger areas, and less than 3 ft in half the area it underlies (Ashley, 1928). In West Virginia, the coal is generally multiple bedded and up to 12 ft thick in the northern part of the state. The main mining area is along the basin's southeast boundary from Preston to Clay counties. In the central and western part of the state it thins, becomes patchy, and commonly loses its multiple-bedded character (Headlee and Nolting, 1940). In Ohio, the Upper Freeport is one of the most valuable coalbeds and is mined extensively in the counties in the southeastern part of the state. The coal thickness is variable and ranges up to 7 ft along the Ohio River adjacent to the West Virginia panhandle but generally averages 2.5 to 3.5 ft over most of southeastern Ohio. Fifty percent of this coal is classified as high-volatile A bituminous; 37%, medium-volatile; and 13%, low-volatile coal (Headlee and Nolting, 1940).

Conemaugh Group

This group decreases in thickness from east to west. Along the eastern basin, it ranges from 500 to 950 ft but thins in the west to 400 ft or less in Ohio. The thickest development lies in Somerset County, Pennsylvania, and the Georges Creek Basin in western Maryland (Ashley, 1928).

Although it may contain up to 30 coalbeds, the group is generally distinguished from the Allegheny Group below and the Monongahela Group above by the relative unimportance of those coals and by an abundance of red shales. In West Virginia, only eight of the coals in this group attain minable thickness and purity. The total original estimated reserve for those coals is slightly less than 3.5 billion tons or about 3% of the state's total coal reserves (Headlee and Nolting, 1940). In Ohio, only four seams are considered in the estimated original reserves of bituminous coals and constitute about 1.3 billion tons or about 2.9% of the state's estimated original reserves. Of that amount, 1.8% is contained in the Anderson (Bakerstown) seam (Brant and DeLong, 1960).

The Conemaugh coals in Pennsylvania are similar to those in Ohio and West Virginia: thin, discontinuous, and shaly with a high ash content. The Bakerstown coal averages 3 to 4 ft thick in Somerset County, Pennsylvania, western Maryland, and eastern West Virginia and has been mined locally. Because they are thin and erratic, these coals will not be considered in determining the total gas content of the basin.

Monongahela Group

The thickness of this group ranges from 200 ft in Ohio, to 400 ft in West Virginia (Ashley, 1928). The major coals in this group are, in ascending order: the Pittsburgh, Redstone, Sewickley, and Uniontown.

Pittsburgh Coal—The Pittsburgh coal is the best known and possibly most valuable coal in this area. In the type locality in Allegheny County, Pennsylvania, the bed is split into a roof and lower bench. Between the two divisions is a clay parting ranging from 1 inch to 3 ft thick, and containing thin bands of coal (Sisler, 1928). The roof division ranges from 2 inches to 8 ft thick and has a high ash content; the lower division ranges from

3.5 to 9 ft thick and is usually divided into benches by the presence of shale and bone partings.

The Pittsburgh coal extends over an area of 6,000 sq mi in West Virginia, Ohio, Maryland, and Pennsylvania and attains its maximum thickness in western Maryland and northeastern West Virginia. In Mineral and Preston counties, West Virginia, thicknesses up to 22 ft thick have been recorded (Averitt, 1975), although the bed averages 7 ft in Pennsylvania and 5 ft in Ohio and West Virginia. This coal has been extensively shaft and strip mined (mainly around the outcrop areas) and has essentially been mined out in Maryland and in Mineral County, West Virginia (Ashley, 1928). As of January 1974, the bed had yielded approximately 9 billion tons of coal (about 35% of the total cumulative production of the Appalachian Basin) and 21% of the total cumulative production of the United States to that date (Averitt, 1975). Of this coal, 97.5% is high-volatile A bituminous and 2% is low-volatile bituminous (Headlee and Nolting, 1940).

Redstone Coal—The Redstone (Pomeroy) coal occurs 30 to 85 ft above the Pittsburgh coal. Although thin and poorly developed over most of the area, it reaches a thickness of 3 to 5 ft in southwestern Pennsylvania, and around 5 ft in West Virginia (Headlee and Nolting, 1940). In Ohio, the coal is of similar thickness, reported to be 3.5 ft thick near Pomeroy (Meigs County), and is relatively uniform through the county. To the southeast, it thins to 2.5 ft, but several thick deposits are known (Brant and DeLong, 1960).

Sewickley Coal—The Sewickley coal (Meigs Creek Seam in Ohio) is better developed than the underlying Redstone coal. Along the northern outcrop in Washington County, Pennsylvania, the Sewickley is thin (4 to 8 inches) but thickens to 5 to 6 ft in southwestern Greene County (Ashley, 1928). In the Wheeling, West Virginia, area, the coal ranges from 3.5 to 5 ft (Cross and Schemel, 1956b). The Sewickley has been commercially mined in Marion, Monongalia, and Marshall counties in West Virginia. In Ohio, the Sewickley (Meigs Creek) is limited in extent, with the bulk of the coal in the counties adjacent to the Ohio River. Coal thickness is variable but reaches 5 ft in several areas. Ninety-three percent of this seam is considered high-volatile A bituminous; 1%, medium-volatile; and 6%, low-volatile (Headlee and Nolting, 1940).

Uniontown Coal—The Uniontown Coal is 3 ft thick or less in Washington and Fayette counties, Pennsylvania, thins to the south and southwestward, and in Greene County rarely exceeds 1 ft. In West Virginia, along the Ohio River in Wetzel and Tyler counties, the coal reaches 2 to 3 ft (Headlee and Nolting, 1940). Across the river in Ohio, the Uniontown is well exposed but is variable in thickness and quality (Brant and DeLong, 1960). The coal presently has no commercial value but may hold future reserves.

Dunkard Group

In the type locality along Dunkard Creek in Greene County, Pennsylvania, this group is 1,100 to 1,200 ft thick and consists of shale, shaly sandstone, limestone, and coal. Most of the coals in this group are thin and erratic (Ashley, 1928), and only two coals are of economic value: the Waynesburg and the Washington. Because the Washington coal lies near the surface, it has probably retained little gas and it will not be considered further in this report.

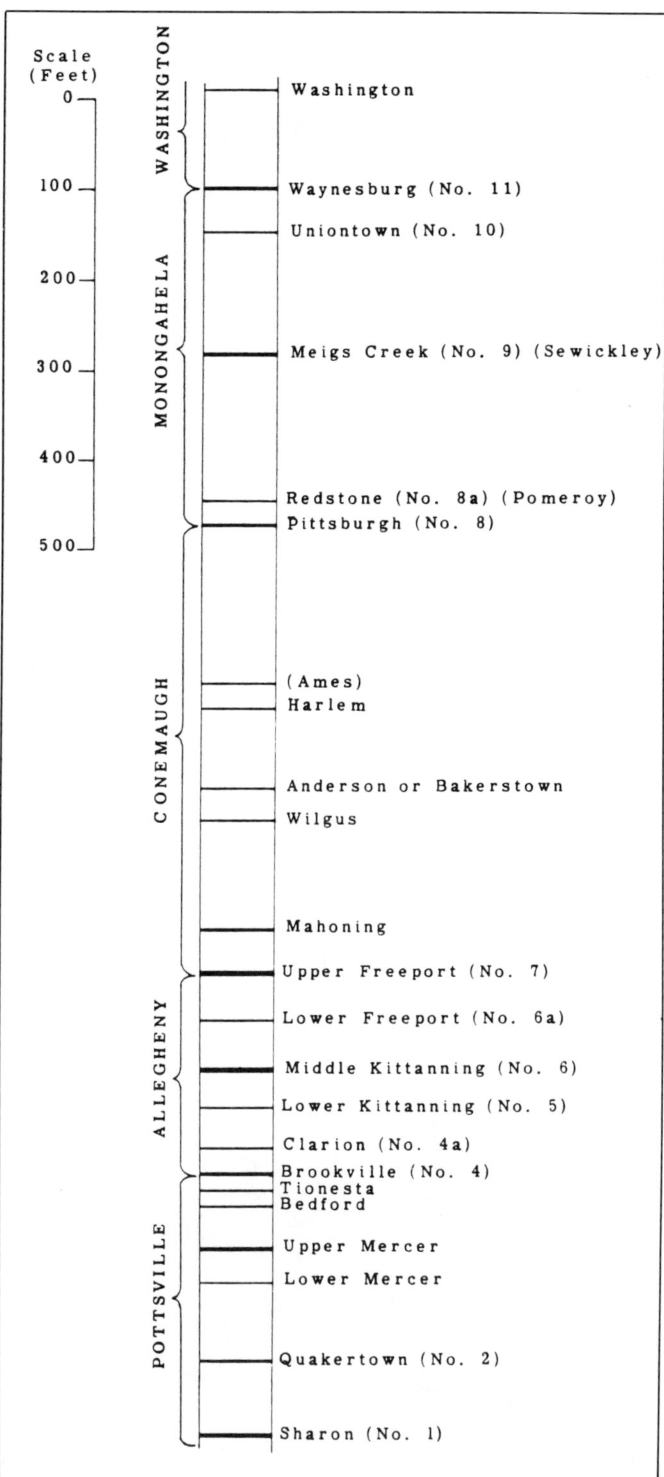

Figure 8—Stratigraphic column of the bituminous coal seams in Ohio. From Brant and DeLong, 1960. Used with the permission of the Ohio Division of Geological Survey.

Waynesburg Coal—The Waynesburg coal is split by a thick parting almost everywhere and is from 5 to 10 ft thick, of which one-fourth to one-half consists of clay partings. In Fayette County, Pennsylvania, the coal has an average thickness of 3.5 ft. To the west, the thickness increases but includes clay partings, and to the south in Monongalia County, West

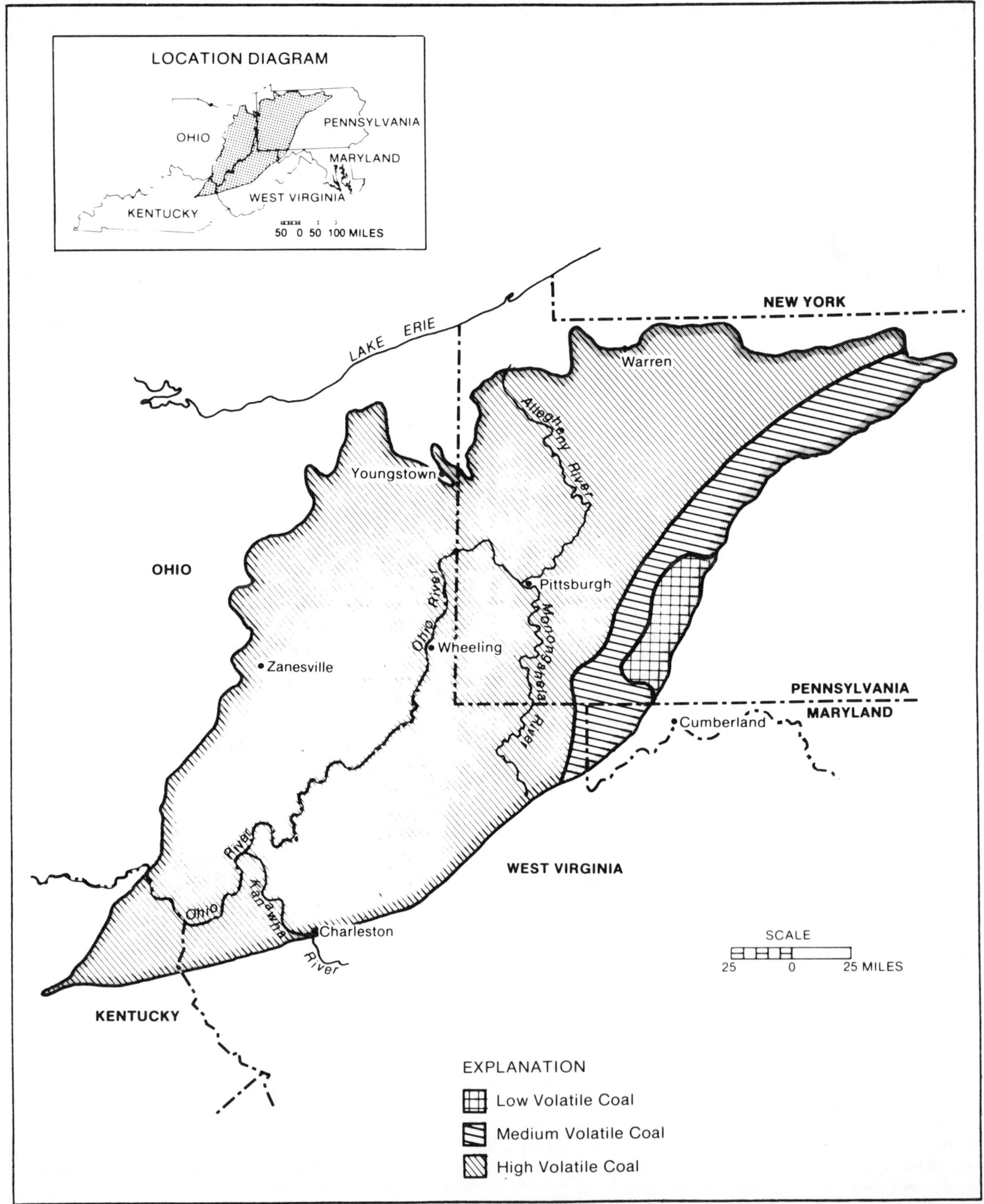

Figure 9—Coal rank map of the Northern Appalachian Coal Basin. From West Virginia Geological and Economic Survey, copyright © 1980, (used with permission) and Pennsylvania Topographic and Geologic Survey, undated.

Virginia, the Waynesburg coal is multiple bedded, ranges from 3 to 12 ft thick, and frequently contains 1 to 36 inches of shale. In Ohio and Brooke counties, West Virginia, the coal thins to 2.5 ft and the parting does not occur (Headlee and Nolting, 1940). In Ohio, the Waynesburg coal lies along the Ohio River from Belmont to Meigs County. In Belmont County, the coal ranges from a few inches to as much as 5 ft but averages 3 to 3.5 ft. Throughout the Ohio counties, the average thickness is about 16 inches (Brant and DeLong, 1960). Based on coal analyses, the seam is classified as high-volatile A bituminous coal (Headlee and Nolting, 1940).

Coal Resource Estimate

The first step for estimating the in-place gas resource of an area is determining the quantity of coal present. Data available in Keystone (1981), Ashley (1928), Brant and DeLong (1960), Headlee and Nolting (1940), and other documents indicate that those authors were mainly concerned with the minable coals or those coals that might be considered reserves for the near future. In West Virginia, in particular, there are few data on the extent of the coals beyond what is minable. The West Virginia Geologic Survey (1982, personal communication) indicated that a lack of data exists for the central part of the Northern Appalachian Coal Basin and that there are no plans to acquire information in that region. The coals are believed to be continuous across the basin (USGS, Coal Branch, 1982, personal communication), but their depth has precluded study, since they are considered unminable by present technical and economic standards.

To use the existing coal resource data would omit a large part of the basin resource and would undoubtedly leave the in-place gas resource estimate inaccurate. The coals are deep, and previous work indicates that coals at depth generally have higher gas contents. It was therefore essential to estimate the coal resource in the central part of the basin to use in calculating the in-place gas resource.

The coal resource estimate was made by grouping major coals in the area rather than examining each bed individually. The groups used are: Brookville-Clarion; Lower, Middle, and Upper Kittanning; Lower and Upper Freeport; Pittsburgh; Redstone-Sewickley; and Waynesburg. In the central part of the basin, the lack of data made it necessary to extrapolate from known areas. Average thickness data for each coalbed were collected and plotted for every county where available. Average thicknesses for the coalbeds in each group were totaled to derive the average coal thickness for that group. The data were then isopached at a 2-ft contour interval (Figs. 10 through 15). The pattern in southeastern Ohio, southwestern Pennsylvania, and northern West Virginia was projected into the areas where few data are available. The area between each contour line was then planimetered to determine area in square miles underlain by the coals of that thickness. Coal reserves were then determined using a value of 1.152×10^6 tons per sq-mi-ft. Results of these calculations are shown in Table 1. For those groups used in the in-place gas resource estimate, the total estimated coal resource was calculated to be 578 billion tons.

COALBED METHANE DATA FROM LITERATURE

Numerous studies have been made of the methane content of coalbeds in the Northern Appalachian Coal Basin. This area was opened early for development of the coal resource because of its proximity to the highly industrialized northeastern United States. With the initial development of the coal resource came an early awareness of the presence of methane in the coals. For years this methane was recognized only as a hazard involved with mining. Only recently has its potential as an energy resource been recognized.

Because of the proximity to the US/DOE Morgantown and Pittsburgh Energy Technology Centers (METC and PETC), USBM, and the Pittsburgh Research Center, much of the preliminary research on coalbed methane was conducted on samples collected in the nearby Northern Applachian Coal Basin and at field sites located within short distances from those facilities. Prior to the onset of the Methane Recovery from Coalbeds Project (MRCP), more was known about this area's coalbed methane resource and its potential producibility than about any other in the conterminous United States.

Systems Studies of Energy Conservation—Methane Produced from Coalbeds

The first comprehensive study of coalbed methane resources for an individual coalbed—the Pittsburgh seam—was conducted in 1976 (TRW, 1977). This study addressed the definition of the methane resource in the Pittsburgh seam; a determination of the technical and economic feasibility of production, collection, conversion, and/or utilization systems; and the synthesis of the basic elements of a field demonstration program plan. This study formed the early basis for the following discussion of the Pittsburgh seam coalbed methane resource.

In 1976, estimates of the coalbed methane resource ranged from 600 billion cubic feet (Bcf) to 4 trillion cubic feet (Tcf) in the Pittsburgh seam alone. These estimates were based on a few coalbed methane desorption tests and a large amount of coal mine methane emission data (Irani et al, 1972; Irani, Jeran, and Deul, 1974).

Mine Emission Data

Additional work published by the USBM in 1977 presents methane emission data from U.S. bituminous coal mines within the conterminous United States (Irani et al, 1977). More than 85% of the approximately 216 MMcfd for the total United States was emitted from Appalachian mines, with the bulk of that from the Northern Appalachian Coal Basin. When mines emitting less than 1 MMcfd are excluded, 31 mines in this coal basin emit a total of 105.6 MMcfd. With very few exceptions, they are all concentrated in ten counties in southwestern Pennsylvania and northern West Virginia in an area encompassing 5,463 sq mi. Emissions for the Freeport coal total 2.1 MMcfd; the Upper Freeport, 3.6 MMcfd; the Upper Kittanning, 1.95 MMcfd; the Lower Kittanning, 7.45 MMcfd; and the Pittsburgh, 90.5 MMcfd.

U.S. Bureau of Mines (USBM) Coalbed Methane Data

The USBM has conducted numerous studies in the area—primarily in the Pittsburgh coalbed—related to safety in dealing with coalbed methane. In addition to utilizing local coal in early studies on the relationship of coal properties and characteristics to the generation and retention of methane in the coalbed, these studies involved directional drilling for coalbed degasification (Diamond, Oyler, and Field, 1977; Oyler, Diamond, and Jeran, 1979), use of coal cleat and other geologic features to predict mine and methane emission problems (McCulloch and Deul, 1973; McCulloch, Deul, and Jeran, 1974; Iannacchione and Puglio, 1979), and a number of other related works (Jeran, Lawhead, and Irani, 1976; Zabetakis et al, 1972).

Recent work in determining the gas content of coalbeds has continued at the USBM. Data for coalbed methane contents for a number of coals in the Northern Appalachians have been reported (Diamond and Levine, 1981). Desorption results are presented for 159 samples from 14 coalbeds and 2 rider seams in 12 counties in the Northern Appalachian Coal Basin (1 in Ohio, 6 in West Virginia, and 5 in Pennsylvania).

Consolidation Coal Company Methane Evaluation

As a part of a coal exploration program in the Northern Appalachian coal field, Consolidation Coal Company (CONSOL) has been involved in a coalbed methane assessment program. Because of legal uncertainties of coalbed methane ownership in this region, CONSOL's interest is primarily from a mine safety rather than a potential resource point of view (Puglio, 1981, personal communication). CONSOL has 16 active and 9 planned underground coal mines in southwestern Pennsylvania and northern West Virginia. More than 58 MMcfd of methane are emitted from CONSOL underground mines in the region with the Loveridge, Humphrey, Blacksville No. 1 and 2, and the Osage mines, among the gassiest in the United States. With operations in these areas, knowledge of coalbed methane distribution would be critical to mine operation, planning, and development.

CONSOL's involvement in coalbed methane drainage began with drilling the underground horizontal methane drainage holes at the Humphrey Mine, Monongalia County, West Virginia, in 1958. Since then, nearly 2.8 mi of horizontal methane drainage boreholes have been drilled in CONSOL mines (Puglio, 1981).

Gas Composition

Studies have been conducted on the chemical composition of gas produced from coalbeds and on gas generation in coalbeds as a function of their thermal maturity or relative time and depth of burial. Studies by Rice et al (1979) have also examined the low-temperature (biogenic) generation of natural gas from coaly material.

Gas samples were collected and analyzed from both vertical and horizontal boreholes in the Pittsburgh and Upper Kittanning coalbeds. Samples from the Pittsburgh were collected over a wide geographic area. Pittsburgh coalbed gas samples ranged from 84 to 96% methane, and the principal contaminant was carbon dioxide (CO_2). Upper Kittanning coalbed gas samples ranged from 95 to 99% methane, and nitrogen (N_2) is the principal contaminant. Heat of combustion values ranged from 900 to 1,059 Btu/cf relative to a "natural gas" heat of combustion of 950 to 1,035 (TRW, 1977).

Vertical Well Artificial Stimulation Studies

Both the DOE and the USBM have conducted studies on the artificial stimulation of coalbeds to enhance production of coalbed methane. The U.S. government, through the USBM and the DOE, sponsored 21 coalbed stimulation treatments in the Northern Appalachian Coal Basin between July 1970 and December 1979. These and other wells are discussed in some detail by Lambert, Trevits, and Steidl (1980) in a publication addressing well completion practices for the removal of methane from coals.

While no prestimulation production tests were conducted on 9 of the 21 wells, the others had improved gas flows ranging from 3.8 to 68 times initial production. The highest observed poststimulation production rate was 100 Mcfd for No. EM-5 well in the Pittsburgh coalbed at the Emerald Mine in Waynesburg, Greene County, Pennsylvania. This well was one of a series of test wells that will be discussed in more detail later in this section.

Gas analyses of samples from ten stimulated coalbed methane wells had methane contents ranging from 92.5 to 98.2%, with CO_2 the principal contaminant in most of the wells ranging from 0.6 to 6.45%. Since seven of the ten sampled wells were stimulated with nitrogen foam, N_2 contamination might be expected. The fact that the highest nitrogen concentrations do not exceed 1.2% of the total gas suggests that the coalbed methane wells clean up well on flowback.

Water chemistry samples collected from stimulated coalbeds in tested wells vary dramatically, as they do for coalbed waters in general. Analyses of six water samples from the Middle Kittanning and the Pittsburgh coalbeds showed total dissolved solids (TDS) ranging from 3,100 to 63,000 ppm; one Middle Kittanning water sample represented the high TDS sample, and five samples from the Pittsburgh coalbed ranged from 3,106 to 17,700 ppm.

Methane Recovery from Coalbeds Project (MRCP) Data

With the inception of the MRCP in 1977, a number of projects were initiated that have provided significant data on the methane content and producibility of various coalbeds throughout the Northern Appalachian Coal Basin. These individual projects include:

- Methane Recovery from Multiple Coal Seams Project, Waynesburg College
- Demonstration of Methane Recovery from Pennsylvania Coal Seams Project, Pennsylvania State University/U.S. Steel/Carnegie Natural Gas
- Space Heating Test Project, Westinghouse Electric Company, Waltz Mill Test
- Kinloch Development, # 1 Murdock Well Test.

These projects will be discussed individually as they pertain to the coalbed methane resource and its producibility.

Methane Recovery from Multiple Coal Seams Project

The Methane Recovery from Multiple Coal Seams Project was conducted near the campus of Waynesburg College in Waynesburg, Greene County, Pennsylvania. The actual test site (located on the flood plain of Purman Run) lay between the Waynesburg Syncline and the Belle Vernon Anticline. The strata at the site were nearly horizontal, dipping approximately 1° northwest.

A corehole was drilled to sample all encountered coal seams to determine the gas content. The well encountered the shallowest coal (the Waynesburg) at 149 ft, and the deepest coal (the Mercer (?)) at 1,407 ft. A cumulative total of 39.9 ft of coal was encountered, and approximately 31 ft were suitable for completion for methane extraction. Gas contents ranged from

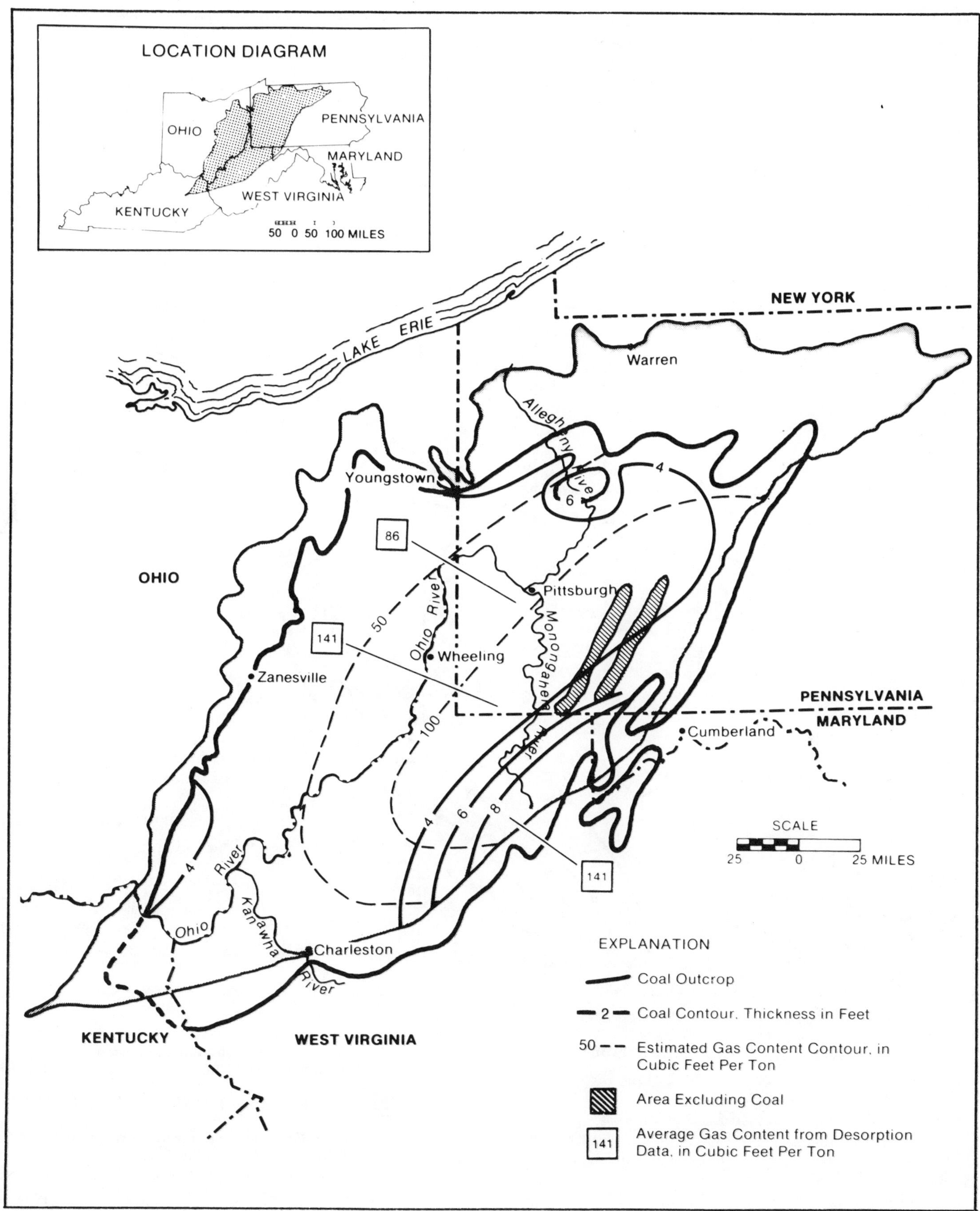

Figure 10—Brookville-Clarion coal isopach and in-place gas distribution map.

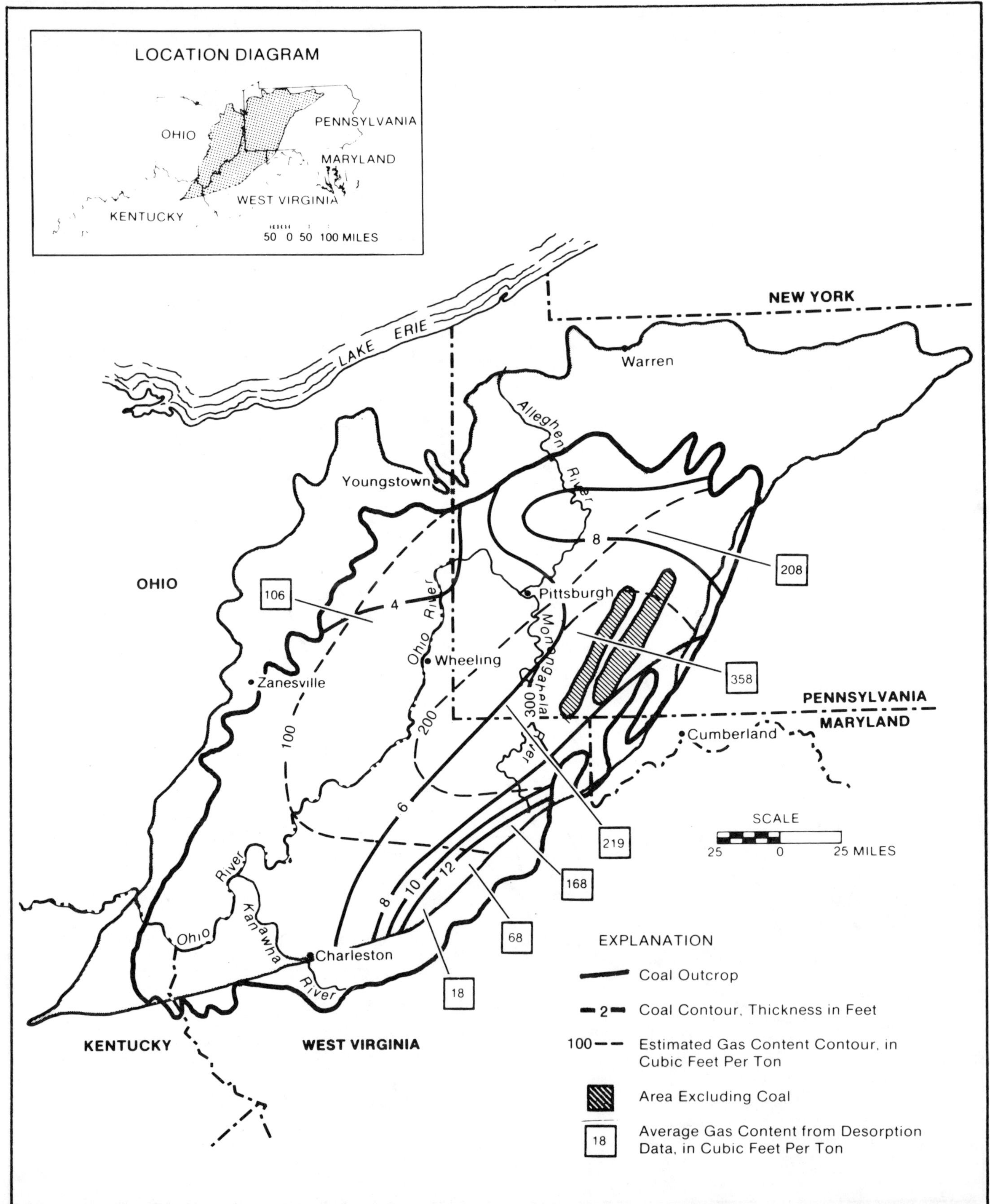

Figure 11—Lower, Middle, and Upper Kittanning coal isopach and in-place gas distribution map.

70 cf/ton (2.2 cc/gm) for the Waynesburg coal to 195 cf/ton (6.1 cc/gm) for the Upper Kittanning coal. The Mercer coal, which was not sampled for desorption testing, was estimated to have a gas content of 141 cf/ton (4.4 cc/gm).

The well was cased, cemented, and perforated in the Pittsburgh, Pittsburgh Rider, Bakerstown, Upper Kittanning, Middle Kittanning, Clarion, Mercer 1, and Mercer 2 coal seams. It was completed and stimulated in three zones:

- No. 3—Pittsburgh Rider
  —Pittsburgh
- No. 2—Bakerstown
  —Upper Kittanning
  —Middle Kittanning
- No. 1—Clarion
  —Mercer # 1
  —Mercer # 2

Together, these have calculated gas reserves of 6.06 MMcf/acre.

The three zones were stimulated using a nitrogen foam frac in late February 1980, and coalbed dewatering began by swabbing between 120 and 230 b/d of water. Gas flow was measured to be 18 Mcfd with 140 b/d water on July 11, and maximum gas flow of 30 Mcfd was observed on August 9, 1980. Gas production from August through December 1980 averaged about 20 Mcfd.

The produced gas meets quality specifications for pipeline injection, having methane content of 94%, and as of February 1981 was being transported via the Equitable Gas Pipeline to the Waynesburg College campus for use in heating buildings at the rate of 20–30,000 cfd during the winter months.

Demonstration of Methane Recovery from Pennsylvania Coal Seams

Pennsylvania State University, Minerals Engineering Department, under the joint sponsorship of the USBM and the Pennsylvania Research Foundation, initiated a methane recovery research project in January 1975 at the Cumberland Coal Operation of U.S. Steel in Wayne Township, Greene County, Pennsylvania. This study continued through 1979 and 1980 with support of the MRCP program.

Three degasification wells were drilled, and two were stimulated and tested at the mine to determine the potential for coalbed degasification using vertical boreholes. Gas production from the Pittsburgh coalbed increased in the stimulated wells as water flow decreased, reaching a maximum of more than 71 Mcfd gas with less than two barrels of water.

One borehole was plugged in April 1980 and mined through shortly thereafter. A study of this mine-through showed that the fracture had been induced from only one side of the well. The frac was observed 60 ft from the wellbore, vertically traversing the entire coalbed, and was open approximately 1/2-inch in the middle of the seam. The fracture curved near the top of the seam. The frac was observed up to 150 ft from the borehole where it was about 1/8-inch wide and contained only a small amount of sand. The report gave no indication that the fracture extends into strata overlying or underlying the coal but does note the presence of several parallel thin cracks at the top of the coal seam (Frantz et al, 1981).

A roof fall occurred in the vicinity of the fracture trace in the crosscut. The true relationship of roof fall to frac trace could not be defined because of shoring for safety reasons, but roof falls had occurred in other areas of this mine where they could not be attributed to stimulation treatments (Frantz et al, 1981).

Space Heating Test Project

Westinghouse Electric Company, Advanced Energy Systems Division, conducted a test to supply gas for space heating in a nearby Westinghouse facility situated on 850 acres overlying virgin coal near Waltz Mill in Westmoreland County, Pennsylvania. Two wells were drilled: one as a production well (Westinghouse Electric Company, FEE # 4); and one as an observation well (Westinghouse Electric Company, FEE # 5). FEE No. 5 was cored through the Sharon coalbed, penetrating a total of 27 ft of coal between 187 and 629 ft. The gas contents of the coal ranged from 13 to 102 cf/ton, with a weighted average of 37 cf/ton. Final USBM desorption values of coal samples indicate an average gas content of 186 cf/ton. The computed coalbed resource in this area is 1.85 MMcf/acre (Westinghouse Electric Company, 1978).

Borehole geophysical logs of FEE # 4 indicated a cumulative thickness of 24 ft of coal above a depth of 760 ft. The well was completed using a Kiel frac in December 1978 into four zones ranging from:

- Zone 4: Surface–348 ft
- Zone 3: 348–470 ft
- Zone 2: 470–583 ft
- Zone 1: 583–757 ft

The Kiel frac procedure involves pumping pulses of water and proppant into the coalbeds at a specified low flow rate. This pulsed pressuring generates fractures in a new direction with each pressure cycle and reportedly creates a dendritic fracture pattern with a significant increase in the interconnection of the face and butt cleat permeability.

No pressure or flow testing was conducted prior to the frac treatment, so the effectiveness of that treatment was difficult to determine. During April 1978, the well produced gas at a rate of 38 Mcfd with about 240 b/d water. By late May, gas flow remained constant and water flow had decreased to approximately 120 b/d. During January 1980, FEE # 4 produced 856 Mcf, with an average production of 28 Mcfd. Maximum production was 43 Mcfd.

Additional analyses of core sample data suggest the in-place resource is 8 MMcf/acre—much higher than the preliminary estimate early in the program.

Kinloch Development, # 1 Murdock Well Test (Site AAA)

As part of the coalbed methane resource assessment in the Northern Appalachian Coal Basin, several wells were drilled and tested by Kinloch Development Corporation in Whiteley Township, Greene County, Pennsylvania. The Harry A. Murdock, Jr., Farm # 1, Kinloch # 1 Well (hereafter called # 1 Murdock Well) was drilled, sidewall cores collected for coal desorption analysis, logged, fraced, and tested for its coalbed methane production potential. Sidewall cores were collected in the Kinloch # 1 Stoner Well for desorption analysis, but no other information is available.

The # 1 Murdock well was located along Frosty Run, approximately 1 mi northwest of Fordyce in Whiteley Township, Greene County, Pennsylvania. It was air-drilled to a total depth of 1,608 ft, and the surface elevation at the well is 1,160 ft. A cumulative total of 61 ft of coal was encountered between 324 and 1,608 ft, of which one-half was encountered in

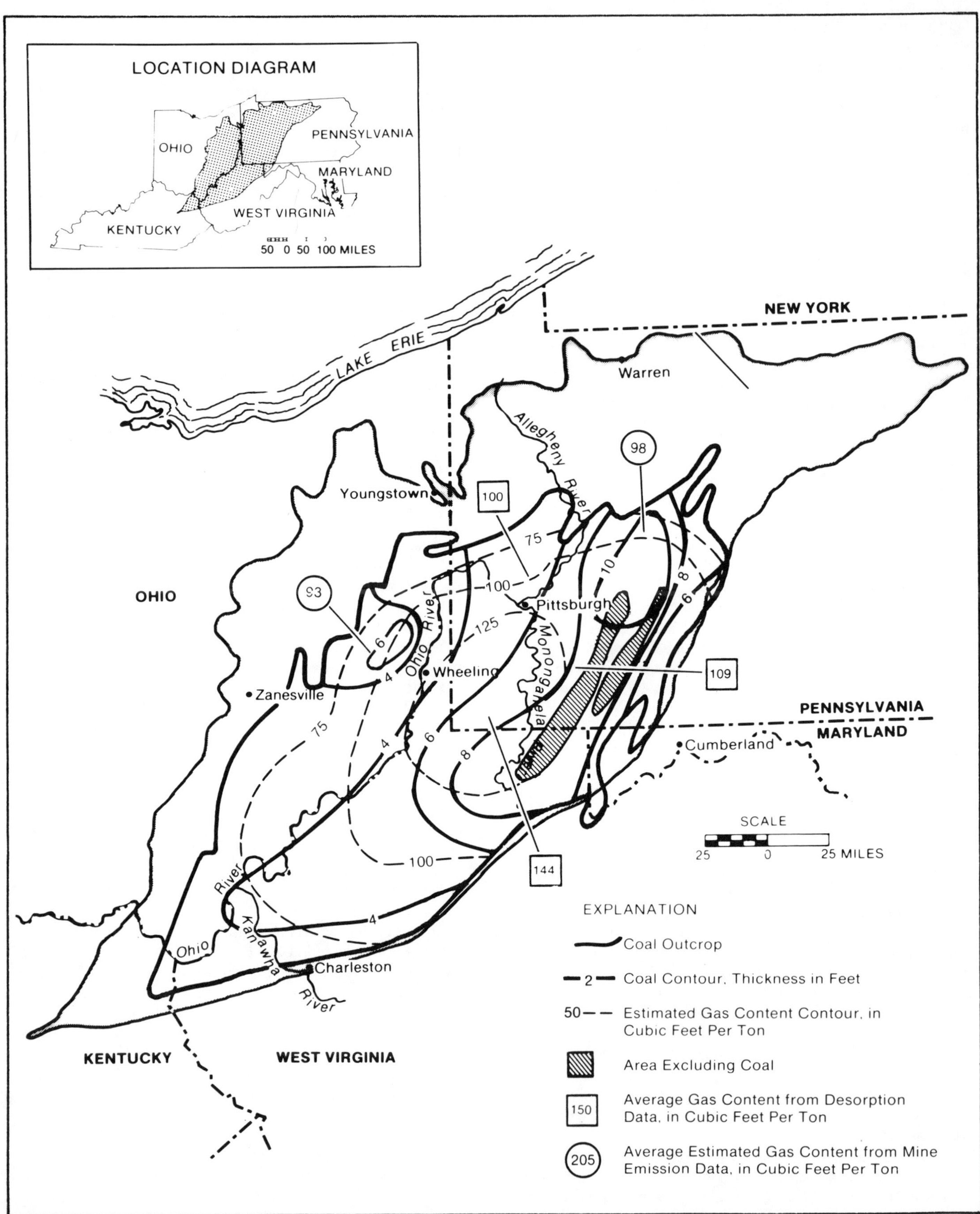

Figure 12—Lower and Upper Freeport coal isopach and in-place gas distribution map.

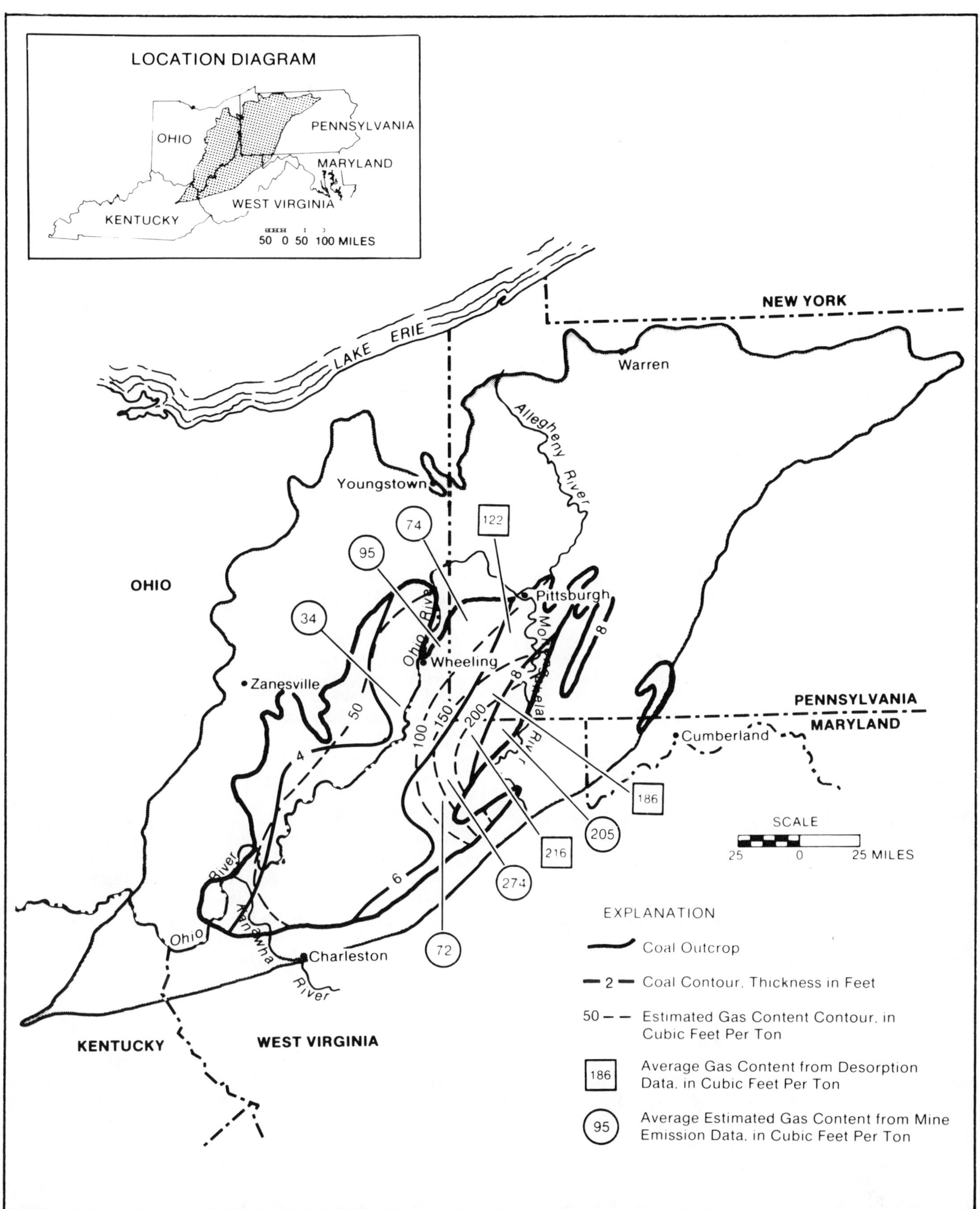

Figure 13—Pittsburgh coal isopach and in-place gas distribution map.

the Pittsburgh Formation (13.5 ft: 9.5-ft Pittsburgh seam; 4-ft Pittsburgh Rider) and the Kittanning Formation (17.5 ft).

The # 1 Murdock was stimulated using an eight-stage Kiel treatment. The high injection pressure required stages 2 and 3 to be shortened, and termination after stage 3. The well flowed back for several days and was cleaned out. It produced approximately 80 b/d water with a show of gas. Poststimulation injection tests indicated that of four zones stimulated, three were positively affected by the stimulation treatment. The well was pumped intermittently for several weeks and gas flow was reported but water or gas production was not measured.

Possible Hydrogeologic Controls on Coalbed Methane Distribution and Potential Production

The geology of the Northern Appalachian Coal Basin strongly influences the local distribution of methane within any individual coalbed. The eastern basin boundary is structurally controlled with the Alleghenian-age coals cropping out on the flanks of southwest-trending folds. The depositional environment suggests that the coals are continuous across the basin, but no data exist at present to confirm this. Limited coalbed methane data exist primarily in southwestern Pennsylvania and northcentral West Virginia, and mine emission data exist for a number of mines in the Pittsburgh coal, but the correlation of emission data to gas-in-place must be considered tenuous at best.

Structural Controls

A cursory examination of the potential structural controls on the distribution and production potential of coalbed methane suggests several features that may be important. Numerous tight to moderate anticlinal and synclinal folds lie within the folded valley and ridge province on the eastern basin boundary. In some instances, these folds bring the coalbeds to the surface with fairly steep dips. Although the coalbeds deepen within a short distance from their outcrops, much of the methane gas may escape by migrating updip over the long periods of time that the seams have been exposed. The Chestnut Ridge and Laurel Hill anticlines located in Fayette, Somerset, and Westmoreland counties are examples of coals with low methane potential because of uplift and erosion.

Further to the west in the basin, the strata flatten and the overburden increases. In the deepest part of the basin in Greene County, Pennsylvania, and Wetzel and Tyler counties, West Virginia, the deepest coals lie beneath 2,300 ft of overburden. This thick overburden and relatively undisturbed coal suggest a potential for large volumes of gas.

Coal rank increases from northwest to southeast across the basin, and some coal adjacent to the southeastern basin boundary is of medium- and even low-volatile bituminous rank. The area underlain by these high-rank, potentially gassy coals is a relatively narrow band approximately 35 mi wide and 250 mi long. The remainder of the basin is underlain by coals of high-volatile A bituminous rank. The higher rank coals lie within the area most strongly influenced by tectonic stresses and therefore exposed to higher pressures and temperatures caused by deeper burial and metamorphic processes.

Coalbeds in the basin exhibit directional permeability owing to the cleat (or fractures) in the coal. The highest permeability is parallel to the face cleat direction, which is shown in Figure 16.

Depositional and Stratigraphic Controls

The Pennsylvanian-age coals in the Northern Appalachian Coal Basin are continuous over broad areas and are uniformly thick. They are cut out in places by channel deposits. The coals of the Allegheny Group—Brookville through Upper Freeport seams—were deposited in a delta plain or alluvial plain environment. This group was overlain by a sequence of marine and nonmarine sediments containing abundant shales and mudstones that provide good seals for trapping gas within the Alleghenian coal-bearing section. The youngest major coal-bearing units—the Monongahela Group—were deposited in a series of lakes and deltaic lake-margin environments. These units consist predominantly of shale, which also provides impermeable barriers to gas migration.

Mica-bearing peridotite dikes are known to penetrate several coalbeds in Greene, Fayette, and Indiana counties in Pennsylvania, indicating the possibility of anomalously high coal rank and gas content resulting from contact metamorphism. Although the regional extent of these dikes is unknown, a high thermal gradient—and, therefore, higher rank and gas content—is possible if they are widespread.

Hydrologic Controls

Throughout the Appalachian region, coalbed methane reservoir pressure is controlled by hydrostatic head. Where the coalbeds crop out or lie above drainage (which is not uncommon for coalbeds isolated on synclinal ridgetops), the methane that was present has been lost to the atmosphere over long periods of exposure.

Where coalbeds are highly permeable, it is possible that large volumes of water must be produced prior to the onset of significant gas flow. Care is required in selecting completion procedures to make certain that aquifers are not penetrated during a stimulation treatment leading to excess water production.

Discussion of Methane Resources

The coalbed methane potential of the Northern Appalachian Coal Basin was assessed on the basis of groups of coalbeds. This approach was useful because the legal status of coal and coalbed methane ownership has been contested in the state of Pennsylvania and because of the paucity of data on some coals. The fact that some coalbeds can be correlated across the basin aided significantly in this approach.

Regional Distribution of Methane in Coals

The coalbeds of interest have been combined into a number of groups containing one, two, or three major coalbeds to allow the use of average gas contents and average cumulative coal thicknesses where those parameters did not exist for individual coalbeds. Average coal thicknesses were assigned for each county where a particular coalbed was reported, and the thicknesses were interpolated between measured points assuming continuity of individual coalbeds or coal zones across the basin. Discussions with coal geologists familiar with this region suggested that this assumption was valid.

Gas contents measured by the USBM direct method were

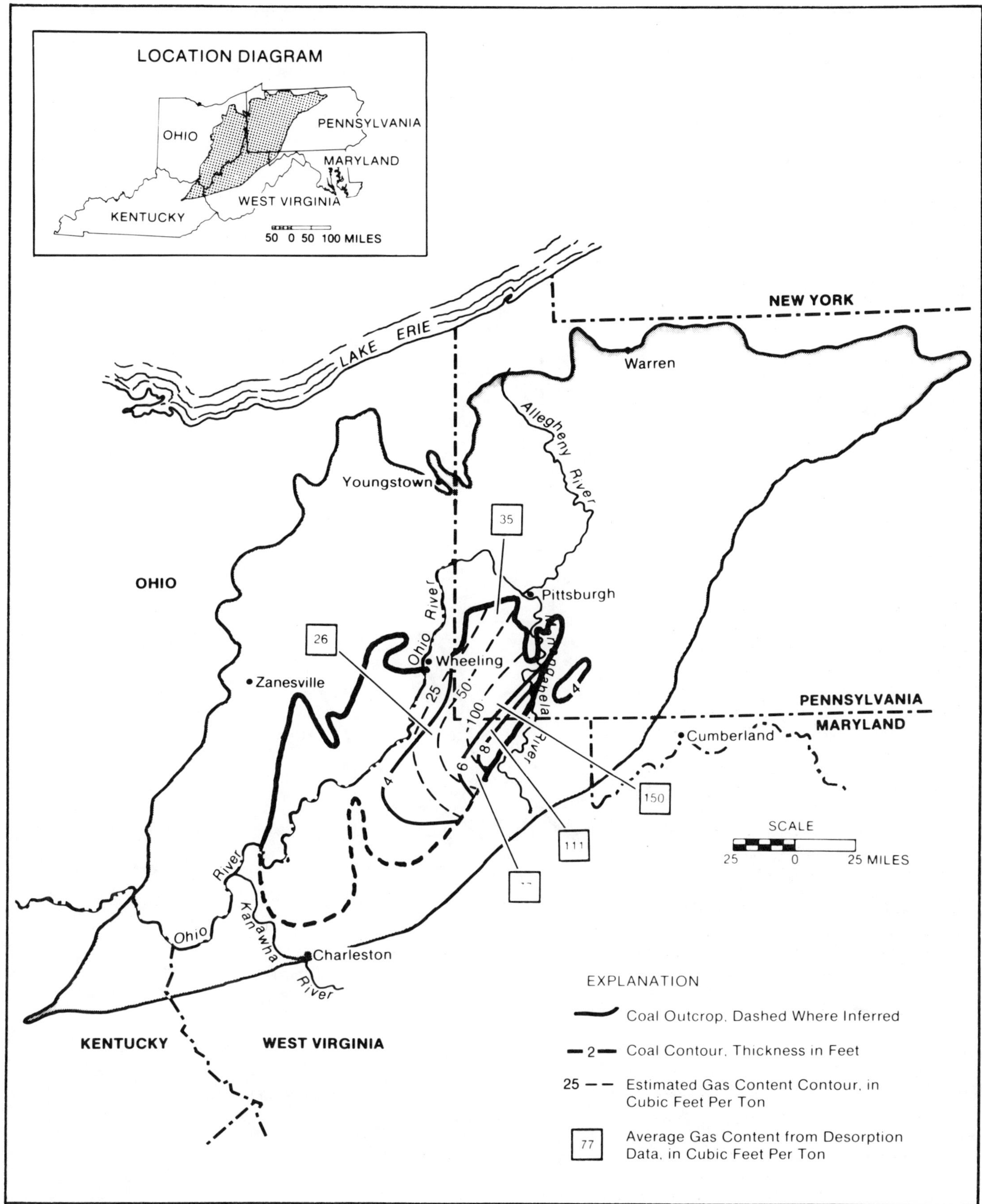

Figure 14—Redstone and Sewickley coal isopach and in-place gas distribution map.

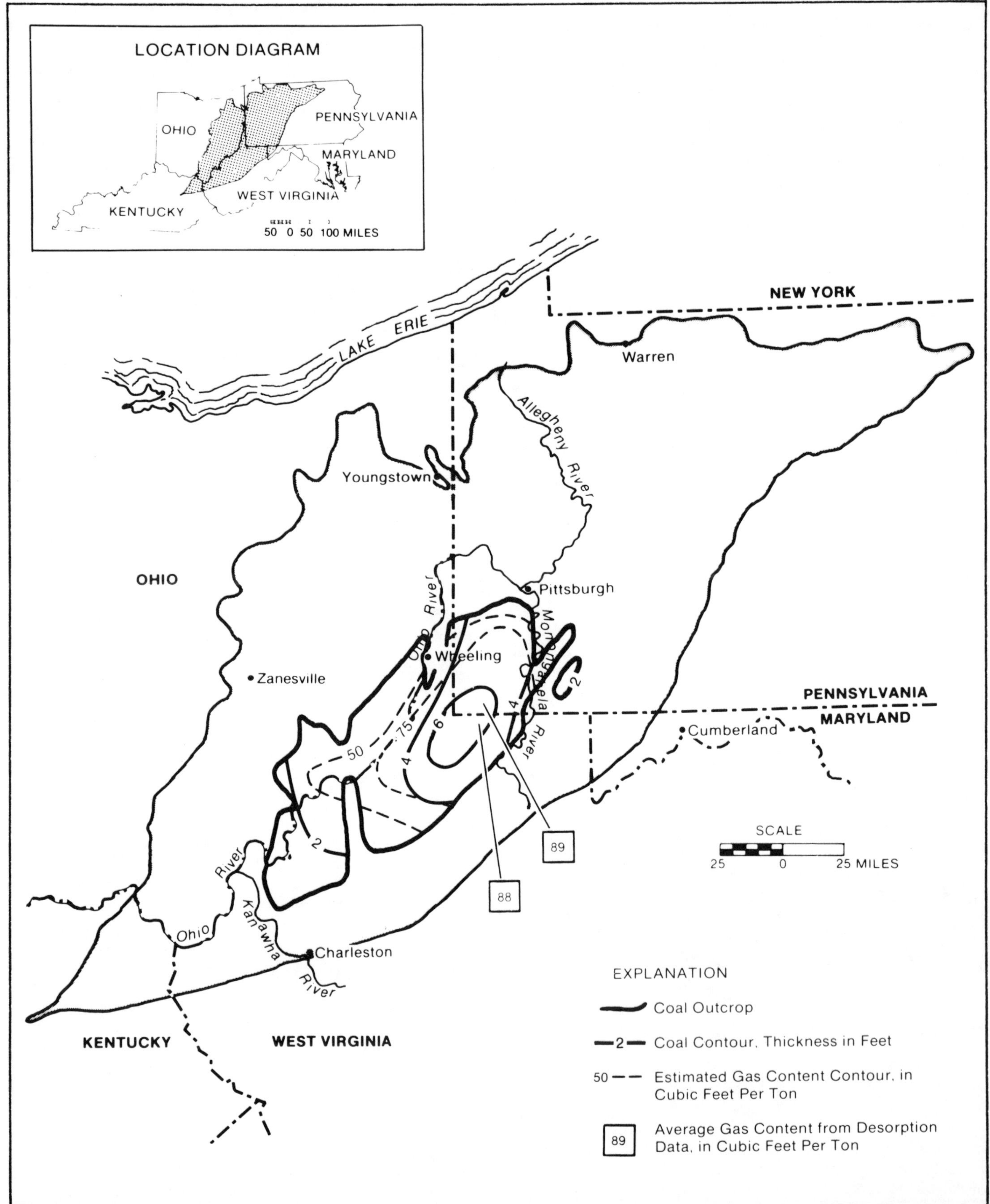

Figure 15—Waynesburg coal isopach and in-place gas distribution map.

Table 1—Estimated coal resource of selected coalbeds in the Northern Appalachian Coal Basin.

Coal Group or Bed	Approximate Area Underlain by Coal Group or Bed (sq. mi.)	Coal Resource (in billions of tons) By Contour Interval (in ft)*						
		<2	2–4	4–6	6–8	8–10	>10	Total
Brookville - Clarion	30,300	—	83.7	23.3	10.7	6.6	—	124.3
Lower, Middle, and Upper Kittanning	28,000	—	5.9	83.4	60.6	23.7	9.6	182.9
Lower and Upper Freeport	22,800	—	33.9	32.8	29.2	27.2	11.8	134.9
Pittsburgh	12,600	—	7.6	39.2	22.6	2.6	—	72.0
Redstone - Sewickley	8,000	—	16.0	15.7	1.7	1.7	—	35.1
Waynesburg	7,000	1.0	12.1	10.4	5.0	—	—	28.5
							Grand Total	577.7

*Calculated by using 1.152×10^6 tons/sq-mi-ft times the average thickness of each interval.

taken as representative of gas-in-place at the locations measured and were averaged by county for the individual coalbeds or groups of coalbeds. Where numerous data are available for a coalbed in the center of the basin, the values are interpolated between known points and between known points and the outcrop based on coal rank and approximate depth. Figures 10 through 15 show the estimated gas content (in cf/ton) for the following coal groups:

- Brookville-Clarion
- Kittanning
- Freeport
- Pittsburgh
- Sewickley-Redstone
- Waynesburg

Average desorption data by individual coalbed, by county, are presented in Table 2.

Where mine emission data are available, they are used to suggest trends in gas-in-place. Using suggestions from USBM personnel, an empirical relationship of gas-in-place equal to 20% of gas emission/coal production ratio is used to approximate the gas within the coalbed. When the derived numbers conflict with direct method values, the latter are used.

Brookville-Clarion Coalbeds

The Brookville-Clarion coalbeds underlie more than 30,000 sq mi in the Northern Appalachian Coal Basin. The outcrop of these beds very closely coincides with the boundaries of the area of study, except in the north and northeast. The coalbed ranges in thickness from about 2 ft in southeastern Ohio to 8 ft along the eastern basin boundary. Average gas contents are reported for Allegheny and Greene counties, Pennsylvania, and for Barbour County, West Virginia. Figure 10 shows the distribution of in-place gas for this unit.

Kittanning Coalbeds

To assess the coalbed methane potential of the Kittanning coals, the combined Upper, Middle, and Lower coalbeds are assumed to underlie approximately 28,000 sq mi of the basin. Average coal thicknesses range from 2 ft in northeastern Ohio (Columbiana County) to 13 to 14 ft in northcentral West Virginia (Barbour and Braxton counties) along the southeastern basin boundary. Average gas contents are presented for four counties in Pennsylvania (Indiana, Westmoreland, Greene, and Allegheny), three counties in West Virginia (Barbour, Upshur, and Braxton), and one county in Ohio (Harrison). Gas content contours indicate that the bulk of the basin is underlain by these coals which have gas contents in excess of 100 cf/ton (Fig. 11). Based on coal rank determinations, a significant part of Fayette, Westmoreland, and Somerset counties in Pennsylvania, and Preston County, West Virginia, are underlain by coalbeds with average methane contents in excess of 300 cf/ton.

Freeport Coals

The Freeport coalbeds here include the Lower and Upper Freeport and the Freeport Coal-undifferentiated. This unit underlies approximately 23,000 sq mi of the basin. Average cumulative coal thicknessess range from approximately 2 ft along the western boundary of the basin to 12 ft in Indiana County, Pennsylvania. Average desorption values are available for Allegheny, Greene, and Westmoreland counties in Pennsylvania and indicate gas contents in excess of 100 cf/ton (Fig. 12). Mine emission data indicate the presence of approximately 100 cf/ton in the Freeport coalbeds mined in Indiana County, Pennsylvania, Harrison County, Ohio.

Pittsburgh Coalbed

The Pittsburgh coalbed and its rider seams underlie approximately 12,500 sq mi of the Northern Appalachian Coal Basin. This coalbed, one of the most important energy resources in the world, accounts for about 85% of the methane emitted by all mining operations in the conterminous United States. The Pittsburgh coalbed ranges in thickness from 3 ft in the southwestern corner of the area to more than 10 ft in northeastern West Virginia. The area of thick coal also corresponds in part to the highest rank coal. Average desorption values are available for Washington and Greene counties, Pennsylvania, and for Marion County, West Virginia (Fig. 13), and range from 122 to 216 cf/ton. Gas contents based on mine emission data are available for those same counties and Monongalia and Harrison counties, West Virginia, and range from 34 to 274 cf/ton. The entire portion of southwestern Pennsylvania and northcentral West Virginia are underlain by the Pittsburgh coalbed, which contains more than 100 cf/ton. Significant portions of this area, however, have been mined out.

Sewickley-Redstone Coalbeds

The combined Sewickley and Redstone coalbeds underlie an area of approximately 8,000 sq mi of the Northern Appalachian Coal Basin. Average cumulative coal thickness ranges from 2 and 3 ft on the southwest and northwest, respectively, to approximately 10 ft in Marion and Monongalia counties in

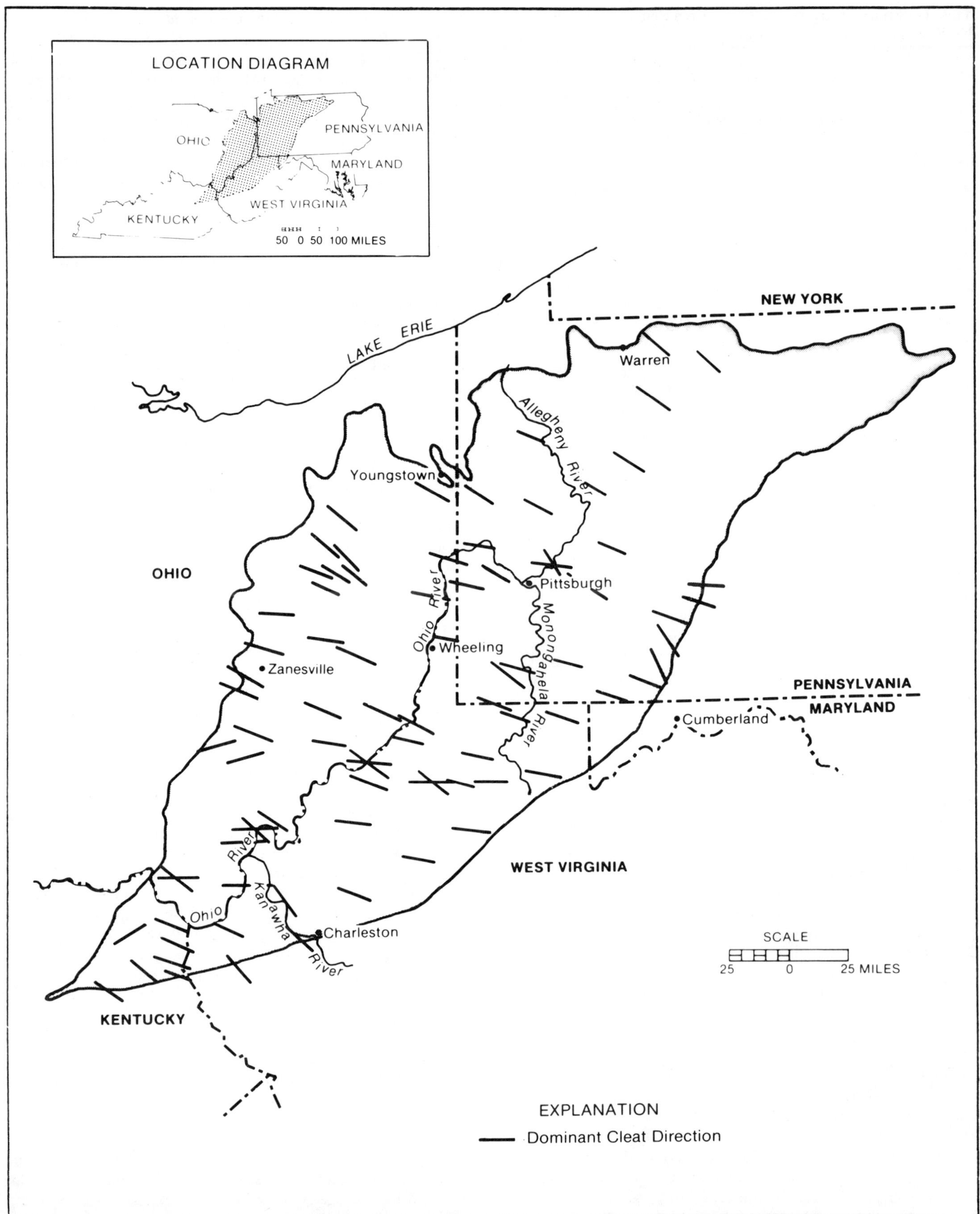

Figure 16—Dominant coal cleat directions. From Colton, Perry, and Mackenzie, 1981; and Kulander and Dean. Copyright © 1980 West Virginia Geological and Economic Survey. Used with permission.

Table 2—Gas content from desorption data by seam.

Coalbed	State	County	No. Samples	Av. Gas Content in cf/ton (range)
Brookville	PA	Allegheny	2	83 (80–86)
Clarion	PA	Allegheny	1	93
		Greene	1	141
	WV	Barbour	2	141 (115–166)
Lower Kittanning	PA	Greene	1	426
		Indiana	3	174 (26–445)
		Westmoreland	1	358
	WV	Barbour	16	177 (45–294)
		Braxton	18	18 (3–32)
Middle Kittanning	PA	Allegheny	1	160
		Greene	1	99
	OH	Harrison	3	106 (90–118)
	WV	Upshur	3	76 (74–80)
Upper Kittanning	PA	Allegheny	3	79 (13–109)
		Greene	2	175 (154–195)
		Indiana	1	311
	WV	Barbour	12	156 (77–230)
		Upshur	1	42
Upper Freeport	PA	Allegheny	11	100 (13–157)
		Greene	7	144 (51–266)
		Westmoreland	1	109
Pittsburgh (including Pittsburgh Rider)	PA	Greene	49	183 (62–266)
		Washington	1	122
	WV	Marion	2	216 (208–224)
Bakerstown	PA	Greene	2	192 (141–243)
Redstone	WV	Marion	1	77
		Monongalia	2	128 (125–131)
		Wetzel	1	26
Sewickley	PA	Greene	10	150 (80–182)
		Washington	1	35
	WV	Monongalia	2	94 (22–166)
Washington	PA	Greene	1	61
Waynesburg	PA	Greene	11	89 (33–144)
	WV	Monongalia	2	88 (86–90)
Mahoning	PA	Allegheny	1	51
Ohio No. 5 or 6	OH	Harrison	3	99 (90–118)
			180	

northcentral West Virginia. Average gas content values are reported for Greene and Washington counties, Pennsylvania, and for Marion, Monongalia, and Wetzel counties, West Virginia (Fig. 14), and indicate that only a relatively small area is underlain by coals with a gas content in excess of 100 cf/ton. Most of these coals are relatively shallow and are not expected to contain large volumes of gas.

Waynesburg Coal

The youngest coalbed of any significance to the coalbed methane resource is the Waynesburg, which underlies approximately 7,000 sq mi of the basin. The coalbed is relatively thin, ranging in thickness from 1.5 ft in Washington County, Ohio, to about 7 ft in northcentral West Virginia. The coal is absent over parts of Wirt, Wood, Ritchie, and Pleasants counties, West Virginia, where it has been eroded on the crest of the Burning Springs anticline (Fig. 15). Average gas contents are available for Greene County, Pennsylvania, and Monongalia County, West Virginia, indicating gas contents of approximately 90 cf/ton. This coal is also generally shallow and not expected to contribute significantly to the producible coalbed methane resource.

Estimated Resource Volume

Based on these data, the coalbed methane resource of the Northern Appalachian Coal Basin was estimated. Figure 17 shows the distribution of that resource in billions of cubic feet of gas per square mile superimposed on the outcrop pattern of the deepest coal group: the Brookville-Clarion. The bulk of the resource is located in the deepest part of the basin, where the coals are thick and of highest rank. Although other, deeper parts of the basin might also be good targets, on the basis of existing data, this area has the greatest potential to contain large volumes of gas. Based on a county-by-county assessment of a coalbed group basis, the total coalbed methane resource in the Northern Appalachian Coal Basin is estimated to be approximately 61 Tcf (Table 3).

CONCLUSIONS

The coalbed methane resource in the Northern Appalachian Coal Basin is estimated here to be 61 Tcf. This basin contains a number of the gassiest mines in the United States, with cumulative methane emissions measured to be more than 180 MMcfd. Thirty-one mines in the northern Appalachians individually emit more than 1 MMcfd.

Desorption data, while limited in their stratigraphic and geographic distribution, indicate the presence of highly gassy coals in certain parts of the basin. This limited information suggests that relatively gassy coals exist along the eastern margin of the basin where high coal rank, significant depth of burial, and tectonic stresses indicate a high potential for coalbed methane production. Measured gas contents range up to 445 cf/ton in an individual Kittanning coal sample.

A high-potential coalbed methane target area has been defined, based on the following parameters:

- Numerous, relatively thick coalbeds
- High or potentially high gas contents
- High-rank coal
- Significant overburden thickness
- High methane mine emission rates
- Significant distance from outcrop
- Producing coalbed methane wells
- Areas of extensive mining.

Figure 18 shows a preliminary target area based on the individual parameters listed above. The target is slightly to the west of some of the highest rank coals because they crop out along several anticlinal structures that trend northeast-southwest through eastern Fayette and Westmoreland counties in Pennsylvania. The coals have been extensively

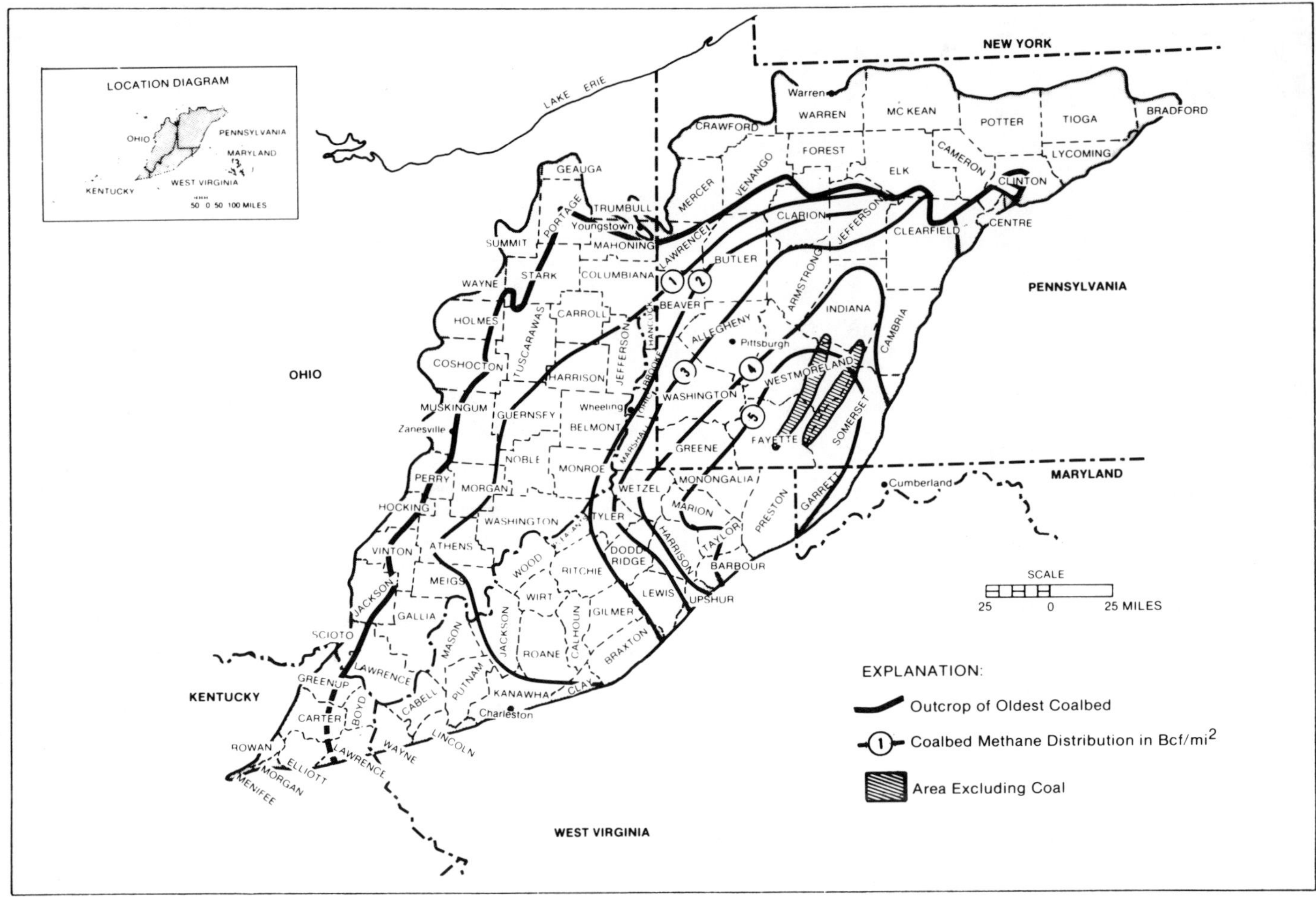

Figure 17—Distribution of the coalbed methane resource in the Northern Appalachian Coal Basin.

Table 3—Coalbed methane resource distribution by coal group.*

Group or Bed	Approx. Area Underlain by Coal Group or Bed (sq mi)	Methane Resource in Coal Group or Bed (Bcf)
Waynesburg	7,000	2,040
Redstone-Sewickley	8,000	1,635
Pittsburgh	12,600	7,130
Freeport Group	22,800	11,690
Kittanning Group	28,000	30,460
Brookville-Clarion	30,300	8,380
Total		61,335

*Calculated using estimated average coal thicknesses and estimated average gas contents by county for counties underlain by the individual coalbeds or groups.

mined along those structures and along the outcrops on the eastern edge of the basin. This target area covers approximately 4,500 sq mi and includes a significant part of northcentral West Virginia purposely limited to the deepest part of the basin northeast of the Burning Springs anticline.

REFERENCES CITED

Adams, M. A., 1982, Geologic overview, coal resources, and potential methane recovery from coalbeds, Central Appalachian Basin, Maryland, West Virginia, Virginia, Kentucky, and Tennessee: TRW Inc., Report to DOE/METC, Contract No. DE-AC21-81 MC 14900.

Arkle, Jr., T., 1974, Stratigraphy of the Pennsylvanian and Permian systems of the Central Appalachians, *in* G. Briggs, ed., Carboniferous of the southeastern United States: U.S. Geol. Sur. Spec. Paper 148, p. 5–29.

——— , et al, 1979, West Virginia and Maryland, *in* J. W. Skehan et al, eds., The Mississippian and Pennsylvanian (Carboniferous) systems in the United States: U.S. Geological Survey Professional Paper 1110-A-L, Sec. D, 35 p.

Ashley, G. H., 1928, Bituminous coal fields of Pennsylvania: Pennsylvania Geological Survey, 4th Series, Bulletin M6, Pt. 1, 241 p.

Averitt, P., 1975, Coal resources of the United States, January 1, 1974: United States Geological Survey Bulletin 1412, 131 p.

Barlow, J. A., (ed.), 1975, The age of the dunkard; Proceedings of the First I. C. White Memorial Symposium, September 25–29, 1972: West Virginia Geological and Economic

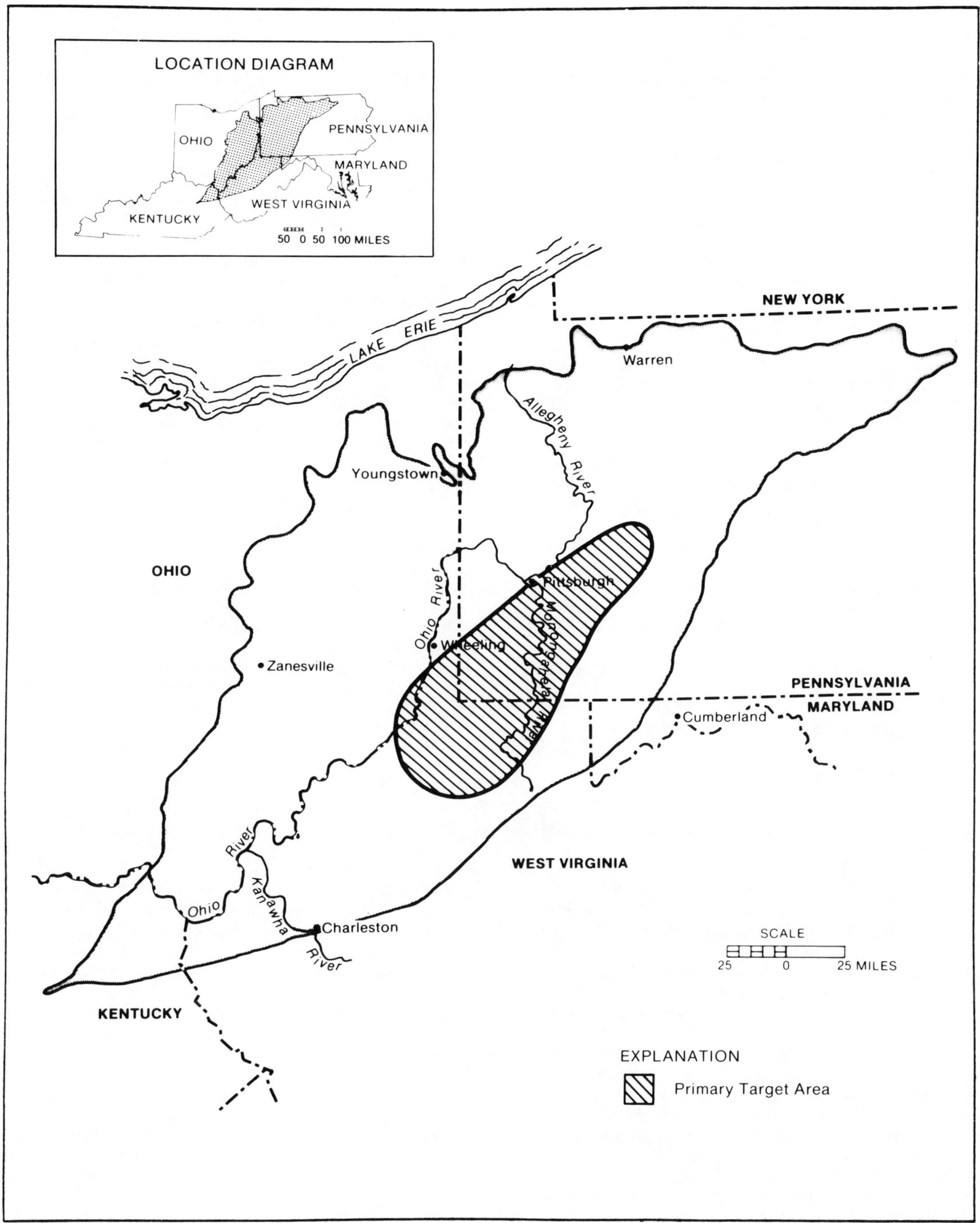

Figure 18—Primary methane target area in the Northern Appalachian Coal Basin.

Survey, 352 p.

Berryhill, Jr., H. L., and V. E. Swanson, 1962, Revised stratigraphic nomenclature for Late Pennsylvanian and Early Permian rocks, Washington County, Pennsylvania, *in* Short papers in geology and hydrology: U.S. Geological Survey Professional Paper 450-C, p. C43–C46.

——, S. E. Schweinfurth, and B. H. Kent, 1971, Coal-bearing Upper Pennsylvanian and Lower Permian rocks, Washington area, Pennsylvania: U.S. Geological Survey Professional Paper 621, 47 p.

Brant, R. A., and R. M. DeLong, 1960, Coal resources of Ohio: Ohio Division of Geological Survey Bulletin 58, 245 p.

Chow, M. M., 1951, The Pennsylvanian Mill Creek limestone in Pennsylvania: Pennsylvania Geological Survey, 4th Section, Bulletin G26, 36 p.

Clendening, J. A., 1969, The base of the Permian in the Appalachian; Ninety years of controversy, *in* A. C. Donaldson, ed., Some Appalachian coals and carbonates, models of ancient shallow-water deposition: Geological Society of America Coal Division, Preconvention Field Trip, November 1969, p. 229–233.

Collins, H. R., 1979, Ohio, *in* J. W. Skehan et al, eds., The Mississippian and Pennsylvanian (Carboniferous) Systems in the United States: U.S. Geological Survey Professional Paper 1110-A-L, 379 p.

Colton, G. W., W. J. Perry, and J. D. Mackenzie, 1981, Joint patterns in coal in the Central and Northern Appalachian Basin: U.S. Geological Survey Open-File Report 81-837.

Cross, A. T., and M. P. Schemel, 1956a, Geology and economic resources of the Ohio River Valley, *in* West Virginia, Pt. 1, Geology: West Virginia Geological and Economic Survey, v. XXII, 149 p.

—— and——, 1956b, Geology and economic resources of the Ohio River Valley, *in* West Virginia, Pt. 2: West Virginia Geological and Economic Survey, v. XXII, 129 p.

Dennison, J. M., 1978, Stratigraphic distribution of decollements in the Appalachian basin: Geological Society of America Abstracts with Programs, v. 10, n. 4, p. 167.

Diamond, W. P., and J. R. Levine, 1981, Direct method determinations of the gas content of coal, procedures and results: U.S. Bureau of Mines Report of Investigations 8515, 36 p.

—— D. C. Oyler, and H. H. Field, 1977, Directionally controlled drilling to horizontally intercept selected strata, Upper Freeport Coalbed, Greene County, Pennsylvania: U.S. Bureau of Mines Report of Investigations 8231, 21 p.

Donaldson, A. C., 1969, Ancient deltaic sedimentation (Pennsylvanian) and its control on the distribution, thickness and quality of coals, *in* Some Appalachian coals and carbonates, models of ancient shallow-water deposition: Geological Society of America Coal Division, Preconvention Field Trip, November, 1969: Morgantown, West Virginia, West Virginia Geological and Economic Survey, p. 357–384.

——, 1974, Pennsylvanian sedimentation of central Appalachians, *in* G. Briggs, ed., Carboniferous of the southeastern United States: Geological Society of America Special Paper 148, p. 47–48.

Edmunds, W. E., and T. M. Berg, 1971, Geology and mineral resources of the southern half of the Penfield 15-minute quadrangle 9 Pennsylvania: Pennsylvania Geological Survey 4th Series, Atlas 74cd, 184, p. 22.

Ferm, J. C., 1974, Carboniferous environmental models in eastern United States and their significance, *in* G. Briggs, ed., Carboniferous of the southeastern United States: Geological Society of America Special Paper 148, p. 79–95.

—— and V. V. Cavaroc, Jr., 1969, A field guide to Allegheny deltaic deposits in the upper Ohio Valley, with a commentary on deltaic aspects of Carboniferous rocks in the northern Appalachian Plateau (1969 Spring Field Trip): Ohio Geological Society and Pittsburgh Geological Society, 19 p.

Frantz, R. L., et al, 1981, Demonstration of methane recovery from Pennsylvania coal seams: prepared for TRW under Subcontract No. H1216 5JJ95, 116 p.

Gillespie, W. H., and J. A. Clendening, 1969, Age of dunkard group, Appalachian Basin: XI Botanical Congress, Seattle, Washington.

Gwinn, V. E., 1964, Thin-skinned tectonics in the Plateau and northeastern Valley and Ridge provinces of the central Appalachians: Geological Society of America Bulletin, v. 75, n. 9, p. 863–900.

Headlee, A. J., and J. P. Nolting, Jr., 1940, Characteristics of minable coals of West Virginia: West Virginia Geological Survey Report 13, 272 p.

Huddle, J. W., et al, 1963, Coal reserves of eastern Kentucky: U.S. Geological Survey Bulletin 1120, 247 p.

Iannacchione, A. T., and D. G. Puglio, 1979, Geology of the Lower Kittanning coalbed and related mining and methane emission problems in Cambria County, Pennsylvania: U.S. Bureau of Mines Report of Investigations 8354, 32 p.

Irani, M. C., P. W. Jeran, and M. Deul, 1974, Methane emission from U.S. coal mines in 1973. A survey—supplement to Information Circular 8558: U.S. Bureau of Mines Information Circular 8659, 47 p.

——, et al, 1972, Methane emissions from U.S. coal mines, a survey: U.S. Bureau of Mines Information Circular 8558, 58 p.

——, et al, 1977, Methane emission from coal mines in 1975, a survey—a supplement to Information Circulars 8558 and 8659, 55 p.

Jeran, P. W., D. H. Lawhead, and M. C. Irani, 1976, Methane emissions from an advancing coal mine section in the Pittsburgh Coalbed: U.S. Bureau of Mines Report of Investigations 8132, 10 p.

Keystone, 1981, Keystone coal industry manual: New York, Mining Informational Services, McGraw-Hill Mining Pub.

Kulander, B. R., and S. L. Dean, 1980, Fracture trends in the Allegheny Plateau of West Virginia: West Virginia Geological and Economic Survey Map WV-11.

Lambert, S. W., M. A. Trevits, and P. F. Steidl, 1980, Vertical borehole design and completion practices to remove methane gas from minable coalbeds: U.S. DOE, DOE/CMTC/TR-80/2, Carbondale Energy Technology Center, 163 p.

Lobeck, A. K., 1948, Physiographic provinces of North America: New York, The Geographical Press, Columbia University, Map.

McCulloch, C. M., and M. Deul, 1973, Geologic factors causing roof instability and methane emission problems, the Lower Kittanning Coalbed, Cambria County, Pennsylvania; U.S. Bureau of Mines Report of Investigations 7768, 25 p.

——, ——, and P. W. Jeran, 1974. Cleat in bituminous coalbeds: U.S. Bureau of Mines Report of Investigations 7910, 25 p.

Meckel, L. D., 1967, Origin of the Pottsville conglomerates (Pennsylvanian) in the central Appalachians: Geological Society of America Bulletin, v. 78, n. 2, p. 223–257.

Oyler, D. C., W. P. Diamond, and P. W. Jeran, 1979, Directional drilling for coalbed degasification: U.S. Bureau of Mines Report of Investigations 8380, 15 p.

Pennsylvanian Topographic and Geologic Survey, undated, Distribution of Pennsylvania coal: Map 11.

Puglio, D. G., 1981, Evaluating geologic conditions and methane contents of coalbeds through exploration programs in the northern Appalachian coalfield: Proceedings, 3rd International Coal Exploration Symposium, 23–26 August, 1981, Calgary, AB, 16 p.

Rice, C. L., and J. H. Smith, 1980, Correlation of coalbeds, coal zones, and key stratigraphic units in the Pennsylvanian rocks of eastern Kentucky: U.S. Geological Survey Miscellaneous Field Studies, Map MF-1188.

——— , et al, 1979, Kentucky, *in* J. W. Skehan et al, eds., The Mississippian and Pennsylvanian (Carboniferous) systems in the United States: U.S. Geological Survey Professional Paper 1110-A-L, 379 p.

Rogers, J., 1963, Mechanics of Appalachian foreland folding in Pennsylvania and West Virginia: Bulletin of the American Association of Petroleum Geologists, v. 47, p. 1527–1536.

Sholes, M. A., and V. W. Skema, 1974, Bituminous coal resources in western Pennsylvania: Pennsylvania Geological Survey, Mineral Resources Report, 68 p.

Sisler, J. D., 1928, Geology of Pennsylvania bituminous coals, *in* Bituminous coal fields of Pennsylvania: Pennsylvania Geological Survey, 4th Series M6, Pt. IV, 268 p.

Stauffer, C. R., and C. R. Schroyer, 1920, The dunkard series of Ohio: Ohio Geological Survey, 4th Series, Bulletin 22, 167 p.

Stout, W., 1946, Mineral resources of Ohio: Geological Survey of Ohio, Fourth Series, Information circular No. 1 (revised).

TRW, 1977, Methane produced from coalbeds, systems studies of energy conservation: U.S. Department of Energy, MERC/CR-77/4, under contract E(46-1)-8042, 2 v.

Wanless, H. R., 1975, Appalachian region, *in* E. D. McKee and E. J. Crosby, coordinators, Introduction and regional analyses of the Pennsylvanian system, Pt. 1, of paleotectonic investigations of the Pennsylvanian system in the United States: U.S. Geological Survey Professional Paper 853, pt. 1, p. 17–62.

Weaver, K. N., J. M. Coffroth, and J. Edwards, Jr. 1976, Coal reserves in Maryland—Potential for future development: Maryland Geological Survey Informational Circular 22, 16 p.

Westinghouse Electric Company, 1978, Coalbed methane extraction and utilization—7th monthly progress report: contract EW-78-C-21-8332, November, 1978, DOE Contract Office, Oak Ridge Ops.

West Virginia Geological and Economic Survey, 1980, coal rank map of the bituminous coals of West Virginia: West Virginia Geological and Economic Survey Plate WV-8.

Williams, E. G., and W. A. Bragonier, 1974, Controls of Early Pennsylvanian sedimentation in Western Pennsylvania, *in* G. Briggs, ed., Carboniferous of the southeastern United States: Geological Society of America Special Paper 148, p. 135–152.

Zabetakis, M. G., et al, 1972, Methane emission in coal mines: Effects of oil and gas wells: U.S. Bureau of Mines Report of Investigations 7658, 9 p.

Geologic Overview, Coal Resources, and Potential Methane Recovery from Coalbeds of the Central Appalachian Basin—Maryland, West Virginia, Virginia, Kentucky, and Tennessee

M. A. Adams

The Central Appalachian Basin occupies an estimated 22,850 sq mi over parts of Maryland, West Virginia, Virginia, Kentucky, and Tennessee. The Mississippian and Pennsylvanian strata that underlie the basin contain numerous coalbeds that have generated significant amounts of methane. Data for analyzing and estimating this resource were obtained from desorption data published by Diamond and Levine (1981); gas data from the USGS/R-9 well in Clay County, Kentucky; data from the horizontal borehole project at Island Creek Coal Company's Virginia Pocahontas No. 5 mine; and the Clinchfield Coal Company's Jawbone coalbed methane drainage project. Mine methane emission studies published by MSHA have been investigated, and these data have indicated that the coalbeds—especially the Pocahontas No. 3 and No. 4 coals—emit substantial quantities of methane. The Beckley mine in Raleigh County, Virginia, was monitored in 1974 to measure methane emission rates during mining.

The methane potential of this area has been estimated from the published gas content information, related projects, mine emission data, and methane-related incidents. It has been estimated that the basin contains a potential in-place methane resource ranging from a low of 10 to a high of 48 trillion cubic feet.

A coalbed methane primary target area for the Central Appalachian Basin has been defined over an area of southwest Virginia, eastern Kentucky, and southeast West Virginia. This area contains the highest measured coalbed gas content values, and other available project data suggest this area contains exceedingly high amounts of methane.

INTRODUCTION

The Central Appalachian Basin incorporates approximately 22,850 sq mi, including parts of eastern Kentucky, western Maryland, eastern Tennessee, southwestern Virginia, and southern West Virginia (Fig. 1). The area covers parts of two physiographic provinces: the Appalachian Plateau, and the Valley and Ridge. The Appalachian Plateau province includes the majority of the Central Appalachian area; the Valley and Ridge covers only the southwest portion of Virginia, a small portion of east-central Tennessee, and Kentucky to the south of the Pine Mountain Fault (USGS and USBM, 1968).

The Appalachian Plateau province, located west of the Valley and Ridge province, is divided into the Allegheny Mountain, Kanawha, Cumberland Mountain, and Cumberland Plateau sections within the basin. The province is a high, horizontal to gently dipping upland, with numerous stream dissections traversing the area. The eastern ridge of the plateau is the highest part, rising from 1,000 to 3,000 ft above the valleys to the east. Along this ridge, the Central Appalachian's altitudes range from 4,800 ft in central West Virginia to 3,000 ft or less in Tennessee. West of the eastern ridge, the altitude decreases to approximately 1,100 to 2,000 ft in central Kentucky and Tennessee. Relief to the west in West Virginia, eastern Kentucky, and northern Tennessee ranges from approximately 500 to 1,500 ft.

The Valley and Ridge province has been characterized as an elongated, narrow band of alternating northeast-southwest–trending mountains and valleys, formed from the highly folded and faulted structure of the region (USGS and USBM, 1968). The area is composed of extensive thrust fault systems, synclines, and anticlines whose trends generally parallel the northeast-southwest trend of the Appalachians.

STRUCTURAL GEOLOGY

The Appalachian Plateau province portion of this basin consists of broad anticlines and synclines from Maryland to Tennessee. Major structures within the province are the Upper Potomac Basin in western Maryland; the Rome Trough in West Virginia; and the Irvine–Paint Creek Fault System, Warfield Fault, Paint Creek Uplift, and Rockcastle Uplift in eastern Kentucky.

The hingeline of the Rome Trough, which divides West Virginia along a northeast-southwest line, separates the northern and southern coal fields in West Virginia and has also been used in this report as the boundary between the Central

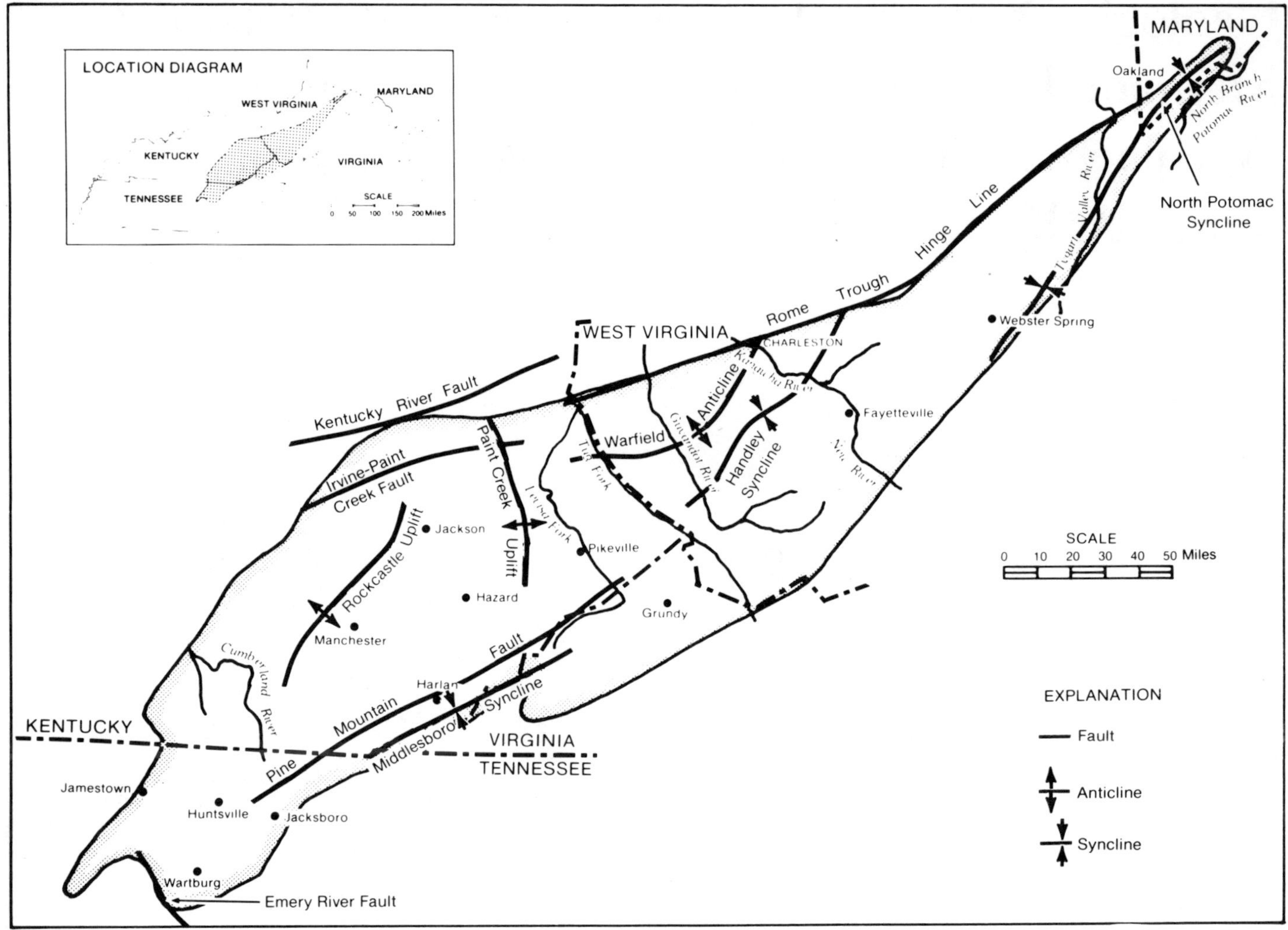

Figure 1—Prominent structural features within and surrounding the Central Appalachian Basin.

Appalachian Basin and the Northern Appalachian Coal Basin in West Virginia. The trough is a graben structure that formed during Lower Cambrian time and remained active throughout the Paleozoic. The trough affected sedimentation patterns in the Devonian, Mississippian, and Pennsylvanian systems.

The boundary separating the Central Appalachian Basin from the Northern Appalachian Coal Basin in eastern Kentucky is located north of the Irvine–Paint Creek Fault system and south of the Kentucky River Fault system. To the north of the Irvine–Paint Creek Fault, the strata's lithologic character correlates to the strata of West Virginia and Ohio in the Northern Appalachian Coal Basin. South of the fault, the rocks thicken and contain more sandstone and conglomerate—more indicative of the Central Appalachian Basin stratigraphy (Huddle et al, 1963).

The Valley and Ridge province is structurally more intense than the Appalachian Plateau province. It contains numerous anticlines, synclines, and overthrust faults with the strata in the area ranging from moderately to steeply dipping. Overturned beds are known to occur in some localities. The largest fault zones in this area are the Pulaski, Russell Fork, Saltville, St. Paul, Clinchport, Jacksboro, Emery River, and Pine Mountain Faults.

The Pine Mountain Fault is the most extensive fault in the basin. It crops out in southeast Kentucky and adjacent Tennessee, and the Pine Mountain overthrust block associated with the fault is bounded on the northwest and southeast by thrust faults and on the northeast and southwest by tear faults (Fig. 2). The Pine Mountain Fault is the northwest boundary of the overthrust block; the Russell Fork Fault marks the northeast; the St. Paul or Hunter Valley Fault bounds the southeast; and to the southwest in Tennessee, the Jacksboro Fault limits the thrust sheet (Miller, 1974). The overthrust block extends 125 mi from southwestern Virginia through southeastern Kentucky to northeastern Tennessee and is approximately 25 mi wide (Rice et al, 1980). The block trends northeast to southwest and parallels Appalachian folding (Miller, 1974). After Pennsylvanian strata were deposited, the block was thrust to the northwest a distance of approximately 10 mi at the southwest end of the block in Tennessee and 5.8 mi at the Tennessee-Virginia border (Wilson and Stearns, 1958).

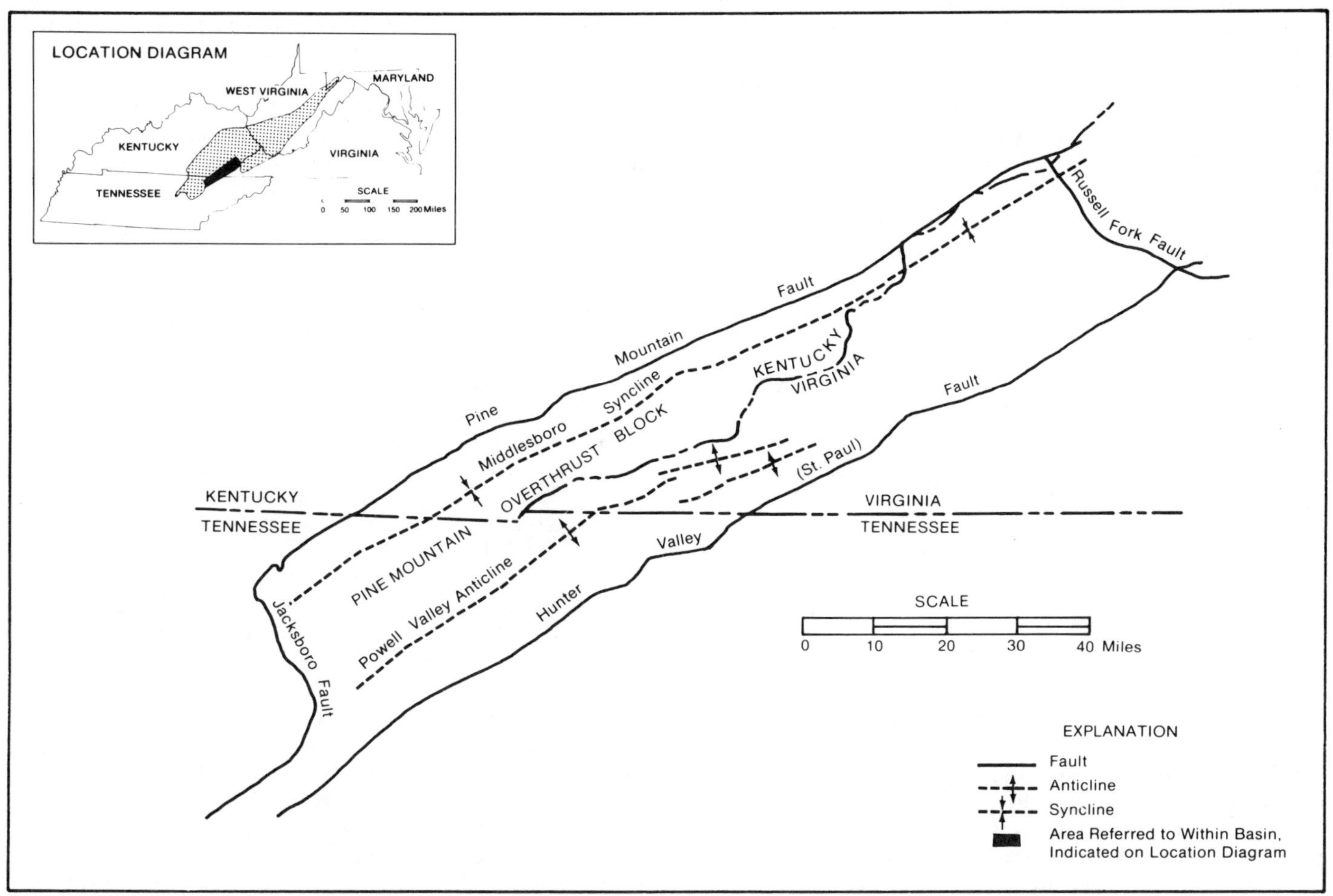

Figure 2—Structure of the Pine Mountain Overthrust Block. From Thornbury. Copyright © 1965. Reprinted by permission of John Wiley and Sons, Inc.

STRATIGRAPHY

Figure 3 is a correlation chart of strata nomenclature for the Appalachian Basin.

Deposition of Mississippian and Pennsylvanian Strata

Depositional environments of Carboniferous strata in the Central Appalachian Basin reflect the fluctuation of marine and continental environments in a shallow, slowly subsiding basin. Regressing and transgressing seas deposited Mississippian strata, with shallow marine seas depositing most of the shale members. Sandstones were deposited in high-energy coastal environments. Mississippian coals and carbonaceous shales with fluvial sandstones indicate that a broad coastal swamp was the dominant depositional environment.

Throughout the basin, deposition was fairly continuous from Late Mississippian through Pennsylvanian time. In a few areas, such as northwest eastern Kentucky, Pennsylvanian sediments were deposited on eroded Mississippian strata.

Pennsylvanian Stratigraphic Units

Western Maryland and Northeast West Virginia Stratigraphy

In western Maryland and northeast West Virginia, the Pennsylvanian formations are, in ascending order: the Pottsville, Allegheny, Conemaugh, and Monongahela. These formations are separated by persistent coalbeds. The Pottsville Formation extends from the top of the Mississippian Mauch Chunk Formation to the Brookville coal. The Allegheny Formation occurs from the top of the Brookville coal to the top of the Upper Freeport coal. The Conemaugh Formation extends from the top of the Upper Freeport coal to the base of the Pittsburgh coal. The Monongahela Formation is not extensive in the Central Appalachian Basin.

Pottsville Formation—The Pottsville Formation averages 400 ft thick and is composed of sandstone with interbedded siltstone and shale. The basal part of the formation is the Shawn Sandstone Member, a thick sequence of sandstone and conglomerate. The member is resistant to erosion and forms ridges and mountains. Above the Shawn Sandstone Member are siltstones, silty clays, shales, and three thin coalbeds (Vokes and Edwards, 1974).

The upper lithology of the Pottsville Formation consists of siltstone and sandstone, including the Upper Connoquenessing and Homewood Sandstones. In western Maryland, the Upper Connoquenessing Sandstone is separated from the uppermost Homewood Sandstone by a thin zone of silty and carbonaceous shale with a lenticular coal, the Mercer (Vokes and Edwards, 1974).

Allegheny Formation—The Allegheny Formation consists of cyclic sequences of sandstone, siltstone, shale, limestone, and coal. It contains more coalbeds than the underlying Pottsville Formation. The sandstones are massive, and vary in grain size from conglomeratic to fine-grained. The formation contains a few nonmarine plant fossils and varies between 275 and 325 ft thick in western Maryland (Vokes and Edwards, 1974).

Conemaugh Formation—The Conemaugh Formation occurs between the top of the Upper Freeport coalbed and the base of the Pittsburgh coalbed. In the Upper Potomac Basin, a complete section ranges from 835 to 925 ft thick, and in West Virginia, from 750 to 850 ft thick. The formation is composed of claystone, shale, sandstone, limestone, red shale, marine shale, and coalbeds. This formation is divided into the Lower and Upper Members (Vokes and Edwards, 1974).

Monongahela Formation—The Monongahela Formation caps one hill in the Upper Potomac Basin in western Maryland and is found in Mineral and Grant Counties in West Virginia. The formation whose base is marked by the Pittsburgh coal is composed of nonmarine cyclic sequences of sandstone, siltstone, shale, limestone, and coal and is approximately 170 ft thick in West Virginia (Cardwell et al, 1968).

Southwest Virginia and Southeast West Virginia Stratigraphy

In southwest Virginia, the Pennsylvanian consists of, in ascending order: the Pocahontas, Lee, Norton, New River, and Kanawha Formations of the Pottsville Group.

Pocahontas Formation—The Pennsylvanian Pocahontas Formation is composed of sandstone, siltstone, shale, coal, and underclay. This formation includes the Pocahontas Nos. 1 through 7 coals. The sandstone is gray, fine to medium grained, poorly sorted, micaceous, feldspathic, and low in quartz. A few quartzose sandstones occur in the formation and are very fine to fine grained, silty, and interbedded with thin dark shales and siltstones. The coals within the formation are medium- to low-volatile bituminous, with low sulfur contents (Miller, 1974).

The intertonguing of the Pocahontas and the underlying Upper Mississippian Bluestone Formations is difficult to trace in the subsurface, but it does not affect any of the major coals. The boundary between the Pocahontas and overlying Lee Formations is distinct to the northeast where the basal white quartzose conglomerate sandstone of the Lee Formation unconformably overlies the dark gray shale, siltstone, and sandstone of the Pocahontas (Miller, 1974).

The Pocahontas Formation is thickest in southeastern Buchanan County and Tazewell County, Virginia, where it ranges from 600 to 750 ft thick. The formation thins to the west and northwest and is absent in northwestern Wise, Buchanan, and Dickenson counties, Virginia (Miller, 1974). In southeast West Virginia, the formation ranges from 450 to 720 ft thick (Cardwell et al, 1968).

Lee Formation—The Lee Formation in southwest Virginia contains three prominent ridge-forming quartz arenite beds. The arenites are fine to coarse and rounded to subangular (Miller, 1974).

The quartz arenites have mostly massive bedding with cross-bedding, cross-lamination, and cross-stratification. The arenites' basal contact is sharp, undulating, and disconformable; while the upper contact is gradational. Thin shale, siltstone, sandstone, coal, and underclay separate the three quartz arenites. To the northwest in southwest Virginia, the arenite beds comprise an estimated one-third of the formation's total thickness. To the southeast and east, the quartz arenites tongue out, leaving gray, micaceous, partly carbonaceous siltstone and shale plus fine- to coarse-grained, poorly sorted, gray sandstone (Miller, 1974).

Norton Formation—The Norton Formation in Virginia is located between the top of the Lee Formation and the base of the Gladeville Sandstone. The formation contains shale and siltstone with lesser amounts of sandstone, coal, and underclay. The shale and siltstone are gray to dark gray, locally carbonaceous, and laminated, and contain plant fossils. The sandstone ranges from light gray to gray, is fine to medium grained, clay-rich, silty, micaceous, and locally feldspathic. The Norton Formation thickens to the east and southeast because the underlying Lee Formation arenites tongue out in those directions (Miller, 1974).

In western Wise, northwest Dickenson, and northwest Buchanan counties where the base of the Norton Formation is limited by the Bee Rock Sandstone, the formation ranges from 500 to 800 ft thick. To the southeast and east of these areas, the Norton is limited by the middle quartz arenite of the Lee Formation (Miller, 1974).

New River Formation—The New River Formation is located in northwest Tazewell County and is correlated to West Virginia strata. The formation includes interbedded sandstone, siltstone, shale, coal, and underclay. The sandstone ranges from light gray to gray and is fine to coarse grained, silty, micaceous, and thinly bedded. The siltstone and shale are gray to dark gray, micaceous, locally carbonaceous, and laminated. The coal and underclays are usually thick and persistent (Miller, 1974).

Kanawha Formation—The Kanawha Formation in West Virginia is composed of sandstone, shale, siltstone, and coal. It extends from the top of the Upper Nuttall Sandstone to the top of the Homewood Sandstone. The formation includes the Douglas, Gilbert, Eagle, Powellton, Campbell Creek, Peerless, Alma, Adar Grove, Williamson, Chilton, Winifrede, Coalburg, and Mercer coals. It ranges from 1,725 to 2,100 ft thick (Cardwell et al, 1968).

Kentucky Stratigraphy

In eastern Kentucky, the Pennsylvanian system includes, in ascending order: part of the Pennington (also top of Mississippian), Lee, Breathitt, Conemaugh, and Monongahela Formations.

Pennington Formation—In southeastern Kentucky, the Pennsylvanian-Mississippian boundary occurs in a gradational sequence between the Mississippian Little Stone Gap Member of the Pennington Formation and the base of the Lee Formation. The Pennington Formation consists of varicolored clay shale, brown-gray siltstone, coarse- to fine-grained orthoquartzite, and fine-grained silty sandstone. Some coals and underclay occur near the top of the Pennington Formation and may be truncated locally by the basal member of the Lee Formation (Rice et al, 1980).

Lee Formation—The Lee Formation is composed of sandstone, conglomerate, siltstone, and shale. The formation is exposed along the western margin of the coal field at the Pottsville Escarpment; in the Irvine–Paint Creek Uplift; at Pine Mountain, where the rocks are upturned by faulting; and along the Cumberland Mountain, at the southeast border of the field (Huddle et al, 1963).

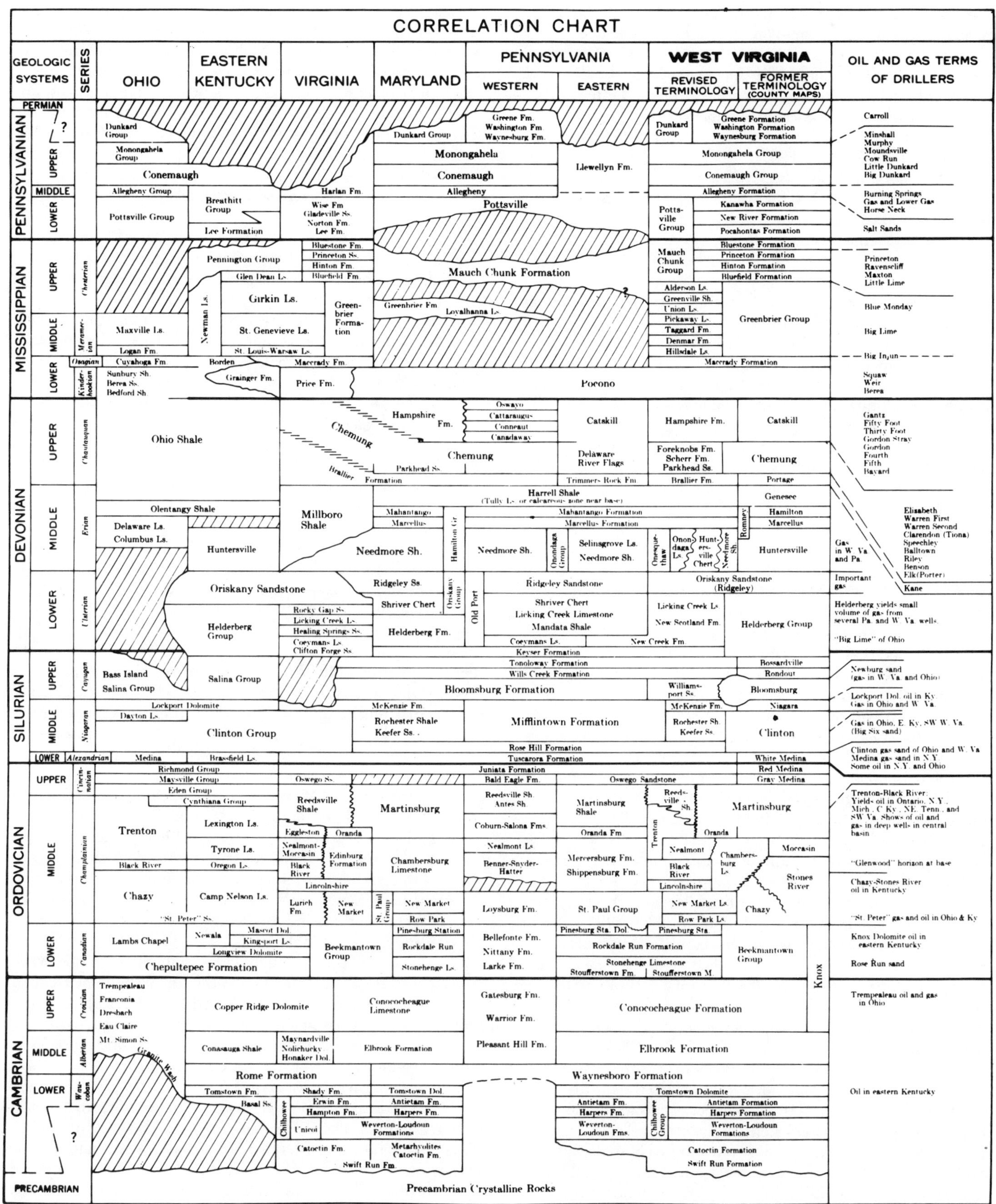

Figure 3—Stratigraphic correlation chart for the Central Appalachian Basin. Cardwell et al. Copyright © 1968 West Virginia Geological and Economic Survey. Used with permission.

In a few areas, a disconformity occurs at the base of the Lee Formation. Near the Irvine–Paint Creek Uplift, the Lee Formation lies disconformably on sandstone, shale, or limestone beds that are Early to Late Mississippian in age. In the Pine and Cumberland Mountains, the formation grades and intertongues with the Mississippian Bluestone Formation. The Lee Formation's contact with the overlying Breathitt Formation is gradational (Huddle et al, 1963).

The formation decreases from 670 ft at the Kentucky-Tennessee state line to 100 ft at the Ohio River in the Northern Appalachian Basin. In the Pine and Cumberland Mountain area, the Lee Formation ranges from 1,000 to 1,500 ft thick. In this area, the formation is composed of sandstone separated by thin zones of shale and coal. This sandstone grades into thinner and fewer shale and siltstone beds to the northwest in Kentucky and intertongues or grades into subgraywacke in southwest Virginia (Huddle et al, 1963).

Breathitt Formation—The Breathitt Formation is Middle Pennsylvanian age and includes strata between the Lee and Conemaugh Formations. Along the Pottsville Escarpment, all Pennsylvanian strata except the named members of the Lee Formation are referred to as the Breathitt Formation (Rice et al, 1980). Over most of eastern Kentucky, the Breathitt Formation is found at the surface, although in a few localities, the entire formation has been removed by erosion (Huddle et al, 1963).

The formation thickens to the south as indicated by the change in strata thickness between the Upper Elkhorn No. 3 and Hazard coals. The formation is about 300 ft thick in the north and reaches 1,300 ft thick south of Pine Mountain (Huddle et al, 1963).

The Breathitt Formation consists of shale, siltstone, sandstone, and coal. The shale is dark or light gray and contains fossils and siderite (also referred to as ironstone). The siltstone and sandstone contain fossils, clay minerals, and rock fragments. The siltstone and clay shale commonly intergrade, are carbonaceous, and contain large argillaceous concretions. The sandstone in the upper half of the formation is feldspathic. Siderite is found as either thin lenses or continuous layers of nodules up to 3 ft thick in shale or claystone. Coals are numerous in the formation (Huddle et al, 1963; Rice et al, 1980).

The Breathitt is divided into units using four marine zones and coalbeds that have distinctive properties and are widespread. The key units are the Fire Clay coal, the Magoffin Member, the Kendrick Shale of Jillson, and Upper Elkhorn No. 3 coal. The Fire Clay coal is hard and contains a flint-clay parting in the lower third of the coalbed, with the coal above or below the parting missing in a few locations. Lithologically, the two marine zones—the Magoffin Member and Kendrick Shale—resemble one another. They generally coarsen upward, are argillaceous, and contain sand (Rice et al, 1980). The Magoffin Member has the wider distribution of the two marine zones but thins and becomes discontinuous to the northeast in Kentucky. The Kendrick Shale Member occurs in the eastern and central parts of eastern Kentucky; south of Pine Mountain, it is not recognized. The member is composed of shale, siltstone, and sandstone plus limestone concretions that average 2 ft in diameter. The Upper Elkhorn No. 3 coal has no distinctive identifications but has been correlated from mine to mine. The Upper Elkhorn No. 3 contains more remaining reserves than any other coal in eastern Kentucky (Huddle et al, 1963).

Conemaugh Formation—The Late Pennsylvanian Conemaugh Formation is found only in Lawrence County, Kentucky, of the Central Appalachian Basin. The Upper Freeport coal of the Breathitt Formation and the Mahoning Sandstone Member of the Conemaugh Formation mark the formation's boundaries. The formation consists of shale, siltstone, and sandstone, with several marine limestones and chert beds. Two insignificant coals are located in the formation (Huddle et al, 1963).

Tennessee Stratigraphy

The Pennsylvanian strata of the Cumberland Plateau in Tennessee consist of sandstone, shale, coal, siltstone, limestone, and siderite. The sandstone ranges from conglomerate with rounded quartz pebbles to a very fine-grained sandstone. The coarse-grained sandstone is well sorted and cross bedded. The finer grained sandstone is not as well sorted and contains clay- to silt-sized particles and shale partings (Wilson, Jewell, and Luther, 1956).

The siltstone generally has gradational lithology ranging from silty or sandy shale to fine-grained shaly sandstone. Shale is the most abundant type of lithology in the Pennsylvanian of Tennessee. The shale is generally silty, sandy, or clayey. The silty shale is gray to brown and ranges from silt- to sand-sized particles. Shale is generally nonfissile and finely and irregularly laminated (Wilson, Jewell, and Luther, 1956).

Wilson, Jewell, and Luther (1956) divided the Pennsylvanian System in Tennessee into the following groups, in ascending order: the Gizzard, Crab Orchard Mountain, Crooked Fork, Slatestone, Indian Bluff, Red Oak Mountain, Vowell Mountain, and Cross Mountain Groups.

In east-central Tennessee, the Gizzard Group includes strata between the top of the Mississippian and the base of the Pennsylvanian Sewanee Conglomerate. The group ranges up to 700 ft thick and is thickest along the southeast border of the Cumberland Plateau. The group consists of three formations: Raccoon Mountain, Warren Point Sandstone, and Signal Point Shale (Wilson, Jewell, and Luther, 1956).

The Crab Orchard Mountain Group includes strata from the top of the Gizzard Group to the top of the Rockcastle Conglomerate. At the group's type locality in Cumberland County, the strata are approximately 640 ft thick. The group forms a wedge that is thickest to the east and is divided into the Sewanee Conglomerate, Whitwell Shale, Newton Sandstone, Vandever Formation, and Rockcastle Conglomerate (Wilson, Jewell, and Luther, 1956).

The Crooked Fork Group includes strata between the top of the Rockcastle Conglomerate and the top of the Poplar Creek coal and averages from 300 to 400 ft thick. The unit has a high concentration of sandstone and occurs mainly in the northern Cumberland Plateau. In some areas, the sandstone becomes a thick conglomerate over the Rockcastle Conglomerate (Wilson, Jewell, and Luther, 1956). The group has alternating sequences of three sandstones and three shales. The units are, in ascending order: Dorton Shale, Crossville Sandstone, Burnt Mill Shale, Coalfield Sandstone, Glenmary Shale, and Warthing Sandstone.

The Poplar Creek coal, which marks the top of the Crooked Fork Group, is generally found within 5 ft of the top of the Wartburg Sandstone. This coal is found in Anderson, Claiborne, and Scott counties and is traceable into Kentucky (Wilson, Jewell, and Luther, 1956).

The Slatestone Group is located between the top of the Poplar Creek coal and the top of the Jellico coal. The group ranges from 420 to 720 ft thick and thins to the northwest. It includes four sandstone units: the Stephens, Petros, Sand Gap, and Newcomb Sandstone, plus the Jellico coal. Separating the sandstones are four shale intervals. In Campbell County, the Jellico coal is found a few feet from the top of the Newcomb Sandstone. This coal is mined in Morgan, Anderson, Campbell, and Scott counties and on the Pine Mountain overthrust block, where it is known as the Log Mountain or Mingo coal (Wilson, Jewell, and Luther, 1956).

The Indian Bluff Group ranges from 150 to 470 ft thick and includes four sandstone units separated by four shale intervals with minor coal and sandstone. The sandstone units are the Seeber Flats, Stockstill, Indian Fork, and Pioneer. In the northeast part of the Cumberland Mountains and on the thrust sheet, both the Stockstill and Indian Fork Sandstones are absent (Wilson, Jewell, and Luther, 1956).

The Graves Gap Group, located between the top of the Pioneer Sandstone and the top of the Windrock coal, averages 200 to 365 ft thick. The group includes the Armes Gap and Roach Creek Sandstones, which are separated by three shale units containing minor sandstone and coal. The Windrock coal marks the top of the Graves Gap Group. The coal can be recognized by an underlying thin flint clay. This flint clay is hard and breaks with a conchoidal fracture (Wilson, Jewell, and Luther, 1956).

The Redoak Mountain Group ranges from 300 to 440 ft thick and includes the strata between the Windrock and Pewee coals. This group consists of three sandstone units—the Caryville, Fodderstack, and Silvery Gap Sandstones—separated by four shale units that contain minor amounts of coal and sandstone (Wilson, Jewell, and Luther, 1956).

The Vowell Mountain Group ranges from 230 to 390 ft thick and includes the Pilot Mountain and Frozen Head Sandstones, which are separated by three intervals of shale with minor sandstone and coal (Wilson, Jewell, and Luther, 1956).

The Cross Mountain Group includes the Low Gap and Tub Spring Sandstones separated by three shale intervals. This is the youngest group of Pennsylvanian rocks in Tennessee. These strata are only found on higher peaks and ridges in the Cumberland Mountains (Wilson, Jewell, and Luther, 1956).

COAL RESOURCES

The sequence of strata of the Mississippian and Pennsylvanian periods was the result of a series of northwest-prograding deltas. Vegetation for the formation of coal flourished in the delta swamps, resulting in a series of coalbeds distributed vertically throughout the sequence. The sizes and shapes of the various deltaic complexes were variable, limiting the areal distribution of some of the coals. Because of this limited distribution, it is difficult to correlate the coals over large areas. Coalbeds with available desorption data and those indicated as key marker beds will be discussed on an individual basis.

The majority of the coal in the basin is high-volatile bituminous with some medium- and low-volatile bituminous. Analyses are available for numerous coalbeds and are listed in Table 1.

Western Maryland and Northeast West Virginia

Coal in northeastern West Virginia and western Maryland occurs in the Allegheny, Conemaugh, and Monongahela Groups (Fig. 4). The Allegheny Formation consists primarily of sandstone and shale, with lesser amounts of clay, coal, and limestone.

The Conemaugh Group, in Mineral and Grant counties, West Virginia, consists mainly of shale, sandstone, and nonmarine limestone, with lesser amounts of marine shale, coal, and clay. There is very little minable coal in the Conemaugh Group.

The Monongahela Group is restricted to the Northern Appalachian Coal Basin, except with erosional remnants in Mineral and Grant counties in West Virginia. The group consists of shale, nonmarine limestone, and sandstone with limited amounts of coal and clay (Keystone, 1980). Coals within this group will not be discussed on an individual basis, owing to their lack of depth and extent. A detailed discussion of the Monongahela Group can be found in the Northern Appalachian report that appears in this volume.

Allegheny Group

Clarion Coalbed—The Clarion coal is multiple bedded, ranging from 2 to 10 ft thick. The coal has an estimated original minable tonnage for West Virginia of 1.3 billion (Keystone, 1980).

Lower Kittanning Coalbed—The Lower Kittanning coal, also referred to as the No. 5 Block coal, is a 2 to 12 ft thick, multiple-bedded coal located 18 to 75 ft above the Clarion coal. It has estimated original minable tonnage in West Virginia of over 10 billion (Keystone, 1980).

Middle Kittanning Coalbed—The Middle Kittanning coal occurs in the lower half of the Allegheny Group. In coalbeds over 27 inches thick, the coal has recoverable reserves of 25.5 million tons in the Upper Potomac Basin in Maryland (Weaver, Coffroth, and Edwards, 1976).

In West Virginia, the Middle Kittanning coal ranges from 2 to 10 ft thick and is found 27 to 75 ft above the Lower Kittanning coal. It is usually multiple bedded and of a fair to poor quality. The coal has an estimated original minable tonnage in West Virginia of 1 billion (Keystone, 1980).

Upper Kittanning Coalbed—The Upper Kittanning coal, also referred to as the Montell coal, is located in the upper half of the Allegheny Group. Recoverable reserves of 35 million tons are estimated to occur in beds greater than 27 inches thick in the Upper Potomac Basin in Maryland (Weaver, Coffroth, and Edwards, 1976).

In West Virginia, the Upper Kittanning coal is double bedded and ranges from 2 to 5 ft thick. It is located 17 to 70 ft above the Middle Kittanning coal. The coal's original minable tonnage for West Virginia is estimated to have been 4 billion (Keystone, 1980).

Lower Freeport Coalbed—In West Virginia, the Lower Freeport coal ranges from 2 to 6 ft thick and is found 65 to 95 ft above the Upper Kittanning coal. This coal usually occurs in multiple beds. The coal is minable in Nicholas, Webster, Braxton, and Preston counties, with original minable tonnage for West Virginia estimated at 700 million (Keystone, 1980).

Upper Freeport Coalbed—The Upper Freeport coalbed, also

Table 1—Proximate analyses of coals in the Central Appalachian Basin.

State	Formation/ Group	Coalbed	Moisture (%)	Volatile Matter (%)	Fixed Carbon (%)	Ash (%)	Sulfur (%)	Btu	Reference*
Western Maryland	Allegheny	Upper Kittanning	1.9–3.3	15.2–16.3	64.5–71.6	9.6–15.5	0.8–3.6	12,510–12,540	1
		Upper Freeport	1.7–5.3	15.4–21.9	63.6–72.8	5.3–14.1	0.8–4.2	12,890–14,245	1
	Conemaugh	Barton	1.3–3.0	14.9–20.3	69.6–73.1	7.8–11.6	0.6–3.0	13,345–14,005	1
		Franklin	1.6–4.8	17.9–19.7	64.8–69.1	9.6–12.8	2.2–3.1	12,780–13,780	1
		Pittsburgh	1.7–4.6	14.0–22.0	61.9–79.3	5.1–11.9	0.6–1.7	13,070–14,480	1
Southeast West Virginia	Pottsville	Pocahontas No. 3	—	12–25	65–82	3–13	0.4–2.0	12,800–15,600	2
		Pocahontas No. 4	—	12–22	70–82	3–9	0.4–1.0	14,000–15,600	2
		Pocahontas No. 6	—	16–27	57–79	2–9	0.4–1.2	13,000–15,400	2
		Fire Creek	—	15–30	57–80	2–15	0.4–1.2	10,600–15,600	2
		Beckley	—	13–25	68–81	2–12	0.4–2.5	13,600–15,400	2
		Welch	—	13–39	57–79	5–13	0.4–1.4	13,000–15,000	2
		Sewell	—	13–34	50–81	2–25	0.4–?	11,200–15,800	2
		Gilbert	—	18–39	54–76	2–17	0.6–6.0	12,800–15,200	2
		Eagle	—	23–41	49–73	2–17	0.6–7.0	13,200–15,400	2
		Powellton	—	29–44	54–66	2–9	0.6–3.2	14,000–15,000	2
		Pearless	—	25–45	45–69	2–15	0.6–3.0	12,800–15,400	2
		Alma	—	29–44	47–65	2–17	0.6–4.0	12,200–15,000	2
		Cedar Grove	—	32–44	50–64	2–14	0.6–3.6	12,800–15,000	2
		Chilton	—	32–41	47–62	2–23	0.6–2.2	11,800–14,800	2
		Winifrede	—	28–43	50–69	2–12	0.6–2.0	12,400–15,200	2
		Coalburg	—	30–42	47–62	3–16	0.6–3.0	12,000–15,200	2
		Stockton–Lewiston	—	29–42	46–61	3–22	0.6–4.6	11,400–14,800	2
	Allegheny	Lower Kittanning	—	16–48	41–72	2–26	0.4–9.0	11,600–15,600	2
		Upper Kittanning	—	16–43	46–73	5–17	0.1–5.6	12,600–14,000	2
		Upper Freeport	—	15–43	43–77	5–20	0.4–9.0	10,200–14,600	2
	Conemaugh	Bakerstown	—	14–37	43–77	5–23	0.4–7.0	11,200–15,000	2
		Elk Lick	—	15–41	45–76	7–18	0.8–5.5	12,800–14,400	2
	Monongahela	Pittsburgh	—	17–45	45–77	4–20	0.6–1.4	11,400–14,800	2
		Redstone	—	32–46	54–66	3–14	1.0–5.5	12,800–14,200	2
Southwest Virginia	Pocahontas	Pocahontas No. 3	0.8–2.3	15.3–22.8	68.9–76.0	3.3–16.0	0.5–0.9	12,840–14,940	2
		Burtons Ford	2.6–3.6	31.7–33.2	60.3–62.8	4.0–7.4	0.5–1.0	—	3
	Lee	Middle Horsepen	3.2	—	—	9.2	0.5	13,730	2
		Tiller	1.8–3.0	26.7–32.5	56.8–63.8	6.1–12.2	0.4–0.7	13,260–14,180	2
		Jawbone	1.2–4.3	17.9–32.9	45.2–71.2	6.5–30.9	0.5–1.2	10,070–14,250	2
		Raven	1.4–5.5	28.9–33.4	55.1–63.9	2.8–12.2	0.4–0.8	12,980–14,940	2
		Kennedy	1.6–4.7	25.0–36.4	43.2–69.0	3.3–27.0	0.5–2.0	10,380–14,790	2
		Lower Banner	1.2–8.1	20.1–36.6	54.6–70.9	5.2–11.2	0.4–1.9	13,030–14.630	2
		Splash Dam	1.0–4.3	25.0–31.3	56.2–64.8	4.4–17.6	0.6–1.6	12,690–14,430	2
		Hagy	2.3	—	—	8.1	1.8	13,880	2
		Norton	2.1–4.1	31.2–33.5	56.1–58.2	6.4–10.2	0.8–1.5	13,390–13,900	2
	Wise	Dorchester	1.2–5.3	25.8–36.1	53.4–63.4	2.2–16.1	0.6–2.9	12,710–14,570	2
		Blair	1.5–11.0	25.9–35.2	50.6–64.2	2.0–20.2	0.5–3.9	11,910–14,490	2
		Clintwood	1.3–3.2	30.5–38.5	56.1–61.9	2.2–8.5	0.7–1.7	12,780–14,530	2
		Imboden	1.8–2.8	33.0–35.7	53.6–61.3	4.0–10.4	0.6–0.9	13,310–14,450	2

continued

Table 1—(Continued)

State	Formation/ Group	Coalbed	Moisture (%)	Volatile Matter (%)	Fixed Carbon (%)	Ash (%)	Sulfur (%)	Btu	Reference*
		Taggart Marker	1.5–64.	33.5–37.5	55.9–59.4	2.5–5.6	0.5–0.8	13,810–14,800	2
		Taggart	1.5–4.3	32.8–38.4	55.4–61.6	1.7–4.6	0.4–0.8	13,720–14,810	2
		Low Splint	2.1–3.7	34.5–37.5	51.3–54.7	5.6–10.5	0.7–1.1	12,790–13,910	2
		Pardee	2.0–5.1	32.4–37.8	48.8–57.5	3.9–12.0	0.7–1.4	12,430–14,240	2
		High Splint	2.7–3.6	30.8–35.3	48.6–60.2	3.6–17.0	0.5–0.8	11,850–14,120	2
Eastern Kentucky	Lee	(Southwest Reserve District)							
		Hudson	3.1–3.8	35.5–39.4	50.0–54.0	5.7–9.9	1.9–3.4	12,870–13,490	4
		Stearns No. 1–1/2	3.7–4.4	35.0–38.2	54.6–55.9	3.5–5.4	0.5–0.7	13,390–13,910	4
		Beaver Creek	2.4–2.7	37.5–40.8	46.7–47.3	10.1–12.5	3.2–3.4	10,380–13,330	4
	Breathitt	(Licking River Reserve District)							
		Zachariah	—	—	—	3.7	1.8–3.1	13,290–13,890	4
		Grassy	—	—	—	2.0	0.7	13,300	4
		Gun Creek	3.4–7.5	38.3–42.7	47.7–48.9	5.3–6.2	0.7–2.0	12,740–13,690	4
		Fire Clay	1.4–5.4	40.0–46.1	34.5–49.0	5.6–21.2	—	11,980–13,090	4
		Prater	4.5	35.6	50.8	9.1	0.8	12,760	4
		Index	7.6	37.4	49.3	5.7	0.7	12,640	4
		Oakley/Nickell	2.3–4.8	37.6–45.8	38.1–51.1	6.5–13.8	0.9–1.6	13,070–13,130	4
		Fugate	5.4–8.2	35.5–38.1	43.1–48.8	8.3–12.9	0.6–2.7	11,360–12,350	4
		(Big Sandy Reserve District)							
		(average)	5.5	—	2–4	6.0	1.3	13,700	4
		Elswick	—	—	—	9.7	2.4	13,600	4
		Upper Banner	—	—	—	8.0	0.7	13,070	4
		Auxier	—	—	—	6.8	1.8	14,000	4
		Millard	—	—	—	8.7	1.3	13,600	4
		Bingham	—	—	—	8.8	1.5	13,400	4
		Lower Elkhorn	—	—	—	4.3	0.6	14,000	4
		Upper Elkhorn No. 1	—	—	—	4.6	0.7	14,200	4
		Upper Elkhorn No. 2	—	—	—	4.3	0.7	14,000	4
		Upper Elkhorn No. 3	—	—	—	3.9	0.9	14,200	4
		Williamson	—	—	—	5.9	2.7	13,750	4
		Fire Clay	—	—	—	11.0	2.4	12,300	4
		Haddix	—	—	—	5.1	1.1	13,010	4
		(Hazard Reserve District)							
		Zachariah	5.2–6.2	40.7–45.1	46.0–50.0	1.8–3.1	3.1–3.7	13,750–3,890	4
		Lower Elkhorn	2.1	36.1	57.6	4.2	0.7	13,790	4
		Upper Elkhorn No. 3	2.1–4.2	33.8–37.6	55.0–59.9	2.4–3.8	0.5–1.1	13,790–14,860	4
		Amburgy	3.2–6.8	33.5–40.7	49.9–55.0	2.0–8.2	0.2–2.9	12,640–13,700	4
		Lower Whitesburg	2.6–3.7	37.4–40.5	49.4–52.9	6.0–7.2	1.3–1.9	13,330–13,820	4
		Fire Clay	2.8–5.6	34.7–39.7	51.7–56.6	3.2–9.5	0.7–1.2	12,880–14,000	4
		Fire Clay Rider	2.8–7.3	35.3–41.2	47.7–53.7	3.8–11.1	0.9–2.5	12,660–13,990	4
		Hazard	3.4–3.6	36.8–38.7	53.9–56.1	3.5–4.6	0.6–0.9	13,710–13,900	4
		Hazard No. 7	4.2–4.8	35.0–37.6	51.1–54.7	5.4–6.5	0.6–0.9	13,100–13,290	4
		Francis	2.8–5.4	36.5–40.1	48.0–51.4	5.3–12.7	0.8–1.8	12,150–13,470	4
		Hindman	3.1–3.9	36.8–37.2	51.7–53.4	5.5–8.4	1.8–2.5	13,130–13,350	4
		Skyline	4.1–5.3	35.8–37.1	50.5–54.3	3.3–9.6	0.8	12,430–13,320	4

continued

Table 1—(Continued)

State	Formation/ Group	Coalbed	Moisture (%)	Volatile Matter (%)	Fixed Carbon (%)	Ash (%)	Sulfur (%)	Btu	Reference*
		(Southwestern Reserve District)							
		Lily	3.0–7.2	33.2–40.0	49.1–54.8	4.2–10.9	0.9–3.3	12,620–13,970	4
		Blue Gem	2.4–6.3	33.7–44.4	48.2–56.2	1.4–8.7	0.7–3.7	12,820–14,170	4
		Jellico	2.0–5.1	34.8–40.5	54.1–58.1	1.2–7.4	0.7–1.7	13,120–14,600	4
		Fire Clay	3.6–7.3	33.8–37.2	51.2–57.5	5.0–8.1	0.6–1.3	12,580–13,480	4
		Fire Clay Rider	1.0–1.2	27.9–33.4	59.7–64.9	5.9–6.0	0.4–0.7	—	4
Eastern Kentucky		(Upper Cumberland Reserve District)							
		Harlan Subdistrict							
		Imboden	1.5	34.0	56.1	8.4	1.3	14,215	4
		Harlan	2.6	36.7	57.2	3.5	0.8	14,130	4
		Darby	2.5–3.6	34.6–37.5	54.3–58.8	2.4–6.0	0.5–1.0	13,640–14,590	4
		Low Splint	3.7	37.8	54.3	4.2	0.7	13,830	4
		Fire Clay	3.1–3.3	37.7–39.1	54.3–54.4	3.3–4.8	0.7–1.0	13,650–13,820	4
		High Splint	4.8	35.8	53.4	6.0	0.7	13,170	4
		Middlesboro Subdistrict							
		Mingo	4.4	36.1	57.1	2.4	0.8	13,940	4
		Sandstone Parting	3.1	35.1	54.9	6.9	1.0	13,460	4
		Poplar Lick	2.8–3.4	36.7–37.9	49.9–53.9	5.0–10.6	1.0–2.8	12,860–13,710	4
		Sterling	2.3	36.7	51.6	9.4	3.1	13,150	4
		Stray	2.4	36.4	50.0	11.2	1.5	12,750	4
		Lower Hignite	3.5	37.0	55.1	4.4	0.8	13,870	4
		Red Spring	4.9	35.9	51.6	7.6	0.6	12,930	4
North Tennessee	Gizzard	Wilder	1.9–4.5	33.7–37.3	49.7–53.0	8.8–11.2	2.2–3.8	12,520–13,310	2
	Slatestone	Coal Creek	1.3–3.7	33.0–41.7	46.3–58.1	2.3–9.1	0.7–4.8	12,780–14,380	2
		Blue Gem	3.4–5.7	36.8–39.8	51.1–56.3	1.2–5.2	0.7–2.2	13,480–14,300	2
		Jellico	2.0–7.6	30.8–39.4	45.3–59.6	1.9–18.2	0.7–3.2	11,690–14,210	2
	Graves Gap	Windrock	1.2–5.7	34.1–40.6	45.1–61.6	3.4–19.3	0.6–6.2	12,230–15,020	2
	Redoak Mountain	Big Mary	0.8–1.5	36.0–38.0	—	9.0–14.0	2.0–4.0	12,000–13,000	2
		Red Ash	3.2–5.2	34.7–39.3	50.4–53.7	4.3–9.7	0.9–1.4	13,120–13,690	2
		Pewee	3.1–4.0	35.2–38.4	54.4–55.3	3.4–6.7	0.6–0.9	13,500–13,880	2

*1. Weaver, Coffroth, and Edwards, 1981.
2. Keystone, 1980.
3. Miller, 1974.
4. Huddle et al, 1963.

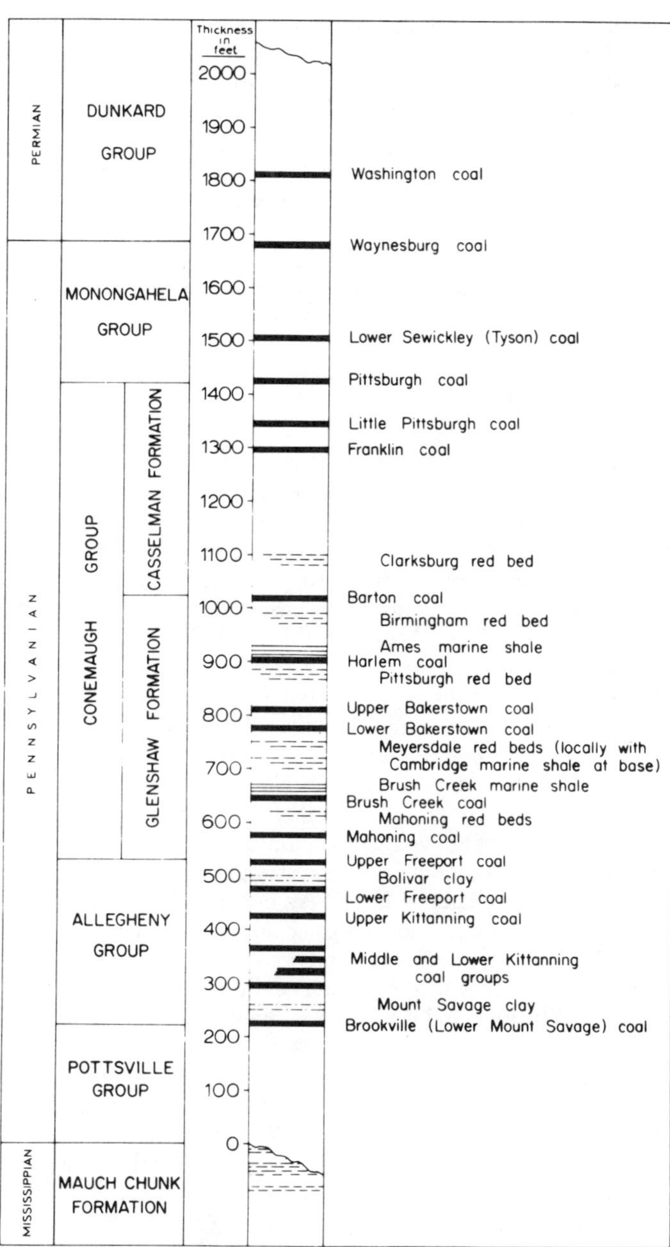

Figure 4—Geologic units in western Maryland with the major coalbeds. Weaver, Coffroth, and Edwards, 1976. Used with the permission of the Maryland Geological Survey.

referred to as the Davis and Split Six coalbeds, occurs in the top of the Allegheny Group in the Upper Potomac Basin and ranges from 48 to 63 inches thick, reaching a maximum thickness of 72 inches (Keystone, 1980). In beds greater than 27 inches thick, the coal has recoverable reserves of 115 million tons in the basin (Weaver, Coffroth, and Edwards, 1976).

In West Virginia, it is a double- or multiple-bedded coal ranging from 2 to 12, and averaging 5, ft thick. The coal is found 30 to 75 ft above the Lower Freeport coal and has estimated original minable tonnage in West Virginia of 3.5 billion (Keystone, 1980).

Conemaugh Group

Mahoning Coalbed—In western Maryland, the Mahoning coalbed (also referred to as the Piedmont or Six-Foot coal) is located near the base of the Conemaugh Group. There is rapid lateral change in thickness for this coal. The coal reaches a maximum thickness of 6 ft and averages 48 to 54 inches (Keystone, 1980). Recoverable reserves in beds greater than 27 inches thick are estimated at 9.5 million tons for this coal (Weaver, Coffroth, and Edwards, 1976).

The double- or multiple-bedded coal averages 3 ft thick and ranges from 2 to 6 ft in West Virginia. The coal is found 24 to 53 ft above the Upper Freeport coal. The original minable tonnage in West Virginia is estimated to have been 230 million (Keystone, 1980).

Brush Creek Coalbed—In West Virginia, the Brush Creek coal's one bench ranges from 2 to 5 ft thick and is of good quality. The coal generally occurs an estimated 60 ft above the Mahoning coal and is minable in Upshur County. Original minable tonnage in West Virginia is estimated at 90 million (Keystone, 1980).

Lower Bakerstown Bed—In western Maryland, the Lower Bakerstown bed is also referred to as the Honeycomb, Thomas, or Three-Foot coalbed. It averages 42 to 48 inches thick in the Upper Potomac Basin (Keystone, 1980). It has recoverable reserves of 33 million tons in beds greater than 27 inches thick (Weaver, Coffroth, and Edwards, 1976).

Bakerstown Coalbed—In West Virginia, the Bakerstown coal is generally located 65 to 100 ft above the Brush Creek coal. The coal ranges from double to multiple bedded and averages 4 ft thick but ranges from 2 to 8 ft. Original estimated minable tonnage in West Virginia is 2 billion (Keystone, 1980).

Upper Bakerstown Coalbed—The Upper Bakerstown coal is found approximately 50 ft above the Bakerstown coal in West Virginia. The coal is generally soft and multiple bedded, ranging from 2 to 5 ft thick. The coal is minable in parts of Mineral County, with original minable tonnage in West Virginia estimated to have been 90 million (Keystone, 1980).

Barton Coalbed—The Barton coalbed, also referred to as the Four-Foot coalbed, occurs in the middle of the Conemaugh Group in western Maryland. In the Upper Potomac Basin, the bed reaches 42 inches thick (Keystone, 1980). Recoverable reserves in beds greater than 27 inches thick are estimated to be 5.5 million tons (Weaver, Coffroth, and Edwards, 1976).

Harlem Coalbed—In West Virginia, the Harlem coal occurs 90 to 150 ft above the Upper Bakerstown coal. The coal is soft, reaching a maximum of 2 ft thick. Original minable tonnage in West Virginia is estimated to have exceeded 270 million (Keystone, 1980).

Elk Lick Coalbed—In West Virginia, the Elk Lick coal usually has two benches but may be multiple bedded. The coal ranges from 1.5 to 8, and averages 3, ft thick. The coal occurs approximately 100 ft above the Harlem coal. The original minable tonnage in West Virginia has been estimated at 650 million (Keystone, 1980).

Franklin Coalbed—The Franklin coalbed, also called the Little Clarksburg or Dirty Nine-Foot coalbed, is located in the upper third of the Conemaugh Group in western Maryland. The coal is found in the northeast part of the Upper Potomac Basin, approximately 130 ft below the Pittsburgh coalbed. The coal is a series of 2- to 4-ft beds, separated by carbonaceous shale and bone averaging 64 inches thick (Keystone, 1980).

In West Virginia, the Little Clarksburg coal is multiple bedded and ranges from 2 to 7 ft thick. The coal (which is not of good quality) is minable only in Mineral County and occurs 200 ft above the Elk Lick coal. The original minable tonnage in West Virginia is estimated to have been 28 million (Keystone, 1980).

Little Pittsburgh Coalbed—In West Virginia, the Little Pittsburgh coal ranges from 3 to 6 ft thick and is not of good quality. The coal occurs 40 to 75 ft above the Little Clarksburg coal and is minable in Tucker County. The original minable tonnage in West Virginia is estimated to have been 8 million (Keystone, 1980).

Southwest Virginia and Southeast West Virginia

The southwest Virginia and southeast West Virginia coal area includes coal ranging from high- to low-volatile bituminous in rank. The coal occurs in sandstone, siltstone, shale, and thin clastic and calcareous marine zones. In southeast West Virginia, the coal field strata expand in thickness from the Rome Trough southeastward. The number of coalbeds also increases to the southeast, indicating deposition in a rapidly subsiding linear belt or trough. The oldest coal-bearing formation is the Pocahontas, which occurs in the southeast part of the Appalachian Plateau (Fig. 5).

The Lee Formation is equivalent to the New River Formation of West Virginia. The New River Formation is approximately 400 ft thick in the Charleston area and thickens to 1,000 ft in McDowell County. The formation includes sandstone with shale, numerous coals, and lesser amounts of clay and limestone (Keystone, 1980).

In southeast West Virginia, the Kanawha Formation overlies the New River Formation. It is 600 ft thick in the Charleston area and 2,100 ft thick in McDowell County. The formation consists of sandstone and gray shale with abundant coals and lesser amounts of clay and limestone (Keystone, 1980).

Above the Lee Formation in southwest Virginia is the Norton Formation, which includes the Kennedy, Big Fork, Lower Banner, Upper Banner, Splash Dam, Hagy, and Norton coals. To the southeast, in southwest Virginia, the formation contains the Tiller, Jawbone, Raven, and Aily coals plus those previously mentioned (Keystone, 1980).

Gas content data are available for the Pocahontas No. 3, Pocahontas No. 4, Beckley, Sewell, and Jawbone coalbeds in this area and will be discussed in a later section of this chapter. These coals will be discussed here on an individual basis.

Pottsville Group

Pocahontas Formation: Pocahontas No. 3 Coalbed—The Pocahontas No. 3 coal ranges from medium- to low-volatile bituminous in rank. This coal is limited in distribution to southwest Virginia and southeast West Virginia, where it is located approximately 70 ft above the Pocahontas No. 2 coal (Keystone, 1980). In southeast West Virginia, the Pocahontas No. 3 coal is fairly continuous over an estimated 650 sq mi. It ranges up to 11 ft thick (Rehbein, Henderson, and Mullenex, 1981). Keystone (1980) estimates the original minable reserves for this coal to be 2.8 billion tons; and Rehbein, Henderson, and Mullenex (1981) indicate that of this original resource, only 0.9 billion tons remain in-place in beds more than 3 ft thick.

For southwest Virginia, distribution and thickness maps for this coal have been developed by Miller (1974) using available data from boreholes in the region. Rehbein, Henderson, and Mullenex (1981) have developed comparable maps for southeast West Virginia.

In southwest Virginia, the Pocahontas No. 3 coal underlies an estimated 750 sq mi. The coal ranges from a maximum thickness of 11 ft in Tazewell County to being absent in northwestern Wise, Dickenson, and Buchanan counties because of truncation by the lowermost arenite of the Lee Formation (Miller, 1974). In these areas, the No. 3 coal ranges from 3 to 4 ft thick. The entire Pocahontas Formation is eroded to the northwest in southwest Virginia (Miller, 1974).

The Pocahontas No. 3 coal has been displaced 4 mi on the northeast by the Russell Fork Fault and 1 mi in the southwest by the Jacksboro Fork Fault. South of the Russell Fork Fault, the Pocahontas No. 3 coal occurs in the subsurface with the coal indicated to exist in southern Dickenson, southeast Wise, and part of Scott and Russell counties (Miller, 1974).

Based on the extent and thickness map developed by Miller (1974) for the Pocahontas No. 3 coal in southwest Virginia, this author estimates that original resources for this coal were between 1.4 and 3.4 billion tons.

Pocahontas Formation: Pocahontas No. 4 Coalbed—In southwest Virginia, the low-sulfur, low-volatile Pocahontas No. 4 coal occurs 30 to 120 ft above the Pocahontas No. 3 coal. The No. 4 coal reaches a maximum thickness of approximately 6 ft in southwestern Buchanan County. An unconformity at the base of the Lee Formation truncates the coal in the northern parts of Wise, Dickenson, and Buchanan counties (Miller, 1974).

In southeast West Virginia, the Pocahontas No. 4 coal ranges from 3 to 6 ft thick and is found approximately 85 ft above the No. 3 coal (Keystone, 1980).

This author estimates that the coal resources in southwest Virginia for the Pocahontas No. 4 coal lie within the range of 0.9 to 1.9 billion tons. For southeast West Virginia, Keystone (1980) estimates the original minable reserves for this coal to have been 800 million tons.

Lee Formation–New River Formation

Beckley Coal—In West Virginia, the Beckley coal, also referred to as the Raleigh or War Creek coal, ranges from 2 to 5 ft thick and is found 80 to 100 ft above the Fire Creek coal. The multiple-bedded coal is minable in parts of McDowell, Wyoming, Raleigh, and Greenbrier counties. The original minable reserve is estimated at 2 billion tons (Keystone, 1980).

Sewell Coalbed—The Sewell coal is a multiple-bedded, columnar coal ranging from 2 to 10 ft thick in southeast West Virginia. The coal is located 30 to 70 ft above the Welch coal. It is minable in parts of McDowell, Raleigh, Wyoming, Fayette, Greenbrier, Nicholas, Pocahontas, Webster, Braxton, Upshur, Randolph, Barbour, and Tucker counties. The original minable tonnage has been estimated at 6.3 billion (Keystone, 1980).

The Sewell "B" coal is a 1- to 5-ft-thick, double-bedded, columnar coal located 85 ft above the Sewell coal. It is minable in Wyoming and McDowell counties, and its original minable reserves are estimated to be 120 million tons (Keystone, 1980).

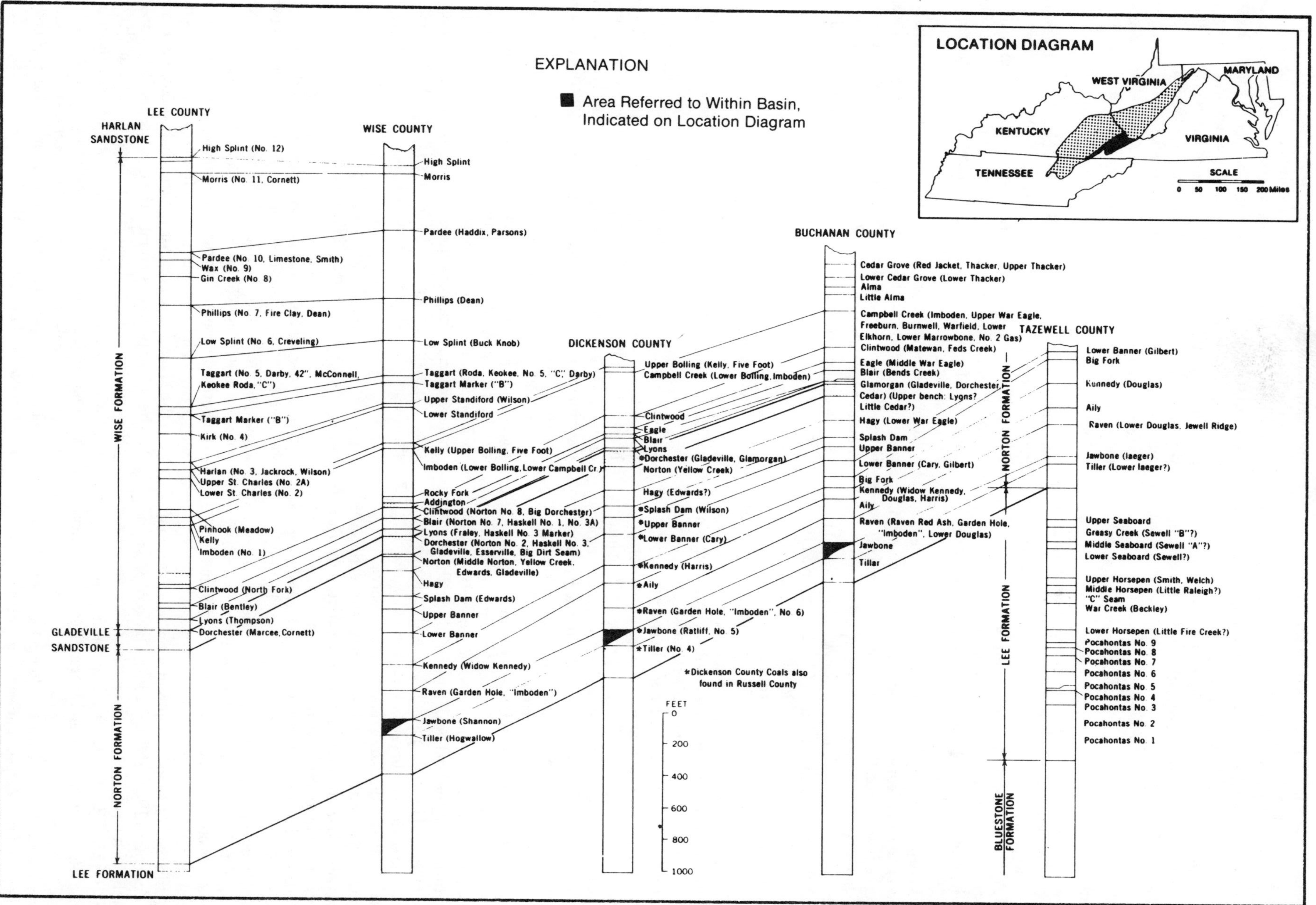

Figure 5—Stratigraphic columns showing correlation of Virginia coalbeds on a county basis. USGS and USBM, 1968.

Jawbone Coalbed—The Jawbone coalbed, in southwest Virginia, is a high-volatile A to low-volatile bituminous coal. This coal occurs up to 100 ft above the Tiller coal, but in some places the Tiller and Jawbone converge and form one coalbed. The Jawbone is found in Buchanan, Dickenson, Tazewell, Russell, and Wise counties. Individually, the Jawbone reaches up to 6.5 ft thick; but when combined with the Tiller coalbed, it can reach 15 ft (Keystone, 1980).

Eastern Kentucky

Huddle et al (1963) divided the eastern Kentucky coal field into six coal reserve districts based on state and county boundaries, geologic features, drainage, and producing areas. The Central Appalachian Basin includes part or all of the following districts: the Princess, Licking River, Big Sandy, Hazard, Southwestern, and Upper Cumberland Reserves (Fig. 6).

In eastern Kentucky, the major coal-bearing strata are the Pennsylvanian Lee and Breathitt Formations. The coals within the Lee Formation are generally thin, discontinuous, and of questionable lateral extent. In eastern Kentucky, the Lee Formation ranges in thickness from 350 ft in Lee County (to the north), to 1,000 to 1,500 ft in the Upper Cumberland Reserve District (to the south) (Huddle et al, 1963).

The Pennsylvanian Breathitt Formation is the major coal-bearing unit in eastern Kentucky. The formation is composed of sandstone, shale, siltstone, coal, and minor underclay and chert. The formation contains 37 identifiable coalbeds and reaches a thickness of 2,500 ft in the Upper Cumberland Reserve District. Coal within this formation ranges from small lenses to 19 ft thick with the thicker coals appearing to be elongated north and east. The coals end horizontally by thinning, grading into other sediments, splitting into numerous beds, or ending rapidly by contact with channel fill deposits or faults. In the Breathitt Formation, the Upper Elkhorn No. 3 and Fire Clay coals are extensive and continuous coal zones (Huddle et al, 1963).

In eastern Kentucky, key beds and sequences are used to correlate coals and related strata. Nine key beds and zones are used in the six reserve districts, three of which are relatively continuous throughout the coal field. The Kendrick Shale of Jillson is characterized by shale and siltstone with calcareous concretions. The shale is an important marker bed in the Princess, Licking River, Big Sandy, and Hazard Reserve Districts. In the Princess Reserve District, the shale overlies the Gun Creek coal; and in the Hazard Reserve District, it overlies the Amburgy or Amburgy Rider coal (Huddle et al, 1963).

The Fire Clay coal is an important marker bed in at least part of all six reserve districts. The coal is characterized by a dark brown to olive-gray flint clay parting, 1 to 8 inches thick (Huddle et al, 1963). Coals stratigraphically above the Fire Clay coal are numerous and will not be discussed on an individual basis.

The Magoffin beds of Morse, an important marker zone, are known to occur 60 to 190 ft above the Fire Clay coal and to be approximately 15 ft thick in the Licking River Reserve District. In addition, the Magoffin is also considered an important marker in the Big Sandy Reserve and Hazard Reserve Districts (Huddle et al, 1963).

In the Princess Reserve District, the "main block ore," the black flint bed, and the Vanport Limestone member are important markers. The "main block ore" occurs 35 to 40 ft above the Princess No. 3 coal and consists of approximately 1 ft of silty limonite with marine fossils. The black flint bed is found approximately 45 ft above the "main block ore" and consists of two units. The lower unit is a few inches thick and consists of a black flint parting with fossils; the upper unit is granular flint with marine fossils from 3 to 4 ft thick. The Vanport Limestone member occurs 20 to 50 ft above the black flint bed and 60 to 75 ft below the Princess No. 7 coal. It is a 2 to 5 ft thick, gray, massive limestone with marine fossils (Huddle et al, 1963).

In the Harland Subdistrict of the Upper Cumberland Reserve District, the Darby coal—one of the major coals mined—is an important key bed. In the Middlesboro Subdistrict, the Sterling coal (also known as the Fire Clay Rider coal) plus the Sandstone Parting coal are important marker beds. The Sterling coal is characterized by pyritized marine fossils in the roof rock. The Sandstone Parting coal occurs 200 to 250 ft below the Poplar Lick coal (Huddle et al, 1963).

Reserve estimates for the Lee and Breathitt Formations have been made for five of the six districts. The Princess, Licking River, Big Sandy, Hazard, and Southwestern Reserve Districts contain approximately 0.6, 2.2, 10.0, 11.1, and 4.6 billion tons of reserves, respectively, totaling 28.5 billion tons (Huddle et al, 1963).

Key Coalbeds in the Breathitt Formation

Zachariah and Zachariah Rider Coalbeds

The Zachariah coal, located at the base of the Breathitt Formation in the Princess and Hazard Reserve Districts, is a high-volatile B bituminous coal. It is equivalent to the Lily coal of the Southwestern Reserve District and possibly correlates to the Bingham coal in the Big Sandy Reserve District (Rice and Smith, 1980).

The thin bed overlying the Zachariah coal is the Zachariah Rider coal. The two beds average 14 to 42 inches thick. To the east in the Licking River Reserve District, the rider bed lies 3.5 to 50 inches above the main bed; and further east, the rider grades into carbonaceous shale. Estimated original reserves for the Zachariah coal are 566 million tons for the four above-mentioned districts (Huddle et al, 1963).

Lower Elkhorn Coalbed

The Lower Elkhorn coal is also called the Pond Creek, Imboden, Mason, Freeburn, Warfield, No. 2 Gas, Campbell Creek, or Shelby Gap coal in the Big Sandy and Hazard Reserve Districts and in the Harlan Subdistrict of the Upper Cumberland Reserve District. The coal ranges from 14 to 93 inches thick and contains partings from 0.5 to 10 inches thick.

In the Hazard Reserve District, the coal outcrops in eastern Letcher County. In this county, the coal occurs 560 to 680 ft below the Fire Clay bed and 200 ft above the Zachariah coal. In the Big Sandy Reserve District, the coal is located 160 ft above the Bingham Coal. In the Harlan Subdistrict, the coal occurs 180 to 200 ft below the Harlan coal. Reserves of 1,759 million tons have been estimated for the coal within the districts mentioned (Huddle et al, 1963).

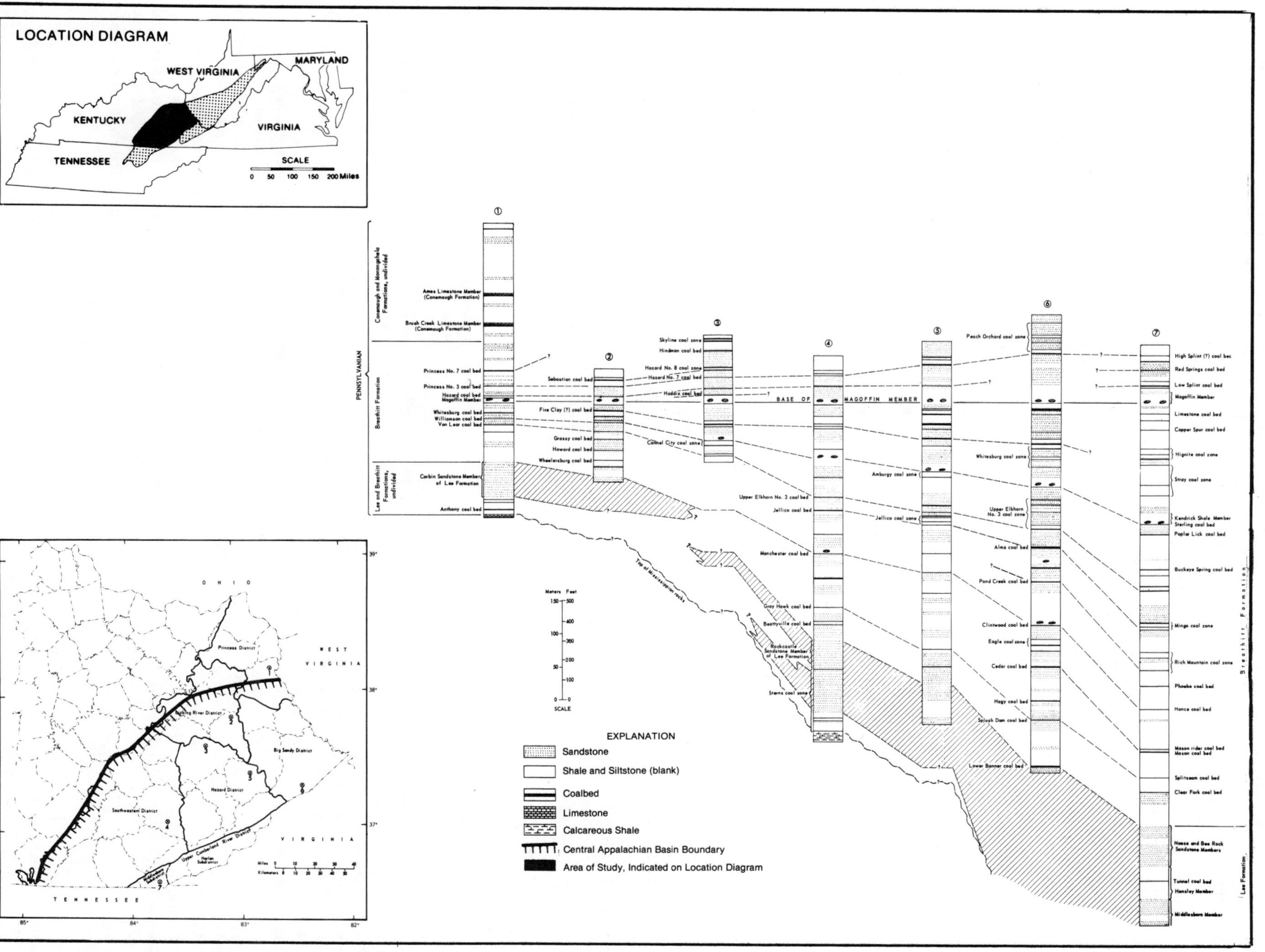

Figure 6—Correlation diagram of Pennsylvanian rocks, eastern Kentucky. Rice and Smith, 1980.

Upper Elkhorn No. 1 Coalbed

The Upper Elkhorn No. 1 coal, also known as the Alma and Wayland coals, occurs at the base of the Upper Elkhorn coal zone and is 260 to 600 ft below the Fire Clay coal in the Hazard Reserve District. In the Big Sandy Reserve District, the coal occurs 100 to 140 ft above the Lower Elkhorn coalbed. The coal crops out in eastern Knott and Letcher counties, with the beds dipping westward. The Upper Elkhorn No. 1 coal ranges from 12 to 55 inches thick, and in some areas the coal contains shale partings up to 16 inches thick. Total estimated reserves for the coal in the district are 1,297.5 million tons (Huddle et al, 1963).

Blue Gem Coalbed

In the Southwestern Reserve District, the Blue Gem coal is correlated to the Upper Elkhorn No. 1 coal, the Lower Howard bed of Russell, and the No. 6 coal of Miller. It is located 30 to 150 ft above the Bacon Creek coal, with the Pine Mountain Fault cutting off the coal to the southeast (Huddle et al, 1963).

Averaging 20 to 25 inches thick, it is generally a bright attrital coal, split by shale partings up to 3 inches thick. The roof rock is shale overlain by a thin to massive sandstone. Original reserves of 470.6 million tons have been estimated for the coal (Huddle et al, 1963).

Upper Elkhorn No. 3 Coalbed

In the Hazard Reserve District, the Upper Elkhorn No. 3 coal ranges from 220 to 450 ft below the Fire Clay coal. The coal, also known as the Cedar Grove or Thacker coal, occurs 20 to 70 ft above the Upper Elkhorn No. 2 coal in the Big Sandy Reserve District. It ranges from 12 to 75 inches thick, with partings that range up to several feet thick, causing multiple bedding. The coal is high-volatile A bituminous in rank and has estimated reserves for the two districts of 3,127.8 million tons (Huddle et al, 1963).

Jellico Coalbed

The Jellico coalbed, in the Southwestern Reserve District (also referred to as the Straight Creek and Beech Creek coal), correlates to the Upper Elkhorn No. 3 coalbed and occurs 60 to 125 ft above the Blue Gem coal. To the west in the district, the coal is absent owing to erosion, and to the southeast the Pine Mountain Fault terminates the coal (Huddle et al, 1963).

The coal averages 28 to 42 inches thick, with 64 inches the maximum recorded thickness. It usually occurs as a single bed but has been split in a few places by clay or shale partings. The roof rock on the coal ranges from sandstone to shale and siltstone. Original reserves of 712.4 million tons have been estimated for the coal (Huddle et al, 1963).

Amburgy Coalbed

The Amburgy coal, also known as the Williamson coal, is a persistent coal ranging from 6 to 78 inches thick, with several shale partings. The coal is located 60 to 225 ft above the Upper Elkhorn No. 3 coal in the Hazard and the Big Sandy Reserve Districts. The Amburgy coal is 120 to 300 ft below the Fire Clay coal. Total estimated reserves for the coal in the two districts are 1,653.1 million tons (Huddle et al, 1963).

Poplar Lick Coalbed

The Poplar Lick coal occurs 200 to 250 ft above the Sandstone Parting bed in the Middlesboro Subdistrict and correlates to the Windrock coal in Tennessee; and to the Dean, Wallis Creek, and Fire Clay beds to the north and east. The coal ranges from 23 to 60 inches thick and in a few areas occurs in two benches. The upper bench averages 24 inches thick, and the lower bench averages 22 inches with a clay or shale parting between 4 and 13 inches thick. The original reserves of the high-volatile A bituminous coal are estimated to have been 102.8 million tons (Huddle et al, 1963).

Whitesburg, Fire Clay, and Fire Clay Rider Coalbeds

The Whitesburg, Fire Clay, and Fire Clay Rider coals all crop out in southern Lawrence County of the Princess Reserve District. The Fire Clay coal is greater than 24 inches thick in the area, with the other two zones being thinner and less persistent over the total district. The total original reserves of the Fire Clay coal are 136.3 million tons. The Whitesburg and Fire Clay Rider coals' total estimated reserves for southern Lawrence County are 13.6 and 18.0 million tons, respectively (Huddle et al, 1963).

Fire Clay Coalbed

The Fire Clay coal is a key stratigraphic marker bed in eastern Kentucky. It contains a characteristic flint clay parting that ranges from 0.25 to 8 inches thick and occurs near the coal's base. The coal generally occurs in irregular bodies that range from 17 to 103 inches thick over all or parts of the Licking River, Southwestern, Hazard, Big Sandy, and Upper Cumberland Reserve Districts. The coal is also referred to as the Hazard No. 4, Dean No. 4, Bevins, Taylor, Chilton, Wallins Creek, or Springfield coal. In the five mentioned districts, the high-volatile A bituminous coal has calculated reserves of 4,051.2 million tons (Huddle et al, 1963).

North Tennessee

The Tennessee coals are of bituminous rank and have the same general character as coals in eastern Kentucky and southwest Virginia. In the Central Appalachian Basin, Tennessee includes all or part of seven mining districts: the Southern Appalachian, Jellico, La Follette, Fork Mountain, Rockwood, Wilde-Fentress, and Monterey.

The principal mined seams within the seven districts are the Blue Gem, Coal Creek, Jellico, Pewee, Red Ash, Windrock, Poplar Creek, Big Mary, Rex, Sewanee, Walnut Mountain, and Bon Air No. 2 coals (Fig. 7). The following coalbed descriptions, in ascending order, include the above-mentioned beds, within their geologic group.

Gizzard Group

Bon Air Coalbed—The Bon Air coal is found in the upper part of the Raccoon Mountain Formation in White County. The

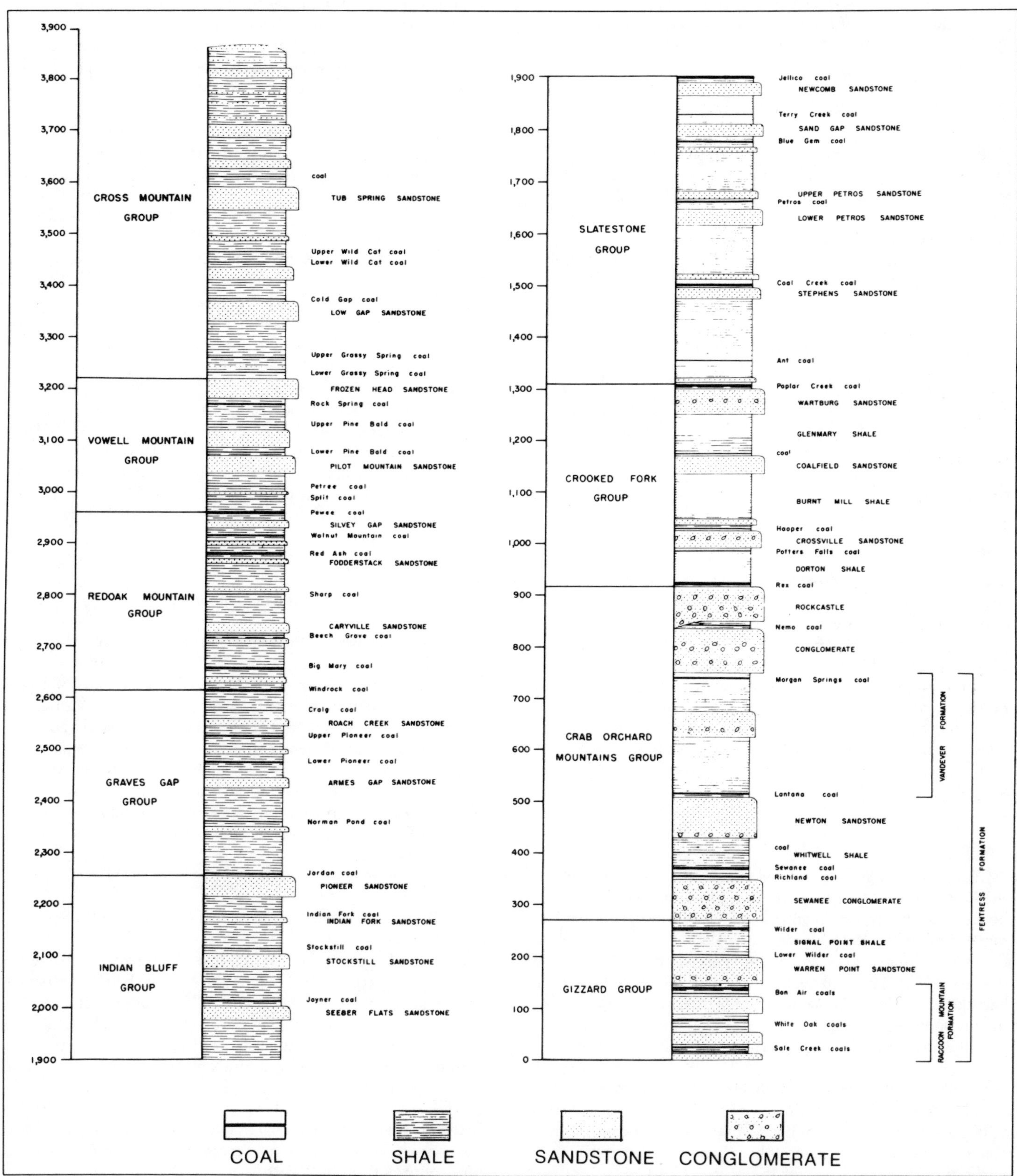

Figure 7—Stratigraphic column of coalbeds within Tennessee strata. Luther, 1959.

majority of this coal is less than 42 inches thick. Known recoverable reserves of 4.1 million tons are estimated in White County (Luther, 1959).

Crab Orchard Mountain Group

Sewanee Coalbed—The Sewanee coal occurs in the Whitwell Shale of the Crab Orchard Mountain Group, approximately 30 to 40 ft above the Sewanee Conglomerate. The coal, also referred to as the Clifty coal, is located 20 to 40 ft above the Wilder coal in Fentress County. Known recoverable reserves of 41.8 million tons are found in Cumberland, Fentress, Morgan, Overton, and White counties (Luther, 1959).

Crooked Fork Group

Rex Coalbed—The Rex coalbed occurs from 20 to 100 ft above the top of the Rockcastle Conglomerate in Campbell, Claiborne, and Morgan counties. The coalbed generally has a slate roof rock and a clay (with minor shale) floor rock. A parting that ranges from 2 to 12 inches is found 12 inches from the top of the coalbed (Keystone, 1980). The coal has known recoverable reserves of 40 million tons in the three counties (Luther, 1959).

Poplar Creek Coalbed—The Poplar Creek coal, also known as the Murray or Swamp Angel coal, is located toward the top of the Crooked Fork Group, a few ft above the Wartburg Sandstone. The coal occurs 170 to 330 ft above the Rex coal in Campbell County and 240 to 300 ft above the Hooper coal in Morgan County (Luther, 1959). The Poplar Creek coal ranges from 18 to 84 inches thick, with an average of 48 to 60 inches. A parting commonly occurs near the top or middle of the coalbed. The coal has low ash and sulfur contents (Keystone, 1980). Known recoverable reserves of 78.6 million tons are found in the four above-mentioned counties (Luther, 1959).

Slatestone Group

Coal Creek Coalbed—The Coal Creek coal is also known as the Kent, Jellico, Bennett Fork, Kings Mountain, Rector, or Turner coal. In Anderson and Campbell counties, the coal ranges from 80 to 260 ft above the Poplar Creek coal and in Claiborne County, from 250 to 340 ft above the Rex coal (Luther, 1959). The coal averages 40 to 48 inches thick but has been known to reach a maximum of 60 inches. If the coal is 42 inches or more, it usually has a parting ranging up to 4 inches thick occurring at the center of the bed (Keystone, 1980). Known recoverable reserves of 145.5 million tons are located in Anderson, Claiborne, and Campbell counties (Luther, 1959).

In Campbell and Claiborne counties, the coal ranges from 36 to 60 inches of solid, bright coal; while in other localities, the coal is separated into numerous benches by clay or shale partings that reach 12 ft thick (Keystone, 1980).

Blue Gem Coalbed—The Blue Gem coal correlates to the Rich Mountain coal and is also known as the Yellow Creek and Billy Goat coal. It is found a few feet below the Sand Gap Sandstone and ranges from 130 to 350 ft above the Coal Creek coal in Campbell and Claiborne counties. The recoverable reserves are known to be 23.1 million tons in Campbell and Claiborne counties (Luther, 1959).

The Blue Gem coal ranges from 20 to 36 inches thick. The coal's roof rock is commonly sandy shale or slate, with the floor rock being fire clay and slate. The coal may be recognized by a thin band of impurities around the center of the bed (Keystone, 1980).

Jellico Coalbed—The Jellico coal is also known as the Big Bushy, Log Mountain, Mingo, and State coalbeds. In Campbell County, the coal is a medium-hard, bright, clean coal with brittle fractures. In this area, the coal ranges from 3 to 7 ft thick and averages 3.5 ft (Keystone, 1980). It occurs 280 to 520 ft above the Coal Creek coal in Anderson County, 100 to 260 ft above the Blue Gem coal in Campbell and Claiborne counties, and 320 to 700 ft above the Poplar Creek coal in Morgan and Scott counties. The known recoverable reserves of 53.9 million tons occur in the five counties mentioned above (Luther, 1959).

The coal usually contains one or two partings. In Claiborne County, the coal is between 48 and 72 inches thick. In Scott County, it ranges from 15 to 36 inches, with a 1- to 2-inch or 6- to 12-inch parting (Keystone, 1980). In Claiborne County, the Mingo coal—which has been correlated to the Jellico coalbed—ranges from 60 to 72 inches thick and is divided by partings into two or more benches. Where the Mingo coal is in two benches, the upper ranges from approximately 24 to 37.5 inches thick, with a clay parting from 2.5 to 31 inches. The bottom bench ranges from 32 to 45.5 inches thick (Keystone, 1980).

Graves Gap Group

Windrock Coalbed—The Windrock coal lies directly below the Big Mary coal and is greater than 28 inches thick in Anderson, Scott, and Campbell counties. The coal ranges from 380 to 630 ft above the Joyner coal in Anderson and Morgan counties and 70 to 215 ft above the Pioneer coal in Campbell and Scott counties. The coal is greater than 42 inches thick in many areas, and known recoverable reserves of 58.7 million tons exist in the four counties (Luther, 1959). It is believed that Windrock coal resources in Claiborne County are essentially depleted by mining. The coal is characterized by a flint-clay underclay (Keystone, 1980).

Redoak Mountain Group

Big Mary Coalbed—The Big Mary coal, also known as Sterling and Klondike coals, occurs commonly in two benches separated by a shale interbed. In a few areas, only the upper bench is evident. The lower bench varies from 10 to 30 inches thick and is found up to 5 ft below the top bench. At places, the two benches combine to form one thick bed (Keystone, 1980). The coal ranges from 10 to 90 ft above the Windrock coal in Anderson, Campbell, Claiborne, and Scott counties, and 200 to 280 ft above the Pioneer coal in Claiborne County. The blocky coal has known recoverable reserves of 78.5 million tons in the four counties (Luther, 1959).

Walnut Mountain Coalbed—The Walnut Mountain coal is generally less than 42 inches thick. The coal ranges from 245 to 320 ft above the Big Mary coal in Anderson and Morgan counties. Known recoverable reserves of 43.9 million tons exist in Anderson, Morgan, Campbell, and Scott counties (Luther, 1959).

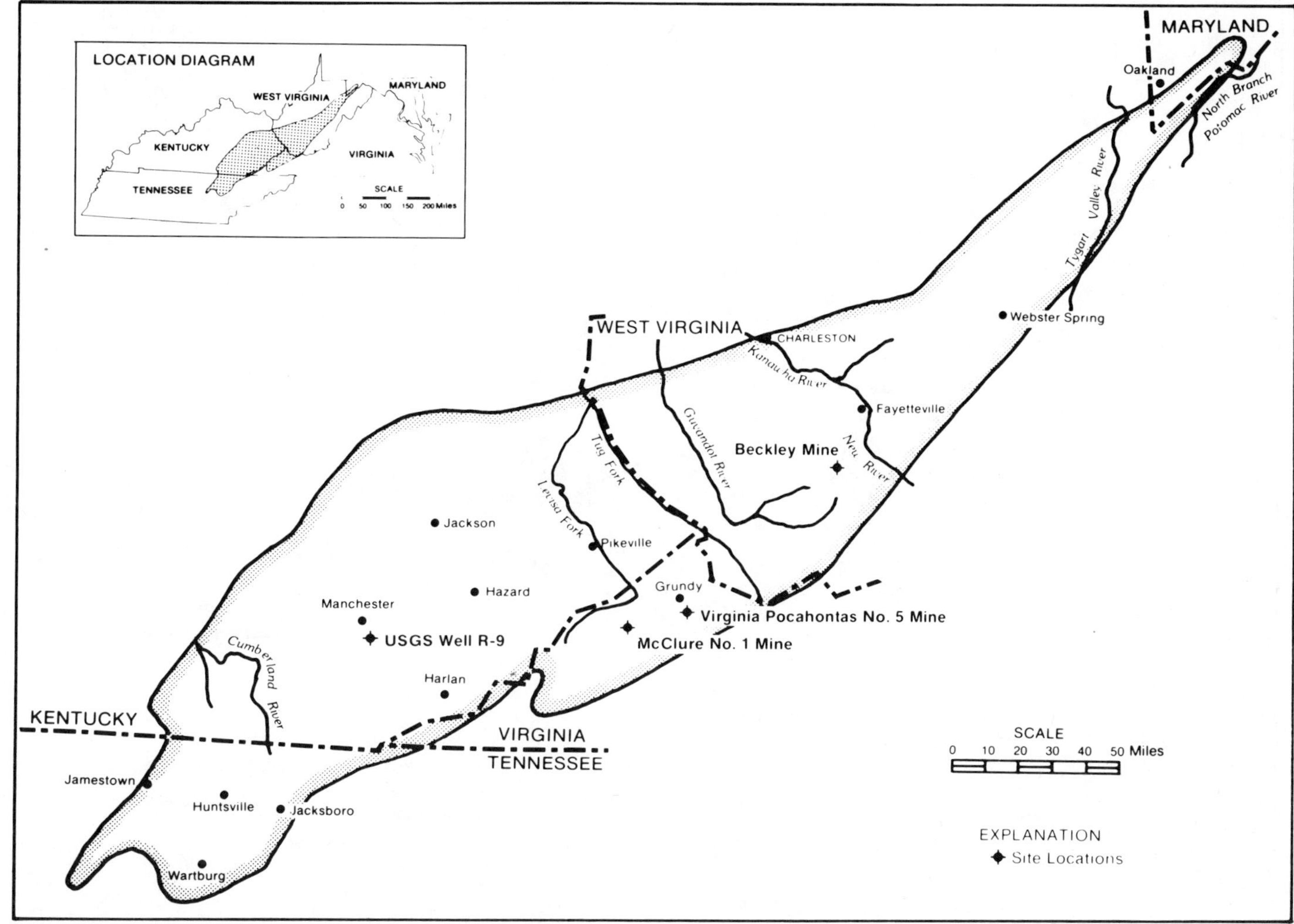

Figure 8—Location of methane projects and wells within the Central Appalachian Basin.

Red Ash Coalbed—The hard, bright Red Ash coal ranges from 36 to 60 inches thick. The coal's roof rock is slate, and the floor rock ranges from shale to sandstone to sandy clay (Keystone, 1980).

The Red Ash coal ranges from 160 to 260 ft above the Big Mary coal in Campbell and Scott counties. The coal has 13.7 million tons of known recoverable reserves in these two counties (Luther, 1959).

Pewee Coalbed—The Pewee coal is also referred to as the Jumbo, Lower Hignite, Merwin, and X beds. In Anderson County, the coal ranges from 42 to 84 inches thick and contains one to three partings ranging from 0.25 to 1.5 inches thick. In Campbell County, the coal is a maximum of 60 inches thick, separated by one to two thin partings. In Morgan County, not much is known about the coalbed's extent and thickness. The coal's roof rock is slate or sandstone, and the floor rock is a hard, sandy underclay (Keystone, 1980). It ranges from 10 to 110 ft above the Walnut Mountain coal in Anderson, Campbell, Morgan, and Scott counties; and from 300 to 340 ft above the Big Mary coal in Claiborne County. Recoverable reserves for this coal of 32.7 million tons are known in the five counties (Luther, 1959).

POTENTIAL METHANE RESOURCE

Methane Data from Published Literature

The coalbed methane potential of the Central Appalachian Basin has been evaluated based on published information and projects directly related to methane extraction from coalbeds. The location of projects is indicated in Figure 8.

An indication of methane gas associated with coalbeds comes from mine methane emission surveys. Data on methane emission from working mines, available from the U.S. Department of Labor Mine Safety and Health Administration (MSHA), have proven that in the basin, the methane emission is substantial. In 1975, individual mines within five counties of southwest Virginia and southeast West Virginia had total daily methane emissions of at least 1 million cubic feet (MMcf) (Fig. 9). Fifty-seven mines in 18 counties recorded average measured methane emission rates of at least 0.1 million cubic feet per day (MMcfd) and at least 10 mines within the basin were producing over 1 MMcfd of methane from the Pocahontas Nos. 3 and 4 coalbeds (Table 2) (Irani et al, 1977). These emission data indicate an abundance of methane within the Pocahontas Nos.

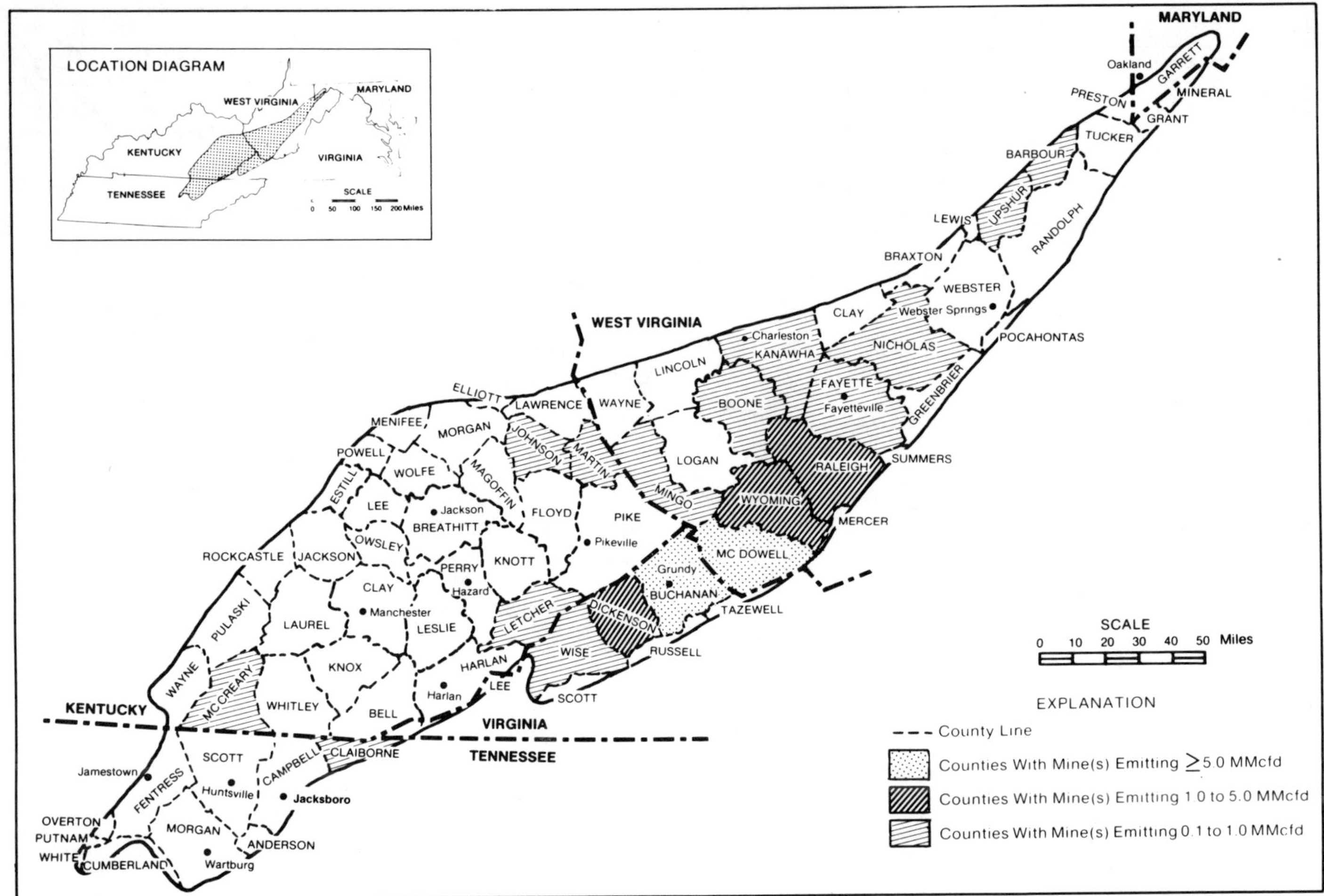

Figure 9—Map showing mine emission data results from 1975. From Irani et al, 1977.

3 and 4 coalbeds in the basin.

Methane emissions from four sections in the Beckley mine in Raleigh County, West Virginia, were monitored from September to December 1974 (Table 3). This project was initiated to study the effects of overburden thickness and daily coal production on methane emissions during mining. Background for the Beckley coalbed indicates that it dips to the northwest and has overburden from 200 to 2,200 ft thick. In the mine area, the roof rock is sandy shale and the floor rock is dark shale. Four sections in the mine were monitored: the North Main; Northwest Main; South Main; and South Right Main. The daily methane emissions ranged from 80 to 450 Mcf for each of the four sections over the monitored period (Jeran, Lawhead, and Irani, 1977).

Methane-related mine accidents are an additional indicator of a significant methane resource. On November 7, 1980, an explosion at the Ferrell No. 17 mine in Boone County, West Virginia, killed five miners. This explosion was reported to have been caused by insufficient mine ventilation, producing an explosive buildup of methane. A more recent methane-related mine incident occurred on December 8, 1981, in Marion County, Tennessee, south of the southern boundary of the Central Appalachian Basin. The No. 21 coal mine had an apparent burnout that killed 13 miners. The burnout occurred in an area where potentially explosive concentrations of methane had recently been detected.

Two projects in southwest Virginia have been developed to analyze the production potential of methane on a commercial basis. In Dickenson County, Virginia, near its McClure No. 1 mine, the Clinchfield Coal Company drilled five vertical boreholes to determine the feasibility of using this system to drain methane gas from the Jawbone coalbed (Manilla, 1980). A second objective of this project was to evaluate the borehole spacing relationship to gas drainage rates.

Desorption tests performed on two cores from the first well indicated a gas content of approximately 280 cf/t. The well was stimulated for production in June 1978. Gas production from this well ranged from a maximum of 43.5 thousand cubic feet per day (Mcfd) in December 1978 to a minimum of 16 Mcfd in March 1980. Total cumulative production for this well from June 1978 to September 1980 was 11.6 MMcf (Manilla, 1980).

The Pocahontas No. 3 coalbed has been described as the second gassiest coal in the United States (Irani et al, 1977). The Island Creek Coal Company has seven active mines in the Pocahontas No. 3 coal in Buchanan County, emitting approximately 20 MMcfd from gob gas vents and ventilation fans (Fig. 10) (Von Schonfeldt, 1981). These seven mines are the Beatrice and the Virginia Pocahontas No. 1 through No. 6 mines.

Occidental Research Corporation, under a cost-sharing

Table 2—Characteristics of central Appalachian coal mines producing over 1 MMcfd of methane in 1975. Irani et al, 1977.

Owner, Name of Mine, and Location	Coalbed	Thickness of Coalbed, Inches	Average Methane Emission, MMcfd	Coal Production, Tons per Day	Gas per Ton of Coal Mined, Cubic Feet	Age of Mine, Years
Allied Chemical Corp. Shannon Branch Mine McDowell County, WV	Pocahontas No. 3 and 4	45 (38–52)*	1.1	1,500	733	70
Beatrice Pocahontas Co. Beatrice Mine Buchanan County, VA	Pocahontas No. 3	54 (48–60)	5.6	3,300	1,697	12
Beckley Coal Co. Beckley Mine Raleigh County, WV	Beckley	66	2.6	4,900	531	4
Clinchfield Coal Co. Moss No. 3 Mine Dickenson County, VA	Tiller	(84–180)	1.4	2,100	666	17
Eastern Associated Coal Corp. Keystone No. 1 Mine McDowell County, WV	Pocahontas No. 3	60	1.4	3,00	467	75
Island Creek Coal Co. Virginia Pocahontas No. 1 Mine Buchanan County, VA	Pocahontas No. 3	50	3.9	3,000	1,300	9
Virginia Pocahontas No. 2 Mine Buchanan County, VA	Pocahontas No. 3	52 (46–58)	3.4	1,100	3,091	8
Virginia Pocahontas No. 3 Mine Buchanan County, VA	Pocahontas No. 3	68	3.3	2,400	1,375	7
Virginia Pocahontas No. 4 Mine Buchanan County, VA	Pocahontas No. 3	50	1.9	850	2,235	4
Itmann Coal Co. Itmann No. 3 Mine Wyoming County, WV	Pocahontas No. 3	48	1.5	5,000	300	16
Olga Coal Co. Olga Mine McDowell County, WV	Pocahontas No. 4	72	5.0	7,000	714	52
United States Steel Corp. Pinnacle Creek No. 50 Mine Wyoming County, WV	Pocahontas	53	1.6	3,000	533	6

*Range of thickness.

Table 3—Methane emission data for the Beckley Mine, Raleigh County, West Virginia. Jeran, Lawhead, and Irani, 1977.

Mine Section	Days Monitored	No. of Production Days	Range of Coal Prod., tons	Average Daily Coal Prod., tons	Range of Methane Emissions, Mcfd	Average Daily Methane Emissions, Mcfd	Average Overburden Over Face, ft	
							Beginning	Ending
North Main	107	43	60 to 500	282	180 to 380	289	650	850
Northwest Main	107	47	50 to 400	224	80 to 200	135	650	700
South Main	107	55	50 to 500	361	180 to 450	304	650	625
South Right Main	86	41	50 to 500	356	120 to 250	158	750	850

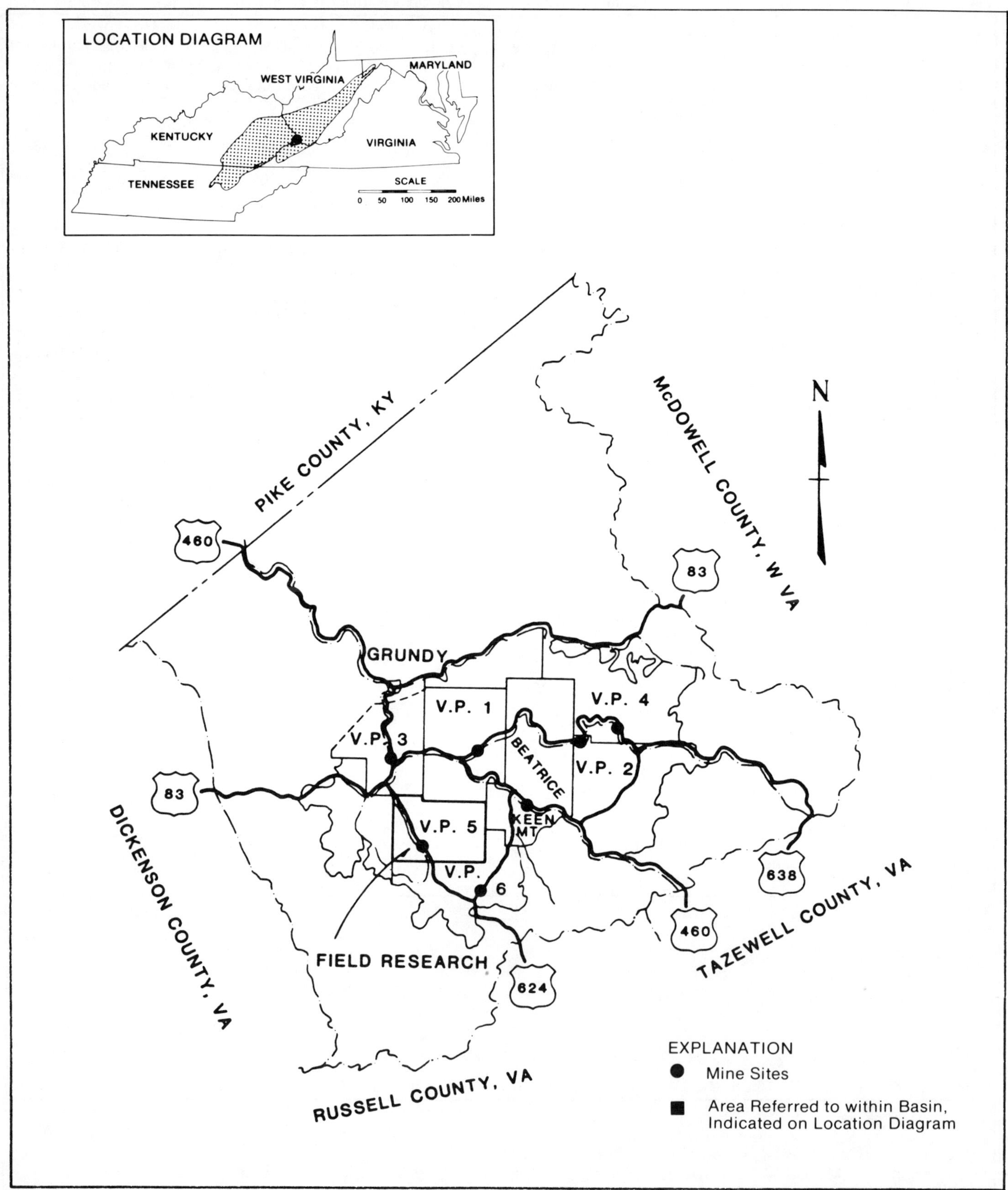

Figure 10—Location of Island Creek Coal Company property in Buchanan County, Virginia. Von Schonfeldt, 1981.

agreement with DOE/MRCP, conducted a project at the Island Creek Coal Company, Virginia Pocahontas No. 5 mine near Grundy, Virginia, in Buchanan County. The objectives of the project were to obtain preliminary data on quality and quantity of methane that can be claimed from a coal seam in advance of mining and to determine the potential effect of methane drainage on overall coalbed gas content. A technique was developed for methane recovery using multiple horizontal boreholes. Twelve boreholes that intersect either the Pocahontas No. 3 coal or the Pocahontas No. 3 and No. 4 coals were drilled by Occidental for methane drainage (Von Schonfeldt, 1981).

The most useful data in calculating the methane resource potential of the Central Appalachian Basin are published desorption data. Diamond and Levine (1981) have published desorption data from 109 samples taken from 12 coalbeds within the basin (Table 4). The gas content data range from a low of 6 cf/t to a high of 573 cf/t. Significant gas was measured in 18 coal samples from the Pocahontas No. 3 coal in Wyoming County, West Virginia, and Buchanan County, Virginia. The gas content ranged from 285 to 573 cf/t from depths between 778 and 2,143 ft. The majority of the samples analyzed had apparent ranks of high-volatile A bituminous, with a few medium-volatile samples (Diamond and Levine, 1981).

In September 1981, TRW participated in the drilling of a well in Clay County, Kentucky, by the USGS Conservation Division. This well was part of a USGS 22-well coal assessment drilling program in eastern Kentucky. Three coal samples were collected for desorption studies. As of January 1982, the gas content of these samples ranged between 25 and 45 cf/t, from depths ranging from 643 to 869 ft. This range of desorption values does not include residual gas determinations.

Potential Methane Resource Calculations

In the Central Appalachian Basin, the potential in-place methane resource has been calculated based on available gas content values from Diamond and Levine (1981) and from the USGS/R-9 well in Clay County, Kentucky.

The Pocahontas No. 3 coal underlies most of southwest Virginia, and part of southeast West Virginia. The compilation of data by Diamond and Levine (1981) suggests that this coal has the highest average gas content value of any of the measured coals within the Central Appalachian Basin.

The Central Appalachian Basin reserves are estimated to be between 80 and 120 billion tons of bituminous coal. This range of approximate reserve estimates is based on information published by Keystone (1980), Huddle et al (1963), USGS and USBM (1968), and other available data. Available gas content data from Diamond and Levine (1981), USGS well R-9, and project data have enabled methane content estimation for the total basinal area. Using an estimated low of 125 cf/t and a high of 400 cf/t, the in-place methane resource for the Central Appalachian Basin has been calculated to range from a minimum of 10 to a high of 48 Tcf. This estimate is based on

Table 4—Desorption data from the U.S. Bureau of Mines. Diamond and Levine, 1981.

Coalbed	State	County	Sample Depth (ft)	Lost Gas (cc)	Desorbed Gas (cc)	Residual Gas (cc/gm)	Total Gas Content		Apparent Rank
							cc/gm	cf/ft	
Beckley	WV	Raleigh	558	32	333	0.1	0.4	13	—
			588	28	3,313	0.3	4.8	154	—
			653	430	7,805	0.8	5.5	176	—
			655	880	14,967	1.8	11.5	368	—
			740	890	16,641	0.6	13.7	438	—
			830	1,660	17,787	0.8	15.3	490	—
			850	1,720	9,630	0.6	9.3	298	—
			852	2,880	16,160	0.8	12.0	384	—
			875	1,880	17,214	0.9	14.4	461	—
			990	640	12,920	0.9	13.1	419	—
			1,198	1,400	14,903	0.1	9.9	317	—
			1,200	1,900	14,016	0.0	10.8	346	—
Cedar Grove (L)	WV	Mingo	684	40	205	0.0	0.2	6	—
			704	78	953	1.2	3.1	99	—
			819	20	142	0.2	0.5	16	—
			833	30	317	0.5	1.1	35	—
			842	50	99	0.5	1.3	42	—
			842	18	27	0.1	0.2	6	hvA
			851	31	60	0.1	0.3	10	—
			862	38	766	1.9	4.5	144	hvA
			878	56	561	0.5	1.3	42	—
			913	—	135	1.4	1.8	58	hvA
			923	17	499	1.3	2.8	90	hvA
			936	36	64	0.1	0.2	6	—
			943	22	100	0.1	0.3	10	—
			949	28	636	2.7	3.7	118	hvA
			996	100	1,150	0.1	1.0	32	hvA
			1,037	42	360	2.7	3.5	112	hvA

continued

Table 4—(Continued)

Coalbed	State	County	Sample Depth (ft)	Lost Gas (cc)	Desorbed Gas (cc)	Residual Gas (cc/gm)	Total Gas Content cc/gm	Total Gas Content cf/ft	Apparent Rank
Clarion	WV	Barbour	819	230	4,951	0.3	5.2	166	hvA
			822	101	3,286	0.3	3.6	115	hvA
Coalburg	WV	Mingo	506	13	32	0.1	0.2	6	—
Elkhorn No. 3	KY	Perry	400	20	734	0.5	1.7	54	—
Jawbone	VA	Dickenson	431	2,225	10,307	0.8	8.0	256	mv
			431	950	5,845	1.2	4.8	154	mv
Kittanning (U)	WV	Barbour	486	198	800	1.9	2.6	83	hvA
			487	216	3,560	2.2	4.4	141	hvA
			489	520	5,395	2.8	6.2	198	hvA
			490	114	1,960	2.7	5.1	163	hvA
			546	396	5,590	1.9	7.3	234	hvA
			547	534	1,530	2.0	3.3	106	hvA
			548	300	3,895	2.5	6.6	211	hvA
			549	168	1,875	0.6	2.4	77	hvA
			610	58	5,175	1.2	5.8	186	hvA
			611	50	3,435	1.7	6.4	205	hvA
			612	46	4,905	1.5	5.8	186	hvA
			708	118	2,389	0.2	2.6	83	hvA
	WV	Upshur	840	43	713	0.5	1.3	42	—
Kittanning (M)	WV	Upshur	909	135	1,578	0.9	2.3	74	—
			911	132	1,646	1.0	2.5	80	—
			912	95	1,529	0.9	2.3	74	—
Kittanning (L)	WV	Braxton	76	48	170	0.3	0.5	16	hvA
			77	104	430	0.4	0.8	26	hvA
			78	67	60	0.6	0.8	26	hvA
			92	60	120	0.3	0.4	13	hvA
			93	59	215	0.7	0.9	29	hvA
			94	77	200	0.7	1.0	32	hvA
			146	56	40	0.0	0.2	6	hvA
			149	45	30	0.0	0.1	3	hvA
			151	79	325	0.0	0.3	10	hvA
			154	89	130	0.4	0.6	19	hvA
			405	13	125	0.1	0.2	6	—
			407	19	180	0.4	0.6	19	mv
			408	12	125	0.3	0.4	13	hvA
			409	18	120	0.1	0.2	6	—
			410	54	395	0.4	0.6	19	hvA
			411	24	270	0.6	0.9	29	hvA
			413	32	313	0.3	0.6	19	hvA
			414	45	390	0.6	0.9	29	hvA
	WV	Barbour	535	342	4,695	1.9	5.9	189	hvA
			536	364	4,990	1.9	5.7	182	hvA
			537	348	5,125	1.3	4.4	141	hvA
			538	207	120	1.6	1.8	58	hvA
			539	225	6,435	1.4	5.8	186	hvA
			540	462	4,970	1.4	5.6	179	hvA
			593	200	95	1.9	2.2	70	hvA
			594	340	5,935	1.7	5.8	186	hvA
			595	540	5,670	1.4	5.7	182	hvA
			596	314	5,335	2.1	9.2	294	hvA
			597	522	1,780	1.9	4.1	131	hvA
			651	102	4,435	1.0	6.2	198	hvA
			652	108	5,272	1.3	7.3	234	hvA
			653	154	6,887	1.2	8.1	259	hvA
			654	118	4,500	1.8	9.1	291	hvA
			806	105	1,504	0.3	1.4	45	—

continued

Table 4—(Continued)

Coalbed	State	County	Sample Depth (ft)	Lost Gas (cc)	Desorbed Gas (cc)	Residual Gas (cc/gm)	Total Gas Content		Apparent Rank
							cc/gm	cf/ft	
Pocahontas No. 3	WV	Wyoming	778	560	12,926	1.3	8.9	285	—
	VA	Buchanan	1,316	1,600	11,558	0.8	12.2	390	—
			1,430	2,087	5,230	0.1	13.6	435	—
			1,518	3,880	22,588	0.2	14.5	464	—
			1,528	2,300	12,011	1.0	15.1	483	—
			1,551	2,150	12,541	1.1	17.3	554	—
			1,554	1,650	11,435	1.1	16.7	534	—
			1,589	1,400	16,475	1.1	16.4	493	—
			1,621	1,700	10,880	0.7	11.5	368	—
			1,621	1,700	11,410	0.8	12.3	394	—
			1,737	1,400	9,685	0.4	10.9	349	—
			1,764	1,350	14,833	1.2	17.9	573	—
			1,845	600	9,530	0.7	11.0	352	—
			1,999	500	10,620	1.0	15.8	506	—
			2,022	1,150	15,778	1.1	16.4	493	—
			2,036	1,350	15,690	1.1	17.6	563	—
			2,108	300	12,460	0.9	13.9	445	—
			2,143	200	8,184	0.7	10.6	339	—
Pond Creek	KY	Pike	125	240	1,210	0.7	2.1	67	—
		Martin	400	51	1,134	0.4	1.8	58	—
		Pike	500	85	1,097	0.3	1.2	38	—
		Mingo	1,070	30	296	2.6	3.2	102	hvA
Sewell	WV	Raleigh	680	960	8,987	0.5	9.3	298	—
			700	220	3,296	1.3	4.1	131	—
		Braxton	981	250	2,780	0.2	2.8	90	—

in-place gas within the coalbeds and does not take into account the influx of gas from strata other than coal.

Coalbeds throughout the basin are numerous and generally discontinuous and are only questionably correlatable. These problems—in addition to limited gas content data—limit establishment of in-place gas estimates for coals on an individual basis. As more exploration is accomplished and more gas content data are acquired within the basin, the methane resource will be defined more accurately.

CONCLUSIONS

The estimated methane content of the coals within the Central Appalachian Basin suggests an opportunity to examine the economic feasibility of recovering and utilizing this energy resource. In analyzing the basin using available data, a primary target area can be delineated, as shown in Figure 11. This area, which covers approximately 4,000 sq mi in southwest Virginia and parts of eastern Kentucky and southeastern West Virginia, appears to have the greatest potential for coalbed methane development in the basin. The target area boundaries were derived using the following criteria:

- Of the coals sampled for desorption purposes, the highest gas yields of 285 to 573 cf/t were obtained from the Pocahontas No. 3 coal at depths ranging from 778 to 2,143 ft.
- Available information from horizontal borehole tests in the Island Creek Coal Company's Virginia Pocahontas No. 5 mine in Buchanan County, Virginia, indicated high methane contents for the Pocahontas No. 3 coal.
- Methane emission data indicated that the area within the target boundaries reflected large amounts of gas released daily from the Pocahontas Nos. 3 and 4 coalbeds.
- Information on the methane content of coals outside the target area is limited. In areas where available, the gas content data indicate significantly lower gas contents than from samples within the target area.

The primary target area, as designated, has undergone extensive investigations by the Occidental Research project and the Clinchfield coal project. These projects investigated three coalbeds within the target area: the Pocahontas Nos. 3 and 4 coalbeds (by the Occidental project), and the Jawbone coalbed (by the Clinchfield project). The target area, which includes parts of eastern Kentucky, southeast West Virginia, and southwest Virginia, contains numerous coals in addition to the three mentioned above, which should be investigated to establish their methane potential.

REFERENCES CITED

Cardwell, D. H., et al, 1968, Geologic map of West Virginia, scale 1:250,000: West Virginia Geological and Economic Survey.

Diamond, W. P., and J. R. Levine, 1981, Direct method determination of the gas content of coal, procedures and

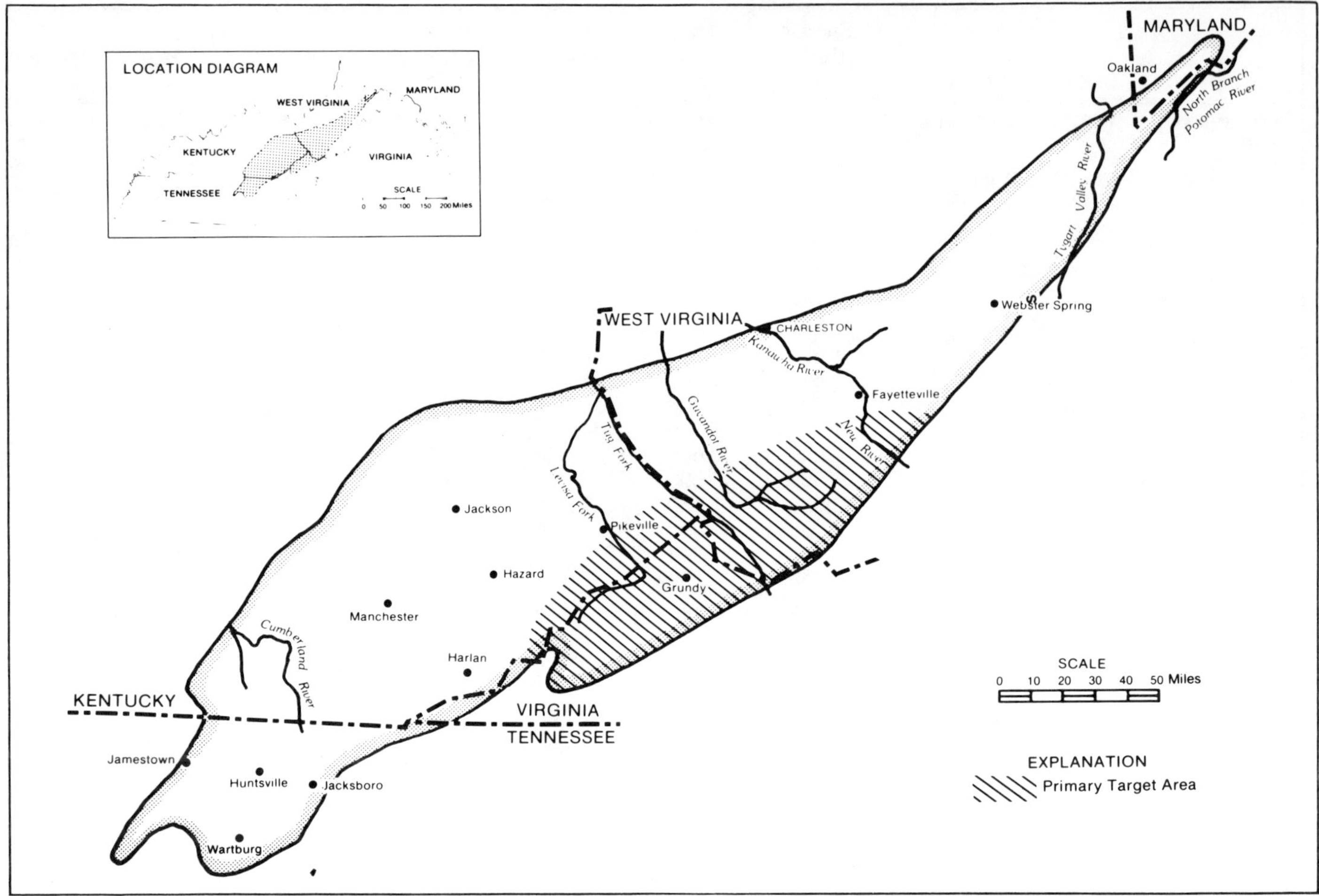

Figure 11—Primary methane target area of the Central Appalachian Basin.

results: U.S. Bureau of Mines Report of Investigations 8515, 36 p.

Huddle, J. W., et al, 1963, Coal reserves of eastern Kentucky: U.S. Geological Survey Bulletin 1120, 247 p.

Irani, M. C., et al, 1977, Methane emissions from U.S. coal mines in 1975, a survey: U.S. Bureau of Mines Information Circular 8733, 55 p.

Jeran, P. W., D. H. Lawhead, and M. C. Irani, 1977, Methane emissions from four working places in the Beckley Mine, Raleigh County, West Virginia: U.S. Bureau of Mines Report of Investigations 8212, 16 p.

Keystone, 1980, Keystone coal industry manual: New York, Mining Informational Services, McGraw-Hill Mining Pub.

Luther, E. T., 1959, The coal reserves of Tennessee: Tennessee Division of Geology Bulletin 63, 294 p.

Manilla, R. D., (ed.), 1980, Semi-annual report for the unconventional gas recovery program: U.S. Department of Energy, Morgantown Energy Technology Center, Morgantown, West Virginia, October, 1979 to March, 1980, p. 195–197.

Miller, M. S., 1974, Stratigraphy and coalbeds of Upper Mississippian and Lower Pennsylvanian rocks in southwestern Virginia: Virginia Division of Mineral Resources, 211 p.

Rehbein, E. A., C. D. Henderson, and R. Mullenex, 1981, No. 3 Pocahontas coal in southern West Virginia—resources and depositional trends: West Virginia Geological and Economic Survey Bulletin B-38, 41 p.

Rice, C. L., and J. H. Smith, 1980, Correlation of coalbeds, coal zones, and key stratigraphic units, Pennsylvanian rocks of eastern Kentucky: U.S. Geological Survey Map MF-1188.

——, et al, 1980, The Mississippian and Pennsylvanian systems in Kentucky: Kentucky Geological Survey Series X, Reprint 5, 32 p.

Thornbury, W. D., 1965, Regional geomorphology of the United States: New York, John Wiley and Sons, Inc., p. 130–151.

U.S. Geological Survey and U.S. Bureau of Mines, 1968, Mineral resources of the Appalachian region: U.S. Geological Survey Professional Paper 580, 492 p.

Vokes, H. E., and J. Edwards, Jr., 1974, Geography and geology of Maryland: Maryland Geological Survey Bulletin 19, 242 p.

Von Schonfeldt, H., 1981, Methane drainage in advance of

mining through horizontal boreholes; Final report for period April, 1979 to December, 1980: Occidental Research Corporation, for TRW Energy Systems Group, Subcontract No. H 15730 JJ9S, 118 p.

Weaver, K. N., J. M. Coffroth, and J. Edwards, Jr., 1976, Coal reserves in Maryland—potential for future development: Maryland Geological Survey Information Circular 22, 16 p. Reprinted 1981.

Wilson, Jr., C. W., and R. G. Stearns, 1958, Structures of the Cumberland Plateau, Tennessee: Tennessee Division of Geology Report 18, 13 p.

——, J. W. Jewell, and E. T. Luther, 1956, Pennsylvanian geology of the Cumberland Plateau: Tennessee Division of Geology, 21 p.

Geologic Overview, Coal, and Coalbed Methane Resources of the Warrior Basin — Alabama and Mississippi

J. L. Hewitt

The Warrior Basin is an area of 35,000 sq mi astride the Alabama–Mississippi state boundaries. This southern extension of the Appalachian basin contains Paleozoic sediments overlain in part by Cretaceous Coastal Plain deposits. The Pennsylvanian System is represented by the Pottsville Formation and contains all the estimated 35 billion tons of coal resources in the basin. Most of the coal is found in 20 seams forming seven coal groups. Approximately 20 billion tons are considered to be present at depths conducive to commercial methane development. Measured methane contents over 500 cubic feet/ton are commonly obtained in several beds, especially in Jefferson and Tuscaloosa counties, Alabama. At present commercial development is limited to those counties and chiefly to the Mary Lee Coal group. The methane resource is estimated to be between 7 and 10 trillion cubic feet.

INTRODUCTION

The Warrior Basin is a triangular area in Mississippi and Alabama, bounded by the Appalachian Mountains on the southeast and the Ouachita front on the southwest. It is a structural basin containing Carboniferous coal-bearing strata partly concealed beneath younger sediments of the Gulf Coastal Plain. The Warrior Basin covers 35,000 sq mi (Mellen, 1947) and is outlined in Figure 1. It takes its name from the Black Warrior River, one of several large, navigable rivers flowing to the west-southwest, which provided an early means of cheap transportation to the mining district. Older reports referred to the basin as the Black Warrior, but, following the current trend, this report will use the shortened name of Warrior Basin (Self and Neathery, 1975).

Throughout most of the Warrior Basin, the rocks are flat or gently inclined. To the east and southeast, the basin is bounded by the steeply overturned beds and thrust faults of the Appalachian system, and to the west and southwest, it is bounded by the deeply buried Ouachita front. The topography in the eastern part of the basin is dominated by the Appalachian structural trend that plunges beneath the Gulf Coastal Plain deposits in eastern Alabama. Two synclinal folds within this trend define the Cahaba and Coosa valleys and their respective coal fields (Fig. 2).

Coal has been mined for over 100 years in Alabama, typically by strip mining near-surface deposits. Some shallow gas fields are actively producing from Mississippian rocks in the Warrior Basin, but the bulk of the oil and gas exploration in the region is focused to the southwest in the younger rocks of the Gulf Coast province. Although no coal is produced in Mississippi, it ranks as the largest mining industry in Alabama, producing an average of 20 million tons annually (Ward and Evans, 1975). Almost all commercial production has come from over 20 seams forming seven minable groups in the Warrior coal field. Most of the coal is high-volatile A bituminous in rank, with low sulfur and variable ash contents. Some higher rank coal of medium- and low-volatile bituminous rank exists along the eastern margin of the basin.

The major coalbeds were formed during the Carboniferous, at a time when the Warrior Basin was one of a number of rapidly subsiding basins in eastern North America (Davis and Ehrlich, 1974). It represented the eastern shoaling of the Ouachita trough (Graham et al, 1976; Thomas, 1974) and was probably continuous with the Arkoma Basin of Arkansas and Oklahoma along the Ouachita structural belt (Davis and Ehrlich, 1974; Graham et al, 1976) (Fig. 3).

All minable coals in the Warrior Basin are Pennsylvanian in age, but some Mississippian coals occur in scattered localities. The Pennsylvanian Pottsville Formation ranges up to 9,000 ft in thickness and overlies a thick sequence of Mississippian deposits that grade from a cherty limestone at the base to become shales and sandstones at the top. A generalized stratigraphic column that will be used in this report is based on the work of Beg et al, (1978), shown in Table 1.

A structural model for the Warrior Basin is important in expanding the search for methane-rich coalbeds, especially in the subsurface, where future exploration will be aimed.

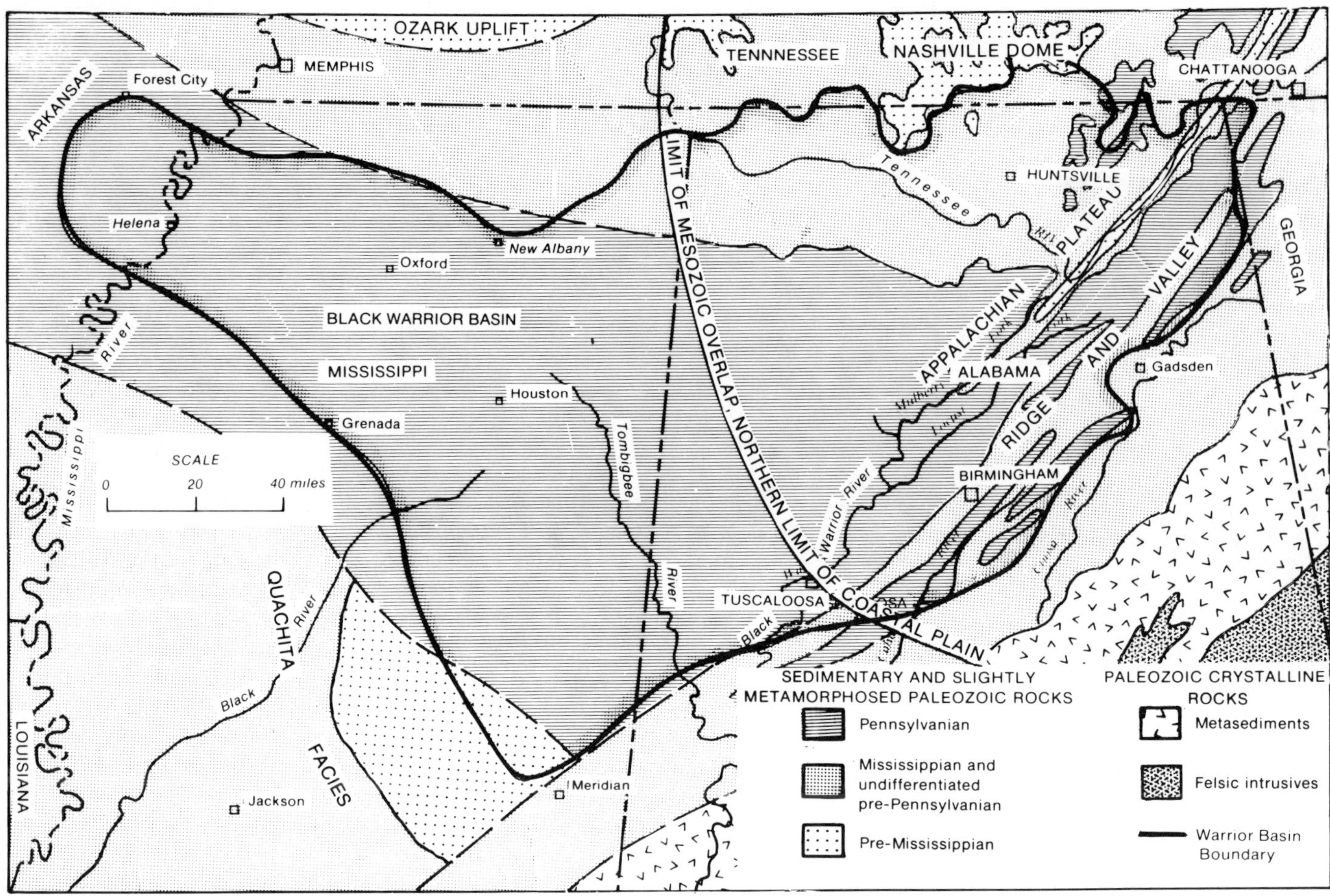

Figure 1—Geologic map of Warrior Basin and surrounding areas. Hobday, 1974. Used with the permission of the author.

Historically, coal production in this area has been limited to seams that can be proven in surface outcrop or at sites where the overburden thickness permits surface mining. Early identification of gas-rich seams and a general knowledge of basin geometry will permit the maximum production from this unconventional gas resource.

STRUCTURE AND STRATIGRAPHY

The roughly triangular shape of the Warrior Basin is formed by the Paleozoic rocks that create its boundaries (Fig. 1). The Ozark Uplift to the northwest and the Nashville Dome to the northeast form the basin's northern boundary. To the east and southeast, the basin is bounded by the southwesterly plunging folds and steeply overturned beds of the Appalachian system. In turn, the deeply buried Ouachita fold and thrust belt defines the basin's western and southwestern limits. The greater part of this triangular area is overlapped by Mesozoic and Tertiary sediments of the Gulf Coastal Plain that thicken from discontinuous outliers capping ridges and hills in central and western Alabama to more than 6,000 ft in thickness in central Mississippi (Self and Neathery, 1975; Kidd, 1976) (Fig. 4).

Rocks of early Paleozoic age are exposed in a few places on the basin's boundaries, and only one well—the F. W. Smith No. 1 in Cullman County, Alabama—has penetrated the basement complex. Precambrian granites have been found at depths of slightly less than 6,000 ft in Tennessee on the Nashville Dome. It is thought that these granites were eroded long ago to produce thick wedges of sandstone lying deep within the Warrior Basin (Mellen, 1971). Where these sandstones are exposed in eastern Alabama and Tennessee they are highly contorted and metamorphosed, but show evidence of having been porous and bituminiferous (Butts, 1926; Mellen, 1947). Several thousand feet of Cambrian and Ordovician carbonates were deposited on a stable shelf environment as the seas transgressed from the southeast; toward the west and northwest, the carbonates are overlain by and interfinger with clastics derived from the west (Boland and Minihan, 1971; Milici et al, 1973). The Silurian in the Warrior Basin is not very thick, but the Red Mountain Formation contains interbedded seams of "red fossil ore" that accumulated through chemical replacement of extensive beds of fossil fragments by iron-rich waters (Butts, 1926). In deeper parts of the basin to the west, porous reef-type accumulations of Silurian limestone are sought as oil and gas reservoirs. The Devonian was a time of erosion or nondeposition throughout the basin and is represented by thin cherts and shales. Only on the eastern margin is evidence of the Acadian orogeny

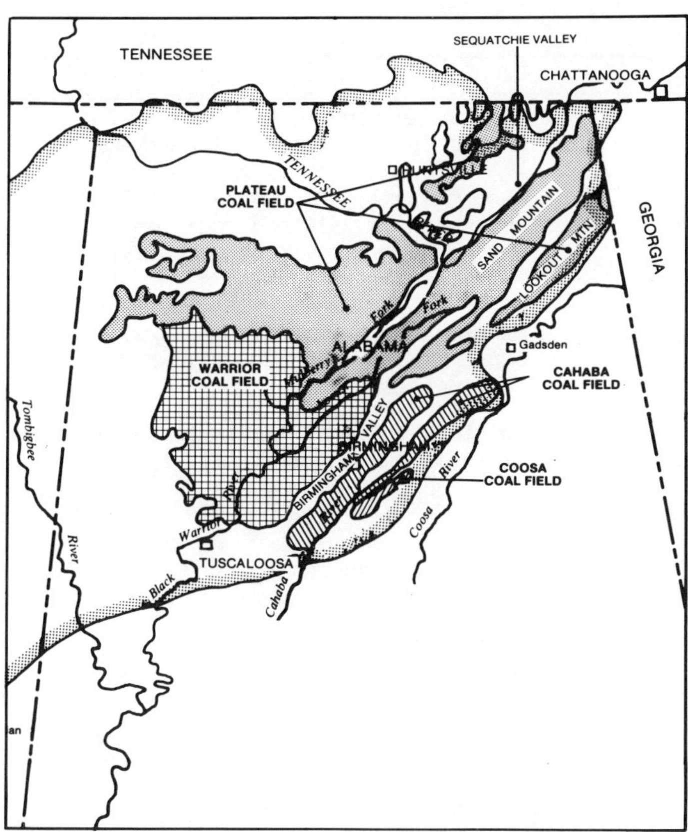

Figure 2—Coal fields of the Warrior Basin. Ward and Evans, 1975. Used with the permission of the Alabama Geological Survey.

preserved as Grenville-age gravels in diamictite and by the stratigraphic position of the Hillabee greenstone volcanic arc complex (Tull, 1979).

In Early Mississippian time, the Warrior Basin included a quiet, tectonically stable platform in the northeast that supported a flourishing carbonate shelf facies. The basin sloped rapidly toward the south and southwest where deep-water clays accumulated.

The Fort Payne Chert is a distinctive lithologic marker at the base of the Mississippian sequence in Alabama and provides a common datum below the carbonates of northeastern Alabama on the platform edge and the clastic facies lying to the southwest (Fig. 5). On the margin of the Warrior platform, the Fort Payne Chert is overlain by the Tuscumbia Limestone (Thomas, 1972). The Tuscumbia formation thins toward the southwest and pinches out entirely in the Appalachian synclines to the southeast as the result either of a facies change from shallow water carbonates to deeper water clastics or as a gradation to the upper Fort Payne. The Tuscumbia is overlain by the Monteagle Limestone in northeastern Alabama and by either of two clastic units; the Floyd Shale or Pride Mountain Formations elsewhere (Fig. 5).

The Monteagle Limestone contains scattered interbeds of dolomitic limestone or dolostone that increase in number to the northeast. A basal layer of blocky chert in northern Alabama appears to be an extension of the Lost River Chert in Tennessee (Ferguson and Stearns, 1967). The Monteagle grades from limestone to interfinger with a time-equivalent shale, the Pride Mountain Formations, or the Hartselle Sandstone in the southwest off the Warrior Platform. To the southeast, carbonates are absent in the Appalachian folds, and shale is found in this interval.

The Pride Mountain Formation is the leading edge of the thick clastic wedge that prograded from the southwest and forms a major tongue of the Floyd Shale. Beyond the southwestern limit of the Hartselle Sandstone, the Pride Mountain is not differentiated from the Floyd Shale. It is the most lithologically variable Carboniferous unit in the Warrior Basin and contains diverse fauna. In the northwest, the Pride Mountain Formation shale grades into limestone, and sandstone is absent; to the southwest, the formation contains one to three units of sandstone, each of which may include interbedded limestones locally. Four sand units exist as widespread horizons within the Floyd Shale. Evidence supports their formation as offshore shallow marine bars and/or intermittent barrier islands developed upon high spots on the East Warrior Platform. Although part of the Pride Mountain and its time equivalent (the Floyd Shale) may be brackish lagoon deposits, much of the sequence is marine or back-barrier marine (Thomas, 1972).

The largest of these sandy units is the Hartselle Sandstone, typically composed of well-sorted light-gray to gray-brown fine quartz sand. Both marine and plant fossil fragments have been found in the Hartselle, and a fossil tree stump was collected in Colbert County (McCalley, 1886). The Hartselle is of variable thickness, with three thickness maxima in southeastward-trending belts spaced about 25 mi apart. The Hartselle Sandstone is thought to represent a barrier island complex whose trend paralleled the edge of the east Warrior Platform (Thomas and Mack, 1982). To the northeast, it prograded over the carbonate bank to interfinger with limestone and pinched out entirely in the back-barrier marine and/or lagoonal clays of the Floyd Shale or Pride Mountain deposits (Fig. 5).

The Bangor Limestone includes a variety of lithologies that lie above the Hartselle Sandstone and below the clastic Pottsville Formation. It formed as a carbonate complex on a

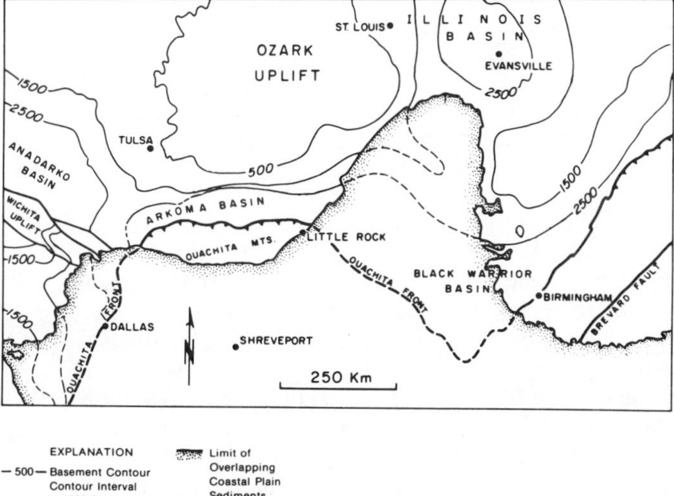

Figure 3—Regional tectonic map showing relationship of the Warrior Basin, Arkoma Basin, and Ouachita Front. Graham et al, 1976.

Table 1—General stratigraphic column for the Warrior coal field. Modified from Beg et al, 1978. Used with the permission of the Alabama Geological Survey.

Era	System	Geologic Unit	Thickness	Lithologic Description
Cenozoic	Quaternary	Alluvium and Terrace deposits	0–50 ft 0–15 m	Clay, sand and gravel, rock fragments, unconsolidated, lenticular, poorly sorted.
Mesozoic	Cretaceous	Eutaw Formation	0–200 ft 0–60 m	Sand, brownish-gray to light tan, unconsolidated, lenses of gravel and varicolored clay; minor medium- to dark-gray carbonaceous clay.
		Tuscaloosa Group	420–850 ft 130–260 m	
Paleozoic	Pennsylvanian	Pottsville Formation	2,500–4,400 ft 760–1,340 m	Sandstone, light brown to medium gray, crossbedded, interbedded with siltstone, shale and coal beds.
	--?--?--?--?	Parkwood Formation	400–1,200+ ft 120–370+ m	Sandstone, medium to light gray, interbedded with shale, minor coal in upper part.
	Mississippian	Floyd Shale	1,000+ ft 300+ m	Shale, dark gray, fissile; thin beds of siltstone locally.
		Fort Payne Chert	90–200 ft 27–60 m	Chert, grayish orange to light gray, fossiliferous; thin- to medium-bedded with minor limestone and shale.
	Devonian	Chattanooga Shale	0–18 ft 0–5 m	Shale, black, highly fissile; absent in places.
		Frog Mtn. Sandstone	0–70 ft 0–21 m	Sandstone, light gray to tan, minor grayish orange chert, absent in places.
	Silurian	Red Mountain Formation	200–500 ft 60–150 m	Sandstone, reddish-brown to olive-gray, siltstone conglomerate and shale; minor limestone and bedded oolitic hematite in sandstone.
	Ordovician	Chickamauga Limestone	200–500 ft 60–150 m	Limestone, medium- to light gray, sublithographic, fossiliferous, medium-bedded; minor bedded shale. Base is marked locally by Attalla member, chert and carbonate conglomerate.
	--?--?--?--?	Knox Group undifferentiated	2,000+ ft 610+ m	Dolomite, light gray, cherty, thick-bedded, interbedded with limestone. Below the Knox Group some chert-free Ketona dolomite may be present.
	Cambrian	Conasauga Formation	1,100–1,900 ft 335–580 m	Limestone, dark gray, sublithographic thin-bedded, minor light olive to medium gray shale and dolomite.
		Chilhowee Group	2,150–2,500+ ft 650–760+ m	Orthoquartzite interbedded with feldspathic quartz arenite; greenish-gray mudstone; minor siltstone and arkosic pebble conglomerate.

very shallow marine shelf that developed along the edge of the east Warrior Platform. To the southwest, it is overlain by the Floyd Shale and Parkwood Formation that were prograding toward the northeast. Upon the platform, the carbonate complex was being overlapped by the marine Pennington clastic sediments prograding from the northeast. Overlying these sediments are the delta and coastal plain deposits of the Pottsville Formation, derived from both northeast and southwest sources.

The Floyd Shale extends into the Warrior Basin below the Parkwood and above the Hartselle Sandstone and represents the early pro-delta clays of an expanding delta complex from a sediment source in the southwest. Fauna indicate that a marine environment existed in the southwest.

The Parkwood Formation is a series of claystone, shale, and sandstone interbeds containing abundant plant fossils and coalbeds near the top of the formation. These show the greatest concentration in the upper 300 ft exposed in the Cahaba syncline (Thomas, 1972). Marine fossils found in sandy and calcareous intervals suggest that the lower Parkwood is Mississippian in age and the upper Parkwood is Pennsylvanian (Butts, 1926). The Parkwood Formation is thought to be the shallow marine clays and distributary front sands of a prograding delta. The southwestward increase in total sand content, abrupt variations in sand body thickness, and a lack of widespread individual sand units suggest a distributary front environment

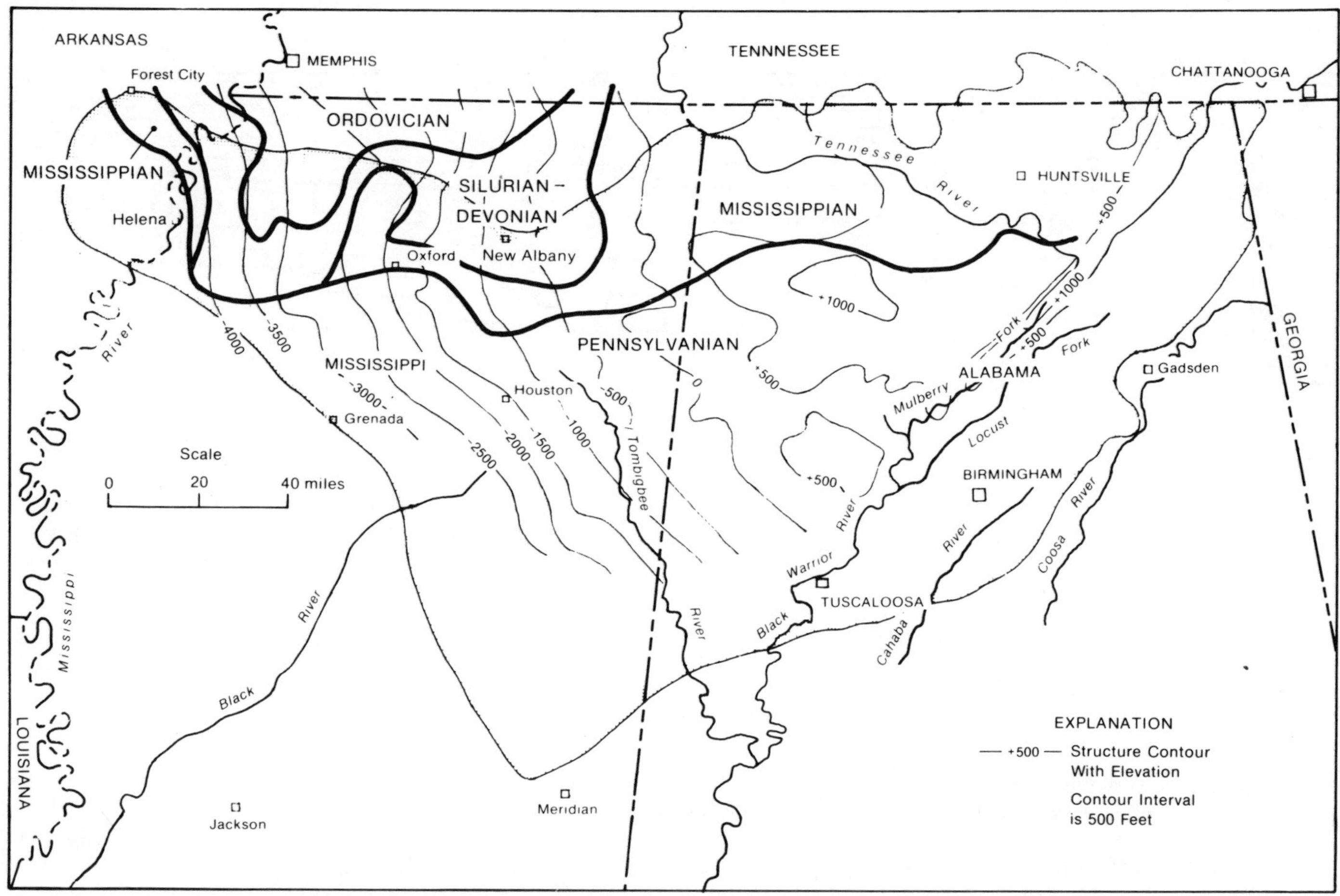

Figure 4—Structure contour map: top of Paleozoic strata. From Boland and Minihan. Copyright © 1971 The Gulf Coast Association of Geological Societies. Used with permission.

that spread northeastward. In the southeast, active subsidence in the Appalachian synclines provided channels for the rapid transport of clastic material around the edge of the East Warrior Platform.

Throughout the Mississippian, the Warrior Basin had filled rapidly with deposits from the southwest, as a smaller delta system developed in the north draining the craton. During the Pennsylvanian, a later clastic wedge derived from the rising Appalachian borderlands spread from the east to merge with the Pottsville.

The Pennington shale represents the northeastern clastic wedge that prograded over the Bangor Limestone to merge with the Floyd-Parkwood clastic wedge advancing from the southwest. The Pennington and Parkwood interfinger with the Bangor Limestone and with each other. Faunal evidence and stratigraphic relationships suggest that the upper parts of these three formations are mutually time-equivalent.

Overlying the Pennington, Parkwood, and Bangor Formations is the Pottsville, containing up to 40 coalbeds of variable thickness and quality. It has been informally divided into two units: a lower barren part of quartz-rich sandstone and quartz pebble conglomerate, and an upper part of sandstone and shale that contains all the commercially productive coal seams in Alabama. It was deposited during rapid progradation of upper delta and sandy coastal sediments across a continental shelf and subsiding clastic wedges. The base of the Pottsville is thought to reflect local topographic conditions at the time (Fig. 6) as a relatively flat, southwestward-sloping surface. In the eastern part of the Warrior Basin, the Sequatchie anticline separates the Coalburg syncline or Pratt Basin from the rest of the Warrior coal field before dying out altogether in northern Tuscaloosa County. This anticline is associated with a major Appalachian thrust fault further to the north in Tennessee. Some small-scale folds such as the Wiley dome in northern Tuscaloosa County have 200 ft of subsurface closure (Culbertson, 1964) and are only revealed by thorough mapping programs. Numerous en echelon faults trend northwest and have been identified as normal faults up to 4 mi in length with up to 300 ft of displacement (Semmes, 1929).

The work of Ehrlich (1964), Ferm et al (1967), and Davis and Ehrlich (1974) presents evidence for a southern source within the Ouachita structural belt for the Pennsylvanian clastics. This model envisions the uplift and subsequent erosion of a southern source terrane composed of low-rank metamorphic rocks laced with volcanics. In this model, Carboniferous sediments are deposited in a prograding delta system building northward—first filling a shallow, subsiding "geosynclinal" portion of the crust and then lapping onto the shelf (Fig. 7). The thick coalbeds and sandstones of the Pottsville Formation reflect the channels, lakes, and swamps of a delta or alluvial plain and the Early Pennsylvanian and Mississippian deposits as delta front and shallow back-barrier marine environments (Fig. 8). The original

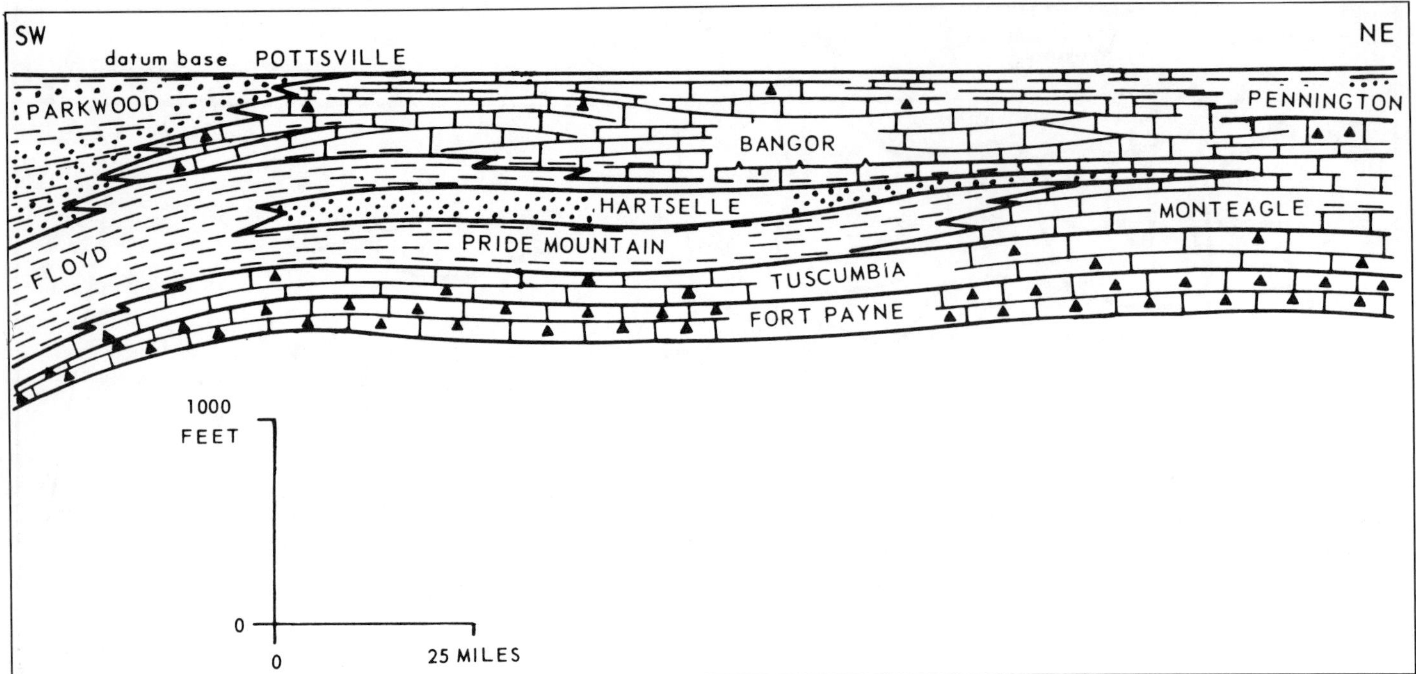

Figure 5—Generalized stratigraphic cross section of northern Alabama. Thomas, 1967. Used with the permission of the Alabama Geological Survey.

southern source area is now buried beneath the Mesozoic coastal plain but was probably an extension of the Ouachita orogenic belt.

Hobday (1969, 1974) examined Pennsylvanian-age sandstone complexes in northern Alabama and established that these sand bodies are similar to modern beach and barrier islands. The pattern of outcrops suggests two different sources of sediment: one from the northeast and one from the south. Further, these two sets of barrier complexes developed more or less synchronously and support a model of the northward spread of a delta complex filling a geosyncline or trough and crossing over a continental shelf (Ferm and Ehrlich, 1967). They concluded that the two sets of mobile barrier bars—one supplied by sediment from the northeast and one fed from the south—merged in north-central Alabama during this time as the sea withdrew to the northwest.

Combining aspects of these models is one by Thomas (1972, 1974, 1977), that envisions the Mississippian carbonate facies lying on the East Warrior shelf in north-central Alabama between two clastic wedges that prograde from the southwest, as in the model of Ferm et al (1967), and from the northeast, as discussed by Hobday (1969, 1974). Close examination of the southwest wedge indicates that it is the eastern shallow-water equivalent of a thick, flysch-like wedge that fills the Ouachita trough. Increasing sediment supply through time, radial transport directions, and a general coarsening-upward sequence suggest a prograding delta complex with several cycles of growth. A tongue of the lowermost cycle or lobe extends across the Warrior Basin to interfinger with the limestone belts that occupy the platform-edge position. From the northeast, another later clastic wedge spread over the carbonate facies, but this was only the distal southwestern fringe of a larger clastic wedge derived from an Appalachian highland source to the northeast, in the direction of the Tennessee salient. Thomas (1974) presents evidence to indicate that the Appalachian synclines that form the Cahaba and Coosa coal fields were active folds during Mississippian time. These provided repositories for the thick accumulations of sediment generated in the southwest and behaved as channels for northeastward transport of material around the shallow southeastern edge of the Warrior platform.

Recent work done by Graham et al (1975, 1976) compared petrographic data from samples collected from the Warrior Basin and Ouachita Mountain systems with paleocurrent data collected by other workers. They found a common sedimentary source for the Warrior Basin-Ouachita deposits and concluded that they were once a linked sediment dispersal system. The Warrior Basin acted as a catchment area for sediment that was then channeled westward to produce the deep-water turbidites of the Ouachita trough.

The Pottsville Formation, which contains the minable coals of the Warrior coal field, has never been subdivided formally, in part because of a lack of extensive stratigraphic marker beds. Informal systems fall into two classes: those that divide the Pottsville into the lower barren and upper productive coal measures, and those that divide the Pottsville into seven intervals based on the presence of major coal groups. For this report, the stratigraphic succession used by Metzger (1965) consisting of seven intervals designated from oldest to youngest as A through G will be adopted. Figure 9 is a generalized stratigraphic column of the Pottsville Formation illustrating the position of the major coal seams, and Figure 10 shows their outcrop pattern in the eastern half of the Warrior Basin.

Stratigraphic Interval A includes all the sediments above the Mississippian and below the Black Creek coalbed. It overlies the Pennington Shale throughout its outcrop area, which forms an east-west band in Alabama (Fig. 10) and has been dated as Pocohontas in age from the Pocohontas coal field region of West Virginia and Virginia. Stratigraphic Interval A thickens in

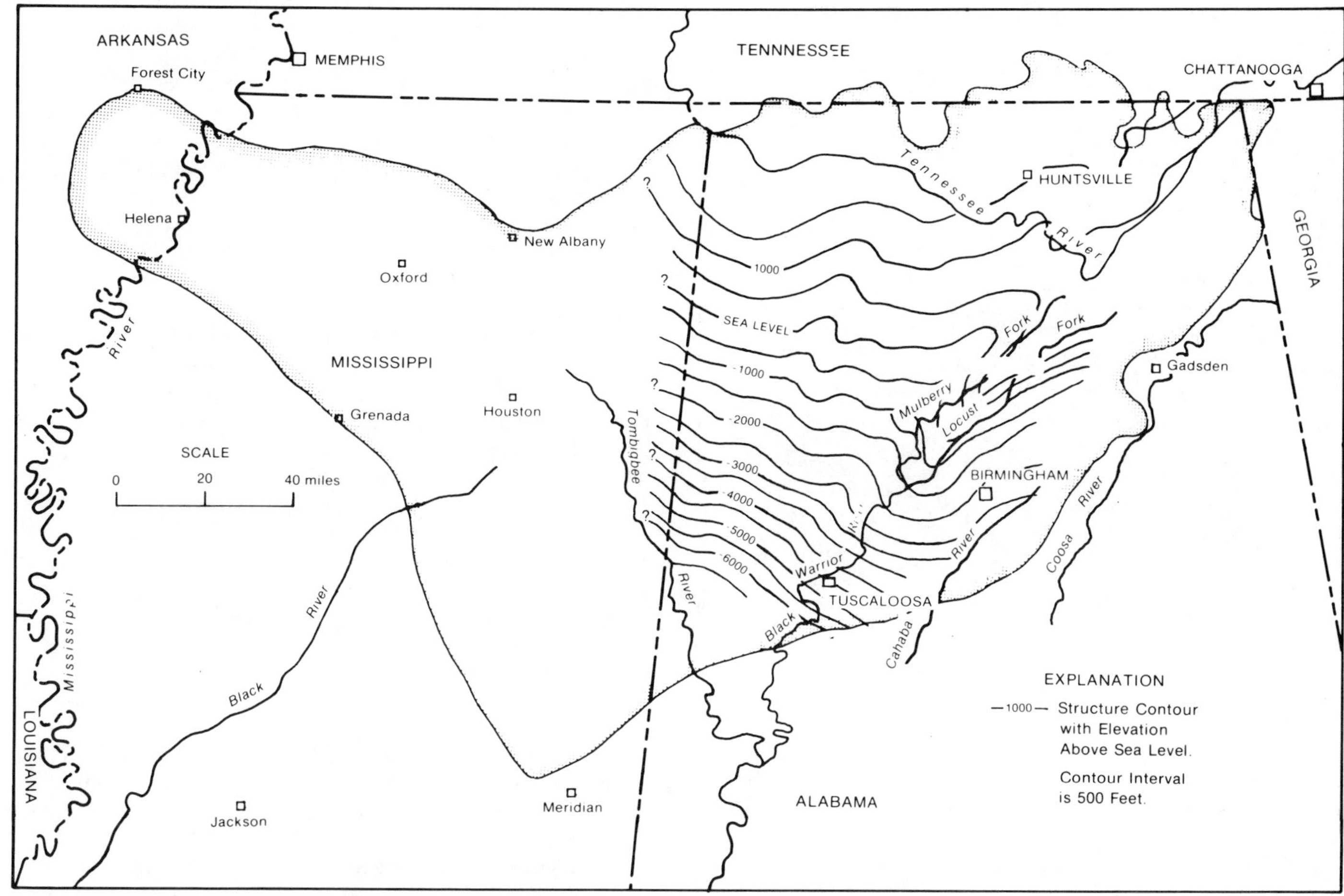

Figure 6—Structure contour map on the top of the Mississippian System in the Eastern Warrior Basin. From Thomas, 1972. Used with the permission of the Alabama Geological Survey.

the subsurface into Mississippi and thins abruptly in the east-central part of the basin overlying the trend of the Sequatchie anticline.

Stratigraphic Interval B is defined as those beds above the Black Creek coal and below the Mary Lee seam. Erosion has produced two unconnected outcrop areas: one in the Appalachian Plateau region in northeastern Alabama and another in the northern basin that has been shown to extend into the subsurface (Fig. 11). This interval is composed of interbedded shales and sandstones that contain a few poorly preserved plant remains. The coals are thin and of poor quality and decrease in thickness and number to the southwest. This unit is 250 ft thick at Warrior in Jefferson County but thickens to over 600 ft at the Mississippi state line.

Stratigraphic Interval C crops out over a wide area in the east-central basin and includes the interval above the Mary Lee coal and below the Pratt. Its thickness averages 400 ft in the outcrop area but increases in the subsurface to the south and west (Fig. 12).

Stratigraphic Interval D extends from the Pratt to the base of the Cobb lower coal. The Pratt coal, also known as the Corona, is overlain by a fossiliferous marine interval about 10 ft above the main seam (Metzger, 1965). Figure 13 shows an isopach map of this interval that thickens uniformly to the south.

Stratigraphic Interval E lies above the Cobb and extends to the base of the Gwin coal. Throughout its outcrop, it is about 200 ft thick but thickens in the subsurface to the southwest (Fig. 14).

Stratigraphic Interval F crops out in the southern part of the basin and was originally mapped by McCalley (1900). Figure 15 shows an isopach map of this unit that thickens to the southwest in the subsurface.

Stratigraphic Interval G is defined as those beds above the Johnson coal extending to the top of the Pennsylvanian sequence. It crops out in isolated topographic high areas and ranges from 60 to 180 ft thick. It includes the Milldale coal lying 35 to 40 ft above the Johnson coal and the Brookwood coal approximately 10 to 30 ft above the Millwood. These are overlain in places by the thin but persistent Guide coal that occurs 5 to 25 ft above the Brookwood where it has not been removed by erosion.

This interval is overlain by Cretaceous and younger deposits that are primarily unconsolidated clay, sand, or coarse gravel. Younger deposits overlap the western boundaries of all these stratigraphic intervals and are shown on a map of the outcrop patterns (Fig. 10).

McCalley (1900) reported that the total thickness of coal in this field ranged from 20 to 120 ft and averaged about 60 ft. Almost all coal production in the Warrior field has come from over 20 seams forming seven minable groups. The lowest of these groups is the Black Creek (Fig. 9) which is located 1,400 ft above the base of the Pottsville Formation.

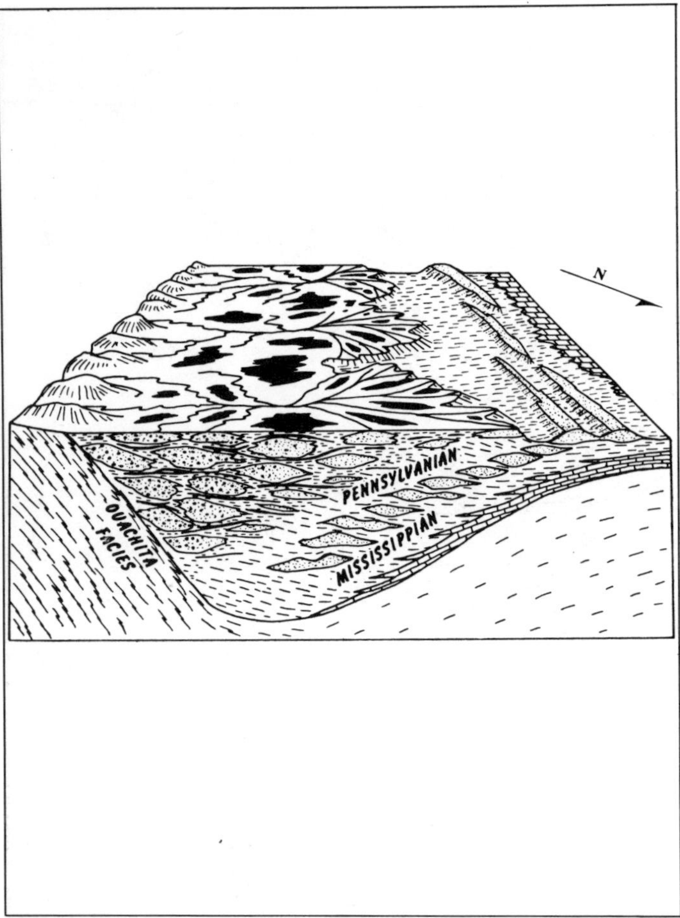

Figure 7—Schematic block diagram of Carboniferous detrital rocks in the Warrior Basin of northern Alabama. Ferm et al, 1967.

COAL RESOURCES

Four coal fields in the Warrior Basin are separated from each other by folding, faulting, and erosion (Fig. 2). The Coosa field lies furthest to the southeast in an elongate syncline along the trend of the Appalachian Mountains. It covers approximately 210 sq mi in Jefferson, Shelby, and St. Clair counties, Alabama. Very little is known of the thickness or extent of coalbeds in this narrow, trough-shaped field, although the total thickness of its coal-bearing strata is estimated to be greater than 7,000 ft. The northwest boundary is formed by the flank of an anticline dipping steeply to the southeast, and the southeast boundary is a highly deformed complex of thin, imbricate thrust sheets of Ordovician- to Mississippian-age rocks. In addition to their trough-like structure, both the Coosa and Cahaba fields are cut by numerous subordinate folds and faults, so that coalbeds generally are steeply inclined and mining conditions more difficult than in the Warrior or Plateau fields.

The Cahaba field lies to the southwest of the Coosa field, separated from it by the Cahaba Valley (Fig. 2). It includes approximately 360 sq mi in parts of Bibb, Shelby, St. Clair, and Jefferson counties, Alabama. The Cahaba coal field lies in a narrow asymmetrical trough that is deepest along the southeastern limb. The boundary on this side is formed by the Helena thrust fault (Butts, 1911, 1940) whose vertical displacement of more than 10,000 ft has brought Pennsylvanian coal-bearing rocks into contact with Cambrian and Ordovician carbonates (Butts, 1906). The base of the Pottsville Formation is placed at the Brock coalbed (Fig. 16), the lowest coalbed exposed in the Cahaba field. The coal-bearing strata thicken toward the southeast, reaching over 9,000 ft in Shelby County (Stearns and Mitchum, 1962) from less than 6,000 ft in the northeast (Butts, 1906).

The coal-bearing interval in the Cahaba field may include as many as 60 coalbeds, but most are too thin to be minable (Tolson et al, 1982). Coalbeds suitable for strip mining are concentrated in the top half of the section and include the Nunnally Group, Harkness, Wadsworth, Youngblood, Clark, Thompson, Yeshic, and Montevallo beds shown in Figure 17 (Shotts, 1971). The Nunnally Group consists of two to six beds throughout an interval of 100 ft or more (Butts, 1911). The third Nunnally coalbed is the most persistent and averages 3 ft in thickness. The Harkness or Big Bone bed averages 6 ft in thickness, but, as the name implies, it is high in ash and uniformly low in grade (Shotts, 1971). It has been correlated with the Black Creek coalbed in the Warrior Basin (Butts, 1926). The Wadsworth coal is defined as two thin but persistent beds, 15 to 20 ft apart, called the Alice and Jones coalbeds by local miners, which lie about 600 ft above the Harkness. The Youngblood is a persistent coalbed with a uniform thickness averaging 3 ft and is locally known as the Coke coalbed because of its excellent coking properties. The Clark bed consistently yields the highest quality coal from the Cahaba field. It is generally over 3 ft in thickness but increases to more than 5 ft near West Blocton in Bibb County where it is known as the Woodstock coalbed. The Thompson coalbed has been the most productive seam in the Cahaba field and contains less sulfur than any other Warrior Basin coal. The Yeshic and Montevallo coalbeds no longer hold significant amounts of strippable reserves and are no longer being mined.

Although this coal field was the site of the first underground mine in the Warrior Basin in 1856 (Squires, 1890, p. 18), current production is only from surface mining activities and constitutes only 10% of Alabama's demonstrated coal reserve base (Tolson et al, 1982). Many published analyses of these coals confirm that they are all of high-volatile A bituminous rank, averaging about 6% ash and less than 1.5% sulfur (Shotts, 1971). Coals in this field are not known to contain significant methane, and Butts stated, "The mines are never dangerously gaseous" (Butts, 1906, p. 115). These seams are of lower rank than many of the coals across the Birmingham Valley in the Warrior coal field to the northwest and most resemble the coals of Walker County west of the Warrior River in rank.

The plateau coal region is composed of several coal-bearing areas in the upland regions of northeastern Alabama (Fig. 2). While it covers a greater area than all the other coal fields combined (4,500 sq mi), its coalbeds are not continuous throughout the field because of erosion, and the same bed may be known by several different names. Over 25 coalbeds in the

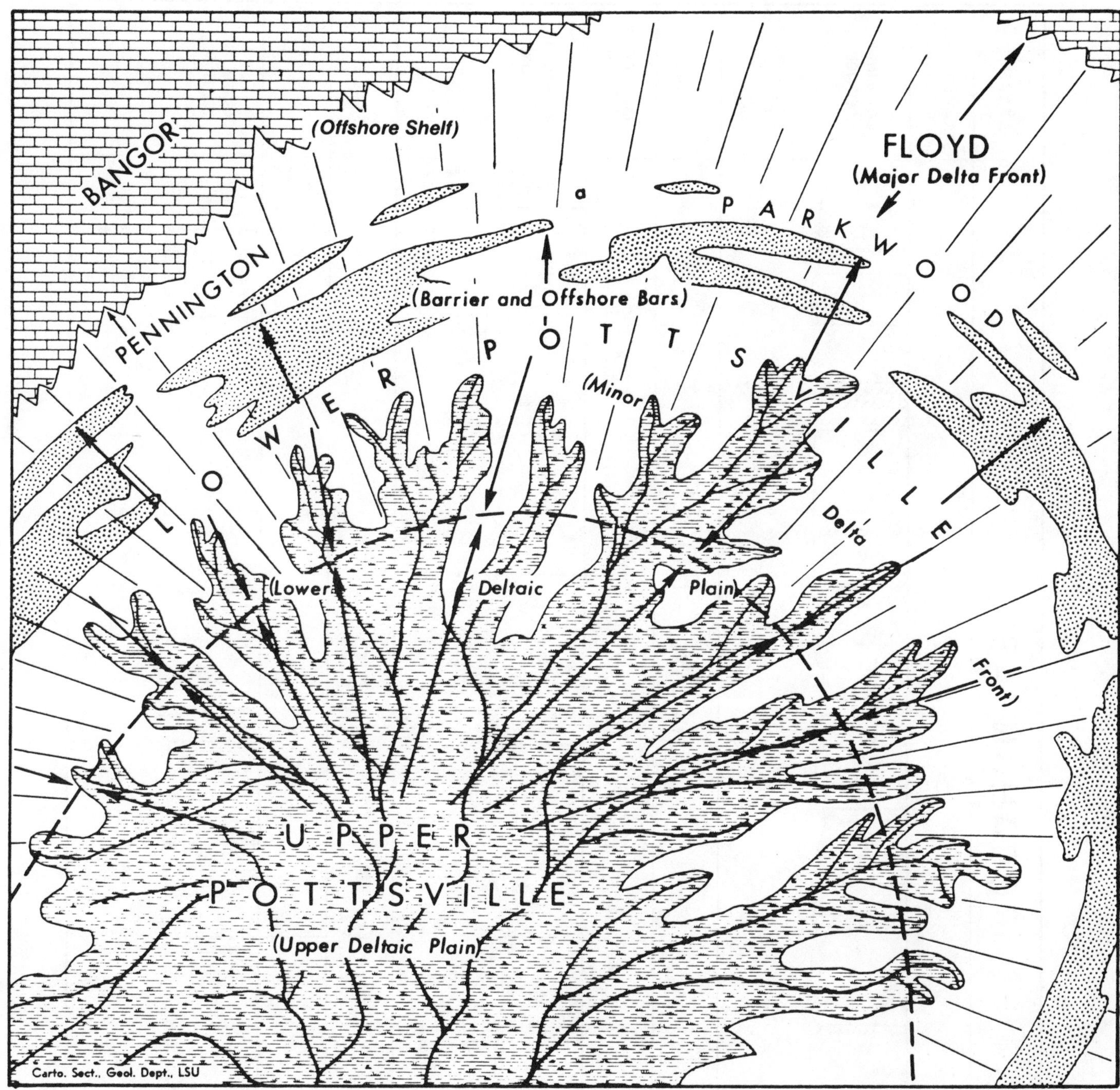

Figure 8—Plan view of generalized delta model for Carboniferous detrital deposits in northern Alabama. Arrows show dominant sediment transport directions. Ferm et al, 1967.

plateau region average 2 to 3 ft in thickness but locally have been reported to be as much as 20 ft thick (Ward and Evans, 1975).

The coal-bearing areas in the plateau region are separated by isolated mesas and flat-topped mountains where resistant sandstone units in the lower Pottsville Formation form the caprocks. The Wills, Sequatchie, and Murphree Valley fields lie within folded and faulted anticlines, and Lookout, Sand, and Blount Mountains are shallow synclines all lying along the edge of the northeast-trending Appalachians (Culbertson, 1964). The Pottsville generally is only 800 ft thick throughout the field, with beds dipping at 30 ft per mi to the southwest. The Cliff bed (McCalley, 1891)—or Castle Rock seam (Coulter, 1947)—averages 18 inches in thickness but may reach 4 ft

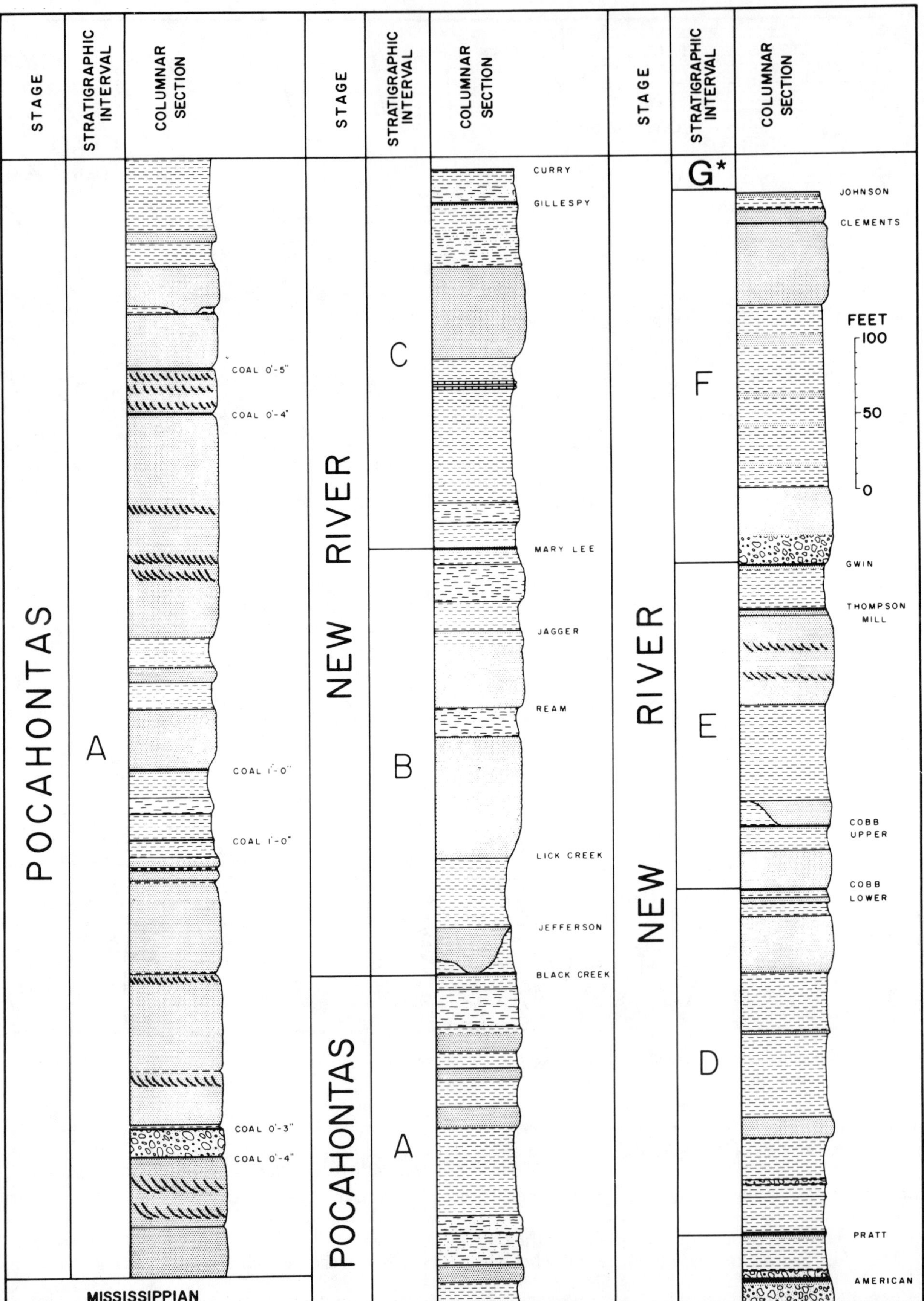

Figure 9—Generalized stratigraphic column of the Pottsville Formation. Metzger, 1965. *Added to Metzger, 1965. Used with the permission of the Alabama Geological Survey.

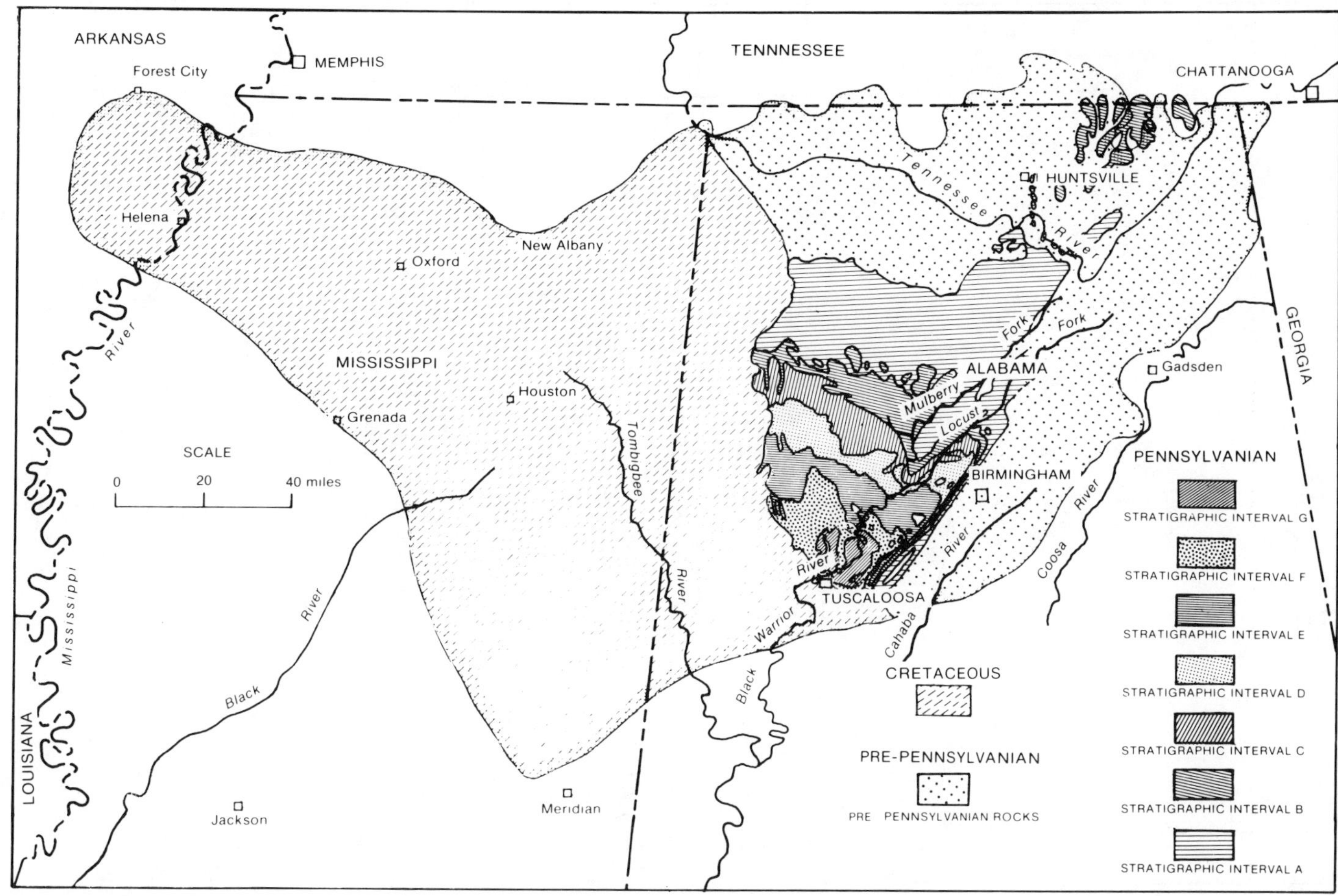

Figure 10—Outcrop map of the Pennsylvanian in the Eastern Warrior Basin. From Metzger, 1965. Used with the permission of the Alabama Geological Survey.

locally. From a high-volatile A bituminous in the northern Sand Mountain field, the coal changes to low-volatile bituminous rank in Lookout Mountain. Although minable in many places, it occurs in lenses and is locally channeled.

In the Sand and Lookout Mountain areas, a coalbed known as the Underwood coal occurs a few feet above the base of the Pottsville Formation (Culbertson, 1964). It ranges from medium- to low-volatile bituminous in rank and generally is high in both ash and sulfur. Above the Underwood at Lookout Mountain, several other coalbeds are found that tend to be thin but persistent, and all have been mined from time to time.

Blount Mountain contains the thickest Pennsylvanian sequence in the plateau field, ranging up to 1,400 ft. Since the 1890s, all recorded production has come from the Swansea and Altoona beds, which probably correlate with the Rosa bed of the Warrior field (Gibson, 1891, 1893) (Fig. 16). Terminating abruptly against the northwest border fault of Blount Mountain, they are high-quality coking coals with low ash and sulfur contents.

The Warrior coal field is the largest and most productive of the four and covers 3,500 sq mi in Tuscaloosa, Jefferson, Lamar, Marion, Winston, Fayette, Cullman, Blount, and Walker counties. The Warrior contains over 20 coalbeds in seven groups, some of which are known to extend into the subsurface of Mississippi (Bicker, 1970).

Physiographically, the Warrior coal field lies within a part of the Cumberland Plateau not covered by coastal plain sediments. To the northeast it merges into the Plateau coal field (Fig. 2) and to the southeast it is bounded by the Opossum Valley thrust fault that passes into a tightly folded anticline in the northeast. Although the strata of the Warrior field thicken and dip regionally to the southwest, they are concealed by overlying Cretaceous and younger sediments that obscure the basin's structure in this direction.

Most of the coals occurring in the Warrior Basin are high-volatile A bituminous in rank, although coal of highest rank is found in eastern Alabama at Lookout Mountain where at least a portion is low-volatile bituminous in rank (Tolson et al, 1982). Elsewhere, the coal's fixed carbon contents average between 55 and 60% and reach a maximum of 64% in the Pratt bed southwest of Birmingham. Volatile matter ranges between 25 and 35%, ash averages about 9%, and moisture about 2.5% throughout Alabama coals (Butts, 1927, p. 213). Most run-of-mine coal samples have calorific values above 13,000 Btu.

Traditionally, surface mining of Warrior Basin coals produces over half the output from that region. Seams were located from their outcrop, mined until the overburden thickness became too great, and then abandoned. As underground mines become deeper, more methane-rich seams are likely to be encountered,

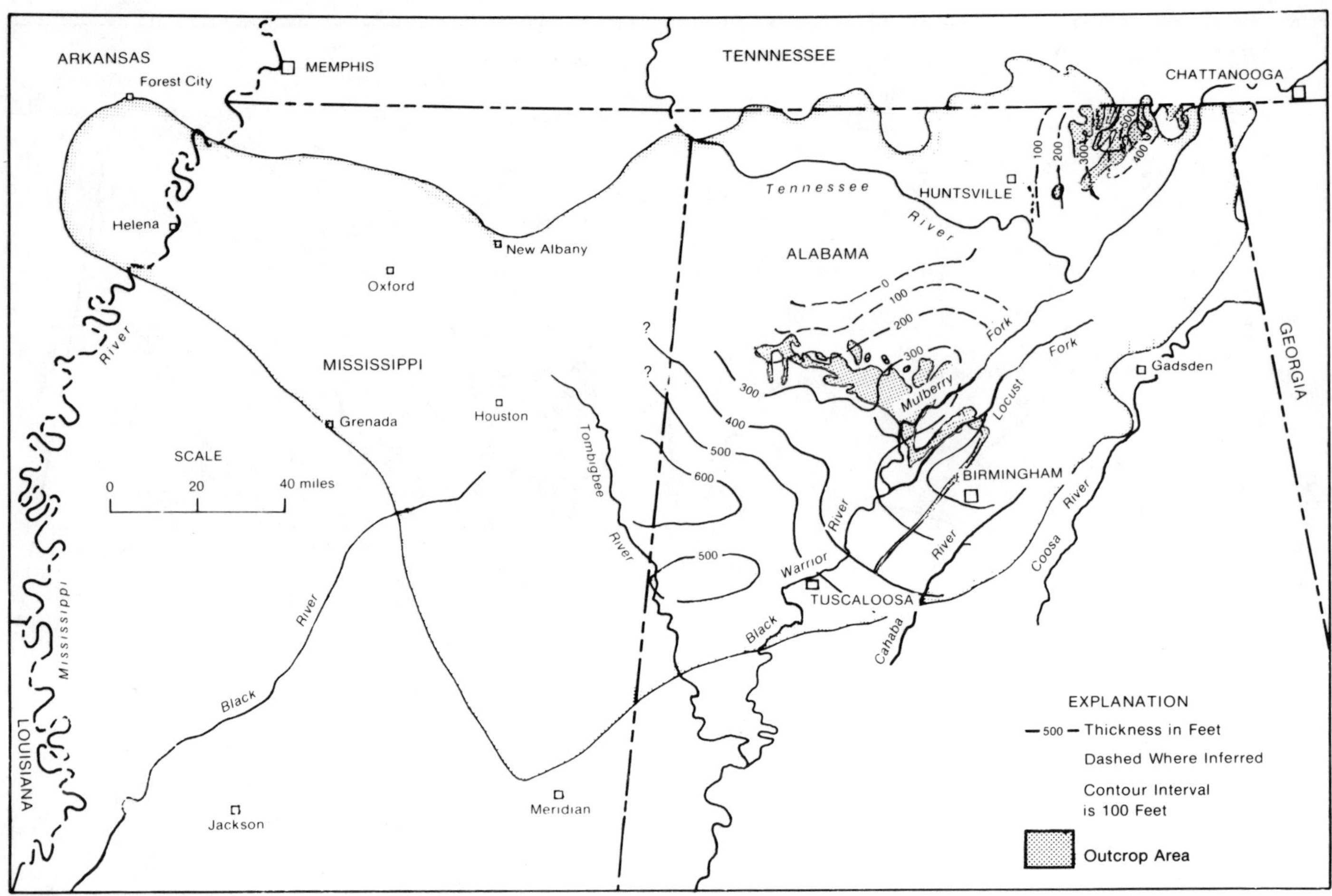

Figure 11—Isopach map of stratigraphic interval B, Pottsville Formation. From Metzger, 1965. Used with the permission of the Alabama Geological Survey.

affording an opportunity to extend our knowledge and understanding of coalbed methane.

The coal resources of the Warrior field are contained primarily in 20 minable seams lying within seven major groups in the lower Pennsylvanian Pottsville Formation (Fig. 18). Average analyses of the major coals in the Warrior Field are given in Table 2. Because many of the coal seams are laterally discontinuous, a number of stratigraphic systems have been applied to the Pottsville Formation in this field.

Black Creek Coal Group

The Black Creek coal group is the lowest consistently workable horizon in the Warrior coal field and underlies a large area in Jefferson, Walker, Cullman, and Blount counties, Alabama (Fig. 18). The Black Creek ranges over a stratigraphic interval of 50 to 150 ft and is composed of three beds which, in ascending order, are the Black Creek, Jefferson, and Lick Creek coals (McCalley, 1900). Because the Lick Creek bed is thin and contains many partings, this group's reserves are wholly contained in the other two seams. The Jefferson bed ranges up to 60 inches in thickness, generally averaging from 10 to 30 inches, and is a high-volatile A bituminous coal with low to moderate ash and variable sulfur contents (Table 2). The Black Creek bed produces a high-quality coal having the lowest ash and sulfur contents of any of the major beds in the Warrior coal field. Like the bulk of the Warrior Basin coals, it is high-volatile bituminous A in rank. It averages between 20 and 36 inches in thickness but thins and grades into carbonaceous shale to the west in northwestern Walker, southern Winston, and southeastern Marion counties.

Mary Lee Coal Group

The Mary Lee coal group contains the largest reserves of the Warrior coal field in five seams which, in ascending order, are the Ream, Jagger, Blue Creek, Mary Lee, and New Castle coals (Fig. 18). McCalley (1900) reported that this group had a combined coal thickness ranging from 3 to 19 ft but averaged between 9 and 10 ft as measured along a line of outcrops. One of the most important and productive coals of the Warrior Basin—the Mary Lee—underlies a large area within the central basin. The lowest coalbed, the Ream, lies some 50 to 200 ft above the Black Creek group and is thin and erratic with many shale partings. The Jagger lies 30 to 65 ft above the Ream and reaches a maximum thickness of 2 ft in the eastern subbasin of

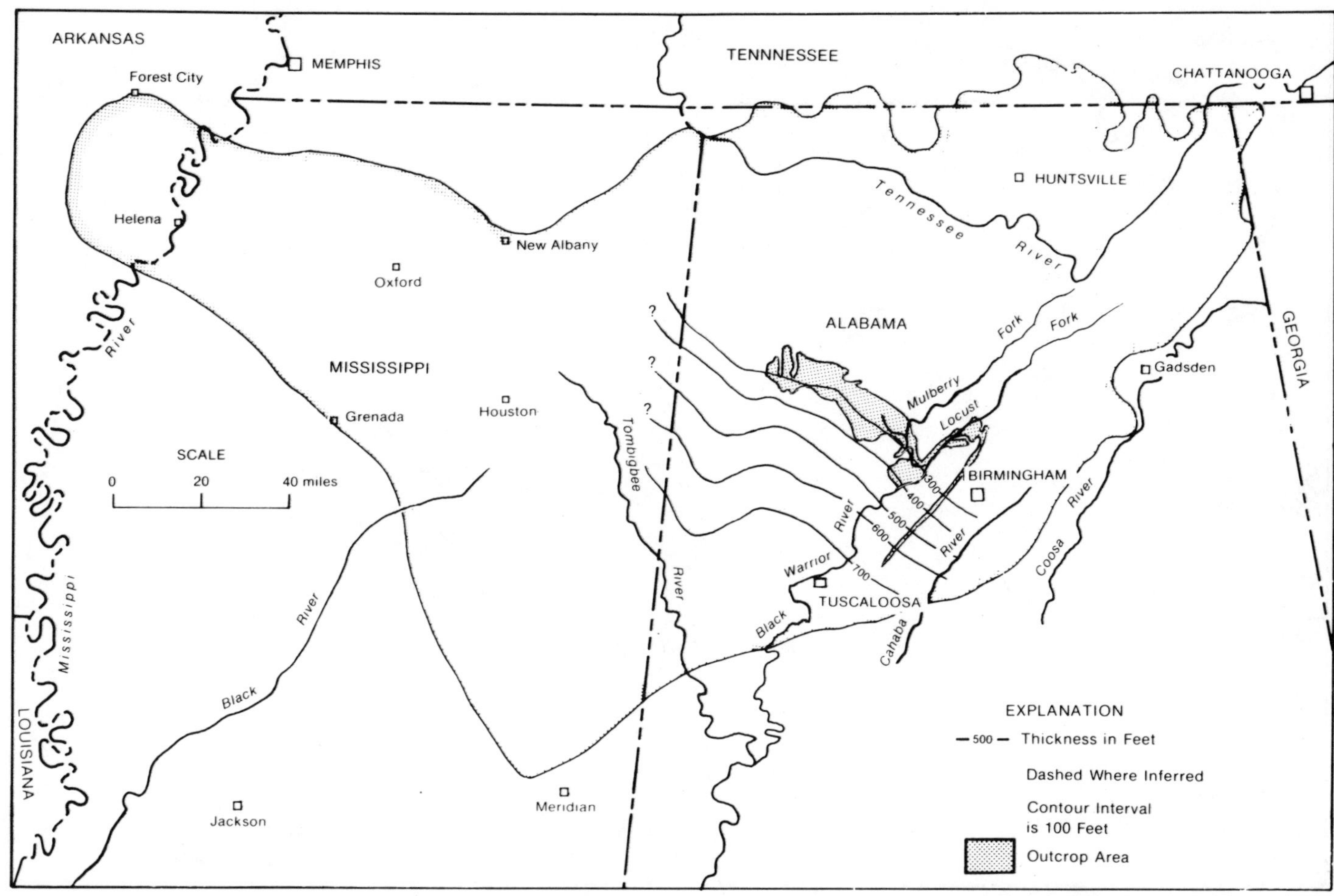

Figure 12—Isopach map of stratigraphic interval C, Pottsville Formation. From Metzger, 1965. Used with the permission of the Alabama Geological Survey.

the Warrior Basin. The Blue Creek coal is mined in the Blue Creek Basin in southern Jefferson County and can be traced northward through logs and boreholes until it coalesces with the top bench of the Mary Lee coal (Murrie et al, 1976). From a maximum of 5 ft, its average thickness lies between 1 and 2 ft and thins in the central part of the field. Because of correlation difficulties and a change in coal rank across the basin, the Blue Creek ranges from low- to medium-volatile in rank with generally low sulfur and high ash contents. The Mary Lee seam is usually reported as two benches that merge to form one seam in the northern part of the field but splits into as many as five benches locally (Murrie et al, 1976). It ranges from 2 to 6 ft thick in the northern basin and from 6 to 8 ft thick in the southwest. Volatility decreases and rank increases along a trend from northwest to southeast. The topmost bed of the Mary Lee group, the New Castle seam, is a thin, discontinuous bed found 15 to 65 ft above the Mary Lee. It may form up to three benches and often is mistaken for the Mary Lee bed.

Pratt Coal Group

The Pratt group consists of five seams which, in ascending order, are the Gillespie, Curry, American, Fire Clay (also called Nickel Plate), and Pratt lying from 400 to 650 ft above the Mary Lee group. The combined thickness of these coals varies from 3 to 16 ft and averages 8 ft. They range over a stratigraphic interval of 100 to 250 ft and vary from three beds in the western part of the field to ten in the east (Fig. 18). Although the lowest two seams, the Gillespie and Curry, thicken generally toward the east, neither reaches a minable thickness of 14 inches. The American bed lies 40 to 100 ft above the Curry and is commercially important in the south-central part of the Warrior field. Throughout this area, it ranges from 30 to 60 inches in thickness but thins and splits into several benches to the west. The Fire Clay or Nickel Plate seam takes its name from a persistent underbed of light-colored, plastic clay and lies 20 to 40 ft above the American coalbed. It reaches a minable thickness only in the eastern part of the coal field, and it merges with the overlying Pratt in Western Walker County. The Pratt is the principal seam in the Pratt group and lies from 20 ft to a few inches above the Fire Clay seam. Within the Coalburg syncline, it ranges from 30 to 75 inches in thickness, contains relatively little ash and sulfur, and is an excellent coking coal. Most of the Pratt coalbed has been mined out in this area. Throughout the rest of the field, the Pratt seam is persistent but less than 36 inches thick.

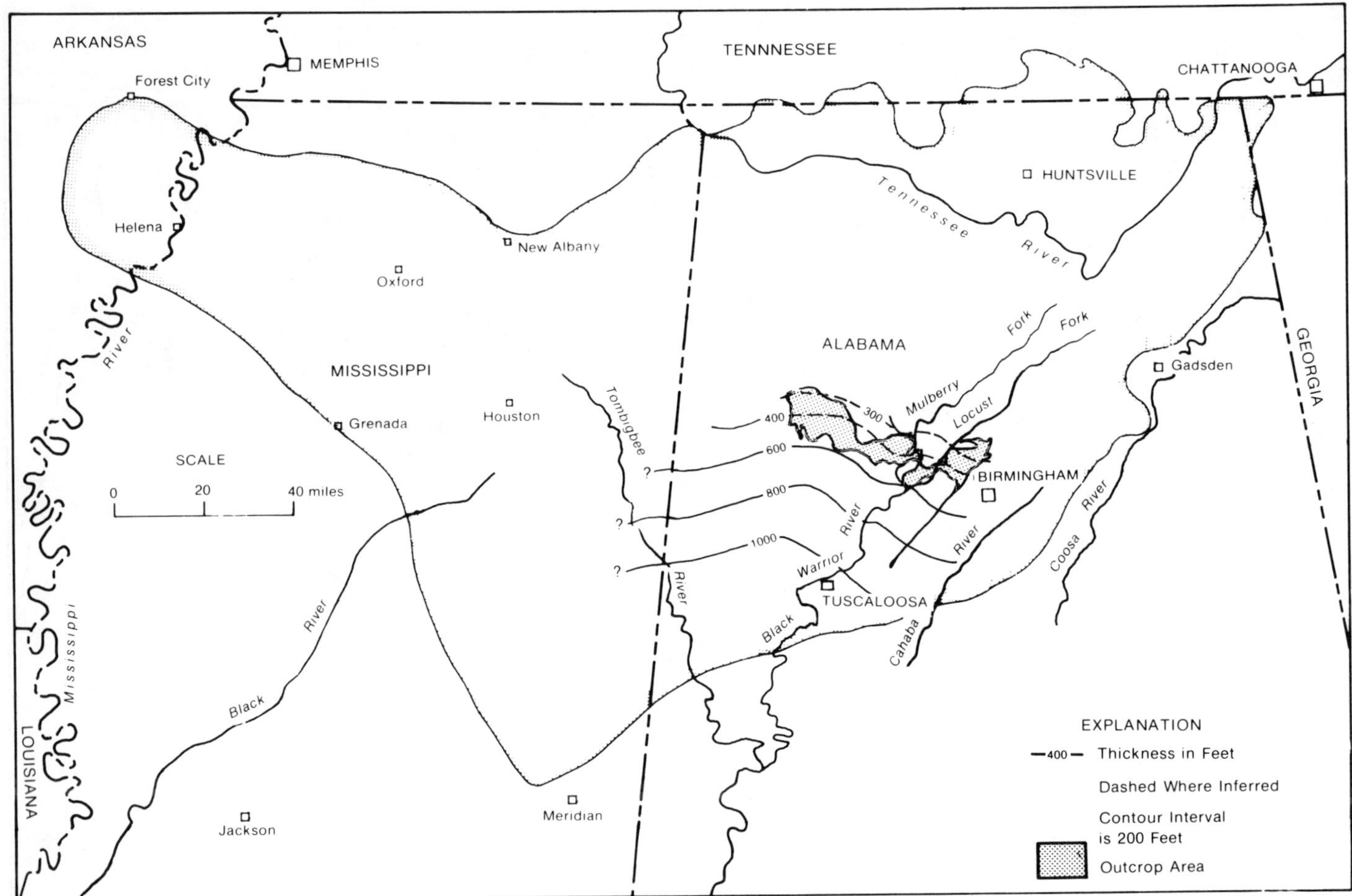

Figure 13—Isopach map of stratigraphic interval D, Pottsville Formation. From Metzger, 1965. Used with the permission of the Alabama Geological Survey.

Cobb Coal Group

The Cobb coal group is composed of up to three beds named, in ascending order, the Thomas, the lower Cobb and the upper Cobb (McCalley, 1900). They lie from 210 to 330 ft above the Pratt group over a stratigraphic interval of 50 ft (Fig. 18). McCalley (1900) reported a total thickness between 3 and 4 ft for this group, but Culbertson (1964) estimated that these beds rarely exceed 2 ft.

Gwin Coal Group

Lying 120 to 160 ft above the Cobb coal group, the Gwin coal group consists of two beds—the Gwin above and the Thompson Mill below. The Thompson Mill is persistent but not thick enough to attract mining interest in the Warrior field. The Gwin bed is found up to 35 ft above the lower bed and ranges from 14 to 52 inches in thickness. Because it is persistent but erratic in thickness, the Gwin coalbed is important only locally.

Utley Coal Group

This group of two to six beds ranges over a vertical interval of 20 to 150 ft and lies from 250 to 320 ft above the Gwin coal group (Fig. 18). In southern Jefferson County, a bed that correlates with the Utley is more than 30 inches thick and medium-volatile bituminous in rank with low sulfur and moderate ash contents.

Brookwood Coal Group

This group lies 200 to 300 ft above the Utley and consists of five seams which, in ascending order, are the Clements, Johnson (or Carter), Milldale, Brookwood, and Guide seams. The lowest, or Clements, bed is thin or absent in many places but has been mined locally where it reaches a thickness of 30 to 36 inches (Fig. 18). Lying 30 to 60 ft above is the Johnson bed, which averages 25 inches in thickness and has been mined only locally. The Milldale is found 30 ft or more above the Johnson coalbed and rarely exceeds 30 inches in thickness. It merges with the overlying Brookwood seam to form a seam 70 to 80 inches thick. The thickest, most extensive, and persistent bed in this group is the Brookwood. It lies 1 to 40 ft above the Milldale, averages 40 to 50 inches in thickness, and has been mined extensively for coking coal. The thin Guide coalbed lies 30 ft above the Brookwood and has been removed by erosion in many places.

Cleat and Joint Orientation

Several studies have been made of the joints and fracture

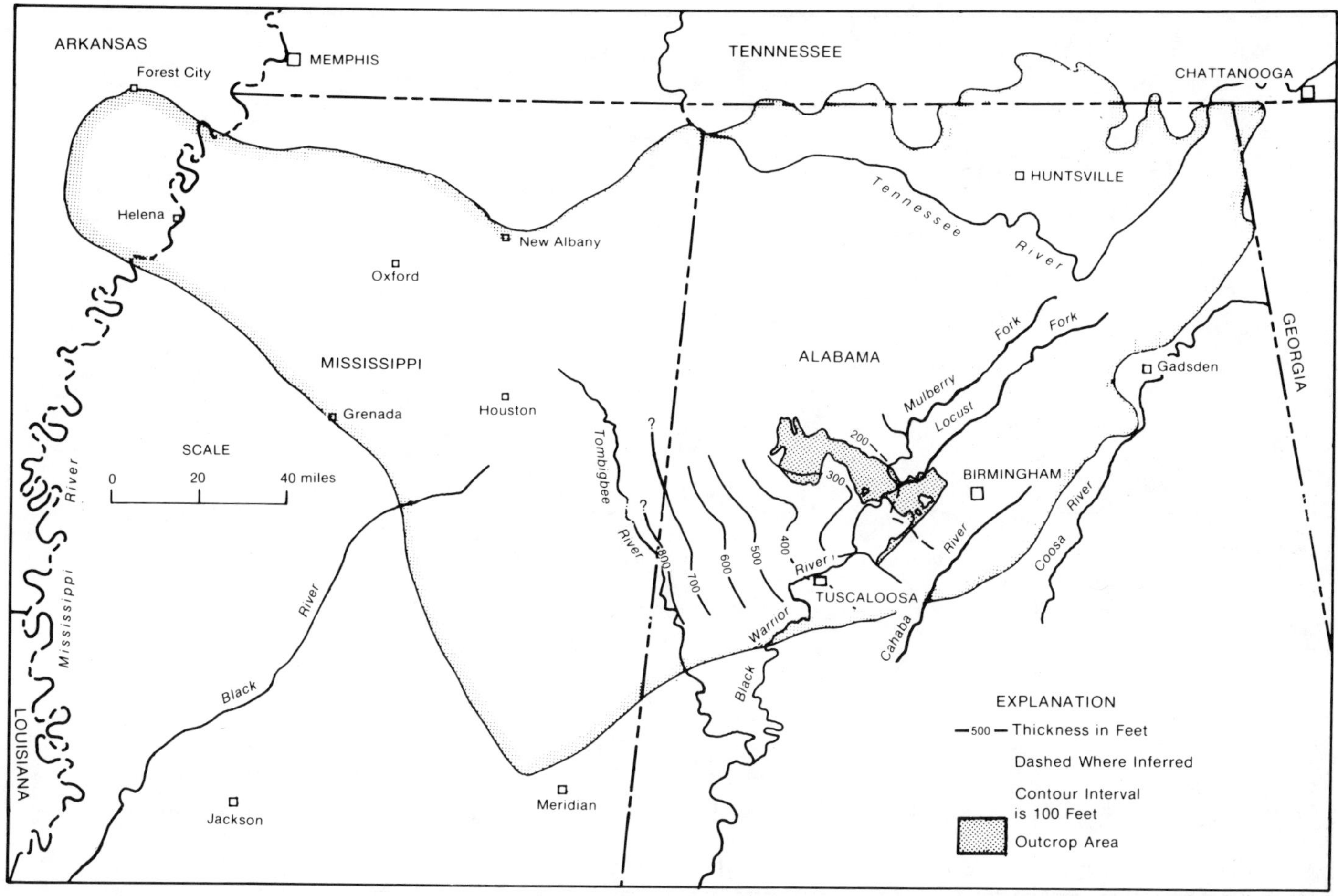

Figure 14—Isopach map of stratigraphic interval E, Pottsville Formation. From Metzger, 1965. Used with the permission of the Alabama Geological Survey.

systems of coal-bearing strata in the Warrior Basin. Murrie et al, (1976) measured surface and underground cleats and inclined fractures in the Mary Lee coal group in and around western Jefferson, southern Walker, and Tuscaloosa counties, Alabama. Fracture directions were quite varied and cleats show numerous trends, especially in the western part of their study area. By constructing a single regional composite rose diagram, they demonstrated that the Mary Lee group has a very pronounced face cleat peak at N 61° E and a bimodal butt cleat peak at N 21° W and N 36° W. They suggested that the stresses which produced the face cleat originated in the Ouachita orogeny to the southwest and were either extensional fractures oriented parallel to the maximum compressive stress or shear fractures. Because the maximum tensional stress in the Warrior Basin would be oriented perpendicular to the northeast-southwest axis of the depositional basin, they concluded that the face cleat was the result of shear.

Ward (1977) studied jointing in an area of the Warrior coal field lying in Walker, Jefferson, and Tuscaloosa counties, Alabama. He found two orthogonal fracture systems: one system lying perpendicular to the Appalachian structural front; the other oriented parallel to the Appalachian front (Fig. 19). He concluded that both were extension-release fracture systems. The latter joint set is prevalent in the western and southwestern parts of the coal field and formed earlier during the northeast-southwest stresses of the Ouachita orogeny. The former cleat system is prevalent only in the eastern and northeastern parts of the coal field and formed later by the northwest-southeast maximum principal stresses of the Appalachian orogeny.

POTENTIAL METHANE RESOURCE

Some coalbeds in the Warrior Basin have a century-old reputation for being rich in methane. McCalley (1886, p. 269) observed that some coals in Jefferson County "are dry and compact coals, though they hold considerable free gas." He reports the existence of natural gas seeps and recounts how a punctured tin can, when placed over one such seep near Village Creek in Jefferson County, was lit and burned steadily with a blue flame.

Butts (1926, p. 11) reported that explosive gas occurrences became more common as mines grew deeper and more extensive in the Warrior and Cahaba fields. Methane was reported in about 20% of the producing mines, which

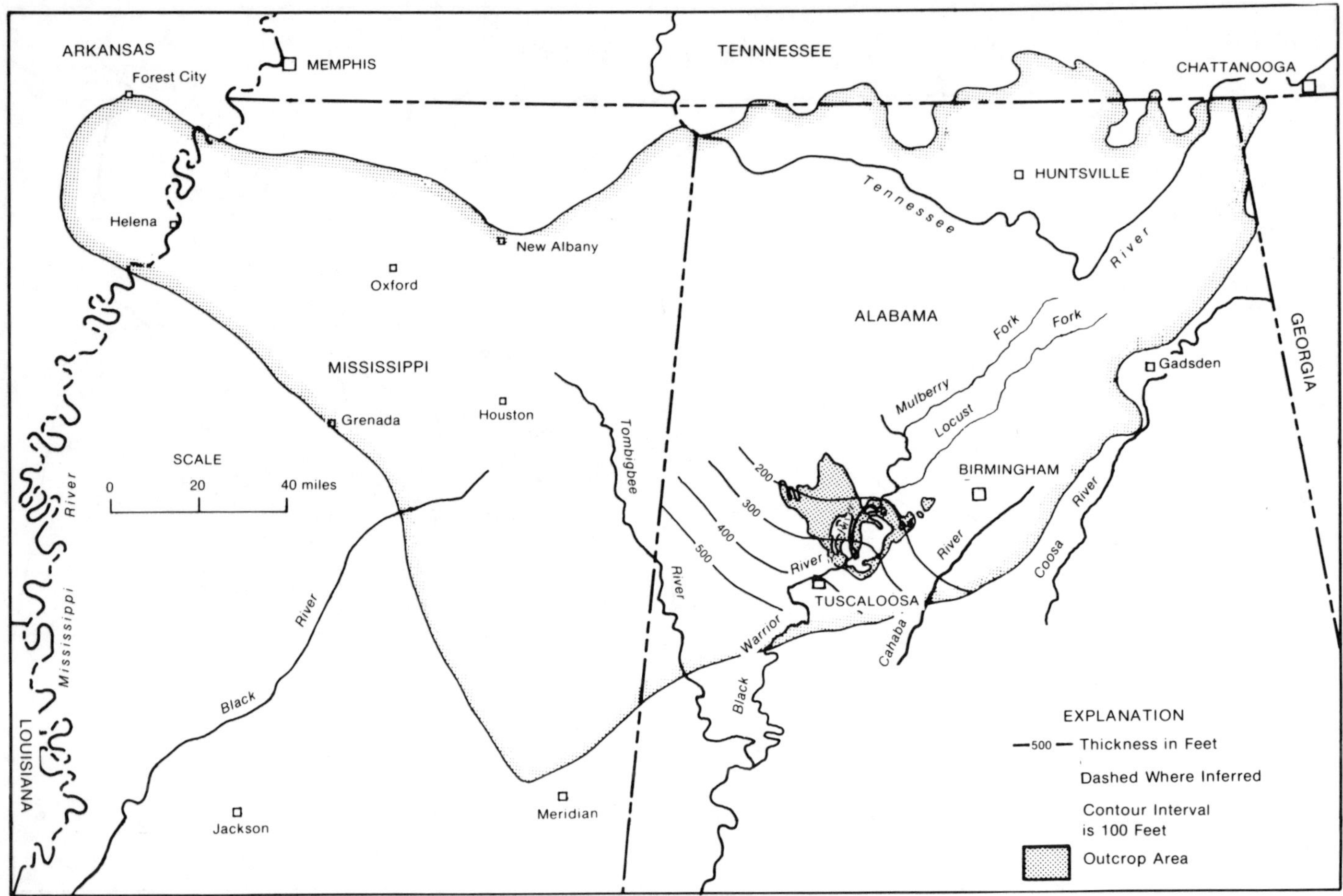

Figure 15—Isopach map of stratigraphic interval F, Pottsville Formation. From Metzger, 1965. Used with the permission of the Alabama Geological Survey.

represented 50 to 55% of the annual output of coal.

Previous studies to determine the methane content of coalbeds in the Warrior Basin have been directed specifically toward problems of methane emissions during underground mining and toward determining the feasibility of degasification in advance of mining. Within the Warrior Basin, the gassy Mary Lee coal group has caused problems during mining but is currently being recovered from depths over 2,100 ft in the deepest coal mine in the United States (Perry et al, 1980). Because early workers were more concerned with mine safety and increased productivity than with methane recovery, most of the older data cluster along a north-south line in western Alabama where overburden thicknesses are too great to allow strip mining, yet low enough to permit commercial underground extraction. The locations of these early projects are shown in Figure 20, and the data are summarized in Table 3; and although they are not directly comparable, some general trends can be seen.

Some of the results of these studies to investigate coalbed methane potential are summarized briefly below.

1. Multi-well fields seemed to attain better production than small fields or single wells and generally improve as more wells are placed on production. Observers report a gradual increase in coalbed permeability with time (Kissell, 1972).
2. Although the consensus favors artificial stimulation of methane wells, there is no agreement as to a best method for coal seams. In the joint DOE/U.S. Steel project, 28 wells were stimulated using different kinds of hydraulic fracturing techniques; the Bureau of Mines stimulated only one well in their five-spot pattern and chose a corner well that had proven to be the best producer. Lambert and Trevits (1980) carefully compared gas and water production in two wells from U.S. Steel's Oak Grove Mine. Despite a number of complications arising during the production test, conventional hydraulic stimulation using sand proppant was more successful in inducing methane drainage than a stimulation test without proppant.
3. There is typically a long period of dewatering before the onset of significant gas production from the Mary Lee seam. Vertical well production data show water rates that decline from initially high values as gas rates begin from zero and increase. This suggests the existence of a complex two-phase relationship between saturation and permeability that changes with time (Cervik, 1969).

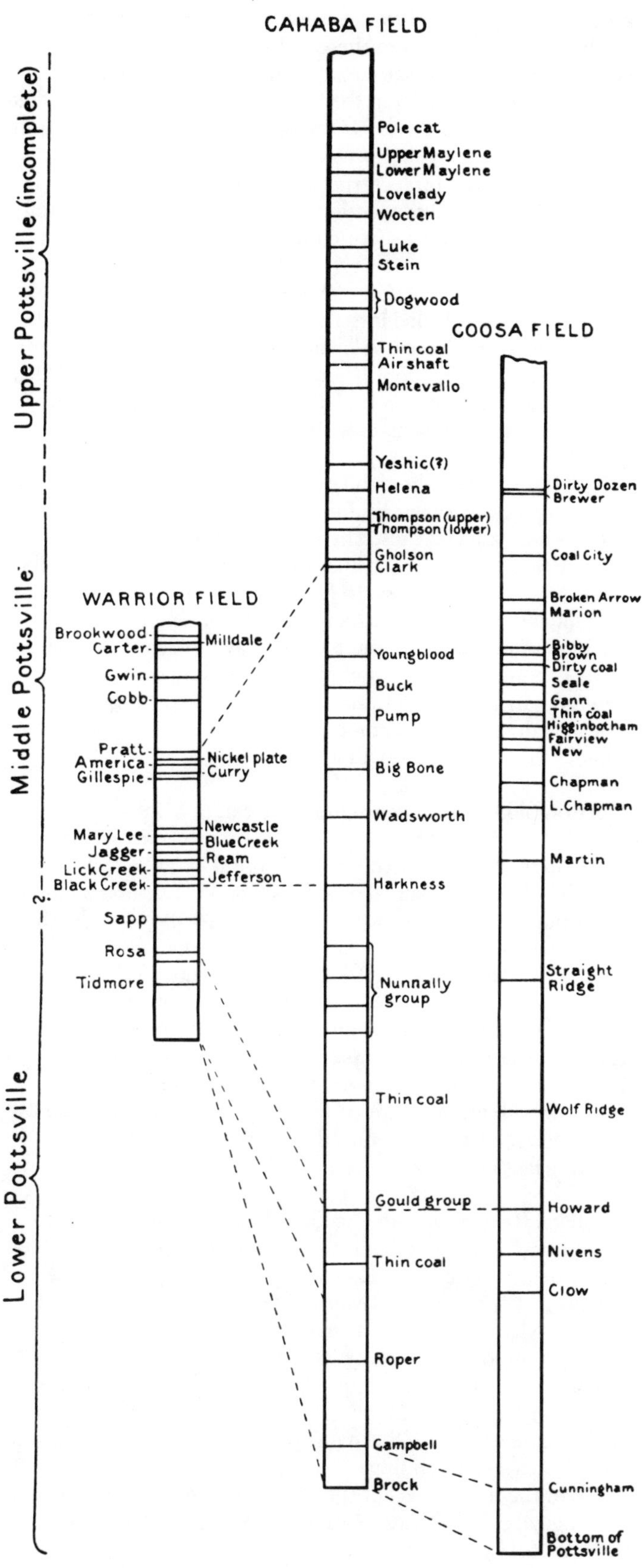

Figure 16—Correlation of coal seams in the Coosa, Cahaba, and Warrior Fields. Butts, 1926.

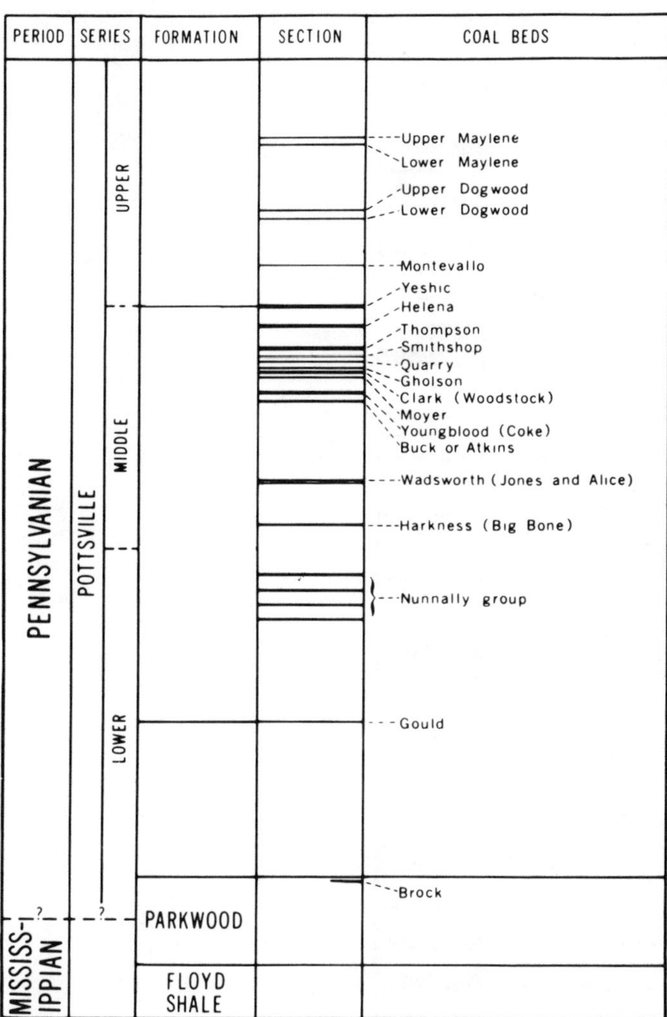

Figure 17—Stratigraphic column of major coal seams in the Cahaba Field. Shotts, 1971. Used with the permission of the Alabama Geological Survey.

DOE/U.S. Steel Oak Grove Project

The DOE/MRCP, in a cost-sharing agreement with U.S. Steel Corporation, tested 28 vertical boreholes near the U.S. Steel Oak Grove Mine in Jefferson County, Alabama. The objectives of this project were to develop the capability for removing coalbed methane using a pattern of vertical boreholes, and to demonstrate this method's compatibility with coal mining. Twenty-three boreholes were drilled on a grid pattern approximately 5 years ahead of active mining, and five boreholes were placed less than 1,000 ft from underground mining operations under the project plan. All 28 wells were completed into the Mary Lee coalbed, stimulated, produced, and mined through (Lambert, 1980, inputs to UGR semiannual report for April–September 1980).

These wells were placed on a 3-by-5 grid pattern of 21.5-acre spacing with two adjacent but outlying boreholes. The wells were originally completed as open holes and then hydraulically stimulated.

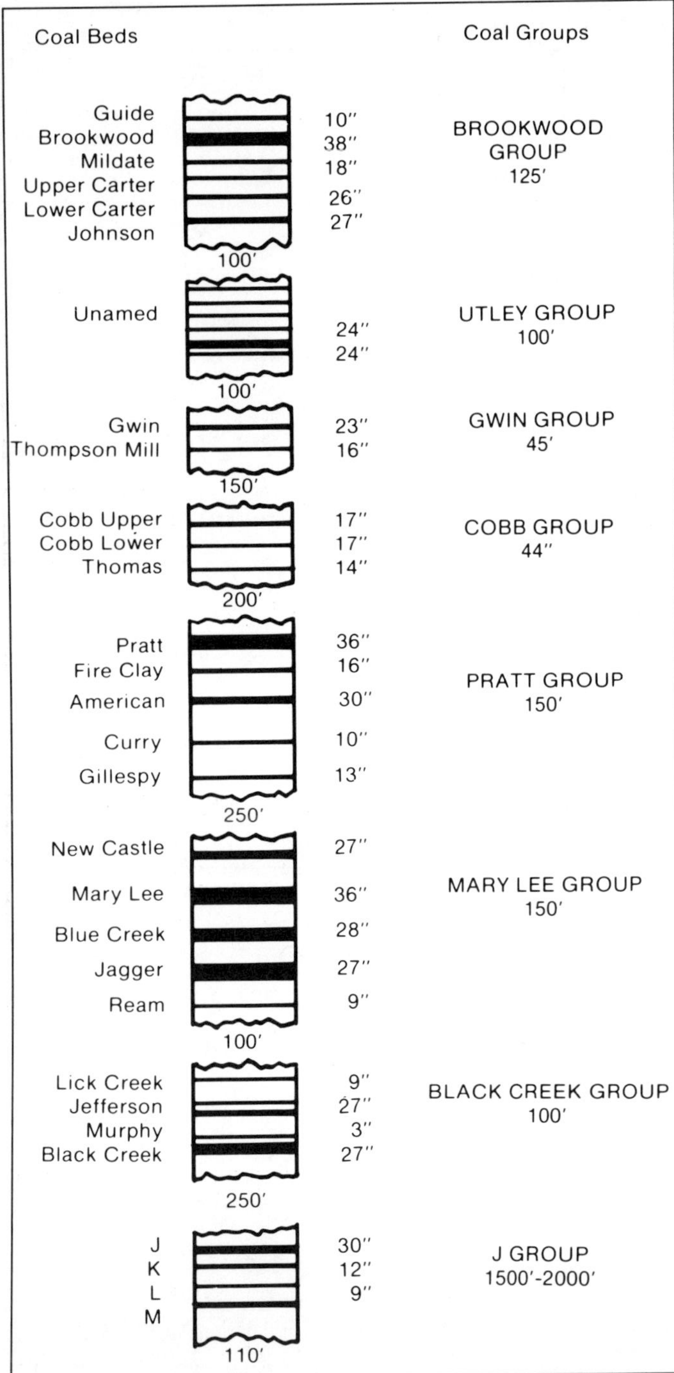

Figure 18—Generalized columnar section for coalbeds in the upper Pottsville Formation in the Warrior Basin.

Production data reveal that increasing volumes of gas were produced from each well as the coal seams were dewatered. For the first 300 days only three widely spaced wells were in operation and produced 150 to 200 barrels of water per day and very little gas. Graphs of the cumulative production records for the first 700 days of operation (Fig. 21) show that as more wells were placed on production, interference between them created areas of pressure drawdown, causing gas volumes to increase.

The data collected from these wells were used by Ancell et al (1980) to construct a mathematical model from which the gas and water production from the initial 17 wells in the field could be predicted. Figure 22 shows the predicted 20-year production for this well field developed by that model.

Since this initial project, the U.S. Steel Oak Grove mine has been on commercial production since December 2, 1981, and is discussed further in the section on commercial development.

Five-Hole Pattern: Bureau of Mines

Five wells were drilled into the Mary Lee coalbed near Oak Grove, Jefferson County, Alabama, to test the feasibility of this method for degasification in advance of mining (Elder and Deul, 1974). Over a 2-month period, four wells were drilled at the corners of a 1,500-ft square, with a fifth well located in the center.

Figures 23 and 24 show the increasing production rates of gas and water for the five wells during 2.5 years. The investigators felt this increase was caused by the long times required to achieve steady-state flow conditions and to dewater the coalbed across an expanded radius. In June 1973, hole 3SW was hydraulically stimulated, which dramatically increased gas production in that well from 5,000 cf/day to 90,000 cf/day in 11 days. The investigators concluded that drilling vertical boreholes in advance of mining is a practical technique for methane drainage but recommended the use of some type of stimulation technique on all wells.

Horizontal Drilling Techniques: Bureau of Mines

To test the feasibility of using this methane drainage method in advance of mining, two long horizontal holes were drilled into the Mary Lee coalbed near Brookwood, Tuscaloosa County, Alabama (Perry et al, 1980). One hole was drilled to a depth of 1,010 ft perpendicular to the face cleat and the second to a depth of 540 ft parallel to the face cleat. Figure 25 shows the flow and pressure data for hole 1, terminating at 68,000 cf/day on the day before it was grouted and mined through. Collapse of hole 2 prevented both completion and gas measurement of this hole. The investigators reported the total gas recovered from hole 1 was 40 million cu ft of 98 to 99% pure methane over 10 months and effectively reduced methane emissions from working faces in the mine.

Methane Recovery from Coalbeds Project (MRCP/DOE)

To investigate the feasibility of recovering and using the methane contained in coal seams, the U.S. Department of Energy in 1978 began the Methane Recovery from Coalbeds Project (MRCP). One of the major foci of this project was to locate the resource, estimate its size, and understand the site characteristics where it occurred. Seven wells in the Warrior Basin were tested for their coal thickness and gas content by TRW under the MRCP. An eighth well was later sampled and tested with TRW funding.

All of these wells except the first one were drilled by the U.S. Geological Survey/Conservation Division prior to a series of federal coal lease sales in the Warrior Basin. Under a cooperative agreement between the Department of Energy and the U.S. Geological Survey, coal seams were cored and

Table 2—Average analyses of coal in the Warrior coal field. Culbertson. 1964.

Coal Bed	Proximate Analysis (%) Moisture	Volatile Matter	Fixed Fixed Carbon	Ash	Ultimate Analysis (%) Hydrogen	Carbon	Nitrogen	Oxygen	Sulfur (%)	Btu	Ash Softening Temperature (°F)	Number of Analyses Averaged
Jefferson County												
Black Creek	3.0	31.6	62.1	3.3	5.3	81.4	1.8	7.5	0.7	14,310	2,500	4
Jefferson	2.3	31.9	58.5	7.3	5.1	76.7	1.6	6.2	3.1	13,780	2,300	3
Blue Creek	3.1	23.4	63.6	9.9	4.9	76.7	1.5	6.2	.8	13,530	2,900	1
Mary Lee	2.4	27.2	58.1	12.3	4.8	73.3	1.6	6.4	1.1	12,840	2,680	10
Newcastle	3.0	30.5	52.7	13.8	5.0	69.9	1.6	7.7	2.0	12,590	2,450	1
American	3.1	30.5	59.6	6.6	5.2	78.1	1.5	7.2	1.2	13,650	2,480	2
Nickel Plate	2.3	29.2	59.2	9.2	4.9	75.5	1.4	6.7	2.2	13,540	2,420	2
Pratt	2.5	29.4	62.6	6.4	5.1	79.4	1.6	5.9	1.4	14,250	2,460	10
Utley[1]	1.2	24.5	65.3	9.0	4.7	77.4	1.4	6.2	1.3	13,360	2,690	2
Marion County												
Jefferson	5.2	37.3	54.1	3.3	5.6	76.0	1.6	12.0	1.3	13,600	2,250	2
Tuscaloosa County												
Jefferson	1.4	32.6	61.4	4.6	—	—	—	—	1.4	14,310	—	1
Blue Creek	1.4	30.5	52.2	15.9	—	—	—	—	2.3	12,400	—	1
Mary Lee	1.0	32.5	51.0	15.5	—	—	—	—	1.6	12,580	—	1
Pratt	1.4	36.6	53.5	8.3	—	—	—	—	2.0	13,650	—	1
Carter	3.2	31.1	59.9	5.8	—	—	—	—	.9	14,020	2,800	1
Milldale	3.8	31.4	59.9	4.9	5.4	78.3	1.4	8.6	1.4	14,030	2,320	3
Brookwood	3.5	28.7	58.0	9.7	5.1	74.5	1.5	8.1	1.0	13,270	2,850	2
Walker County												
Black Creek	3.0	36.3	58.1	2.5	5.6	80.4	1.8	8.8	0.9	14,280	2,460	5
Jefferson	4.1	36.7	55.0	4.2	5.5	76.3	1.7	10.8	1.5	13,590	—	1
Jagger	3.8	33.1	50.4	12.5	5.1	67.9	1.6	11.4	1.2	12,210	2,730	2
Mary Lee	3.3	30.5	53.5	12.7	5.0	70.4	1.6	9.7	.7	12,280	2,740	12
American	2.3	32.5	52.9	12.3	5.0	71.5	1.5	8.0	1.6	12,530	2,720	2
Pratt	2.0	34.8	54.9	8.3	5.3	75.4	1.7	7.2	2.1	13,420	2,360	3
Corona[2]	2.4	38.9	49.0	9.7	5.4	71.3	1.7	9.6	2.3	12,880	2,400	1

[1]This coal bed is called Clements, where sampled.
[2]The Corona coal bed at the western edge of Walker County is equivalent to the Pratt coal bed.

collected, and the amount of gas emitted was carefully measured for some period of time. These data and the well locations are listed in Table 4 and will soon be published as a series of USGS open file reports.

Tuscoal Project

The Tuscoal Project was the first recognized attempt at large-scale commercial gas production from coalbeds in the Warrior Basin. Texas Eastern Transmission Corporation, Texas Energy Services, Inc., the Public Service Electric and Gas Company, and Sun Gas were partners in this project, which acquired oil and gas leases for 23,000 acres in Tuscaloosa County, Alabama. If commercial potential was indicated from four test wells, an additional 92 wells would be drilled in patterns of 16-well modules and a gas pipeline constructed.

The initial feasibility study indicated that the particular coalbeds involved in this project had potentially recoverable reserves of 100 billion cubic feet (Bcf) of natural gas that could be produced over a 20-year period. Project plans estimated that methane could be successfully and economically produced at a cost of $1.75/thousand cubic feet (Mcf) (U.S. Federal Energy Regulatory Commission, 1980). Results from two wells drilled to 2,965 ft and 3,130 ft in Section 15, T20S, R10W indicated that less coal was present than originally projected, however, and the project was abandoned in late 1982.

Commercial Production of Coalbed Methane

The American Public Gas Association (APGA) and the DOE have initiated several projects to determine the technical and commercial feasibility of gas production from coal in the Warrior Basin.

One of the first wells was drilled to a total depth of 1,831 ft in T16S, R2W near New Castle, in Jefferson County, Alabama, in mid-1980. A cumulative coal thickness of 19.5 ft was encountered, which included a 4.4-ft New Castle seam lying at 441 ft and a 1.5-ft Jane "B" seam at 1,764 ft. Gas contents in these seams were generally low, however, and production was poor. The well was finally plugged and abandoned in March 1983.

Three wells were drilled in T18S, R4W in the Pleasant Grove Coal Degasification field in Jefferson County in 1980, with depths averaging about 1,550 ft. These wells encountered

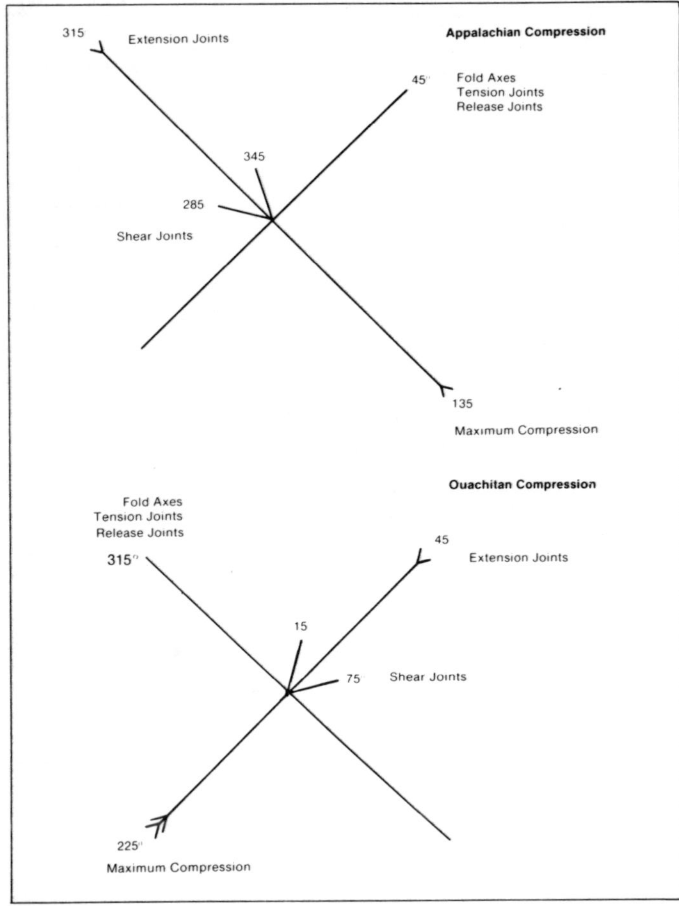

Figure 19—Theoretical joint orientations in the Warrior Basin. Ward, 1977. Used with the permission of the author.

cumulative coal thicknesses of about 10 ft in the Mary Lee Group (1,040 to 1,090 ft), 4 ft in an Unnamed Group (1,250 to 1,260 ft), and over 4 ft in the Black Creek Group (1,415 to 1,460 ft). Measured gas contents ranged from 9.1 cc/gm (290 cf/ton) to 15.6 cc/gm (500 cf/ton) with an average of 11.3 cc/gm (360 cf/ton) for six samples. Commercial production from these three wells has been reported from April 1981 to the present. Initial gas flow from the three wells averaged 1,400 Mcf/month with recent production at 1,800 Mcf/month. Cumulative field production over 23 months is 45,747 Mcf.

Three test wells were drilled in and near Tuscaloosa, Alabama, with the assistance of the U.S. DOE to determine the feasibility of gas production from coal seams. The Kaulton 34-10 No. 1 was drilled to a depth of 2,882 ft late in 1981 in T21S, R10W by the City of Tuscaloosa. This well penetrated a cumulative coal thickness of 34 ft from the Utley group encountered at 388 ft below the surface through the Black Creek group encountered at 2,765 ft. Gas contents of samples taken from this corehole ranged from 1.89 cc/gm (60 cf/ton) for the shallow Utley coals to 9.18 cc/gm (294 cf/ton) for the Black Creek (City of Tuscaloosa, Alabama, and School of Mines and Energy Development, University of Alabama, 1982). B.F. Goodrich drilled the Goodrich-Warrior 29-13 No. 1 in the same township to a depth of 3,429 ft to investigate potential gas production for use in their manufacturing plant nearby. This well penetrated a cumulative coal thickness of 28.6 ft from the Brookwood group through the Black Creek group. Gas contents measured from these core samples ranged from 1.47 cc/gm (47 cf/ton) for the Brookwood coal samples to 9.14 cc/gm (293 cf/ton) in the Mary Lee group (City of Tuscaloosa, Alabama, and School of Mines and Energy Development, University of Alabama, 1982). Similarly, the Tuscaloosa County Industrial Development Authority drilled a 2,905-ft well in T21S, R10W to test the potential for gas production in an industrial complex near the Tuscaloosa airport. A cumulative coal thickness of 26.2 ft of coal was encountered, despite the absence of the coal group that had been faulted out. Gas content values were still being determined at the time this report is being written (City of Tuscaloosa, Alabama, and School of Mines and Energy Development, University of Alabama, 1982). Drilled as research coreholes to gather stratigraphic and gas content data, all three wells were subsequently plugged and abandoned.

The University of Alabama is attempting to utilize the gas contained in coals lying beneath their Tuscaloosa campus with the help of the APGA. Their first well, the East Campus 24-2 No. 1, was drilled to 2,289 ft in T21S, R10W and is still in production testing but penetrated 31.3 ft (cumulative) of coal. This ranged from only 2.0 ft in the Cobb group, whose gas content was measured to contain 2.2 cc/gm (70 cf/ton), to 14.1 ft in the Mary Lee group, which was measured to contain 9.4 cc/gm (301 cf/ton). The Black Creek group contained 6.3 ft of coal in this corehole, and samples indicated its gas content to be 9.6 cc/gm (307 cf/ton) (City of Tuscaloosa, Alabama, and School of Mines and Energy Development, University of Alabama, 1982). A second well in the same township was drilled to 2,766 ft in June 1983. If successful, the gas will be used by the University to defray energy costs on the campus.

To the east along the Black Warrior River in T21S, R9W, two coalbed methane wells have been drilled by Coal Tech, Inc., for the Reichhold Chemical Company. The first well, the Reichhold Chemical 3-6 No. 1, was drilled to 2,606 ft early in 1982 and supplies gas to the plant on an as-needed basis. A second well, the Reichhold Chemical 3-11 No. 2, was spudded in May of 1983 with a projected depth of 2,500 ft. Gas in these wells will be recovered from several seams including over 5.0 ft contained in the Pratt group at 1,340 ft to more than 6 ft in the Black Creek coals at 2,550 ft.

From its inception as a cost-sharing research agreement between the U.S. Bureau of Mines and U.S. Steel Co., the methane recovery project from the Oak Grove mines has grown into the commercial production of coal seam gas. The Mary Lee/Blue Creek seam at that location lies between 1,000 and 1,100 ft and contains an average 300 cf/ton of gas. Wells placed 5 years in advance of the active mine are completed as open holes, stimulated, and pumped to produce water and gas. A gathering system connects the wells and produced gas is compressed, dried, and injected into a nearby utility company pipeline.

A field of 38 wells is currently producing an average of about 70 Mmcf/month, with two wells shut in. In 15 months of commercial production, the field has produced 715 Mmcf, with new wells continually being drilled and brought on production.

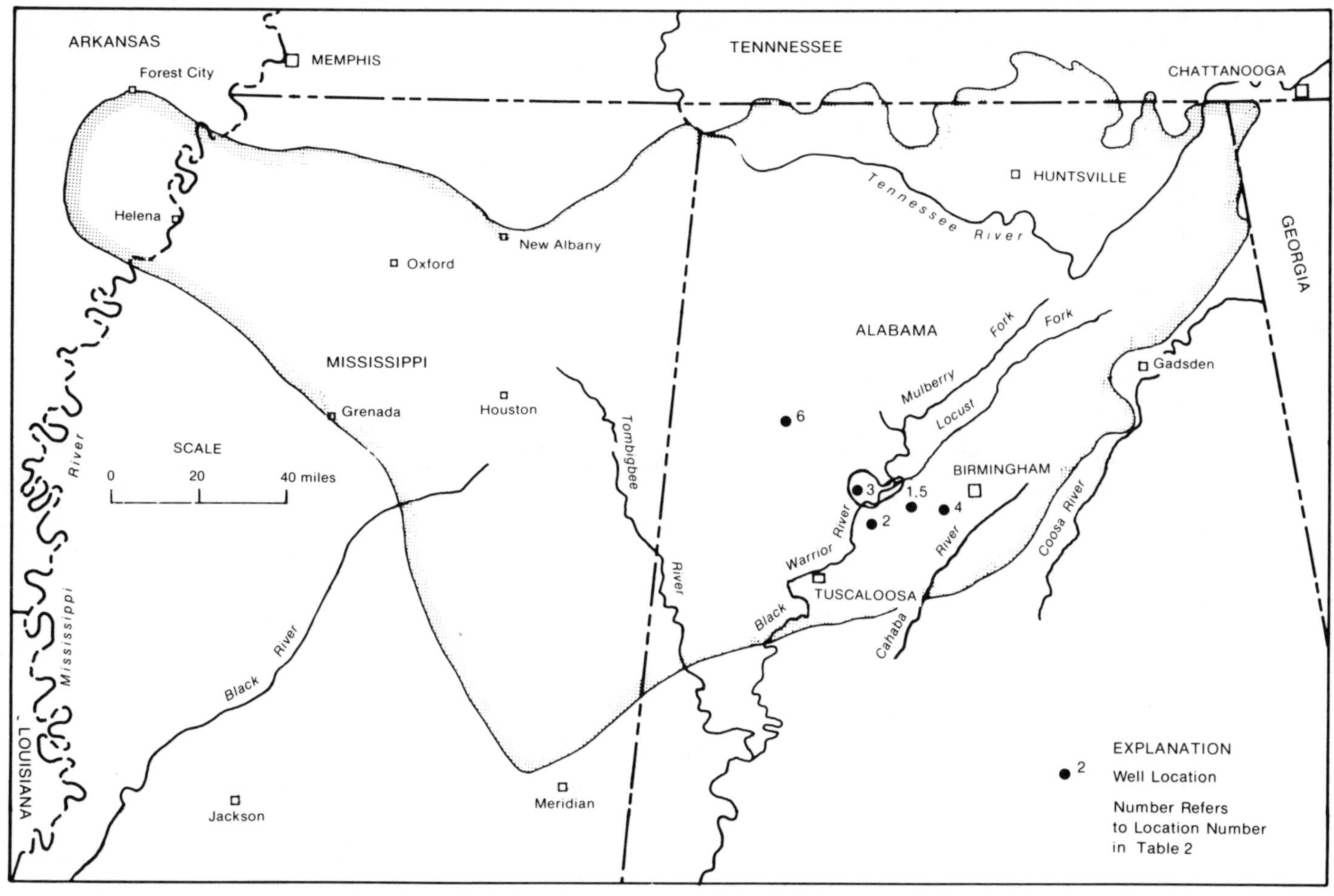

Figure 20—Coalbed methane project locations in the Warrior Basin.

About 20 mi east of Tuscaloosa at the Jim Walters' No. 4 Mine, Enhanced Energy Resources, Inc., has established the Brookwood Coal Degasification field. Although the Mary Lee/Blue Creek seam is mined from 2,100 ft, gas is recovered from both the Mary Lee and Pratt coals at 1,500 ft. Since beginning commercial production in February of 1982, the Brookwood field has produced 1,490 Mmcf in 13 months from 31 wells with one well shut in. Plans to drill up to 700 wells when the field is completely developed requires an active drilling schedule. The produced gas is collected in a gathering system, compressed, dried, and sold to a local pipeline company.

Other operators in the Warrior Basin have drilled and tested coalbed methane wells and plan to begin commercial sales in the future. The Alabama Methane Production Company has drilled five research wells in T21S, R8W, but a lack of local markets has delayed commercial development. TRW Inc. has begun developing a 3,000-acre tract in T20S, R9W north of the Black Warrior River. The first well (the Ramsey-McCormick 35-3 No. 1), drilled to a depth of 2,813 ft, is presently being tested, and plans call for 30 to 40 development wells upon completion of the field. Norris and Beard have drilled four wells in T20S, R8W to depths varying from 800 to 2,250 ft, and they are currently being tested. 4J Exploration has drilled or staked five wells in T21S, R9W for the purposes of coalbed methane production.

WELL TEST DATA

Figure 26 shows a plot of total gas content (both free and adsorbed) vs. overburden thickness for more than 50 coal samples from the Mary Lee group in the Warrior Basin. These data were compiled from the MRCP and a later publication (Diamond and Levine, 1981) and illustrate the general relationship between gas content and depth. Table 4 summarizes some recently published methane content data for a number of Warrior Basin coals.

ESTIMATED RESOURCE VOLUME

The most important factor for successful recovery of coalbed methane is the amount of coal in place. Estimating the total volume of coal in a largely unexplored basin requires certain

Table 3—Methane well data for Warrior Basin wells. Compiled from Lambert, 1980; Perry et al, 1980; Diamond et al, 1976; Lambert and Trevits, 1980.

	Location 1	Location 2	Location 3	Location 4	Location 5	Location 6
Owner/Performer	U.S. Steel/INTERCOMP and Dept. of Energy	Jim Walter Resources Inc. No. 4 Mine/Bureau of Mines	Several/Bureau of Mines	Several/Bureau of Mines	U.S. Steel/Dept. of Energy	Grace Petroleum Corp./Alabama Methane
Date of Publication and Reference	1980 Lambert	1980 Perry et al	1976 Diamond et al	1974 Elder and Deul	1980 Lambert and Trevits	1979 MRCP Well Test Pkg.
No. of Wells in Field	17	2	7	5	2	1
Well Completion Data	Open-hole completion; hydraulic stimulation	Slotted pipe completion; not stimulated; 2 horizontal holes drilled parallel and perpendicular to face cleat	Unspecified completion; some wells stimulated	Open-hole completion; hydraulic stimulation	Open-hole completion; no-proppant hydraulic stimulation	Cased to core point; not stimulated
Depth to Coalbed	1,100 ft	2 holes: 1,010 and 540 ft	From 600 to 2,200 ft	1,090 ft	1,150 ft	From 593 to 1,880 ft
Coalbed Thickness and Type	5.2 ft of Mary Lee Group coal	6′9″ Mary Lee Group	From 1 to 9 ft thick over 3 counties	5 ft of Mary Lee Group	5.5 ft of Mary Lee Group	10 ft total in 4 seams mostly below Mary Lee Group
Reservoir Pressure	421 psi	70 psi initially	Not given	390 psi	Not given	> 160 psi
Production Data: 1) Average 2) Maximum 3) Trends	1) 500,000 cf/day for 17 wells 2) 1.3 million cf/day for 17 wells 3) 400,000 cf/day predicted for future decade	1) 75,000 cf/day 2) 200,000 cf/day 3) Pressure and flow rate declined with time but effectively reduced methane problems in mine	1) Not given 2) Not given 3) 90% of in situ methane is at depths over 1,000 ft	1) 5,000 cf/day before stimulation; 50,000–60,000 cf/day after stimulation 2) 90,000 cf/day 3) Low initial production but increased gas flow with time as steady-state conditions achieved	1) 60–80,000 cf/day 2) 150,000 cf/day 3) Adjacent wells in field give widely varying results	Not given

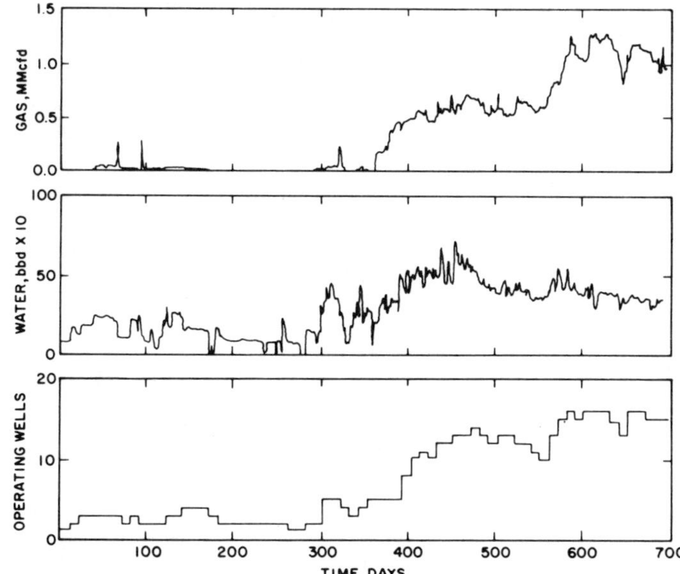

Figure 21—Production record for 17-well degasification test. Lambert, 1980.

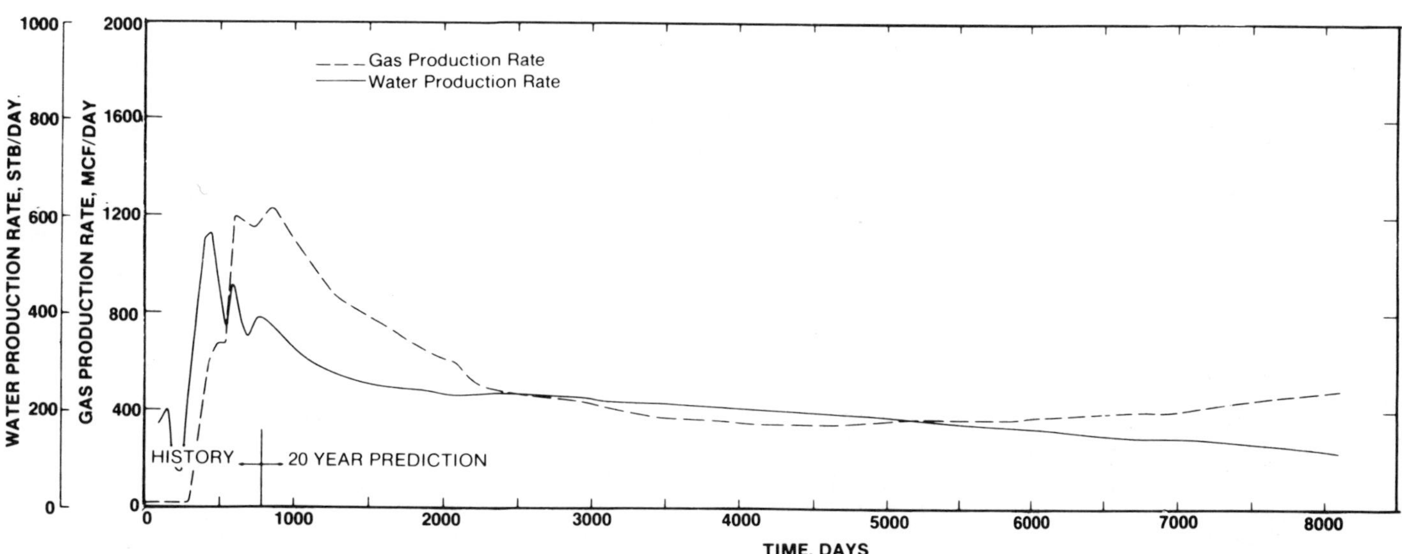

Figure 22—Twenty-year production prediction for 17-well degasification test. Ancell et al, 1980.

assumptions and restrictions that give rise to a range of values. One of the most carefully researched studies was that of Culbertson (1964), who estimated coal reserves in Alabama to be 13.8 billion tons, of which 86% or 11.9 billion tons is contained by the Warrior field. He restricted his estimate to those coalbeds more than 14 inches thick and shallower than 3,000 ft. He terminated his investigations west of an imaginary line running through the Alabama cities of Tuscaloosa and Fayette because of limited core data and the overlying Upper Cretaceous rocks that thicken rapidly to the southwest. He also excluded an unknown number of coalbeds lying between 1,500 and 3,000 ft in the southeastern Warrior field and those in the Parkwood and lower Pottsville Formations because of a lack of data. He speculated that either more coalbeds are present in the

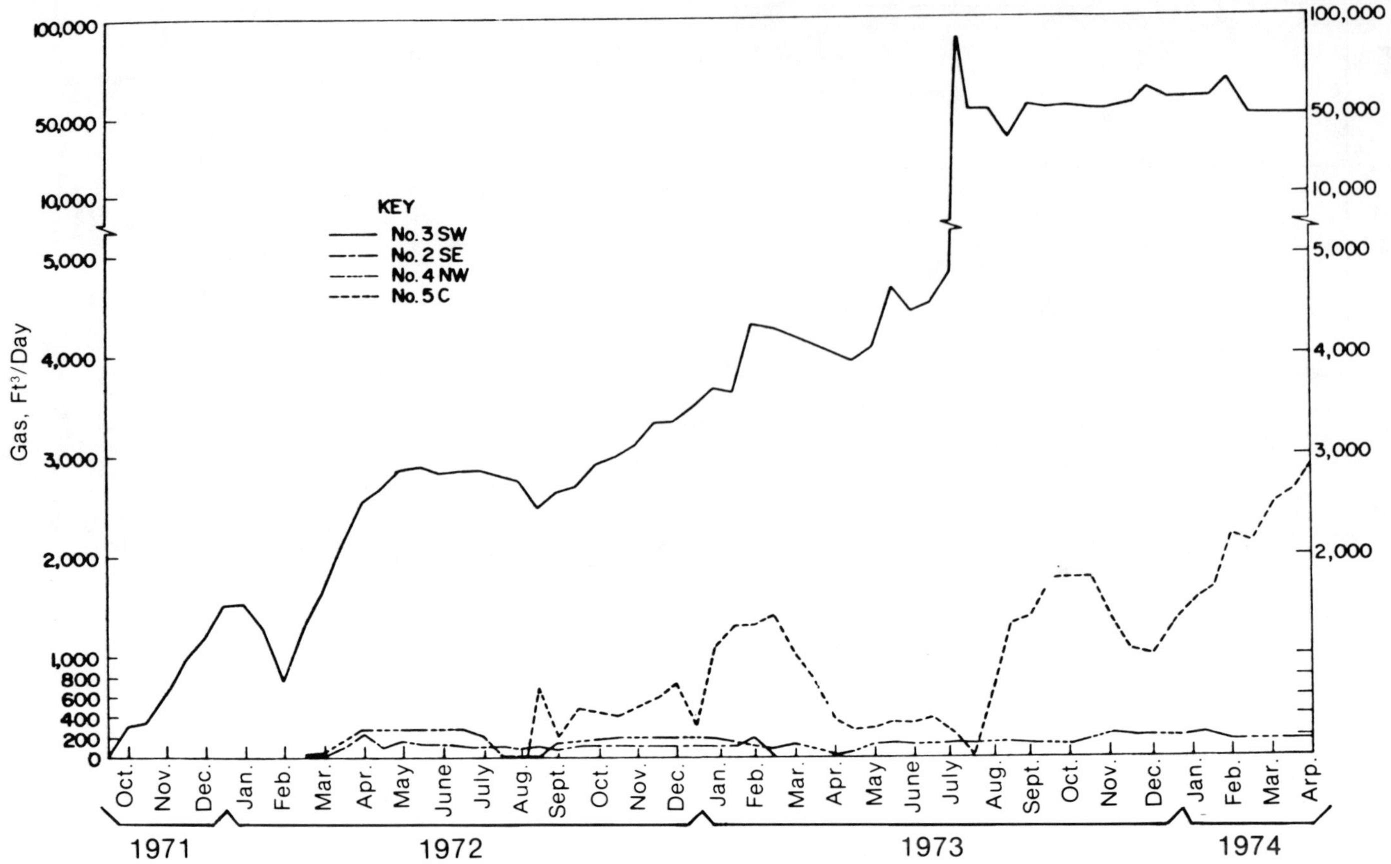

Figure 23—Gas production for five-hole degasification test. Elder and Deul, 1974.

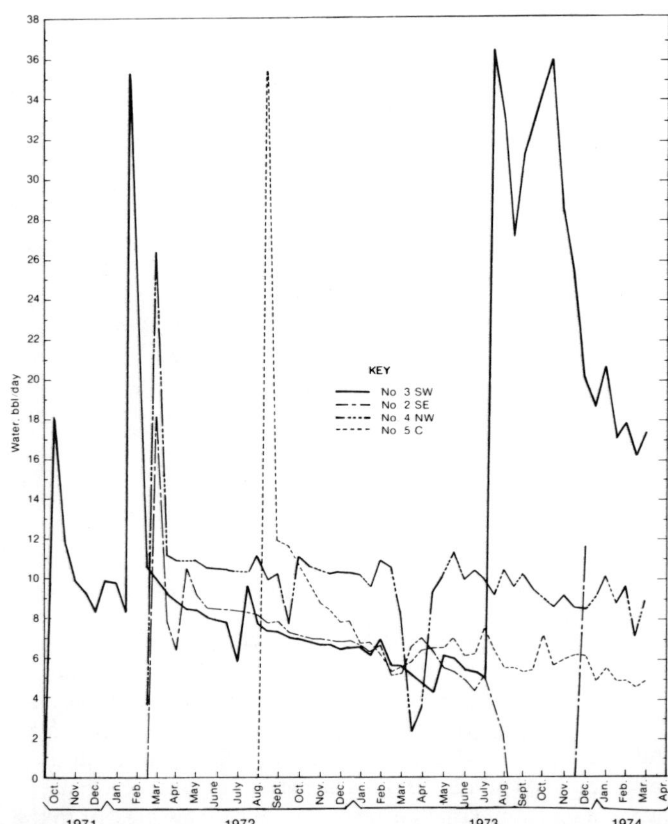

Figure 24—Water production for five-hole degasification test. Elder and Deul, 1974.

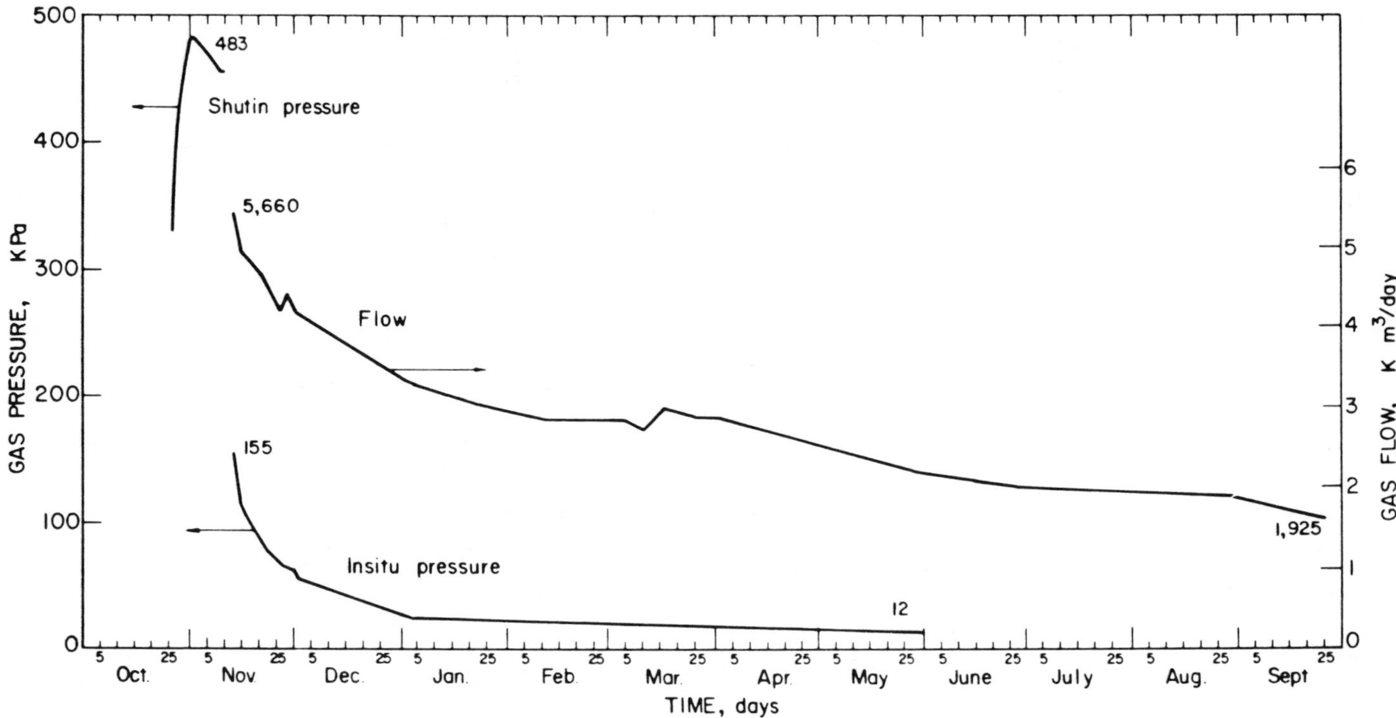

Figure 25—Flow and pressure data for horizontal methane drainage well. Perry et al, 1980.

southwestern part of the Warrior field, or the Pottsville grades into a thick, sandy sequence containing few thin coals as suggested by trends observed in the southwestern Coosa field. He estimated that removing these restrictions would add only several hundred million tons, and that the potential coal resource in beds more than 14 inches thick lying at less than 3,000 ft in his excluded geographical areas probably would total 20 billion tons (Culbertson, 1964, p. B68). This is in close agreement with Averitt's later estimate of 35.3 billion tons for the same conditions in the same area (Averitt, 1975, p. 14). For the purposes of this report, 35 billion tons will be used as a general estimate of the coal resource within the defined basin.

Diamond et al (1976) estimated the methane content at various depths in the Mary Lee group of coalbeds that varies from 150 cf/ton at 1,000 ft to 550 cf/ton at 2,100 ft. These estimates were made for a selected area in Jefferson, eastern Tuscaloosa, and southern Walker counties of north-central Alabama. They withheld exact locations and coal isopach maps because of the confidentiality of the data but estimated that their study area contained one trillion cubic feet (Tcf) of methane, of which 70% or 700 Bcf was contained in the Mary Lee group (Table 5). From their data, a coal resource estimate of 6.27 billion tons can be calculated. They add that although their study did not include the Mary Lee group in the deeper part of the Warrior Basin, the methane potential there is enormous. Each additional sq mi of coal 6 ft thick lying at a depth of 2,000 ft would contribute an additional 3.3 Bcf of gas to the total (Diamond et al, 1976, p. 7).

Additional published data provide hints that unexplored parts of the basin may contain an abundance of coal and methane. In a report on Pennsylvanian coal in the subsurface of Mississippi, Bicker observes that coalbeds as much as 2 ft thick have been reported at depths greater than 1,000 ft below the surface but that these beds are found nearest the surface in Chickasaw, Clay, and Monroe counties, Mississippi (Bicker, 1970, p. 65). Combining the methane content estimate of Diamond et al (1976)–that each additional sq mi of coal 6 ft thick lying at a depth of 2,000 ft would contribute another 3.3 Bcf of gas–and the total basin resource estimate of this report would indicate an approximate methane potential of 16.5 Tcf for the entire basin. Because this includes a great deal of the proven coal resources that we know contain little methane, this value represents a maximum for the Warrior Basin. Applying the methane content figure to the estimated 20 billion tons of coal resources lying in the deeper part of the basin gives an estimate of 10 Tcf as a realistic maximum. Diamond et al (1976) estimated 7 Tcf for a study area in Alabama. The assumptions and estimates discussed in this section are summarized in Table 6.

Analyses of gas composition indicate that Warrior Basin coal fields contain high-quality methane. These analyses are shown in Table 7.

CONCLUSIONS

On the basis of these estimates and assumptions, 35 billion tons of coal and 10 Tcf of methane exist within the basin defined in this report. Early mapping studies indicate that although many seams exist within the four coal fields, most are too thin, too ashy, or lie too deep to be of commercial mining interest. Consequently, some of the most promising areas for coalbed methane are still largely unexplored, and this lack of data—especially in the deeper parts of the basin to the west—is the most serious handicap for methane recovery in the region.

Table 4—Gas content data collected during the methane recovery from coalbeds project.

Location	USGS Well Number	MRCP Well Name	Date Sampled	Group/Seam	Depth (ft)	Thickness (ft)	Sample Weight (gm)	Lost Gas (cc)	Desorbed Gas (cc)	Residual Gas (cc)	Gas Content (cc/gm)	Gas Content (cf/ton)
Grace, Grimsley 35-15 #1	N/A	CAA	7/12/79	Blue Creek	593	1.0	1,789	715.6	3,757	358	2.7	86
SW/4-SE/4 Sec. 34, T14S, R11W								394.2	1,511	197	3.2	102
Fayette Co., AL												
	N/A	CAA	7/13/79	Unnamed H	1,102	2.0	657	119.8	240	240	0.5	16
			7/16/79	Rosa	1,675	2.5	1,198	111.3	1,336	1,002	2.2	70
			7/18/79	Tidmore B	1,879	3.0	1,113					
SW/4 Sec. 19, T14S, R9W	57	CAB	10/16/79	Mary Lee/Mary Lee	578	1.17	No Samples Collected					
Walker Co., AL	KRCRA-2		10/21/79									
SW/4-SW/4 Sec. 18, T18S, R9W	37	CAC	2/5/81	Mary Lee/Blue Creek	1,502	2.6	566.5	0	6,875	385	12.8	410
Tuscaloosa Co., AL	KRCRA-2											
SE/4-SE/4 Sec. 4, T17S, R9W	49	CAD	2/20/81	Mary Lee/Blue Creek	1,314	6.6	1,157	125	10,457	575	9.6	309
Tuscaloosa Co., AL	KRCRA-2											
SE/4-SE/4 Sec. 35, T17S, R10W	1	CAE	5/5/81	Mary Lee/New Castle	1,147	1.3	437	91.5	4,312	445	11.1	355
Tuscaloosa Co., AL	KRCRA-2			Mary Lee/Mary Lee	1,171	3.6	570	114	5,444	300	10.3	329
NW/4-SE/4 Sec. 28 T18S, R9W	5	CAF	5/5/81	Mary Lee/Mary Lee	1,588	2.7	384	71.6	4,300	235	12.0	384
Tuscaloosa Co., AL	KRCRA-2		5/5/81	Mary Lee/Mary Lee	1,589	2.7	408	42.4	5,185	390	13.8	441
			5/11/81	Black Creek/Jefferson	1,871	2.35	609	−7.03	1,995	test not perf.	3.3*	105*
SW/4-NE/4 Sec. 3, T17S, R10W	3	CAG	5/14/81	Black Creek/Jefferson	1,487	3.83	278	−24.1	1,776	340	7.5	240.8
Tuscaloosa Co., AL	KRCRA-2			Black Creek/Jefferson	1,487	3.83	459	27.8	2,877	410	7.2	231.1

continued

Table 4—(Continued)

Location	USGS Well Number	MRCP Well Name	Date Sampled	Group/Seam	Depth (ft)	Thickness (ft)	Sample Weight (gm)	Lost Gas (cc)	Desorbed Gas (cc)	Residual Gas (cc)	Gas Content (cc/gm)	Gas Content (cf/ton)
SE/4-NW/4 Sec. 28, T18S, R2W Tuscaloosa Co., AL	9	CAH	2/5/83	Cobb/Cobb	769	0.14	137	32.01	1,115	N/D	8.4	268
			2/5/82	Cobb/Cobb	780	1.0	862	33.5	4,500	N/D	5.3	168
			2/8/82	Cobb/Cobb	793	0.52	232	11.9	1,838	N/D	8.0	255
			2/8/82	Cobb/Cobb	793	0.52	301	16.1	1,783	N/D	6.0	191
			2/11/81	Pratt/Pratt	1,061	2.52	1,494	158.3	12,643	111	8.6	274
			2/11/82	Pratt/Pratt	1,063	2.52	509	150.5	6,074	N/D	11.9	391
			2/11/82	Pratt/Nickel Plate (?)	1,074	0.60	776	49.3	9,568	N/D	12.4	397
			2/12/82	Pratt/America	1,128	1.42	N/D	N/A	6,458	N/D	9.6	307
			2/12/82	Pratt/America	1,129	1.42	616	88.9	8,375	N/D	13.7	440
			3/18/82	Mary Lee/New Castle	1,486	3.1	1,377	555.0	14,417	144	10.9	348
			3/18/82	Mary Lee/New Castle	1,486	3.1	1,211	341.3	12,513	282	10.6	340
			3/19/82	Mary Lee/Mary Lee-Blue Creek	1,522	5.4	1,487	726.0	19,992	71	13.9	446
			3/19/82	Mary Lee/Mary Lee-Blue Creek	1,522	5.4	1,375	594.0	19,132	352	14.3	459
			3/1/82	Mary Lee/Ream	1,696	0.36	309	32.5	3,152	15	10.3	330
			3/1/82	Mary Lee/Ream	1,715	0.28	187	47.6	2,664	N/D	14.3	467
			3/3/82	Black Creek/Lick Creek	1,778	0.43	410	18.1	2,241	41	5.5	176
			3/3/82	Black Creek/Lick Creek	1,797	0.49	470	17.2	2,653	N/D	5.7	182
			3/4/82	Black Creek/Jefferson	1,831	0.40	348	35.6	1,736	N/D	5.0	163
			3/5/82	Black Creek/Jefferson	1,850	1.52	678	709.9	10,386	39	16.4	524
			3/5/82	Black Creek/Jefferson	1,851	1.52	684	505.3	6,406	123	10.1	323
			3/5/82	Black Creek/Black Creek	1,879	2.93	1,659	701.0	25,158	122	15.6	499

*Does not include residual gas volumes.

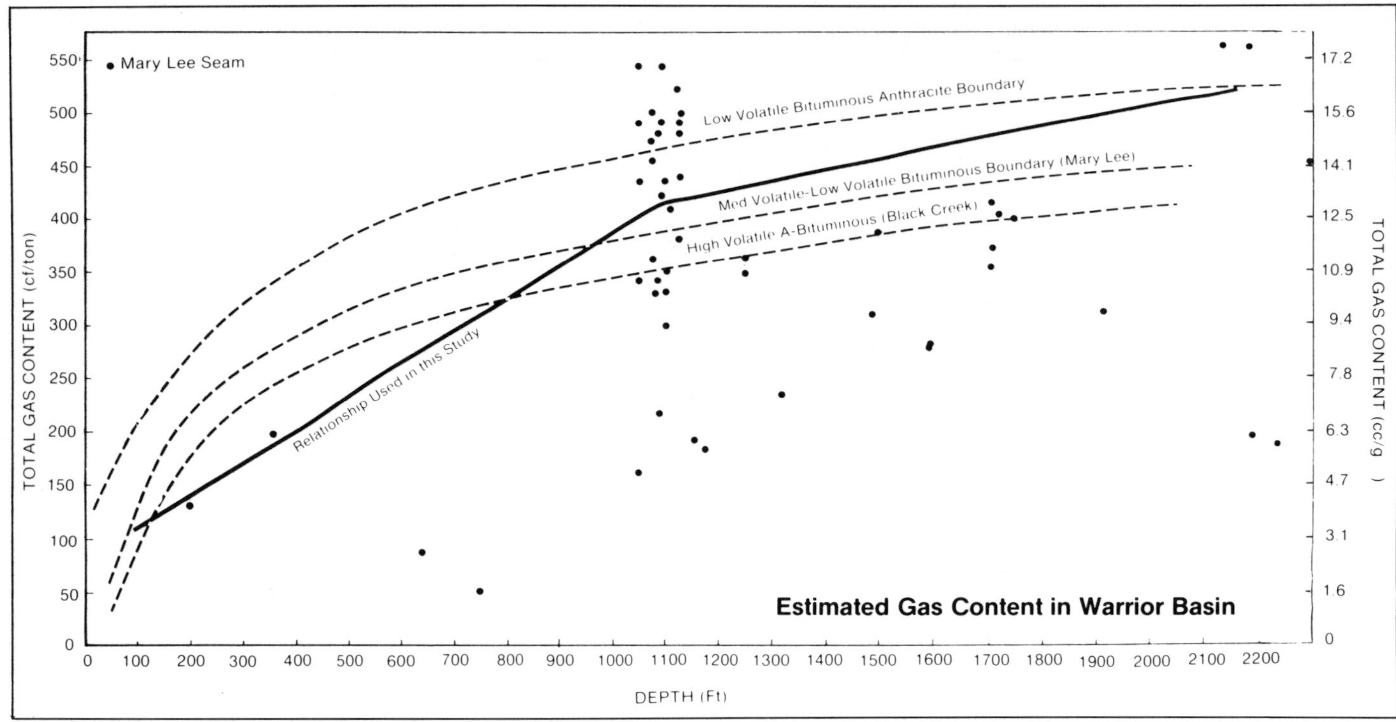

Figure 26—Estimated gas content in Warrior Basin.

Table 5—Methane content data for study area in north-central Alabama. From Diamond et al, 1976.

Overburden, ft	Average Methane Content		Methane Potential, cu ft		Methane Potential, %	
	cc/gm	cf/ton	Mary Lee	Mary Lee Group	Mary Lee	Mary Lee Group
0–500	0.25	7.5	0.09×10^{11}	0.13×10^{11}	1	1
500–1,000	2.00	64.0	0.90×10^{11}	1.15×10^{11}	13	10
1,000–1,500	7.50	240.0	2.65×10^{11}	3.76×10^{11}	37	36
1,500–2,000	13.10	419.0	2.30×10^{11}	3.54×10^{11}	32	34
2,000–2,500	18.60	595.2	1.23×10^{11}	1.96×10^{11}	17	19
Total	—	—	7.17×10^{11}	10.54×10^{11}		

No mining activity exists to the west of a north-south line where the overburden thickness becomes greater than 2,000 ft, so few data are available for the deep basin. The work of Diamond et al (1976), however, indicates that the greatest potential for coalbed methane lies at depths of 1,000 ft or more (see Table 5). The primary target area shown in Figure 27 was selected on the basis of the following criteria:

- Available data such as early maps, exploratory boreholes, or basin trends support the existence of coal seams in this area.
- Emission data from underground mines indicate that methane-rich coal seams lie at depth.
- Overburden thickness trends for coals that have proven to be rich in methane, such as the Mary Lee group, suggest this area is a likely prospect for coalbed methane.

REFERENCES CITED

Ancell, K. L., S. Lambert, and F. S. Johnson, 1980, Analysis of the coalbed degasification process at a seventeen well pattern in the Warrior Basin of Alabama, *in* Proceedings of the First Annual Symposium on Unconventional Gas Recovery, Pittsburgh, PA: Society of Petroleum Engineers/U.S. Department of Energy, May 18–20, 1980, p. 355–370.

Averitt, P., 1975, Coal resources of the United States, January 1, 1974: U.S. Geological Survey Bulletin 1412, 131 p.

Beg, M. A., et al, 1978, Mineral resources of Tuscaloosa County, Alabama: State Oil and Gas Board, Geological Survey of Alabama, Map 185.

Bicker, Jr., A. R., 1970, Economic minerals of Mississippi: Mississippi Geologic, Economic and Topographic Survey

Table 6—Coal reserves and methane estimates in the Warrior Basin.

	Culbertson, 1964	Averitt, 1974	Diamond et al, 1976	Tuscoal Project, 1980	This Study 1980
Area considered	Warrior Basin coal fields	Warrior Basin coal fields	Selected area of Mary Lee group coals in Jefferson, Tuscaloosa, and Walker counties, Alabama	23,000 acres in unspecified location in Tuscaloosa County, Alabama	Warrior Basin coal fields
Coalbed and areas excluded	All areas west of a north-south line through Tuscaloosa and Rayette, Alabama; coalbeds between 1,500′–3,000′ in southwestern Warior field; all seams <14″ thick; all coals >3,000′ deep	Coals less than 14″ thick; coal 6,000′ deep	All coals not mined within study area	All areas outside study area	All areas outside of study area
Coal resource estimate	33.8 billion tons	41.3 billion tons	6.3 billion tons	Not given	35 billion tons
Methane estimate	Not given	Not given	1 Tcf; 70% in Mary Lee group	100 Bcf over 20 years	10 Tcf

Table 7—Gas analysis from five-well degasification test of Mary Lee coalbed, Jefferson County, Alabama. Elder and Deul, 1974 and TRW's Ramsey McCormick 35-3 well (in mole %).

Sample	Bore-Hole	CO_2	CH_4	C_2H_6	C_3H_8	C_4H_{10}	C_5H_{12}	H_2	He	O_2	Inerts N_2 and A
1	3SW	0.1	96.2	0.01	—	—	—	0.01	0.26	0.0	3.4
2	3SW	0.1	95.9	0.01	—	—	—	0.01	0.28	0.1	3.6
3	3SW	2.9	88.6	0.08	0.018	0.002	0.0003	3.9	—	0.2	4.3
4	3SW	2.6	89.1	0.08	0.019	0.002	0.0003	3.9	—	0.2	4.1
5	3SW	0	96.7	0.01	Trace	—	—	0	0.22	0	3.0
1	TRW	0.1	96.8	0.18	0.01	<0.01	<0.01	NA	NA	0.38	2.45

Bulletin 112, 80 p.

Boland, L. F., and E. D. Minihan, 1971, Petroleum potential of the Black Warrior Basin: Gulf Coast Association of Geological Societies Transactions, v. 21, p. 139–158.

Butts, C., 1906, The northern part of the Cahaba coal field, Alabama: U.S. Geological Survey Bulletin 316, p. 76–115.

———, 1911, The southern part of the Cahaba coal field, Alabama: U.S. Geological Survey Bulletin 431, p. 89–146.

———, 1926, The Paleozoic rocks, *in* G. I. Adams, C. Butts, L. W. Stephenson, and W. Cooke, Geology of Alabama: Geological Survey of Alabama, Special Report No. 14, p. 40-223.

———, 1927, Bessemer and Vandiver Quadrangles, Alabama: U.S. Geological Survey Geologic Atlas, Folio 221.

———, 1940, Description of the Montevallo and Columbiana Quadrangles: U.S. Geological Survey Geologic Atlas, Folio 226.

Cervik, J., 1969, Behavior of coal-gas reservoirs: U.S. Bureau of Mines Technical Progress Report 10, 10 p.

City of Tuscaloosa, Alabama, and University of Alabama School of Mines and Energy Development, 1982, Utilizing the unconventional gas resources of the Pottsville Formation coals in Tuscaloosa County, Alabama: Feasibility Study, 88 p.

Coulter, D. M., 1947, Coking coal deposits on Lookout Mountain, DeKalb and Cherokee counties, Alabama: U.S. Bureau of Mines Report of Investigations 4030, 89 p.

Culbertson, W. C., 1964, Geology and coal resources of the coal-bearing rocks of Alabama: U.S. Geological Survey Bulletin 1182-B, 79 p.

Davis, M. W., and R. Ehrlich, 1974, Late Paleozoic crustal composition and dynamics in the southeastern United States, *in* G. Briggs, ed., Carboniferous of the southeastern United States: Geological Society of America Special Paper 148, p. 171–185.

Diamond, W. P., and J. R. Levine, 1981, Direct method determinations of the gas content of coal, procedures and results: U.S. Bureau of Mines, Report of Investigations 8515, 36 p.

———, G. W. Murrie, and C. M. McCulloch, 1976, Methane gas content of the Mary Lee Group of coalbeds, Jefferson, Tuscaloosa, and Walker counties, Alabama: U.S. Bureau of Mines Report of Investigations 8117, 9 p.

Ehrlich, R., 1964, Evidence on relative ages of the Appalachian and Ouachita structural trends: Geological Society of America Special Paper 82, 299 p.

Elder, C. H., and M. Deul, 1974, Degasification of the Mary Lee coalbed near Oak Grove, Jefferson County, Alabama, by vertical borehole in advance of mining: U.S. Bureau of Mines Report of Investigations 7968, 21 p.

Ferguson, C. C., and R. G. Stearns, 1967, Stratigraphy and petrology of the Upper Mississippian of southernmost Tennessee, *in* A field guide to Mississippian sediments in northern Alabama and south-central Tennessee: Alabama Geological Society 5th Annual Field Trip Guidebook, p. 53–60.

Ferm, J. C. and R. Ehrlich, 1967, Petrology and stratigraphy of the Alabama coal fields, *in* J. C. Ferm, R. Ehrlich, and T. L. Neathery, eds., A field guide to Carboniferous detrital rocks in northern Alabama: Geological Society of America, Coal Division, and Alabama Geological Society Field Trip Guidebook, p. 11–15.

———, R. Ehrlich, and T. L. Neathery, 1967, A field guide to Carboniferous detrital rocks in northern Alabama: Geological Society of America, Coal Division, Field Trip Guidebook and Alabama Geological Society , 101 p.

Gibson, A. M., 1891, Report on the coal measures of Blount Mountain: Geological Survey of Alabama, Special Report No. 3, 80 p.

———, 1893, Geologic structure of Murphree's Valley, its minerals and other materials of economic value: Geological Survey of Alabama, Special Report No. 4, 132 p.

Graham, S. A., W. R. Dickinson, and R. V. Ingersoll, 1975, Himalayan-Bengal model for flysch dispersal in the Appalachian-Ouachita system: Geological Society of America Bulletin, v. 86, p. 273–286.

———, R. V. Ingersoll, and W. R. Dickinson, 1976, Common provenance for lithic grains in Carboniferous sandstones from Ouachita Mountains and Black Warrior Basin: Journal of Sedimentary Petrology, v. 46, n. 3, p. 620–632.

Hobday, D. K., 1969, Upper Carboniferous shoreline systems in northern Alabama: Doctoral Dissertation, Baton Rouge, Louisiana State University, 75 p.

———, 1974, Beach- and barrier-island facies in the Upper Carboniferous of northern Alabama, *in* G. Briggs, ed., Carboniferous of the southeastern United States: Geological Society of America Special Paper 148, p. 209–223.

Kidd, J. T., 1976, Configuration of the top of the Pottsville Formation in west-central Alabama: State Oil and Gas Board, Geological Survey of Alabama, Map 1.

Kissell, F. N., 1972, Methane migration characteristics of the Pocahontas No. 3 coalbed: U.S. Bureau of Mines Report of Investigations 7649, 19 p.

Lambert, S. W., 1980, Technical progress reports, monthly reports, 1977 to 1980: Contract No. ET-75-C-01-9027, U.S. Department of Energy, Morgantown Energy Technology Center, Morgantown, West Virginia.

———, and M. A. Trevits, 1980, The feasibility of no-proppant stimulation to enhance removal of methane from the Mary

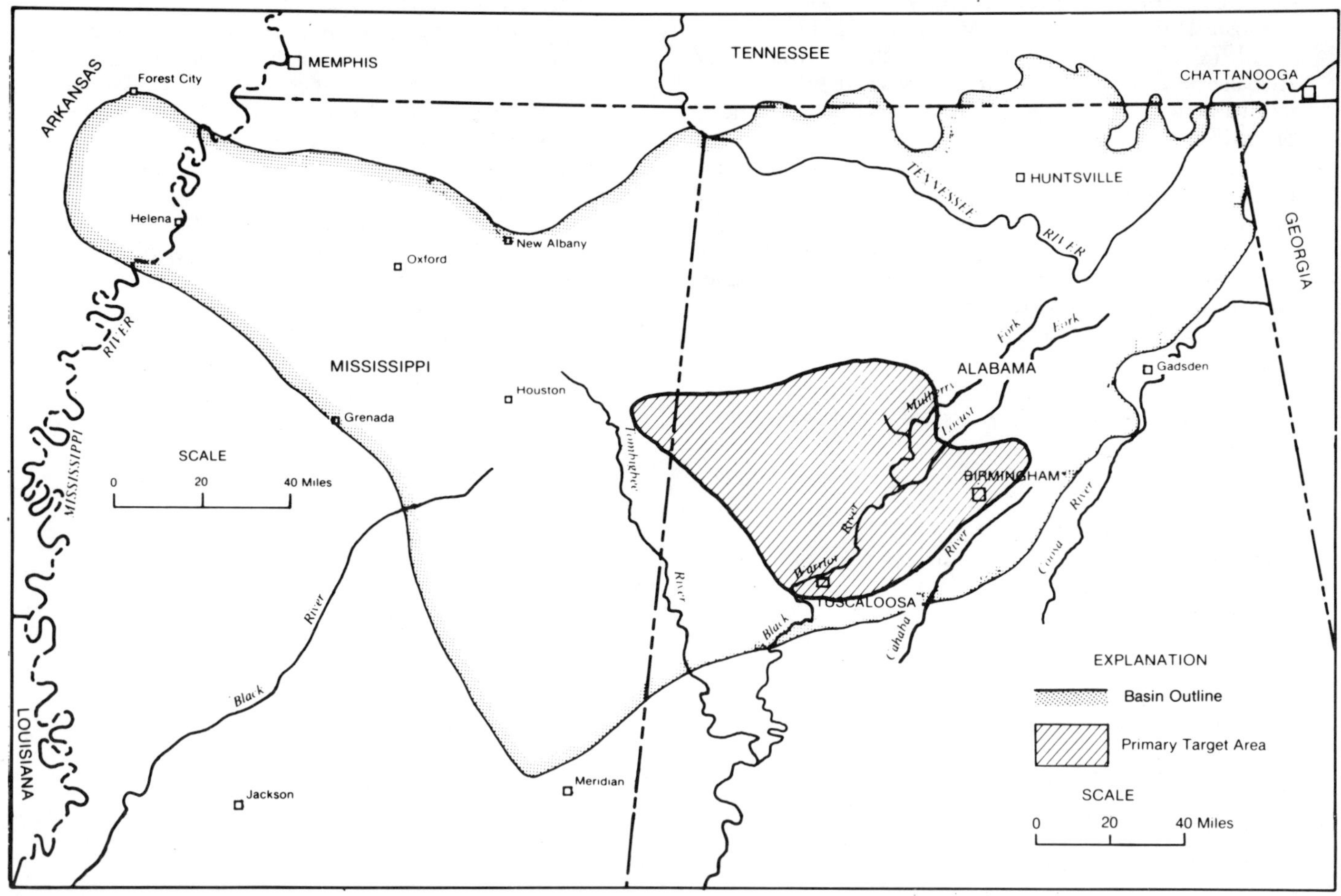

Figure 27—Primary target area of the Warrior Basin.

Lee coalbed: U.S. Department of Energy Report of Investigation CMTC-TR80-1, April, 1980, 18 p.

McCalley, H., 1886, On the Warrior coal field: Geological Survey of Alabama Special Report No. 1, 571 p.

——— , 1891, Report on the coal measures of the Plateau Region of Alabama: Geological Survey of Alabama, Special Report No. 3, 238 p.

——— , 1900, Report on the Warrior coal basin: Geological Survey of Alabama, Special Report No. 10, 217 p.

Mellen, F. F., 1947, Black Warrior Basin, Alabama and Mississippi: Bulletin of the American Association of Petroleum Geologists, v. 31, n. 10, p. 1801–1816.

——— , 1971, What's ahead for the Black Warrior Basin?: World Oil, v. 173, n. 2, p. 45–53.

Metzger, W. J., 1965, Pennsylvanian stratigraphy of the Warrior Basin, Alabama: Geological Survey of Alabama, Circular 30, 80 p.

Milici, R. C., W. B. Brent, and K. R. Walker, 1973, Depositional environments in upper Conasauga lagoon-fill sequence, *in* Geology of Knox County, Tennessee: Tennessee Division Geology Bulletin 70, p. 138-148.

Murrie, G. W., W. P. Diamond, and S. W. Lambert, 1976, Geology of the Mary Lee Group of coalbeds, Black Warrior coal basin, Alabama: U.S. Bureau of Mines Report of Investigations 8189, 49 p.

Perry, J. H., L. J. Prosser, Jr., and J. Cervik, 1980, Methane drainage from the Mary Lee coalbed, Alabama, using horizontal drilling techniques, *in* Proceedings of the First Annual Symposium on Unconventional Gas Recovery, Pittsburgh, Pennsylvania: Society of Petroleum Engineers/U.S. Department of Energy, May 18–20, 1980, p. 335–340.

Self, D. L. and T. L. Neathery, 1975, Lexicon of structural features in Alabama: University of Alabama, Geological Survey of Alabama and U.S. Geological Survey, 496 p.

Semmes, D. R., 1929, Oil and gas in Alabama: Geological Survey of Alabama, Special Report No. 15, 408 p.

Shotts, R. Q., 1971, Coal reserves of Bibb County, Alabama: Geological Survey of Alabama Circular 72, 50 p.

Squires, J., 1890, Report on the Cahaba Coal Field: Geological Survey of Alabama, Special Report No. 2, 189 p.

Stearns, R. G., and R. M. Mitchum, Jr., 1962, Pennsylvanian rocks of the southern Appalachians, *in* Pennsylvanian system of the United States: American Association of Petroleum Geologists, p. 74–96.

Thomas, W.A., 1967, Mississippian stratigraphy of the Tennessee Valley, Alabama, *in* E. Smith, ed., A field guide to Mississippian sediments in northern Alabama and southcentral Tennessee: Alabama Geological Society Guidebook, 5th Annual Field Trip, p. 5–9.

——— , 1972, Mississippian stratigraphy of Alabama: Geological Survey of Alabama, Monograph 12, 121 p.

——— , 1974, Converging clastic wedges in the Mississippian of Alabama, *in* G. Briggs, ed., Carboniferous of the southeastern United States: Geological Society of America Special Paper 148, p. 187–207.

——— , 1977, Evolution of Appalachian-Ouachita salients and recesses from reentrants and promontories in the continental margin: American Journal of Science, v. 277, p. 1233–1278.

——— , and G. H. Mack, 1982, Paleogeographic relationship of a Mississippian barrier-island and shelf-bar system (Hartselle Sandstone) in Alabama to the Appalachian-Ouachita orogenic belt: Geological Society of America Bulletin, v. 93, p. 6–19.

Tolson, J. S., C. G. Musgrove, and P. K. Sokalosky, 1982, Alabama coal data for 1980, Geological Survey of Alabama Information Series 58B, 96 p.

Tull, J. F., 1979, Stratigraphic and structural relationships of the eastern Talledega Slate Belt of Alabama, *in* J. F. Tull, and S. H. Stow, eds., The Hillabee metavolcanic complex and associated rock sequences: Alabama Geological Society Guidebook, 17th Annual Field Trip, p. 3–13.

U. S. Federal Energy Regulatory Commission, 1980, Pipeline rates: Order Denying Petition for Advance Approval and Granting Interventions, February 28, 1980.

Ward, W. E., II, 1977, Jointing in a selected area of the Warrior coal field, Alabama: Unpublished Master's Thesis, University of Alabama, 61 p.

Ward, W. E., and F. E. Evans, Jr., 1975, Coal—its importance in Alabama: Geological Survey of Alabama Information Series 53, University of Alabama, 27 p.

Pennsylvanian Geology, Coal, and Coalbed Methane Resources of the Illinois Basin — Illinois, Indiana, and Kentucky

P. L. Archer
J. N. Kirr

The Illinois Basin is a broad, spoon-shaped structural depression encompassing approximately 53,000 sq mi in Illinois, southwestern Indiana, and western Kentucky. The U.S. Geological Survey has estimated the bituminous coal reserves in Pennsylvanian rocks to be as much as 365 billion tons. More than 75 individual coal seams have been identified in this area, 20 of which are mined. The majority of the coals are not continuous and do not maintain constant thicknesses. Individual seams range from a few inches to 15 ft in thickness over large areas. The coals outcrop at the basin's periphery and dip gently towards its deeper portions in southeastern Illinois and western Kentucky. Lower and Upper Pennsylvanian coals are thin and discontinuous while the Middle Pennsylvanian coals are thick, generally continuous, and provide the major reserves of the basin. Thin Lower and Upper Pennsylvanian coals have not been studied in much detail and are not as well correlated as the thicker coalbeds of the Middle Pennsylvanian. The greatest cumulative thickness of coal seams presumably occurs in the southeastern portion of the basin (near the tristate boundary) where the thickest Pennsylvanian section occurs. All Illinois basin coal seams are covered by less than 3,000 ft of overburden, and the major coals are within 1,500 ft of the surface.

The Springfield-Harrisburg (No. 5) coals in Illinois and their correlatives, Springfield V in Indiana and No. 9 coals in Kentucky, are the most extensive and uniformly thick coals in the Illinois basin; estimated coal reserves are over 67 billion short tons. The Herrin (No. 6) coal is also thick and extensive in Illinois and contains estimated coal reserves of over 77 billion short tons. Some deeper coals, the Colchester (No. 2), which is uniformly present over the entire basin, and the Davis (No. 6) and Mannington (No. 4) occurring primarily in Kentucky, contain combined reserves estimated at over 39 billion short tons.

The potential methane resource from selected coalbeds in the Illinois basin can be estimated from desorption data generated by the U.S. Bureau of Mines (USBM), the Illinois and Indiana State Surveys, and field tests performed under the MRCP. The Herrin (No. 6) and the Springfield-Harrisburg (No. 5) coals and their correlatives in Indiana have been sampled the most often. The estimated gas content of the Herrin (No. 6) ranges from 32 to 125 cubic feet per ton (cf/ton). A similar range of values, 32 to 147, has been calculated for the Springfield-Harrisburg (No. 5) coals. This variability cannot be directly related to sample depth, since some of the gassier coals were from relatively shallow horizons—the opposite of what might have been expected—but is related to other geologic controls on the gas content, to analytical errors, or to the method used to determine the "remaining" gas.

MRCP data on the methane gas content of Illinois basin coals is presently limited to coring and well testing in Clay and Marion counties of Illinois Posey County, Indiana, and Webster County, Kentucky. In Clay County, the Danville (No. 7), Herrin (No. 6), Briar Hill (No. 5a), Harrisburg (No. 5), and Seelyville coals were cored between depths of 994 and 1,352 ft and samples were desorbed for approximately 5 months using the USBM's direct method. Total gas contents ranged between 32 and 48 cf/ton. In Marion County, the same coal seams, excluding the Seelyville, were cored and sampled at depths between 664 and 736 ft. Gas contents ranged from 22 to 35 cf/ton. All major coals in Posey County, Indiana, were sampled and showed a gas content from 26 to 144 cf/ton. Results from Webster County, Kentucky, showed a gas content of 45 cf/ton for the Nos. 13 and 9 seams.

Investigations of data have shown that the gas content of the coals in the Illinois basin is generally low, ranging from less than 40 to 150 cf/ton. Based on the limited data available, ranges for the maximum and minimum expected in-place gas resource have been made for the Danville, Herrin, Springfield-Harrisburg, and their equivalent coals. The Danville coals are anticipated to have a minimum of approximately 500 billion and maximum of nearly 1.7 trillion cubic feet (Tcf) of in-place gas. The Herrin, likewise, is estimated to contain 2.5 and 9.6 Tcf of gas and the Springfield-Harrisburg, 2.2 and 9.9 Tcf of gas. The minimum in-place gas resource for these three seams totals over 5 Tcf. It is reasonable to assume that the methane contained in major deeper coals (Colchester, Davis, etc.) may add significantly to this figure. It should be noted that although the specific gas content of coals in the Illinois basin is quite low, the simple magnitude of the coal resource produces large in-place gas resource estimates.

The gas content of coals in the Illinois basin is generally higher towards the southeastern portion of the basin and initial target area defined by the MRCP. On the basis of information gathered in this report, two target areas (totaling approximately 4,300 sq mi) hold the greatest probability for early commercial gas production. Target Area A, located in western Kentucky, contains a thick section of deep coals in a highly disturbed structural belt. Target Area B, in southeastern Illinois and southwestern Indiana, contains previously reported gassy coals and thick coal sections at considerable depths.

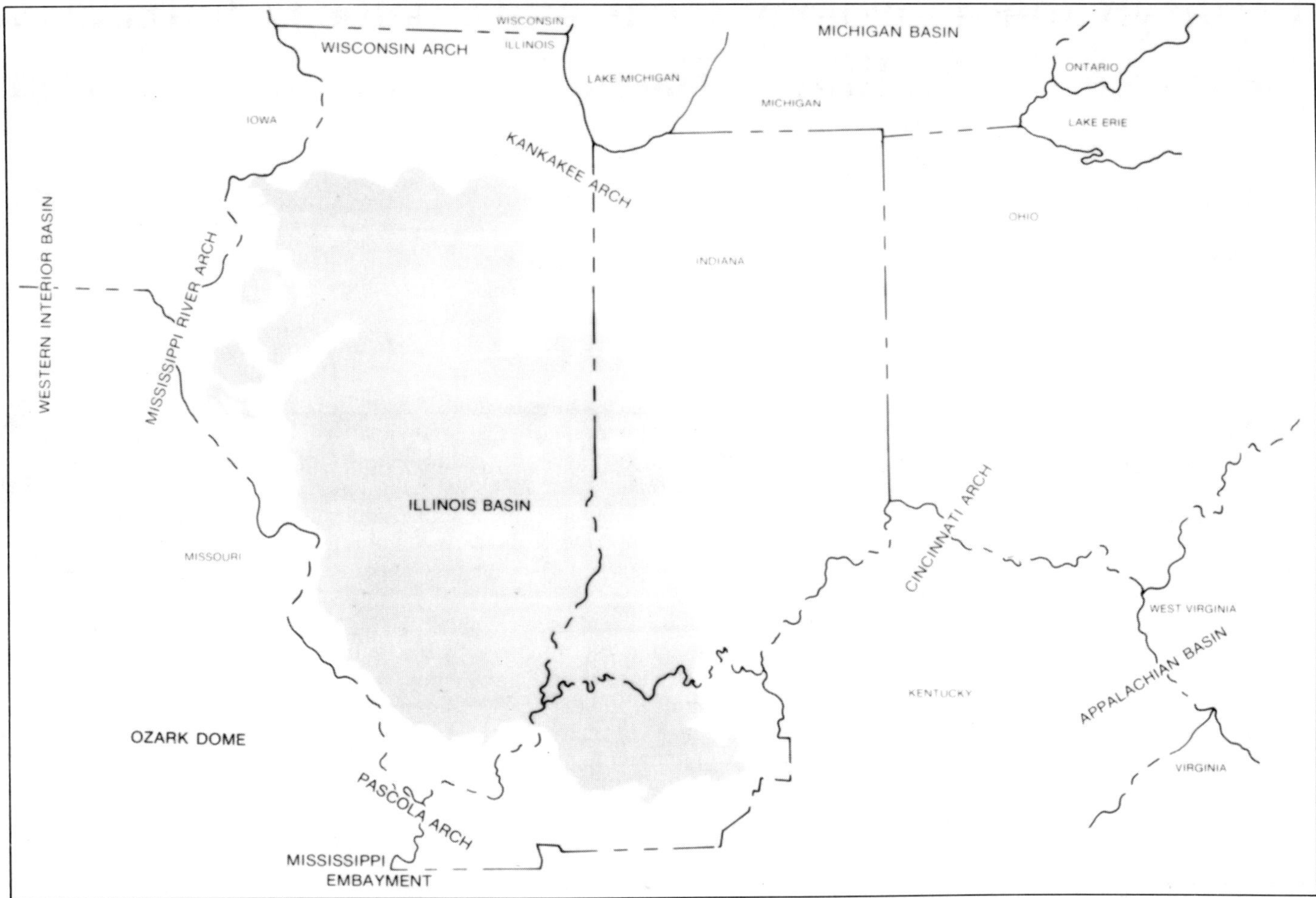

Figure 1—Regional setting of the Illinois Basin.

INTRODUCTION

Situated in the east-central United States, the Illinois Basin (Fig. 1) covers a major portion of Illinois and extends into southwestern Indiana and western Kentucky. For the purpose of this report, the Illinois Basin's areal extent, as defined by the cross-hatched region in Figure 1, is that area underlain by Pennsylvanian-age rocks encompassing approximately 50,000 sq mi.

Physiographically, a major portion of the Illinois Basin lies in the Central Lowland province and is characterized by low relief and gently rolling topography. Much of this lowland is a glacial till plain presently covered by loess, lacustrine, and alluvial deposits. Southernmost Illinois, south-central Indiana, and the Kentucky portion of the basin lie within the Interior Low Plateaus province. The plateaus province, contrasting with the lowland province, is typified by rolling uplands with moderate relief (400 to 500 ft above sea level) and is dissected by numerous entrenched meandering streams. Total relief in the Illinois Basin is never over 1,000 ft and, except for the southern part of the basin, seldom over 200 ft. The entire basin area is drained by the Mississippi River and its major tributaries: the Ohio, Wabash, and Illinois Rivers.

STRUCTURAL GEOLOGY

As part of a central stable platform bordering the Canadian Shield, this region has seen little structural deformation since Precambrian time other than regional warping and differential sinking (Swann, 1968; Atherton, 1971). The Illinois Basin itself is a broad, north-northwest/south-southeast–trending, spoon-shaped synclinal feature, surrounded on all sides by positive structural arches and domes. The position of regional structural features relative to the basin is shown in Figure 1. Adjacent coal-bearing basins having similar characteristics and age—the Appalachian, Michigan, Forest City (Western Interior), and the Black Warrior Basins—are isolated from the Illinois Basin by the Cincinnati, Kankakee, Mississipppi River, and Pascola arches, respectively. Other prominent surrounding features include the Wisconsin arch to the north, the Ozark dome to the southwest, and the Mississippi embayment area to the south.

The major structural features within the Illinois Basin (Fig. 2) are expressions of relief caused by high-angle faulting. These include the La Salle anticlinal belt, a series of anticlines and westward-dipping monoclines; the Du Quoin monocline, dipping east; and the Cottage Grove and Shawneetown-Rough

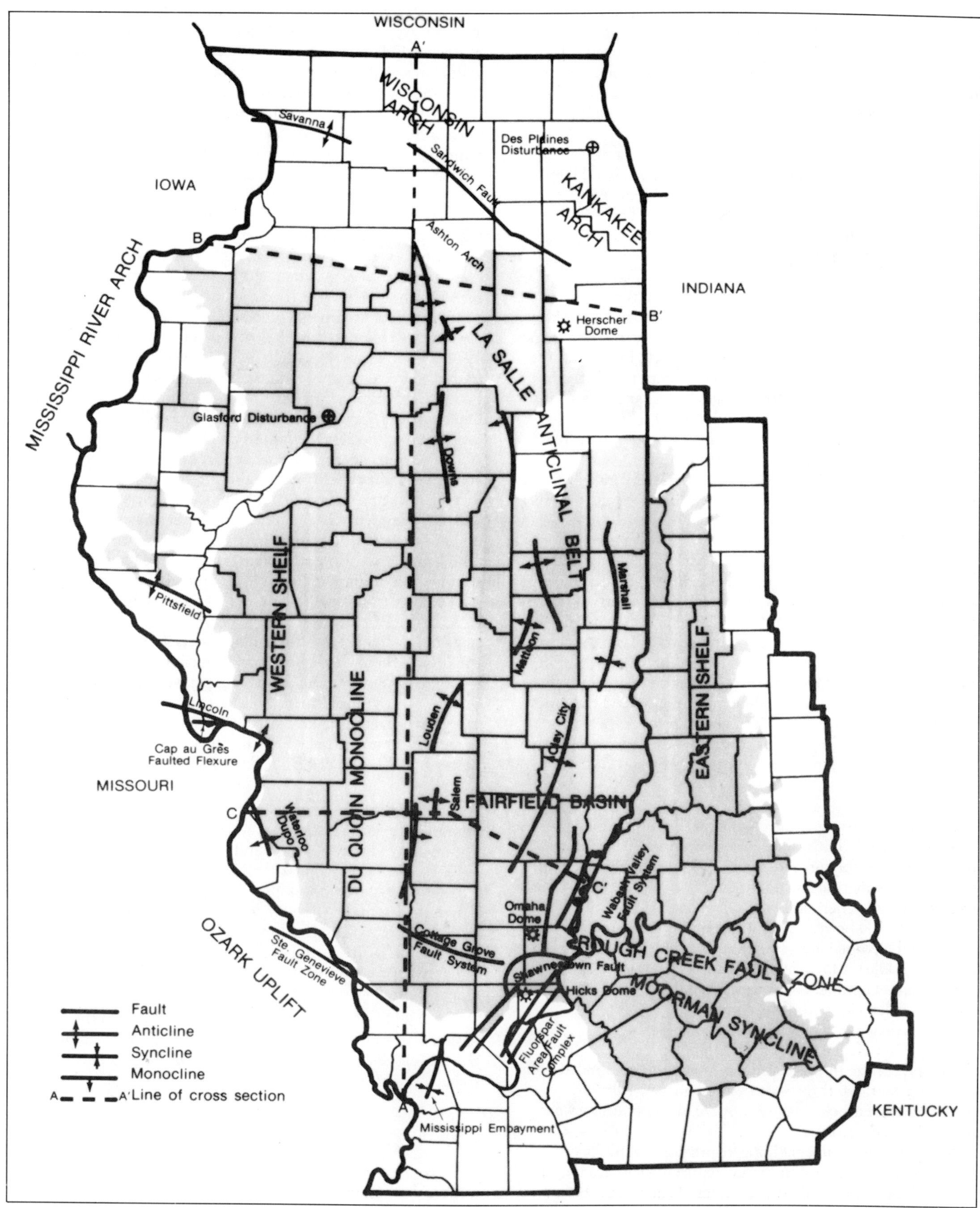

Figure 2—Major structural features of the Illinois Basin. After Willman et al, 1975. Used with the permission of the Illinois State Geological Survey.

Creek Fault systems, trending east-west. Each of these uplifted areas rises from a deep central interior basin: the Fairfield. South of the Shawneetown-Rough Creek fault zone is the Moorman syncline, structurally the deepest portion of the Illinois Basin, with Precambrian rocks at about 15,000 ft (Willman et al, 1975). The thickest Pennsylvanian section also is found in this region.

Although numerous folds, faults, and domes occur throughout the basin (Buschbach and Ryan, 1963; Buschbach and Heim, 1972; Willman et al, 1975), only those faults mentioned above have significant displacement or lateral extent. The Shawneetown-Rough Creek fault zone is composed of a complex pattern of high-angle basement faults with more than 3,000 ft of vertical displacement in some areas (Davies, 1973). Major faults are subparallel in an east-west direction and frequently are intersected by smaller cross faults. In a few places, limestone of Late Mississippian age is brought to the surface and is in contact with Late Pennsylvanian-age rock. In areas where this intense faulting occurs (in the southeastern part of the Illinois basin), there are abrupt changes in rock units vertically and laterally; locally, the bedrock dips steeply. The Fluorspar District, an extensively faulted, mineralized, and mined area, is located at the southern boundary of the basin.

Near-surface bedrock of the Illinois Basin is of Pennsylvanian and Upper Mississippian age. The generalized geologic map (Fig. 3) shows their distribution. Rocks surrounding the Illinois Basin are generally older except to the south, where Tertiary and Cretaceous sediments of the northern Mississippi embayment overlap Ordovician and Mississippian sediments. Paleozoic units dip gently from all edges toward the deeper parts of the basin, and progressively younger rocks are exposed toward the basin's center. The sedimentary layering of the basin rocks has been likened to a nest of graduated measuring spoons (Swann, 1968). Exaggerated cross sections through the basin illustrate this configuration (Fig. 4). As shown, the majority of the units thicken toward the basin's structural center in southeastern Illinois. Detailed geologic cross sections of the Paleozoic section in the Illinois Basin have been prepared by Swann (1968).

Stratigraphy

A generalized stratigraphic column of the Illinois Basin is shown in Figure 5. In the deeper portions of the basin, the thickness of this column exceeds 14,000 ft. Paleozoic rocks are predominantly marine and dominated by dolomite, limestone, shale, sandstone, chert, anhydrite, and coal, in that order (Swann, 1968). The entire basin was emergent numerous times during the Paleozoic; major regional unconformities occur beneath Middle Ordovician, Middle Devonian, and Pennsylvanian sediments (Swann, 1968; Buschback, 1971). Since the Paleozoic, with the exception of periods of Pleistocene glaciation, the area has been a site of nondeposition. Because very little bedrock is exposed, the stratigraphy of the Illinois basin has been largely compiled from subsurface data. All three state geologic surveys (Illinois, Indiana, and Kentucky) have extensive files containing well logs, cores, and cuttings.

Although a few thin coal horizons are present in Upper Mississippian-age rocks of Indiana, Pennsylvanian-age rocks contain the primary coal-bearing seams of the Illinois Basin. Pennsylvanian rocks were deposited unconformably on older Paleozoic rocks following a major period of uplift and erosion at the end of the Mississippian. The region consisted of a southwesterly inclined coastal plain that received sediment from the Appalachian and Canadian Shield areas to the northeast. Subareal erosion of the coastal plain produced river valleys in a well-developed linear drainage pattern. Alluvial sands and muds first filled these valleys, and as the seas gradually migrated northward, clastic marine (deltaic) sedimentation filled and covered the erosion surface (Pryor and Sable, 1974). Although the original areal extent of Pennsylvanian rocks is not known, they were deposited over a much larger area and subsequently eroded (Wanless, 1962). Oldest Pennsylvanian rocks were deposited only in the southern portions of the region; and in most places, progressively younger Pennsylvanian rocks lie upon older Paleozoic rocks in northern Illinois (Willman et al, 1975). The majority of Middle and Upper Pennsylvanian deposition occurred in transitional and continental environments such as alluvial and delta plains, distributary channels, marshes, and swamps (Pryor and Sable, 1974). Several times, the delta platform was covered by fresh-water coal swamps, and, as sea level rose, the swamps were flooded and black, pyrite-bearing shales with beds of fossiliferous limestone were deposited. At least 51 repeated cycles of deposition (cyclothems: Wanless and Weller, 1932) have been identified in the Pennsylvanian sequence. Each cyclothem consists of sandstone, shale, limestone, and coal units arranged in a regular sequence (Fig. 6). The repeated initiation, growth, and abandonment of delta lobes, combined with their lateral migration, led to the complex vertical and horizontal arrangement of these cyclothems now observed in the stratigraphic column of Pennsylvanian rocks.

The present stratigraphic nomenclature of Pennsylvanian rocks in the Illinois basin area is shown in Figure 7. The sequence is divided into three major rock-stratigraphic units by the Illinois State Geological Survey: a basal McCormick Group, a middle Kewanee Group, and an upper McLeansboro Group. Within these groups, the Illinois State Geological Survey recognizes seven formations; the Indiana State Geologic Survey names ten formations; and the Kentucky Geological Survey names five. More specifically, Pryor and Sable (1974) report that:

- McCormick Group sediments and their equivalents are mostly fluviatile sandstones and mudstones, the coalbeds are thin and of limited extent, and limestones are rare or absent.
- Kewanee Group sediments and their equivalents mark the beginning of well-developed cyclic deposits with shales, limestones, and coals being much more common.
- McLeansboro Group sediments and their equivalents are also cyclic but more marine in character than earlier Pennsylvanian units, with thicker, more numerous limestones and thinner coals.

The thickest Pennsylvanian sections are found in the Fairfield basin (> 2,400 ft) and Moorman syncline (> 3,300 ft) areas of the Illinois Basin (Fig. 8). These areas also are generally believed to have the greatest cumulative thicknesses of coal.

COAL RESOURCE

Coal deposits of the Illinois Basin are primarily confined to rocks of the Pennsylvanian system, although thin coalbeds of limited extent have been found in Upper Mississippian rocks. The extent of the coal-bearing Pennsylvanian rocks is depicted

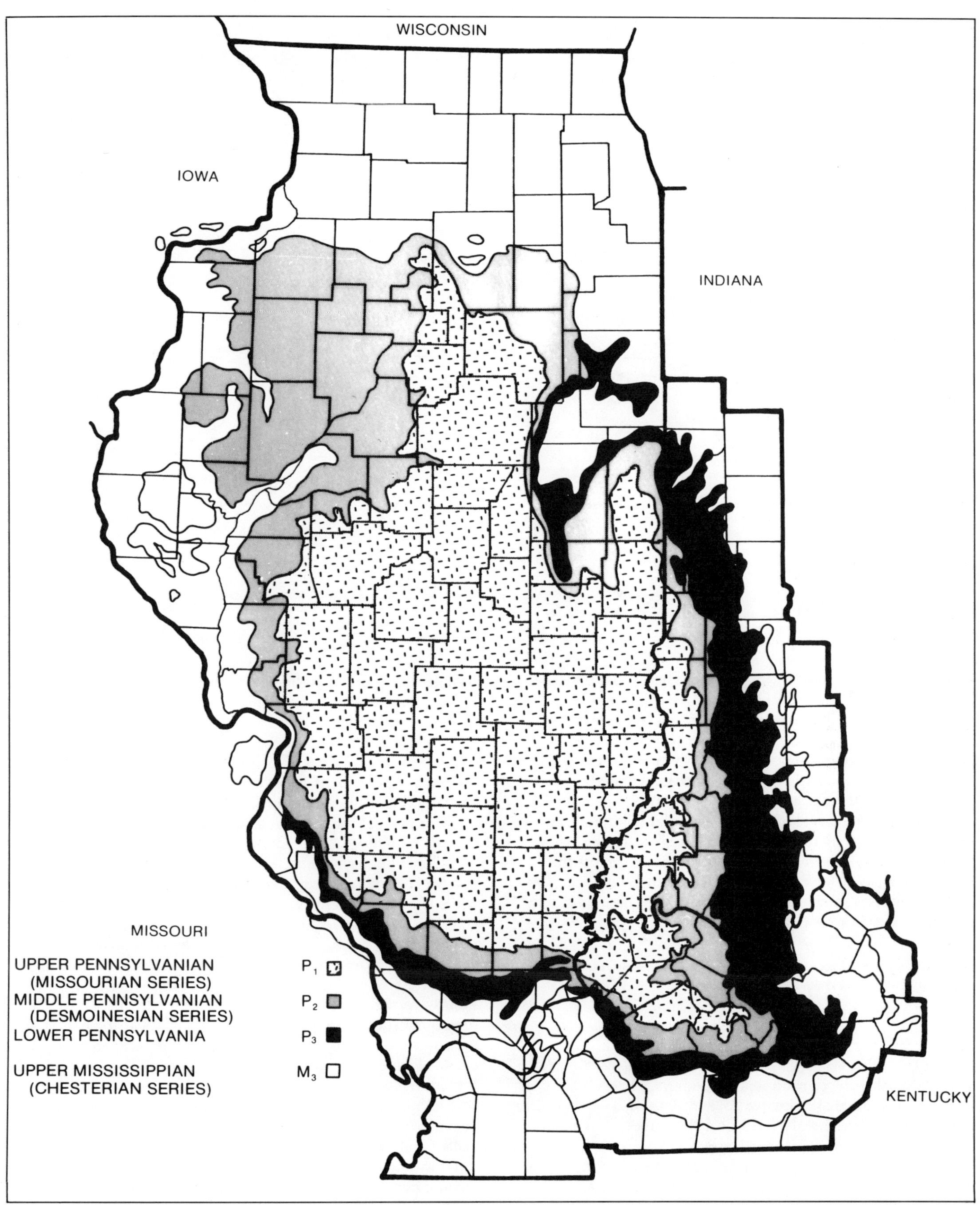

Figure 3—Generalized geologic map of the Illinois Basin. From Geologic Map of the United States, 1:2,500,000, USGS, 1974.

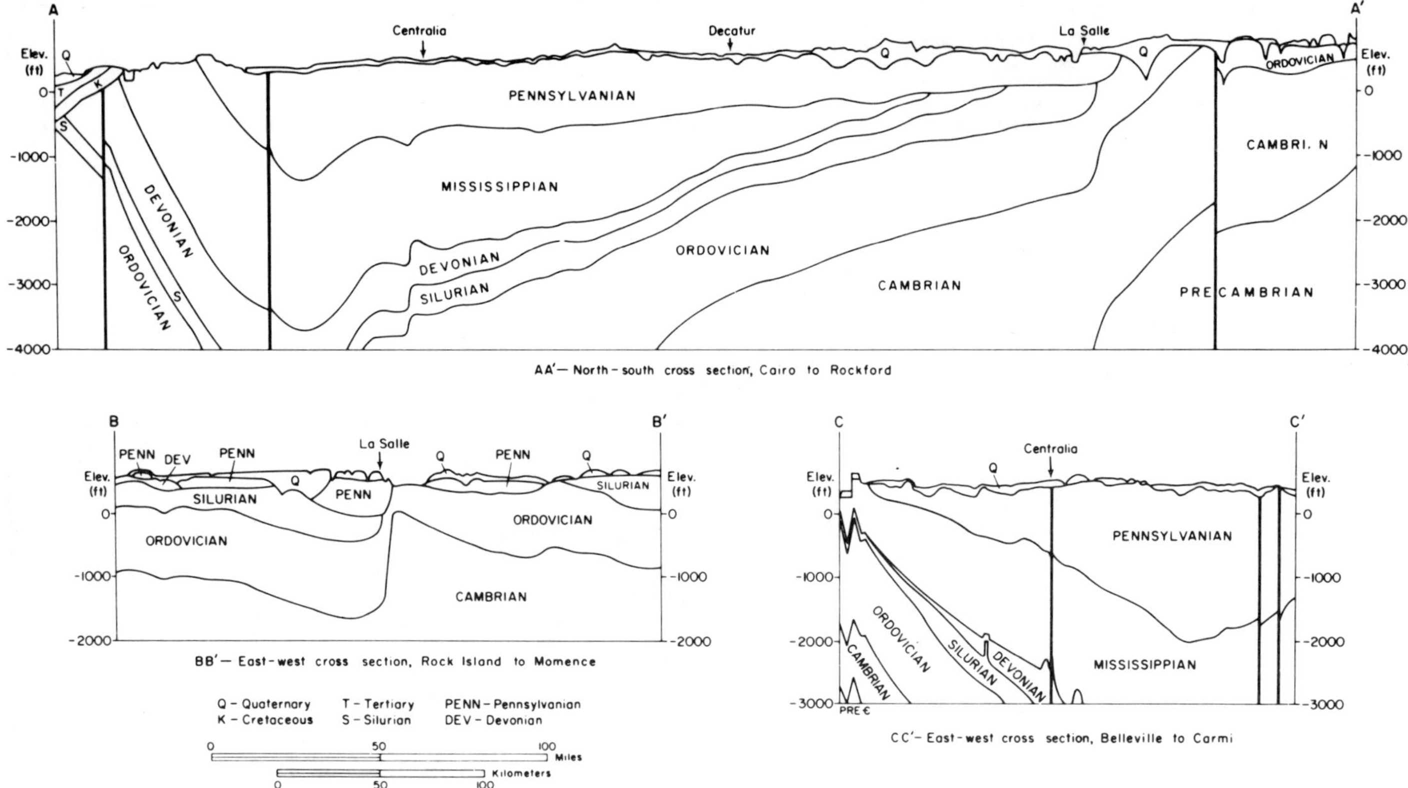

Figure 4—Geologic cross sections along the lines shown in Figure 2. Willman et al, 1975. Used with the permission of the Illinois State Geological Survey.

by the Illinois Basin outline in Figure 1. These rocks underlie approximately 36,900 sq mi (102 counties) of Illinois, 6,500 sq mi (25 counties) of Indiana, and 6,400 sq mi (14 counties) of Kentucky. Coals of the Illinois Basin are thought to have accumulated in fresh-water swamp environments on broad delta plains (Willman et al, 1975). Bright- and dull-banded coals, cannel coals, and paper coals have been found in the basin. These coals cover the full range of high-volatile bituminous, with a progressive increase in rank from high-volatile C to high-volatile A coals in a southeasterly direction across the basin (Fig. 9), as observed and documented by Damberger (1971). Analyses show that coals of the Illinois Basin generally fall within the following range of values:

Moisture (%)	5–20
Ash (%)	6–14
Sulfur (%)	2–5
Btu/lb	10,000–13,000

The principal authigenic minerals found in these coals are kaolinite, calcite, marcasite, pyrite, and gypsum. The chemistry of the authigenic minerals in the coal is commonly related to the mineralogy of the overlying strata.

Detailed stratigraphic columns of the Pennsylvanian system in Illinois and Indiana are shown in Figures 10 and 11, respectively; none is available for western Kentucky. Although more than 75 individual coal seams have been identified (of which 20 are mined), the coals make up less than 3% of the total section. The majority of Illinois Basin coals are not continuous and do not maintain constant thicknesses; some peat accumulated only in small isolated pools. Individual seams range in thickness from a few inches to 15 ft, and some average 4 to 6 ft over large areas. Thicknesses often are controlled by the presence of prominent structural features (Fig. 2) and proximity to major drainage systems in swamps (Smith and Stall, 1975). Little or no coal was deposited on topographic highs, and thick coal accumulated in the vicinity of principal channels. Lower and Upper Pennsylvanian coals (McCormick, lower Kewanee, and McLeansboro Group coals) are thin and discontinuous, while the Middle Pennsylvanian coals (upper Kewanee Group coals) are thick, generally continuous, and provide the major reserves of the basin. Thin Lower and Upper Pennsylvanian coals have not been studied in as much detail and are not as well correlated as the thicker coalbeds of the Middle Pennsylvanian. The greatest cumulative thickness of coal seams presumably occurs in the southeastern portion of the basin (near the tristate boundary) where the thickest Pennsylvanian section occurs. Figure 12 shows the stratigraphic correlation of important coal seams in the Illinois Basin.

Overburden thickness varies greatly throughout the basin and depends upon local topography and location within the basin itself. Coal seams outcrop on the basin's periphery and gently dip toward its center. In southwestern Indiana, the coal seams dip at a rate of 25 to 30 ft per mi toward the deeper parts of the basin in Illinois and Kentucky. Approximate depths to individual coalbeds can be determined by studying isopach maps of the Pennsylvanian strata and Quaternary deposits overlying

ERA, ERATHEM	PERIOD, SYSTEM	ORIGIN AND CHARACTER	GREATEST THICKNESS (ft)	EPOCH, SERIES		
				ILLINOIS	INDIANA	KENTUCKY
CENOZOIC	QUATERNARY	Continental — glacial, river and stream, wind, lake, swamp, and colluvial deposits and soils	600	PLEISTOCENE	PLEISTOCENE	PLEISTOCENE
		Major unconformity				
	TERTIARY	Continental — river deposits, mostly gravel, some sand	50	PLIOCENE	PLIOCENE	PLIOCENE
		Major unconformity				
		Deltaic — mostly sand, some silt	300	EOCENE	EOCENE	EOCENE
		Marine — mostly clay, some sand	150	PALEOCENE	PALEOCENE	PALEOCENE
		Unconformity				
MESOZOIC	CRETACEOUS	Deltaic and nearshore marine — sand, some silt and clay, locally lignitic	500	GULFIAN	GULFIAN	GULFIAN
		Major unconformity				
PALEOZOIC	PENNSYLVANIAN	Marine, deltaic, continental — cyclical deposits, mostly shale, sandstone, and siltstone with some limestone, coal, clay, black sheety shale; sandstone dominant in lower part, shale above; coal most prominent in middle part, limestone in upper part	3000	VIRGILIAN MISSOURIAN DESMOINESIAN ATOKAN MORROWAN	CONEMAUGHIAN ALLEGHENIAN POTTSVILLIAN	UPPER MIDDLE LOWER
		Major unconformity				
	MISSISSIPPIAN	Marine, deltaic — cyclical deposits of limestone, sandstone, shale	1400	CHESTERIAN	CHESTERIAN VALMEYERAN KINDERHOOKIAN	CHESTERIAN MERAMECIAN OSAGIAN KINDERHOOKIAN
		Marine, deltaic — limestone, siltstone, shale, chert, sandstone	2000	VALMEYERAN		
		Marine — shale, limestone, siltstone	150			
	DEVONIAN	Marine — shale, limestone	300	UPPER	SENECAN AND CHAUTAUQUAN ERIAN ULSTERIAN	UPPER MIDDLE LOWER
		Unconformity				
		Marine — largely limestone, some shale	450	MIDDLE		
		Major unconformity				
		Marine — cherty limestone, chert	1300	LOWER		
	SILURIAN	Marine—shale, siltstone, limestone	100	CAYUGAN	CAYUGAN NIAGARAN ALEXANDRIAN	CAYUGAN NIAGARAN ALEXANDRIAN
		Marine — dolomite, limestone, shale, local reefs	1000	NIAGARAN		
		Marine — dolomite, limestone, shale	150	ALEXANDRIAN		
		Unconformity				
	ORDOVICIAN	Marine—shale, limestone, siltstone, dolomite	300	CINCINNATIAN	CINCINNATIAN CHAMPLAINIAN CANADIAN	CINCINNATIAN CHAMPLAINIAN CANADIAN
		Unconformity				
		Marine — limestone, dolomite, sandstone	1400	CHAMPLAINIAN		
		Major unconformity				
		Marine — dolomite, sandstone	1000	CANADIAN		
	CAMBRIAN	Marine—sandstone, dolomite, shale	4000	CROIXAN	ST. CROIXAN	ST. CROIXAN
		Major unconformity				
PRECAMBRIAN		Intrusive igneous rocks — mostly granite				

Figure 5—Generalized stratigraphic column of the Illinois Basin. From Willman et al, 1975. Used with the permission of the Illinois State Geological Survey.

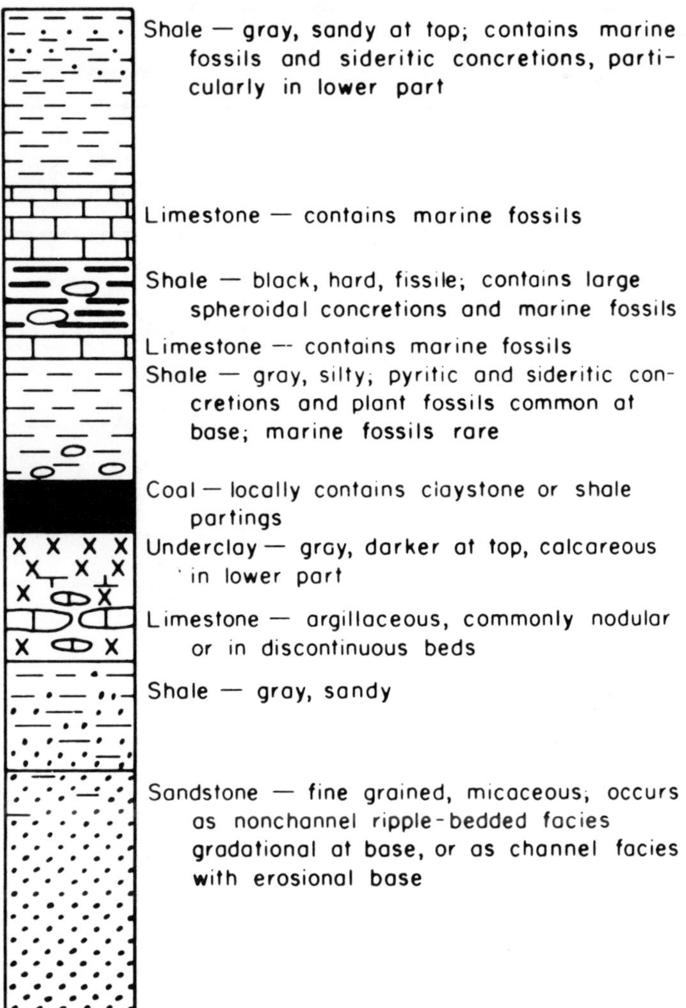

Figure 6—Sequence of lithologic units in a cyclothem. Willman et al, 1975. Used with the permission of the Illinois State Geological Survey.

ILLINOIS		INDIANA		KENTUCKY	
GROUP	FORMATION	GROUP	FORMATION	GROUP	FORMATION
MCLEANSBORO	MATTOON	MCLEANSBORO	MATTOON	MCLEANSBORO	(HENSHAW)
	BOND		BOND		STURGIS
	MODESTO		PATOKA		
			SHELBURN		(LISMAN)
KEWANEE	CARBONDALE	CARBONDALE	DUGGER		CARBONDALE
			PETERSBURG		
			LINTON		
	SPOON	RACCOON CREEK	STAUNTON		TRADEWATER
			BRAZIL		
MCCORMICK	ABBOTT		MANSFIELD		
	CASEYVILLE				CASEYVILLE

Figure 7—Stratigraphic nomenclature of the Pennsylvanian system by state. Used with the permission of the Illinois State Geological Survey.

the coal seam of interest. All Illinois Basin coal seams are covered by less than 3,000 ft of overburden; the major coals are everywhere within 1,500 ft of the surface.

Illinois Basin coals have been mined by both underground and surface methods. The majority of the mining activities are near the basin's perimeter, where the coal crops out or is shallow, as in the southeastern portion of the basin, where local vertical structural displacements have brought the coal to the surface. Estimated resources (those coals 18 or more inches thick and 1,000 or less ft in depth) on a county basis are shown in Figure 13. As shown, the southeastern and west-central basin portions have the greatest accumulation of remaining coal resources. Certain portions of the basin (Fairfield Basin and Moorman Syncline) have significant coal reserves at depths exceeding 1,000 ft.

The Herrin (No. 6) and the Springfield-Harrisburg (No. 5) coals are the most commercially important in Illinois. The Springfield (V) coal is the most important commercial coal in Indiana. The Nos. 11, 9, and 6 coals are commercially most important in the western Kentucky coal field. Total coal resources by seam on a state basis are given in Table 1. The U.S. Geological Survey (USGS) has estimated that the total in-place resource (hypothetical and identified) is 365 billion tons (Averitt, 1975).

Rocks of the Pennsylvania system have been classified in three groups (the McCormick, the Kewanee, and the McLeansboro) by the Illinois State Geological Survey (Fig. 7). Available information concerning these groups varies significantly and is reflected in the discussions below. For simplicity and continuity, the names of coal members discussed in the following sections are those used by the Illinois State Geological Survey, and stratigraphically equivalent coals in Indiana and Kentucky are discussed under the same names. Figures 7 and 12 serve as keys to these sections of the basin report. The coal seams discussed below are marked with an asterisk(*) in Figure 12.

McCormick Group Coals

The McCormick Group coals are the lowest in the section, comprised of coals in the Caseyville and Abbott Formations in Illinois and their correlatives in the Mansfield and Brazil Formations in Indiana and the Caseyville and Tradewater Formations in Kentucky (Fig. 12).

Numerous coal members have been identified in the McCormick Group sediments, the lowest members being characterized by:

- Lack of lateral persistence
- Greatly varying thickness within seams (few inches to 4 ft)

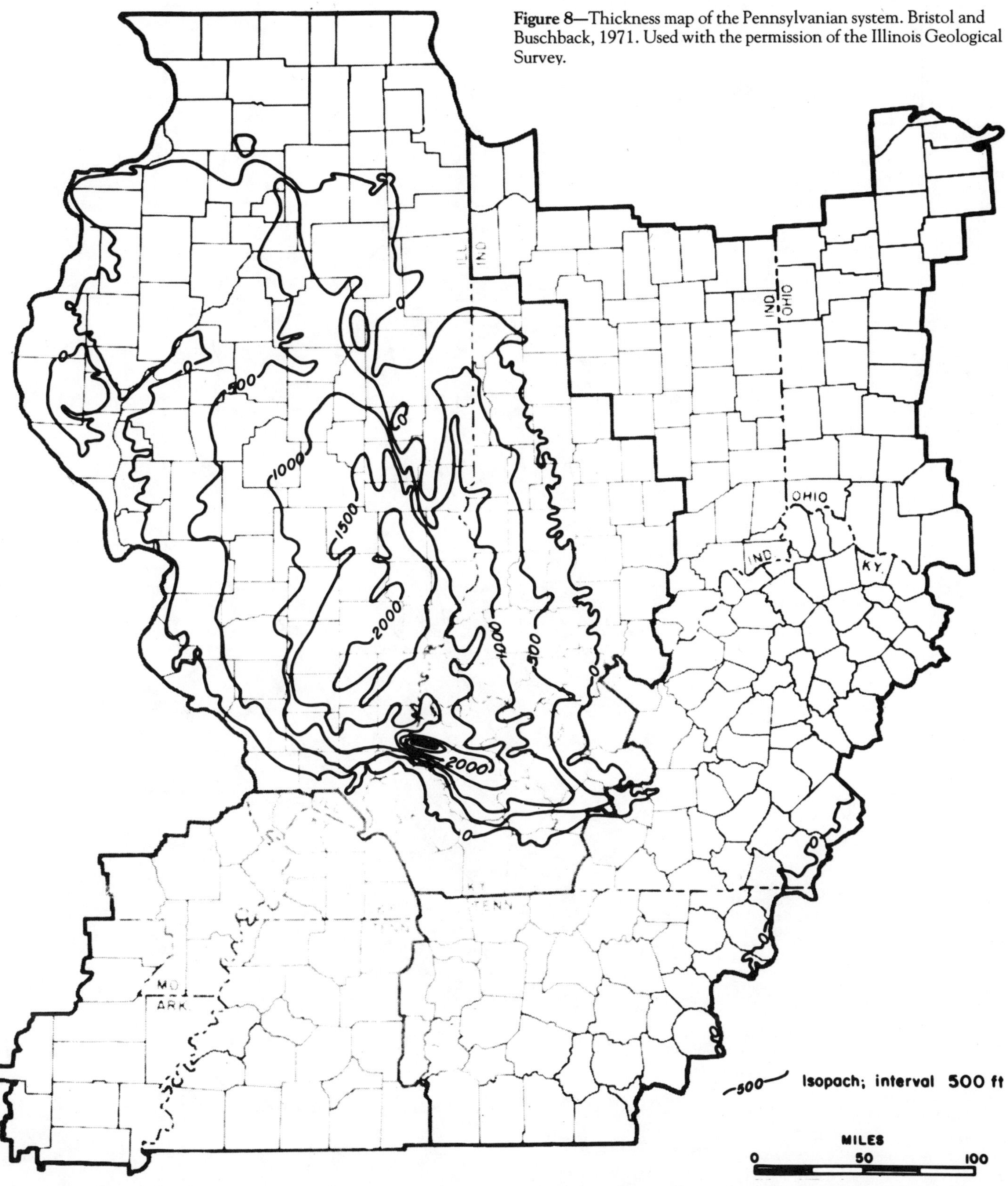

Figure 8—Thickness map of the Pennsylvanian system. Bristol and Buschback, 1971. Used with the permission of the Illinois Geological Survey.

- Occurrence restricted to Indiana, Kentucky, and southeastern Illinois
- Mining on a local basis
- High moisture and ash but low sulfur content.

Higher in the McCormick section, the coals are more widespread and generally thicker. However, these coals (the Willis and the Lower and Upper Block members) are still quite areally restricted and require detailed mapping before mining or other commercial utilization can be attempted. The Lower and Upper Block coals are distinct in that a semisplint variety of bituminous coal splits along two well-developed slip patterns at 90° to each other, oriented at approximately N 20° W and N

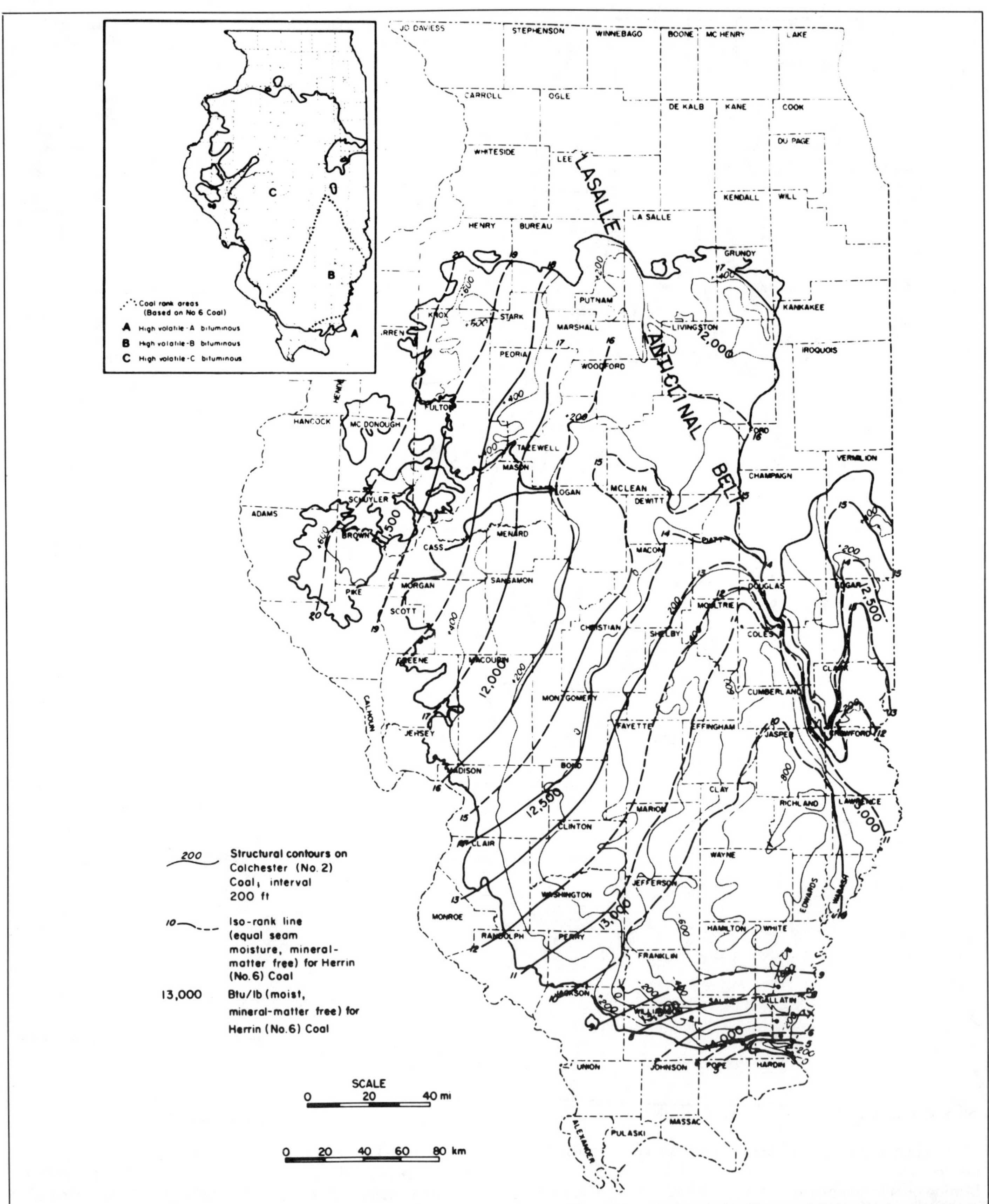

Figure 9—Coalification pattern in Illinois shows progressive increase in rank in the Herrin coal southward.

70° E. These coals are dull, banded, low in sulfur, and the Upper Block has a low ash content. Many of the McCormick Group coals outcrop in the extreme eastern portions of the basin and are locally strippable.

Spoon Formation Coals of the Kewanee Group

The Spoon Formation coals of Illinois are part of the lower Kewanee Group. The uppermost Brazil and all of the Staunton Formation in Indiana correlate with the Spoon Formation, as do the upper part of the Tradewater Formation and the lower part of the Carbondale Formation in western Kentucky. These formations contain the oldest widespread coalbeds in the basin; the coals are thicker and more extensive than coals of the McCormick Group and thinner than those of younger Pennsylvanian rocks of the Illinois Basin.

Rock Island (No. 1) Coal Member

The Rock Island coal is the basal member of the Spoon Formation and is the oldest coal in western Illinois. It occupies a series of linear troughs up to 4 mi wide, trending east-west and northeast-southwest in western Illinois. The coal is typically 4 ft thick but varies from 8 ft to a few inches at its margins. The roof rock of the coal is typically a fossiliferous limestone (the Seville), and the floor rock is composed of the Bernadotte Sandstone. Wanless et al (1969) report that the Rock Island was deposited in estuarine valleys prior to their inundation by the sea. It is absent over large areas in Illinois but has been correlated with the Minshall coal of Indiana and the Mannington No. 4 coal of Kentucky. The Mannington, or Mining City No. 4 coal, is the most extensive of the major coals in western Kentucky. Throughout, it is at least 14 inches thick: typically 42 to 56 inches thick in large areas of Hopkins and Muhlenburg counties and up to 70 inches thick near Greenville, Kentucky. The No. 4 coal is generally 200 to 250 ft below the Kentucky No. 6 coal. The following is a summary of selected analyses of these coals:

	Rock Island No. 1 (Illinois)	Minshall (Indiana)	No. 4 (Kentucky)
Moisture (%)	14–18	11	10
Volatile Matter (%)	35–40	41	36
Fixed Carbon (%)	36–42	37–43	48
Ash (%)	7–10	11	5.8
Sulfur (%)	3–6	3–4	2.4
Btu/lb	10,400–11,200	11,100–11,400	12,300

Davis Coal Member

The Davis coal member lies approximately 200 ft above the Rock Island coal, averages 4 ft thick in southern Illinois, and thins to the north and west. It is called the No. 6 coal in Kentucky, where it thickens to 56 inches, and is present over nearly all of the western Kentucky coal field region and parts of southwestern Indiana. Much of the thicker coal (42 to 56 inches) is found in Union and Henderson counties of Kentucky. The roof is typically a black marine shale overlain by limestone. The No. 6 coal varies from 265 to 230 ft below the Kentucky No. 9 coal. The Davis coal is generally too deep for strip mining but has been mined locally by surface methods. Although relatively little is known about the chemical character of the No. 6 coal in Kentucky, it apparently has a low sulfur content.

Selected analyses show the following:

	Davis (Illinois)	No. 6 (Kentucky)
Moisture (%)	5–7	5–12
Volatile Matter (%)	35–37	36–43
Fixed Carbon (%)	46–48	49–50
Ash (%)	8–10	8–13
Sulfur (%)	3–4	2–3
Btu/lb	12,500–12,800	11,100–12,800

De Koven Coal Member

The De Koven coal member is present in southernmost Illinois, Indiana, and Kentucky. It occurs from a few inches to 40 ft above the Davis and averages from 3 to 3.5 ft thick in Williamson, Saline, and Gallatin counties in Illinois. To the north and west, the coal is either absent or thin. Its easternmost extent is in Union and northwestern Henderson counties of Kentucky. Analysis of the coal is shown below:

	De Koven (Saline Co., Illinois)
Moisture (%)	5–7
Volatile Matter (%)	35–37
Fixed Carbon (%)	46–48
Ash (%)	8–13
Sulfur (%)	3–5
Btu/lb	11,900–12,700

Seelyville Coal Member

The Seelyville coal member is among the uppermost coal members of the Spoon Formation and is an important minable coalbed in the eastern part of the Illinois Basin. Its occurrence in Indiana is widespread, averaging 6 ft in thickness and increasing to as much as 10 ft. The Seelyville, or Indiana coal III, was 6.7 and 11.5 ft thick in two MRCP wells drilled in Posey County, Indiana. The latter is believed to be the thickest Seelyville coal ever measured in the Illinois Basin. It also occurs up to 8 ft thick in a 1,200-sq-mi area in Edgar, Clark, Crawford, Cumberland, Jasper, and Lawrence counties of Illinois. The Seelyville is a highly banded coal that in some places has numerous shale partings that result in locally high ash contents. An analysis of the coal is shown below:

	Seelyville (Illinois)
Moisture (%)	11
Volatile Matter (%)	36–40
Fixed Carbon (%)	38–39
Ash (%)	11–15
Sulfur (%)	3–6
Btu/lb	10,500–11,100

Carbondale Formation Coals

The Carbondale Formation, from the basal Colchester (No. 2) coal member through the Danville (No. 7) coal member,

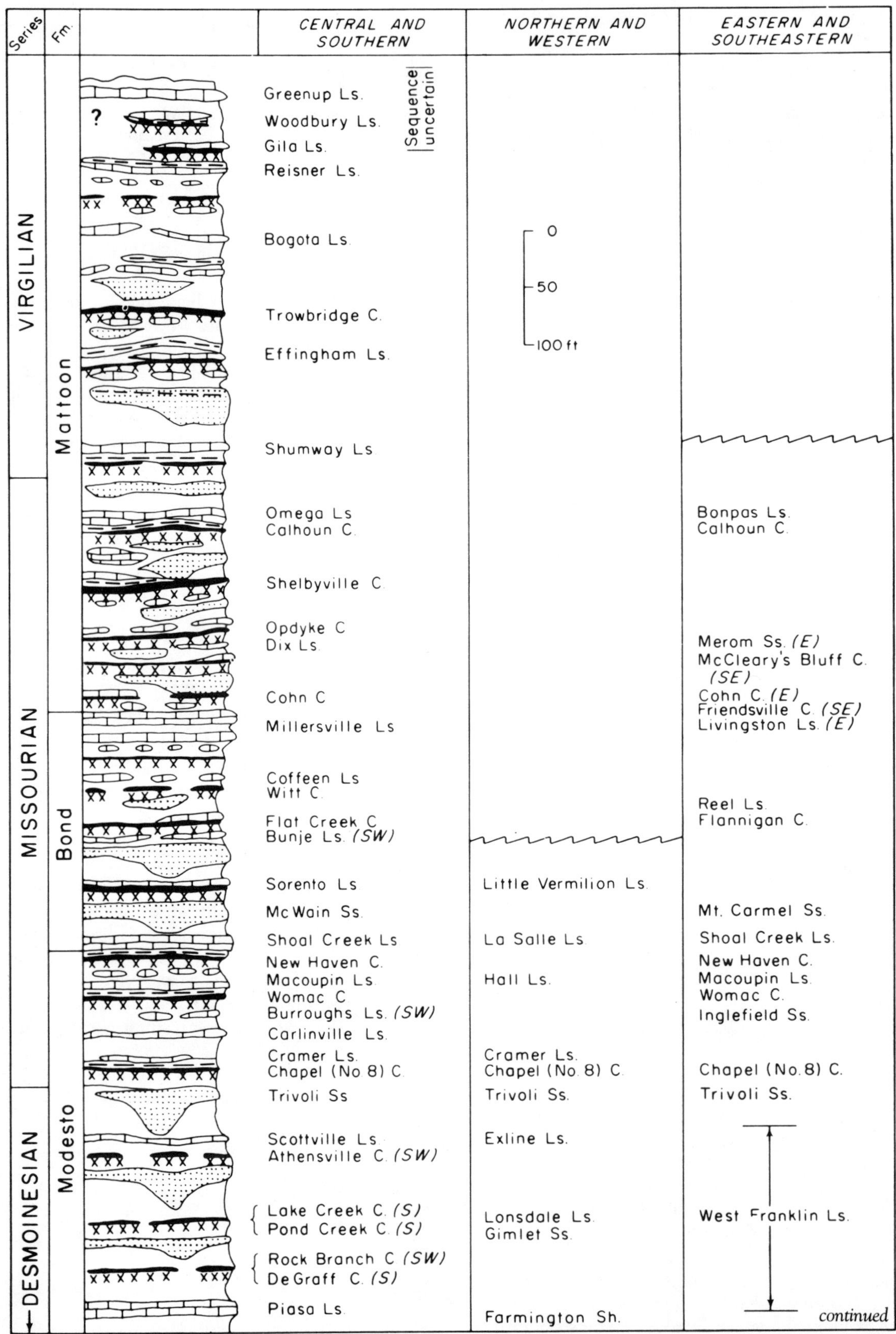

continued

Figure 10—Generalized stratigraphic column of rock units in the Pennsylvanian system as named by the Illinois Geological Survey. Willman et al, 1975. Used with the permission of the Illinois State Geological Survey.

Kewanee Group

Series	Fm.	SOUTHWESTERN AND SOUTHEASTERN	NORTHERN AND WESTERN	EASTERN
DESMOINESIAN	Carbondale	Danville (No.7) C.	Danville (No.7) C.	Danville (No.7) C.
		Galum Ls.		
		Allenby C.		
		Bankston Fork Ls.		Bankston Fork Ls.
		Anvil Rock Ss.	Copperas Creek Ss.	
			Lawson Sh.	
		Conant Ls.		Conant Ls.
		Jamestown C.		Jamestown C.
		Brereton Ls.	Brereton Ls.	Brereton Ls.
		Anna Sh.	Anna Sh.	Anna Sh.
		Herrin (No.6) C.	Herrin (No.6) C.	Herrin (No. 6) C.
			Spring Lake C.	
			Big Creek Sh.	
			Vermilionville Ss.	
		Briar Hill (No 5A) C.		Briar Hill (No. 5A) C.
		Canton Sh.	Canton Sh.	Canton Sh.
		St. David Ls.	St. David Ls.	St. David Ls.
		Dykersburg Sh.		
		Harrisburg (No 5) C.	Springfield (No.5) C.	Harrisburg (No.5) C.
			Covel Cgl.	
		Hanover Ls.	Hanover Ls.	
		Excello Sh.	Excello Sh.	Excello Sh.
		Summum (No.4) C.	Summum (No.4) C.	Summum (No.4) C.
			Breezy Hill Ls.	
		Roodhouse C.	Kerton Creek C.	
		Pleasantview Ss.	Pleasantview Ss.	Pleasantview Ss.
			Purington Sh.	
		Shawneetown C.	Lowell C.	Shawneetown C.
		Oak Grove Ls.	Oak Grove Ls.	
		Mecca Quarry Sh.	Mecca Quarry Sh.	Mecca Quarry Sh.
			Jake Creek Ss.	
			Francis Creek Sh.	
			Cardiff C.	
		Colchester (No 2) C	Colchester (No.2) C.	Colchester (No.2) C.
	Spoon		Browning Ss.	
			Abingdon C.	
		Palzo Ss.	Isabel Ss.	
		Seelyville C.		Seelyville C.
		De Koven C.	Greenbush C.	
		Davis C.	Wiley C.	
		Seahorne Ls.	Seahorne Ls.	
		Vergennes Ss.		
		Stonefort Ls.		
		Wise Ridge C.		
		Mt. Rorah C.	De Long C.	
		Creal Springs		
		Murphysboro C.		
		Granger Ss.		
		New Burnside C.	Brush C.	
		Bidwell, O'Nan C.		
			Hermon C.	
		Curlew Ls.	Seville Ls.	
		Litchfield, Assumption C.	Rock Island (No.1) C.	

Cheltenham Clay (Browning Ss. through Brush C.)

0 — 50 — 100 ft.

Rocks present only in subsurface.
Local oil-field names have been used informally.

continued

Figure 10—(Continued)

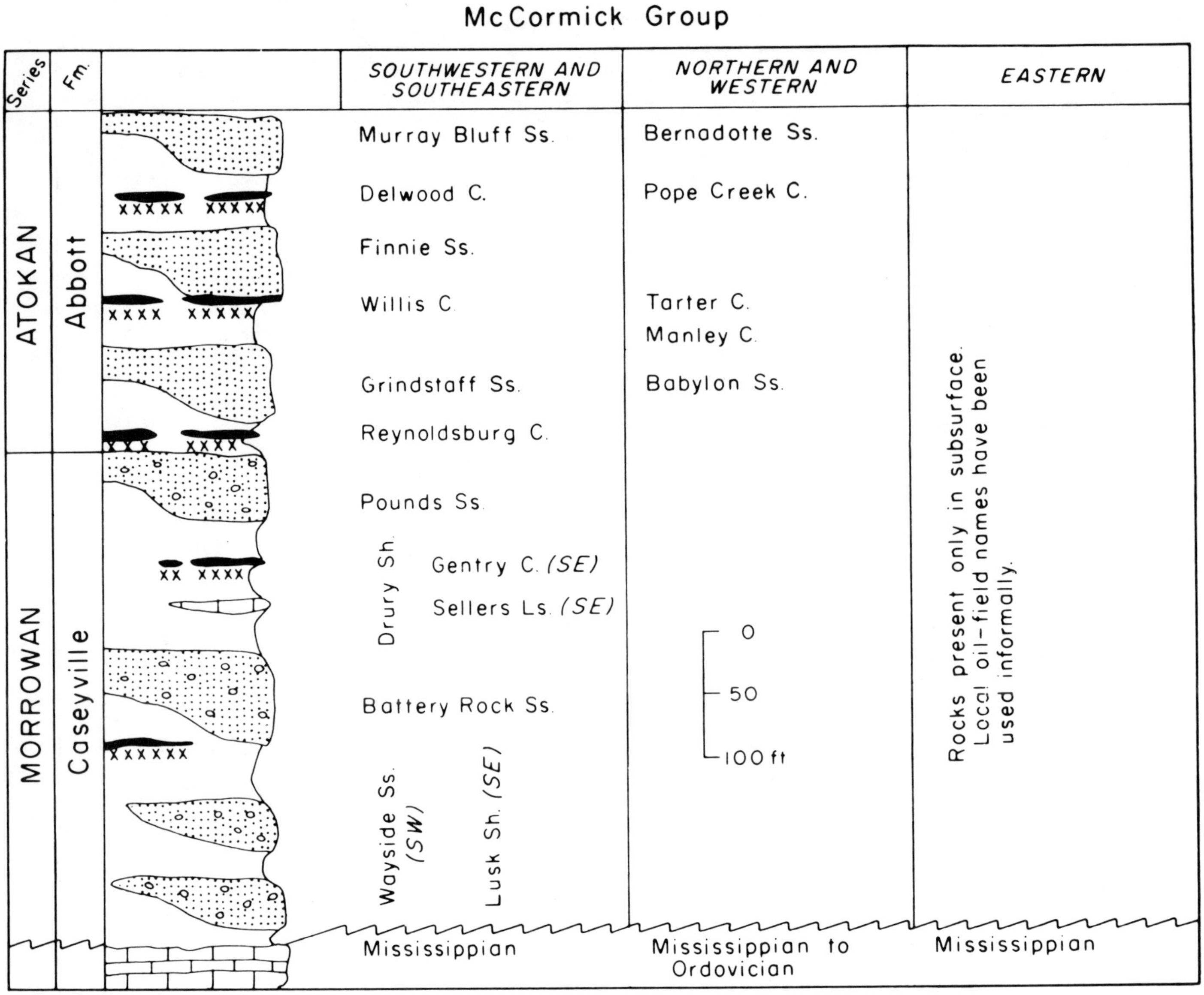

McCormick Group

Series	Fm.	SOUTHWESTERN AND SOUTHEASTERN	NORTHERN AND WESTERN	EASTERN
ATOKAN	Abbott	Murray Bluff Ss.	Bernadotte Ss.	Rocks present only in subsurface. Local oil-field names have been used informally.
		Delwood C.	Pope Creek C.	
		Finnie Ss.		
		Willis C.	Tarter C.	
			Manley C.	
		Grindstaff Ss.	Babylon Ss.	
		Reynoldsburg C.		
MORROWAN	Caseyville	Pounds Ss.		
		Drury Sh.: Gentry C. *(SE)*		
		Drury Sh.: Sellers Ls. *(SE)*		
		Battery Rock Ss.		
		Wayside Ss. *(SW)*; Lusk Sh. *(SE)*		
		Mississippian	Mississippian to Ordovician	Mississippian

Scale: 0 — 50 — 100 ft

Figure 10—(Continued)

includes the principal economic coals of the Illinois Basin. The Carbondale Formation in Illinois correlates with the Linton, Petersburg, and Dugger Formations, or approximately the Carbondale Group in Indiana; and with the upper portion of the Carbondale Formation and the lower part of the Sturgis (Lisman) Formation in Kentucky (see Fig. 7). The coals of these formations are the thickest and most widespread of any coals in the Pennsylvanian section in the Illinois Basin.

Colchester (No. 2) Coal Member

The Colchester coal (No. 2) is the lowest coal member of the Carbondale Formation in Illinois and correlates with the Colchester (IIIa) coal in Indiana and the Schultztown coal in Kentucky. This coal is believed to be one of the most widespread coalbeds of the Illinois Basin and the United States. It is absent in less than 5% of its outcrop area in the Illinois Basin and has been correlated with coals in Iowa, Missouri, Oklahoma, and Pennsylvania. The Colchester (No. 2) is thickest (about 3.5 ft) in the northern and western parts of the basin. It uniformly thins to a few inches in southern Illinois, Indiana, and western Kentucky, where it is useful as a marker bed for correlation and mapping. The roof rock is a marine black shale, except in parts of western and most of northern Illinois, where the overlying strata are gray shales. The floor rock is a well-developed underclay. The range of typical analyses in northern and western Illinois is:

	Colchester (No. 2) (Illinois)
Moisture (%)	10–19
Volatile Matter (%)	31–45
Fixed Carbon (%)	35–48
Ash (%)	3–11
Sulfur (%)	1–5
Btu/lb	10,400–11,700

SYSTEM	SERIES		GROUP	FORMATION, MEMBER, AND BED		MISCELLANEOUS UNOFFICIAL NAMES CITED IN TEXT	
	APPALACHIAN	MIDCONTINENT					
PENNSYLVANIAN	Monongahelan	Virgilian	McLeansboro	Mattoon Fm.	Merom Ss. Mbr.		Grayville Ls. Mbr., McCleary's Bluff Coal Mbr., Friendsville Coal Mbr.
					Cohn Coal Mbr.		
		Missourian		Bond Fm.	Livingston Ls. Mbr.	Merom Gr.	
	Conemaughian				Riverview Ls. Mbr.		
					Fairbanks Coal Mbr.		
					St. Wendel Ss. Mbr.		
					Shoal Creek Ls. Mbr.		Hayden Branch Fm., New Haven Ls. Mbr., Parker(s) Ls.
				Patoka Fm.	Parker Coal Mbr.		
					Rabens Branch Mbr.		
					Dicksburg Hills Ss. Mbr.		Murphys Bluff Fm.
					Vigo Ls. Mbr.		
					Hazelton Bridge Coal Mbr.		
					Inglefield Ss. Mbr.		
					Ditney Coal Mbr.		
		? Desmoinesian		Shelburn Fm.	West Franklin Ls. Mbr.		Maria Creek Ls., Somerville Fm., Ls.
					Pirtle Coal Mbr.	Millersburg Fm.	Coal VIIa
					Busseron Ss. Mbr.		
	Alleghenian		Carbondale	Dugger Fm.	Danville Coal Mbr. (VII)		Coal VII, Little Newburg Coal, Millersburg Coal, Upper Millersburg Coal
					Universal Ls. Mbr.		
					Hymera Coal Mbr. (VI)		Coal VI, Lower Millersburg Coal
					Providence Ls. Mbr.		Main Newburg Ls.
					Herrin Coal Mbr.		
					Bucktown Coal Mbr. (Vb)		Coal Vb
					Antioch Ls. Mbr.		Upper Alum Cave Ls.; Coal Va
					Alum Cave Ls. Mbr.		Arthur Ls.
				Petersburg Fm.	Springfield Coal Mbr. (V)		Coal V, Coal VII, coal at Alum Cave, Main Newburg Coal, Petersburg Coal
					Stendal Ls. Mbr.		Houchin Creek cap. Ls.
					Houchin Creek Coal Mbr. (IVa)		Coal IVa
				Linton Fm.	Survant Coal Mbr. (IV)		Coal IV
					Velpen Ls. Mbr.		Mecca Quarry Sh. Mbr.
					Colchester Coal Mbr. (IIIa)		Coal IIIa, Velpen Coal
					Coxville Ss. Mbr.		
			Raccoon Creek	Staunton Fm.	Seelyville Coal Mbr. (III)		Coal III, Coal VI, Lower Hanging Rock Coal, Rock Creek Coal, Staunton Coal
					Silverwood Ls. Mbr.	Silverwood Cyclothem	
					Holland Ls. Mbr.		Upper Huntingburg Chert; Holland Coal; Coal II
	Pottsvillian				Perth Ls. Mbr.		Minshall Ls.
		? Atokan		Brazil Fm.	Minshall and Buffaloville Coal Mbrs.		
					Upper Block Coal Mbr.		
					Lower Block Coal Mbr.		
				Mansfield Fm.	Shady Lane Coal Mbr.	Cannelton, Bloomfield and Shoals Lithofacies; Coal I, Kirksville Coal; Shoals Coal	Coal I, Peacock coal
		? Morrowan			Ferdinand Ls. Mbr.		Grandview Ls.
					Fulda Ls. Mbr.		
					Mariah Hill Coal Bed		Lower Huntingburg Coal, Upper Mariah Hill Coal
					Blue Creek Coal Mbr.		
					Pinnick Coal Mbr.		Cannelton Coal, Ss.; Lower and Upper Cannelton Coals; Troy and Upper Troy Coals
					St. Meinrad Coal Bed		
					French Lick Coal Mbr.		Hindostan Whetstone Beds; Coal I

Figure 11—Generalized stratigraphic column of rock units in the Pennsylvanian system as named by the Indiana Geological Survey.

Illinois		Indiana		W. Kentucky	
Modest Fm.		Shelburn Fm.		Sturgis Fm.	No. 14 Coiltown No. 13 Baker No. 12 Paradise
Carbondale Fm.	Danville (No. 7)*	Dugger Fm.	Danville (VII)		
	Jamestown*		Hymera (VI)		
	Herrin (No. 6)*		Herrin	Carbondale Fm.	No. 11 Herrin
	Briar Hill (No. 5A)*	Petersburg Fm.	Coal (Va)		No. 10 Briar Hill
	Springfield-Harrisburg (No. 5)*		Springfield (V)		No. 9 Mulford
					No. 8b Upper Well
	Summum (No. 4)*	Linton Fm.	Houchin Creek (IVa)		No. 8 Well
	Shawneetown Coal* (Lowell)		Survant (IV)		
	Colchester (No. 2)*		Colchester (IIIa)		Schultztown
Spoon Fm.	Seelyville*	Staunton Fm.	Seelyville (III)		
	DeKoven*				No. 7 DeKoven
	Davis*				No. 6 Davis
			Buffaloville	Tradewater Fm.	
	Murphysboro	Brazil Fm.	Minshall		
	Rock Island (No. 1)*				No. 4 Mining City Mannington
Abbott Fm.	Willis		Upper Block Lower Block		
		Mansfield Fm.	Mariah Hill Blue Creek		
			St. Meinrad		No. 1b Bell
	Reynoldsburg				
Caseyville Fm.	Gentry			Caseyville Fm.	Main Nolin
			Pinnick French Lick		

* Coal seams discussed in text.

Figure 12—Stratigraphic correlation of principal coal seams in the Illinois Basin. From Malhotra, 1977. Used with the permission of the Illinois State Geological Survey.

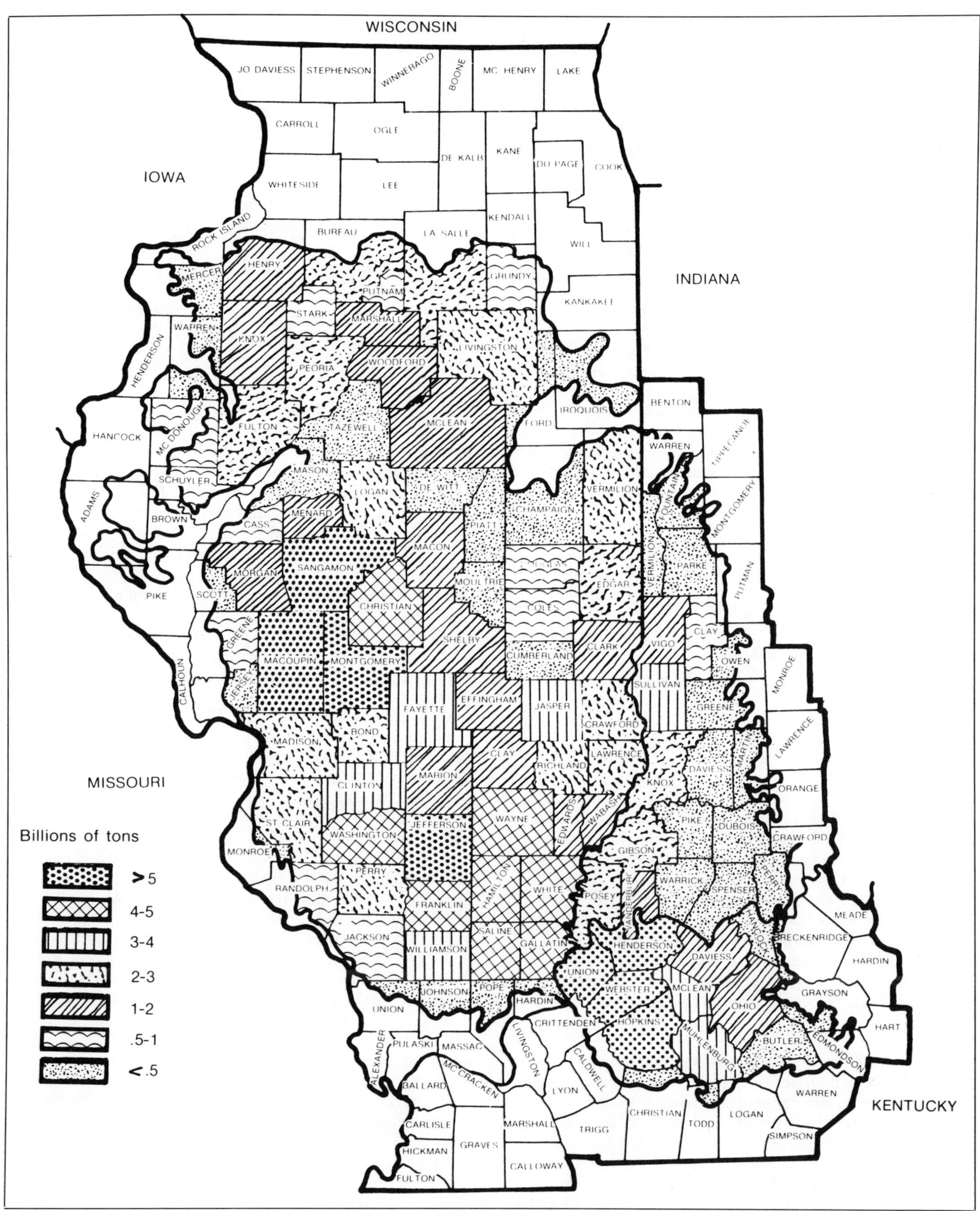

Figure 13—Remaining coal resources of the Illinois Basin by county. Data from Illinois, Indiana, and Kentucky Geological Surveys.

Shawneetown Coal Member

The Shawneetown coal member is a thin persistent coal occurring in southern and eastern Illinois. It is a relatively unimportant coal in Illinois, though it has been locally reported to occur as thick as 8 ft. It is correlated with the Survant coal (IV) (an extensively mined coal in Indiana) and the No. 8 coal (a relatively unimportant unit in Kentucky). The Shawneetown is overlain by a dark shale and underlain by underclay. A shale parting divides the Survant coal into two benches. The upper bench is somewhat blocky, and the lower bench is more friable. Where mined, the Survant coal averages 4 ft in thickness, if the shale parting is thin and the sulfur and ash contents are relatively low.

Summum (No. 4) Coal Member

The Summum coal is a laterally persistent coal member throughout Illinois. It can be correlated with the Houchin Creek (IVa) coal of Indiana and the No. 8b coal of western Kentucky. The coal reaches a maximum thickness of 1 to 2 ft in the southern and eastern portions of the basin. It is overlain by a black shale and underlain by underclay. It normally lies within 25 ft of the overlying Springfield-Harrisburg (No. 5) and 50 ft of the underlying Shawneetown coal. Analyses of this coal in northwestern Illinois show:

	Summum (No. 4) (Illinois)
Moisture (%)	14–16
Volatile Matter (%)	37–39
Fixed Carbon (%)	38–40
Ash (%)	7–9
Sulfur (%)	3–4
Btu/lb	10,800–11,300

Springfield-Harrisburg (No. 5) Coal Members

The Springfield-Harrisburg coal members are the same unit, but the name varies with location in Illinois. They are known as the Springfield (V) in Indiana, and the No. 9 in Kentucky. Together, they are commercially the most important coals in the Illinois Basin. The Springfield-Harrisburg is present everywhere in the basin except the extreme northern portions and a few localities where eroded or not deposited. Typical thicknesses of the coals are 4 to 8 ft in Illinois, 4 to 7 ft in Indiana, and 5 ft in Kentucky. It thins northward at a rate of 1 inch per mi in western Illinois and is known to be as thick as 13 ft in Clay County, Indiana. The floor rock is underclay, and the coals are usually overlain by a black fissile shale less than 3 ft thick, which in turn is overlain by marine limestone generally less than 2 ft thick. Where the coal is thick enough, it may be distinguished by a positive resistivity response on electric logs bracketed between two negative peaks corresponding to the roof and floor rock. The thickness and quality of the coal are largely controlled by its proximity to a major channel that extends toward the northeast direction from Saline County, Illinois, to Pike County, Indiana. Coal thickens approaching the channel but may thin abruptly. It is commonly split by shale partings and is overlain by the gray silty Dykersburg Shale. Coal that is overlain by 20 ft or more of this shale has a low sulfur content (<3%). In addition, coal in parts of western Illinois, and Sangamon, Logan, and Menard counties is frequently cut by claystone dikes.

The following are typical analyses of the Springfield-Harrisburg coals across the basin. They show a systematic decrease in volatile matter and an increase in fixed-carbon percent and Btu/lb in a southeasterly direction across the basin.

	Harrisburg-Springfield (No. 5) (Illinois)	Springfield (V) (Indiana)	No. 9 (Kentucky)
Moisture (%)	4–18	8–10	5
Volatile Matter (%)	33–40	41	40
Fixed Carbon (%)	34–53	39–41	50
Ash (%)	8–12	9–11	11
Sulfur (%)	2–5	3–6	3
Btu/lb	10,100–12,700	11,600–11,700	12,900

Herrin (No. 6) Coal Member

The Herrin coal member is the most extensively mined coal in Illinois. It is correlated with the Herrin Coal Member in Indiana and the No. 11 coal in Kentucky. The Herrin, present throughout much of the Illinois Basin, is typically more than 6—and up to 15—ft thick, although it averages approximately 1 ft thicker than the Springfield-Harrisburg (No. 5) across the entire basin. It thins in parts of central and southeastern Illinois and is probably not greater than 5 ft thick in Indiana. The Du Quoin monocline apparently affected its deposition; the eastern boundary of thicker Herrin (No. 6) coal is adjacent to the Salem and Loudon anticlines. The Herrin (No. 6) coal is separated from the Springfield-Harrisburg (No. 5) coal by only 30 ft in the northern part of the basin and more than 130 ft in the southeastern portion of the basin. The Herrin is characterized by a persistent claystone (shale) parting of 1 to 3 inches in the lower part of the bed. The roof rock is typically composed of the black fissile Anna Shale (<2 ft thick), overlain by the marine Brereton limestone (3 to 4 ft thick). In any given location, either or both of these units may be missing or cut-out and replaced by a gray silty shale. Both contemporaneous and erosional channels cut through the Herrin (No. 6) coal in Illinois. The main erosional channel sand—the Anvil Sandstone—cuts an east-west path across Perry, Washington, Jefferson, Wayne, and Edwards counties and continues on a southern and southwesterly course through White and Gallatin counties. The presence of this channel deposit does not affect the thickness of the coal adjacent to it. The Walshville Channel trends in a north-south direction. It was deposited at the same time as the coal and is associated with relatively thick coal, coal split with partings, and gray roof shales representing crevasse splays, natural levees, and other nonmarine fluvial processes. Where these gray shales (the Energy Shale) are thick (>20 ft), the Herrin coal has a low-sulfur content. An area of highest quality Herrin coal in Franklin, Williamson, and Jefferson counties of Illinois has been extensively mined. The Fairfield basin area of southeastern Illinois probably has the largest reserves of the Herrin (No. 6) coal; a detailed report, including thickness and structure maps of the Herrin in this region, was compiled by Allgaier and Hopkins (1975). The Herrin coal is similar to the Springfield-Harrisburg coals in that there is a general systematic decrease in volatile matter and an increase in fixed-carbon and Btu/lb from north to south across the basin (see Fig. 9).

Analyses across Illinois show the following range (as-received basis):

	Illinois
Moisture (%)	3–31
Volatile Matter (%)	13–40
Fixed Carbon (%)	35–50
Ash (%)	7–13
Sulfur (%)	1–5
Btu/lb	9,700–13,700

Jamestown Coal Member

The Jamestown coal member is a widespread, thin, and relatively unimportant coalbed occurring in southern Illinois, correlating with the commercially important Hymera (VI) and No. 12 coals of Indiana and Kentucky, respectively. It averages 5 ft but occurs up to 8 ft thick in underground mines in Indiana. The coal varies greatly in thickness over the entire western Kentucky coal field. It is generally thin (less than 14 inches) near the Illinois boundary, in Union and Henderson counties, but thickens through central Hopkins, Webster, and northern Muhlenburg counties, where it may occur more than 70 inches thick.

In Indiana, the coal is stratigraphically close to the overlying Danville and is mined in conjunction with it. In Kentucky, the No. 12 coal is often mined in conjunction with the underlying No. 11 coal. Analyses show the Hymera coals to have ash content similar to the underlying Springfield coals but to be considerably lower in sulfur content. A typical range of analyses in Indiana is shown below:

	Hymera VI (Indiana)
Moisture (%)	8–13
Volatile Matter (%)	32–36
Fixed Carbon (%)	41
Ash (%)	10–18
Sulfur (%)	3–4
Btu/lb	10,400–11,100

Danville (No. 7) Coal Member

The Danville coal member is the uppermost member of the Carbondale Formation and the youngest commercial coalbed of the Illinois Basin. It is extensive but generally of minable thickness only in the northern portion of the basin and east of the LaSalle anticline. The Danville (No. 7) is known as the Danville (VII) in Indiana; its relationship to coals in western Kentucky is presently unknown. The coals are 2.5 to 6 ft thick where mined and generally a few inches to 3 ft thick in other areas.

The overlying strata are usually gray shales, and the underlying stratum is a thick (50 to 100 ft), uniform underclay. These coal horizons are quite easily found on geophysical logs, thus serving as excellent marker beds. In contrast with the underlying Hymera coal, the Danville has relatively few partings and is usually lower in ash content.

	Danville (No. 7) (Illinois)	Danville (VII) (Indiana)
Moisture (%)	13–19	11–13
Volatile Matter (%)	33–39	35
Fixed Carbon (%)	34–40	40–42
Ash (%)	9–15	11–12
Sulfur (%)	3–4	2
Btu/lb	9,600–11,300	10,300–10,900

McLeansboro Group Coals

The McLeansboro Group of the Upper Pennsylvanian sequence contains at least 15 coal horizons within the Illinois Basin. These coalbeds are generally quite thin (less than 1 ft), of extremely variable quality, and of limited or unknown lateral extent. Only rarely has a coalbed occurred in minable thickness, but one example is the Fairbanks coal in western Sullivan County, Indiana, where the coal has been mined along its outcrop. Coals of the McLeansboro Group are reported to have an average ash content of 28% and sulfur content of 2%.

Structural Character of the Coals

The structural character of coals (the fracture, cleat density, or orientation) has not been systematically investigated on a basin-wide scale. Comparison of local or county-wide studies is difficult, because there is no consensus among coal workers on definition of cleat and fracture or of what tectonic forces or conditions are responsible for producing them. From the available data, however, the following generalizations can be drawn:

- Major cleat orientation often is not apparent when investigations are made on a small scale (outcrop, small section of mine, etc.).
- No areas in the basin have been reported with higher densities of cleats or fractures relative to other areas.
- The southeastern portion of the basin is intensively faulted and folded in comparison with the remainder of the basin, suggesting the presence of at least local areas with high fracture density.
- Fractures in the coal typically are mineralized, thus having low permeabilities.

POTENTIAL METHANE RESOURCE

Overview

Generally, all coal contains methane as either free gas in fissures (cleats) or adsorbed gas on coal surfaces or in pore spaces. This gas is trapped in the coal during either breakdown of organic matter during coalification or migration of gases from other source rocks. Coals of the Illinois basin are not "gassy" relative to many coals in other parts of the country. Because of this, few tests of the gas content of coals in this region were made until recently.

Studies involving desorption of methane gas from coal in the Illinois Basin have been performed by the coal industry, the Illinois and Indiana State Geological Surveys, and the U.S. Bureau of Mines (USBM), and under the auspices of the

Table 1—Illinois Basin coal resource by seam (thousands of short tons).

Illinois		Indiana		Kentucky		Total
Danville (No. 7)	7,791,547	Danville VII	5,539,698	No. 14	1,242,932	14,574,177
Jamestown		Hymera VI	5,643,829	No. 12		5,643,829
Herrin (No. 6)	68,747,199	Herrin		No. 11	8,366,246*	77,113,445
Springfield-Harrisburg (No. 5)	50,620,791	Springfield V	7,432,426	No. 9	9,382,424	67,435,641
Shawneetown		Survant IV	2,986,880			2,986,880
Colchester (No. 2)	20,837,085	Colchester IIIa	4,270	Schultztown		20,841,355
Seelyville		Seelyville III	6,034,830			6,034,830
Davis	3,410,355	—		No. 6 (Davis)	7,474,404	10,884,759
Rock Island (No. 1)	1,575,194	Minshall	185,105	No. 4 (Mannington)	6,526,852	8,287,151
	152,982,171		27,827,038		32,992,858	213,802,067

Illinois Data:	Includes strippable, all coals indicated and hypothetical, 28″ thick, less than 1000′ (Illinois Geological Survey, oral communication)
Indiana Data:	Total tons (January 1, 1965) measured, indicated, and inferred, includes strippable (Indiana Geological Survey, 1979, written communication)
Kentucky Data:	Coal Resources of western Kentucky, includes strippable (Kentucky Geological Survey Open File Report)

*Includes Kentucky No. 12 resources.

Department of Energy's Methane Recovery from Coalbeds Project (MRCP). Desorption data from these tests are compiled in Table 2. Location of these data, by county, is shown on the map in Figure 14. The Herrin (No. 6) and the Springfield-Harrisburg (No. 5) coals and their correlatives in Indiana have been sampled most often. Very few data exist on the methane content of major coal seams such as the Colchester No. 2 coal of Illinois and the No. 6 Davis and Mannington No. 4 coals of Kentucky.

The estimated gas content of the Herrin (No. 6) *and equivalent coals* ranges from 32 to 125 cubic feet per ton (cf/ton). A similar range of values—32 to 147 cf/ton—has been estimated for the Springfield-Harrisburg (No. 5) coals. This variability cannot be directly related to sample depth alone (since the gassier coals were from the shallower horizons, contrary to what might be expected) but is related to other geologic controls, analytical errors, or the method used in determining the "remaining" (residual) gas. The graphical method of determining residual gas in coal samples (McCulloch et al, 1975) has been found by the USBM to be invalid. Aside from the problem of ascertaining whether a particular coal sample should be classified as blocky or friable, the graph used for estimating residual gas was constructed using data obtained by mechanically crushing samples in a "crushing box." Data generated through use of the crushing box are now disputed because of confirmed leakage of gas from the box. Thus, the data in Table 2 generated by use of the graphical method, denoted by (G), should be considered questionable and used with care. Although recent tests have been performed under generally uniform conditions, a major problem still exists in trying to recognize analytically bad data points and eliminating them, so that evaluation and estimates of gas contents are not biased.

A gross indication of "gassy" areas within the basin comes from mine ventilation records made available by the U.S. Department of Labor Mine Safety and Health Administration (MSHA). Methane gas emission data, by coal mine, are provided in Table 3. The quantity of gas released per ton of coal mined cannot be used directly to estimate the gas content of a coalbed because once mining has occurred, a significant portion of the methane emitted into a mine may come from adjacent strata. However, correlation can be made between the amount of gas emitted from coal mines and the gas content predicted

Table 2—Methane desorption data from the Illinois Basin.

Coalbed	State	County	Depth (ft)	Desorbed Gas[2] (cc/gm)	Remaining Gas[3] (cc/gm)	Total (cc/gm)	Gas Content (cf/ton)
Shelbyville (?)	Illinois	Coles	504	0.1	0.1 (BM)	0.2	8
Danville (No. 7)	″	Clay	994	0.9	0.4 (BM)	1.3[1]	40[4]
″	″	Coles	963	2.0	0.7 (BM)	2.7	87
″	″	Marion	664	0.7	0.1 (BM)	0.8	26[4]
Herrin (No. 6)	″	Clay	1,035	0.6	0.4 (BM)	1.0[1]	32[4]
″	″	Coles	1,067	1.0	0.5 (BM)	1.5	48
″	″	Franklin	850	1.7	—	1.7	53
″	″	Franklin	670	2.3	—	2.3	72
″	″	Franklin	670	2.2	—	2.2	69
″	″	Jefferson	733	1.8	0.1 (CB)	1.9	61
″	″	Marion	698	0.9	0.2 (BM)	1.1	35[4]
″	″	Wayne	900	1.2	0.7 (G)	1.9	61
″	″	Wayne	969	1.6	1.8 (G)	3.4	109
″	″	White	781	3.5	0.4 (BM)	3.9	125

continued

Table 2—(Continued)

Coalbed	State	County	Depth (ft)	Desorbed Gas[2] (cc/gm)	Remaining Gas[3] (cc/gm)	Total (cc/gm)	Gas Content (cf/ton)
Briar Hill (No. 5A)	"	Clay	1,075	0.5	0.5 (BM)	1.0	32[4]
"	"	Marion	727	0.4	0.3 (BM)	0.7	22[4]
Harrisburg (No. 5)	"	Clay	1,090	0.9	0.3 (BM)	1.2	38[4]
"	"	Coles	1,092	0.8	1.0 (BM)	1.8[1]	58
"	"	Franklin	916	1.2	—	1.2	38
"	"	Franklin	715	2.2	—	2.2	70
"	"	Franklin	733	1.9	—	1.9	62
"	"	Jefferson	793	0.8	0.2 (CB)	1.0	32
"	"	Marion	732	0.8	0.1 (BM)	0.9[1]	29[4]
"	"	Wayne	1,010	2.4	1.3 (G)	3.7	118
"	"	Wayne	1,066	1.4	1.3 (G)	2.7	86
"	"	White	908	2.4	0.5 (BM)	2.9	93
Colchester (No. 2)	"	Peoria	133	0.6	0.5 (G)	1.1	35
Seelyville	"	Clay	1,352	1.1	0.4 (BM)	1.5	48[4]
"	"	Wayne	1,287	1.3	0.7 (G)	2.0	64
"	"	Wayne	1,290	1.5	1.6 (G)	3.1	99
Danville (VII)	Indiana	Knox	339	1.8	1.5 (G)	3.3	106
"	"	Knox	413	2.2	1.4 (G)	3.6	116
"	"	Posey	467	1.0	0.1 (BM)	1.1	35
"	"	Posey	506	2.0	0.2 (BM)	2.2	70
"	"	Sullivan	145	0.7	0.2 (BM)	0.9	29
Herrin	"	Gibson	580	1.8	1.1 (G)	2.9	93
"	"	Posey	562	2.5	0.2 (BM)	2.7	87
Hymera (VI)	"	Knox	361	3.0	0.8 (G)	3.8	122
"	"	Knox	442	2.2	1.4 (G)	3.6	116
"	"	Sullivan	178	1.1	0.2 (BM)	1.3	42
Coal Va	"	Sullivan	238	1.8	0.3 (BM)	2.1	67
Coal Vb	"	Knox	522	1.8	1.2 (G)	3.0	96
Springfield (V)	"	Gibson	665	2.7	1.8 (G)	4.5	144
"	"	Knox	420	2.7	1.9 (G)	4.6	147
"	"	Knox	536	2.5	1.7 (G)	4.2	134
"	"	Posey	616	0.4	0.4 (BM)	0.8	26[4]
"	"	Posey	665	1.8	0.3 (BM)	2.1[1]	67
"	"	Sullivan	261	2.1	0.3 (BM)	2.4	77
Houchin Creek (IVa)	"	Posey	728	1.4	0.4 (BM)	1.8	58
"	"	Posey	772	1.9	0.5 (BM)	2.4	77
Survant (IV)	"	Knox	695	2.8	1.9 (G)	4.7	149
"	"	Posey	787	1.5	0.5 (BM)	2.1	67
"	"	Posey	827	2.8	0.4 (BM)	3.2	103
Colchester (IIIa)	"	Posey	907	3.2	0.7 (BM)	3.9[1]	124
Seelyville (III)	"	Gibson	943	1.3	0.9 (G)	2.2	70
"	"	Posey	879	0.3	0.5 (BM)	0.8	26
"	"	Posey	889	0.0	0.4 (BM)	0.4	13
"	"	Sullivan	430	2.2	0.3 (BM)	2.5	80
Coal #13	Kentucky	Webster	1,200	0.7	0.7 (BM)	1.4[1]	45[4]
Coal #9	"	Webster	1,305	0.6	0.8 (BM)	1.4[1]	45[4]

[1]Data is average of two or more samples.
[2]Desorbed gas includes estimated "lost" gas.
[3]Method of determination is indicated: (BM)—Gas released by crushing sample in ball mill.
(G) —Graphical method as in USBM R1 8043.
(CB)—Gas released in crusting box.
[4]MRCP data.

from the "direct method" of coal core desorption for large deep mines that have been producing for a few years up to levels of at least a few thousand tons a day (Kissell et al, 1973; McCulloch et al, 1975). In the Illinois Basin area, total gas emitted from "mature" mines during operation is typically on the order of four to seven times greater than the gas indicated to be present in the coal from USBM "direct method" determinations.

Additional information concerning the potential methane resource of coals comes from predrainage or degasification tests prior to coal mining operations. Again, the relatively low gas content of Illinois basin coals has precluded the requirement for these operations in this region. However, as mining activity

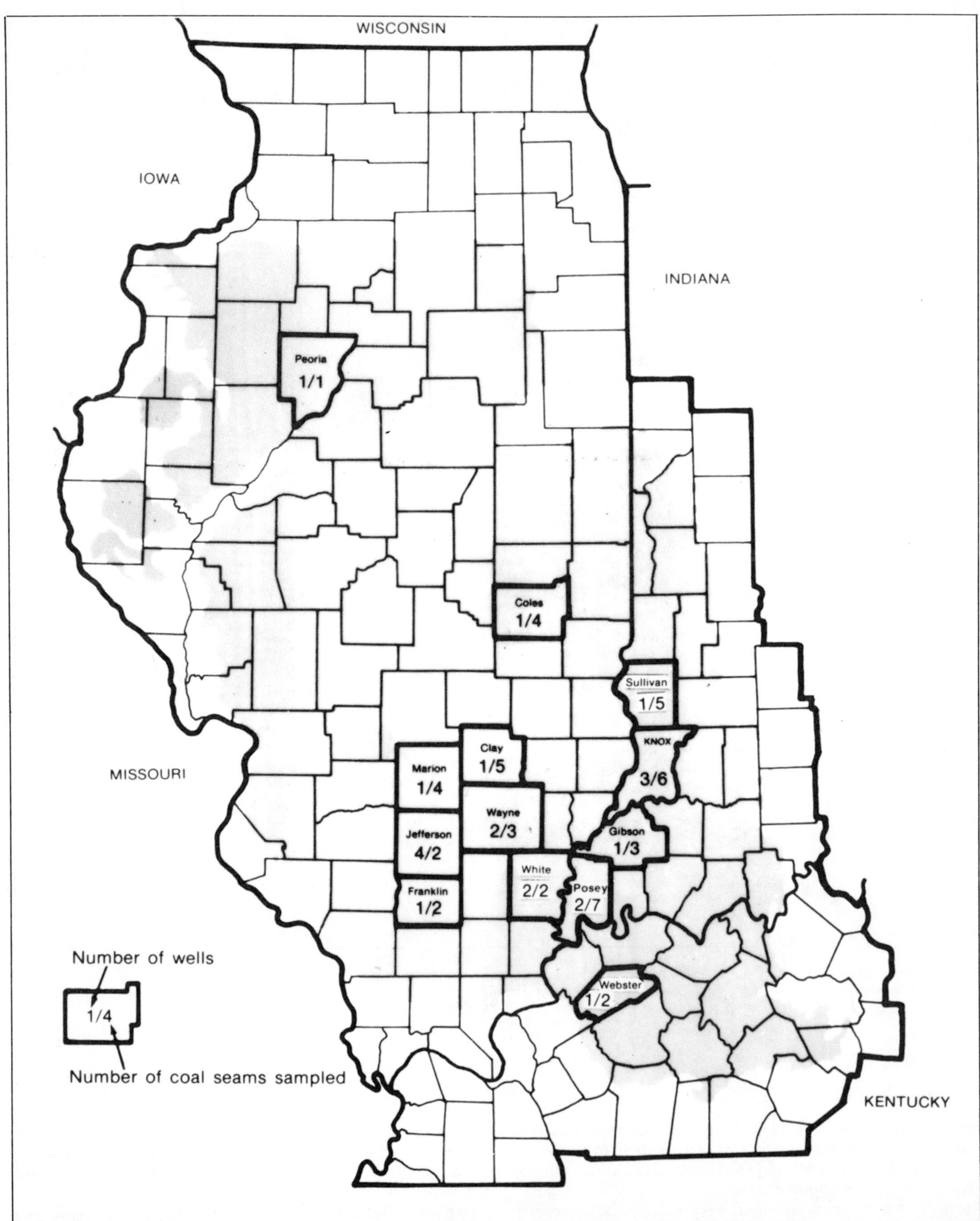

Figure 14—Illinois Basin map locating counties for which methane desorption data are available.

Table 3. Gas emission data from underground mines in the Illinois Basin.

Mine Name	Total Air Circulated (cfm)	Cubic Feet of Methane/ 24-hr Period	Daily Production (tons)	Coal Seam	Depth of Shaft/ Slope (ft)	Age of Mine	Size of Mine (No. of Units)
			Illinois				
4th Vein Mine	28,400	0	400	No. 4 Seam	200 (Drift)	9 months	1
J & R Mine	116,250	17,000	850	No. 6 Seam	130 (Drift)	3 years	2
Spur Underground	63,900	18,000	1,250	No. 5 Seam	120 (Drift)	6 years	2
R & H Mine	29,057	13,900	300	No. 5 Seam	90	12 years	1
Peabody No. 10	664,053	1,330,000	12,000	Illinois No. 6	350	29 years	10
Murdock Mine	206,240	316,000	5,500	Illinois No. 5	200	34.5 years	4
Hillsboro Mine	680,920	160,000	7,929	Illinois No. 6	540	16 years	11
Monterey No. 1	1,097,521	1,199,000	10,785	Illinois No. 6	300	10 years	11
Zeigler No. 5	283,300	615,000	5,500	Illinois No. 6	200	7 years	5
Crown No. 2	757,172	864,000	6,000	Illinois No. 6	331	4 years	12
Calefy Mine	34,000	4,900	300	Grape Creek No. 5	146 (Drift)	1.5 years	1
Wabash Mine	1,000,000	1,500,000	9,000	Illinois No. 5	850	7 years	13
Morris No. 5	153,000	22,000	1,700	Illinois No. 5	100 (Drift)	3 years	4
Brushy Creek	200,000	0	700	Illinois No. 6	300	6 months	1
Orient No. 3	825,000	1,500,000	7,200	Illinois No. 6	750	32 years	16
Orient No. 4	240,000	138,000	4,000	Illinois No. 6	350	25 years	6
Orient No. 6	640,000	1,600,000	3,000	Illinois No. 6	700	8 years	6
Old Ben 21	797,085	1,197,000	8,000	Illinois No. 6	750	22 years	13
Old Ben 24	870,000	1,250,000	6,130	Illinois No. 6	700	20 years	13
Old Ben 25	555,000	1,000,000	5,500	Illinois No. 6	750	2 years	10
Old Ben 26	827,560	1,230,000	7,612	Illinois No. 6	750	15 years	13
Old Ben 27	346,525	249,000	3,600	Illinois No. 6	750	1.5 years	10
Inland No. 1	665,000	1,000,000	8,300	Illinois No. 6	780	12 years	11
Inland No. 2	161,728	47,000	925	Illinois No. 5	950	1 year	5
Ziegler No. 4	240,000	275,000	1,500	Illinois No. 6	350	36 years	2
Sahara No. 20	230,000	200,000	1,800	Illinois No. 5	250	11 years	4
Sahara No. 21	386,000	278,000	2,000	Illinois No. 5	200	10 years	3
Sahara No. 22	50,000	0	400	Illinois No. 5	70 (Drift)	6 months	1
Eagle No. 2	400,000	600,000	4,900	Illinois No. 5	300	12 years	7
Baldwin Underground	470,000	183,000	10,000	Illinois No. 6	138	8 years	7
Monterey No. 2	790,000	466,000	13,400	Illinois No. 6	320	2 years	13
Marissa Mine	146,500	38,000	2,200	Illinois No. 6	160	1.5 years	2
River King Underground	436,760	129,000	8,000	Illinois No. 6	170	10 years	8
Spartan	170,000	52,000	1,900	Illinois No. 6	230	28 years	4
Zeigler No. 11	225,300	98,000	3,300	Illinois No. 6	220	3 years	6
			Western Kentucky				
Busick	145,000	2,100	2,500	Kentucky No. 9	200	4 years	1
Camp 1	503,300	288,000	7,000	Kentucky No. 9	400	10 years	41
Camp 2	353,260	305,000	6,500	Kentucky No. 9	400	10 years	47
Camp 11	235,000	34,000	2,100	Kentucky No. 11	270	3 years	16
Dotiki	360,900	416,000	6,000	Kentucky No. 9	800	13 years	69
Drake No. 4	238,000	343,000	2,500	Kentucky No. 14	300	6 years	137
Graham Hill	57,000	0	1,200	Kentucky No. 9	150	4 years	0
Hamilton No. 1	737,000	213,000	7,000	Kentucky No. 9	600	12 years	30
Hamilton No. 2	329,000	47,000	4,200	Kentucky No. 9	250	9 years	11
Ken No. 4	183,500	31,300	3,000	Kentucky No. 9	260	13 years	10
No. 2 Mine	247,000	214,000	3,000	Kentucky No. 9	325	12 years	71
No. 9 Mine	307,000	309,000	3,000	Kentucky No. 9	350	9 years	103
Pride Mine	160,000	92,000	2,400	Kentucky No. 6	600	3 years	38
Providence No. 1	265,000	191,000	5,000	Kentucky No. 9	450	12 years	38
Pyro No. 6	139,600	101,000	3,200	Kentucky No. 6	600	3 years	32
Pyro No. 11	115,000	153,000	1,400	Kentucky No. 11	300	3 years	109
Retiki Mine	103,800	45,000	2,500	Kentucky No. 9	230	?	18
River Queen UG	132,000	16,000	2,000	Kentucky No. 9	220	14 years	8
Sinclair UG Mine No. 2	330,000	285,000	4,200	Kentucky No. 9	320	6 years	68
Star North	288,000	207,000	5,300	Kentucky No. 9	220	10 years	39
Star South	62,000	8,900	1,800	Kentucky No. 9	220	2 years	5
Walton Creek	72,000	0	1,800	Kentucky No. 9	175	2 years	0
Wheatcroft	143,000	206,000	1,200	Kentucky No. 9	300	2 years	172

moves toward areas of deeper virgin coal, methane emission is more of a problem and predrainage may become a more viable option. The only predrainage or degasification test to date in the Illinois Basin was performed by the USBM in Jefferson County, Illinois. Five vertical boreholes were drilled and stimulated at an average depth of 733 ft in the Herrin (No. 6) coalbed. Although a drop in methane concentration was observed when the borehole was passed during mining, the holes produced an insignificant amount of gas. According to a USBM report (Elder, 1977), the gas flow increased from an initial 10 to 4,300 cubic feet per day (cf/day). Such low flow rates raise questions as to the potential for producing the methane resource via vertical wells in this tectonically undisturbed portion of the basin.

Well Tests

Summaries are given below of recent coring and well testing operations performed to delineate the methane resource of coals in the Illinois Basin. These include MRCP well tests (see Fig. 15) performed in Clay and Marion counties, Illinois, Webster County, Kentucky, and Posey County, Indiana; and an Illinois Geological Survey test in Coles County, Illinois. The gas-in-place (GIP) estimates* made for these tests are based on coal analyses only and do not take into account gas that might be available in surrounding strata. More detail concerning the MRCP well tests is available in Preliminary Well Tests Reports prepared by TRW for each well. The holes in Webster and Posey counties were drilled subsequent to preparation of the first draft of this report and are located in areas A and B (Fig. 16), respectively, outlined in the recommendations of that report.

MRCP Site AA, Clay County, Illinois

This well—Hagen Oil, Henderson No. 2, Section 19, T3N, R8E—tested coals in one of the deeper untested portions of the Illinois Basin. Eight coal seams were penetrated between depths of 990 and 1,400 ft. Five of the seams—the Danville (No. 7), Herrin (No. 6), Briar Hill (No. 5A), Harrisburg (No. 5), and Seelyville coals—were sampled for desorption testing. Core measurements indicated the total thickness of these five seams to be 8.0 ft; however, geophysical logs indicate the coals total 14.6 ft. Each of the coals is high-volatile B bituminous.

Successful drill stem tests (DST) were performed in the Seelyville and Briar Hill (No. 5A) coal intervals. The Seelyville is believed to have a "relatively high degree of permeability" (TRW, 1979); however, it was not possible to calculate meaningful values of transmissibility or permeability. The DSTs indicated that coals were at pressures close to the normal hydrostatic pressure at those depths. Porosity and permeability tests of coal core samples indicated that at reservoir conditions, the Danville (No. 7) coal has a porosity of 11.4% and permeability of 0.15 md; the Herrin (No. 6) coal has a porosity of 6.4% and a similar permeability.

The gas content of the coals ranged from 32 to greater than 48 cf/ton. The deepest coal had the highest gas content. Using an average of 38 cu ft of gas per ton of coal, a coalbed thickness of 14.6 ft, and a drainage area of 80 acres, the total GIP for these five seams is estimated at 77 MMcf/80 acres.

MRCP Site AB, Marion County, Illinois

At this site—GeoWest, VIICA 722-ICD (Section 7, T2N, R2E)—the Danville (No. 7), Herrin (No. 6), Briar Hill (No. 5A), and Harrisburg (No. 5) coals were penetrated between 652 and 740 ft and were sampled for desorption testing. These coals were shallower and had lower rank (high-volatile C bituminous) than those at the MRCP AA site. Gas content of the coals ranged from 22 to 35 cf/ton. Using an average content of 28 cf/ton and a coalbed thickness of 12.8 ft, the GIP for the four seams sampled is estimated to be 50 MMcf/80 acres.

MRCP Site AC, Posey County, Indiana

In cooperation with the MRCP, the Indiana State Geological Survey wireline-cored an interval between 465 and 895 ft in well SDH 301, Section 33, T6S, R13W, in Posey County, Indiana. The Indiana VII, Herrin, V, IVa, IV, and III coals were sampled for desorption testing. The total coal section was 19.4 ft thick. Gas contents ranged from 26 to 83 cf/ton. These estimated gas contents were a little lower than expected based on previous investigations by the Indiana Survey in southwestern Indiana; however, they were still higher than those results obtained from previous MRCP tests. The gas content generally increased as the overburden thickness increased, although the upper split of coal III had an abnormally low gas content value. Of the 19.4 ft of coal, 17.4 ft would probably be targeted in the event of attempted production of gas. Using an average of 50 cu ft of gas per ton of coal and a coalbed thickness of 17.4 ft, the GIP for this location is estimated to be 122 MMcf/80 acres.

MRCP Site AD, Posey County, Indiana

The Indiana Survey wireline-cored a second hole, SDH 302, in Posey County, approximately 15 mi north of the previous hole. This well (in Section 26, T4S, R13W) has shown the greatest potential for the production of coalbed methane of all wells drilled in the Illinois Basin during the MRCP. The coal section between 506 and 945 ft was 28.9 ft thick. All major seams (the Indiana VII, Herrin, IVa, V, IV, IIIa, and III coals) were sampled for desorption testing. Only one seam, coal IIIa, was so thin (0.7 ft) that it should probably be ignored in making resource estimates. Indiana coal III was 11.5 ft thick, including a 2-ft shale binder. This is the thickest section of coal III ever reported in Indiana. Gas samples from coal III were analyzed as being approximately 61% CH_4, 1.2% CO_2, 2.5% O_2, and 34% N_2 and having very few heavier hydrocarbons. Isotopic analyses indicated that the gas was very "light" ($\delta^{13}C = -61.3$) but typical of other coal gas samples analyzed in the Illinois Basin (Coleman, personal communication). The gas content of the coals from this site was generally the gassiest tested to date in this basin by the MRCP. Gas content again basically increased with overburden thickness and ranged from 45 to 144 cf/ton. Using an average content of 88 cf/ton and a coalbed thickness of 28.2 ft, the GIP is estimated to be 347 MMcf/80 acres.

MRCP Site AE, Webster County, Kentucky

A core hole, R. W. Beeson, Coy Vandygriff strat test, was drilled on a downthrown fault block in the Moorman Syncline of western Kentucky. This hole, at Carter coordinates 4 L24 in

---

*The gas-in-place is calculated from the following equation:

- GIP = (drainage area) × (methane content per cubic foot) × (thickness).
- The drainage area in square feet equals: acres × 43,560.
- The methane content per cubic foot equals: (cubic feet/ton × 80 lb/cf) ÷ 2000 lb/ton = cubic feet of gas per cubic foot of coal).

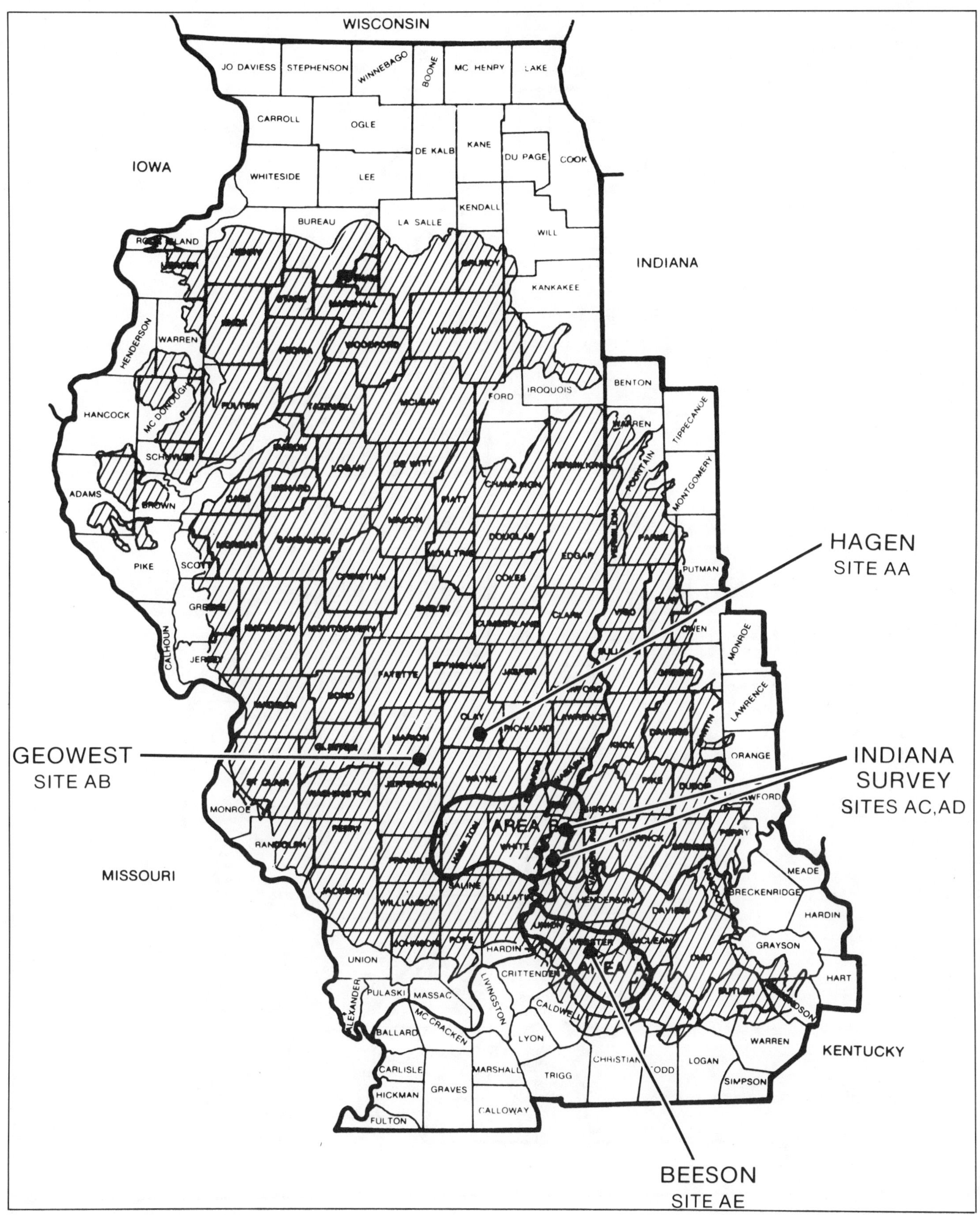

Figure 15—Location of MRCP wells.

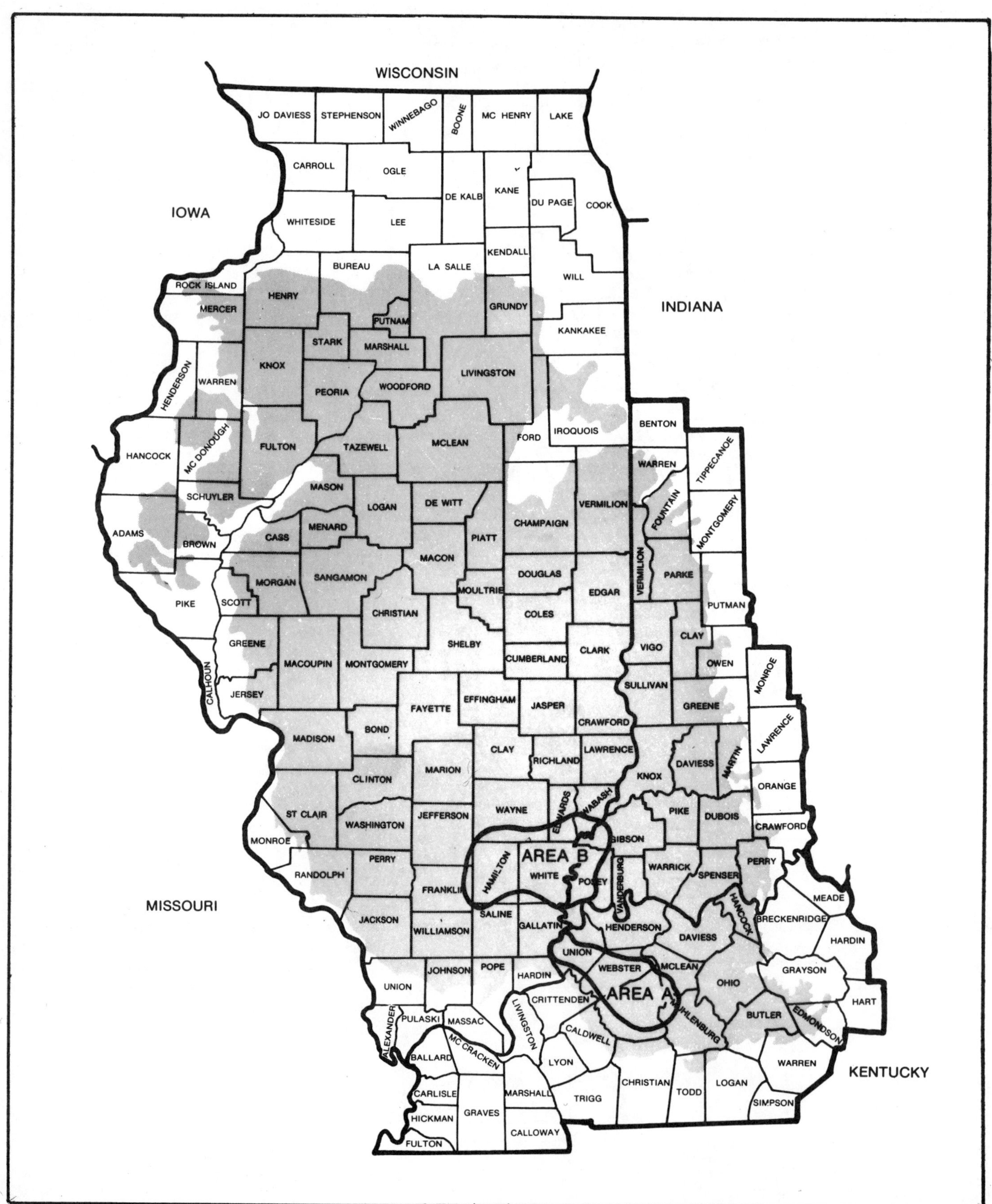

Figure 16—Redefined Illinois Basin coalbed methane target areas.

Webster County, was to test the gas potential of the Kentucky No. 13, No. 11 (Herrin), and No. 9 (Springfield-Harrisburg) coals in a region where no coal gas content data existed. The Kentucky No. 11 seam was missing; however, the Nos. 13 and 9 seams were 5.7 and 5.0 ft thick, respectively, in the interval between 1,198 and 1,310 ft. The rank of these coals was determined to be high-volatile A bituminous. Geophysical logs indicated that 5—and possibly as much as 16—ft of coal exists in shallower horizons. DSTs indicated the coals were at hydrostatic pressures; very little flow was measured, and the formations were considered very tight. Permeabilities could not be calculated. It was somewhat surprising that the coals sampled were as hard and unfractured as observed, considering the complex structural setting from which they were taken. Gas analyses indicated that the CH_4 percentage was low; however, this is attributed to sampling problems and is not believed inherent to the desorbed gas. Hydrocarbons measured were essentially all (99%) CH_4; stable isotopic analyses were not performed on these samples. The estimated gas contents of the Nos. 11 and 9 coals were 45 cf/ton. The superior thickness of each of the seams makes them attractive from the stimulation and production standpoint; however, the low gas contents will probably preclude any production from these seams in the near future. Based on a thickness of 10.7 ft and a gas content of 45 cf/ton, the estimated GIP for these two seams is 67 MMcf/80 acres.

Charleston Drilling Project, Coles County, Illinois (Popp et al, 1979)

The Illinois Geological Survey recently drilled a hole near Charleston, Illinois, for the purpose of studying the concentration and composition of gas in coals at that site. They measured the gas content of four seams sampled between 504 and 1,097 ft: the Shelbyville (?), the Danville (No. 7), the Herrin (No. 6), and the Springfield (No. 5) coals. Gas analyses indicated the desorbed gas to have a composition of approximately 65% methane, 10% other hydrocarbons, 5% CO_2, and 20% nitrogen. Isotopic compositions of the gases varied with the coal seams, suggesting that some gases had different origins and/or histories. The survey found that the Herrin (No. 6) coal released its gas most readily and suggested that although it was not the gassiest coal, it might produce gas most rapidly in the wellbore. Gas concentrations ranged from a low of 8 cf/ton for the Shelbyville (?) coal to 87 cf/ton for the Danville (No. 7) coal. The Shelbyville (?) coal has such a low gas content that it will not be considered in making resource estimates. The remaining coal section is only 9.4 ft thick. Thus, although the average gas content (63 cf/ton) of the coals is relatively high compared with other tests in the Illinois basin, the total GIP is still low at 83 MMcf/80 acres and similar to the estimate for the MRCP Site AA in Clay County, Illinois.

Commercial Gas Production from Coalbeds

Virgin coal in the Illinois Basin has never been a target to which an operation has drilled for the purpose of commercial gas production. To date, coalbed gas production has been limited to mined-out (gob) areas in Saline County, Illinois. Three companies (Petro-Search Associates of Harrisburg, Illinois; RE Company; and WOERA of Eldorado, Illinois) produced a combined 400 Mcf/day from three wells in 1980. Gas production was at low pressures. Methane content of the gas has been between 90 and 92%. Information regarding these wells is given in Table 4. Low pressures at WOERA's Farris well have been attributed to the fact that a mine has been open and venting gas from the coalbed continuously since 1954. Daily production from RE's Adams #1 well is limited by the demand of the low-pressure system to which it is connected. Cumulative production over a 2-year period from the three wells approaches 250 MMcf of gas. An undetermined amount of gas was previously produced from the Adams #1 well by the Jade Oil Company between 1960 and 1963. Gas reserves from the Dering and Southern counties No. 1 Mines are estimated to be in excess of 1 billion cubic feet (Bcf).

Gas from these wells is used locally (after compression to about 25 psi) by the United Cities Gas Corporation, and, in the case of the Farris well, the gas is compressed and added to a high-pressure transport system. It should be noted that the wells were placed into relatively old and apparently well sealed-off mines. The RE Company believes additional gob gas wells may be economically viable at this time; however, the absence of nearby high-volume-demand markets somewhat restricts rapid development of this resource.

As of this writing, no oil and gas operator plans to fracture virgin coal and attempt to produce methane gas in the Illinois Basin.

Estimated Resource Volume

The variability and absence of data on the methane content of coals in some settings in the Illinois basin make it difficult to estimate the resource accurately. A detailed resource estimate will require better distribution of data points throughout the geologic section and the basin. In addition, we would optimally have precise knowledge of each coal's physical and chemical character (including rank, overburden thickness, porosity, etc.) and the geologic history of specific regions, combined with an understanding of how these parameters affect the methane content of the coal.

Based on the limited data in Table 2, it is possible to present reasonable ranges for the maximum and minimum expected in-place gas resource of several coals in the Illinois Basin. These estimates have been made on a simple volumetric basis over the entire basin for the Danville, Herrin, Springfield-Harrisburg, and their equivalent coals, (Table 5). Estimated total coal

Table 4. "Gob" gas production data (1980).

Well Name (Company)	Abandoned Mine Works (Date)	Daily Production (Ave.)	Cumulative Production	Pressures	
				Initial	Present
Farris (WOERA)	Dering (1954)	180 Mcf	88 MMcf	22 oz.	2–7 oz.
Adams #1 (RE)	Southern Counties No. 1 (1929)	155 Mcf	62 MMcf	17.5 lbs.	8.5 lbs.
(Petro-Search)	Ogera No. 8	60 Mcf	~100 MMcf	?	1 oz.

resource for the Danville (No. 7) coalbed and its equivalents is over 14.5 billion short tons. Given a range of gas contents from 40 to 116 cf/ton, minimum and maximum gas resource estimates total 0.5 and 1.7 trillion cubic feet (Tcf). The total Herrin (No. 6) and equivalent coal resource is estimated at over 77.1 billion tons. Given a range of gas contents from 32 to 125 cf/ton, minimum gas resource estimates are 2.5 and 9.6 Tcf. Likewise, the Springfield-Harrisburg (No. 5) and equivalent coals, with total resource of over 67.4 billion tons and gas contents of 32 to 147 cf/ton, have gas resource estimates of 2.2 and 9.9 Tcf.

The minimum in-place gas resource for these three seams totals over 5 Tcf. It is reasonable to assume that the methane resource contained in other major coals (Colchester, Seelyville, Davis, etc.) may add significantly to this figure. It should be noted again that even though the specific gas content of coals in the Illinois Basin is quite low, the magnitude of the coal resource in the Illinois Basin has produced relatively large in-place gas resource estimates. The volume of gas that could actually be produced is of course unknown at this time. Productivity will, by necessity, be controlled by the reservoir and economic conditions existing at any specific location.

Estimates of the gas resource in the redefined target areas of the previous draft report were made using 90 cf/ton and 70 cf/ton a specific gas content of the coals, respectively, in areas A and B (Fig. 16). Recent tests indicate that the specific gas contents will probably not be this high. The typical coal section thickness, however, will probably be greater than previously assumed. Given proper evaluation and exploration ahead of drilling, it should be reasonable to expect to penetrate coal sections greater than 20 ft thick. An 80-acre location with a 20-ft coal section that has an average gas content of 65 cf/ton would contain 181 MMcf of gas-in-place.

Additional Observations

Various sources have indicated that the gassiest coal mines in the Illinois Basin have always been those near sandstone channel cut-outs and with gray shale or silty roofs. Further investigation is necessary before this observation can be documented; however, the following rationale suggests the validity of such an occurrence. A vast majority of gas in coal mines is often attributed to flow of gas into a mine from surrounding roof and floor rock. Silty gray shale roofs are generally more porous and are expected to contain more gas than dense, black, organic-rich shales. Porous sandstone channels that are not water saturated likewise serve as potential gas reservoirs. As mining approaches sandstone channels, or encounters gray shale roofs, the accompanying pressure drop causes gas to flow from these "reservoirs" into the mine, creating unusually high gas concentrations for a given coalbed. Another alternative explanation is that the gray shale roof is a better cap rock, preventing vertical immigrations of coal gas. This gray shale causes coal in this environment to contain more gas. Likewise, it has been suggested that coal adjacent to coal swamp channels, which was more rapidly buried as a result of sudden and frequent floodings, is likely to better retain gas generated in the early stages of coalification. These solutions are not as attractive because gray shale does not appear to be less permeable than black shale, and more than enough gas is generated during coalification to offset any loss prior to complete burial by sediment. In any case, the empirical observation has been made that coal mines with gray shale roofs appear to be gassier than those with other types of roofs. Whether the coal in these environments has a higher or lower than expected specific gas content remains to be determined.

Table 5. Estimated in-place coalbed gas resource (1 March 1980).

Coal Seam	Gas Content cf/ton	Gas Resource cu ft	
		Minimum	Maximum
Danville (and equivalents)	40–116	5.83×10^{11}	1.69×10^{12}
Herrin (and equivalents)	32–125	2.47×10^{12}	9.64×10^{12}
Springfield-Harrisburg (and equivalents)	32–147	2.16×10^{12}	9.91×10^{12}
		5.21×10^{12}	21.24×10^{12}

A number of isotopic analyses of gas from coal and coal mines have raised the question as to the source and origin of the methane in coal of the Illinois Basin. The ratio in methane of the ^{13}C isotope to the ^{12}C isotope has been related to the mode and conditions of formation of the gas and is used to "fingerprint" the origin of gases (Fuex, 1977; Stahl, 1977). Stahl (1977) indicates that coal gas derived from a humic source material such as predominates in Illinois coals should produce methane having a $\delta\,^{13}C$ value in the range of -25 to -40%. Analyses of methane gas from Illinois Basin coals and coal mines have invariably resulted in $\delta^{13}C$ values below -50% (Coleman, 1979, personal communication; Popp et al, 1979). Such light gas is usually derived from material attributed to forming petroleum or from material that has a higher thermal maturity than present in the Illinois Basin coals. These data suggest that the gas associated with coals in the Illinois Basin was formed (wholly or in part) in surrounding shales and has migrated into the coals. Similar conclusions were reached by investigations of coalbed gas in the Arkoma Basin (Iannacchione and Puglio, 1979).

CONCLUSIONS

Although the data are derived from a relatively small data base, it appears that the gas content of coals in the Illinois basin is low: 30 to 150 cf/ton. Additionally, there seems to be a general increase in gas content toward the southeastern portion of the basin as a whole and the initial target area defined by the MRCP.

On the basis of information gathered in this report, it is possible to define those areas of the basin that hold the greatest probability for early commercial coalbed methane gas production. The criteria used to select areas with the greatest potential methane resource are:

- Exceptionally gassy coals
- Thickest cumulative section of coal
- Highest coal rank
- Major coal units at the greatest depths
- Some structural deformation having potentially enhanced coal porosity and permeability.

Two target areas that meet these criteria and thus have the highest potential for early commercialization of coalbed gas in the Illinois Basin are outlined in Figure 16.

Target Area A covers approximately 1,100 sq mi. The total estimated coal resource in the area is near 15.5 billion short tons. Although no coals have been sampled for desorption testing in this area, a conservative estimate of the gas content based on the depth, rank, and reported gassy nature of the coals is 90 cf/ton. The resulting estimated in-place resource is 1.4 Tcf of gas or 2,000 Mcf/acre. The cumulative thickness of major seams in this area is typically 15 ft; under these conditions, the specific in-place gas resource is on the order of 133 Mcf/acre-ft.

Target Area B includes approximately 3,200 sq mi. The total estimated coal resource of this area is over 18.6 billion short tons. Of this, the Herrin (No. 6) and Springfield-Harrisburg (No. 5) coals contain 5.5 and 7.6 billion tons, respectively. Based on desorption data presently available, a modest estimate of 70 cu ft of gas per ton of coal can be used to provide in-place gas resource estimates. The total in-place gas resource is approximately 1.3 Tcf, which converts to 635 Mcf/acre. The Herrin (No. 6) coal is estimated to contain 400 Bcf of gas, or 195 Mcf/acre. The Springfield-Harrisburg (No. 5) is estimated to contain 500 Bcf of gas, or 244 Mcf/acre. The cumulative coal thickness in this area is typically 10 ft; under these conditions, the specific in-place gas resources are estimated to be near 64 Mcf/acre-ft for the total coal section.

ACKNOWLEDGMENTS

TRW gratefully acknowledges the assistance of those organizations and persons who provided information, data, and advice during preparation of this report. Among these, special thanks are due to John Popp (Illinois State Geological Survey), Donald Carr (Indiana State Geological Survey), Russell Brant (Kentucky Geological Survey), and Pat Diamond (U.S. Bureau of Mines) for their invaluable support.

In addition, TRW greatly appreciates the cooperation of the Illinois State Geological Survey, American Association of Petroleum Geologists, McGraw-Hill Publications, and W. R. Freeman and Company of San Francisco in permitting the reproduction of various figures and tables contained in the text.

REFERENCES CITED

Allgaier, G. J., and M. E. Hopkins, 1975, Reserves of the Herrin (No. 6) coal in the Fairfield basin in southeastern Illinois: Illinois State Geological Survey Circular 489, 31 p.

Atherton, E., 1971, Tectonic development of the eastern interior region of the United States: Illinois State Geological Survey Petroleum 96, p. 29–43.

Averitt, P., 1975, Coal resources of the United States, January 1, 1974: U.S. Geological Survey Bulletin 1412, 131 p.

Bristol, H. M., and T. C. Buschbach, 1971, Structural features of the eastern interior region of the United States: Illinois State Geological Survey Petroleum 96, p. 21–28.

Buschbach, T. C., 1971, Stratigraphic setting of the eastern interior region of the United States: Illinois State Geological Survey Petroleum 96, p. 2–20.

——, and G. E. Heim, 1972, Preliminary geologic investigations of rock tunnel sites for flood and pollution control in the greater Chicago Area: Illinois Geological Survey, Environmental Geology Note 52, 35 p.

——, and R. Ryan, 1963, Ordovician explosion structure at Glasford, Illinois: Bulletin of the American Association of Petroleum Geologists, v. 47, p. 2015–2022.

Damberger, H. H., 1971, Coalification pattern of the Illinois basin: Economic Geology, v. 66, no. 3, 488–494.

Davies, R. E., 1973, Structural features of southern Illinois, in depositional environments of selected lower Pennsylvanian and upper Mississippian sequences of southern Illinois: 37th Annual Tri-State Field Conference, Southern Illinois University, p. 3–10.

Elder, C. H., 1977, Effects of hydraulic stimulation on coalbeds and associated strata: U.S. Bureau of Mines Report of Investigations 8260, 19 p.

Fuex, A. N., 1977, The use of stable carbon isotopes in hydrocarbon exploration: Journal of Geochemical Exploration, v. 7, p. 155–188.

Iannacchione, A. T., and D. G. Puglio, 1979, Geological association of coalbed gas and natural gas from the Hartshorne Formation in Haskell and LeFlore counties, Oklahoma: Comptes Rendus, IX International Congress of Carbon Stratigraphy and Geology, Urbana, Illinois.

King, P. B., and H. M. Beikman, (compilers), 1974, Geologic map of the United States, scale 1:2,500,000: U.S. Geological Survey.

Kissell, F. N., C. M. McCulloch, and C. H. Elder, 1973, The direct method of determining methane content of coalbeds for ventilation design: U.S. Bureau of Mines Report of Investigations 7767, 17 p.

Malhotra, R., 1977, Market potential for coals of the Illinois basin: Illinois State Geological Survey Illinois Industrial Minerals Note 67, 60 p.

McCulloch, C. M., J. R. Levine, F. N. Kissell, and M. Deul, 1975, Measuring the methane content of bituminous coal beds: U.S. Bureau of Mines Report of Investigations 8043, 21 p.

Popp, J. T., D. D. Coleman, and R. A. Keogh, 1979, Investigation of the gas content of coal seams in the vicinity of Charleston, Illinois: Illinois Institute of Natural Resources Document 79-38.

Pryor, W. A., and E. G. Sable, 1974, Carboniferous of the eastern interior basin, *in* G. Briggs, ed., Carboniferous of the southeastern United States: Geological Society of America Special Paper 148, p. 281–313.

Smith, W. H., and J. B. Stall, 1975, Coal and water resources for coal conversion in Illinois: Illinois State Geological and Water Surveys Cooperative Resources Report 4, 79 p.

Stahl, W. J., 1977, Carbon and nitrogen isotopes in hydrocarbon research and exploration: Chemical Geology, v. 20, p. 121–149.

Swann, D. H., 1968, A summary geologic history of the Illinois basin, in geology and petroleum production of the Illinois basin: Illinois-Indiana-Kentucky Geological Societies, Schulze Printing Co., p. 3–21.

TRW, 1979, Preliminary well test package DAA, Hagen Oil, Henderson No. 2 well, Clay County, Illinois: U.S. Department of Energy, Morgantown Energy Technology Center Open File Report, UGR-384.

Wanless, H. R., 1962, Pennsylvanian rocks of the eastern interior coal basin, *in* Pennsylvanian system in the United States, symposium: Tulsa, OK, American Association of Petroleum Geologists, p. 4–59.

——, and J. M. Weller, 1932, Correlation and extent of

Pennsylvanian cyclothems: Geological Society of American Bulletin, v. 43, p. 1003–1016.
——, J. R. Baroffio, and P. C. Trescott, 1969, Conditions of deposition of Pennsylvanian coal beds, *in* Environments of coal deposition: Geological Society of America Paper 114, p. 105–142.
Willman, H. C., et al, 1975, Handbook of Illinois stratigraphy: Illinois State Geological Survey Bulletin 95, 261 p.

Geologic Overview, Coal, and Coalbed Methane Resources of the Arkoma Basin — Arkansas and Oklahoma

H. H. Rieke
J. N. Kirr

The Arkoma Basin, an elongated, linear east-west–trending sedimentary basin, encompasses an area of approximately 13,488 sq mi in east-central Oklahoma and west-central Arkansas. Within this basin, extensive reserves of bituminous and some semianthracite coals are contained in Pennsylvanian age rocks. The Oklahoma Geological Survey and the U.S. Geological Survey have estimated that the total coal resource of this basin might be 7.89 billion tons.

More than ten individual coal seams have been identified in this area, eight of which have been mined or are being mined presently. The majority of the major coalbeds are continuous but do not maintain constant thickness. Individual beds range from a few inches to 7 ft in thickness. The coalbeds outcrop along the periphery of the synclines and dip steeply (up to 90°) toward the axis. Lower (Atokan) and Upper (Missourian) Pennsylvanian coals are thin and discontinuous, whereas the Middle Pennsylvanian (Desmoinesian) coals are thick, generally continuous, and provide the basin's major proven reserves. The thin Lower Pennsylvanian coals have not been studied in as much detail. The Hartshorne, the McAlester (Stigler), and the Secor coalbeds are the most extensive and uniformly thick coals in the Arkoma Basin; estimated coal reserves are over 7.4 billion tons, constituting 93% of the basin's total coal resource.

Methane Recovery from Coalbed Project (MRCP) data on the methane gas content of Arkoma Basin coals is presently limited to coring and well testing in Pittsburg, Haskell, and Le Flore counties, Oklahoma. In Pittsburg County, the Upper Hartshorne, Upper Booch, and McAlester coals were cored in two wells between the depths of 1,905 and 3,650 ft and samples were desorbed for approximately 5 months using the USBM's direct method. Total gas contents ranged between 73 and 211 cubic feet per ton (cf/ton). In Le Flore County, Oklahoma, the Upper Hartshorne was sampled at a depth of 192 ft and found to contain 310 cf/ton. Variability in gas content cannot always be directly related to coalbed depth, inasmuch as some of the gassier coals were from shallower horizons—the opposite of what might have been expected—but is related to other geologic controls on the gas content, analytical errors, or to the method used in determining the "remaining" gas content.

The Hartshorne coals in Oklahoma have been the most frequently sampled coalbeds. Estimated gas content of the Hartshorne ranges from 73 to 570 cf/ton. Other measured values are 200–211 cf/ton for the Booch coals and 131 cf/ton for the McAlester coal. The latter two estimates are from coals located in the subshelf part of the basin where the coals are thinner and not as well developed.

Methane liberated from mines during operation also provides an indication of the potential methane resource in an area. The correlation between the amount of gas emitted from coal mines and that predicted to be there by desorption testing is not obvious. Experience in the Arkoma Basin at the Howe mine in Le Flore County, Oklahoma, shows that about seven times more gas is liberated from a mature mining operation than what was measured by desorption testing.

Based on the limited data available, ranges for the maximum and minimum expected in-place gas resource have been made for all the major coalbeds. The Hartshorne coals are anticipated to have a minimum of about 5.4 billion short tons and a maximum of nearly 2.45 trillion cubic feet (Tcf) of in-place gas. The McAlester (Stigler) Coal, likewise, is estimated to contain a maximum of 0.69 (Tcf) of gas. The minimum in-place gas resource for the Hartshorne and McAlester coals totals about 1.4 Tcf.

It is reasonable to assume that the methane contained in the deeper parts of the various major synclines may add significantly to this figure. The gas content of the coals in the Arkoma Basin is generally higher towards the east. Several target areas totaling approximately 3,600 sq mi have been identified as having the greatest probability for early commercialization of gas. The main target area is located in the deepest portion of the Cavanal syncline in Le Flore County, Oklahoma. Another principal target is located in the Lehigh syncline in Coal and Atoka counties, Oklahoma. An alternate target would be the Cavanal syncline area in the western part of Sebastian County, Arkansas. These areas contained reasonably thick coal sections at considerable depths.

INTRODUCTION

The Arkoma Basin has been referred to as the Arkansas-Oklahoma Coal Basin and is known in Arkansas as the Arkansas Valley Basin and in Oklahoma as the McAlester Basin (Fig. 1). The basin extends for approximately 250 mi in an east-west direction and is situated in east-central Oklahoma and west-central Arkansas. This long, arcuate trough is 20 to 50 mi wide from north to south and is bounded along the southern side by the very complex Ouachita overthrust belt. The deepest part of this basin is adjacent to the Ouachita Mountains system, where the sedimentary column is estimated to be 30,000 ft thick. In central Arkansas, the Arkoma Basin plunges beneath the overlapping Tertiary and Cretaceous strata of the Mississippi Embayment, where its nature becomes obscure. The basin is bounded on the north by the Boston Mountains and the Ozark dome, and its northwest boundary is the Central Oklahoma platform. On the west is the Seminole uplift, or Hunton arch. The basin terminates abruptly on the southwest against the Tishomingo anticline in Atoka County, Oklahoma, northeast of the Arbuckle Mountains. An estimated 40,000 cu mi of sediment fills this basin (Branan, 1968).

The Arkoma Basin is located between two major physiographic provinces, the Ouachita Province and the Ozark Plateaus, and is a subset of the Ouachita Province. Both of these major provinces resemble the Appalachian Provinces and Interior Low Plateaus in many ways. The east-west–trending Ouachita Mountains are fold mountains of thick Paleozoic formations. The Ozark Plateaus lie north of the basin, forming a broad upwarp that exposes lower Paleozoic formations similar to those of the Cincinnati Arch.

Mesas, buttes, cuestas, recent fault-line scarps, anticlinal ridges, and synclinal valleys are the most common erosional forms in the eastern Arkansas part of the basin. Folding is present everywhere in eastern Oklahoma and increases to the south, resulting in sandstone-capped, isolated synclinal mountains that rise to 2,000 ft above the wide hilly plains. Relief increases toward the east, reaching a maximum on either side of the Arkansas-Oklahoma state line. In the western portion of the basin, the relief is about 350 ft; whereas near the eastern boundary of the basin, the relief is about 600 ft. Folding decreases in eastern Oklahoma northward to less prominent east-west monoclinal ridges with smooth continuous uplands and distinct scarps that are commonly precipitous (Fig. 2). This southern portion has been classified by Curtis and Ham (1972) as the McAlester Marginal Hills Belt. Folding continues to decline northward, resulting in broad, gently rolling lowlands of shale lying between sandstone ridges and hills that rise less than 300 ft above the plains. Curtis and Ham (1972) have classified this northern portion of the Arkoma Basin as the Arkansas Hill and Valley Belt. The boundary separating the Arkansas Hill and Valley Belt from the McAlester Marginal Hills Belt trends east-west near the center of the Panama Quadrangle, Oklahoma.

The Arkoma Basin is one of seven similar basins that lie along the fronts of the Ouachita and Appalachian mountain systems. Extending eastward from west Texas, they are the Marfa, Val Verde, Kerr, and Fort Worth Basins in Texas; the Arkoma Basin in Oklahoma and Arkansas; and the Black Warrior Basin at the intersection of the Appalachian and Ouachita Mountain systems in northern Mississippi and northwestern Alabama. The Appalachian Basin extends northeastward into New York State. These basins are closely related, both stratigraphically and tectonically, and all were part of the gigantic Ouachita geosyncline of the eastern and southeastern United States (Branan, 1968).

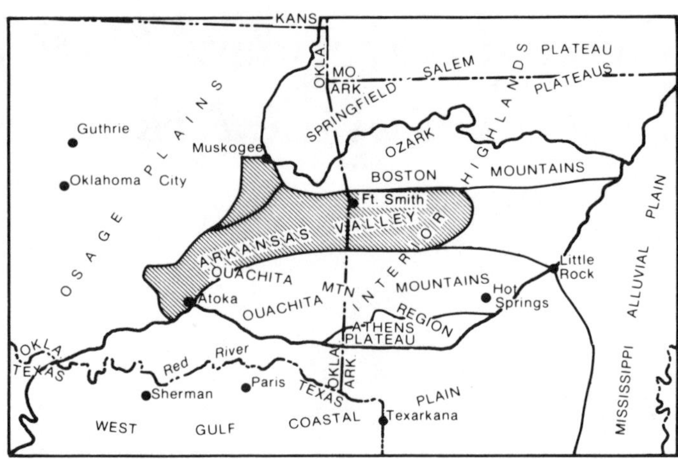

Figure 1—Map showing location of the Arkansas-Oklahoma coal basin (shaded area) and physiographic provinces. From Hendricks et al, 1936, p. 1342.

An ancient Oklahoma embayment was the predominant sedimentary feature of Oklahoma and Arkansas until the post-Hunton (Devonian) orogeny changed sedimentary patterns. Two shallow basins occupied the Arkansas Valley region during Ordovician time, one being generally east and the other west of the Pope-Conway County area (Fig. 2) (Caplan, 1957). Both basins persisted into Atoka time (Table 1). Formations older than Mississippian in the Arkoma area thicken southwestward into the Oklahoma embayment. Above this unconformity, these deposits thicken toward the south-southeast into the Ouachita geosyncline.

At the beginning of Mississippian time, a large part of what now is the trough of the Arkoma Basin was a high area. This is evidenced by the presence of the Silurian-Devonian limestone of the Hunton Group on the basin shelf; whereas in much of the trough area, the Hunton is either thinner than on the shelf or absent by truncation below Mississippian strata (Table 1) (Branan, 1968). The elevated arch began to subside during Mississippian time; maximum subsidence was in the Early and Middle Pennsylvanian in the Atokan (Pottsville) and the early Desmoinesian.

Pennsylvanian-age strata were deposited in deltaic fashion in a rapidly subsiding environment. These clastic sediments probably were derived primarily from the eroding highlands located on the northeast, north, and northwest sides of the basin. Paleocurrent azimuths in the deeper Atoka strata show that the direction of flow was southward (Briggs and Cline, 1967).

Sedimentation continued until Late Pennsylvanian time, when the Arbuckle orogeny of southern Oklahoma took place (Fig. 2). This Late Pennsylvanian orogeny uplifted the Tishomingo anticline in south-central Oklahoma, thereby separating the Arkoma depression from the Fort Worth Basin south of the uplift. During Early Permian time, the violent Ouachita orogeny occurred.

The Ouachita Mountains area south of the basin was an incipient uplift from late Atokan time until the major Permian

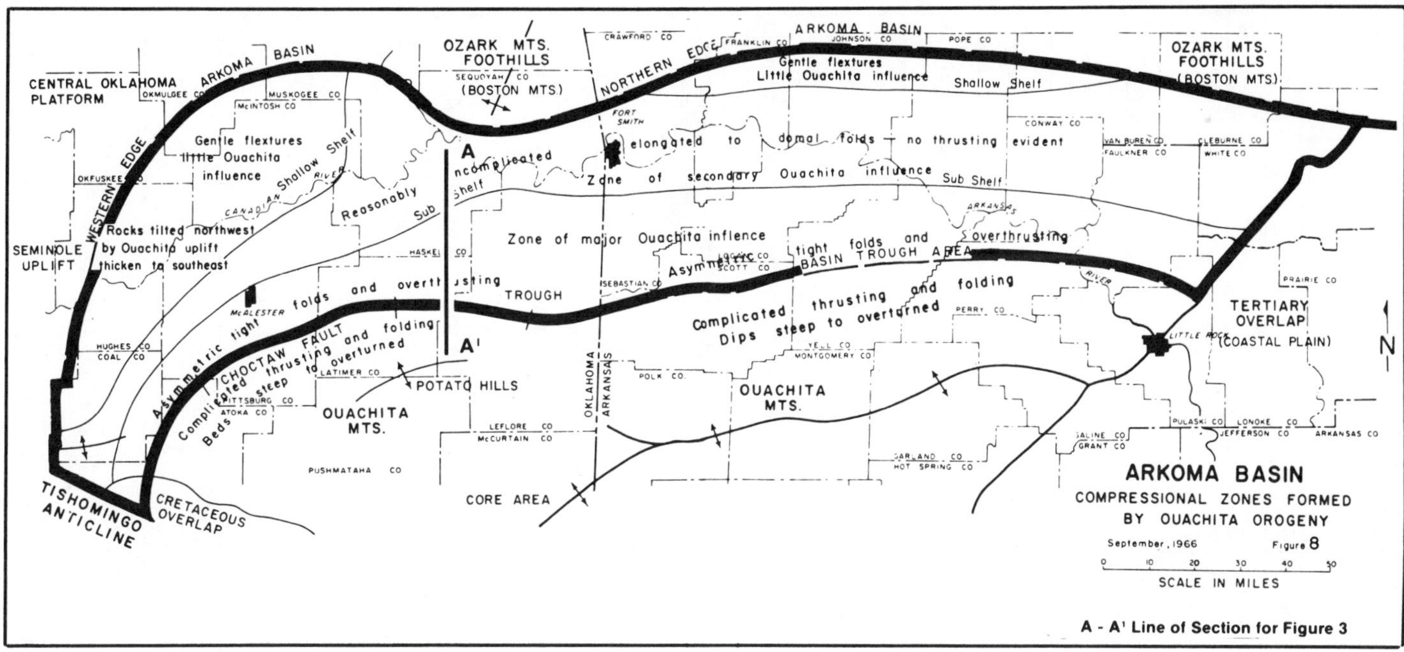

Figure 2—Regional setting of the Arkoma Basin. From Branan, 1968.

orogeny. During this period, it had an effect on the sedimentary patterns of the Middle and Late Pennsylvanian strata but little effect on the structural framework of the basin. Pulsations of the Ouachita uplift began in late Atokan time and continued throughout Pennsylvanian time. These pulsations that preceded the major Early Permian orogeny had a dramatic effect on Pennsylvanian sedimentary patterns over the entire mid-continent region of the United States (Branan, 1968).

During the Desmoinesian epoch of Middle Pennsylvanian time, access to the open sea on the southeast was cut off periodically by Ouachita pulsating uplifts.

The Permian movement was intense and modified the entire province. In the Arkoma Basin, this great force from the south caused compaction of sediments, crustal shortening through folding, and thrust faulting (Branan, 1968). At nearly the same period, the Ozark dome was reelevated, causing beds to tilt and block faulting to occur in that area.

The entire Arkoma Basin was tilted westward subsequent to the Ouachita folding, presumably by the later Appalachian movements (Branan, 1968). There also were movements in Late Permian time and during the Cretaceous Laramide orogeny that had a minor influence on the basin. As evidenced by earthquakes, movements still occur today in the Ouachita Mountains.

From Mesozoic to Cenozoic time, heat was introduced into the Arkoma Basin from igneous intrusions in the eastern Ouachita Mountains and the adjacent areas of the Mississippi Embayment (Craney, 1978). It is believed that this igneous influence resulted in the Arkoma Basin coal's eastward increase in rank.

STRUCTURAL GEOLOGY

Rocks and coalbeds in the Arkoma Basin were highly deformed by a combination of forces. In parts of the basin, the Precambrian/Cambrian basement rocks were depressed to depths greater than 25,000 ft. Influence of the Ozark uplift and the Ouachita orogenic province has caused many unique structural conditions in the Arkoma Basin. This has given rise to the unique physiography where the synclines form the mountains and the anticlines are expressed as valleys.

In order to explain the tectonics, Branan (1968) divided the Arkoma Basin into three structural provinces, ranked in order of Ouachita deformation (Fig. 2).

Zone 1 is at the forefront of the Ouachita Mountains in the *trough* area. Steep dips are common. The rocks are folded into a series of long, narrow anticlines, generally overthrust on the north.

Zone 2 is an area of elongate and domal features, with little or no evidence of thrusting. This is known as the *subshelf* area.

Zone 3 is the basin shelf area where Ouachita influence was slight, and only gentle folds are found. In Zone 3 in Oklahoma, the outcropping beds are Middle and Upper Pennsylvanian. These rock units thicken southeastward; however, they were elevated by the Ouachita force and now dip northwestward.

There are two basic structural patterns affecting the area of interest in the basin: 1) folding and northward overthrusting; and 2) block faulting. Stability of the Ozark Plateau on the north during basin subsidence caused tensional forces to develop, resulting in the evolution of major block faulting in the basin (Fig. 3). Evidence indicates that some of these faults were growing contemporaneously with deposition of the Lower Pennsylvanian beds.

Early Permian (Ouachita) mountain building on the south compressed basin sediments into a series of long, narrow, east-west anticlinal and synclinal folds. Overthrusting along anticlinal axes near the mountain front is common. Some of the folds have surface expressions extending 75 mi or more (Fig. 2). Disney (1960) reported that the anticlines occupy approximately 35% of the total basin area. The synclines' being wider accounts for approximately 65% of the basin area. Indications are that these superficial folds diminish with depth. A high anticline on the surface may not be expressed as an

Table 1—General geologic section of Oklahoma oil- and gas-producing areas. Dott, 1945. Used with the permission of the author.

System	Series	Group	Group or Formation	Description
Cretaceous	U. Cretaceous		Colorado gp.	Clay, Some Limestone.
			Dakota-Woodbine ass.	Sandstone
	L. Cretaceous		Washita gp.	Clays, Some ls. in Southeast Okla., Cheyenne ss. and Klowa sh. in Clmarron Co.
			Fredericksburg gp.	Represented by Goodland ls., Southeast Okla.
			Trinity gp.	Sandstone, Southeastern Okla. Madill sd., Madill Pool, Marshall County.
Triassic			Morrison fm. Exeter ss.	Variegated sh. and Marl: Some ss. Jurassic Present Only in Panhandle.
Triassic			Dockum gp.	Red Beds Present Only in Panhandle.
Permian	Quadalupian		Quartermaster fm.	Red Shale and Sandstone.
			Cloud Chief fm.	Red Shale, Local Lenses Gypsum and Limestone.
		Whitehorse	Rush Springs ss.	Massive Sandstone.
			Marlow fm.	Shale, Some ss., Thin Gypsums. Contains Verden Sandy Dolomite.
	Wolfcampian	Leonardian	El Reno gp.	Duncan-Chickasha, Grating Northward into Flowerpot, sh., Illalene gyp., and Dog Creek Shale.
			Hennessey sh. Garber ss. Wellington fm.	Shale and ss., Probably Contains Chickasha Gas Sand, and Other Horizons in Southwestern Oklahoma.
	Virgilian	Pontotoc	Stratford fm.	Equivalent to Stillwater fm. of North-Central Oklahoma. Contains Hoy, Hotson, and Other Sds., in North-Central Okla.: Gas-Bearing Dolomite and ls. in Texas Co., and Probably Producing Horizons in Southwestern Okla.
			Vanoss fm.	Sandstone, Shale, Limestone, Conglomerate: Produces in Ardmore Basin, Cement, and Elsewhere in Southwestern Okla.: Referred to as "Pontotoc"; Contains Smith and Brown sds. of Italnola Pool, Stephens Co.
			Pawhuska fm.	Shale and ls., Some ss.
			Elgin ss.	Sandstone and sh. ss. at Base = Carmichael sd., Higher sss. = Hoover sds. of North-Central Okla.
			Nelagoney-Vamoosa	Sandstone and sh. Bigheart and Fourmile sss. at Base = Tonkawa-Stainaker sds. in North-Central Okla.: Lovell ls. About Middle; Wynona ss = Endicott sd. Above Middle: Oread ls. at Top.
	Missourian	Hoxbar-Canyon (Ardmore Basin): Ochelata Group	Wildhorse ls.	Dolomitic ls. } = "Avant" ls. of Some Geologist; Above Perry Gas sd. in North-Central Okla. Sandstone and South Southward.
			Shale	Calcareous sh.
			Okesa ss.	Calcareous ss.
			Wann fm.	Shale and Lenticular sss., Contains Perry Gas sd.
			Avant ls.	Limestone, Somewhat Dolomitic = Oil City Lime in Southern Osage Co.
			Chanute sh.	Shale and Sandstone, Including Cottage Grove ss. = So-Called "Layton" sd. of Osage Co. (not the Laten sd. of Creek Co.), = "Musselem" and Peoples sds. of Osage Co. (not Musselem sd. of Oilton Pool, Creek Co.).
		Hoxbar-Canyon (Ardmore Basin): Skintook Group	Dewey ls.	Probably Equivalent to Belle City Limestone in Central Oklahoma.
		Skintook Group: Francis fm	Nellie Bly fm.	Shale and Variable Sandstones: Thins Westward in Subsurface.
			Hogshooter ls.	Limestone, Layton Lime of Creek County.
			Coffeyville fm.	Shale and Sandstone Dodds Creek ss. = Layton Sand (of Creek County) at Top: Checkerboard Limestone at Base.
			Seminole fm.	Two Sandstones and Middle Shale Containing Dawson Coal. Equivalent to Upper and Lower Cleveland Sands, Jones and Dillard Sands.

continued

Table 1— (Continued)

System	Series	Group or Formation			Description
		H'ville		Lenapah ls.	Limestone and Shale Overlayed by Memorial Shale of Des Moines Age, Near Tulsa.
				Nowata Shale	Shale, Some Sandstone, Contains Wayside Sand.
Pennsylvanian	Desmoinesian	Deese Formation (Ardmore Basin)	Wewoka	Oologah fm.	Limestone the "Big Lime": Splits into Pawnee Limestone, Bandera Shale, and Altamont Limestone ls. Northern Rogers and Nowata Counties.
				Labette sh.	Shale, Contains Sandstone Equivalent to Peru Sand.
			Wetumka	St. Scott ls.	Limestone Oswego Lime of North-Central Oklahoma: Oolitic Phase in Cushing Field, etc., Called Wheeler "sd."
			Cherokee	Upper Calvin	Sandstone, Upper Calvin = Pure Sand.
				Middle Calvin	Shale, Contains Thin lss. Lower ls. Probably = Verdigris Limestone.
				Lower Calvin Senora fm.	Lower Calvin Is Sandstone: Senora Formation Is Shale and Sandstone, Skinner Sand Probably = L. Calvin and Sandstone of Senora, Zone also Includes Allen Sand.
				Stuart Shale	Pink Lime of Subsurface Probably in Struart Shale.
				Thurman ss. Boggy fm.	Red Ford-Burbank-Earlsboro sd. Zone Probably Included in Thurman and Upper Boggy, Bartlesville-Glenn-Salt sds. (= Bluejacket ss.) in Lower Boggy, Inola Lime is Between Bartsville and Red Fork sds.
				Savanna fm.	Shale, ss, and Few Thin lss., Including the "Brown Limes" (Sam Creek and Spaniard).
		U. Dornick Hills fm		McAlester fm.	Shale and Sandstone. One or Another ss. and in Some Places the Underlying Hartshorne ss as have Been Correlated With Booch (Taneha) sd.
				Hartshorne ss.	Sandstone. Some Shale, Contains Hartshorne Coals, Produces Gas in Eastern Oklahoma.
				Atoka fm.	Thick Succession of Shale and Sandstone: Contains in Lower Part the Following: Timber Ridge, Muskogee, and Boynton sds. in Muskogee Co.: Butcher (= Boynton), and Dilerease-Deaner, in Central Oklahoma; Other Horizons Produce Gas in Le Flore, Sequoyah, Haskell, and Pittsburg Counties.
	Morrowan			Wapanucka fm. Union Valley fm.	Wapanueka fm. Contains ls., Sandy in Relatively Pure; Shale, and Sandstone. Union Valley fm. Contains ss. With ls. at Top, and is Correlated With Primrose ss of Ardmore Basin and Hale fm. of Ozark Area. Both Wapanueka and Union Valley fms. Produce Oil in East-Central Oklahoma. Lyons Lime and Sand of Okmulgee Dist., Cromwell, Quinn, Smith, and Sykes sds. of Seminole, Hughes, and Okfuskee Counties are of Morrowan Age. Probably = Union Valley fm. Lower Dornick Hills fm. Represents Morrowan Series in Ardmore Basin.
	Springeran				Dark Grey to Black Shale Called "Pennsylvanian Caney," Contains Several sss. in Ardmore Basin, Productive in Some Oil Fields in Southern Oklahoma. Outcrop Names From Base Upwards: Rod Club, Overbrook, and Lake Ardmore sss. Black sh. Between Rod Club ss. and Top of Mississippian Caney.
Mississippian	Chesterian	Caney sh.		Pitkin ls.	Fine-Grained, Drab ls. Present Only in Northeastern Oklahoma; Produces Locally.
	Meramecian			Fayetteville sh.	Black Shale, Caney Contains Black, Granular, ls. ("False Stayes") in Lower Part Locally in Central Oklahoma.

continued

Table 1— (Continued)

System	Series	Group or Formation		Description
Mississippian	Osagian ("Boone")	Mayes = Sycamore ls.	Mayes ls. Mississippi Lime	"Mississippi Lime" (White, Cherty ls.) in Northern Okla. in Lincoln Co. Grades into Seminole Mayes" (Black, Granular ls.), Probably = Sycamore ls. Bed of Glauconite at Base of Mayes. Burgess sv. in Northeastern Oklahoma Lies at Contact of Basal Pennsylvanian (Cherokee) and Weathered Surface of "Mississippi Lime": Exact Age of Overlying Beds Varies.
	Kinderhookian	Woodford sh.	Chatanooga sh.	Black Shale; Woodford is Cherty in Vicinity of Arbuckle Mts. and Subsurface of Central and Southern Okla. Chattanooga Black Shale in Northern and Northwestern Okla., has at Base the Lenticular Sylamore ss. = Misemer sd.
Devonian		Hunton gp	Frisco ls. Bois d'Arc ls. Harakan Snarl	Limestone, Cherty in Part; Marly. Considerable Production Especially at West Edmond and Elsewhere in Central and South-Central Oklahoma. Subdivided in Arbuckle Mts.
Silurian			Henryhouse sh. Chimneyhill ls.	
Ordovician	Cincinnatian		Sylvan Shale	Grey, Grey Shale, Green in Upper Part: Black in Middle Part Locally; Locally Dolomitic in Lower Part in Southern Oklahoma.
		Viola ls.	Fernvale ls.	White, Crystalline Limestone; Called "Buttermilk Lime," Most Widely Used Datum for Subsurface Structure Mapping.
	Mohawkian		Trenton ls.	Fine-Grained, Dolomitic ls. Produces Oil in Many Areas: "Arst" or "Seminole Wilcox" Sand Probably Belongs in Trenton.
	Chanyan	Simpson Group	Bromida fm. Tulip Cr. fm. McLish fm. Oil Creek fm. Joins fm.	Ss., ls., Dolomite, and Green Shale; Contains Wilcox sd. Which Probably Represents Different Formations in Different Areas. Contains "Lower Simpson" sds. at Oklahoma City, etc. Hominy or Burgen, Tyner, and Wilcox in Northern Oklahoma Probably Belongs in Tulip cr. and Bromide fms. "Tucker" sd of Cushing Field = Wilcox. Wilcox sd. Gives Largest Area-Yields of State's Producing Formations.
	Beekmantown		Arbuckle group	Known as "Siliceous Lime": Porous Dolomite in Top of Arbuckle has Been Turkey Mt. "sd." Production in Northern Oklahoma; Discovery Horizon at Oklahoma City.
Cambrian	U. Cambrian		Timbared Hills gp.	Contains Arkosic Reagan ss. at Base, and Honey Creek fm. No Known Oil Production in Oklahoma.
Pre-Cambrian				Granite etc.

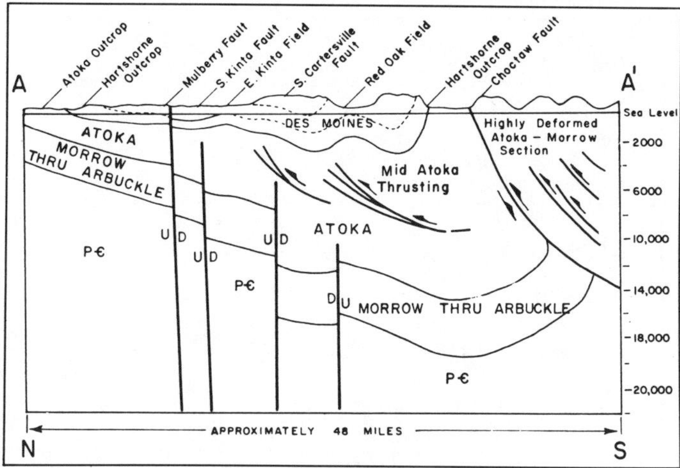

Figure 3—North-south cross section showing general structural relationships in the basin. From Branan, 1968.

anticline in the subsurface below the Atoka.

These features are much more complex than simple anticlines. In many cases, low-angle sole thrusts in the Atoka section break into imbricate thrusts as they rise to the surface, causing much repetition of section (Fig. 3). This has created the unique situation, in parts of the basin, where shallower beds and surface rocks were folded and thrusted, whereas deeper beds were only block-faulted.

There were unusual structural developments as Ouachita tectonic pressures were applied to the zones of major down-to-the-basin block faulting. When the force came from the south, the faces of these large block faults became buttresses that, in some cases, caused the thicker section of the Atoka Formation on the south side of the fault to be shoved and thrust upward. The final effect is an anticline on the surface that is actually on the downthrown side of a major block fault in the deeper beds (Branan, 1968).

Many types of fault systems are found in the basin. Understanding the nature of the faulting in a particular area of the basin might be of great value in developing a coalbed methane reservoir. Block faulting took place as the basin was subsiding in Pennsylvanian time and being filled with Atoka sediments. As a result, deep in the basin, major zones of faulting are commonly obscure or absent on the surface, because enough Atoka age shale was deposited on the downthrown side to compensate for the throw. Later Permian uplift on the south and consequent folding and overthrusting of the beds in the basin produced some complex structural situations.

Near the forefront of the mountains is a system of either block faulting or high-angle thrusting, in the deeper and older strata (Fig. 3). These faults are upthrown on the south and were probably created by Ouachita pressure. The faults appear unrelated to the system of major down-to-the-basin faults previously described.

Throws on the faults in the Cartersville-Kinta-Red Oak area range from 3,000 to 6,000 ft. The throw is generally to the south, or basinward. However, there are notable exceptions where the throw is reversed. The faults downthrown on the north generally are not of the magnitude of the down-to-the-basin faults (Fig. 3).

STRATIGRAPHY

Stratigraphic relationships are important in reconstructing the geological depositional environments of the coal and associated sediments. Table 1 presents the general geologic section for oil- and gas-producing areas in eastern Oklahoma. Table 2 presents a nomenclature chart for both the McAlester and the Arkansas Valley regions, prepared by Branan (1968). The stratigraphic terminology in Arkansas differs from that in Oklahoma, and adoption of a single set of names for the numerous formations and members by industry and government is improbable. This conflict in terminology has arisen between surface geology and subsurface geology interpretations of the depositional patterns in the basin. Beds above and below the Atoka correlate well across the basin, and although different names are used, the rocks are essentially alike lithologically (Branan, 1968). However, the Atoka series thickens so markedly on the south in parts of the basin that it becomes impossible to correlate updip sandstone beds with those within the thicker Atoka units (Branan, 1968). As the formation thickens, sandstone bodies become more numerous.

The coal-bearing strata of Middle and Late Pennsylvanian age in eastern Oklahoma occur in an area of 14,500 sq mi. At least 24 coals have been recognized as Desmoinesian and Early Missourian (Pennsylvanian) in eastern Oklahoma. This area is located in the southern part of the western region of the Interior Coal Province of the United States (Campbell, 1917). Some 8,000 sq mi in Oklahoma are known to contain coalbeds 1 to 7 ft thick that are presently of commercial value, have been in the past 100 years, or may be in the future (Fig. 4). The rocks of the McAlester region are chiefly of Pennsylvanian age, with some tentatively classified as Quaternary (Hendricks et al, 1939). The rocks tentatively referred to as Quaternary consist of sand and gravel that cover stream terraces.

The Pennsylvanian rocks in the basin contain these valuable coalbeds and have been subdivided into the following formations, listed in order of age, with the oldest first: Atoka Formation, Hartshorne Sandstone, McAlester Shale, Savanna Sandstone, Boggy Shale, Thurman Sandstone, Stuart Shale, and Senora Formation (Fig. 4). Each of these formations consists of alternating sandstone and shale, and all except the Atoka and Thurman include beds of commercially exploitable coal. The shales and sandstones at some horizons are of continental origin, and those at others are marine. Some continental beds grade laterally into marine beds. All of these formations were deposited in a shallow water environment. No marked unconformity in the Pennsylvanian rocks can be recognized at any one horizon over the entire basin, but locally between the Savanna and McAlester Formations, there is an unconformity that probably extends over the entire area without being recognizable at all places. Minor breaks within the formations are indicated by considerable irregularity in the bedding of the rocks (Hendricks et al, 1939).

The Quaternary age Gerty Sand consists of unconsolidated sand, gravel, and clay up to 50 ft thick and derived from a western source of Cretaceous age. These sediments are widely distributed over the north half of the McAlester district and are found along the former course of the Canadian River in parts of the Lehigh District and the Howe-Wilburton District. Other terrace deposits of the same age as the Gerty Sand have an eastern source in the Ouachita Mountains. Alluvium of recent

Table 2—Correlation chart, Arkoma Basin. Branan, 1968. Used with the permission of the Oklahoma Geological Survey.

System	Series (Group)	Oklahoma	Arkansas	
Pennsylvanian	Desmoinesian (Krebbs)	Boggy Shale Savanna Sandstone McAlester Shale[1] Hartshorne Sandstone[1]	Boggy Shale Savanna Sandstone McAlester Shale Hartshorne Sandstone	
	Atokan	Webbers Falls Sand Gilcrease Sands[1] Fanshawe Sand[1] Red Oak Sand[1]	Carpenter "A" Sand[1] Upper Alma Sand[1] Middle Alma Sand[1] Lower Alma Sand[1] Carpenter "B" Sand[1] Morris-Tackett Sand[1] Self-Areci Sand[1] Freiburg-Henson-Pearson Sand[1] Casey-Vernon-Hudson Sand[1]	
			Cecil Series[1]	Dunn "A"-Sells-McGuire Sand Ralph Barton Sand Dunn "B" Sand Dunn "C"-Dawson "A" Sand Paul Barton-Dawson "B" Sand Cecil Spiro-Hamm Sand
		Basal Atoka-Spiro Sand[1]	Spiro-Orr-Kelly-Barton Sand[1]	
	Morrowan	Wapanucka Limestone[1] Union Valley Formation Cromwell Sand[1] Jefferson Sand	Bloyd Shale[1] Kessler Limestone Member Brentwood Limestone Member (Cline) Hale Formation[1] Prairie Grove Member Cane Hill Member	
Mississippian		Pitkin Limestone Caney Shale Ada Mayes Member Welden Limestone	Pitkin Limestone Fayetteville Shale Wedington Sandstone Member[1] Batesville Sandstone Moorefield Formation Boone Formation[1]	
		Woodford Shale Misener Sand	Chattanooga Shale Sylamore Sandstone Member	
Sil.-Dev.	(Hunton)[1]	Haragan Shale Chimneyhill Limestone	Penters Chert[1] Lafferty Limestone St. Clair Limestone[1] Brassfield Limestone	
Ordovician		Sylvan Shale	Cason Shale	
		Fernvale Limestone Viola Limestone[1] 2nd Wilcox Sand	Fernvale Limestone Kimmswick Limestone	
	(Simpson)	McLish Formation[1] Oil Creek Formation Joins Formation	Plattin Limestone Joachim Dolomite St. Peter Sandstone Everton Formation	
Ordovician / Cambrian	(Arbuckle)	Dolomite	Powell Dolomite Cotter Dolomite Jefferson City Dolomite Roubidoux Formation Gasconade Dolomite Eminence Dolomite Potosi Dolomite Derby Dolomite Davis Formation Bonneterre Dolomite	
Cambrian		Reagan Sandstone	Lamotte Sandstone	
Pre-C		Granite	Granite	

age forms mainly the floors of the larger valleys and consists of an ash-gray silt. Classical stratigraphic descriptions for the McAlester region of the Arkoma Basin are as follows.

McAlester Region

Atokan Series

Atoka Formation—The formation consists mainly of gray to black sandy shale with thick sandstone units interbedded at widely spaced intervals. Maximum thickness is probably greater than 9,000 ft. The shale units are brown to black, micaceous, and contain lenses and nodules of siderite. Knechtel (1949) stated that these shale units in the Atoka Formation are more silty and less clayey than those of the younger, overlying Pennsylvanian formations. The contact between the Atoka Formation and the overlying Hartshorne Sandstone needs better definition but is thought by Oakes and Knechtel (1948) to be at the base of the massive Lower Hartshorne Sandstone, some 42 ft above the thin laterally discontinuous sandstone used by Hendricks.

Desmoinesian Series

Hartshorne Formation—The Hartshorne Sandstone is the basal unit of the Desmoinesian series. This formation is conformably underlain by the Atoka Formation and conformably overlain by the McAlester Formation and is a clastic unit containing no pure carbonate rocks.

McAlester Formation—Taff (1899) described the McAlester Formation that conformably overlies the Hartshorne Formation and is named after McAlester, Pittsburg County, Oklahoma. The contact is normally placed at the top of the Hartshorne coal (Fig. 4). The formation consists mostly of gray shale and siltstones. The Booch sandstones are named after the Booch Farm, Morris Field, Okmulgee County, Oklahoma (Jordan, 1957), and are stratigraphically equal to the Warner, Cameron, and Tamaha sandstones.

Savanna Formation—The formation consists of gray shale units with interbedded black carbonaceous shales and fine-grained, silty to coarse-grained sandstones of variable thickness. The lower boundary is at the base of the lowermost sandstone. The Savanna Sandstone overlies the McAlester Formation with an irregular contact, although at some places, the transition from one formation to another appears to be gradational (Hendricks et al, 1939). In the northeast part of the McAlester Basin, the Savanna Formation changes facies into the Spaniard Limestone (Knechtel, 1949). The lower boundary of the Savanna Formation has been placed by Knechtel (1949) at the base of the Spaniard Limestone in the northeastern shelf area (Fig. 2). The upper boundary of the Savanna Formation is at the base of the Bluejacket Sandstone member of the overlying Boggy Formation.

Boggy Formation—The Boggy Shale, as it is commonly called, consists of alternating gray shales and sandstones. The Thurman Sandstone member ranges from 250 to 335 ft thick in the Lehigh and Quinton-Scipio Districts. A basal bed of chert-pebble conglomerate is rather widespread (Dane et al, 1938). The basal conglomerate rests with a sharp contact and local erosional irregularity on the underlying Boggy Shale. The Stuart Shale consists of dark laminated shales, which sometimes carry ferruginous or calcareous nodules and concretionary beds.

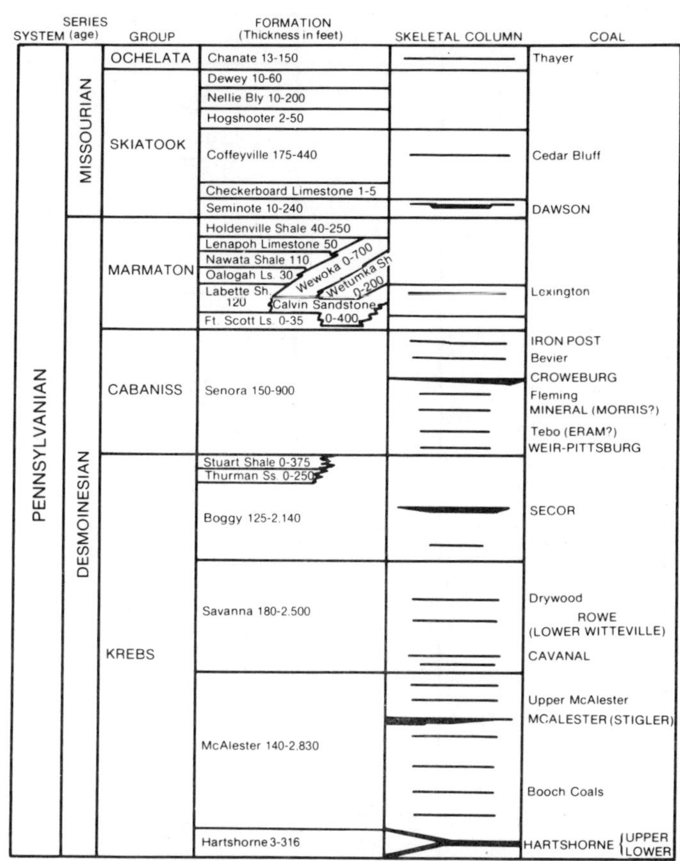

Figure 4—Generalized geologic column showing the sequence of coalbeds of Pennsylvanian age in Oklahoma. Friedman, 1974. Used with the permission of the Oklahoma Geological Survey.

Senora Formation—The Senora exists only as a thin formation in the Quinton-Scipio District and consists of sandstone with some shale beds.

Of the 19 or 20 coalbeds in the Arkansas Valley investigated by state and federal agencies, only four have been considered to be of economic importance by the U.S. Geological Survey (USGS). These are the lower Hartshorne, upper Hartshorne, Charleston, and Paris coalbeds (Fig. 5).

The stratigraphic boundaries of the post-Atoka rocks established by Hendricks and Parks (1950) are used in this report. The stratigraphic nomenclature of these rocks as reported in Hendricks and Parks (1950) was modified by Merewether and Haley (1961), who adopted Miser's (1954) terminology for equivalent rocks in Oklahoma of the McAlester, Savanna, and Boggy Formations, and Oakes' (1953) classification of the Krebs Group, which includes the Hartshorne Sandstone and the McAlester, Savanna, and Boggy Formations. Haley (1961) moves the base of the Boggy Formation stratigraphically upward, thereby conforming to the definition established by Miser (1954). The formations as described here for the Arkansas Valley—with the exception of the Hartshorne Sandstone—are stratigraphically equivalent to the formations in the type areas in Oklahoma (Haley and Hendricks, 1972). The stratigraphic nomenclature and boundaries of Atoka rocks introduced in this section conform, in general, to those used elsewhere in Arkansas.

Rocks of the Pennsylvanian and Quaternary systems are exposed or have been penetrated by wells drilled for gas in the

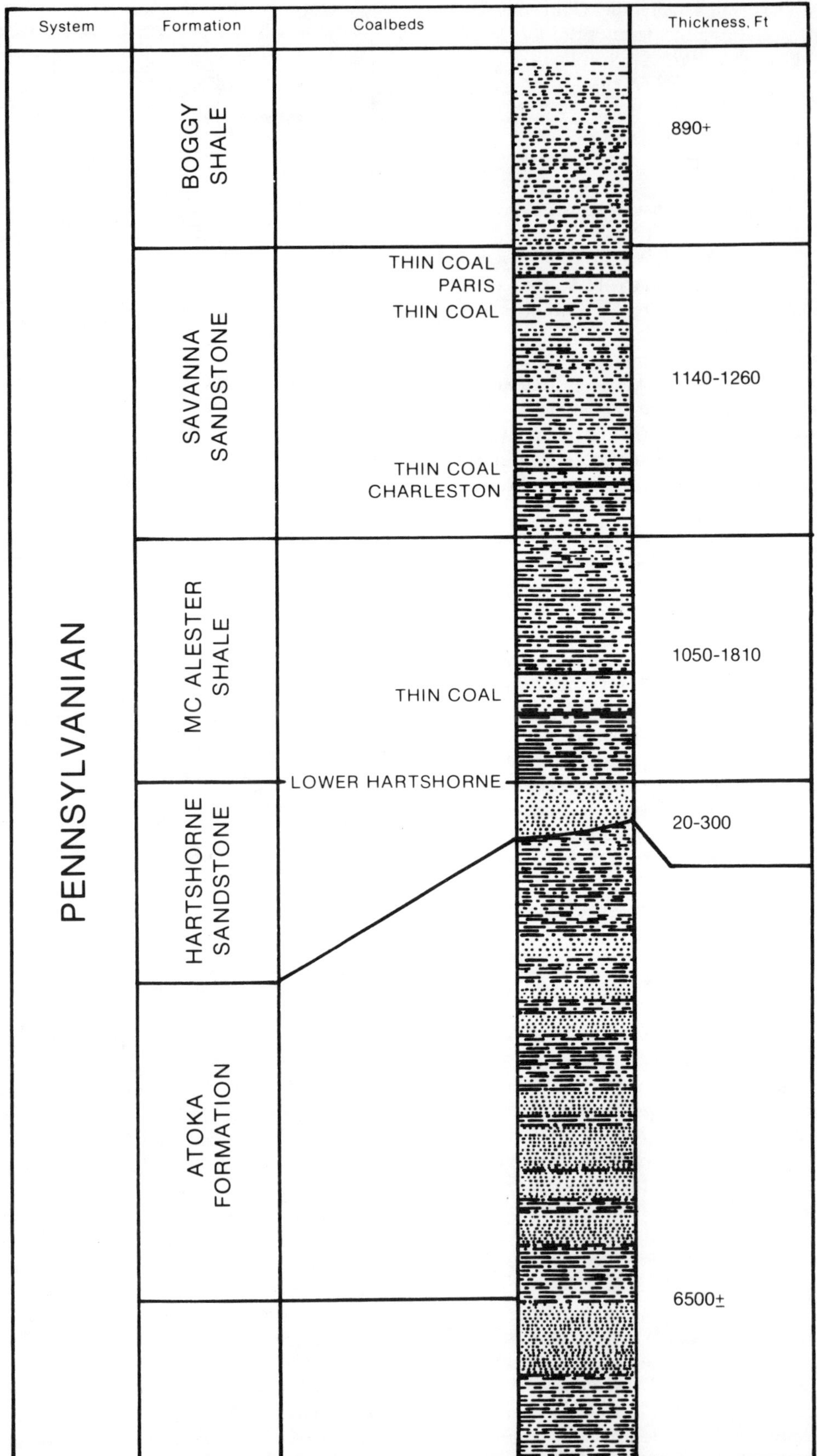

Figure 5—Generalized geologic column showing the sequence of coalbeds of Pennsylvanian age in Arkansas. From the Keystone Coal Industry Manual. Copyright © 1979 McGraw-Hill Publishing Co. Used with permission.

Arkansas Valley. The Pennsylvanian rocks, from older to younger, are the Atoka Series and the Krebs Group. The Quaternary rocks are stream and river terrace deposits of Pleistocene age (Hendricks and Parks, 1950) and stream and river alluvium of recent age. The Hartshorne Sandstone in the western part of the Arkansas Valley is equivalent to the lowermost sandstone in the Hartshorne Sandstone in Oklahoma. Stratigraphic descriptions, along with subsurface interpretations for the Arkansas Valley region of the Arkoma Basin, are as follows.

Arkansas Valley Region

Atokan Series

Atoka Formation—Henbest (1953) described the contact between the Atoka Formation and the Morrow Series and named the basal sandstone of the Atoka the Greenland Sandstone member of the Atoka Formation. According to Henbest, the Greenland Sandstone member consists of silty, ripple-marked, flaggy sandstone with shaley partings and locally interfingering marine quartz-gravel conglomerate.

This unit of sandstone (or a lithologically similar unit) is at the base of the Atoka Formation throughout most of northwestern Arkansas. In the Arkansas Valley region, the contact between the Atoka Formation and the Morrow is not exposed but has been penetrated by wells drilled for gas or oil (Haley and Hendricks, 1972). In preparing a stratigraphic log of the rock cuttings from any one of these wells, the contact is placed at the base of the first sandstone unit overlying a unit of rocks similar to those of the Morrow (shale and limey sandstone, sandy limestone, and limestone). The lithologic character of this basal sandstone unit of the Atoka Formation, as identified from the well cuttings in the Arkansas Valley region and in the Delaware quadrangle, is similar to those of the Greenland Sandstone member as described by Henbest. However, the regional decrease in amount of carbonate and increase in amount of sand in the rocks of the upper part of the Morrow Series make them increasingly similar to the rocks in the lower part of the Atoka Formation south and east of Washington County, Arkansas; thus, the two are not easily separated south, southeast, and east of the Delaware quadrangle.

Hendricks and Parks (1950) reported a minor unconformity between the Atoka Formation and the overlying Hartshorne Sandstone. The basal sandstone bed of the Hartshorne Sandstone overlies the uppermost bed (generally dark-gray shale or silty shale) of the Atoka Formation with a channel-type relationship that Hendricks and Parks (1950) have interpreted as a minor unconformity. The stratigraphic sections show a convergence of the base of the Hartshorne Sandstone and the top of the youngest sandstone of the Atoka in the western Arkansas Valley.

Desmoinesian Series

Hartshorne Sandstone—The Hartshorne Sandstone of Hendricks and Parks (1950) is equivalent to the lowermost sandstone in the Hartshorne Sandstone underlying the Lower Hartshorne coalbed in the type area near Hartshorne, Oklahoma. The minor unconformity between the Hartshorne Sandstone and the Atoka Formation is considered present in the Arkansas Valley region. The Hartshorne Sandstone is conformably overlain by the McAlester Formation.

The Hartshorne Sandstone is one of the most persistent sandstone units in Arkansas. As a sandstone unit, it has several lithologic characteristics that tend to set it apart from most of the sandstone units in the overlying McAlester Formation or in the underlying Atoka Formation. In general, the Hartshorne Sandstone is lighter in color (grayish white to light gray), coarser in grain size (very fine to medium), less silty or clayey, and more widespread. Pebbles of shale and siltstone are present near the base of some sandstone beds (Haley and Hendricks, 1972).

McAlester Formation—The McAlester conformably overlies the Hartshorne Sandstone and is overlain by the Savanna Formation with a contact that, according to Hendricks and Parks (1950), represents a minor unconformity. Haley (1961) reported that the McAlester and Savanna Formations in the vicinity of Paris, Arkansas, have an interfingering relationship. Merewether and Haley (1969) stated that the formations interfinger in the vicinity of Clarksville, Arkansas. In the southeastern part of the Van Buren quadrangle where the basal sandstone of the Savanna is in channel cuts in shale of the McAlester, the contact appears unconformable.

The McAlester Formation is mostly shale, with minor amounts of siltstone and sandstone. The shale is dark gray, nonsilty, and fissile bedded for the most part. Pyrite and ironstone concretions are common in all the shale.

Savanna Formation—The sandstone in the Savanna is finer grained, siltier, and more argillaceous than the sandstone in the McAlester, Hartshorne, or Atoka. In some places, the sandstone and siltstone weather yellowish or greenish brown, whereas the sandstone and siltstone in the other three formations generally weather light gray or grayish white. Thin coalbeds are common in the lower part of the formation. The Charleston coalbed is near the base of the formation, and the Paris coalbed is in the upper part.

The lower part of the Savanna Formation interfingers with the upper part of the McAlester Formation in the eastern part of the Paris quadrangle (Haley, 1961). The same relation between the two formations exists in the Scranton and New Blaine quadrangles. The shale is mostly dark gray, micaceous, and slightly silty to silty. Some of the shale is light to medium gray, and from the appearance of the rock samples from shallow exploration wells, it is nonbedded. Thus, it could be either claystone or underclay. Ironstone concretions are common in the shale, and pyrite is present in some places.

Boggy Formation—There are a few remnants of the Boggy Formation in the Arkansas Valley region located in Logan County, Arkansas. Part of the basal sandstone unit caps Short and Little Short Mountains.

Figure 6 illustrates the relation between general structure and stratigraphy for the Greenwood quadrangle in the western part of the Arkansas Valley region.

COAL RESOURCES

Accurate coalbed correlation is an essential prerequisite to determination of coal resources and reserves, especially particularly gassy beds. The bituminous coals of Arkoma Basin are shown in sequence in generalized geologic columns (Figs. 4, 5). Although some minor coals have been noted or described in the geologic literature, they are not shown in the geologic column, having been omitted from the present coalbed

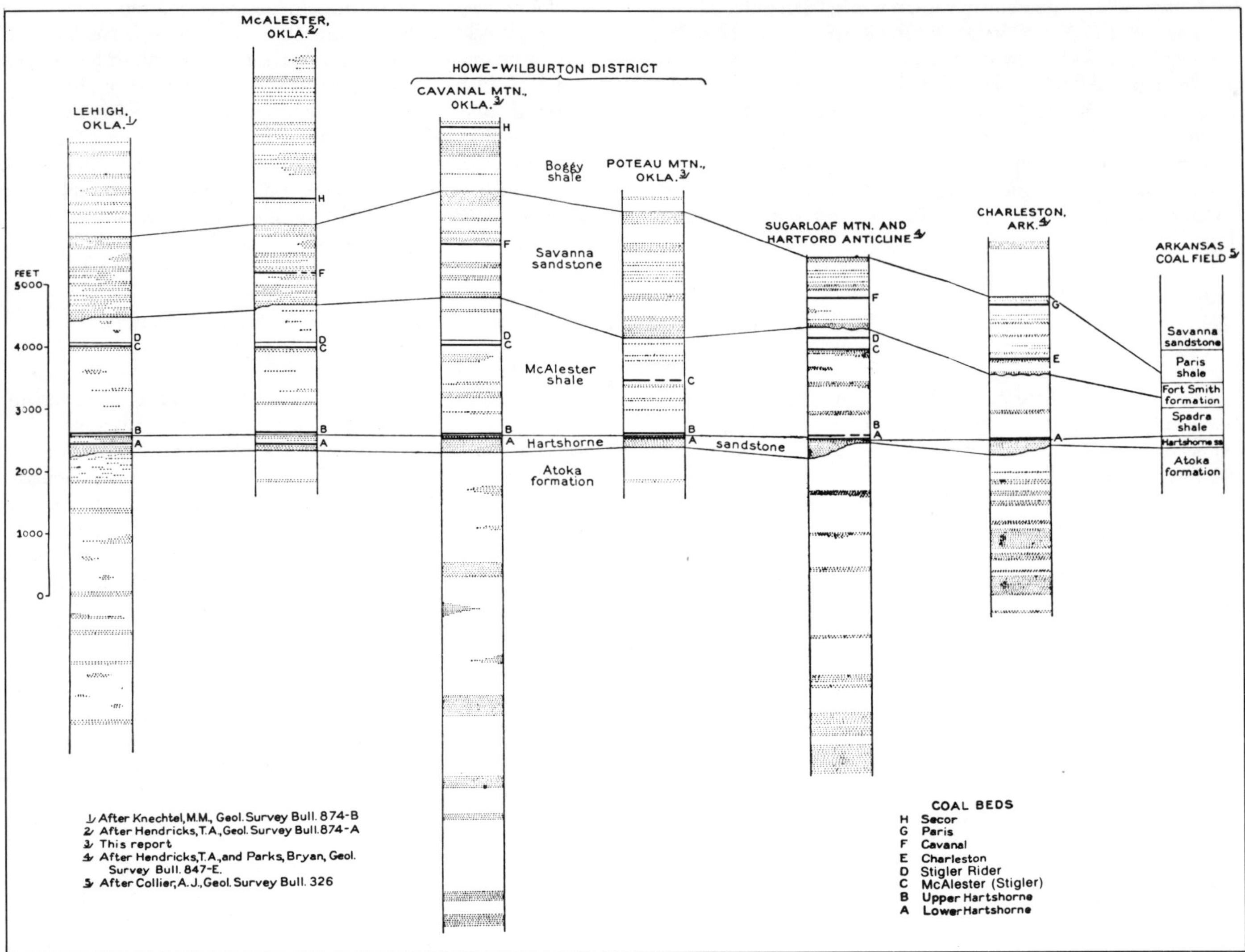

Figure 6—A generalized correlation of Pennsylvanian formations in the Howe-Wilburton district with those of the McAlester and Lehigh Districts, Oklahoma, and the Arkansas Districts. From Hendricks et al, 1939.

methane investigation because of insufficient information. A generalized correlation of the Pennsylvanian age formations of the Lehigh District in the western part of the Arkoma Basin with those in Arkansas is shown in Figure 6 (after Hendricks et al, 1939).

Friedman (1974) noted that there are certain stratigraphic sequences and correlation uncertainties in the McAlester region of the Arkoma Basin, such as the four coals present in the McAlester Formation above the Hartshorne coal and beneath the McAlester, Stigler coal. The McAlester coal appears to correlate with the Stigler. The upper McAlester coal appears to correlate with a "rider" coal above the Stigler. At least one additional coal is present above this rider coal in the McAlester Formation in Le Flore County.

As many as four coals are present in the "Cavanal coal zone" in the lower part of the Savanna Formation in Le Flore County. A previously undetected coal occurs about midway between the Secor (Upper Witteville) and the Lower Witteville coals in Cavanal Mountain, Le Flore County. This coal may be the one present 30 to 45 ft below the Secor in Pittsburg County in places where the Lower Witteville cannot be identified (Friedman, 1974).

An unnamed coal lies some 30 to 50 ft above the Secor coal in Pittsburg County, and it may or may not be equivalent to the Mayberry coal in Cavanal Mountain in Le Flore County. Another unnamed coal lies at least 100 ft stratigraphically above this "rider" of the Secor coal, and if it is in the Boggy Formation, it cannot be equivalent to the Weir-Pittsburg coal of the Northeast Oklahoma shelf area.

Strictly speaking, the Hartshorne Sandstone forms the base of the commercial coal-bearing portion of the Pennsylvanian rock sequence in the coal basin. Coalbeds are present in the Atoka Formation of Arkansas (Haley and Hendricks, 1972). The uppermost known coal occurs in the basal portion of the Boggy Shale in the McAlester District, to the northeast, but in the Lehigh District, no beds of coal are known to occur in rocks younger than the McAlester Shale.

Because of the deltaic environment in which the coal was deposited, available information concerning these stratigraphic units varies significantly. This is reflected in the following discussion.

Atoka Coals

All coalbeds in the Atoka Formation are generally thin, of poor quality, and of local extent (Haley, 1966). Three coalbeds (2, 1, and 1 ft thick) were penetrated by the Hambee Well No. 1, T8N, R26W, Sec. 13, Logan County, Arkansas, at depths of 1,800, 2,305, and 2,977 ft, respectively. The 2 ft thick bed is an exception, inasmuch as the Atoka coals encountered are generally less than 15 inches thick. Sulfur content ranges from 0.6 to 5.7% in Logan, Johnston, and Pope counties, Arkansas (Haley, 1977).

Donica (1978) reported that two thin coals, each 6 inches thick, occur 10 and 40 ft, respectively, below the top of the Atoka Formation in parts of Le Flore County, Oklahoma (Fig. 7). The exact stratigraphic position of these coals is undecided (Hendricks et al, 1939; Friedman, 1978).

Hartshorne Coals

At the southern edge of the coal region in Oklahoma, the Hartshorne coal commonly is split into two beds by shale and sandstone 5 to 100 ft thick (Friedman, 1974). These beds are known as the Upper and Lower Hartshorne coals and have been mined extensively. Oakes and Knechtel (1948) and Knechtel (1949) recognized a convergence between the Upper and Lower Hartshorne coals and redefined the Hartshorne Formation to include both coals. Until this redefinition, the Upper Hartshorne coal had been placed in the overlying McAlester Formation (Fig. 8). Merewether and Haley (1969) reported that the thickness of the split in Johnson County, Arkansas, ranges from 1 inch to about 20 ft and is less than 12 inches in much of this coal-bearing area. The split apparently changes rapidly and without uniformity in either rate or direction of thickening. In the Greenwood quadrangle (Sebastian County, Arkansas), the Upper Hartshorne is present only in the southern part of the quadrangle. The interval between the Upper and Lower coals is from 60 to 90 ft (Haley and Hendricks, 1968). Northward, in the middle of the Arkoma Basin, the Hartshorne coal "undivided" is a single bed containing a persistent black-shale parting about 1 inch thick. In parts of the basin, three coals occur within the Hartshorne Formation. This third unit is known as the Middle Hartshorne coal and is laterally discontinuous in Le Flore County, Oklahoma, where it is present. In the McAlester District, the Hartshorne coal can exist as three separate beds as interpreted from geophysical well logs. Recent core drilling has indicated significant underground coal resources in the Hartshorne coal in areas where it is 3 to 5 ft thick and of low- or medium-volatile rank. The Hartshorne has the qualities of a good coking coal with a high, free-swelling index. In the vicinity of the Arkansas River in northern Le Flore County, it is essentially noncaking. The Hartshorne coalbed contains 0.5 to 6.0% sulfur and averages 1.8% (raw). This coal requires cleaning for use in coke manufacture and metallurgical processes.

The Lower Hartshorne coal is 0.7 to 7.0 ft thick and occurs about 60 ft above the base of the Hartshorne Formation within a gray shale interval that separates the Upper and Lower Hartshorne Sandstone units. It has been mined recently in Le Flore County in underground mines, where it is 3.0 to 3.7 ft thick. This bed is the thickest, most extensive, and most economically important coal in the basin. The bed is underlain by a thin underclay. The Lower Hartshorne coalbed contains 0.4 to 5.1% sulfur and averages 1.0% (raw) in Oklahoma. The range in sulfur content in Arkansas is 0.4 to 4.6%. The Lower Hartshorne coalbed contains one to four irregular shale partings that persist locally. The partings consist mainly of black carbonaceous shale and range in thickness from 1/32 to 1/4 inch.

The Upper Hartshorne coal has been extensively mined from slopes and drifts in Haskell, Latimer, and Le Flore counties. It is 2 to 4 ft thick and is low- or medium-volatile in rank on the east side of the Arkoma Basin and high-volatile on the west side. The upper coalbed is relatively free of shale and bone partings. It contains 0.8 to 2.6% sulfur and averages 1.6% (Friedman, 1974).

McAlester Coals

The McAlester coal was extensively mined by underground methods at McAlester in central Pittsburg County and in southeastern Coal County, where it had been called the Lower McAlester (Friedman, 1974). Significant resources of this coal remain in these areas and are amenable to underground mining. The McAlester coalbed is 1.5 to 5.0 ft thick and is mostly high-volatile in rank. It is not mined at present but is suitable for use in electric power generation; for blending with higher rank coal for coke manufacture; and for gasification and liquefaction. The McAlester coalbed contains 0.8 to 4.8% sulfur and averages 2.1%.

A correlative of the McAlester coal, the Stigler coal, has been mined by surface methods in Haskell, Le Flore, Muskogee, and Sequoyah counties, Oklahoma. Hendricks' (1937) suggestion that the coal above the Lower McAlester coal may be correlative with the Stigler coal of northern Le Flore and Haskell counties has not yet been determined (Donica, 1978). The Stigler coal may correlate with the Lower McAlester coal. The upper coal, named tentatively by Hendricks in 1939 as being the Stigler, may be more appropriately named the Stigler rider coal (Friedman, 1979, personal communication) (Fig. 6). Hendricks (1939) reported that four thin coalbeds (1 ft or less in thickness) are present at least locally between the Stigler (?, upper coal) and the top of the McAlester Shale in Le Flore County (see Donica, 1978). In the McAlester District, numerous thin coalbeds are present a short distance above the McAlester coal, and some of these are present over considerable areas (Hendricks, 1939). Craney (1978) reported that in the Panama quadrangle, the McCurtain Shale contains three unnamed thin coals underlain by sandstones.

There are several unnamed coalbeds in the McAlester Formation in the Arkansas Valley region (Haley, 1968). Haley and Hendricks (1968) reported that in the Greenwood quadrangle, Sebastian County, Arkansas, and eastern Le Flore County, Oklahoma, the Stigler coal was near the top of the McAlester Formation and the McAlester coalbed was near the middle. Haley and Hendricks misinterpreted the upper coal as being the Stigler; it is probably the Stigler Rider coal (see Friedman, 1974).

In the McAlester District, the McAlester coalbed contains more partings and bands of impurities than the Lower and Upper Hartshorne beds—as many as seven partings have been recorded (Hendricks, 1937). Partings consist of pyrite.

Of low- and medium-volatile rank, the Stigler coal is used in coke manufacture. The Stigler coal contains 0.4 to 5.2% sulfur and averages 1.5% in Oklahoma.

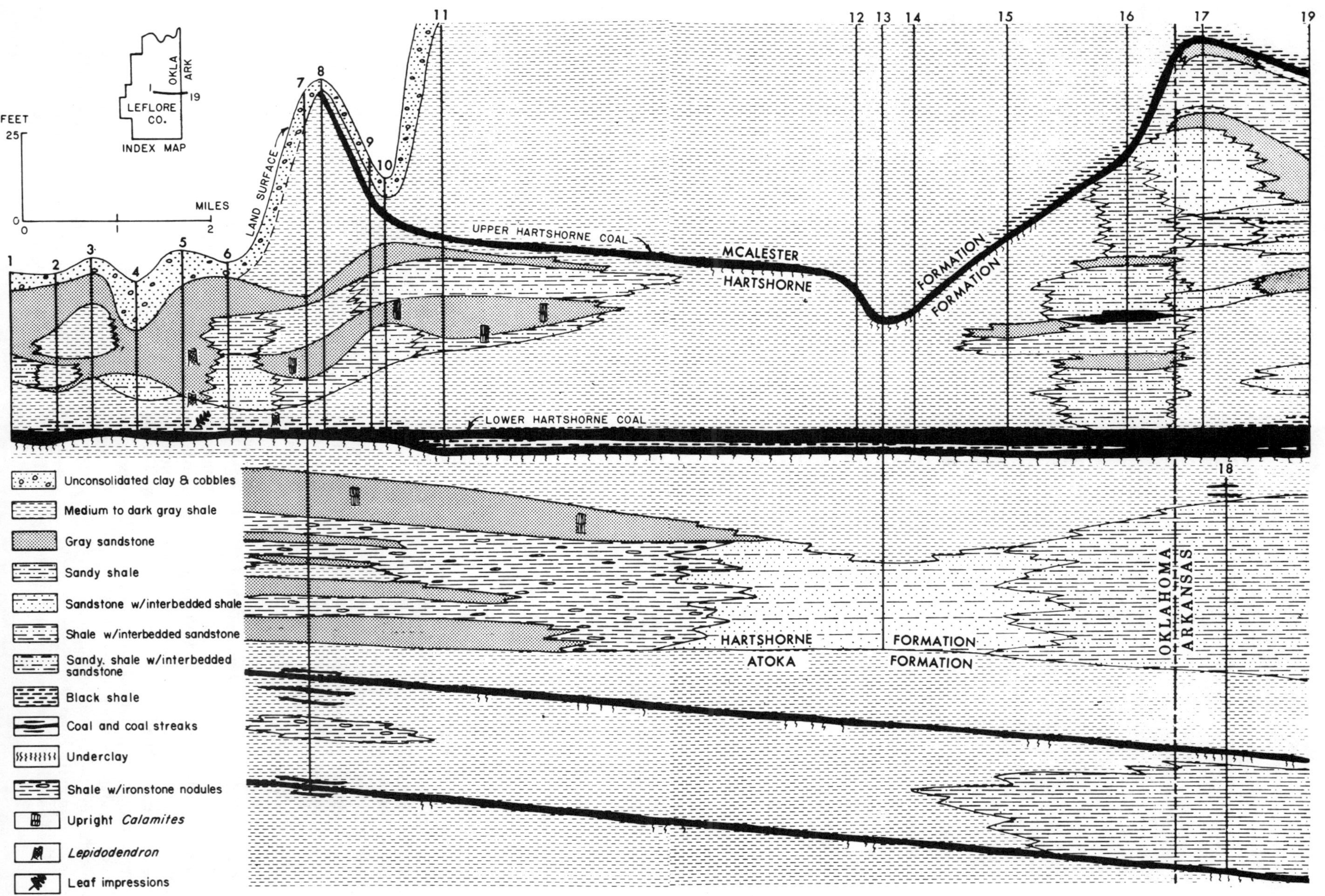

Figure 7—Cross section showing stratigraphic relations of coals and sandstones in the Hartshorne and Atoka Formations, Le Flore County, Oklahoma, and Sebastian County, Arkansas. From Friedman, 1978.

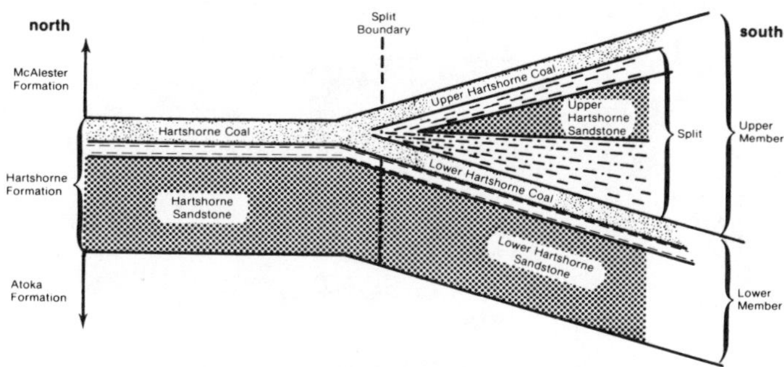

Figure 8—Diagram showing the definition of the Hartshorne Formation in this study. Modified from Oakes and Knechtel, 1948, in Trumbull, 1957. Used with the permission of the Oklahoma Geological Survey.

Savanna Coals

There are numerous local coalbeds in the Savanna Sandstone, but only two could be of importance in coalbed methane determinations.

The Cavanal coal, named after Cavanal Station in Le Flore County, Oklahoma, is 1.2 to 3.5 ft thick, has a high (+100) Hardgrove grindability index, and is mostly medium-volatile and in part high-volatile in rank. It is overlain by 20 to 50 ft of shale and sandstone, at the top of which another, thinner coal is present at some places (Friedman, 1974). The Cavanal coal contains 2.1 to 4.8% sulfur and averages 3.3%.

Dane et al (1938) did not report the existence of the Cavanal coal in the Quinton-Scipio District. The Cavanal coal is present in the Howe-Wilburton District. Donica (1978) reported an Upper Cavanal coal lying 20 to 50 ft above the Cavanal coal in the Heavener area, Oklahoma. Hendricks (1937) reported a coalbed about 2 ft thick and younger than the Cavanal coal, just south of Krebs in the McAlester District.

The Lower Witteville coal is high ash, high sulfur, and medium volatile, and is present in Cavanal Mountain, Le Flore County, where it has been mined underground. It is 3 to 4 ft thick, with one or more clay-stone partings, and the underlying shale contains numerous coal stringers. The Lower Witteville coal contains 4.4% sulfur (Friedman, 1974).

The Rowe coal is a high-volatile coal of the Northeast Oklahoma shelf area (Fig. 2), believed to be a correlative of the Lower Witteville. The Rowe coalbed is 1.0 to 2.5 ft thick in Craig, Mayes, Muskogee, Rogers, and Wagoner counties. The Rowe coal may be suitable for gasification and liquefaction, although it was considered of only marginal value, owing to its high sulfur and ash contents and its dull appearance. The Rowe coal contains more than 13,000 Btu in mine-run condition. The sulfur content is 2.8 to 3.4%, and averages 3.1% (mine run) (Friedman, 1974).

The Savanna coalbeds in Arkansas are poorly developed, but only the Charleston and the Paris are sufficiently thick and extensive to be of any economic importance. The coalbed mined as the Charleston coal probably is a single continuous bed in Arkansas; but like most of the coalbeds in the Savanna, it is probably lenticular and of local extent (Haley and Hendricks, 1972).

The Charleston coal is near the base of the Savanna (possibly correlative to the Cavanal) and is not known to be more than 2 ft thick. Sulfur content ranges from 0.9 to 3.8%.

The Paris coalbed is the thickest and most extensive coalbed in the Savanna Formation in Arkansas. In the Paris quadrangle, it ranges in thickness from 19 to 32 inches and is covered by less than 250 ft of overburden (Haley, 1961). There is a thin coal rider above the Paris coal, approximately 50 ft above the Paris in Sebastian County, that ranges in thickness from 4 to 9 inches. Sulfur content ranges from 0.6 to 3.3% (Haley, 1977).

Boggy Coals

Various coals exist in the Boggy Formation, above and below the Secor, and are very lenticular. No Boggy coals are known to exist in the Arkansas Valley region.

The Secor coal is 1.5 to 4.3 ft thick in places where it has been mined. It is a medium-volatile bituminous coal in Le Flore County and a high-volatile coal in Haskell, Pittsburg, Muskogee, and McIntosh counties. This coalbed commonly contains one or two shale partings and is high in ash and sulfur contents. Hendricks (1937) measured sections of the Secor coal in the McAlester District. The Secor is well-developed in the Quinton-Scipio District (Dane et al, 1938). In places where it is of high-volatile rank, the Secor coal probably is suitable for use in gasification and liquefaction processes (Friedman, 1974).

Recent exploratory drilling indicates that in Pittsburg County, the Secor (3 ft thick) can be mined by surface and underground methods and that additional recoverable reserves in Le Flore County are amenable to underground mining. The coal contains 3.5 to 6.6% sulfur and averages 4.9% (raw).

Senora Coals

No Senora coals are known to exist in the Arkoma Basin, as defined by Figure 1. Although the Eram, Morris, and Croweburg coals abut the northwestern edge of the subshelf portion of the basin in Okmulgee and Hughes counties, Oklahoma, they have been excluded from this discussion owing to the lack of burial depth and evidence of a gassy nature.

Coal Quality

Analyses show that coals of the Arkoma Basin generally fall within the following range of values, although some compositions are known to be both higher and lower.

Moisture (%)	1.0–7.0
Ash (%)	4–11
Sulfur (%)	1–5
Btu/lb	11,500–15,000

Friedman (1974) integrated data from more than 600 logs of boreholes drilled for coal during the past 20 years with some 200 proximate, sulfur, Btu, and ash-fusion analyses provided by private parties involved in coal mining and/or exploration. In addition, analyses that appeared in various publications were incorporated. The average analyses of the Krebs Group of coals in eastern Oklahoma are presented in Table 3. Similar results for the Arkansas Valley are given in Table 4 (Haley, 1977).

The sulfur content of most of the coal resources in the Arkoma Basin area is low compared with those resources of the northeast Oklahoma shelf area. For example, the original and remaining coal resources of Haskell, Pittsburg, Latimer, Le Flore, and Sequoyah counties in the Arkoma Basin average less than 2% sulfur. Exceptions are Atoka and Coal counties, whose

Table 3—On a county basis, average analyses of coals in eastern Oklahoma. From Friedman, 1974. Used with the permission of the Oklahoma Geological Survey.

		Typical Proximate Analyses (as received, %)								
County	Coalbed	Moisture	Volatile	Fixed	Ash	Sulfur (%)	Btu	Ash Softening Temp. (°F)	No. of Anal.	Data
Haskell	Hartshorne	3.1	22.0	68.2	6.7	0.9	13,960	2,200	23	Pre-1953
Haskell	Hartshorne	Dry	22.3	69.1	8.6	1.8	14,346	—	64	1958-1968
Le Flore	Hartshorne	2.4	20.6	71.4	5.6	1.0	14,190	2,250	5	Pre-1953
Sequoyah	Hartshorne	5.3	16.8	72.4	5.5	1.7	13,980	2,170	1	Pre-1953
Coal	Lower Hartshorne	5.9	35.7	50.5	7.9	1.4	13,782	—	1	1915
Haskell	Lower Hartshorne	0.2	21.3	72.1	6.4	0.8	14,233	—	1	1965
Latimer	Lower Hartshorne	4.7	35.9	53.5	5.6	1.4	13,450	2,030	21	Pre-1953
Le Flore	Lower Hartshorne	2.9	17.8	72.3	7.1	0.8	14,000	2,100	23	Pre-1953
Le Flore	Lower Hartshorne	Dry	17.3	73.5	9.1	0.8	—	—	47	1957-1968
Pittsburg	Lower Hartshorne	3.6	37.2	52.8	6.5	1.5	13,490	2,080	39	Pre-1953
Haskell	Upper Hartshorne	0.04	21.0	71.9	7.0	1.3	13,969	—	1	1965
Latimer	Upper Hartshorne	3.4	37.1	55.0	5.3	1.0	13,590	2,190	11	Pre-1953
Le Flore	Upper Hartshorne	2.5	21.3	66.3	10.1	4.1	13,702	2,340	2	Pre-1953
Pittsburg	Upper McAlester	2.3	38.9	50.4	8.3	4.1	13,327	—	1	—
Pittsburg	Upper Hartshorne	4.5	35.4	53.7	6.5	1.5	13,230	2,000	6	Pre-1953
Coal	McAlester	6.8	38.6	44.3	10.4	3.6	11,590	2,180	23	Pre-1953
Latimer	McAlester	2.5	31.6	54.2	11.7	3.2	—	—	1	Pre-1953
Pittsburg	McAlester	3.3	36.1	55.1	5.6	0.9	13,640	2,210	59	Pre-1953
Haskell	Stigler	3.1	25.4	67.3	4.2	1.2	14,430	2,180	13	Pre-1953
Haskell	Stigler	5.0	27.4	63.2	4.4	1.1	13,869	—	3	1972
Muskogee	Stigler	2.3	30.5	60.6	6.6	3.7	14,110	2,340	1	Pre-1953
Le Flore	Cavanal	2.3	22.0	66.3	9.4	3.4	13,840	2,410	4	Pre-1953
Le Flore	Cavanal	2.0	20.9	66.2	10.9	4.7	13,580	—	1	1969
Le Flore	Lower Witteville	1.7	22.1	63.0	13.3	4.3	13,180	—	5	Pre-1953
Rogers	Rowe	1.9	35.6	54.6	7.9	2.3	13,995	—	3	1971
Rogers	Rowe	0.7	38.3	47.9	13.2	3.4	13,348	—	1	1973
Le Flore	Secor	2.4	18.3	56.7	22.6	4.1	11,711	—	1	1972
Pittsburg	Secor	2.5	37.2	48.4	11.8	5.4	12,880	2,240	3	Pre-1953
Pittsburg	Secor	3.0	34.5	46.2	16.5	5.9	11,065	—	2	1973
Okmulgee	Croweburg	7.1	34.5	52.8	5.6	1.6	12,910	2,020	25	1973
Rogers	Croweburg	7.1	34.7	51.9	6.3	2.0	12,780	2,280	17	1973
Rogers	Croweburg	Dry	33.3	59.7	7.0	0.4	12,690	—	1	1965

Table 4—Average analysis of Arkansas coals on an as-received basis. Haley, 1977; Keystone Industry Coal Manual. Copyright © 1979 McGraw-Hill Publishing Co. Used with permission.

County	Coal Seam	No. of Samples	% Moisture	% Vol. Matter	% Fixed Carbon	% Ash	% Sulfur
Johnson	Lower Hartshorne	30	2.5	12.6	75.4	9.6	2.6
Logan	Lower Hartshorne	6	3.6	14.4	78.1	9.5	1.35
Pope	Lower Hartshorne	6	2.1	12.3	75.1	10.6	2.4
Scott	Lower Hartshorne	19	4.05	23.6	67.4	10.5	1.3
Sebastian	Lower Hartshorne	72	2.4	17.5	72.8	7.3	1.1
Franklin	Charleston	12	2.1	19.4	76.1	5.7	3.5
Sebastian	Charleston	2	3.8	16.3	77.3	4.5	2.2
Franklin	Paris	3	1.0	20.5	76.1	4.7	2.4
Logan	Paris	63	1.8	68.9	11.5	11.5	2.0

coal resources average 4.1 and 5.0% sulfur, respectively. Coal resources of the counties in the shelf area (Fig. 2) average more than 3.0%. An exception to this is the Croweburg coal (Henryetta coal) in Okmulgee, Okfuskee, and Rogers counties. Here, the Croweburg averages 2.2, 2.3, and 0.8% sulfur, respectively.

The sulfur content of coal is influenced by the depositional origin of the coal, as well as by the diagenetic and postdiagenetic changes affecting the coalbeds. In addition, deep-lying bituminous coalbeds tend to contain less sulfur than shallow-lying coalbeds. Friedman (1974) stated that the low- and medium-volatile coals of Le Flore and Haskell counties tend to contain less sulfur than the high-volatile coals of the northeast Oklahoma shelf area, with two exceptions: The Secor coal of the Arkoma Basin is high in sulfur content, and the Croweburg coal of the shelf is low in sulfur content. Variations in percent of sulfur for different coalbeds on an average county basis are given in Tables 3 and 4.

In coalbed methane studies, Btu content of the coalbed is extremely important in evaluation of the resource and variations in Btu content are noted for various coalbeds on a county basis. These analyses are given in Tables 3 and 4. It is interesting to note that the carbon/oxygen (C/O) atom density ratio in coal is related to the Btu content, as indicated in Figure 9, and a C/O well logging tool would be useful in evaluating coals.

For quite some time, it has been known that coal rank increases with depth in many basins. This can be observed using different rank parameters obtained from core samples selected at different depths. The basic assumption, however, is that the stratigraphic column tested is essentially the same as the stratigraphic column at the time the rocks were most deeply buried and that subsequently they were not markedly disturbed. Thus, if the rocks were subsequently faulted or folded, as in the Arkoma Basin, the relationship of rank versus depth of burial may not apply.

Coal rank in the Arkoma Basin increases from west to east (Fig. 10). These coals cover a wide range in rank, from high-volatile bituminous B to semianthracite. There is a progressive increase in rank from high-volatile in the western part of the basin to semianthracite in Johnson, Logan, and Pope counties, Arkansas.

Hendricks et al (1939) prepared an isocarb map based on analyses of the coals ranging in age from the Lower Hartshorne to the Croweburg (Henryetta) and covering both the McAlester and Arkansas Valley regions, as shown in Figure 11. Most of the figures for fixed carbon used in the preparation of the map represent an average of several analyses. In general, there is an increase in fixed carbon in the coals from west to east across the Arkoma Basin—from 51.2% near Atoka, Oklahoma, to 88.4% near Russellville, Arkansas. The Lower Hartshorne coal, however, does not have an apparent increase in fixed carbon over the Lower Witteville coal, although it is 4,800 ft stratigraphically lower (Hendricks, 1935)

Fuller (1920) presented data showing the relationship of fixed carbon to structural deformation. Table 5 illustrates that folding of the rocks in the Arkoma Basin resulted in more fixed carbon than was produced by faulting and also shows comparisons of the fixed carbon in the coals within the folded structures. Coals in the anticlines have a higher percentage of fixed carbon than those in the adjacent synclines (Craney, 1978). The fixed carbon content of the coals in the basin probably is related to structural deformation.

Examination of detailed isocarb maps of the Hartshorne coals reveals some regional trends. The Lower Hartshorne coal in the Heavener area is low- and medium-volatile bituminous in rank (Fig. 10) (Donica, 1978). The boundary between the two ranks (78% isocarb contour) is accentuated on the isocarb map (Fig. 11) and trends east-west through the area. The fixed-carbon percentage (mmf) ranges from 70 to 84 in the Heavener area and fairly constantly decreases northeast to south (Donica, 1978).

Iannacchione and Puglio (1979) reported on cleat directions measured at strip mines and surface exposures of coal in Le Flore County, Oklahoma. They then analyzed cleat data using a method devised by Diamond et al (1976). Results of these analyses show that the direction of face cleat varies from N 32° W to N 17° W, perpendicular to structural trends in the area of Le Flore County. Butt cleat directions vary from N 52° E to N 77° E and are generally parallel to structural trends in the area.

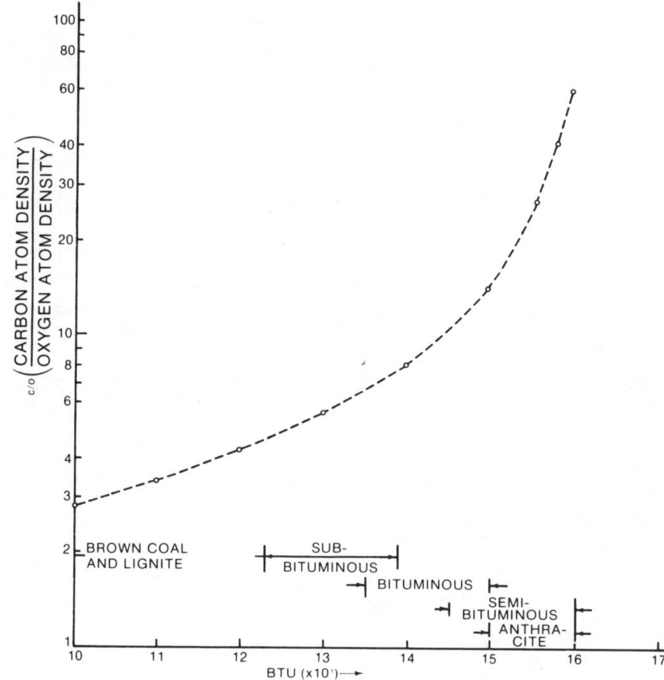

Figure 9—General relation between C/O ratio and Btu content of coals. U.S. Patent 3,849,646.

The friability of the Hartshorne coalbed is due to close spacing of cleats and frequent occurrence of shear fractures with dips of 45° to 55° within the coalbed. Directional permeability characteristics of coalbeds are generally dependent on the orientation of these cleat fractures.

At the Howe Coal Company mine, the Lower Hartshorne coalbed, located 6 mi south of Poteau, Le Flore County, Oklahoma, on the southern border of the Arkoma Basin, the strike of the butt cleat is N 74° E, subparallel to the local axial trend (McCulloch et al, 1974). The strike of the nonsystematic rock joints in this area is N 72° E, also subparallel to the regional trend. The face cleat in the coal and the systematic rock joint both strike N 15° W—subperpendicular to the trend and, therefore, parallel to the axis of compression.

The face cleats tend to be perpendicular to the axial trend of the folds and probably formed as extension fractures. The butt cleats tend to be parallel to the axial trend and formed after compression forces were released.

Cleats in the McAlester (Stigler) coal are so closely spaced (1/8 to 1/2 inch) that they contribute to the friability and high Hardgrove grindability index of the coal at the Garland Coal and Mining Company Stigler No. 9 strip mine (R21E, T9N, Sec. 4), Haskell County, Oklahoma.

The cleat directions in the Croweburg coal (on the shelf) measured by Friedman (1978) at pit 3 of P&K Company, Ltd., near Henryetta, Okmulgee County, Oklahoma, are N 32° E and N 45° W.

Prominent cleat directions in the Secor coal at the Burdett mine 5 mi east of Checotak, McIntosh County, Oklahoma, were determined by Friedman (1978) to be N 45° W and N 40° E.

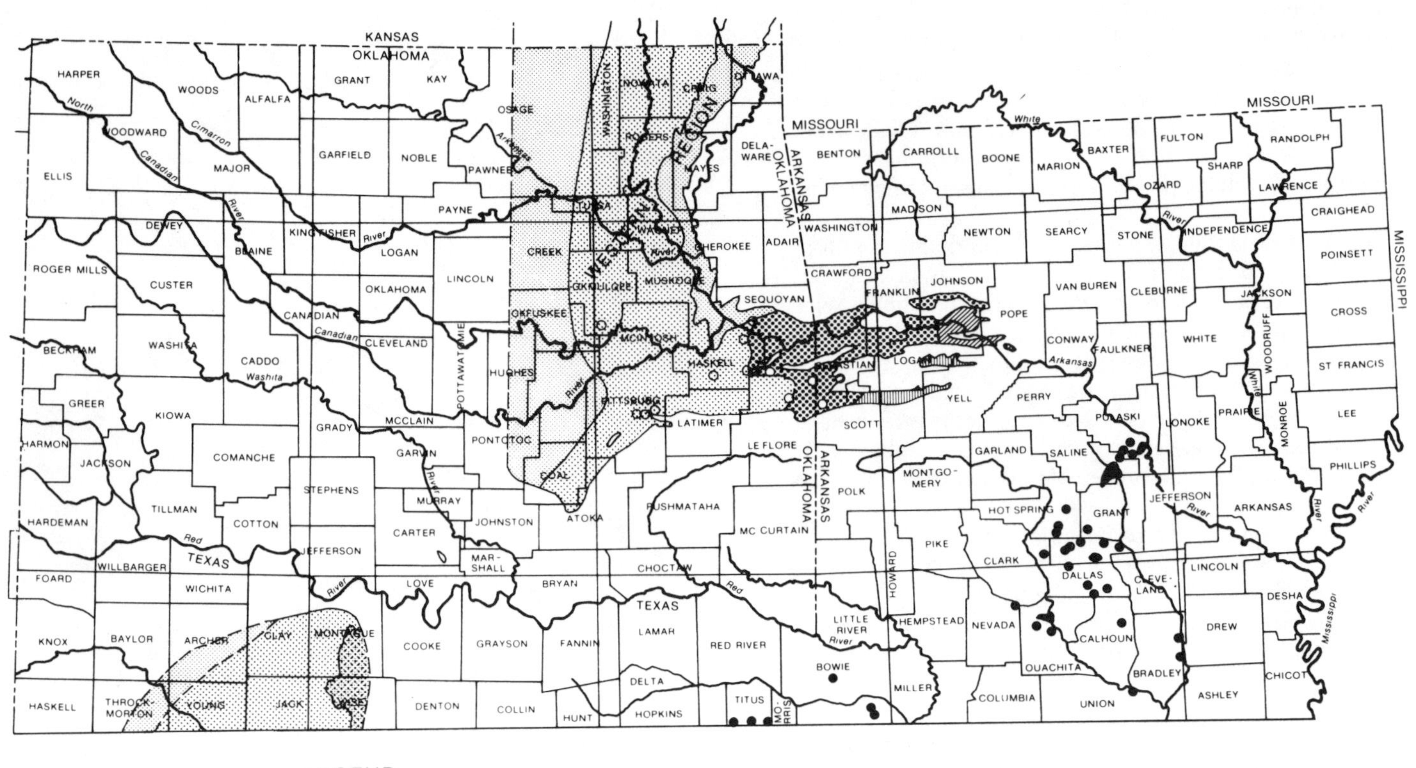

LEGEND

Anthracite, Semianthracite, and Meta-Anthracite

Low-Volatile Bituminous Coal

Medium- and High-Volatile Bituminous Coal

Lignite and Brown Coal

Coking Coal (Coal Coked at the Present Time, or Coal That Has Been Coked in the Past)

Deep color represents areas known to contain coal beds that are of commercial value at the present time or that may be of value in the future. In general the minimum thicknesses included are 14 inches for anthracite and bituminous coal, and 30 inches for subbituminous coal and lignite.

Light color represents areas of doubtful value for coal. These may be divided into three classes — (1) areas containing thin or irregular beds, which generally have little or no value, but which locally may be thick enough to mine, (2) areas in which the coal is poor in quality, and (3) areas where information on the thickness and qualify of coal beds is meager or lacking.

Light color and stippling denote that the coal-bearing formations are under cover which may range from a few hundred to several thousand feet.

Figure 10—Map showing the general rank of the Pennsylvanian coals in the Arkoma Basin. From Trumbull, 1960.

POTENTIAL METHANE RESOURCE

Previous studies involving determination of methane gas in coal in the Arkoma Basin were performed by the USBM and consist of specific degasification and desorption tests.

Gas emission data exist for two mine locations in Oklahoma: Choctaw Coal Facility, Kerr-McGee Coal Corporation, Haskell County, has an average methane emission per day of 0.4 million cubic feet (MMcf), whereas the Howe No. 1 mine, Howe Coal Company, Le Flore County, emits 1.6 million cubic feet/day (MMcfd). Both emissions are from the Hartshorne coal.

In Figure 12, cu ft of methane emitted per ton of coal mined is plotted versus the methane content of the sample for several mines. The resulting correlation is good for mines that are large and deep, have a sustained coal production of at least several thousand tons a day, and have been in operation for a number of years. New mines emit less methane per ton of coal mined than older mines with extensive old working and gob areas that still bleed gas. Hence, an estimate using Figure 12 may be too high for a new mine, but after the mine has been worked for some time, the emission will approach the relationship shown in Figure 12 (McCulloch et al, 1975). The Howe mine in Le Flore County, Oklahoma, is the only representative mine from the Arkoma Basin on the graph.

A total of 16 Hartshorne coal samples have been desorbed by the USBM from Haskell and Le Flore counties, Oklahoma. The data set has not been published by USBM. Figure 13 shows the location of counties in which methane desorption data are available.

Methane content of the Hartshorne coalbed increases with depth but at a decreasing logarithmic rate, so that the maximum content probably will not exceed approximately 700 cubic feet/ton (cf/ton) at 3,000 ft of overburden. According to Iannacchione and Puglio (1979), gas content values in cf/ton range from approximately 160 at 200 ft buried depth to a value of 576 at 1,600 ft. The distribution of methane in the Hartshorne coalbed in Haskell and Le Flore counties, Oklahoma, is given in Table 6.

Well Test Data

Methane Recovery from Coalbeds Project data on the methane gas content of Arkoma Basin coals are presently limited to coring and well testing in Pittsburg, Haskell, and Le Flore counties of Oklahoma (Fig. 14).

In Pittsburg County, two wells were cored in cooperation with private industry:

- Brown Estate Well Number 1-2, Arkla Exploration Company

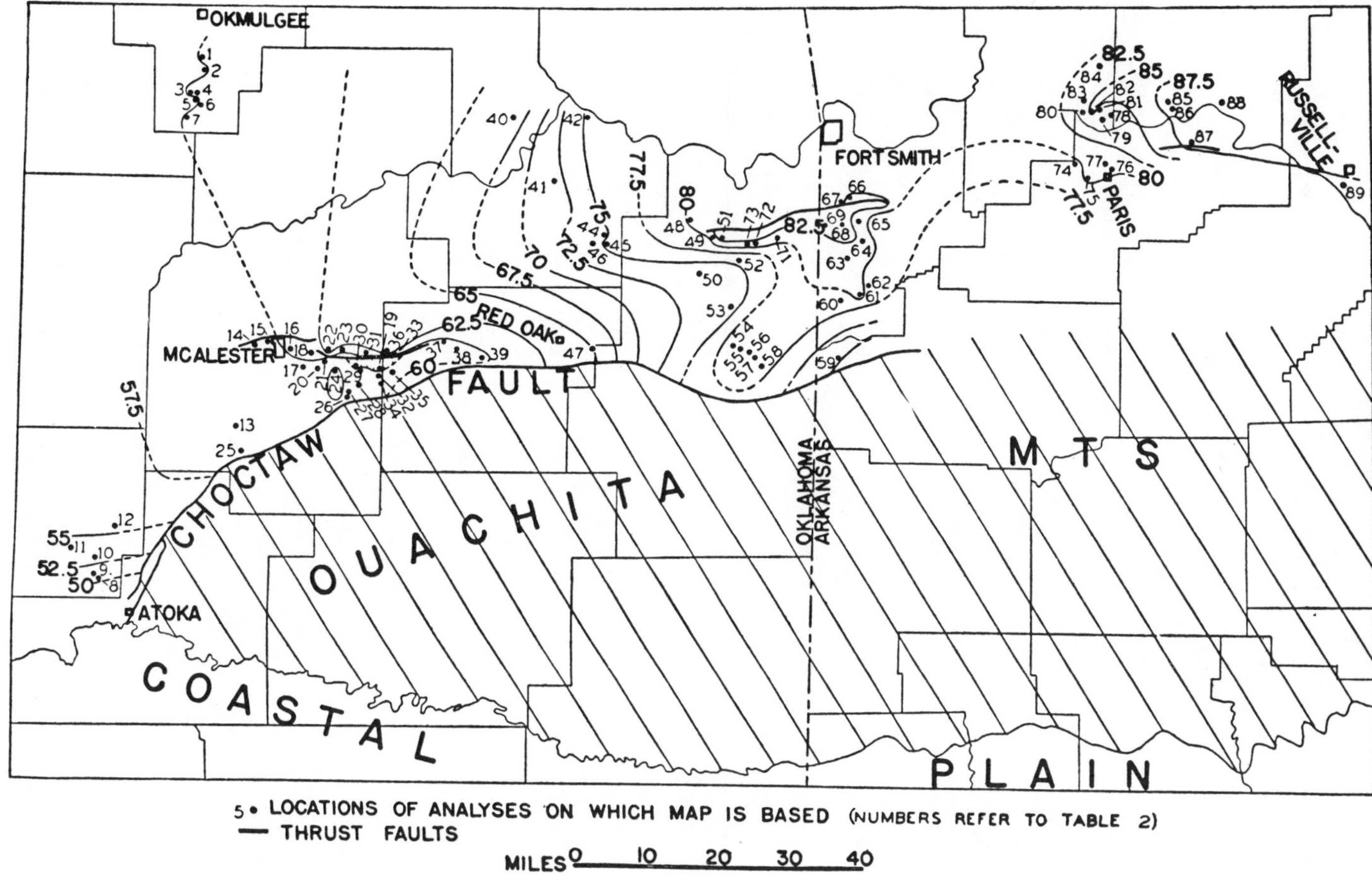

Figure 11—Isocarb values in the Arkoma Basin. From Hendricks et al, 1939.

- Barringer Well Number 1-11, Mustang Production Company.

In Haskell County, one well was cored in cooperation with private industry:

- Day Well Number 1-14, Mustang Production Company.

In Le Flore County, one well was cored in the Rock Island area near the Arkansas-Oklahoma state line in cooperation with the U.S. Bureau of Land Management, Oklahoma City, Oklahoma:

- DH-A17 Well, USBM/USBR.

MRCP desorption test results are presented in Table 7. Using weighted average values for gas content measured in the Brown Estate Number 1-2 well (T6N, R13E, Sec. 2), the gas-in-place is estimated at 1.4 billion cubic feet (Bcf)/640 acres. Using maximum permeability of 1.0 millidarcy (md) and maximum relative permeability curve and minimal permeability of 0.1 md and minimum relative permeability curve, a range of discounted gas deliverability of 40,000 thousand cubic feet (Mcf) and 3500 Mcf, respectively, was calculated.

A DST was run on the Hartshorne coal in the Brown Estate well. The test was conducted over the interval 2,700 to 2,740 ft, and showed flow of 9.0 barrels of water per day at a shut-in pressure of 716 psig. The average permeability calculated over the interval was 4.5 md. Table 8 is a summary of the bottomhole pressure and time data for the successful mechanical test.

Table 5—Relationship of fixed carbon to fault and fold structures and to adjacent structures in the Arkoma Basin. From Fuller, 1920; Wilson, 1961. Used with the permission of the Tulsa Geological Society.

Group	Locality	Ratio	Adjacent Structures
I	Edwards	55	Choctaw Fault
	Pittsburg	57	Choctaw Fault
	Savanna	61	Savanna Anticline
II	Wilburton	59	Near Choctaw Fault
	Lutie	61	Near Choctaw Fault
	Hughes	63	Cavanal Syncline

Locality	Adjacent Structure	Fixed C-%
Savanna	Savanna Anticline	61
Chambers	Krebs Syncline	54
Craig	Flank of Anticline	65
Dow and Coleman	Kiowa Syncline	61
McAlester	McAlester Anticline	63
So. McAlester	Krebs Syncline	59
Coalgate	Coalgate Anticline	58
Lehigh	Lehigh Syncline	54

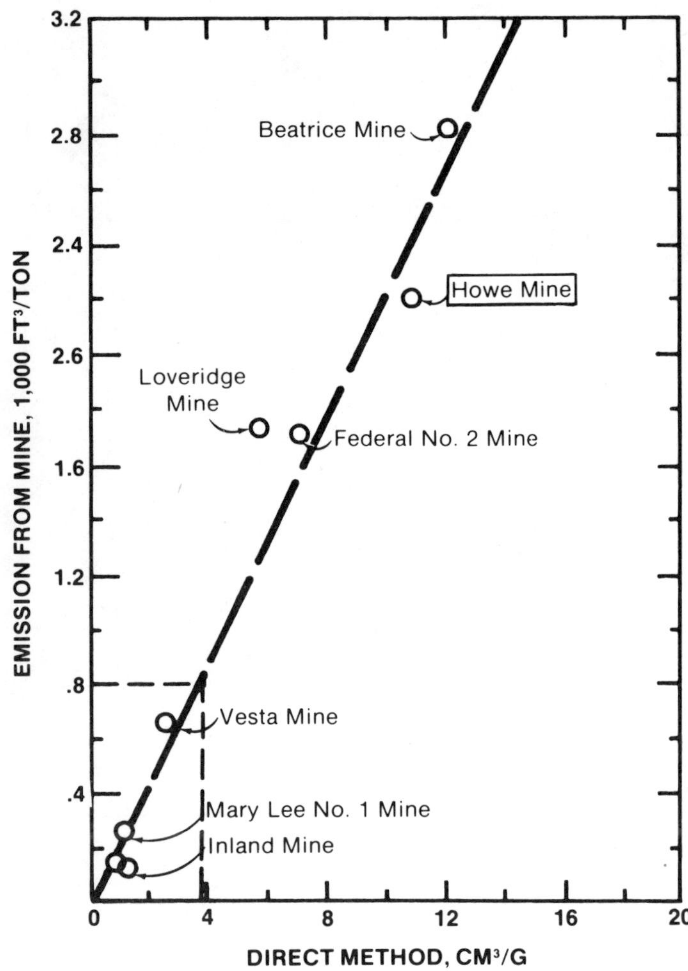

Figure 12—Gas content of coal versus actual mine emission. From Kissell et al, 1973.

In the Barringer Well Number 1-11 (T4N, R15E, Sec. 11), three zones were conventionally cored. Only one coal core was successfully recovered from the Upper Booch coal. No DST were run in the wells. Planned well flow tests are pre- and postfracture. These will be conducted on the Hartshorne coal (4,628 to 4,632 ft) and the McAlester coal (3,212 to 3,218 ft).

No coal cores were obtained from the Day Well Number 1-14 (T7N, R20E, Sec. 14).

The DH-A17 well (T8N, R26E, Sec. 14) penetrated a total of 2.9 ft of Hartshorne coal. Initial desorption results indicate that the Hartshorne coalbed at this location contains a moderate amount of gas (Table 7).

Estimated Resource Volume

The very limited data available on the methane content of the different coalbeds in the Arkoma Basin make it difficult to construct an accurate resource estimate. However, based on the data in Tables 7 and 9, it is possible to present reasonable ranges for the maximum and minimum expected in-place gas resource of the coals in the Arkoma Basin. These estimates have been made on county and bed bases for the Arkoma Basin.

Iannacchione and Puglio (1979) calculated the total volume of methane contained within the Upper and Lower Hartshorne coals for an area within Haskell and Le Flore counties, Oklahoma. Table 10 presents the methane gas content versus the overburden for the Hartshorne undivided, Upper Hartshorne, and the Lower Hartshorne. Within their study area as shown, the methane content of these coals was calculated to be between 1.1 and 1.5 trillion cubic feet (Tcf) of gas. This estimate was based on a reserve of 2,330 to 3,120 million short tons of coal-in-place (Table 9). The estimate of total coal-in-place was made using over 900 coal data points.

The volumetric assessment of the coalbed methane resource in the Arkoma Basin is based on Friedman's 1974 study for the Ozarks Regional Commission, which provides the data for the bituminous coal resources and recoverable reserves of Oklahoma. The coal resource data for Arkansas is from Haley's 1977 study. The purpose of these coal investigation studies was to obtain, evaluate, and provide basic information pertaining to the extent, thickness, depth of burial, and quality of the coals. The coal resources for Oklahoma include those greater than 3,000 ft deep and coalbeds greater than 12 inches thick, whereas the Arkansas data were not specified. Neither data set separated out the shallow (strippable) from the deep (nonstrippable) coals.

The most recent comprehensive coal investigation in Oklahoma was completed in 1952 by Trumbull (1957). He reported 3.25 billion short tons of remaining bituminous coal resources in Oklahoma from an estimated 3.67 billion short tons of original resources.

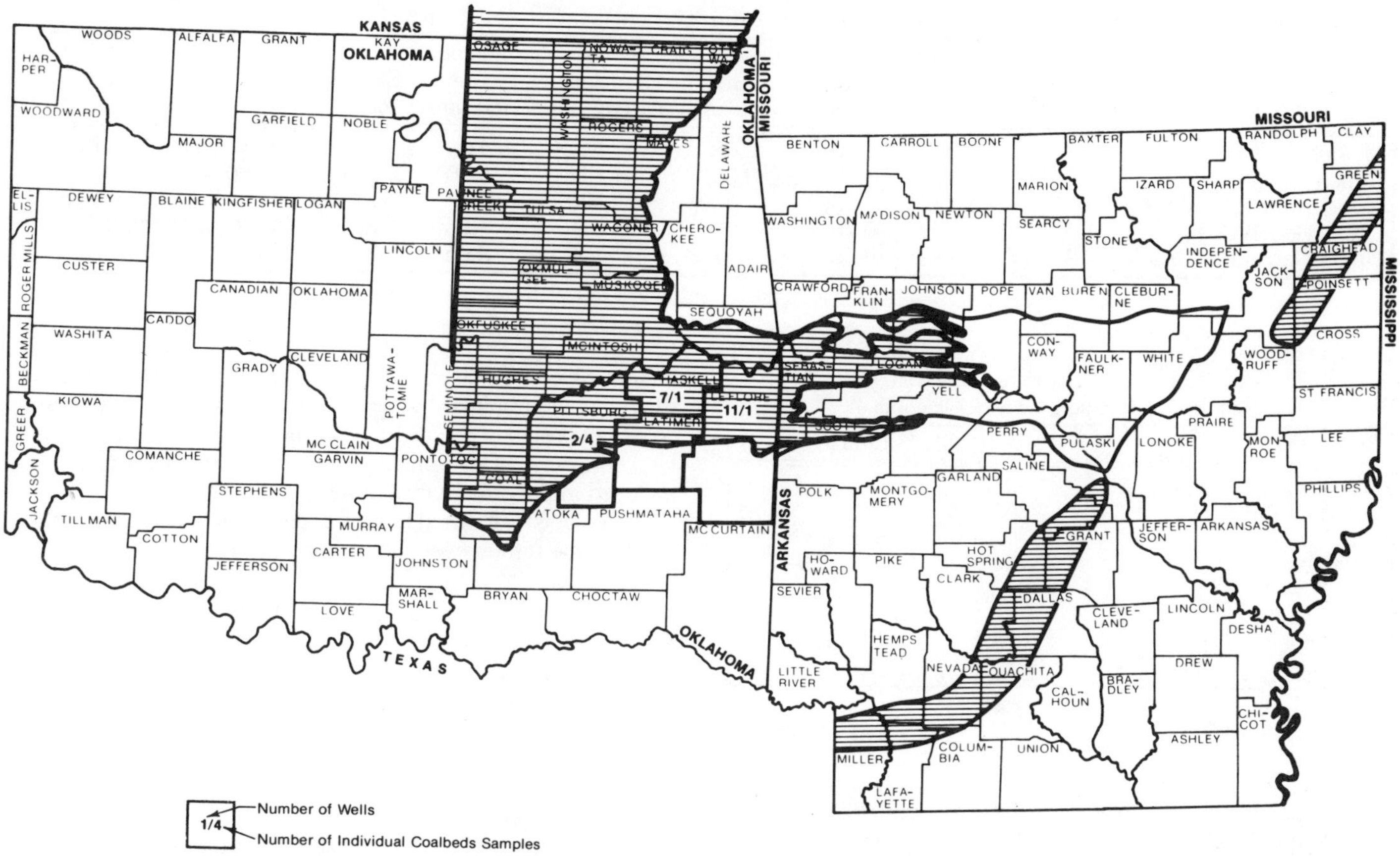

Number of Wells
1/4
Number of Individual Coalbeds Samples

Figure 13—Arkoma Basin map locating counties in which methane desorption data are available.

Table 6—Distribution of methane in the Hartshorne coalbed. From Iannacchione and Puglio, 1979.

Overburden (ft)	Methane Distribution (%)			
	Undivided Hartshorne	Upper Hartshorne	Lower Hartshorne	Total
0–500	6	15	9	9
500–1,000	31	28	23	27
1,000–1,500	38	24	24	31
1,500–2,000	18	14	22	19
2,000–3,000	6	19	22	4

The coal resource is broken down by rank. Trumbull (1957) reported that of the remaining reserves in Oklahoma,

- 65% was high-volatile-bituminous
- 13% was medium-volatile-bituminous
- 22% was low-volatile-bituminous

For Oklahoma, Friedman plotted coal datum points on base maps compiled from 7.5-minute topographic-quadrangle maps at a scale of 1:24,000, and on county road maps and geologic maps at a scale of 1:63,680. The datum points were obtained by

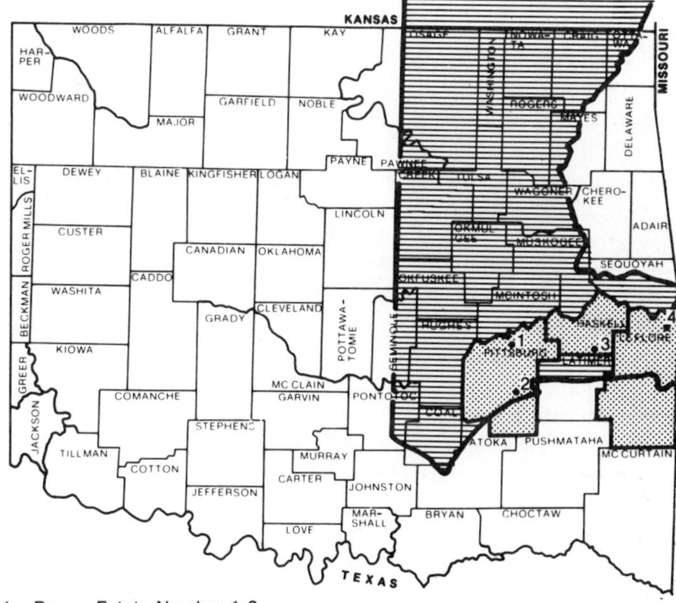

Figure 14—MRCP well locations in the Arkoma Basin.

Table 7—MRCP methane data from coals sampled in the Arkoma Basin of Oklahoma.

Coalbed	Thickness	Depth	Lost Gas (cc/gm)	Desorbed Gas (cc/gm)	Remaining Gas (cc/gm)	Total Gas (cc/gm)	Content (cf/ton)	Well
McAlester Coal(?)	2.0(?)	1,905	1.407	2.683	Nil	4.09	131	Brown Estate Number 1-2
Booch Coal(?)	1.0(?)	2,130	3.514	3.087	Nil	6.60	211	Brown Estate Number 1-2
Upper Hartshorne Coal	4.0(?)	2,733	0.630	1.639	Nil	2.27	73	Brown Estate Number 1-2
Upper Booch Coal	1.0	3,650	0.76	5.60	0.9	7.25	232	Barringer Number 1-11
Upper Hartshorne Coal	2.9	192	0.08	8.37	1.2	9.65	309	DH-A17

Table 8—Drill stem test data, Upper Hartshorne coal, Arkla Well, Brown Estate Number 1-2, Pittsburg County, Oklahoma.

Description	Pressure (psi)	Time (minutes)
Initial hydrostatic mud	1,437.9	
Initial flow (1)	350.6	
Initial flow (2)	361.0	6
Initial shut-in	700.8	90
Final flow (1)	370.5	
Final flow (2)	377.1	31
Final shut-in	678.1	120
Final hydrostatic mud	1,616.1	

Depth internal: 2,700–2,740 ft
Bottom hole temperature: 102°F

Friedman from available sources, excluding geophysical well logs. Arcs or circles were drawn around all datum points, delineating the measured, indicated, and inferred resources and reserves. Although coal deposits were judged by Friedman to exist in geologic continuity with adjacent resources, these deposits were not included in the tonnage figures of resources if they were more than 2 mi from a datum point. This was done to adhere as closely as possible to the criteria established by the USGS and used conservatively by Trumbull (1957) in maintaining the standard method of determining coal resources and reserves (Friedman, 1974).

Arkoma Basin Coalbed Methane Resource Base

Table 11 presents the minimum in-place gas resource of 1.58 Tcf, based on a minimum average methane content of 200 cf/ton. The following coalbeds were included:

- Lower Hartshorne
- Upper Hartshorne
- Hartshorne Undivided
- McAlester (Stigler)
- McAlester (Stigler) Rider
- Charleston
- Cavanal
- Paris
- Lower Witteville
- Secor.

It is reasonable to assume that the methane resource contained in these coalbeds should be much higher, since no depth scaling for the gas content was made in this study such as that done for the USBM study (Iannacchione and Puglio, 1979). An average gas content of 450 cf/ton would give a methane gas resource of about 3.55 Tcf. Additional drilling in the Arkoma Basin also has delineated areas that contain more coal than previously thought, as in the case of the Day Well in Haskell County, Oklahoma. These new reserves were not included in the present evaluation.

The coalbeds are listed in rank order of the magnitude of the methane gas resource for the Arkoma Basin.

	200 cf/ton (Tcf)	450 cf/ton (Tcf)
• Lower Hartshorne	0.636	1.431
• Hartshorne Undivided	0.314	0.707
• McAlester (Stigler)	0.306	0.688
• Upper Hartshorne	0.138	0.311

Gas Composition

Iannacchione and Puglio (1979) estimated that the Hartshorne coalbed in Haskell and Le Flore counties, Oklahoma, contains between 1.1 and 1.5 Tcf of gas. In the same area, the Hartshorne Sandstone produces natural gas from the Quinton, Poteau-Gilmore, and Camaron gas fields. These small gas fields are producing from combination traps (structural and stratigraphic). Gas samples were collected from nine producing wells in the three Oklahoma gas fields, from several gas drainage holes in the Hartshorne coalbed, and from two Hartshorne coal cores (Iannacchione and Puglio, 1980). The ^{13}C value of the methane, ranging from -34 to -44 per mil, and the amount of ethane, propane, and butane were found to be dependent on the rank of the coal associated with each gas field. Coalbed gas sampled from a mine near the Quinton field contained more methane and less ethane, propane, and butane than natural gas from the Quinton field. This may be due to the high sorption properties of coal. The heavier hydrocarbon gases remain adsorbed on the surface of the coalbed micropore structures, with higher proportions of methane released into coalbed macropore structures. Data collected so far indicate that the gas from the Poteau-Gilmore and Camaron fields is derived, at least partially, from associated coalbeds; while the gas from the Quinton field may have another origin. Gas sample analysis from the Hartshorne Formation shows gas composition variation in the western part of the Arkoma Basin (Moore, 1977). The Btu content of the Hartshorne gas ranges from 993 to 1,062.

Analysis of the composition of gases by Iannacchione and Puglio from the Hartshorne coalbed and sandstone has given some insight into the generation, migration, and retention of gas within the Hartshorne Formation. Samples were collected from nine production wells (C-1, 2, 3; PG-1, 2, 3; Q-1, 2, 3) in the Hartshorne Sandstone; eight horizontal degasification holes

Table 9—Methane content versus overburden, Haskell and Le Flore counties, Oklahoma. Iannacchione and Puglio, 1979.

Overburden (ft)	Average Methane Content (cc/gm)	(cf/ton)	Estimated Methane Content (cf ft) Undivided Hartshorne	Upper Hartshorne	Lower Hartshorne
0–500	6.6	211	Min 32 × 10^9 Max 47 × 10^9	Min 26 × 10^9 Max 37 × 10^9	Min 37 × 10^9 Max 50 × 10^9
500–1,000	13.5	432	Min 159 × 10^9 Max 210 × 10^9	Min 47 × 10^9 Max 71 × 10^9	Min 99 × 10^9 Max 125 × 10^9
1,000–1,500	16.7	534	Min 195 × 10^9 Max 253 × 10^9	Min 40 × 10^9 Max 64 × 10^9	Min 106 × 10^9 Max 135 × 10^9
1,500–2,000	18.8	602	Min 95 × 10^9 Max 123 × 10^9	Min 23 × 10^9 Max 36 × 10^9	Min 96 × 10^9 Max 119 × 10^9
2,000–3,000	21.0	672	Min 33 × 10^9 Max 41 × 10^9	Min 32 × 10^9 Max 51 × 10^9	Min 95 × 10^9 Max 119 × 10^9

(HC-1 to 8) in the Choctaw mine (Hartshorne coalbed); and two coal cores collected from the Hartshorne coalbed (HC-9 and HC-10, south of Poteau-Gilmore gas field). Isotopic and chemical data showed variations in the composition of the gases. Carbon isotope values given in ^{13}C per mil express parts per thousand deviation from PDB standard (Table 12). Compositions differ with changing rank of associated coalbeds, different sorption properties of the respective reservoir rocks, and type of source material from which the gases were generated (Iannacchione and Puglio, 1980).

Relation of Gas Composition to Coalbed Rank

Both the ^{13}C value of the methane and the amount of heavier hydrocarbons (ethane, propane, and butane) from gas produced from the nine gas wells sampled in the three different gas fields are correlated to the rank of the coal associated with these gas fields. Chemical analysis of coalbed gas also shows a slight increase in the percentage of methane with increasing rank (Table 13).

Recent work has shown that the composition of gas generated is generally dependent upon the degree of thermal maturation of the organic matter in the associated sedimentary rocks (Rice, 1975; Von Stahl, 1975). Different proportions of methane, ethane, propane, and butane are generated at different stages of metamorphism. An excellent indicator of the degree of thermal maturation is rank of coal. With increasing rank (metamorphism) of the coal associated with gas reservoir rocks, the methane becomes isotopically heavier. Percentages of methane and associated heavier hydrocarbons generated are also affected by this increased metamorphism.

The ^{13}C content of the methane from coalbed gas was determined at only one location (HC-1 to 8, T8N, R21E south of Stigler, Haskell County, Oklahoma), where the Hartshorne coalbed is a medium-volatile bituminous coal. The ^{13}C content of this coalbed gas is similar to the ^{13}C content of the natural gas from the nearby Quinton gas field, where the associated Hartshorne coalbed is high-volatile A to medium-volatile bituminous rank. Therefore, rank of the coal, which reflects the temperature and pressure history of the Hartshorne Formation, can give some insight into the expected thermal and chemical composition of the gases generated from associated rocks.

Table 10—Total Hartshorne coal-in-place, Haskell and Le Flore counties, Oklahoma. Iannacchione and Puglio, 1979.

Overburden (ft)	Hartshorne (Undivided)	Millions of Short Tons Upper Hartshorne	Lower Hartshorne	Percent of Total Coal
0–500	Min 149 Max 221	Min 122 Max 178	Min 174 Max 238	19
500–1,000	Min 368 Max 487	Min 108 Max 165	Min 228 Max 290	29
1,000–1,500	Min 365 Max 475	Min 76 Max 120	Min 198 Max 252	27
1,500–2,000	Min 158 Max 204	Min 37 Max 60	Min 160 Max 198	15
2,000–3,000	Min 49 Max 61	Min 48 Max 75	Min 141 Max 176	10
Total	Min 1,089 Max 1,448	Min 391 Max 598	Min 901 Max 1,154	100

Evidence of Coalbed Gas Fractionation

There is some evidence that physical fractionation of coalbed gas occurs after prolonged production from the coalbed reservoir. Physical fractionation refers to the process of separating different hydrocarbon gases according to their differences in physical properties of weight and size. Eight coalbed gas samples were collected from a horizontal degasification system in the Hartshorne coal.

Seven samples (HC-1 to 7, Table 13) were collected from production holes drilled 3 to 14 months before sampling. The maximum percentages of hydrocarbons from these samples were 97.8% methane, 0.86% ethane, 0.0016% propane, and 0.0005% butane. The eighth sample (HC-8) was collected from a horizontal hole drilled 31 months before sampling. This sample contained 94% methane, 4.2% ethane, 0.0045% propane, and 0.0049% butane. All holes were drilled in the same general area; however, the oldest hole (HC-8) has less methane and more ethane, propane, and butane (Iannacchione and Puglio, 1980).

Table 11—Total remaining bituminous coal and coalbed methane on a county basis, Arkoma Basin, in Arkansas and Oklahoma.

County	State	Remaining Coal (1,000 Short Tons)	Average Gas Content (Bcf) 200 cf/ton (6.25 cc/gm)	450 cf/ton (14 cc/gm)
Atoka	Oklahoma[1]	29,619	5.92	13.33
Coal		292,875	58.58	131.79
Haskell		1,513,681	302.74	681.16
Hughes		Not Kown	—	—
Latimer		841,968	168.39	378.89
Le Flore		1,973,362	394.67	888.01
Muskogee		61,199	12.24	27.53
Pittsburg		1,383,832	276.77	622.72
Sequoyah		27,146	5.43	12.22
Subtotal	Oklahoma	6,123,682	1,224.74	2,755.65
Crawford	Arkansas[2]	289,900	57.98	130.46
Franklin		212,400	42.48	95.58
Johnson		59,400	11.88	26.73
Logan		41,400	8.28	18.63
Scott		104,200	20.84	46.89
Sebastian		1,063,000	212.60	478.35
Subtotal	Arkansas	1,770,300	354.06	796.29
Total	Arkoma Basin	7,893,982	1,578.80	3,552.29

[1]As of January 1, 1974.
[2]As of January 1, 1977.

Table 12—Composition of natural gas from the Hartshorne Sandstone in Haskell and Le Flore counties, Oklahoma. Iannacchione and Puglio, 1980.

Location	Well No.	$\delta^{13}C$ Methane per Mil	Percent Methane	Ethane	Propane	Butane	CO^2	O^2+N	Rank of Coal Associated with Gasfield
Camaron Gasfield	C-1	−35	95.85	0.41	0.02	1,400 ppm[1]	0.08	2.90	Low-volatile bituminous to semianthracite
	C-2	−34	98.20	0.61	0.04	2,000 ppm	1.00	0.11	
	C-3	−35	89.15	0.61	0.05	1,800 ppm	0.77	9.40	
Poteau-Gilmore Gasfield	PG-1	−36	98.00	0.75	0.03	5,000 ppm	0.87	0.27	Semianthracite coal
	PG-2	−36	94.50	1.10	0.06	1,000 ppm	0.44	3.83	
	PG-3	−36	98.15	1.10	0.07	8,000 ppm	0.45	0.13	
Quinton	Q-1	−43	92.85	5.15	0.52	0.16	1.00	0.23	High-volatile A to medium-volatile bituminous coal
	Q-2	−43	93.35	4.85	0.83	0.22	0.44	0.24	
	Q-3	−42	94.85	3.50	0.61	0.14	0.53	0.31	

[1]ppm = parts per million.

It appears that coalbed gas does fractionate with lowering of reservoir pressure and time. This drop in pressure eventually weakens the physical bonds of the adsorbed gas, allowing the gas to desorb and migrate.

Relation of Gas Composition to Geology

Closely associated with this coalbed is the Hartshorne Sandstone, which has produced natural gas from three small gas fields for approximately 65 years. Natural gas has accumulated in commercial quantities in the Hartshorne Sandstone in at least three fields within Haskell and Le Flore counties. These small gas fields are located along the axes of anticlines, where the Hartshorne Sandstone is locally thickly developed. This indicates that a combination of structural and stratigraphic trapping mechanisms were operative. Production from all three gas fields began between 1910 and 1915. Average depth to reservoir of all three fields is approximately 1,500 ft, with initial bottomhole pressures ranging from 205 to 600 psi (Dane et al, 1938). The range of initial production rates is from 250 to 16,000 Mcfd.

CONCLUSIONS

Although the methane data are derived from a relatively small data base, it appears that the methane content of the coalbeds in the Arkoma Basin is low to moderate (73 to 211 cf/ton) in the northwest subshelf area; moderate to high (200 to 700 cf/ton) in the central trough; and unknown in Arkansas. There appears to be a general increase in the gas content within the trough portion of the basin toward the east.

Table 13—Chemical composition of Hartshorne coalbed gas. Iannacchione and Puglio, 1980.

Location	Well No.	$\delta^{13}C$ Content per Mil	Percent				Rank of Coal
			Methane	Heavier Hydrocarbons[1]	CO^2	O^2+N	
Horizontal degasification system, Choctaw Mine Haskell County	HC-1	−44	97.65	0.05	1.5	0.36	Medium-volatile bituminous coal
	HC-2	−44	97.75	0.49	1.6	0.18	
	HC-3	ND[2]	97.80	0.50	1.5	0.19	
	HC-4	−43	97.40	0.81	1.6	0.18	
	HC-5	−44	97.35	0.86	1.6	0.19	
	HC-6	−43	97.85	0.50	1.2	0.44	
	HC-7	−43	97.35	0.85	1.6	0.20	
	HC-8	−45	94.20	4.21	1.3	0.28	
Coal core, Le Flore County	HC-9	ND	98.50	0.07	0.08	1.35	Low-volatile bituminous coal
	HC-10	ND	99.25	0.02	0.10	0.63	

[1]Predominantly ethane with small amounts of propane and butane.
[2]Not determined.

Data world-wide tend to indicate that petroleum and natural gas generation and entrapment are enhanced in basins of high heat flow. Iannacchione and Puglio (1980) worked on the premise that the methane gas was generated with increased thermal maturation (coalification). Kim (1978) pointed out that the gas content of a coalbed depends primarily upon rank, temperature, and pressure. However, it must be noted that coals of the same rank may exhibit a 10,000-fold difference in gas content (Kim, 1978). This is an indication that present gas content of a coalbed is related not only to gas formation during coalification but also to the post-deposition geologic history of the coalbed. The degree of fracturing, distance to outcrop, depositional environment, depth of burial, bed thickness, rate of burial, and permeability of adjacent strata could be important. Keeping this information in mind, the following criteria were developed for use in defining areas with the greatest potential methane resource.

- Thick coalbeds have higher methane contents.
- The higher the coal rank, the greater the amount of thermally generated methane present in the coalbed.
- The lower the fixed carbon values for bituminous coal, the higher the remaining methane content in the Arkoma Basin coalbeds.
- Fixed carbon values in the Arkoma Basin coals are related to structural deformation.
- Coals in anticlines have a higher percentage of fixed carbon than those in adjacent synclines.
- The anticlines in the basin are narrower than the synclines—indicating more deformation.
- More deformation—more fractures—more gas loss to surrounding sediments.
- Anticlines occupy approximately 35 % of the total basin area.
- Synclines occupy 65 % of the total basin area.
- Methane content in the Hartshorne coalbeds increases with depth of burial.
- Thick Hartshorne sand development; thin lower Hartshorne coalbed.
- Optimum conditions. Look for:
  broad synclines
  thick coalbeds
  deeply buried coalbeds
  rank greater than Hvab; less than anthracite.

The target area that best meets these criteria is shown in Figure 15 and covers about 3,600 sq mi.

ACKNOWLEDGMENTS

The author gratefully acknowledges the assistance of those organizations and persons who provided information, data, and advice during preparation of this report. Among these, special thanks are due to Fred E. Galliers (Dakota-Ohio Company), Charles W. Cargile (consulting geologist), S. A. Freedman (Oklahoma State Geological Survey), Walter H. Fertl (Dresser Atlas), Anthony T. Iannacchione (U.S. Bureau of Mines), Robert C. Brown (U.S. Bureau of Land Management), Boyd R. Haley (U.S. Geological Survey), and William V. Bush (Arkansas Geological Commission).

Also greatly appreciated is the cooperation of Mustang Production Company, Arkla Exploration Company, American Association of Petroleum Geologists, McGraw-Hill Publications, Oklahoma State Geological Survey, the Arkansas Geological Commission, U.S. Bureau of Land Management, U.S. Bureau of Mines, and U.S. Geological Survey, in permitting the reproduction of various figures and tables and data contained in the text.

REFERENCES CITED

Branan, Jr., C. B., 1968, Natural gas in Arkoma Basin of Oklahoma and Arkansas: American Association of Petroleum Geologists Memoir 9, p. 1616–1635.

Briggs, G., and L. M. Cline, 1967, Paleocurrents and source areas of late Paleozoic sediments of the Ouachita Mountains, southeastern Oklahoma; Journal of Sedimentary Petrology, v. 37, n. 4, p. 985–1000.

Campbell, M. R., 1917, The coal fields of the United States: U.S. Geological Survey Professional Paper 100-A, p. 1–33.

Caplan, W. M., 1957, Subsurface geology of northwestern Arkansas: Little Rock, Arkansas, Arkansas Geological and Conversation Commission Information Circular 19, 14 p.

Craney, D. L., 1978, Distribution, structure, origin, and resources of the Hartshorne coals in the Panama Quadrangle, Le Flore County, Oklahoma; M.S. Thesis, University of Oklahoma, Norman, Oklahoma, 126 p.

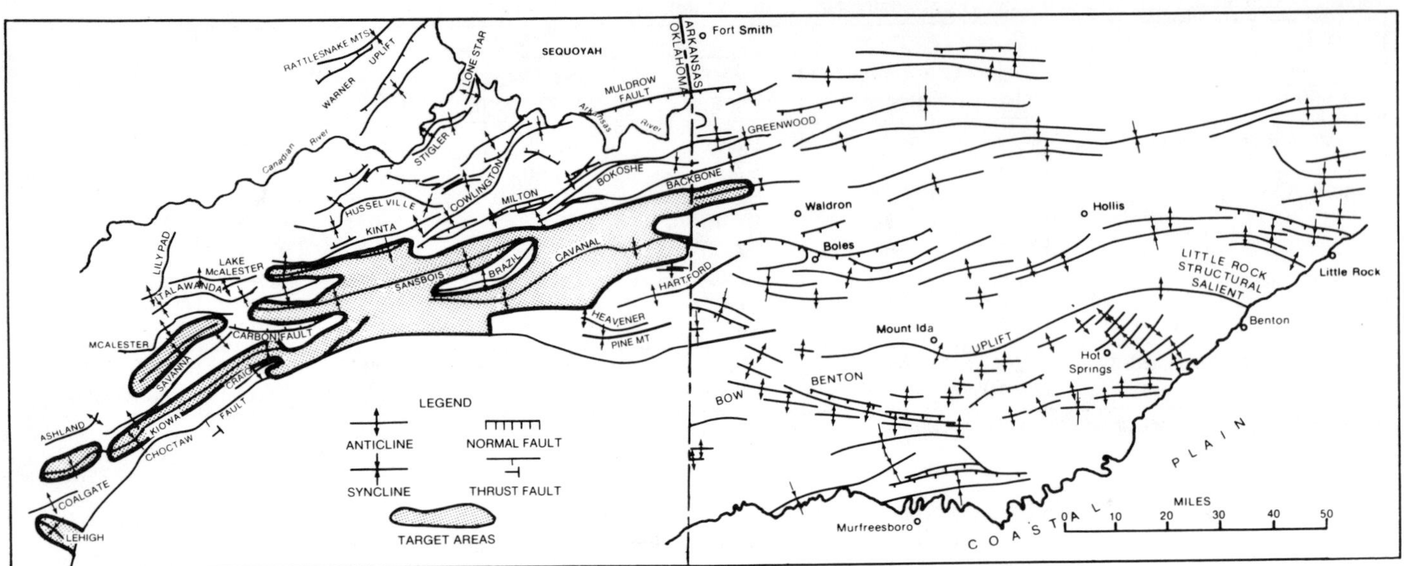

Figure 15—Redefined target areas in the Arkoma Basin.

Curtis, N. M., and W. E. Ham, 1972, Geomorphic provinces of Oklahoma, *in* Geology and Earth resources of Oklahoma, atlas of maps and cross-section: Oklahoma Geological Survey Education Publication 1, 8 p.

Dane, C. H., H. E. Rothrock, and J. E. Williams, 1938, Geology and fuel resources of the southern part of the Oklahoma coal field, pt. 3, The Quinton-Scipio district, Pittsburg, Haskell, and Latimer counties: U.S. Geological Survey Bulletin 874-C, p. 151–253.

Diamond, W. P., C. M. McCulloch, and B. M. Bench, 1976, Use of surface joint and photolinear data for predicting subsurface coal cleat orientation: U.S. Bureau of Mines Report of Investigations 8120, 13 p.

Disney, R. W., 1960, Subsurface geology of the McAlester Basin, Oklahoma: Doctoral Dissertation, University of Oklahoma, 116 p.

Donica, D. R., 1978, The geology of the Hartshorne coals (Desmoinesian) in parts of the Heavener 15' Quadrangle, Le Flore County, Oklahoma: M.S. Thesis, University of Oklahoma, 128 p.

Dott, R. H., 1945, General geologic section of Oklahoma oil and gas producing areas: Oklahoma Geological Survey.

Friedman, S.A., 1974, Investigation of the coal reserves in the Ozarks section of Oklahoma and their potential uses: Oklahoma Geological Survey Special Publication 74-2, 117 p.

———, 1978, Demoinesian coal deposits in part of the Arkoma Basin, eastern Oklahoma: American Association of Petroleum Geologists Field Guidebook, Oklahoma City Geological Society, Oklahoma City, Oklahoma, 62 p.

Fuller, M. L., 1920, Carbon ratios in carboniferous coal in Oklahoma, and their relation to petroleum: Economic Geology, v. 15, n. 3, p. 225–235.

Haley, B. R., 1961, Geology of Paris Quadrangle, Logan County, Arkansas: Arkansas Geological Commission Information Circular 20-B, 40 p.

———, 1966, Geology of the Barber Quadrangle, Sebastian County and vicinity, Arkansas: Arkansas Geological Commission Information Circular 20-C, 76 p.

———, 1968, Geology of the Scranton and New Blaine Quadrangles, Logan and Johnson counties, Arkansas: U.S. Geological Survey Professional Paper 536-B, 10 p.

———, 1977, Low-volatile bituminous coal and semianthracite in the Arkansas Valley coal field: Arkansas Geological Commission Information Circular 20-K, 26 p.

———, and T.A. Hendricks, 1968, Geology of the Greenwood Quadrangle, Arkansas-Oklahoma: U.S. Geological Survey Professional Paper 536-A, 15 p.

———, 1972, Geology of the VanBuren and Lavaca Quadrangles, Arkansas-Oklahoma: U.S. Geological Survey Professional Paper 657-A, 41 p.

Henbest, L. G., 1953, Morrow group and lower Atoka Formation of Arkansas: Bulletin of the American Association of Petroleum Geologists, v. 37, no. 8, p. 1935–1953.

Hendricks, T. A., 1935, Carbon ratios in part of the Arkansas Oklahoma coal field: Bulletin of the American Association of Petroleum Geologists, v. 19, p. 937–947.

———, 1937, Geology and fuel resources of the southern part of the Oklahoma coal field, pt. 1, The McAlester District, Pittsburg, Atoka, and Latimer counties: U.S. Geological Survey Bulletin 874-A, 90 p.

———, 1939, Geology and fuel resources of the southern part of the Oklahoma coal field pt. 4, The Howe-Wilburton District, Latimer and Le Flore counties: U.S. Geological Survey Bulletin 874-D, p. 255–300.

———, and B. Parks, 1950, Geology of the Fort Smith District, Arkansas: U.S. Geological Survey Professional Paper 221-E, p. 67–94.

———, C. H. Dane, and M. M. Knechtel, 1936, Stratigraphy of Arkansas-Oklahoma coal basin: Bulletin of the American Association of Petroleum Geologists, v. 20, p. 1342-1356.

———, M. M. Knechtel, C. H. Dane, H. E. Rothrock, and J. S. Williams, 1939, Geology and fuel resources of the Oklahoma coal field: U.S. Geological Survey Bulletin 874, 300 p.

Iannacchione, A. T., and D. G. Puglio, 1979, Geological

association of coalbed gas and natural gas from the Hartshorne Formation in Haskell and Le Flore counties, Oklahoma: Comptes Rendus, IX International Congress of Carbon Stratigraphy and Geology, Urbana, Illinois.

——, 1980, Methane content and geology of the Hartshorne coalbed in Haskell and Le Flore counties, Oklahoma: U.S. Bureau of Mines Report of Investigations 8407, 14 p.

Jordan, L., 1957, Subsurface stratigraphic names of Oklahoma: Oklahoma Geological Survey Guidebook VI, 220 p.

Keystone, 1979, Keystone coal industry manual: New York, Mining Informational Services, McGraw-Hill Mining Pub., 1311 p.

Kim, A., 1978, Experimental studies on the origin and accumulation of coalbed gas: U.S. Bureau of Mines Report of Investigations 8317, 18 p.

Kissell, F. N., C. M. McCulloch, and C. H. Elder, 1973, The direct method of determining methane content of coalbeds for ventilation design: U.S. Bureau of Mines Report of Investigations 7767, 17 p.

Knechtel, M. M., 1949, Geology and coal and natural gas resources of northern Le Flore County, Oklahoma: Oklahoma Geological Survey Bulletin 68, 76 p.

McCulloch, C. M., M. Deul, and P. W. Jeran, 1974, Cleat in bituminous coalbeds: U.S. Bureau of Mines Report of Investigations 7910, 25 p.

——, J. R. Levine, F. N. Kissel, and M. Deul, 1975, Measuring the methane content of bituminous coalbeds, U.S. Bureau of Mines Report of Investigations 8043, 22 p.

Merewether, E. A., and B. R. Haley, 1961, Geology of Delaware Quadrangle, Logan County and vicinity, Arkansas: Little Rock, Arkansas, Arkansas Geological and Conservation Commission Information Circular 20-A, 30 p.

——, 1969, Geology of the Coal Hill, Hartman and Clarksville Quadrangles, Johnson County and vicinity, Arkansas: Little Rock, Arkansas, Arkansas Geological Commission Information Circular 20-H, 27 p.

Miser, H. D, 1954, Geologic map of Oklahoma: U.S. Geological Survey Map (1967-G68100) 1:500,000.

Moore, B. J., 1976, Analyses of natural gases: U.S. Bureau of Mines Information Circular 8749, 94 p.

Oakes, M. C., 1953, Krebs and Cabaniss groups, of Pennsylvanian age, in Oklahoma: Bulletin of the American Association of Petroleum Geologists, v. 37, n. 6, p. 1523–1526.

——, and M. M. Knechtel, 1948, Geology and mineral resources of Haskell County, Oklahoma: Oklahoma Geological Survey Bulletin 67, 134 p.

Rice, D. D., 1975, Origin and significance of natural gases of Montana: U.S. Geological Survey Open-File Report 75-188, 13 p.

Taff, J. A., 1899, Geology of McAlester-Lehigh coal field, Indian Territory: U.S. Geological Survey 19th Annual Report, pt. 3, p. 423–456.

Trumbull, J. V. A., 1957, Coal resources of Oklahoma: U.S. Geological Survey Bulletin 1042-J, p. 307–382.

——, 1960, Coal fields of the United States: U.S. Geological Survey Map (1967-G67306) 1:5,000,000.

Von Stahl, W. J., 1975, Kohlenstoff-Isotopenverhaltnisse von Erdgasen-Reifekkennzeichenihrer Muttersubstanzon: Erdol und Kohl, v. 28, p. 188–191.

Wilson, L. R., 1961, Palynological fossil response to low-grade metamorphism in the Arkoma Basin: Tulsa Geological Society Digest, v. 29, p. 131–140.

Geologic Overview, Coal, and Coalbed Methane Resources of Raton Mesa Region — Colorado and New Mexico

D. Jurich
M.A. Adams

The Raton Mesa coal region, a north-south–trending basin occupying 2,200 sq mi in southeastern Colorado and northeastern New Mexico, contains an in-place coal reserve of more than 17 billion tons. Of this reserve, more than 3 billion tons are coking coal. Test data from methane desorption show these coals contain between 25 and 490 cu ft of gas per ton of coal. When used with the coal resource estimate, a potential of at least 8.0 and possibly as much as 18.4 trillion cu ft of coalbed methane may be present in the Raton Mesa Region.

An initial target area for exploitation of this resource covers a vast majority of the basin. However, an area of about 310 sq mi in the western part of the basin has been identified as having the highest potential for development. Coals in this area are generally higher in rank and found at greater depths. Desorption data for coal cores from this area average about 250 cu ft of gas per ton of coal (288 MMcf/ton coal per sq mi).

INTRODUCTION

The Raton Mesa Region, herein defined as that part of the Raton Mesa Basin underlain by Upper Cretaceous and Paleocene coal-bearing strata, covers approximately 2,200 sq mi in southeastern Colorado and northeastern New Mexico (Fig. 1). The region extends 175 mi north to south and is a maximum of 65 mi wide near the Colorado-New Mexico state line. It includes parts of Huerfano and Las Animas counties of Colorado and Colfax County of New Mexico.

The Raton Mesa Region is in the westernmost portion of the Great Plains Province where the coal-bearing Upper Cretaceous and Paleocene formations form an intermediate plateau between the mountains of the Rocky Mountain Province to the west and the lowlands of the Great Plains Province to the east. The plateau has been dissected by erosion into a complex system of deep canyons, prominent ridges, and flat-topped mesas, although toward the northern part of the field, the topography gradually changes to rolling hills.

Regional elevations vary from about 6,100 ft along the eastern margin to approximately 10,000 ft above sea level along the southwestern margin. Two conical mountains, East Spanish Peak (with an elevation of 12,669 ft) and West Spanish Peak (at 13,610 ft), rise abruptly in the northern half of the region and dominate the surrounding country. Igneous dikes crisscross the region and, where exposed by erosion, stand as vertical walls up to 100 ft high. The eastern edge of the region is characterized by steep escarpments ranging from 500 to more than 2,000 ft high above the surrounding plains.

GEOLOGIC STRUCTURE

The Raton Mesa Region is situated in the northern half of the Raton Mesa Basin, southernmost of the Laramide basins along the eastern margin of the Rocky Mountains. The Raton Mesa Basin is bordered on the west by the Sangre de Cristo Uplift and merges to the east with the Sierra Grande-Las Animas Arch. The Wet Mountains Uplift and its southeastern continuation—the Apishapa Arch—make up the northern boundary.

Structural elements of the Raton Basin were formed mainly during Early Pennsylvanian and Permian time, when several thousand feet of sediments were deposited in the Central Colorado Basin, a geosynclinal structure that trended northwestward between the ancestral Wet Mountains and the Apishapa Uplift on the east and the San Luis Uplift on the west. The Central Colorado Basin probably was connected to the Rowe-Moro Basin to the south, but during Middle and Late Pennsylvanian time, the ancestral Cimarron Arch rose as part of the Sierra Grande Uplift and partially separated the two basins (Blatz, 1965). The Raton Mesa Basin is an asymmetrical trough that trends north in New Mexico and north-northwest in Colorado. It is characterized by a steep western limb, a gently sloping eastern limb, and a broad central portion in which the beds are essentially horizontal (Figs. 2, 3). The beds at the western edge of the coal region dip steeply to the east and in places are vertical to overturned. Toward the axis of the basin, the dip of the beds gradually decreases. The eastern limb of the trough has a gentle dip of 1 to 10° to the west. The Colorado

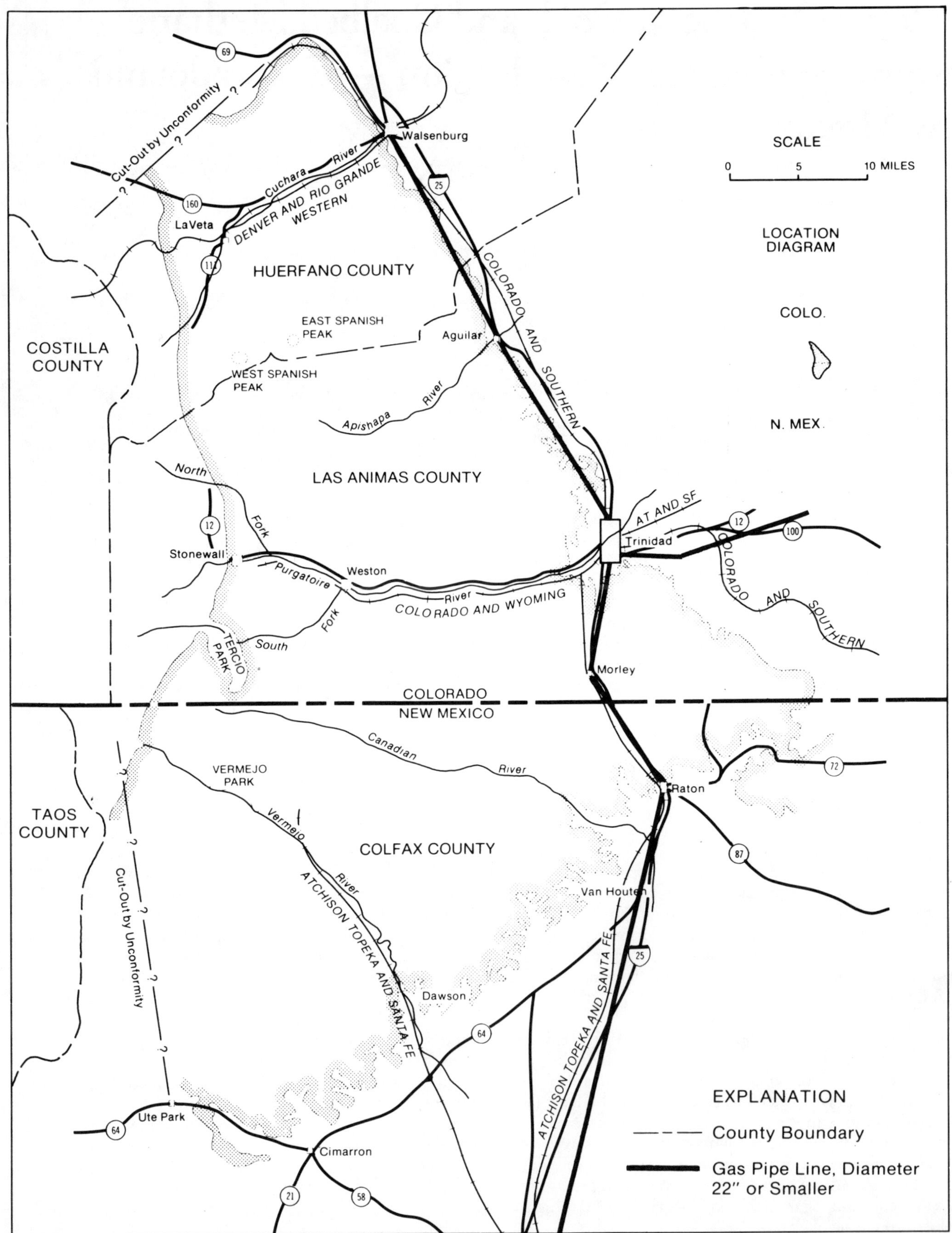

Figure 1—Raton Mesa Region of Colorado and New Mexico.

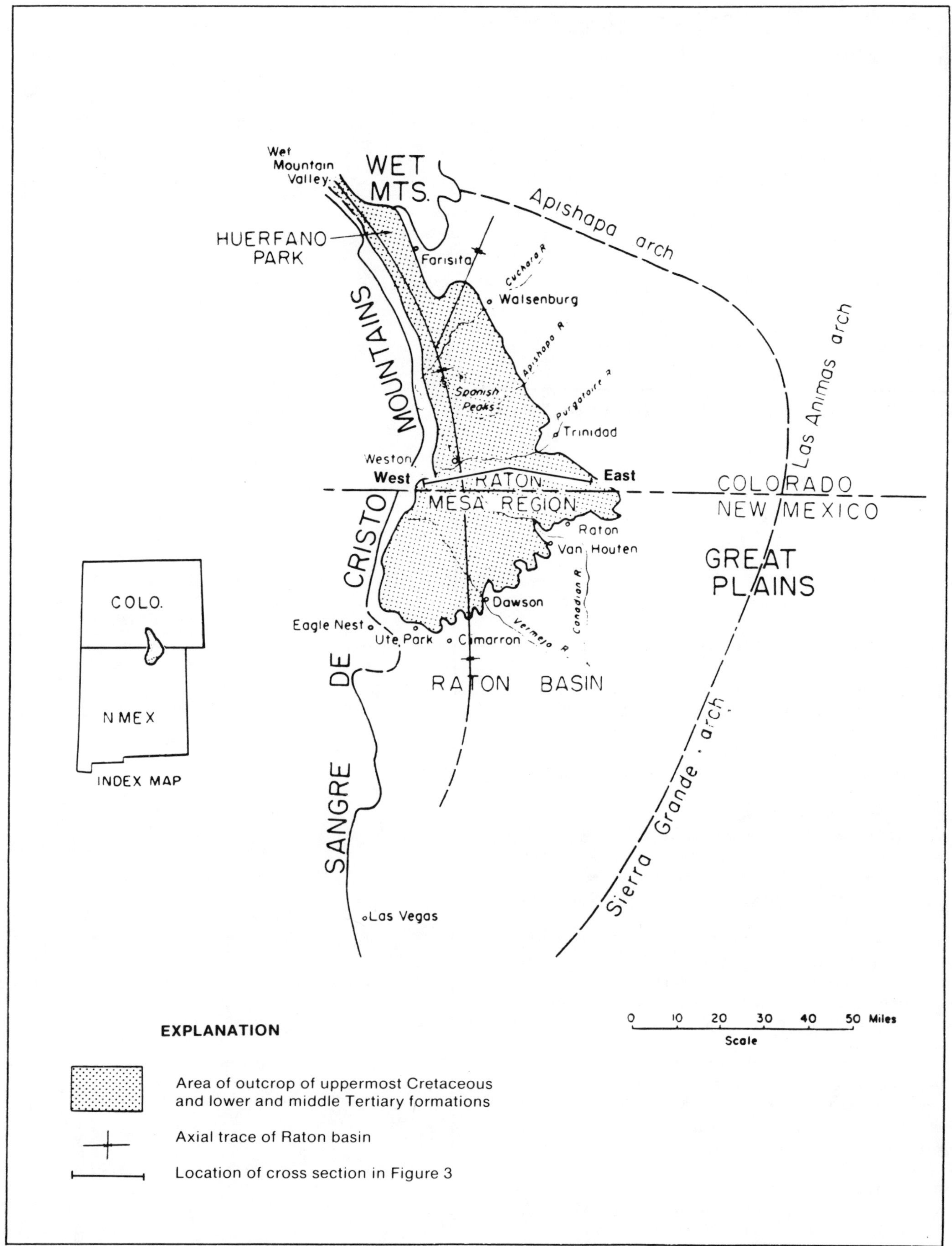

Figure 2—Map of the structural Raton Basin of Colorado and New Mexico. Johnson and Wood, 1956. Courtesy of RMAG.

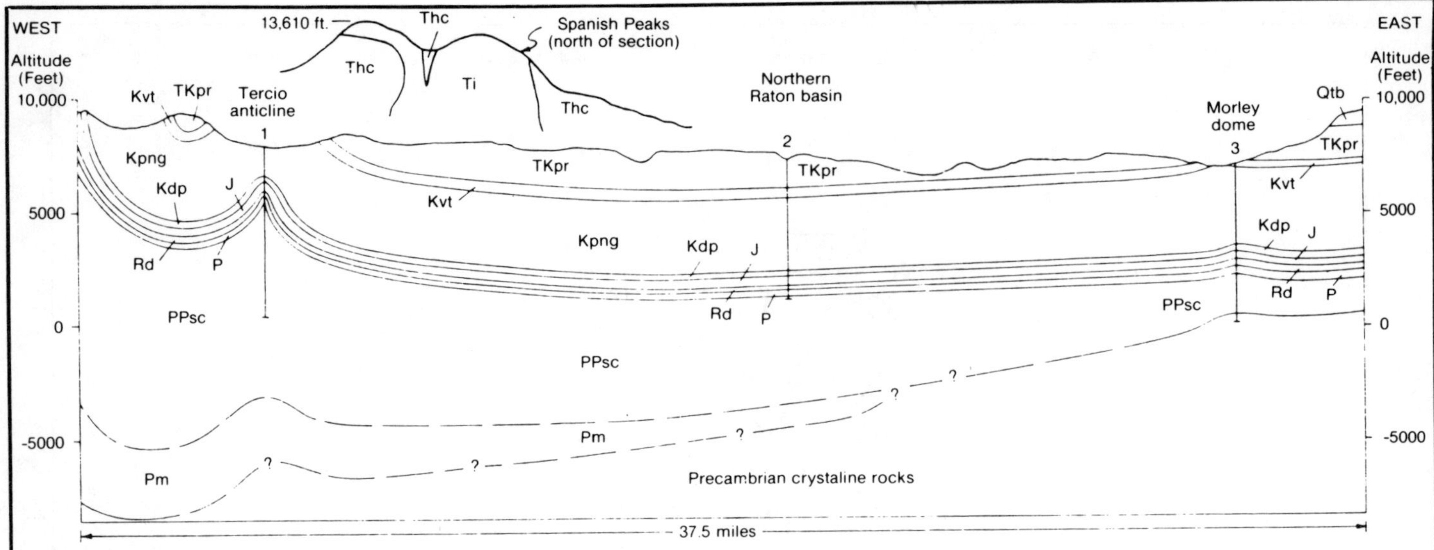

Figure 3—Diagrammatic east-west cross section of the Raton Basin.

portion of the basin is known as the La Veta Syncline (Johnson and Stephens, 1954).

The flanks of the Raton Mesa Basin are characterized by uniform dips and a relatively simple structure, although faults and folds do complicate the structural trends of the region. Several long, narrow, irregular folds of low structural relief occur south of the Spanish Peaks. These folds, which have no preferred orientation, appear to have been formed at the same time as the basin (Johnson, 1961).

Within the Raton Mesa Region, the Vermejo Park Anticline is one of the most prominent structural features. This feature exhibits more than 2,500 ft of structural relief across a distance of 4 mi on its east flank and has more than 500 ft of closure (Pillmore, 1969a).

Middle Tertiary intrusive activity related to the Spanish Peak's stocks and the associated radial dike and sill complexes appears to be concordant with the regional structure of associated sedimentary beds. It is possible that one or more of the anticlinal structures is a laccolith or dome formed over an underlying stock.

Faulting, although rare, is present in the region. In the western margin of the basin, one to three thrust faults parallel the east front of the Sangre de Cristo Mountains. One thrust sheet is complicated by a series of overturned structures that apparently resulted from compression associated with the thrusting.

Isolated groups of normal faults are found in the region. Several small normal faults occur northeast of Weston, Colorado. These faults trend north, east, northeast, and northwest and seem related to a small anticline or dome. Most of the faults near Weston have a displacement of less than 50 ft

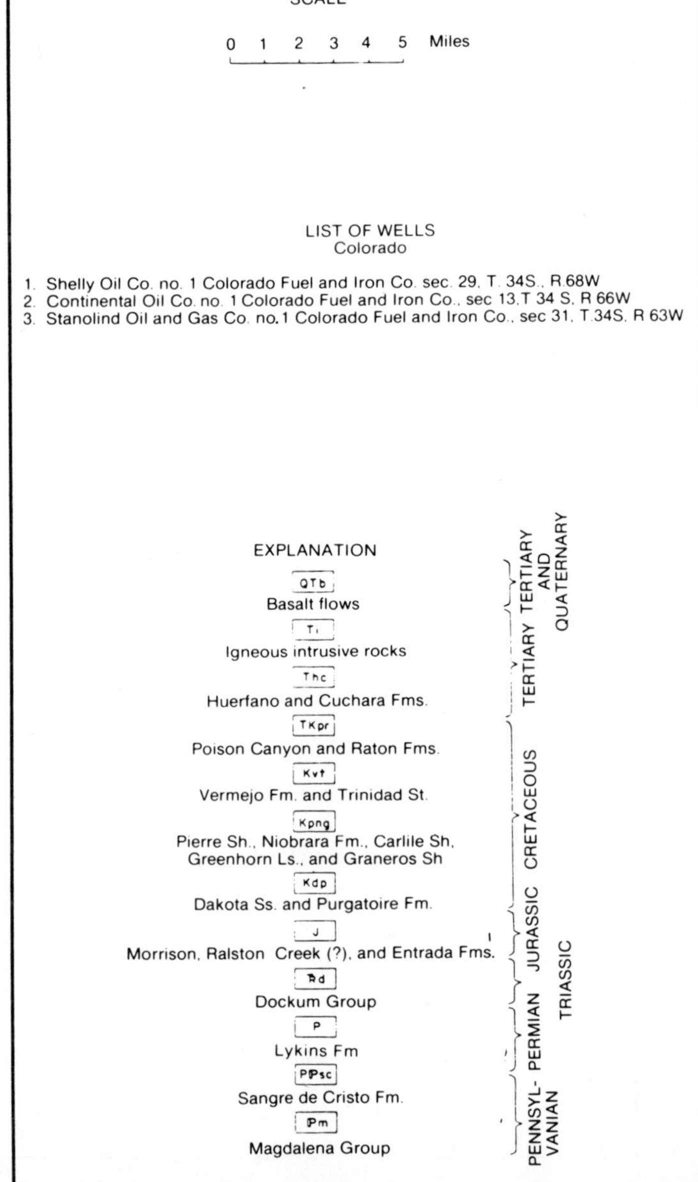

(Johnson, 1961).

The sedimentary rocks between the Spanish Peaks appear to have been brought from depth by the intrusion of the East Spanish Peak magma. The faults associated with this intrusion are normal, with a vertical displacement of 6,000 ft (Johnson, 1961). On the western margin of the basin, Wanek (1963) mapped several enechelon faults that displace the Poison Canyon Formation approximately 1,500 ft.

STRATIGRAPHY

During Precambrian time, several thousand feet of sediments were deposited in the Raton Mesa Region. These sediments were subsequently deformed, intruded by several igneous bodies, and metamorphosed into gneiss, quartzite, and schist. Later, these rocks were subjected to an extended period of erosion, lasting perhaps until Late Mississippian time in some areas of the region.

During much of Paleozoic time, the Raton Mesa Basin probably was part of a broad, southwest-trending, shelf-like "continental backbone" (King, 1951) on which sedimentary rocks older than Devonian are not present. It is not known whether pre-Devonian sediments were deposited and subsequently eroded away to the Precambrian crystalline basement. Maher and Collins (1949) suggested that the thick sequences of Paleozoic rocks of Kansas, Oklahoma, and eastern Colorado that thin toward the Sierra Grande Uplift indicate that it had a long history as a nondepositional area.

The oldest sedimentary rocks present in the Raton Mesa Basin are marine sandstone and dolomitic limestone of the Espiritu Santo Formation of Devonian (?) age (Fig. 4). This unit ranges in thickness from a few to 80 ft. The Tererro Formation of Mississippian age rests unconformably on the Espiritu Santo Formation. This unit is composed of limestone breccia and marine conglomerate, clastic and crystalline limestone, and siltstone varying in thickness from a few to more than 100 ft.

The Magdalena Group of Pennsylvanian age and the Sangre de Cristo Formation of Pennsylvanian and Permian age record the first stages of the formation of the Raton Mesa Basin when the orogenic forces that produced the Ancestral Rockies began to downwarp the Central Colorado and Rowe-Mora Basins. These complex suites of sedimentary rocks, whose vertical and lateral variations reflect marine transgression, regression, and eventual filling of the geosynclinal basin, have a cumulative thickness in excess of 10,000 ft (Blatz, 1965). In the waning phase of the Pennsylvanian-Permian orogeny, the sea returned to cover the Raton Mesa Basin, and several hundred feet of sediments were deposited. In the Raton Mesa Basin, these sediments are referred to as the Lykins Formation and an unnamed unit.

Following an extended period of nondeposition and minor amounts of erosion, terrestrial rocks of Late Triassic age were deposited as fluvial sandstones and shales over much of the Raton Mesa Basin region. These sediments, known as the Dockum Group, are approximately 300 to 400 ft thick on the eastern limb of the basin and thin northward and westward, eventually wedging out in the subsurface. In Late Jurassic time, an extensive series of shallow marine and terrestrial deposits was laid down over the entire Raton Mesa Basin. These beds range in thickness from about 100 to 600 ft; the variations in thickness are attributed to an erosion surface at the top of the Jurassic rocks.

The Purgatoire Formation of Early Cretaceous age rests unconformably on the Late Jurassic Morrison Formation and is present in most of the Raton Mesa Basin and the Sierra Grande and Apishapa Arches. The Purgatoire consists of a lower conglomeratic sandstone member and an upper member composed of gray carbonaceous to coaly shale and interbedded thin sandstone 100 to 150 ft thick. The Dakota Sandstone Group, which lies conformably on the Purgatoire Formation, is generally about 100 ft thick over much of the Raton Mesa Basin and consists of several thin to thick even-bedded sandstones interbedded with thin gray shales. In general, the sandstone beds thin eastward and the proportion of shale increases slightly eastward. The Dakota Sandstone Group represents the initial dune-beach-shallow marine and transgressional phases of the Greenhorn marine cycle (Kauffman et al, 1969). The rest of the Greenhorn marine cycle is recorded in the members of the Benton Formation, which includes the Graneros Shale, the Greenhorn Limestone, and the Carlile Shale up through the Codell Sandstone. A second half-cycle, the transgression of the Niobrara cycle, is recorded in the overlying Juana Lopez Member of the Carlile Shale through the Fort Hayes Limestone of the Niobrara Formation.

The youngest beds of the Pierre Shale record the beginning of the retreat of the marine sea and the rise of the Laramide Rocky Mountains. As the sea retreated from the basin, the sediments of the Pierre Shale became less clayey and more sandy. The Pierre Shale locally intertongues with the overlying Trinidad Sandstone through a transition zone 20 to 50 ft thick (Johnson and Wood, 1956) and represents the epeirogenic movement west of the basin. The resulting uplift forced the strandline to retreat to the northeast. As the strandline retreated, the Trinidad Sandstone accumulated in the sea and on its margins as regressive beach and offshore deposits. The Trinidad Sandstone is thickest near the axis of the Raton Mesa Basin, where it ranges in thickness from 140 to about 300 ft. More recent work (Billingsley, 1977,; Manzolillo, 1976) divides the sandstone into an upper fluvial zone and a lower delta front sandstone. The Trinidad is believed to be a slightly younger representative of the same littoral-paralic regressive sequence represented by the gas-productive Pictured Cliffs Sandstone of the San Juan Basin (Weimer, 1960).

The Trinidad Sandstone is overlain by and intertongues with the Upper Cretaceous Vermejo Formation (Lee, 1917). The Vermejo Formation is composed of up to 550 ft of varied proportions of buff to gray shale, carbonaceous shale, coal, and slightly arkosic fine- to medium-grained sandstones deposited in swamps and on flood plains near the coast of the retreating sea. The Vermejo is believed to be the slightly younger equivalent of an identical lithofacies unit represented by the coal-bearing Fruitland Formation of the San Juan Basin.

Coalbeds in the Vermejo Formation range in thickness from a few inches bo several feet, and are present throughout most of the Raton Mesa Region. The most persistent and widely mined bed in the Raton coal field is the Raton coalbed, normally found within a few feet of the base of the Vermejo Coalbed. This coalbed appears to be restricted to the western part of the region, where the formation is more than 250 ft thick (Pillmore, 1969b).

The Raton Formation, overlying the Vermejo Formation, is the thickest and most widely distributed of the coal-bearing

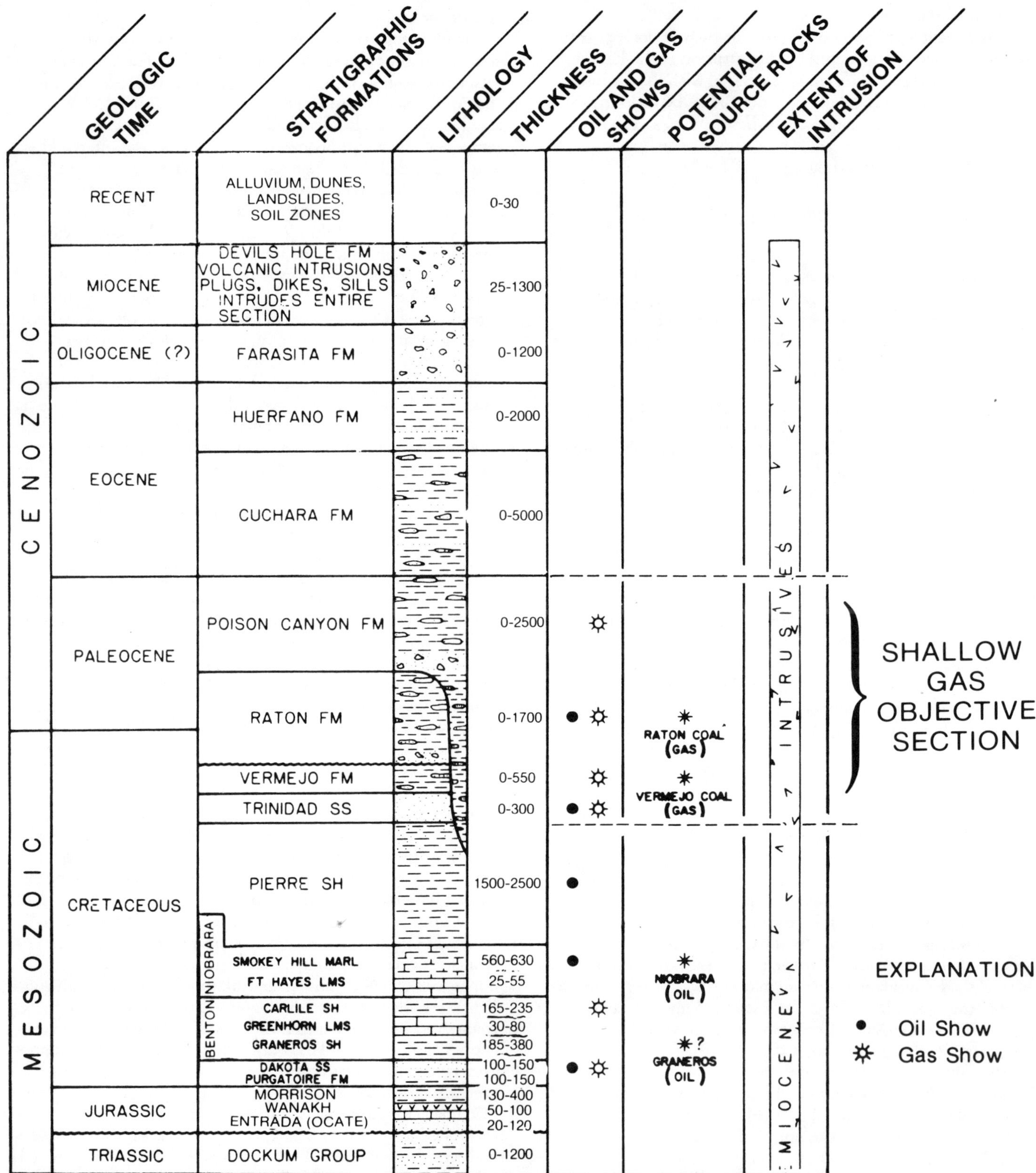

Figure 4—Generalized stratigraphic column of Mesozoic and Cenozoic era strata in the Raton Mesa Region. From Dolly and Meissner, 1977. Used with the permission of the Rocky Mountain Association of Geologists.

units in the region. The Raton Formation, as well as the Vermejo Formation, Trinidad Sandstone, and Pierre Shale, is truncated by the erosional surface at the base of the Poison Canyon Formation. The contact between the Raton and the Poison Canyon Formations is generally indefinite and gradational through a transition zone up to 150 ft thick. The contact is mapped between the highest coal or carbonaceous zone and the bottom of a sandstone unit that contains a basal conglomerate zone.

The Raton Formation is comprised of three generally recognizable field divisions: a basal sandstone, conglomeratic throughout most of the western part of the field; a lower zone, predominantly sandstone, siltstone, and mudstone; and an upper coal-bearing zone, consisting of sandstone, siltstone, mudstone, and beds of coal. These zones vary greatly in lithology and thickness throughout the basin. Total thickness of the Raton Formation ranges from 0 to 1,700 ft. The thickest coal-bearing zone of the Raton Formation ranges from 0 ft in the western part of the basin to over 1,000 ft in the central part. All commercial coalbeds in the Raton Formation occur in this zone (Lee, 1917).

The Raton Formation is of Upper Cretaceous-Lower Tertiary (Paleocene) age and contains the Mesozoic-Cenozoic time boundary. The basal conglomerate of the Raton Formation records the beginning of the Laramide orogenic event. The lithology of the rapidly deposited basal conglomerate indicates Precambrian terrains north and west of the basin as the source areas of the sediment. Depositional conditions remained similar to that of the Vermejo Formation; subsequent sediments of the Raton Formation accumulated on flood plains and in swamps at the same time that the Poison Canyon Formation sediments were deposited on piedmont surfaces further west. Intermittent minor tectonic disturbances continued in the uplifted areas west and north of the basin.

The Poison Canyon Formation of Paleocene age ranges in thickness from 0 to 2,500 ft (Hills, 1888). It consists of massive, lenticular, coarse-grained to conglomeratic, arkosic sandstone and thin, yellow shale derived mainly from Precambrian terrains to the west. Locally thin, irregular, impure coalbeds occur near the base of the Poison Canyon Formation. During middle Paleocene, while the Upper Poison Canyon Formation was deposited in the southern part of the basin, the northwestern part of the basin was uplifted and older formations were tilted, folded, and eroded. Sediments derived from this erosion cycle were deposited in the rest of the basin as layers of pebbles, cobbles, and coarse sand and formed the uppermost beds of the Poison Canyon Formation. In late Paleocene or early Eocene, the mountains to the west and north were again uplifted and the Raton Mesa Region was tilted and folded. The Poison Canyon Formation was eroded from the uplifted areas to the west and north of the Raton Basin, and sediments of the Cuchara Formations (Hills, 1891) of Eocene age were deposited on piedmonts and flood plains in the northern part of the Raton Mesa Region. These sediments are composed of thin to thick beds of red, pink, and white sandstone and thin beds of red and tan shale. Where present, they measure up to 5,000 ft in thickness.

The late Eocene Huerfano Formation rests on the Cuchara Formation (Hills, 1888). The Huerfano Formation appears to lie conformably on the Cuchara Formation on the north and east flanks of West Spanish Peak but unconformably on the south and west flanks of the peak (Johnson, 1961). The Huerfano Formation consists of interbedded arkose and graywacke conglomerate, conglomeratic sandstone, and siltstone, with a few beds of claystone. The formation becomes more conglomeratic upward; the top of the formation contains boulders up to 10 ft in diameter. These sediments were deposited as the result of another episode of uplifting in the regions west and north of the Raton Mesa Basin.

In late Eocene or early Oligocene time, extensive major thrusting, normal faulting, and folding occurred throughout the present mountainous areas of southeastern Colorado and northeastern New Mexico and resulted in formation of the present-day Sangre de Cristo Mountains, Wet Mountains, and Raton Mesa Basin. During this last pulse of the Laramide orogeny, sedimentary rocks of the Raton Mesa Basin were intruded by numerous sills, dikes, plugs, stocks, laccoliths, and sole injections of igneous rocks.

During this time, volcanic stocks, which form the present-day Spanish Peaks, were emplaced near the deepest part of the Raton Mesa Basin. Related sills and plugs are widespread. These intrusions had an intense metamorphic effect, limited to relatively minor contact zones immediately adjacent to the intrusives. Regional low-level metamorphism is believed responsible for the anomalously high rank of coals found in the basin (Dolly and Meissner, 1977). Today, the Raton Mesa Region is an area of anomalously high terrestrial heat flow (Fig. 5). This "hot spot" is probably due to the intrusion of the more deeply seated igneous rocks.

The dikes of the Raton Mesa Basin range from basic to silicic in composition. Most of the dikes in the region are vertical or near vertical and range in thickness from a few inches to more than 100 ft. Most of the dikes belong to two systems. The first system radiates from the Spanish Peaks area, and, along with some of the sills and small plugs, appears to be related to intrusive bodies that form the Spanish Peaks. The second, more extensive system consists of subparallel dikes present throughout the entire region. The strike of this system is from N 60° E in the northern part of the region to N 86° E in the southern part. The dikes trend perpendicular to the axis of Raton Mesa Basin and were probably intruded along fractures resulting from tension during the folding of the basin. Other small localized swarms of dikes are present throughout the region (Johnson, 1958, 1961). Figure 6 illustrates the distribution of coalbeds altered or destroyed by igneous intrusions in the Raton Mesa Region.

After the last period of tectonic activity, uplifted areas were subjected to a period of intense erosion. The only area with a preserved record of this erosion is north and west of the basin, in Huerfano Park, a northern extension of the Raton Mesa Basin. Here, the Oligocene Farasita Conglomerate and Miocene Devil's Hole Formation composed of 25 to 2,500 ft of pebbles, cobbles, boulders, sandstone, water-laid tuff, and volcanic conglomerate rest unconformably on rocks dating from Precambrian to Oligocene.

During Quaternary time, volcanoes in the eastern and southern parts of the Raton Mesa Basin extruded basaltic lava over large areas of it. Remnants of these flows are preserved on

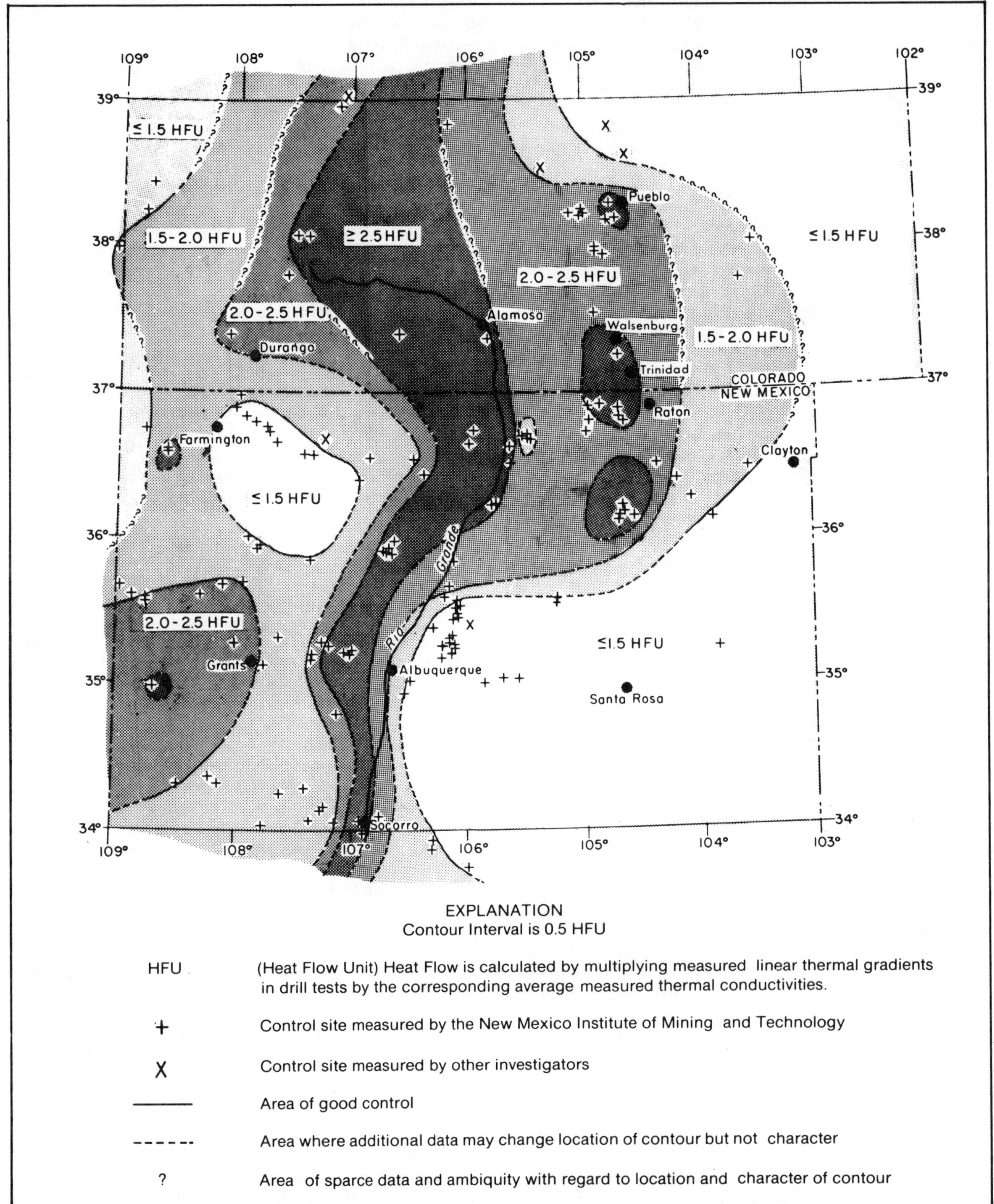

Figure 5—Terrestrial heat-flow contour map of northern New Mexico and southern Colorado. From Edwards et al, 1978.

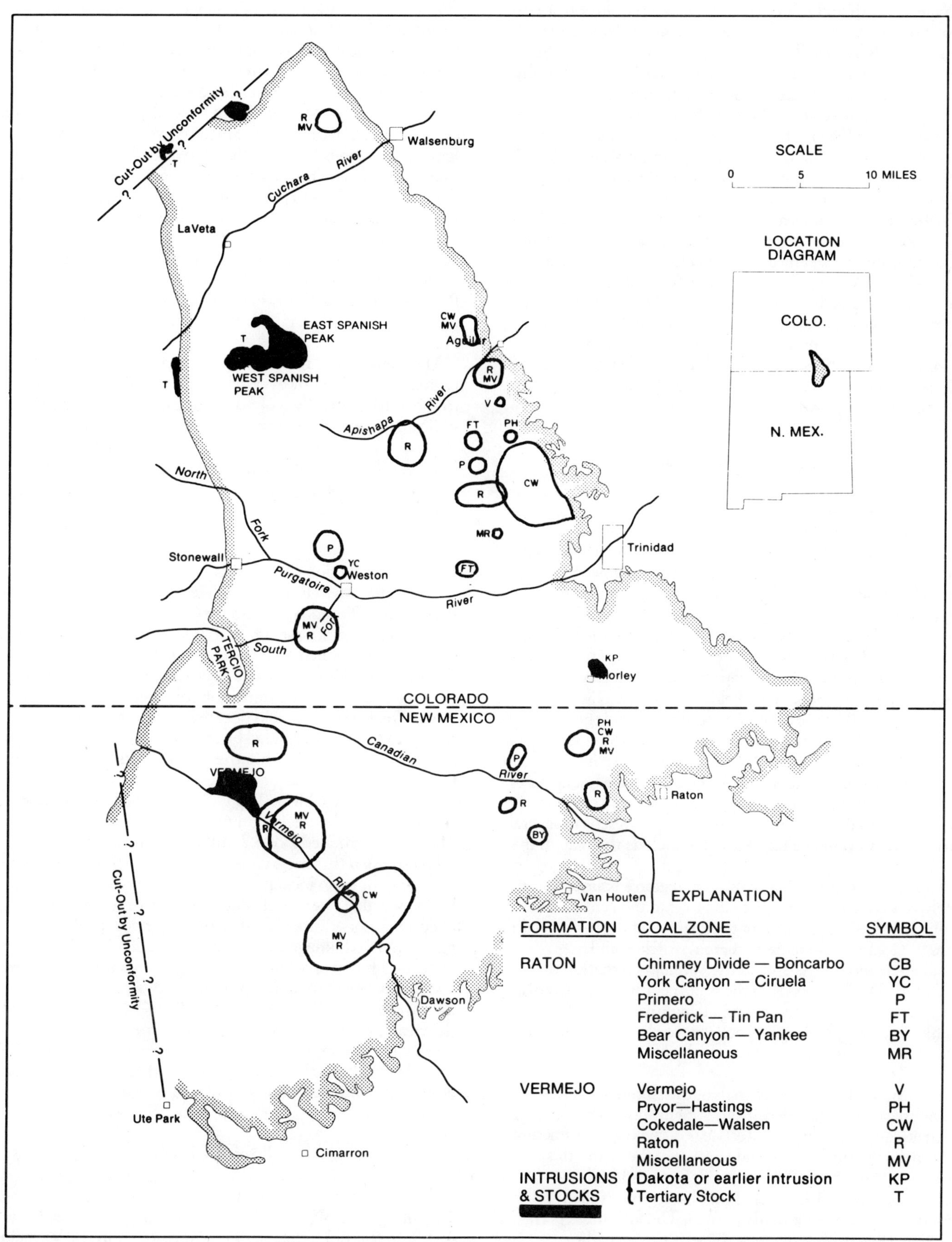

Figure 6—Areas where coal has been destroyed or altered to natural coke by igneous intrusions. Amuedo and Bryson, 1977. Used with the permission of the Colorado Geological Survey.

high mesas north and east of the city of Raton (Johnson, 1959). Also during this time, sills were extruded extensively into and along coalbeds, especially into the Vermejo Formation. The intrusions usually altered or destroyed the coal (Fig. 6). Locally, prismatic coke was formed by the sills; and near Raton, potentially commercial graphite was formed by metamorphism of the Raton coalbed in the Vermejo Formation during intrusion (Lee, 1917). Erosion has continued uninterrupted since the last volcanic episode, and, as a result, only small remnants of the younger Cenozoic formations are preserved.

Quaternary alluvium deposits are found throughout the basin. Gravel and sand deposits (up to 30 ft in thickness) are present along many major present-day stream channels such as the Purgatoire and Vermejo Rivers. Landslide debris and talus cover many of the mountain slopes, and alluvial fans are found along the base of many mountains. Coarse morainal deposits are present at the lower termination of the Cirque Basin on the north side of West Spanish Peak (Johnson, 1961). Soil and pediment deposits cover much of the basin. Quaternary deposits generally are poorly sorted and unconsolidated.

COAL RESOURCE

Regional Character

Coal in the Raton Mesa Basin was deposited during the Cretaceous and Tertiary periods. During the Cretaceous, coal swamps developed along the margins of the shallow seaway. The Tertiary coal swamps developed within the intermontane Raton Mesa Basin.

It is difficult to correlate single coalbeds over long distances, because their discontinuous nature is pronounced in this region. Therefore, this report will use correlation of coal zones in the Trinidad and Walsenburg coal fields, as presented by Boreck and Murray (1979). Because of their discontinuity and the inconsistency in their naming, several coalbeds are identified as occurring in more than one coal zone. In the descriptions of the coal zones, coalbeds that appear in the zone name but are not discussed either are of minor importance or have been mined out. Because of the large number of coalbeds present in several of the zones, it would be impractical to include them all in the zone name.

Coalbeds are exposed along the coal region's boundaries and in canyons cut by streams. They dip toward the center of the basin, so that a large portion of the coal lies at depths exceeding 1,000 ft. Coalbeds are more numerous, thicker, and considerably more extensive in the Vermejo Formation than in the Raton. Thick coalbeds in the Raton Formation generally are lenticular, local in extent, and found mainly in the lower part. Roehler and Danilchik (1980) believe the discontinuous nature of the Raton Formation coalbeds resulted from compaction faults and fluvial scouring.

The structural character of coals of the Raton Mesa Region, such as fracture or cleat density and orientation, has not been systematically investigated. Most coal in the region is fractured (cubically or prismatically), and in some of the mines this cleating enables the coal to be readily removed from the working faces (Lee, 1924). Coal that has been intruded and metamorphosed is columnarly jointed. The jointing is polygonal and is developed at right angles to the intrusive bodies.

One report of cleat direction in coalbeds was by Pillmore (1969b). He recorded a well-developed face cleat system oriented N 70° W in an unnamed coalbed of the Vermejo Formation, located throughout the Vermejo Park area. The axis of the Vermejo Park Anticline, north to N 20° W, is essentially perpendicular to the face cleat direction.

Coals of the Raton Mesa Basin generally range in rank from high-volatile C bituminous to medium-volatile bituminous. Igneous intrusions have locally altered coalbeds to anthracite. In other areas, coalbeds have been metamorphosed to prismatic coke; and near Raton, potentially commercial graphite was formed by the metamorphism of the Raton coalbed in the Vermejo Formation. The thin, relatively rare coalbeds of the Poison Canyon Formation are lignitic in rank. Tables 1 and 2 show the range of analyses of coalbeds in the Vermejo and Raton Formations. Coal that has been locally altered by igneous intrusives generally has high ash and lower calorific and volatile matter values.

The Raton and Trinidad coal fields yield coal that forms a high-quality coke; however, coal from the Walsenburg coal field to the north generally is of noncoking quality. Coal analysis indicates a general and continuous increase in rank from the Walsenburg coal field southward to the Trinidad coal field. The Huerfano-Las Animas County line is considered only an approximate boundary between coking and noncoking coal (Goolsby and Reade, 1979).

Raton Mesa Region Coal Fields

The Raton Mesa Region is divided into three coal fields: the Walsenburg, Trinidad, and Raton (see Fig. 7). The Walsenburg coal field (containing steam coal) is located in the northern part of the region. Immediately south is the Trinidad coal field (containing coking coal). The approximate boundary between the two fields is also the transition zone between the two coal types. The Raton coal field (located in New Mexico) is separated from the Trinidad by the Colorado/New Mexico state line. Mining districts within these three fields are shown on Figure 7, and estimated original coal reserves within the districts are indicated on Table 3. The U.S. Geological Survey (USGS) estimated that the original in-place coal reserve of the Raton Mesa Region totaled 2,304.2 million short tons.

Overburden thickness varies greatly throughout the basin and is dependent upon local topography and position within the basin. Minable reserves of coal occur in the vicinity of the region's periphery, in a few structural uplifts, and along the major stream drainages. Coalbeds with individual thicknesses up to 14.5 ft and an estimated average cumulative thickness of 15 ft are believed to underlie about 2,100 sq mi in the basin, from the outcrops to maximum depths of 3,000 to 4,000 ft at the basin axis near the Spanish Peaks. Only the southwesternmost part, which totals approximately 80 to 100 sq mi, is barren. Sills that intruded into and along the coalbeds destroyed hundreds of millions of tons of coals (Pillmore, 1969b).

Raton Coal Field

Total resource estimates for the Raton coal field range from 1.5 billion tons (Wanek, 1963) to 4.8 billion tons (Read et al, 1950). In this field, the major coals are found in the Vermejo and Raton Formations.

Table 1—Range of typical analyses of coalbeds in the Raton coal field.

Formation	Coalbed	Moisture*	Volatile Matter	Fixed Carbon	Ash	Sulfur	Btu	Rank
Vermejo	Raton	1.0–9.2	30.1–9.2	43.5–50.9	6.7–22.8	0.4–0.8	10,380–12,990	hva–hvb
	Vermejo**	—	28.2	53.7	18.0	0.5	10,030	hvb
	Sugarite	2.9–3.8	38.1–39.7	48.7–49.1	8.7–8.9	0.5–0.6	13,000–13,300	hva
Raton	Potato Canyon**	1.5	35.1	53.6	9.8	0.5	13,410	hva
	Tin Pan***	2.6	36.1	45.3	16.1	0.6	12,150	hva
	Yankee	5.0–6.1	35.2–37.4	42.1–48.3	9.5–16.6	0.5–1.0	12,000–12,700	hva
	Chimney Divide** (Ridge)	—	38.4	47.6	13.4	0.5	13,980	—
	York Canyon	1.7–2.2	33.3–36.1	45.3–54.0	8.0–19.2	0.4–0.6	11,810–13,550	hva
	Ancho Canyon**	—	36.6	54.0	8.6	0.5	13,840	—
	Upper Left Fork	1.5–2.1	36.1–36.7	50.8–54.4	7.5–10.4	0.4–0.6	12,830–13,740	hva

*Unless otherwise stated, all analyses are on an as-received basis.
**Representative moisture-free analyses of coalbeds.
***Only one sample analysis available.

Table 2—Range of typical analyses of coalbeds in the Trinidad and Walsenburg coal fields.

Formation	Coalbed/ Coal Zone	Moisture*	Volatile Matter	Fixed Carbon	Ash	Sulfur	Btu	Rank
Vermejo	Cameron, Lower Bunker	1.9–7.5	31.4–39.9	45.1–57.5	8.1–14.5	0.5–1.0	11,280–13,510	hva–hvb
	Berwind, Upper Bunker	1.7–5.9	31.4–33.6	51.8–54.4	10.4–15.4	0.6–0.9	11,990–13,240	hva–hvb
	Majestic, Mammoth, Piedmont, Starkville, Walsen	1.7–10.2	28.7–41.8	44.2–58.0	4.8–19.2	0.4–1.1	10,740–13,520	hva–hvb
	Empire, Upper and Lower Ludlow, Majestic, Pryor	2.2–6.1	29.1–37.0	48.3–54.8	8.2–18.0	0.6–1.9	11,390–13,420	hva
	Hastings and Robinson	1.5–7.3	28.2–41.5	48.6–56.9	8.1–17.7	0.5–0.8	12,050–13,660	hva
	Cokedale, Kebler, Occidental, Rapson, Thompson	2.3–5.2	25.8–38.6	49.4–57.3	10.2–17.7	0.5–0.6	11,810–12,850	hva
	Gem and Sopris	1.9–2.1	28.2–35.5	51.9–56.9	10.7–17.7	0.7	12,360–13,110	hva
	Apache**	3.1	34.6	54.8	10.6	0.4	13,440	hva
Raton	Alfreda, Bear Canyon, Cass, Frederick	1.8–4.4	30.3–40.7	48.1–58.3	6.1–16.4	0.4–0.8	12,520–13,550	hva
	Delagua and Peacock	1.9–4.3	36.2–39.1	47.7–52.0	7.2–15.9	0.5–0.8	12,110–13,160	hva
	Primero	1.4–4.4	30.8–40.8	44.1–58.0	7.5–21.4	0.5–1.5	11,260–13,850	hva
	Boncarbo	1.4–4.4	30.8–40.8	44.1–58.0	7.5–21.4	0.5–1.5	11,260–13,850	hva

*Unless otherwise stated, all analyses are on an as-received basis.
**Only one sample analysis available.

Vermejo Formation—The Vermejo Formation contains up to three coalbeds greater than 14 inches thick in most of the coal field. These coalbeds are lenticular, irregular in thickness, and interbedded with shale and siltstone. Total coal thickness in the Vermejo Formation is indicated on Figure 8. The coal is generally brittle and friable, with a bright luster. It has prismatic or cubic cleating and platy cleavage. The coal's floor and roof rock are carbonaceous siltstone, sandstone, and conglomeratic sandstone (Lee, 1924). The major coals within the formation are the Raton, Vermejo, and Sugarite.

The Raton coalbed is the name commonly applied to commercial beds found at or near the base of the Vermejo Formation throughout the Raton coal field, regardless of whether the beds can be traced back to the Raton area. In areas where two or more beds are present, the thickest bed usually is called the Raton. In some areas, it may be as thick as 14.5 ft (Lee, 1924) and is the most extensive and valuable bed in the Raton Field. The beds are not continuous, although they lie at the base of the Vermejo Formation. The beds become progressively older to the west. The coalbeds are grouped in large, elongate, pod-shaped deposits, some of which have been intruded by igneous sills and partially destroyed.

Early mining of the Raton bed was concentrated along the Dawson, Koehler-Van Houten, and Gardner trends. The Raton coalbed probably correlates with those found in the Cameron, Lower Bunker coal zone of Boreck and Murray (1979).

The Vermejo is a thick coalbed near the top of the Vermejo Formation at Vermejo Park and in the upper part of the formation in the Castle Rock District. It is more irregular in thickness and distribution than the Raton coalbed. It is generally restricted in occurrence to areas where the Vermejo Formation is thicker than 250 ft.

Mined primarily in the vicinity of Raton and Sugarite, New Mexico, the Sugarite coalbed is believed to be located in the lower part of the Vermejo Formation, approximately 85 ft above the top of the Trinidad Sandstone. It is not located in the normal stratigraphic interval of the Raton or Vermejo coalbeds; and it has not been determined whether the Sugarite bed is in

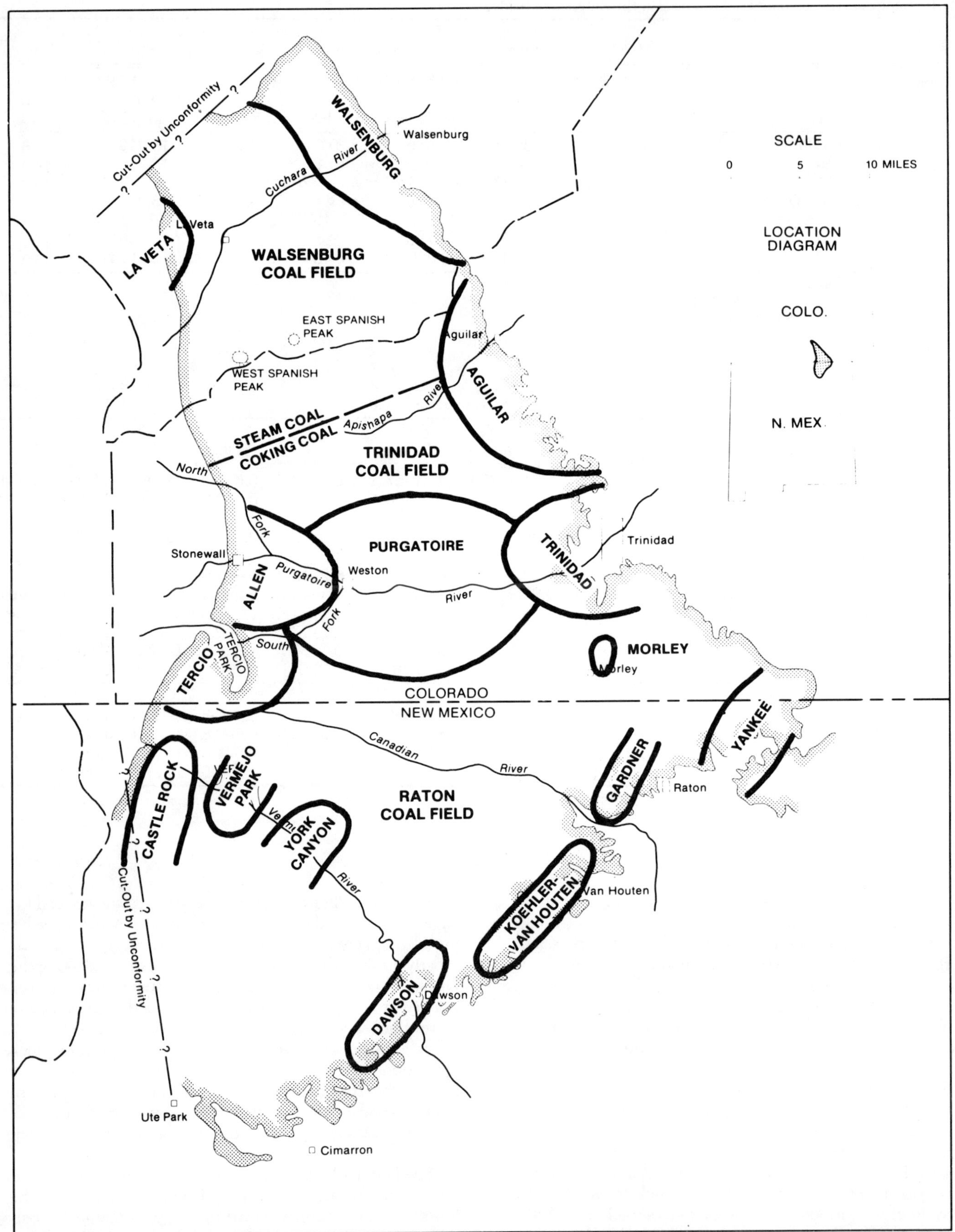

Figure 7—Mining districts of the Raton Mesa Region. Amuedo and Bryson, 1977. Used with the permission of the Colorado Geological Survey.

Table 3—Total estimated original coal resources of coalbeds at least 14 inches thick, with less than 3,000 ft of overburden. Johnson, 1961; Pillmore, 1969b.

State	Area	Coalbed Basis*	Coal Zone Basis**	Total
Colorado	La Veta	120.9	201.9	322.8
	Walsenburg	667.5	974.4	1,641.9
	Spanish Peaks	106.8	386.6	493.4
	Trinidad-Aguilar	1,215.6	1,869.7	3,085.3
	Guinare, Cuchara Pass, and Stonewall	524.1	1,755.9	2,280.1
	Cuarto	262.1	768.0	1,030.1
	Stonewall-Tercio	491.0	1,895.1	2,386.1
	Starkville-Weston	880.3	4,364.7	5,244.5
			Subtotal	16,484.2
New Mexico	Castle Rock	400.0	—	400.0
	Gardner	184.0	—	184.0
	Vermejo Park	24.7	—	24.7
	York Canyon	241.9	—	241.9
	Subtotal	850.6		
	Grand Total	17,334.8		

*All figures are in thousands of short tons.

**Additional estimates of reserves on a zone basis are based on the assumption that the coal-bearing rocks contain as much coal at depth as at the outcrop.

the Raton Formation and equivalent to the Vermejo bed or whether it is an intermediate bed in the Vermejo Formation.

Raton Formation—Coalbeds of the Raton Formation are thinner, more lenticular, irregular in thickness, and more widely spaced than coalbeds of the Vermejo Formation. The roof and floor rocks are generally shale and bony coal, although at some localities the roof is a thick sandstone bed (Lee, 1924). Major coals within this formation are the Potato Canyon, Tin Pan, Yankee, Left Fork, Cottonwood Canyon, Ancho Canyon, York Canyon, and Chimney Divide beds.

The Potato Canyon coalbed ranges up to 8 ft in thickness, is characterized by numerous shale partings, and lies from 100 to 150 ft above the Tin Pan coalbed.

The Tin Pan coalbed underlies an area of 40 sq mi west of Raton and ranges in thickness from a streak to more than 8 ft. It usually contains one or more partings as thick as 4 ft. The Tin Pan coal lies 700 ft stratigraphically above the Trinidad Sandstone and crops out in many canyons along the Canadian River. Even though the Tin Pan coalbed has been mined for several years, substantial coal resources still remain in the ground.

The Yankee coalbed is thinner and less extensive than the Tin Pan and is characterized by shale partings. It is approximately 300 to 400 ft stratigraphically above the Trinidad Sandstone. The Yankee and Tin Pan coals are possibly equivalent stratigraphically, indicating a thinning of the lower zone of the Raton Formation.

The Left Fork coals are two beds about 50 ft apart, exposed on the steeply dipping east flank of the Vermejo Anticline in the Left Fork of York Canyon northeast of Yermejo Park. The lower coalbed is about 1,250 ft above the top of the Trinidad Sandstone. The upper bed is thicker and more persistent and ranges in thickness from 5 to nearly 8 ft throughout much of its known extent. Estimated resources in the Left Fork upper bed, where it is greater than 3.5 ft thick, total almost 25 million tons (13.3 measured, 5.5 indicated, 5.9 inferred) (Pillmore, 1969b).

The Cottonwood Canyon coal zone underlies an area of 10 sq mi in Cottonwood and Caliente Canyons. The Cottonwood Canyon bed ranges in thickness from a few inches to 7 aggregate ft of coal in a 9-ft zone and is characterized by several shale partings. An estimated 51 million tons of coal (19 measured, 23 indicated, 9 inferred) are contained in the Cottonwood Canyon bed (Pillmore, 1969b).

The Ancho Canyon coalbed crops out in an area of 16 sq mi in the vicinity of Ancho, Salyers, Cachupin, and Vermejo River Canyons. The bed generally consists of two layers of coal, each 2 ft thick, separated by a carbonaceous shale 0.5 to 2 ft thick. Coal resources for the Ancho Canyon bed are an estimated 38.4 million tons (21.5 measured, 6.4 indicated, 10.5 inferred), with more than 25 million tons lying beneath less than 100 ft of overburden (Pillmore, 1969b).

The York Canyon coalbed underlies an area of 12 sq mi in York and Road Canyons. It varies from a bed containing many partings and only a few inches of coal in the southeastern part of the area to a single coalbed more than 10 ft thick near York Canyon. A major parting rapidly thickens westward to 30 ft and divides the bed into upper and lower benches along the area's western margin. The York Canyon coalbed is 1,520 ft stratigraphically above the Trinidad Sandstone. Measured coal resources in the York Canyon area total 33.4 million tons in beds at least 4 ft thick (Pillmore, 1969b).

The Chimney Divide coalbed outcrops through an area of 50 sq mi and commonly consists of two benches, each 2 ft thick, separated by a carbonaceous shale parting from 6 inches to 1 ft thick. In some areas, it may consist of a series of lenses at approximately the same stratigraphic position rather than a single continuous bed. North of the Canadian River, the coal becomes dirty and splits into many layers of dirty coal and shale. Resources are estimated to be 97.5 million tons (48.3 measured, 29.3 indicated, 19.9 inferred) in beds at least 28 inches thick (Pillmore, 1969b). An estimated 22 million tons of the resources are in beds at least 42 inches thick.

Trinidad and Walsenburg Coal Fields

Boreck and Murray (1979) report that as of 1977 the remaining reserves of the Trinidad and Walsenburg coal fields were approximately 671.5 million short tons. The major coals in

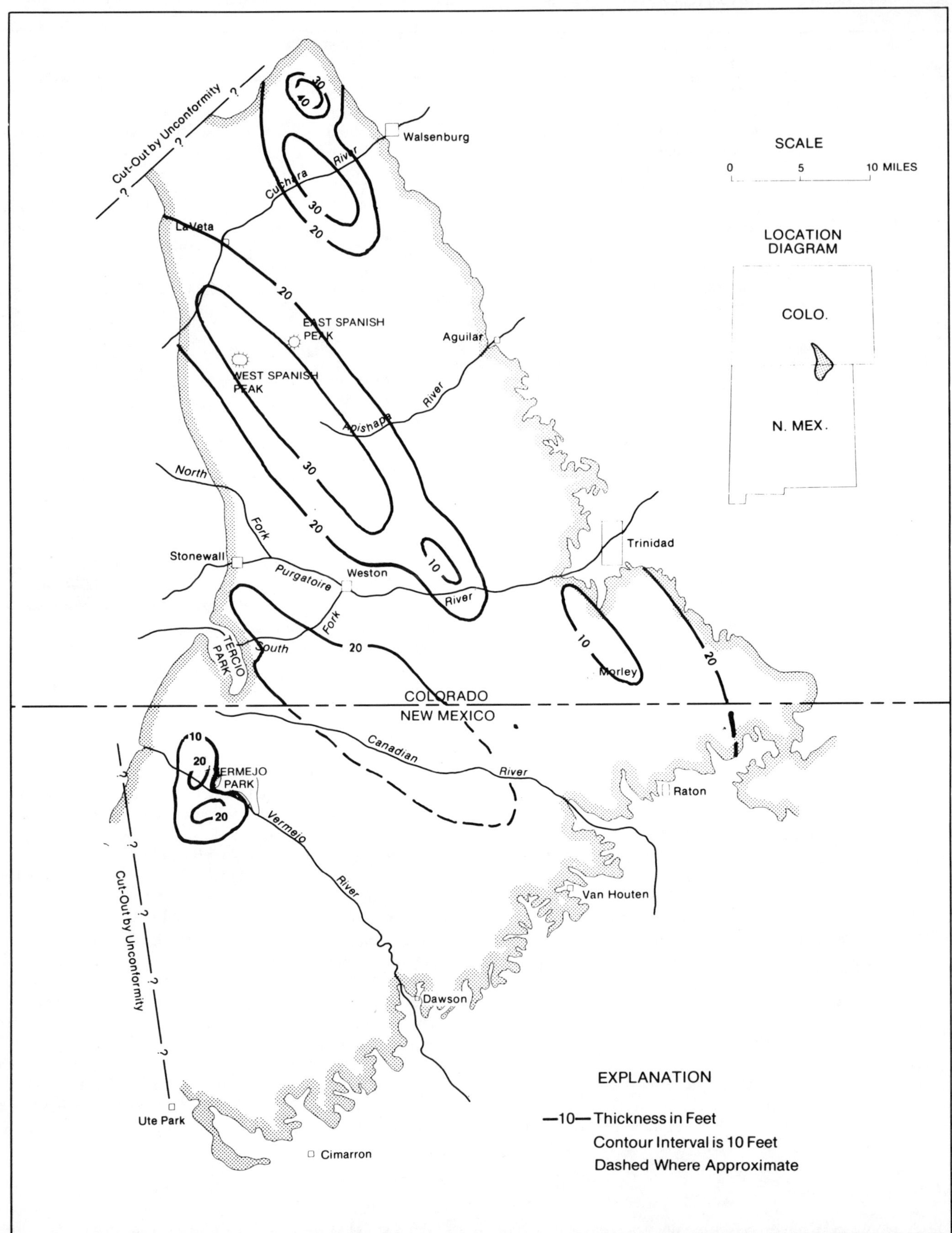

Figure 8—Isopach map of the coalbeds in the Vermejo Formation. From Tremain, 1980; used with the permission of the Colorado Geological Survey; and Johnson and Wood, 1956.

these coal fields also occur in the Vermejo and Raton Formations.

Vermejo Formation—The Vermejo Formation contains 3 to 14 coalbeds greater than 14 inches thick in the Trinidad and Walsenburg coal fields. These coalbeds are lenticular and irregular in thickness. Roof and floor rocks of coalbeds in the formation are usually carbonaceous shale and claystone but locally may be carbonaceous siltstone or sandstone. Bony coal and shale form partings in the coalbeds (Johnson, 1961). The coal generally is of the same character as the Vermejo Formation coalbeds in the Raton coal field: brittle, friable with bright luster, and prismatic or cubic cleavage. Some of the coalbeds yield spherical bodies of coal up to 2 ft in diameter, evidently caused by squeezing within the beds. Coals in these fields are divided into eight major coal zones and beds.

Coalbeds mined in the Cameron, Lower Bunker coal zone include the Cameron, Lower Alamo, Lower Bunker, Lower Piedmont, Maitland, and Rouse (Fig. 9). In the Walsenburg field, the only bed in this zone with significant reserves is the Cameron. It averages approximately 3 ft in thickness and ranges from 2 to a maximum of 5.8 ft, with virtually no partings. The Cameron is usually less than 500 ft stratigraphically above the base of the Vermejo Formation. The Lower Bunker Hill coalbed has also been mined in the Walsenburg field but does not contain significant reserves. This bed is usually less than 2 ft thick and has been naturally coked by igneous intrusions in the southern half of the field. Locally it is dirty and may contain partings of bony coal or shale. In the Trinidad coal field, the Lower Piedmont coalbed is the only bed of this zone having significant reserves. In this field, the Lower Piedmont underlies 12 sq mi along the Purgatoire River and ranges from 1 to 3.5 ft thick, averaging 2.5 ft.

The Berwind, Upper Bunker coal zone contains three beds that have significant reserves in both the Walsenburg and Trinidad fields. These beds—the Rainbow, Upper Bunker Hill, and Berwind—are very similar in occurrence, each averaging approximately 2.5 ft of coal, with thicknesses ranging from a trace to more than 7 ft. Two of the beds have been intruded and coked and each contains local deposits of dirty coal and bony partings.

Two beds of the Majestic, Mammoth, Piedmont, Starkville, Walsen coal zones—the Lenox and the Walsen—underlie 30 sq mi of the Walsenburg field and comprise a significant coal reserve. The most continuous of the two, the Lenox, averages 2.5 ft in thickness and locally contains bony coal and shale partings. In the Trinidad field, two coalbeds of this zone are present: the Piedmont and the Lower Starkville. The Piedmont is more extensive and averages 5 ft of coal, usually with a single parting that averages 1 ft in thickness.

The Pryor, Empire, and Majestic coalbeds occur in the Empire, Lower and Upper Ludlow, Majestic, Pryor coal zone, and underlie 80 sq mi of both the Walsenburg and Trinidad coal fields. The most extensive coalbed—the Pryor—averages 3 ft in thickness, with some local occurrences of bony coal and shale partings. The Majestic (thickest of the three) averages 7 ft and has been coked by intrusions along its southern extension. The Empire coalbed has an average aggregate thickness of 2 ft near the middle of the coal zone. The Empire has also been intruded, destroying most of the coalbed's northern portion. The Lower and Upper Ludlow coalbeds underlie an area of 10 sq mi, 15 mi northwest of Trinidad. These coalbeds are not very extensive and locally have been intruded and coked. The Upper and

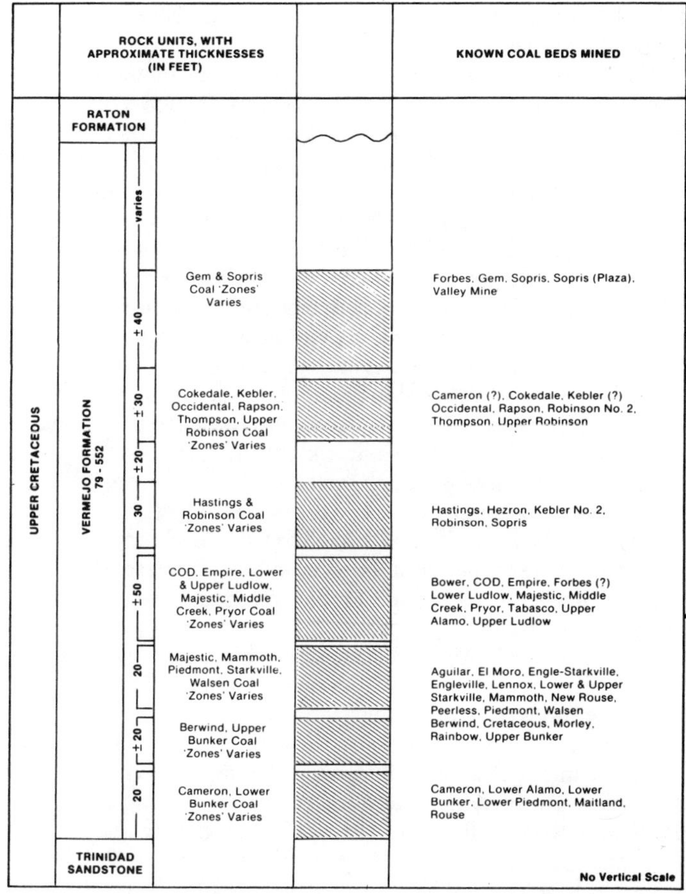

Figure 9—Generalized columnar section of coal-bearing rocks in the Vermejo Formation, Raton Mesa Region, Colorado. Boreck and Murray, 1979. Used with the permission of the Colorado Geological Survey.

Lower Ludlow coalbeds have an average thickness of 5 ft.

The Lower Robinson coalbed in the Walsenburg field is the most extensive in the Hastings and Robinson coal zone. It averages 2.5 ft in thickness and is a relatively clean coal. The less extensive but thicker Hastings coalbed is found in the Trinidad coal field south of the town of Aquilar. Here, the coalbed averages 6 ft thick, with a maximum of 8 ft of coal. The southern portion of the coalbed has been coked by igneous intrusion.

The Rapson coalbed has minable reserves in both the Walsenburg and Trinidad coal fields in the Cokedale, Kebler, Occidental, Rapson, Thompson coal zone. It is not very extensive and contains local zones of dirty or bony coal. It averages 2.5 ft in total thickness, with at least one shale parting that may be as thick as 3 ft. The Upper Robinson coalbed of the Walsenburg field is discontinuous in nature and is truncated by an erosional unconformity at its northern limit. The Upper Robinson averages 4 ft of coal in aggregate thickness and normally has a shale parting 6 inches thick near its base. The Cokedale coalbed underlies the Trinidad field and has an average thickness of 2 ft, which thins to the north and south, where its average thickness drops to 1 ft.

The Gem coalbed of the Gem and Sopris coal zone, mined in both the Trinidad and Walsenburg coal fields, is a small bed of minimal extent. It averages 2.5 ft in thickness and is found at relatively shallow depths. The Supris coalbed of the Trinidad coal field is a discontinuous deposit averaging 4 ft in thickness. It has been coked in the western portion of the deposit by igneous intrusives.

The newly designated Apache coalbed of the Vermejo Formation is presently mined at the recently opened Maxwell Mine, 3 mi southeast of Stonewall, Colorado. The Apache coalbed has an average thickness of 5 ft and is overlain by 410 to 1,400 ft of overburden.

Raton Formation—Coalbeds of the Raton Formation are generally thinner, more lenticular, irregular in thickness, and more widely spaced than coalbeds of the Vermejo Formation. The coal is brittle and friable, with a bright to dull luster, and has cubic or prismatic cleats and platy cleavage. At some locations, conchoidal fractures and spheroidal coal are common. A few coalbeds along the Purgatoire River have been destroyed by the intrusion of igneous sills. Roof and floor rocks are generally carbonaceous shale and siltstone; however, at many places, the roof is a thick sandstone bed (Johnson, 1961). The formation contains six major coal zones or beds.

The Alfreda, Bear Canyon, Cass, Frederick coal zone is the lowermost zone of the Raton Formation in Colorado and consists of several coalbeds that range from a trace to greater than 10 ft in thickness (Fig. 10).

These beds are lenticular and have not been correlated the length of the Trinidad and Walsenburg coal fields. The Frederick bed is the thickest and most continuous of coalbeds in this zone. It averages 4 ft in aggregate thickness, with one or more partings of shale and siltstone averaging 18 inches total thickness. The Frederick was mined extensively along the Purgatoire River in the Trinidad coal field. Other coalbeds of this zone include the Upper Rugby coalbed, averaging 2.5 ft of coal split by a 6-inch shale parting; the Lower Rugby coalbed, averaging 4 ft of coal with local zones of dirty or bony coal; the Alfreda coalbed, averaging 2.5 ft of coal; the Cass coalbed, averaging 3.5 ft of coal with one or more shale partings from 2 to 8 inches thick near the top of the bed; and the extensive Bear Canyon coalbed, which averages 2 ft of coal and is characterized by several thin shale partings. All of the above beds, in addition to the Brodhead Number 4, the Martinez, the Primrose Number 2, the Rugby Number 3, and the Upper Series Number 3 coalbeds, occur within an interval of 150 ft located about 350 ft above the base of the Raton Formation.

The Delagua and Peacock coal zone is composed primarily of the Delagua coalbeds mined along the eastern margin of the basin. Here, the Delagua Number 1 coalbed ranges from 1.5 to 6 ft in thickness. In its northern and southern extremities, the coalbed is split by a shale parting up to 32 ft thick. This bed has been correlated for a lateral distance of about 15 mi. It lies 700 ft above the base of the Raton Formation.

The Primero coalbed of the Primero coal zone has been mined along the Purgatoire River in the Trinidad coal field. Along with the Frederick coalbed, it underlies as much as 75 sq mi in the above-mentioned area. It ranges in thickness from a few inches to greater than 6 ft and averages approximately 3 ft. The Allen coalbed of this zone is presently mined at the Allen Mine, located along the Purgatoire River near the town of Stonewall. This bed averages 5 ft in thickness and is overlain by 100 to 2,500 ft of overburden. It is a high-volatile B bituminous coal.

The Boncarbo coalbed ranges in thickness from 5 to 7 ft, commonly with two or more partings composed of shale and sandstone with an aggregate thickness of 6 to 12 inches.

The Ciruela coalbed, mined primarily along the Purgatoire River, appears to be quite lenticular and locally may be absent. It averages 1.5 ft thick and has a maximum of 2.5 ft.

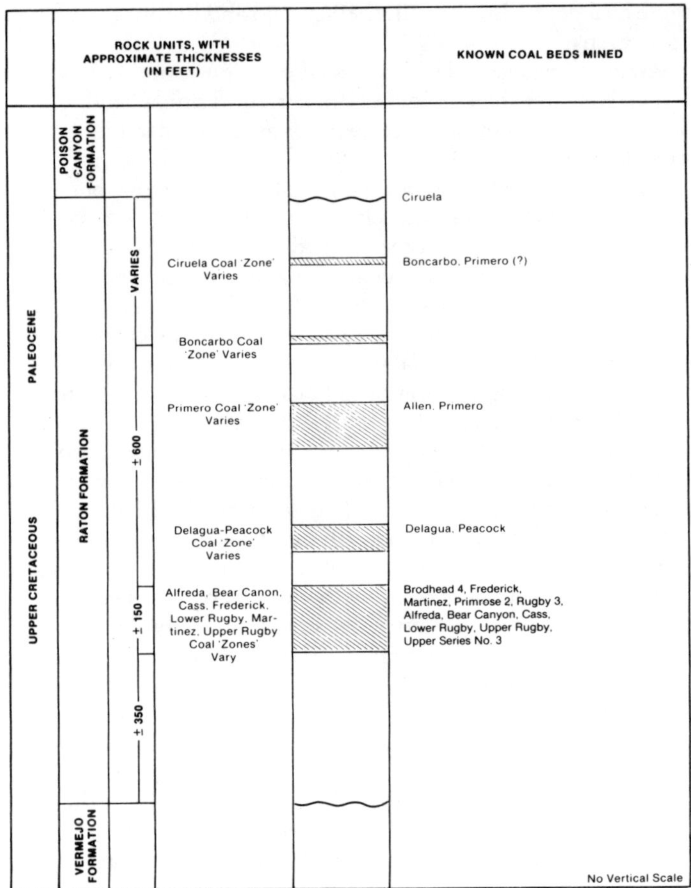

Figure 10—Generalized columnar section of coal-bearing rocks in the Raton Mesa Region, Colorado. Boreck and Murray, 1979. Used with the permission of Colorado Geological Survey.

Along the eastern margin of the Walsenburg coal field, two coalbeds—termed the "A" and "B" coalbeds (Johnson, 1958)—occur at or near the base of the Raton Formation. These beds have been correlated by measured coal sections at outcrops and by drill hole sections. The lower "A" coalbed, usually located within 5 ft of the base of the Raton Formation, is the more continuous and averages 2.5 ft in thickness. Both beds pinch out 6 mi south of Walsenburg; the "B" coalbed pinches out about 9 mi north of Walsenburg.

POTENTIAL METHANE RESOURCE

Studies investigating methane gas content of coal in the Raton Mesa Region are limited to four test holes drilled by the Colorado Geological Survey and the USGS along the Purgatoire River; two production holes drilled by the U.S. Bureau of Mines (USBM) near the town of Morley, Colorado; and three holes drilled by the American Public Gas Association (APGA) near Weston, Colorado. Figure 11 shows the location of the well test sites. Gas content of the coalbeds tested at the USGS sites ranged from 23 to 193 cubic feet/ton (cf/ton) for the Raton Formation coalbeds and 115 to 492 cf/ton for coalbeds of the Vermejo Formation. These tests indicate that the Vermejo Formation, at depths greater than 700 ft, has the greatest

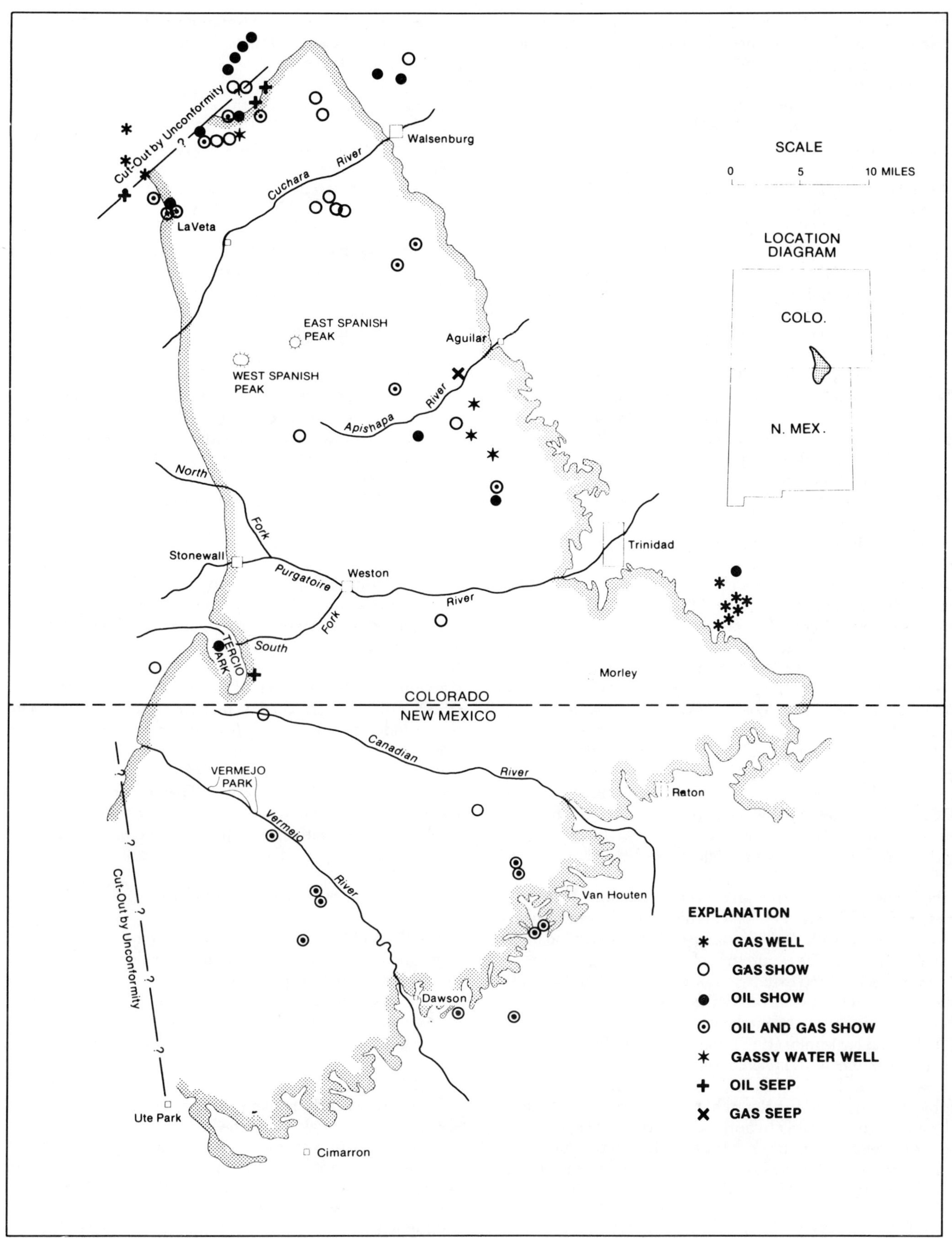

Figure 11—Location of well test sites, gassy mines, and oil and gas shows in the Raton Mesa Region.

potential for coalbed methane development. The USBM completed a drilling program near the Morley Dome, aimed at producing methane from coalbeds (Bench, 1979). Results of this drilling and testing indicate low methane contents south of Trinidad, Colorado. Table 4 summarizes the results of coalbed methane desorption work completed by the USGS and USBM in the Raton Mesa Region.

The city of Trinidad has investigated the possibility of producing methane gas from the Raton coalbed of the Vermejo Formation for use as a municipal natural gas supply. Three wells were drilled by the APGA in the Raton Mesa Region, west of Trinidad. The first hole (MGP # 1) was drilled as a control hole to a total depth (TD) of 1,605 ft and no coal samples were taken. The second hole (MGP # 2) was drilled approximately 500 ft north-northwest of the first hole to a TD of approximately 1,449 ft (Gustavson Associates, Inc., 1980). Southeast of MGP # 2 and northeast of MGP # 1, well MGP # 3 was drilled to a TD of 1,444 ft. Seven samples of coal, carbonaceous shale, and sandstone samples were collected for desorption from MGP # 2 and six samples of coal and carbonaceous mudstone were collected from MGP # 3. Of these 13 samples, 6 were coal. As of June 15, 1981, the gas content of these coals ranged from 300 to 510 cf/ton, from depths ranging from 1,092.6 to 1,209.0 ft (Tremain, 1980). Gas content data from all 13 samples are located in Table 5.

A second indication of methane gas associated with coalbeds comes from mine methane emission surveys. Methane emission reports are available from the U.S. Department of Labor, Mine Safety and Health Administration (MSHA) indicating that more than 2 million cubic feet per day (MMcfd) are vented daily from three mines in the region.

Fender and Murray (1978) recorded 32 mines in the Colorado portion of the region with reported occurrences of gas. Figure 11 shows the location of gassy coal mines and gas shows in exploratory and water wells. Other indications of the occurrence of coalbed methane in the region include two exploration wells that reported 100% methane background gas throughout the coal interval (Energetics Healy No. 13-8, Sec. 8, T31S, R65W), and the recovery of some burnable gas after fracturing a 5-ft coalbed in the Raton Formation (Filon Exploration Corporation No.1—Zeles Hope Sec. 31, T31S, R65W).

The Raton Mesa Region is an area of anomalously high terrestrial heat flow (see Fig. 5). This high heat flow probably is related to the Spanish Peaks intrusive and associated underlying magmatic activity. The occurrence of an area of high heat flow, igneous intrusives, and extensive low volatile-matter coalbeds in the Raton Mesa Region indicates the potential generation of substantial amounts of dry methane gas.

Choate and Rightmire (1982) studied the San Juan Mountain geothermal anomaly and related Tertiary igneous events in southwestern Colorado and their influence on the coalbed methane potential. In the Raton Mesa Basin, coalbeds in areas of igneous intrusions alter to anthracite, natural coke, or graphite. These authors have stated that these heat sources are, in part, responsible for the high rank coals and thus the high gas contents in the western Raton Mesa Basin.

Estimated Resource

Dolly and Meissner (1977) calculated that approximately 23 trillion cubic feet (Tcf) of gas has been generated by the coalbeds of the Vermejo and Raton Formations in the Trinidad and Walsenburg coal fields of Colorado. Their calculations were based on the following:

1. Average aggregate coal thickness—15 ft
2. Areal extent of coal—1,200 sq mi
3. Coal density—1,800 tons per acre ft
4. Weight of coal
   Thickness × Areal extent × Density = $15 \times 1{,}200 \times 640 \times 1{,}800 = 20.74 \times 10^9$ tons
5. Average coal rank is approximately a medium-volatile bituminous coal containing 26% volatile matter and has generated 1,121 cu ft of gas per ton.
6. Volume of gas generated
   Weight of coal × Volume of gas per ton = $20.74 \times 10^9 \times 1{,}121 = 23.2 \times 10^{12}$ cu ft.

If Dolly and Meissner's (1977) calculations are expanded to include the Raton coal field with its 800 sq mi, an estimated 38.7 Tcf of gas has been generated by the Vermejo and Raton Formations. Assuming that approximately 50% of the original gas generated is still confined within the coalbeds, approximately 18.4 Tcf of gas is still in place in the Raton Mesa Region.

Danilchik et al (1979) estimated that approximately 84 billion cubic feet (Bcf) of methane gas is present in coalbeds of the Vermejo Formation underlying an area of 25 sq mi located along the Purgatoire River. Their estimate is based on results of the coal desorption work completed by the USGS and the assumption that a 7-ft thickness of coal is present throughout the entire area at depths ranging from 1,500 to 2,000 ft and containing at least 416 cu ft of gas per ton of coal. Tremain's (1980) estimate of the coalbed methane resource for 54 sq mi in the Trinidad coal field totals 311 Bcf for a 10 ft thick coalbed.

If Danilchik's estimate is expanded to include the entire Raton Mesa Region, approximately 8 Tcf of methane gas is present in-place in the coalbeds. If Tremain's resource estimate is expanded accordingly, there is approximately 12 Tcf of in-place coalbed methane gas in the Raton Mesa Region.

CONCLUSIONS

The Raton Mesa Region contains substantial resources of high- and medium-volatile bituminous coals that extend from outcrops along the periphery of the region to depths of at least 3,000 ft in the deepest parts of the region. Potentially, methane gas associated with the deeper of these coalbeds is very substantial. Test results indicate substantial methane content in higher rank coalbeds of the Vermejo Formation in the western part of the region.

Based on presently available data, an area (Fig. 12) that encompasses the majority of the Raton Mesa Region can be designated as a primary target for methane production. The target area does not include the periphery of the region. Here, where the coal is not mined out, it is either very shallow or outcropping, thus allowing methane gas to escape into the atmosphere. The target also does not include areas where there has been extensive igneous intrusive activity. In those areas, significant amounts of coal have been intruded and destroyed. The northern and eastern portions of the region also have been

Table 4—Desorption results for the Raton Mesa Region. Tremain, 1980. Used with the permission of the Colorado Geological Survey.

Field[1]	Drill Hole	Test No.	Formation	Depth to Bed (m)	Depth to Bed (ft)	Sample Weight (gm)	Desorbed Gas (cc)	Lost Gas (cc)	Residual Gas (cc/gm)	Total Gas (cc/gm)	Total Gas (cf/ton)	Apparent Rank of Coal	% Methane in Gas (air free)	Heating Value of Gas (kcal/cu m)	Heating Value of Gas (Btu/cf)
Trinidad		1	Raton	246.89	810.0	2,319	3,086	370	0.10	1.59	51	an[1]	66.78	6,282	698
		2	Raton	252.37	828.0	1,098	755	130	0.0	0.81	26	hvCb[2]	—	—	—
	USGSDH78-1	3	Raton	321.11	1,053.5	2,308	4,731	480	0.0	2.26	74	hvBb[3]	91.75	8,233	925
		4	Raton	324.03	1,063.1	1,710	9,441	850	0.01	6.03	193	mvb[4]	83.34	7,485	841
		5	Vermejo	515.48	1,691.2	1,600	14,255	3,400	0.04	11.07	350	mvb	46.14	4,139	465
		6	Vermejo	546.20	1,792.0	1,724	18,098	8,300	0.06	15.37	492	mvb	97.16	8,722	980
		7	Raton	94.24	309.2	1,461	3,390	170	0.16	2.60	83	1vb[5]	—	—	—
	USGSDH78-2	8	Raton	146.97	482.2	1,057	2,031	890	0.0	2.76	88.4	mvb	—	—	—
		9	Raton	152.31	499.7	767	2,719	1,110	0.0	4.99	160	mvb	—	—	—
	USGSDH78-3	10	Vermejo	222.32	729.4	1,768	10,176	3,300	0.33	7.95	254	mvb	98.98	8,873	997
	USGMDH-1	11	Vermejo	30.63	100.5	808	158	170	0.30	0.71	23	mvb	—	—	—
		12	Vermejo	51.18	167.9	553	1,057	800	0.20	3.56	115	mvb	—	—	—
		13	Vermejo	218.11	715.6	876	88	—	1.35	1.45	46	hvAb[6]	—	—	—
		14	Vermejo	246.89	810.0	1,051	74	115	0.0	0.18	6	hvAb	—	—	—
	USGSDH78-4	15	Vermejo	247.65	812.5	1,657	56	65	0.0	0.07	2	hvAb	—	—	—
		16	Vermejo	261.37	857.5	1,107	4,409	280	0.60	4,84	155	hvAb	81.62	7,316	822
		17	Vermejo	265.02	869.5	1,661	6,589	360	0.40	4.58	147	hvAb	81.49	7,307	821
		18	Vermejo	266.70	875.0	1,223	3,183	100	0.50	3.20	102	hvAb	—	—	—
		19	Vermejo	264.57	868.0	1,035	505	70	0.60	1.16	37	hvAb	—	—	—
		20	Vermejo	265.94	872.5	1,122	220	10	0.14	0.40	13	hvAb	—	—	—
		21	Vermejo	293.13	961.7	753	260	130	0.61	1.13	36	hvAb	—	—	—
	USBMDH-2	22	Vermejo	293.80	963.9	1,104	270	70	0.69	1.03	33	hvAb	—	—	—
		23	Vermejo	306.48	1,005.5	1,152	745	130	0.44	1.20	38	hvAb	—	—	—
		24	Vermejo	308.70	1,102.8	796	445	170	1.90	2.70	88	hvAb	—	—	—
		25	Vermejo	313.67	1,029.1	809	320	90	1.20	1.71	55	hvAb	—	—	—
		26	Vermejo	313.94	1,030.0	938	335	160	1.11	1.64	52	hvAb	—	—	—
		27	Vermejo	313.94	1,030.0	478	847	200	0.65	2.84	91	hvAb	—	—	—
Walsenburg		28	Vermejo	33.83	111.0	1,049	51	75	0.36	0.93	30	hvCb	—	—	—
		29	Vermejo	47.24	155.0	1,211	82	241	0.41	1.08	35	hvCb	—	—	—
		30	Raton	205.65	674.7	315	185	300	0.10	1.60	53	hvAb	—	—	—
		31	Raton	273.10	896.0	352	157	370	0.0	1.50	48	hvAb	—	—	—
		32	Vermejo	346.47	1,136.7	549	134	550	0.80	2.05	66	hvAb	—	—	—
		33	Vermejo	306.81	1,006.6	369	32	300	0.0	0.90	29	hvAb	—	—	—
		34	Vermejo	308.64	1,012.6	584	54	30	0.0	0.14	5	hvAb	—	—	—
		35	Vermejo	327.20	1,073.5	257	48	90	0.0	0.54	17	hvAb	—	—	—

[1]Anthracite.
[2]High-volatile C bituminous coal.
[3]High-volatile B bituminous coal.
[4]Medium-volatile bituminous coal.
[5]Low-volatile bituminous coal.
[6]High-volatile A bituminous coal.

Table 5—Gas data from MGP #2 and #3.

Well	CGS Number	Date Collected	Lithology	Sample Interval (ft)	Sample Weight (gm)	Gas Content As of 8-14-80 cc/gm	Gas Content As of 8-14-80 cf/ton	Gas Content As of 6-15-81 cc/gm	Gas Content As of 6-15-81 cf/ton
MGP #2	170	6-22-80	Carb. SH, SS, Coal	1,184–1,185	1,930	3.78	121	4.10	131
	171	6-27-80	Carb. SH	1,190–1,190.75	1,644	8.12	260	8.48	271
	172	6-27-80	Carb. SH	1,190	2,161	4.39	141	4.80	154
	173	6-27-80	Coal	1,205–1,209	1,017	9.05	290	10.59	339
	174	6-28-80	Carb. SH	1,218–1,219	2,356	0.67	22	0.78	25
	175	6-28-80	Carb. SH	1,219–1,219.5	1,805	0.72	23	0.82	26
	176	6-28-80	Carb. SH	1,234–1,235	2,748	0.27	8	0.37	12
MGP #3	177	7-23-80	Coal	1,092.6–1,094.6	1,475	8.70	278	9.43	302
	178	7-23-80	Coal	1,093.9–1,094.6	1,647	8.11	260	9.78	313
	179	7-23-80	Coal	1,099.3–1,100.3	1,620	8.62	276	10.30	330
	180	7-23-80	Coal	1,108–1,109	1,667	11.90	381	12.65	405
	181	7-25-80	Coal	1,157–1,158	1,670	11.23	359	15.95	510
	182	7-27-80	Carb. Mudstone	1,179–1,180	996	1.81	58	1.97	61

excluded from the primary target area. Desorption data from coal cores taken in these parts of the region indicate low gas contents, generally less than 50 cf/ton.

In the western part of the target area, 310 sq mi have been identified as having the greatest potential for initial development of coalbed methane gas. Coal found in this area generally is higher in rank and found at greater depths (1,000 to 2,000 ft) than is coal to the north, south, and west. Desorption data for coal cores taken from this area indicate methane contents ranging from 75 to 510 cf/ton and averaging approximately 250 cf/ton.

REFERENCES CITED

Amuedo, C. L., and R. S. Bryson, 1977, Trinidad-Raton Basins (Colorado - New Mexico): A model coal resource evaluation program, *in* D. K. Murray, ed., Geology of Rocky Mountain coal, Proceedings of the 1976 Rocky Mountain Coal Symposium: Colorado Geological Survey Resource Ser. 1, p. 45–60.

Bench, B. M., 1979, Drilling of methane gas in the Fishers Peak Area, Las Animas County, Colorado: U.S. Bureau of Mines Information Circular, 26 p. (unpublished).

Billingsley, L. T., 1977, Stratigraphy and clay mineralogy of the Trinidad Sandstone and associated formations (Upper Cretaceous), Walsenburg Area, Colorado: M.S. Thesis, Colorado School of Mines, 105 p.

Blatz, E. H., 1965, Stratigraphy and history of Raton Basin and notes on San Luis Basin, Colorado–New Mexico: Bulletin of the American Association of Petroleum Geologists, v. 49, n. 11, p. 2041–2075.

Boreck, D. L., and D. K. Murray, 1979, Colorado coal reserves depletion data and coal mine summaries: Colorado Geological Survey Open-File Report 79-1, p. 47–50.

Choate, R., and C. T. Rightmire, 1982, Influence of the San Juan Mountain geothermal anomaly and other Tertiary igneous events on the coalbed methane potential in the Piceance, San Juan, and Raton Basins, Colorado and New Mexico; Proceedings of the Unconventional Gas Recovery Symposium, May 16–18, 1982, Pittsburgh, Pennsylvania: Society of Petroleum Engineers/U.S. Department of Energy, 10805, p. 151–164.

Danilchik, W., J. E. Schultz, and C. M. Tremain, 1979, Methane from coal cores taken from four U.S. Geological Survey coreholes drilled during 1978 in Las Animas County, Colorado: U.S. Geological Survey Open-File Report 70-762.

Dolly, E. D., and F. F. Meissner, 1977, Geology and gas exploration potential, Upper Cretaceous and Lower Tertiary strata, northern Raton basin, Colorado, *in* Exploration frontiers of the central and southern Rockies: Rocky Mountain Association of Geologists Field Conference Guidebook, p. 247–270.

Edwards, C. G., M. Reiter, C. Shearer, and W. Young, 1978, Terrestrial heat flow and crustal radioactivity in northeastern New Mexico and southeastern Colorado: Geological Society of America Bulletin, v. 89, p. 1341–1350.

Fender, H. B., and D. K. Murray, 1978, Data accumulation on the methane potential of the coal beds of Colorado: Colorado Geological Survey Open-File Report 78-2, 25 p.

Goolsby, S. M., and N. S. Reade, 1979, Evaluation of coking coals in Colorado: Colorado Geological Survey Resource Series 7, 72 p.

Gustavson Associates, Inc., 1980, Geologic and drilling report (phase 1): Trinidad, Colorado, American Public Gas Association Test Well.

Hills, R. C., 1888, The recently discovered Tertiary beds of the Huerfano River Basin, Colorado: Proceedings of the Colorado Scientific Society, v. 3, p. 148–164.

——, 1891, Additional note on the Huerfano beds: Proceedings of the Colorado Scientific Society, v. 3, p. 217–223.

Johnson, R. B., 1958, Geology and coal resources of the Walsenburg area, Huerfano County, Colorado: U.S. Geological Survey Bulletin 1042-0, p. 557–592.

——, 1959, Geology of the Huerfano Park area, Huerfano and Custer counties, Colorado: U.S. Geological Survey Bulletin 1071-D, p. 87–119.

——, 1961, Coal resources of the Trinidad coal field in Huerfano and Las Animas counties, Colorado: U.S. Geological Survey Bulletin 1112-E, p. 129–180.

——, and J. G. Stephens, 1954, Coal resources of the La Veta

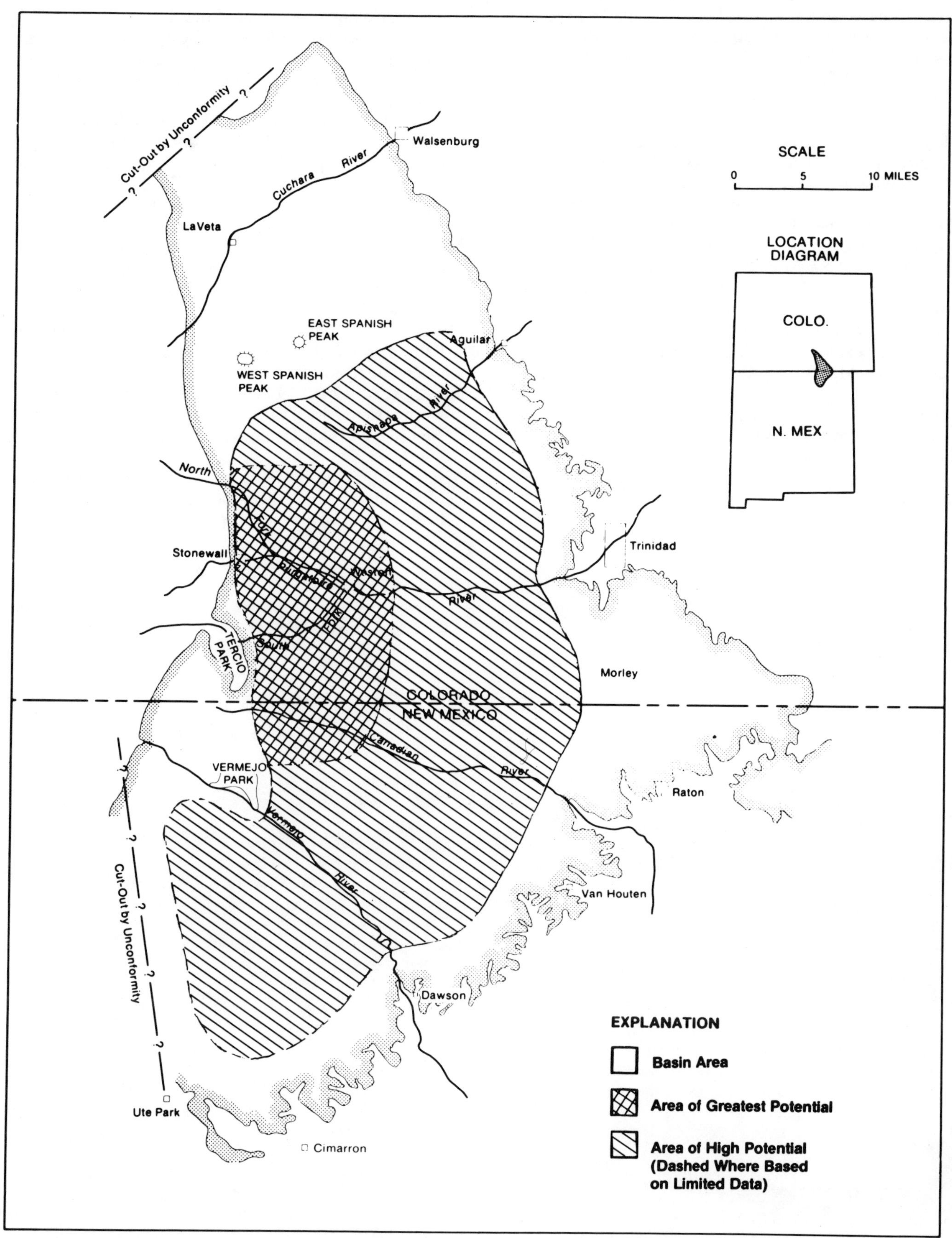

Figure 12—Map showing the location of the Raton Mesa Region and area of greatest coalbed methane potential.

area, Huerfano County, Colorado: U.S. Geological Survey Coal Investigations Map C-20.

——, and G. H. Wood, Jr., 1956, Stratigraphy of Upper Cretaceous and Tertiary rocks of Raton basin, Colorado and New Mexico: Bulletin of the American Association of Petroleum Geologists, v. 40, p. 707–721.

Kauffman, E. G., J. D. Powell, and D. E. Hatten, 1969, Cenomanian-Turonian facies across the Raton basin: The Mountain Geologist, v. 6, n. 3, p. 93–118.

King, P. B., 1951, The tectonics of middle North America: Princeton University Press, 203 p.

Lee, W. T., 1917, Geology of the Raton Mesa and other regions in Colorado and New Mexico, *in* W. T. Lee and F. H. Knowlton, Geology and paleontology of Raton Mesa and other regions in Colorado and New Mexico: U.S. Geological Survey Professional Paper 101, p. 9–211.

——, 1924, Coal resources of the Raton coal field, Colfax County, New Mexico U.S. Geological Survey Bulletin 752, 254 p.

Maher, J. C., and J. B. Collins, 1949, Pre-Pennsylvanian geology of southwestern Kansas, southeastern Colorado and the Oklahoma panhandle: U.S. Geological Survey Oil and Gas Investigations Preliminary Map 101, 4 sheets.

Manzolillo, C. D., 1976, Stratigraphy and depositional environments of the Upper Cretaceous Trinidad Sandstone, Trinidad Aquilar Area, Las Animas County, Colorado: M.S. Thesis, Colorado School of Mines, 147 p.

Pillmore, C. O., 1969a, Geologic map of the Casa Grande Quadrangle, Colfax County, New Mexico and Las Animas County, Colorado: U.S. Geological Survey Geological Quadrangle Map GQ-823.

——, 1969b, Geology and coal deposits of the Raton coal field, Colfax County, New Mexico: The Mountain Geologist, v. 6, p. 125–142.

Read, C. B., R. T. Duffney, G. H. Wood, Jr., and A. D. Zapp, 1950, Coal resources of New Mexico: U.S. Geological Survey Circular 89, 24 p.

Roehler, H. W., and W. Danilchik, 1980, Relation of intraformational faults of fluvial sedimentation and coal-bed discontinuity, Raton Formation, Colorado and New Mexico: Colorado Geological Survey Resource Series 10, 13 p.

Tremain, C. M., 1980, The coal bed methane potential of the Raton Basin, Colorado: Colorado Geological Survey Open-File Report 80-4.

Wanek, A. A., 1963, Geology and fuel resources of the southwestern part of the Raton coal field, Colfax County, New Mexico: U.S. Geological Survey Coal Investigations Map C-45.

Weimer, R. J., 1960, Upper Cretaceous stratigraphy, Rocky Mountain area: Bulletin of the American Association of Petroleum Geologists, v. 40, p. 1–20.

Upper Cretaceous Geology, Coal, and the Potential for Methane Recovery from Coalbeds in San Juan Basin—Colorado and New Mexico

R. Choate
J. Lent
C. T. Rightmire

The San Juan Basin is a northwest-southeast–trending asymmetrical basin located in northwestern New Mexico and southwestern Colorado. The portion of the basin discussed in this report is underlain predominantly by the Upper Cretaceous Fruitland Formation and covers approximately 7,500 sq mi in area.

While coal is present in several formations (particularly the Dakota Sandstone, Gallup Sandstone, Crevasse Canyon Formation, Menefee Formation, and Fruitland Formation), the bulk of more than 200 billion tons of coal resource is present within the Fruitland Formation.

Coals in the Fruitland Formation are generally thicker, of higher rank, and found at greater depths of burial in the northern part of the basin. Gas content reports for Fruitland coals in parts of the basin exceed 500 cubic feet/ton (cf/ton). The total coalbed methane resource in the Fruitland Formation coals is estimated at 31 trillion cubic feet (Tcf), with the bulk of that in a crescent-shaped, high-potential area along the north and east-central parts of the basin.

INTRODUCTION

This report is an introduction to the San Juan Basin of northwestern New Mexico and southwestern Colorado. It provides a regional framework for interpreting and extrapolating methane recovery test data from sites throughout the basin.

It briefly describes the geologic, hydrologic, physical, cultural, and economic context of the basin. The next section characterizes the several coal-bearing formations in the basin. A preliminary estimate of the coalbed methane resource appears at the end of the discussion on previous and ongoing coal methane studies and presents available data.

Largely because of energy resource interest, the San Juan Basin has been heavily explored and studied relative to other areas in New Mexico; this work has produced an abundance of published data. While most of these reports focus in detail on a specific portion of the basin or its resources, several overview studies are available and furnished many of the illustrations used in this report.

BASIN SETTING

Geography/Physiography

The San Juan Basin (as discussed in this report) occupies the portion of southwestern Colorado and northwestern New Mexico (Fig. 1) defined by surface outcrops of the Fruitland Formation. Basin boundaries would be irregularly extended if the basin is defined in terms of the deeper lying Dakota Formation, which contains the deepest coal. In Colorado, the San Juan Basin area includes parts of La Plata and Archuleta counties; and in New Mexico, parts of Rio Arriba, San Juan, Sandoval, and McKinley counties. Parts of the Southern Ute, Jicarilla Apache, and Navajo Indian reservations also are located within the San Juan Basin. The basin is the largest coal-bearing area in New Mexico.

Elliptical in shape, the San Juan Basin is roughly 100 mi long (in the north-south direction) and about 90 mi wide. The basin covers an area of about 7,500 sq mi.

The Continental Divide trends north-south along the east side of the San Juan Basin. The San Juan River drainage lies on the west side of the divide; the Rio Grande drainage on the east. In the basin, the major stream is the San Juan River, which flows in a southwesterly direction through the Colorado portion of the basin into New Mexico, to the town of Blanco; then west across the basin, joining the Colorado River in Utah. The Animas River flows south through Durango, Colorado, to join the San Juan River at Farmington, New Mexico. Near the west rim of the basin, the La Plata River flows south to join the San Juan River about 2 mi downstream from the mouth of the Animas River.

The San Juan Basin is part of the Navajo physiographic section of the Colorado Plateau province. Land surface elevations within the basin range from a low of about 5,100 ft on the west side, where the San Juan River exits the basin, to more than 8,000 ft in the northern portion, for a total topographic relief of about 3,000 ft. Topographic relief is related to the moderate dissection by stream erosion of the relatively

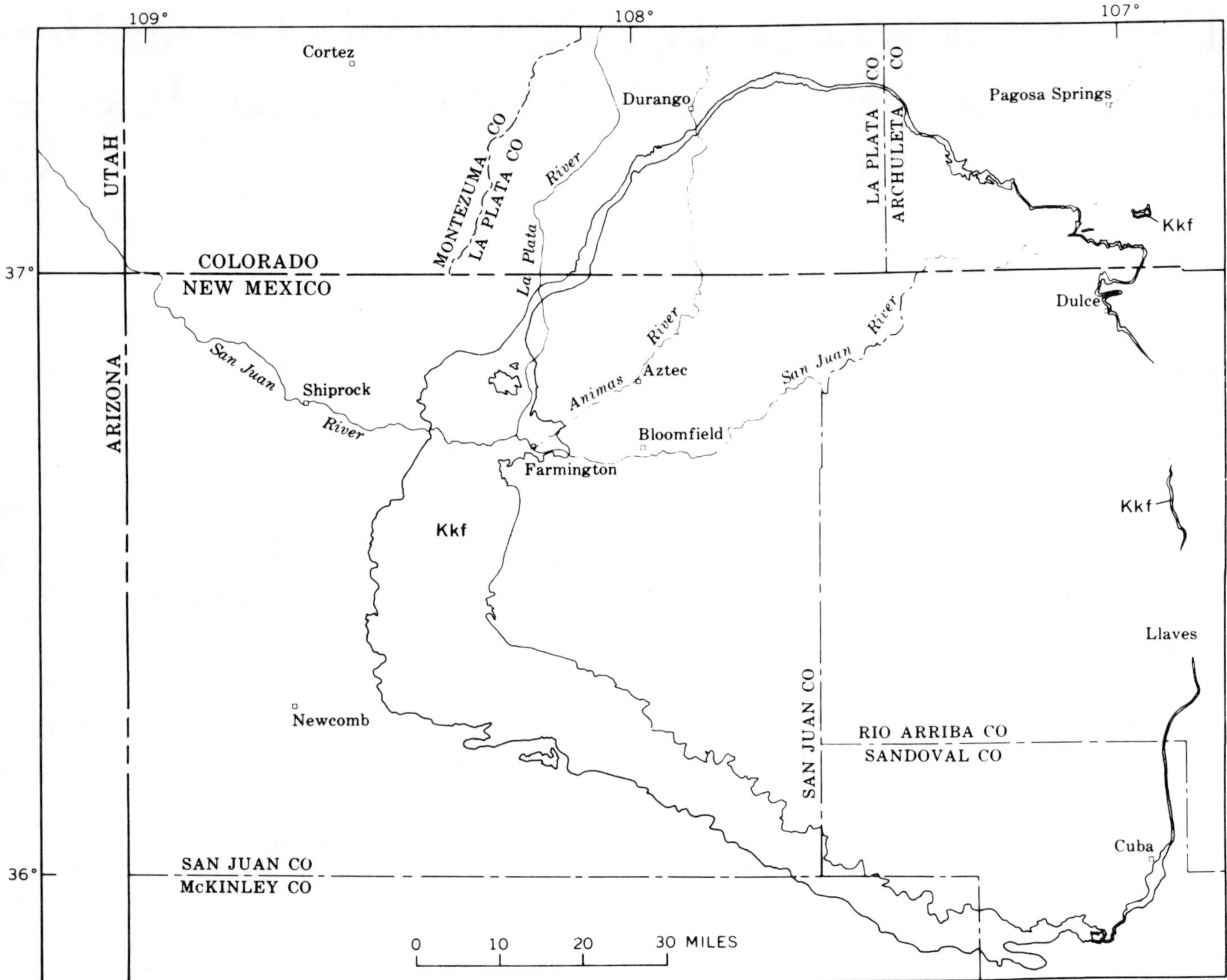

Figure 1—Geographic location of the San Juan Basin. Fassett and Hinds, 1971.

flat-lying Tertiary and Upper Cretaceous sandstone and shales that cover most of the basin. The land surface is characterized by broad plateaus and mesas cut by steep-sided canyons.

Geology

Regional Geologic Setting

The Four Corners area of Utah, Colorado, Arizona, and New Mexico contains three well-defined structural basins: the San Juan, Black Mesa, and Paradox. Located in the southeast part of the area, the asymmetric structural depression that is the San Juan Basin contains sediments ranging in age from Cambrian to Recent. One well drilled into Precambrian rocks penetrated 14,030 ft of sediments.

The general geologic framework is shown in Figure 2. The San Juan Basin is bounded by the San Juan and La Plata Mountains on the north, the Archuleta and Nacimiento Uplifts on the east, the Zuni Uplift on the southwest, and the Four Corners Platform on the northwest. The Chaco slope is the wide structural shelf at the south that slopes gently northeastward into the basin. The deepest part of the basin lies in the northeast, near and parallel to the northern and northeastern boundary, as shown in Figure 3. The background geology has been well summarized in numerous articles, particularly Peterson et al (1965), and will be addressed here only as it influences the coal or coalbed methane resource.

Structure

Tectonic Evolution of the Basin—In general, the San Juan Basin is the result of geologically slow growth along mobile belts. Although basin formation began and uplift and minor deformation recurred at various places and times during the Mesozoic era, main structural features of the basin began to take shape during Late Cretaceous and Early Tertiary time, contemporaneous with the Laramide orogeny. The Hogback Monocline along the northwest flank was developed and large

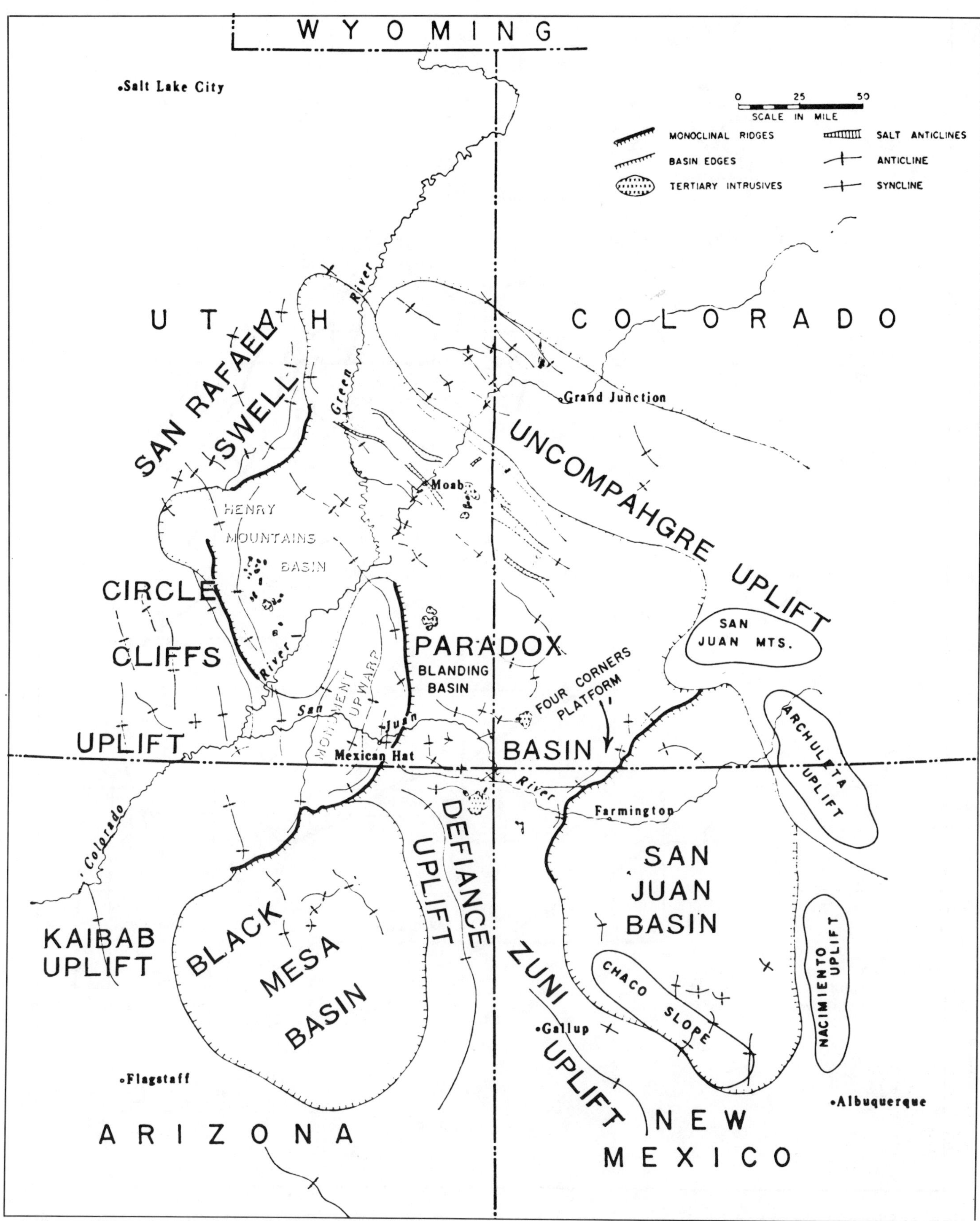

Figure 2—Four Corners area major uplifts and basins. Peterson et al, 1965.

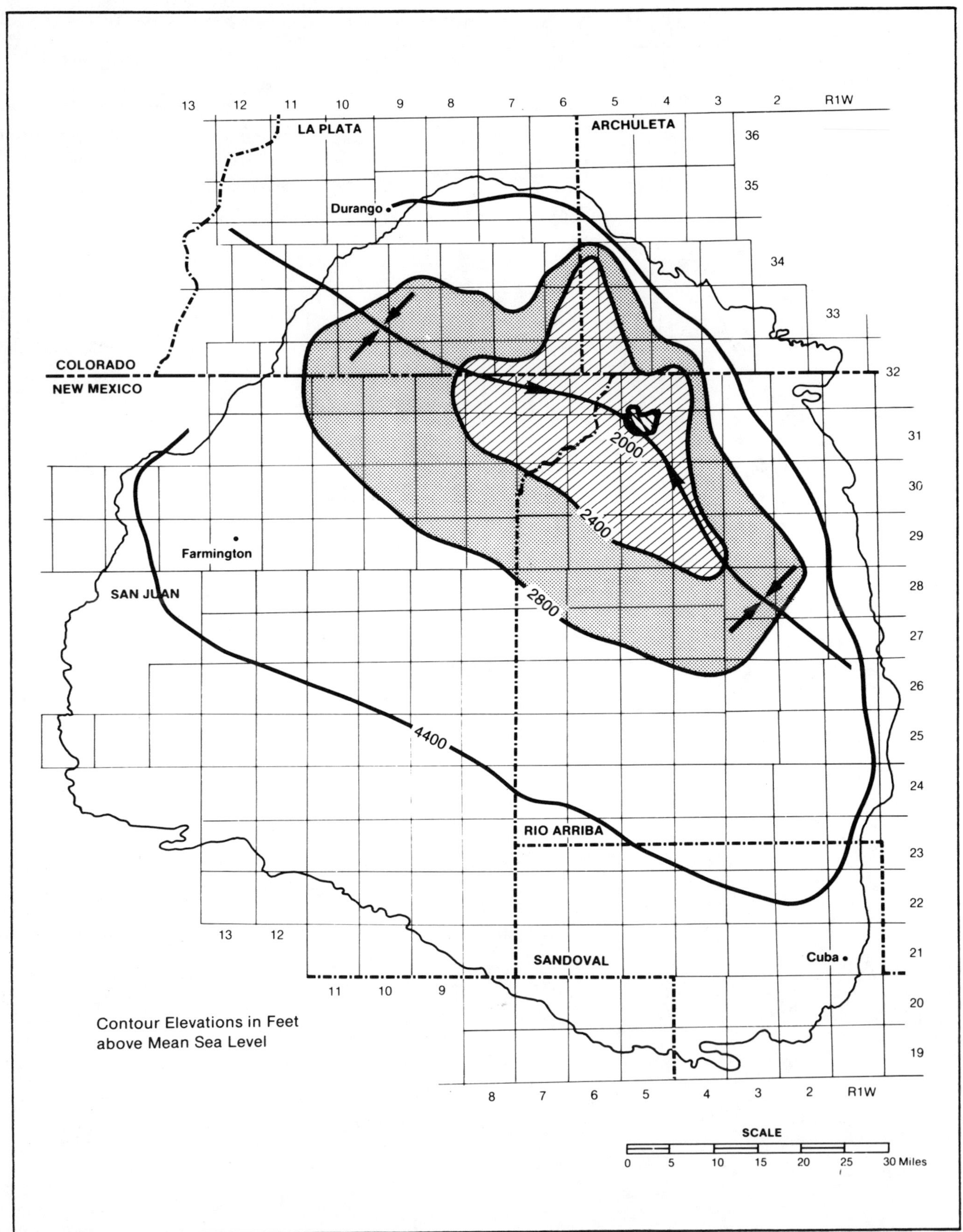

Figure 3—Simplified contour map of the Huerfonito Bentonite Bed, in the Lewis Shale, showing the shape and location of the deeper central portion of San Juan Basin. Modified from Fassett and Hinds, 1971.

bordering structural uplifts accentuated at this time. It is thought that deformation gradually spread from the outer rim to the inner basin (Kelley, 1951) as the deforming mobile basin rims widened and encroached on the depressed, undeformed remainder. Uplift and erosion were pronounced during the Early Tertiary; episodes of basin downwarping and localized uplift characterized the Middle Tertiary. By the late Cenozoic, the entire basin (along with the Colorado Plateau) was widely elevated, accompanied by extensional faulting and a further rise of at least one major uplift. These events were related to the overriding of a subduction zone by the North American crustal plate and subsequent development of the Rio Grande rift.

General characteristics of local structures in the basin include the following:

- Fold intensity may increase with depth.
- Dominant fold axis trend is northwest and north.
- Folds northwest of the Hogback Monocline are apt to be tangential and result in doubly plunging anticlines where they intersect radial axes.
- Folds basinward from the Hogback Monocline are typically radial, resulting in anticlinal noses.
- Folds may diminish in intensity with increasing distance basinward from the Hogback Monocline.

Stratigraphy

The general stratigraphy and stratigraphic relationships of the San Juan Basin are shown in Figure 4. A summary description of the stratigraphy by geological period is given here; specific formations are not addressed.

Precambrian—The Precambrian basement complex has been penetrated in several wells drilled in the basin, but little is known about Precambrian paleogeography except that during this era, the region was part of a geosyncline into which poured debris from rising adjacent rock masses. Regional metamorphism associated with a widespread Precambrian orogeny then produced a sequence of quartzite, slate, schist, and phyllite several thousand feet thick. The basement complex was locally intruded by Precambrian granite.

Schists, gneisses, greenstones, and quartzites compose the Precambrian rocks in the largest areal exposure, which is in the San Juan Mountains on the northern border of the basin.

Cambrian—Cambrian rocks have been described only in the northern part of the basin. Here, Upper Cambrian sandstone and quartzite reach a thickness of about 150 ft, thin rapidly to the south, and are not present in the south half of the basin. On the Chaco slope and Four Corners Platform, depth to Cambrian or Precambrian rock ranges from 4,500 to 8,500 ft.

Ordovician-Silurian—Neither Ordovician- nor Silurian-age sediments have been found in the basin. Generally, either Upper Devonian or Mississippian sediments rest disconformably on the Cambrian or Precambrian surface.

Devonian—About 300 ft of Late Devonian sediments have been described in the Four Corners area. Thinning southward, these sediments pinch out in the north-central part of the basin. Of marine origin, the Upper Devonian sediments consist of sandy dolomite, some thin sandstones, and green and red beds of shale.

Mississippian—In the northern part of the basin, the Mississippian rocks are of Kinderhook and Osage age. In the southeastern part of the basin, thin erosional remnants of Meramec-age sediments rest on Precambrian rocks.

A 150 to 200 ft thick Early Mississippian carbonate section in the northwestern part of the basin contains a 50- to 75-ft interval of permeable dolomite, reported to yield large quantities of salt water. Accumulation of carbon dioxide gas (some with a low percentage of helium) has been reported in several areas. The Early Mississippian carbonate section is thin in the northern part of the basin.

Pennsylvanian—Rocks of Pennsylvanian age are exposed in the mountain ranges along the eastern and northern basin borders and have been discovered in a number of deep boreholes in the basin interior. Composed of a complex sequence of continental and marine strata representing a major sedimentary cycle of marine transgression and regression, the Pennsylvanian strata exhibit lateral facies and thickness changes. Their geometry and composition have been explained in terms of deposition in a region of positive land masses paralleled by localized basins and geosynclinal depressions. Rock units in areas dominated by limestone tend to be thinner than those in areas where clastic beds predominate.

Permian—Sedimentary Permian rocks are largely red clastics—the colorful "red beds" that crop out only on major uplifts at basin margins.

The Permian system changes from a thick arkosic sandstone wedge in the northeast to thinner, finer grained clastic strata overlain by a heavily cross-stratified aeolian sandstone and marine formations toward the south and west. The basal red-bed–bearing formations locally attain a maximum thickness of more than 2,000 ft and are generally thicker than 1,000 ft.

Triassic—The Triassic rock formations in and adjacent to the San Juan Basin are significant as a source of mineral deposits: oil, helium gas, uranium, vanadium, and copper. Triassic rocks crop out in the uplifts forming the basin margins and are present throughout the basin, thickening from northeast to southwest to 1,000 ft and more over broad areas. Typically, they are bounded at the top and bottom by unconformities.

Lower and Middle Triassic rocks comprise red-bed siltstone and sandstone, with some coarse conglomerate. Upper Triassic rocks are exclusively nonmarine, predominantly shales and siltstones, in formations each several hundreds of feet thick.

Jurassic—On the south and west sides of the basin, Jurassic formations form a complex system of sediments, nearly all of which are nonmarine and widespread in the Colorado Plateau region outside the basin. Rocks are largely sandstones formed in similar continental depositional environments.

Cretaceous—Rocks of the Cretaceous period are the most important from the standpoint of coal content. Cretaceous-age strata consist largely of intertonguing sandstone, coal, and shale. Rocks associated with the coal are of marine and nonmarine origin. Stratigraphic relationships of the various units have been complicated by the nature of their deposition: during advances and retreats of a shallow epicontinental sea, and in some cases, by local deformation. Thickness of a single unit can vary significantly.

From lower to upper, the Cretaceous-age stratigraphic sequence generally consists of the: Dakota Sandstone—a lenticular formation in which sandstone usually predominates, containing coal and ranging from 54 ft to a little more than 200 ft thick; Mancos Shale—a 400 to 2,000 ft thick carbonaceous marine shale; an intertonguing, coal-bearing massive sandstone

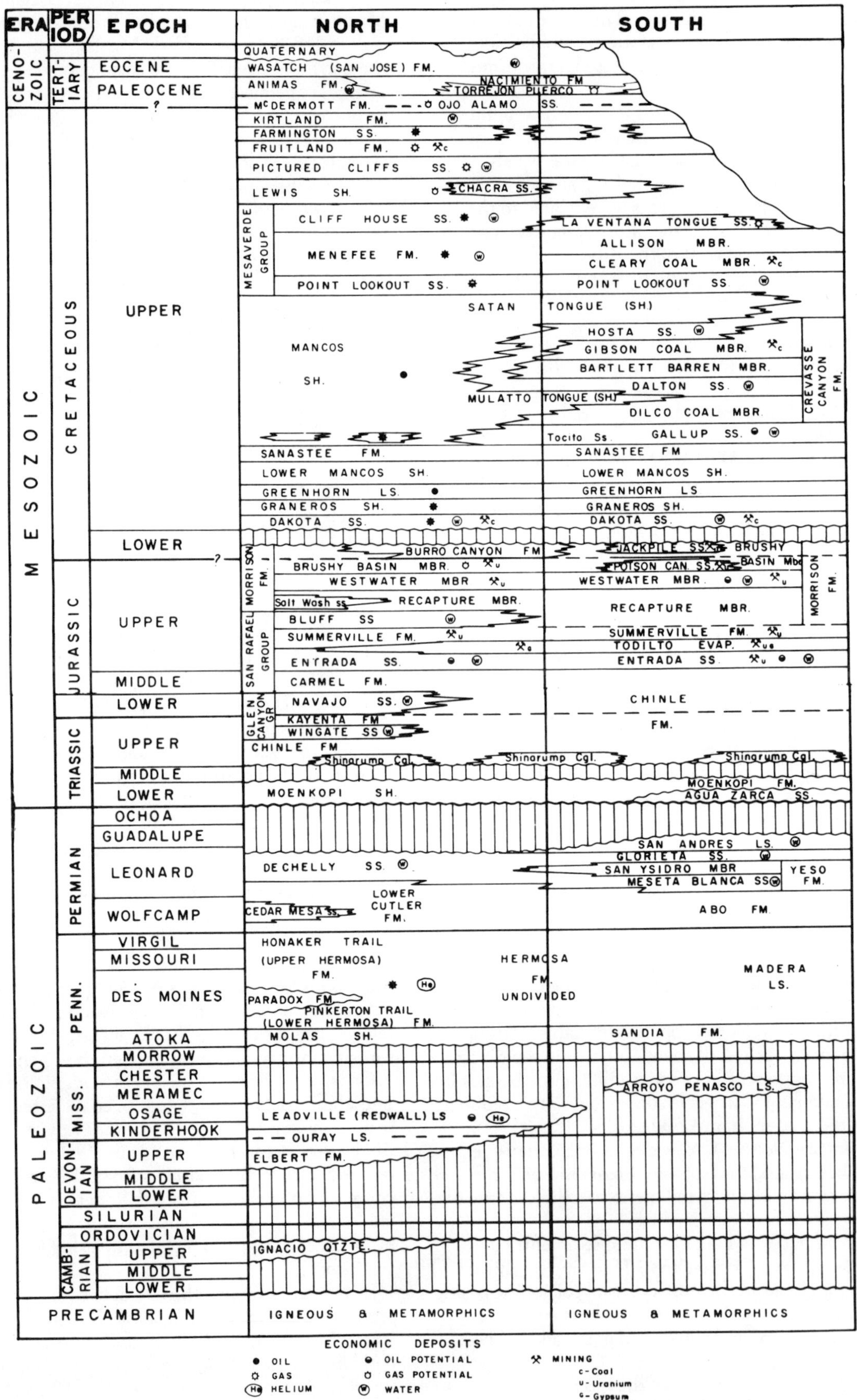

Figure 4—Stratigraphic correlation chart of San Juan Basin. Peterson et al, 1965.

from 180 to 250 ft thick; Crevasse Canyon Formation—a coal-rich formation consisting largely of lenticular sandstone; Point Lookout Sandstone—a localized sandstone formation split into two members, one about 200 ft thick and the other, 100 to 300 ft thick; overlying Menefee Formation—split into several thick members and tongues comprised primarily of shale and sandstone (some containing coal); transition-zone lenticular sandstone deposits; Pictured Cliffs Sandstone—a fine-grained marine sandstone 50 to 400 ft thick; and Fruitland Formation—a 200 to 300 ft thick coal-bearing sequence of interbedded sandstone, shale, carbonaceous sedimentary rocks, and limestone.

Geologic events during the Cretaceous period relate almost exclusively to movements of the sea. Some formations represent landward and seaward equivalents of a particular environment, beach sands and flood-plain deposits, for example; while others such as the Dakota Sandstone (which contains the oldest rocks in the sequence) comprise a complex assemblage of marine and nonmarine deposits.

While coal is present in zones throughout the Cretaceous strata, the thickest and best coal development is in the Mesaverde Group and Fruitland Formation. The coal resource is described in detail later in this paper.

Cenozoic—Cenozoic-era rocks of the San Juan Basin are typically basin-fill deposits formed during a late Laramide mountain-building episode. One persistent formation consists of at least 2,500 ft of alternating sandstone and variegated shale. Erosion has removed much of the Tertiary geologic record in the southern part of the basin, where an erosional unconformity separates Quaternary deposits from the Tertiary section. Tertiary and Quaternary rocks reach a maximum total thickness of more than 3,900 ft.

Surface Geology

Basin asymmetry is expressed by outcrops at the surface. Broad exposures of Cretaceous and Tertiary rocks characterize the west and southwest side of the "central" basin (defined in Fig. 1), where basin contours and dips are relatively gentle. These same rocks have very narrow outcrops on the steeply dipping northern and eastern basin margins.

Rocks exposed within the central basin are exclusively sedimentary. Jurassic and Triassic rocks crop out along the extreme northern and eastern margins, as well as near the McKinley County line, outside the central portion. The basin interior contains shale, sandstone, and greywacke of the Eocene San Jose Formation.

Crystalline rocks consist of Quaternary basalts on the remote fringes of the basin; Tertiary volcanic necks; plugs; dikes; and basalt flows (restricted to an area about 600 mi square in the extreme southeast; the Mount Taylor Volcanic field); minor Tertiary intrusives; and Precambrian granite, gneiss, schist, and quartzite. These last are exposed on the major uplifts on the east, extreme south, and north.

Hydrology

Water Supply and Quantity—In addition to the river systems previously mentioned, water is obtained in the central part of the basin from Tertiary sandstones and Quaternary gravels and sands, from which wells yield 100 to 300 gallons per minute (gpm) of relatively fresh water. Sandstone aquifers in northwestern New Mexico differ with respect to grain size and clay and silt-to-sand ratio; and this affects yield substantially. One aquifer—the Ojo Alamo Sandstone—yields as much as 30 gpm of potable water (Hale et al, 1965). Cretaceous sandstones are an important source of water on the basin's periphery. In the southern part of the basin, along the northeast border of the Zuni Uplift, water is produced from older sandstones and a Permian-age limestone.

Energy and Minerals

Exploration for and development of energy sources and mineral resources constitutes a large part of the livelihood and activity in the basin. Uranium mining and gas exploration and production are major energy-related activities. In addition to coal, mining activities in the basin include gold and silver, pumice, fluorite, and perlite.

Oil and Gas

In the San Juan Basin, petroleum has been commercially developed from sedimentary formations with ages starting as early as the mid-Paleozoic era (Devonian), continuing through the Mississippian and Pennsylvanian, and climaxing in the Mesozoic Late Cretaceous. Table 1 presents a summary of hydrocarbon production by geologic formation or period for the San Juan Basin.

The Blanco gas field, underlying about 1,340,000 acres of the central San Juan Basin, is the second largest producing gas field in the conterminous United States. Ultimate recovery is estimated at about 23 Tcf of gas and additional significant amounts of condensate. Total gas production in 1974 was about 556 billion cubic feet (Bcf). Major Cretaceous gas-producing zones are, in ascending order, the Dakota, Point Lookout, Cliff House, and Pictured Cliffs. Pennsylvanian exploration has begun only recently in the basin, but the Mesozoic section is known to contain essentially proven but undeveloped resources that fall within known field areas. In the structurally low part of the basin, accumulation is determined largely by hydrodynamics.

Oil is produced from the Dakota in several small fields on the basin flanks, mainly on the west side of the basin at Table Mesa and Rattlesnake, and in the Hogback Field of the Chaco Slope, which produces 5 to 7 million barrels. Total estimated ultimate recovery is about 150 million barrels, from about two dozen fields or pools.

COAL RESOURCE

Coal is nearly ubiquitous through the Upper Cretaceous strata of the San Juan Basin. In part because of heavy petroleum and mineral exploration and development activity (more than 14,000 logs have been run in wells drilled in the basin), the coal-bearing rocks have been the subject of numerous investigations and reports, and their lithology and occurrence are well known, compared with other coal-rich areas in the West.

Since coal-bearing rocks are essentially continuous throughout the basin, the entire area has been considered as a

Table 1—Summary of hydrocarbon production by formation/period.

Formation/Period	Liquids		Gas		Ratios	
	Oil (barrels)	Condensate (barrels)	Casinghead (Mcf)	Dry (Mcf)	Oil/Casinghead Gas (gallons/Mcf)	Condensate/Dry Gas (gallons/Mcf)
Farmington	69,536	0	656	993,382		
Fruitland		21,092		37,816,812		0.023
Fruitland/Pictured Cliffs		0		1,653,920		
Pictured Cliffs	89,622	367,496	248,138	2,217,088,455	15.17	0.007
Chacra		13,374		88,656,567		0.006
Mesaverde	648,286	19,127,223	319,512	4,681,247,230	85.05	0.171
Gallup	120,211,950	241,572	260,499,289	58,954,082	19.38	0.172
Tocito	3,888,826		7,832,963		20.86	
Sanastee	29,882		0			
Mancos	11,185,930		6,463,067		72.82	
Greenhorn	1,126	0	165	276,868		
Graneros	5,797		0			
Gallup-Dakota	1,384,166		9,474,151			
Mancos-Dakota	39,892		3,065			
Dakota	15,667,886	28,204,554	12,572,388	2,926,858,007	52.33	0.405
Entrada	2,090,164		0			
Pennsylvanian	12,762,086	0	23,321,876	152,824,302	23.07	
Mississippian	9,229	828	0	2,436,017		
Devonian	18,438		0			
	168,102,816	47,976,139	320,735,270	10,168,822,932		

single field. However, for purposes of discussion, the literature divides the San Juan Basin into several coal fields or districts (Fig. 5) whose boundaries are seldom clearly defined (Shomaker et al, 1971).

Table 2 summarizes the distribution of strippable coal within the basin to provide information on rank, thickness, and bed continuity on the basin perimeter to aid in predicting the coal distribution with depth and geography across the entire basin.

A total of 5,909.4 million short tons of coal was estimated as "original" reserves of strippable coal—without deductions for mining—by Shomaker et al (1971).

These coal fields and areas are briefly described in the following sections.

Regional Character

The single most important factor determining the regional character of San Juan Basin coal is its depositional environment: the mainly freshwater swamps of flood plains along the migrating margins of a shallow continental sea that repeatedly encroached on and retreated (transgressed and regressed) from a broad southwestern land mass. Four or five major transgressions, and many minor ones, have been documented. A relatively stable climate, constant shoreline trend, and uniform source of sediment supply produced coalbeds of comparable quality with a depositional strike of N 50° W to N 60° W, contained in strata whose composition is predictable according to their formation during a particular stage of the transgressive-regressive cycle.

Since the depositional setting favoring the accumulation and preservation of the peat layers that would eventually become coal was restricted to a narrow zone paralleling the strandline, the lenses and pods of coal typical in the basin are similarly defined and restricted. Two related characteristics are apparent. First, the smaller dimension of a pod or lens is likely to be oriented normal to the shoreline or parallel to the direction of movement of the sea margin. Second, the age of such sediments tends to decrease in the direction of shoreline movement. The rate of shoreline shift also had an effect on the swamp, or paludal, deposits: For zones representing rapid shifts, the coal is thinner and more elongated and discontinuous than for a gradual change.

Coal Quality and Type—Coal in the basin is of bituminous and subbituminous rank; the higher rank correlates with high thermal gradient owing to depth of burial or anomalous heat sources.

San Juan Basin coals have a high ash content, possibly attributable to a large surface drainage volume of fine detrital matter over the region of deposition. Ash content is generally higher in the east than in the west.

Heating values for economically significant San Juan Basin coals tend to increase from southwest to northwest (depending on the formation), from about 9,000 to about 12,000 or 15,000 Btu per pound. In some cases, zones of similar heating value closely relate to conditions of deposition.

Data from cores and strip mine samples show sulfur content of San Juan Basin coals to be uniformly less than 1%, except on

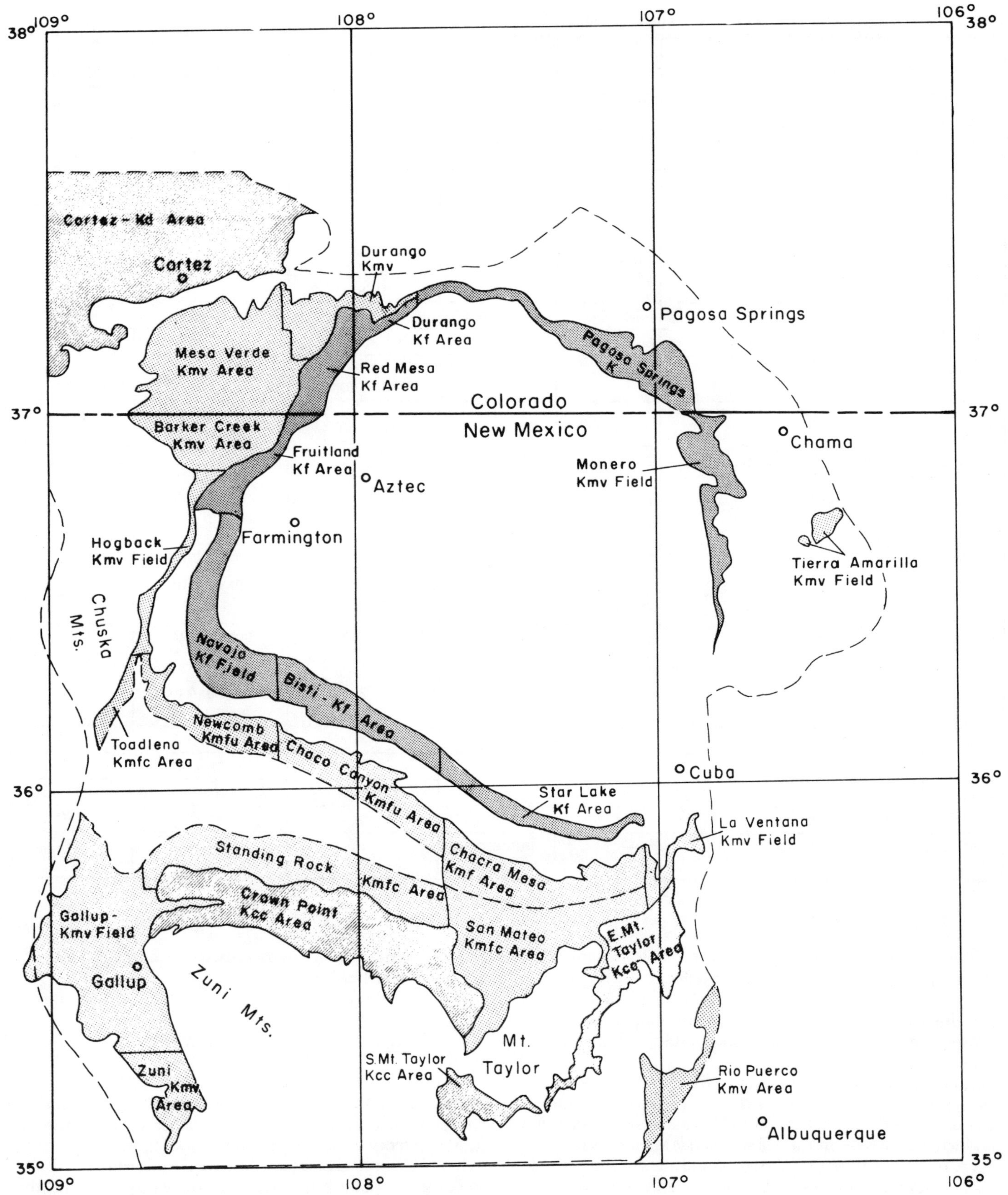

Figure 5—Coal fields and areas in the San Juan Basin. Shomaker et al, 1971. Used with the permission of the New Mexico Bureau of Mines and Mineral Resources.

Table 2—Summary of coal fields and areas containing strippable coal.

Coal Field	Formation	Member	Number of Beds	Individual Coalbed Thickness (ft)	Rank	Original Strippable Reserves (million short tons)
Cortez	Dakota	—	—	3–9	hi-vol A,B,&C	157.8
Gallup	Gallup Sandstone	Myers	3	3–9	subbit.	358.0
	Crevasse Canyon	Dilco	9	4–7	subbit.	
		Gibson	—	4–10	subbit.	
	Menefee	Cleary	—	4–10	subbit.	
Newcomb	Menefee	—	numerous	irregular	subbit. A & B	84.8
Chaco Canyon	Menefee	Allison	numerous	<3	subbit. A & B	31.0
San Mateo	Menefee	Cleary	?	<6	—	21.2
Standing Rock	Menefee	Cleary	?	thin, discontinuous	—	138.5
Zuni	Gallup Ss	—	—	—	—	6.2
	Crevasse Canyon	Dilco	numerous	5	—	
Crownpoint	Gallup Ss	—	numerous	irregular	subbit.	15.0
	Crevasse Canyon	Dilco	numerous	thin	subbit.	
		Gibson	numerous	thin	subbit.	
South Mt. Taylor	Crevasse Canyon	Dilco	—	—	hi-vol bit.	1.4
		Gibson	5	7	hi-vol bit.	
	Menefee	Cleary	numerous	<2	—	
LaVentana	Menefee	Cleary	?	?	subbit. A to hi-vol C bit.	15.0
		Allison	?	?	subbit. A to hi-vol C bit.	
Red Mesa	Fruitland	—	—	20	hi-vol C bit.	40.0
Fruitland	Fruitland	—	2	—	hi-vol C bit.	158.0
Navajo	Fruitland	—	—	—	subbit.	2,377.5
Bisti	Fruitland	—	numerous	—	subbit.	1,870.0
Star Lake	Fruitland	—	numerous	16	subbit. A to hi-vol C bit.	635.0

the surface and at very shallow depths in drillholes, where significant amounts of sulfur occur as gypsum.

Coal Occurrence—Coal occurs in the following Cretaceous units, listed in ascending order: Gallup Sandstone, Crevasse Canyon Formation, Menefee Formation, and the Fruitland Formation.

Figure 6, a time-stratigraphic section through the Upper Cretaceous rocks of the basin, illustrates the paleogeographic relationship of the different units and assigns ages to them, some of which remain to be further substantiated. Marine offshore and coastal deposits, nonmarine rocks, and areas of erosion and nondeposition are designated. This figure shows that numerous transgressive and regressive cycles occurred during the Upper Cretaceous, accounting for the presence of several coal-bearing sequences in the section. Table 3 summarizes the coals identified in the basin. With the exception of the Fruitland Formation, the coal resource—and, hence, the coalbed methane resource—is not of major impact; therefore, only the Fruitland will be addressed in detail.

Fruitland Formation

The Fruitland Formation contains the largest coal resource and has been described in greatest detail because of its strippable depths, extensive outcrop, and importance to the oil and gas industry. Composed of interbedded sandstone, siltstone, shale, carbonaceous shale, carbonaceous siltstone and sandstone, thin limestone, and coal, the Fruitland changes in lithology vertically in a fairly consistent way: Thick coal beds are always in the lower third or fifth of the formation; limestone occurs at the base; sandstone content is greater in the lower part; and siltstone and shale predominate in the upper part.

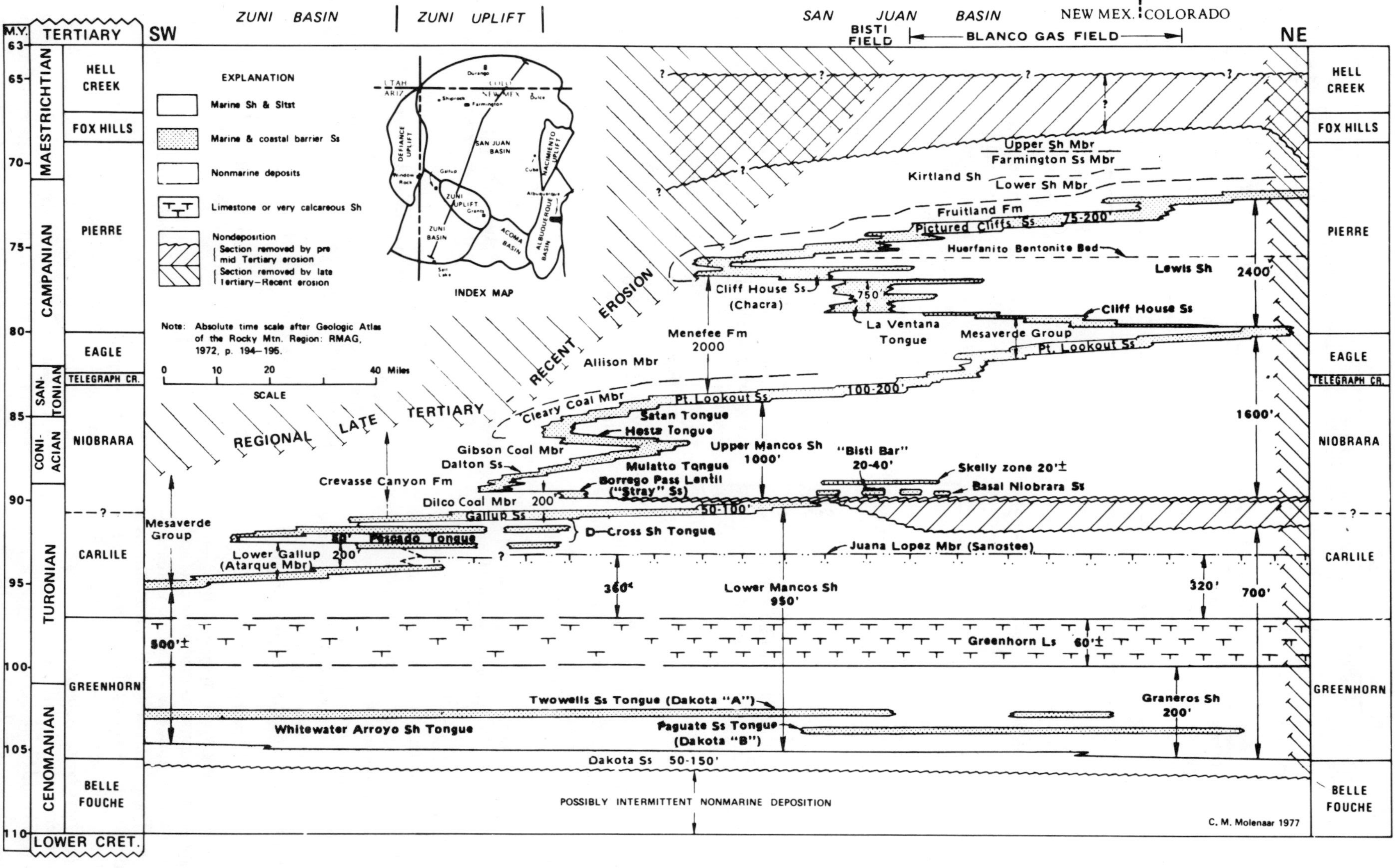

Figure 6—Upper Cretaceous time stratigraphic sequence. Molenaar, 1977. Used with the permission of the New Mexico Geological Society.

Table 3—Summary of coal distribution in the San Juan Basin.

Formation	Coalbed Name	Cumulative Coal Thickness	Distribution	Rank
Fruitland	—	up to 70+ ft		subbit. to med-vol bit.
Menefee	Allison	up to 30 ft	lenticular	subbit. A to hi-vol A bit.
	Cleary	few inches to >11 ft	lenticular	
Crevasse Canyon	Gibson	<3 ft	lenticular	hi-vol bit.
	Dilco	14 inches to 3 ft	lenticular	hi-vol bit.
Gallup Sandstone	—	thin to 4 ft	discontinuous	
Dakota Sandstone	—	3 inches to 4 ft	lenticular	

Table 4—Coal resources of the Fruitland Formation. Shomaker and Whyte, 1977. Used with the permission of the New Mexico Bureau of Mines and Mineral Resources.

Overburden (ft)	Total coal in beds of indicated thickness (millions of short tons)			Total
	2–5 ft	5–10 ft	10 ft	
0–500	4,021.1	4,888.3	5,728.9	14,638.3
500–1,000	3,583.2	4,780.0	5,505.0	13,868.2
1,000–2,000	8,468.3	9,809.1	9,660.0	27,937.4
2,000–3,000	11,736.5	14,759.7	32,312.0	58,808.2
3,000–4,000	14,032.1	17,291.8	51,500.2	82,824.1
4,000+	501.4	594.5	1,964,9	3,060.8
Total	42,342.6	52,123.4	106,671.0	201,137.0

Most of the rock units identified above as comprising the Fruitland Formation are discontinuous, pinching out laterally within a fairly short distance. Coalbeds are the exception, however, and, in places, Fruitland coals have been traced for several miles.

Fruitland coals represent the swamp deposits laid down shoreward of the Upper Cretaceous sea, as it retreated for the last time from the basin, depositing the Pictured Cliffs sediments. The Fruitland overlies the Pictured Cliffs on the north, west, and south sides of the basin, but on the southeast, overlies the Lewis Shale. The formations intertongue locally on the basin's eastern edge, with accompanying changes in thickness. Formational contacts are disconformable in places (Fassett and Hinds, 1971). Conventional usage designates the top of the uppermost coal or carbonaceous shale bed as the contact between the Fruitland and the overlying Kirtland Shale.

Quality—Coal of the Fruitland Formation is black, hard, and brittle, and decomposes readily when exposed to the atmosphere. It commonly contains charcoal fragments and lumps of resin as well as iron sulfide, as grains and veinlets of pyrite and marcasite. The coal is considered subbituminous in New Mexico (Shomaker and Whyte, 1977) but does range to medium-volatile bituminous in Colorado (Kelso et al, 1980). No definite zones of rank have been established.

Abundant data on Fruitland Formation coals are reported by Fassett and Hinds (1971) and Shomaker and Whyte (1977). Absolute Btu values vary between 9,000 in the southwest and 15,720 in the northwest. Moisture content of fresh coal ranges from 2 to 5%. Sulfur content is usually less than 1%.

Three general zones of ash distribution occur in Fruitland Formation coal across the basin: an irregular area in the west-central part with an ash content of 8 to 15%; a north-south zone near the basin's center with a 15 to 20% ash content; and an eastern zone where ash exceeds 20%.

Quantity—The Fruitland contains about 200 billion tons of coal between its outcrop and its deepest point of about 4,000 ft below surface (Fassett and Hinds, 1971), based on a subbituminous rank and an average weight of 1,770 tons per acre-ft. This is recognized to be a conservative estimate based on rank distribution. Three categories of bed thickness and six overburden categories were used to compile the resource estimate shown in Table 4. Approximately 14 billion short tons of Fruitland coal may be strippable, another 14 billion is from 500 to 1,000 ft deep, and about 28 billion tons occur between 1,000 and 2,000 ft in depth.

Depth and Thickness—Estimates of Fruitland coal quantity were based on thickness and depth. Figure 7 shows an asymmetrical basin with a relatively regular dip northward from the outcrop to an area about the size of two townships, where depth to the Fruitland exceeds 4,000 ft. Trending northeast from the west side, deepening is more gradual as well as more complex. Steepening of strata is apparent along the northern and eastern margins.

An isopach map of the cumulative Fruitland coal (Fig. 8) outlines areal changes in total coal thickness. It shows that the largest amount of total coal—more than 70 ft—occurs slightly northwest of where the Fruitland is deepest. Total coal is thickest in the central part of the basin trending southeast-northwest. Thickest portions generally contain several thick beds (most often near the bottom of the formation) in multiple zones, plus a few thin beds. The thickest individual coalbeds (those in which all partings are less than 3 ft) exceed 40 ft in a small area but commonly exceed 20 ft.

Menefee Formation

Menefee Formation coals have been difficult to correlate because of their thinness and lenticularity and because they have not been penetrated as densely or studied as thoroughly as the Fruitland. The Menefee depositional environment was similar to that of the Fruitland Formation: nonmarine swamps and alluvial-plains landward from the Point Lookout and Cliff

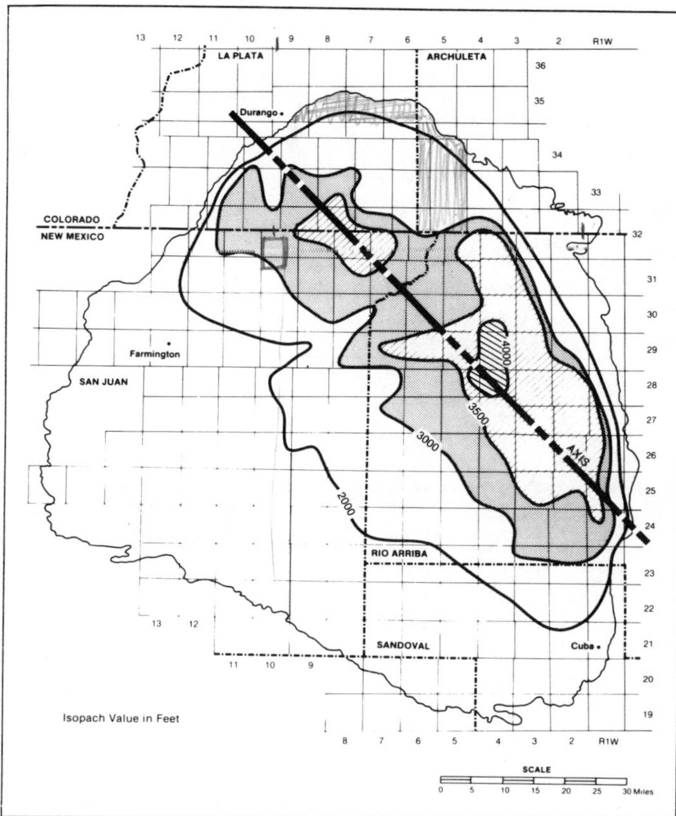

Figure 7—Simplified isopach map of average thickness of overburden above Fruitland Formation coal deposits. Modified from Fassett and Hinds, 1971.

House shorelines. Paludal carbonaceous shales, fluvial sandstones, and flood-plain shales compose the noncoal portions of the formation.

Quality—Menefee Formation coal is of subbituminous A rank in the southwest (near Standing Rock), increasing to high-volatile C bituminous in the northeast. In the north, near Monero, this coal is primarily high-volatile A bituminous, but the southern boundary of the bituminous area has not been established.

Heating values obtained from Menefee coal samples range from 9,550 to 14,940 Btu per pound. The indicated average for the southwest area is about 9,860 Btu per pound; the average increases to the east and north.

Sulfur content averages 1.5% for the basin as a whole (slightly higher than for the Fruitland) and is lowest (less than 1%) in the southern basin. Coal contains 3.5% sulfur in the north, near Monero.

The Menefee coals have more ash than Fruitland coals. Near Standing Rock, Menefee ash content averages 12%; in the northeast, a single sample had an ash content of 6.3%.

Quantity—Shomaker and Whyte (1977) estimated resources for 42 townships (1,512 sq mi) for which drillhole data were available. Estimates are divided into northern and southern portions at about T23N, where depth to coal increases markedly. The tables show an estimated 1,001.7 million tons of coal, of which 660.9 are deeper than 2,000 ft, and 282.8 are less than 500 ft deep. About 54 million tons of the shallowest coal exist between 200 and 300 ft in the Standing Rock area east of Crown Point, and an additional 39.4 million tons occur at less than 500 ft north of this area.

Thickness—Unlike the Fruitland, the deepest part of the Menefee Formation—in the northeastern section of the basin—contains some of the thinnest coal. Total coal thickness is greatest (20 to 30 ft) near the southern end of the "central" basin, at depths to the first coal of about 2,000 to 4,000 ft.

Dakota Sandstone

Quantity—The Dakota Sandstone coal crops out in Montezuma County, Colorado, between Cortez and Mancos, where coal has been mined on a small scale for many years. In a reserve estimate made by Shomaker et al (1971), the Cortez Dakota totaled 158.8 million tons, with 148.8 in a 30-sq-mi area between the two towns. Of the 148.8 million tons, 119.3 are under less than 150 ft of overburden. This estimate is believed to be low, however, because data are scarce.

Quality—Dakota coal quality is fairly well represented by samples from about 3.5 mi east of Cortez. These possess a heating value from 11,580 to 13,650 Btu per pound, an ash content ranging from 4.8 to 16.2%, and a sulfur content ranging from 0.5 to 0.7%.

POTENTIAL METHANE RESOURCES

History of Gas Production from San Juan Basin Coal-Bearing Formations

Overview of San Juan Basin Coalbed Methane Potential

Although San Juan Basin is but one of a number of coal-bearing basins in the Rocky Mountain intermontane west, it is key to understanding western coalbed methane: its

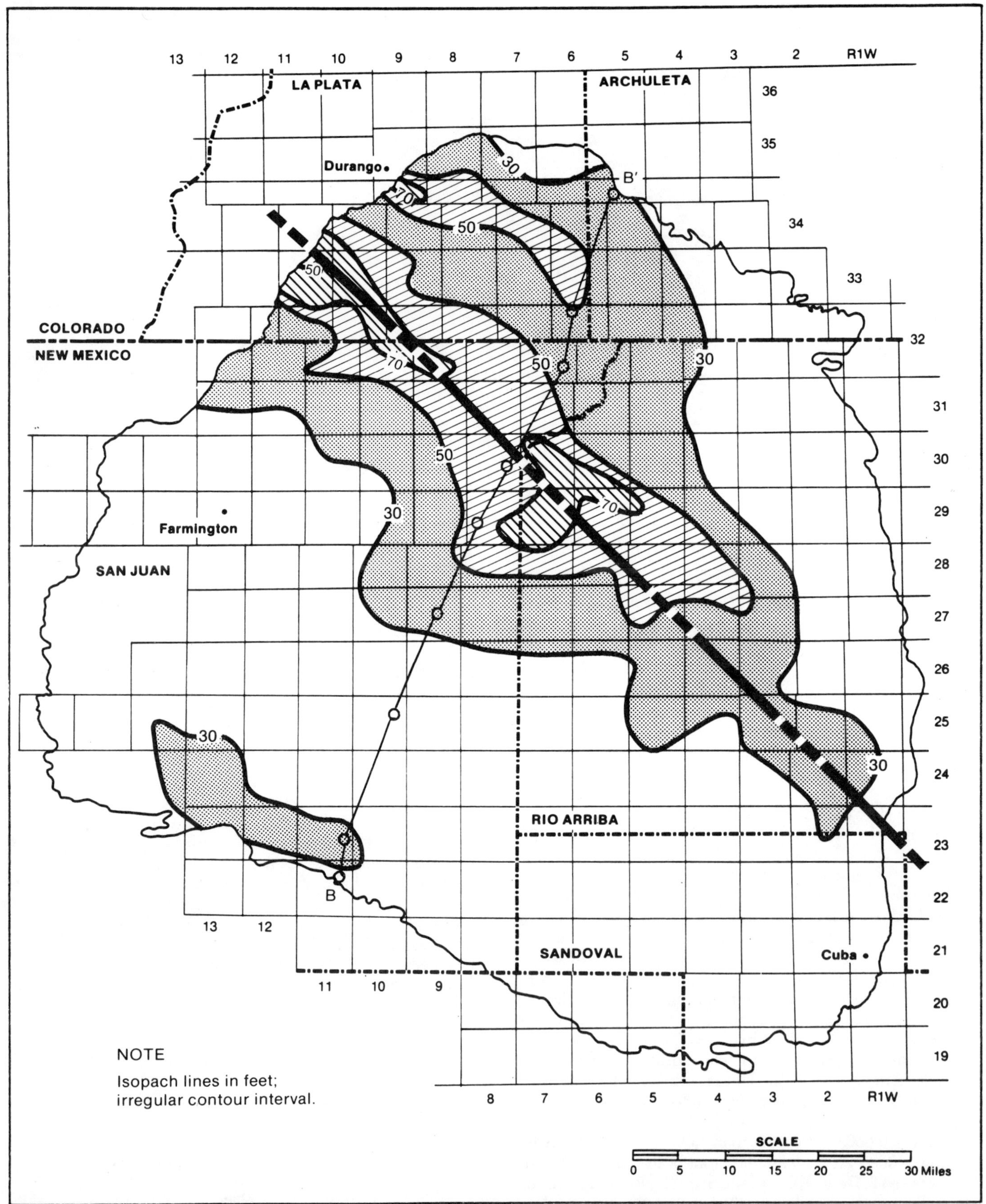

Figure 8—Simplified isopach map of total coal thickness in the Fruitland Formation for areas where total coal thickness exceeds 30 ft. Contours selectively picked to define areas of thickest coal accumulation. Modified from Fassett and Hinds, 1971.

formation, the geologic factors controlling its entrapment, and the production techniques necessary for its economic recovery. San Juan Basin contains all the critical geologic factors observed individually in the other basins; nowhere else are these factors as clearly displayed.

Generation of methane during the coalification process is due to three possible heating processes, including: 1) burial of the organic materials to increasingly greater depths—accompanied by increasing temperature—with total heat exposure dependent on the time/depth-of-burial relationship; 2) emplacement of large igneous bodies causing heating on a regional scale because of very large heat sources that cool slowly over very long periods; and 3) local heating caused by small heat sources such as stocks or sills, or geothermal sources—especially those involving circulation of heated juvenile or meteoric waters along deep-seated fault zones. An example of such a zone adjacent to San Juan Basin is the Rio Grande Rift.

A second major factor for consideration in the study of gas generation in western basins is the history of sedimentation and host-rock development through geologic time.

A third major factor complicating interpretation of gas-formation history in the San Juan Basin is the intimate interfingering in space and time of marine and continental (nonmarine) source beds. The oldest deposits—in the Paleozoic Devonian source beds—are entirely of marine origin. In the youngest deposits, however, as in the Upper Cretaceous Fruitland Formation, source beds are dominantly nonmarine. Through most of Late Cretaceous times, petroleum source beds developed from both marine and nonmarine origins, with the inland seaway western strandline being the interface between the coal-forming swamps just behind the barrier sand beaches to the west and the marine shales developed just to the east on the shallow sea floor. This strandline shifted in geologic time both westward and eastward with each transgression and regression of the sea. The percentage of organic source material of continental origin steadily increased in the Late Cretaceous, starting with deposition of the Dakota Sandstone. With the Dakota, swamp development behind the beaches was minor, and only small amounts of organic material accumulated to act as source beds for oil or gas generation. But as the formations become younger, organic accumulations in the backbeach swamps become increasingly more dominant over organic accumulation on the adjacent shallow sea floor.

A fourth major factor in evaluating San Juan Basin and the origin of its oil and gas deposits is the temperature regime and maturity cycle of these deposits. In San Juan Basin, there is a very distinct geographic factor in the location of the oil and gas deposits. Most of San Juan Basin strata had been buried deep enough that the petroleum deposits have gone beyond the mature stage of development. The post-mature stage has developed to the point where most of the central portion of San Juan Basin contains only gas, with the oil fields lying only along the basin margin close to where their host rocks crop out. Specifically, the oil field pattern forms an open horseshoe pointing north and dipping down into the basin, with the oil deposits lining the northwestern and southwestern sides, and to a lesser extent, possibly the eastern side of the basin. Within the central and northern portions of the basin, only gas deposits are present. The older a formation, the further it is necessary to look on the rim of the basin to find oil deposits. As formations become younger, the horseshoe pattern shifts inward—and becomes shallower—toward the center of the basin. This forms a series of open rings stacked on echelon where it is likely to find oil pools.

A fifth factor is the very large number of oil and gas wells—about 15,000—that have been drilled throughout San Juan Basin. Well spacing is such that, from geophysical logs, detailed sections can be constructed everywhere and compared to oil and gas production records with regard to chemical composition variations and production volumes. These data are probably more extensive for San Juan Basin than for any other basin in the intermontane west.

A sixth major factor is that San Juan Basin is depositionally and structurally very simple. Within the basin, there is very little folding or faulting to complicate the geology and therefore make its history difficult to decipher.

San Juan Basin, then, is key in understanding the origins of coalbed methane in the west and separating this resource from conventional marine-origin oil and gas deposits.

Comparison of Oil and Gas Production from San Juan Basin Strata

Table 1 (presented previously) summarizes oil and gas production from San Juan Basin, starting with Devonian rocks and progressing upward stratigraphically to the Farmington Sandstone—the youngest formation with a record of production. For dry gas production, the Mesaverde Group leads, with 4.7 Tcf. Next are the Dakota Sandstone, with 2.9 Tcf, and the Pictured Cliffs Sandstone, with 2.2 Tcf. Of special coalbed methane interest is the Fruitland Formation, with 37 Bcf of gas produced.

Shown in Table 1 are ratios for the two pairs of petroleum resources produced from the basin: 1) oil (in gallons) versus casinghead gas (in thousand cubic feet [Mcf]); and 2) condensate (in gallons) versus dry gas (in Mcf). These ratios are given for the major producing formations, extending from the mid-Paleozoic to the end of the Mesozoic Late Cretaceous. Of particular interest are the very large differences in the ratios of liquids to gases for the two types of petroleum products. For oil, the ratio of liquid to casinghead gas ranges from 15.0 to 72.0 (gal/Mcf). But for liquid-as-condensate/dry gas, the ratio ranges from only 0.006 to 0.405. Of particular interest in this chart is the fact that ratios for condensate versus dry gas become smaller as the geologic formations become younger. For the Dakota Sandstone, the value is 0.405 (gal/Mcf); for the Gallup Sandstone and the Mesaverde Group, the values are 0.170. Higher in the stratigraphic column, however, as in the Pictured Cliffs Sandstone and the Fruitland Formation, the values are much smaller: 0.007 and .023, respectively.

This trend possibly indicates differences in origin for gases produced from the various geologic formations. With the Dakota Sandstone, gas should be mostly of marine origin; in the Gallup Sandstone and Mesaverde Group, gas is probably of mixed marine and nonmarine origin; and in the Pictured Cliffs Sandstone and Fruitland Formation, gas probably is partly of marine, but mostly of nonmarine, origin.

Almost all desorption data collected within the inner basin have pertained to samples from 26 Fruitland Formation wells. Although coalbed methane has undoubtedly been produced from formations deposited throughout Upper Cretaceous time, the remainder of this section will be concerned principally with methane contained in and produced from coalbeds within the

Fruitland Formation—first with a summary of actual gas production from the Fruitland; then a summary of current coalbed methane exploration by industry within the basin. This is followed by a discussion of the Methane Recovery from Coalbeds Project (MRCP), including data collected by TRW, the University of Mexico, the U.S. Geological Survey (USGS), and the Colorado Geological Survey; methane resources; and an estimate of resource volume.

Exploration and Production of Probable Coalbed Methane Resources

Summary of Fruitland Formation Gas Production—All available data indicate that Fruitland Formation coalbeds have been not only the source beds for the gas produced from the Fruitland Formation itself, but probably also the major source of the gas produced from the underlying Pictured Cliffs Sandstone. In certain areas of the basin, it is generally recognized by the oil industry that Fruitland Formation coal zones are commonly highly overpressured. Because of this overpressured state, it can be expected that gas generated in the coalbeds will not only migrate upward but, following pressure gradients, may also migrate downward into the Pictured Cliffs Sandstone, if that formation is at hydrostatic pressure. As indicated in Table 1, production for both the Fruitland Formation and the Pictured Cliffs Sandstone is comprised mostly of dry gas and associated condensate. Analysis of petroleum products produced from the Fruitland Formation and Pictured Cliffs Sandstone indicates a very high probability that the coalbeds are the dominant source of gas produced from the two formations. Fruitland Formation production is completely restricted to dry gas, with total production of 37.8 Bcf and 21,000 barrels of associated condensate. Commingled Fruitland and Pictured Cliffs production is 1.6 Bcf of dry gas. Only in the Pictured Cliffs is there any production of oil: 89.6 thousand barrels of oil and 248 Mcf of associated casinghead gas. Contrasted to the relatively minor amount of oil, however, is a very large amount of dry gas: 2.2 Tcf. Associated with this gas production are 367 thousand barrels of condensate.

Gas production for all formations within San Juan Basin covers approximately two-thirds of the entire basin and is centered on the basin's center. The long axis of the producing area trends northwest-southeast, generally following the ancient beach strandlines. Figure 9 shows that Fruitland Formation gas production is restricted to about 20 to 25% of the basin and is presently confined to its northwestern portion. A total of 18 defined gas fields exists within the Fruitland Formation in New Mexico. These are, in alphabetical order indicating number of Fruitland wells: the Aztec (60); Aztec North, Blanco (13); Conner, Flora Vista, Gallegos, Gallegos South (20); Harper Hill, Jasis Canyon, Kutz (19); Kutz West, La Jara, Los Pinos North, Los Pinos South (23); Mt. Nebo, Ojo (15); Pinyon (13); and Pinyon North, Pump Mesa, and WAW (157) fields. Colorado contains one major field, the Ignacio Blanco (58). As indicated in the records of the New Mexico Oil and Gas Commission (monthly summary, September, 1981), there are 385 producing wells within the 18 fields in New Mexico. According to the records of the Oil and Gas Commission for the State of Colorado, as of January 1, 1981, there were 58 producing wells for the Ignacio Blanco Field in Colorado.

Of the 19 fields in both states, five have wells that coproduce from the Fruitland Formation and the Pictured Cliffs Sandstone. They comprise the Ignacio Blanco field in Colorado and the WAW, Ojo, Harper Hill, and Los Pinos South gas fields in New Mexico.

The boundaries of the Fruitland gas pools, shown in Figure 9, are not the official boundaries of the gas fields as defined by the Oil and Gas Commission of New Mexico. Rather, the boundaries shown are those indicated by producing Fruitland Formation gas wells.

Figure 10 is a map showing locations of the principal gas fields in the Pictured Cliffs Sandstone. Boundaries shown on this map are as defined by the New Mexico Oil and Gas Commission. The fields shown are the Aztec, Ballard, Blanco, Blanco South, and Blanco East. The area covered by Pictured Cliffs Sandstone production is probably twice that of the Fruitland Formation. Also different from the Fruitland Formation is the apparent trend of these producing fields. The Pictured Cliffs Sandstone stretches from northwest to southeast across the basin and almost exactly crosses its center. The Pictured Cliffs gas fields thus form a linear band about three townships wide—i.e., 18 to 20 mi—and approximately 80 mi long. For dry gas production, the Pictured Cliffs thus is third for all formations within San Juan Basin and follows only the production from the Mesaverde Group and the Dakota Sandstone. Figure 9 shows an apparent northeast-southwest trend of Fruitland Formation gas fields demonstrated by alignments within the WAW, Gallegos, Aztec, Blanco, and Los Pinos South fields, which parallels the northeast-southwest trend of the Hogback Monocline that forms the northwestern boundary of the inner portion of San Juan Basin. This implies that there may be a structural control within the Fruitland Formation that may not be present in, or may be different from, any structural controls in the Pictured Cliffs Sandstone. In the Pictured Cliffs Sandstone, the trapping mechanism seems stratigraphic, representing the sedimentation that occurred along the ancient northwest-southeast–trending beaches. In the Fruitland Formation, however, there appears to be a northeast-southwest–trending structural control perpendicular to, and superimposed on, the expected stratigraphic control, based on the northwest-southeast–trending strandlines behind which the coal-forming swamps developed.

Current Coalbed Methane Exploration by Industry—The previous section discussed evidence pointing toward a coalbed origin for the gas deposits within the Fruitland Formation and Pictured Cliff Sandstone, as indicated by the history of production, pool orientations, age of deposits, and contrast of dry gas versus oil production, all on a historical basis. Direct evidence will be presented here to show that gas being produced is indeed coming from the coalbeds themselves. Evaluation will be made of the location of gas produced stratigraphically from the Fruitland Formation.

Figure 11 is a generic geophysical log from a representative Fruitland Formation gas well. This log covers the entire Fruitland Formation and includes both the base of the overlying Kirtland Shale and the upper portion of the underlying Pictured Cliffs Sandstone. In most of the basin, coalbeds occur throughout the entire stratigraphic section within the Fruitland Formation; however, coalbeds normally are concentrated in two major zones. In general, the thickest coal zone is the lower, which lies immediately above the Pictured Cliffs Sandstone. Within this coal zone are several thick coalbeds, commonly with two to four thick beds present, in addition to thinner ones.

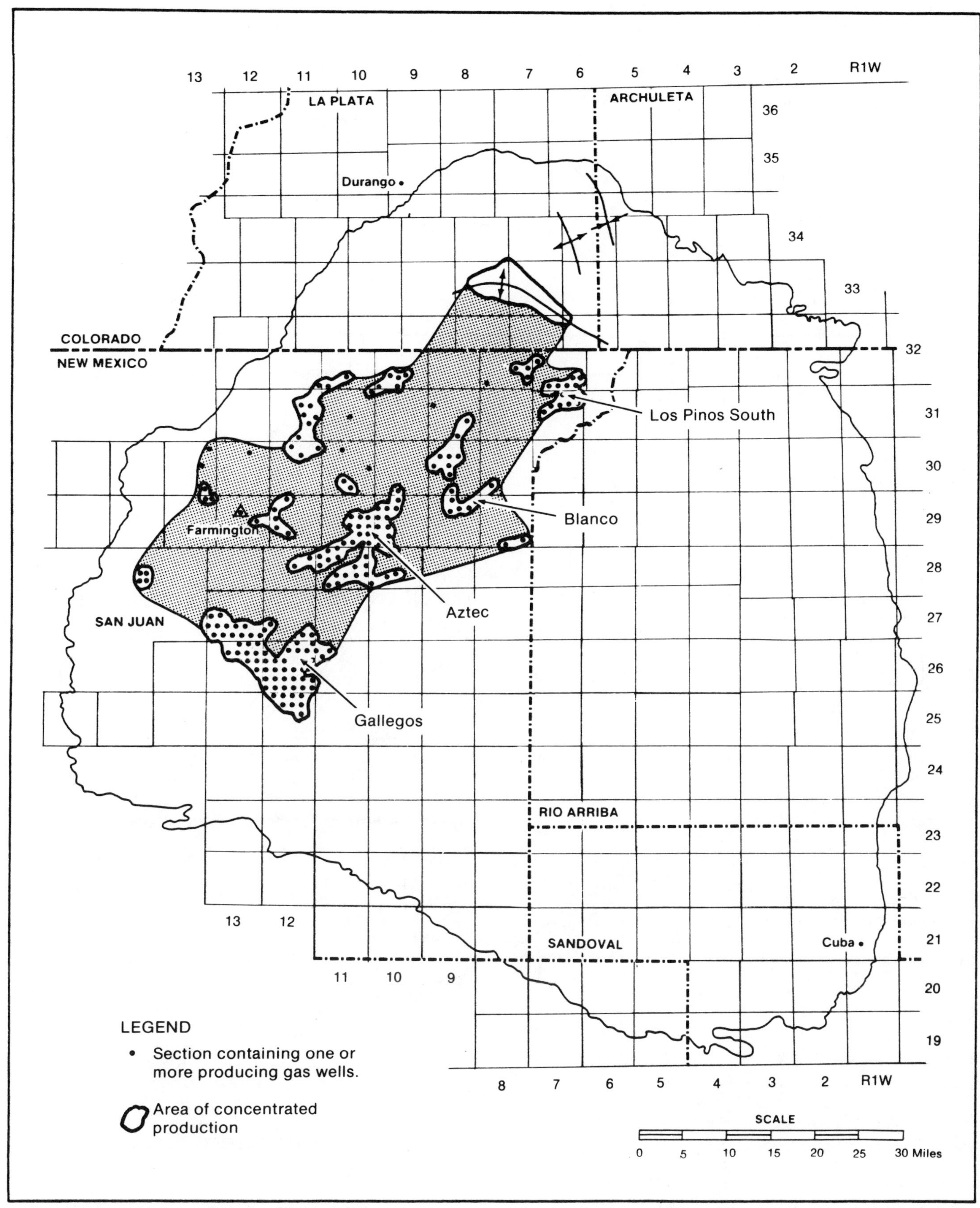

Figure 9—Areas of current production from Fruitland or Fruitland-Pictured Cliffs.

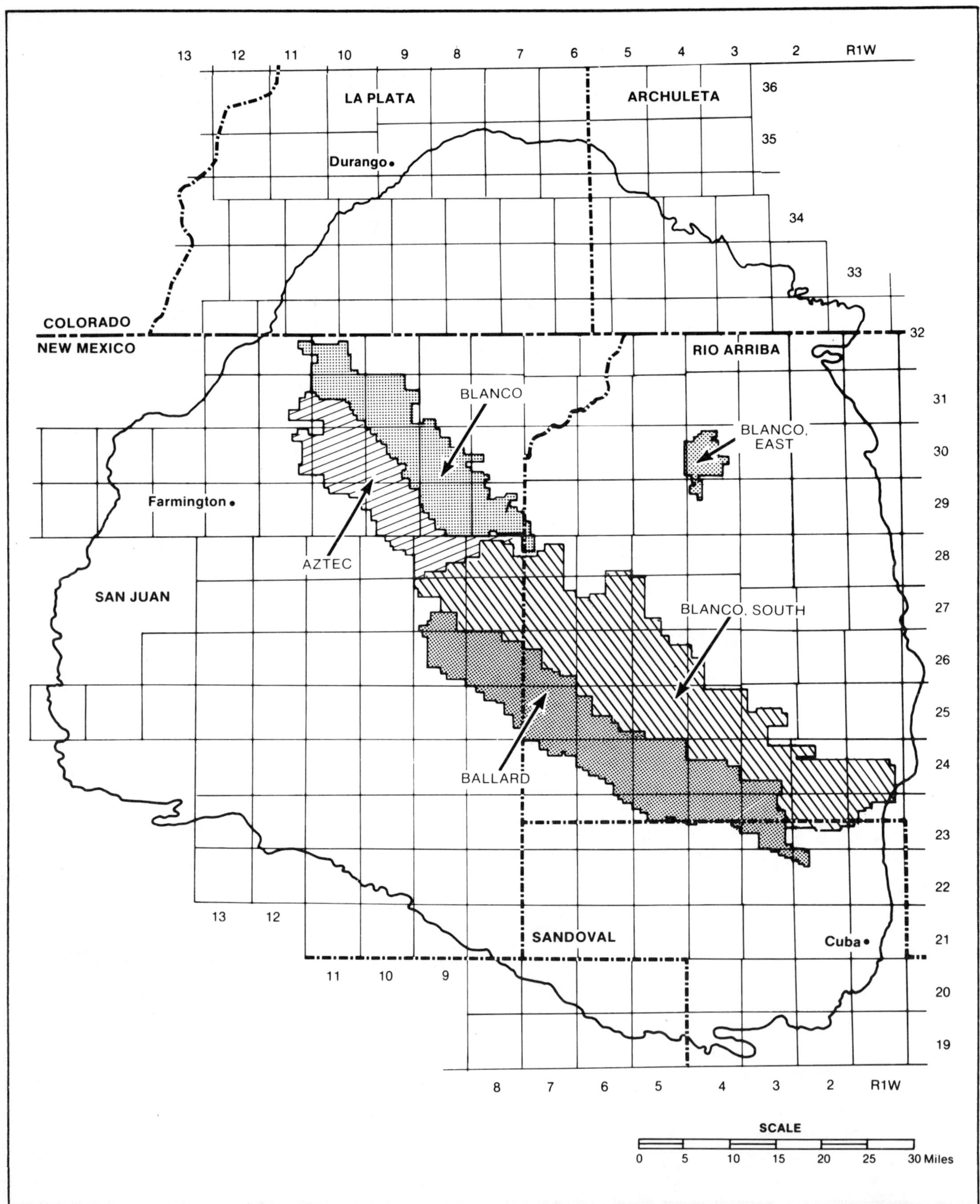

Figure 10—Pictured Cliffs gas fields.

The second most important zone normally is near the top of the Fruitland Formation. Again, this zone contains several individual beds. The occurrence of both lower and upper coal zones within the Fruitland is demonstrated in Figure 11. It is of extreme interest that in 11 of these 18 fields, the zone of perforation is shown quite clearly to be restricted within coalbeds as interpreted by the very large excursions on the electrical resistivity logs. These 11 fields are the Aztec, Aztec North, Blanco, Flora Vista, Harper Hill, Jasis Canyon, Kutz, La Jara, Mt. Nebo, Ojo, and Pinyon North. In several of the other fields, the completion zone occurs across the entire stratigraphic section in which coalbeds are prevalent. The completion is not shown to be as clearly limited within individual coalbed boundaries as it is in the 11 fields previously mentioned. Based on logs selected by Fassett et al (1978) as being representative of Fruitland Formation gas production, it is considered that a large portion of the gas actually being produced from the Fruitland Formation does, in fact, come from the coalbeds and represents coalbed methane gas production. In some cases, it is probable that the companies producing this gas have not been consciously exploring for coalbed methane; rather, simply producing gas wherever they find it. It is the authors' opinion, however, that some companies are quite aware of the origin of this gas and are actively exploring for it. Prominent among these companies is Amoco Production, which has a very active coalbed methane exploration program. A second company well aware of the probable origin of the gas within the Fruitland Formation and actively exploring for and producing from the coal zones within the Fruitland Formation is Dugan Production Corporation of Farmington, New Mexico.

In general, two basically different drilling and completion techniques are currently employed in successful San Juan Basin wells economically producing coalbed methane today.

The first technique incorporates an open-hole completion over the planned producing zone. The basic technique was initially used early in the production of gas from the Fruitland Formation but, with time, has become fairly sophisticated in its use. Today, the designs of successful wells are carefully tailored to the intricate and differing sedimentary complexes and subtle geologic controls normally encountered at each different site in the basin. The first requirement is to identify the zone most likely to be productive within the Fruitland Formation. Within the northern San Juan Basin, in many cases that is the lower coal zone, which lies immediately above the Pictured Cliffs Sandstone. An open-hole completion is performed on this zone. If control is good, the well is drilled to the top of the coal zone or, specifically, to the top of the section to be left open-hole. Production casing is then installed and cemented. After cementing, the well is drilled through the bottom of the casing to TD. This open-hole section may represent 100 ft or more of coalbeds with interbedded sandstones and shales within the Fruitland, or it may represent a single bed of coal. In drilling the open-hole section, where practical, air or air mist is used to keep all liquids off the coalbeds. In general, fluids with any acids are especially avoided. In certain coals, there is a tendency for the acid to soften the coal around the wellbore, making future gas production difficult. In the nominal case, no stimulation of the coalbeds is required.

The second technique involves a more conventional completion; i.e., drilling through a coal zone, setting production casing, cementing, then perforating the zone's most promising beds—identified from a suite of detailed geophysical logs, including, particularly, a density log. Those zones that have high gas potential within a coalbed or adjacent sandstone are perforated and a very small frac performed. In general, the volume of the frac is only large enough to break down the formation. It may involve a few hundred, or a few thousand, gallons of fluid.

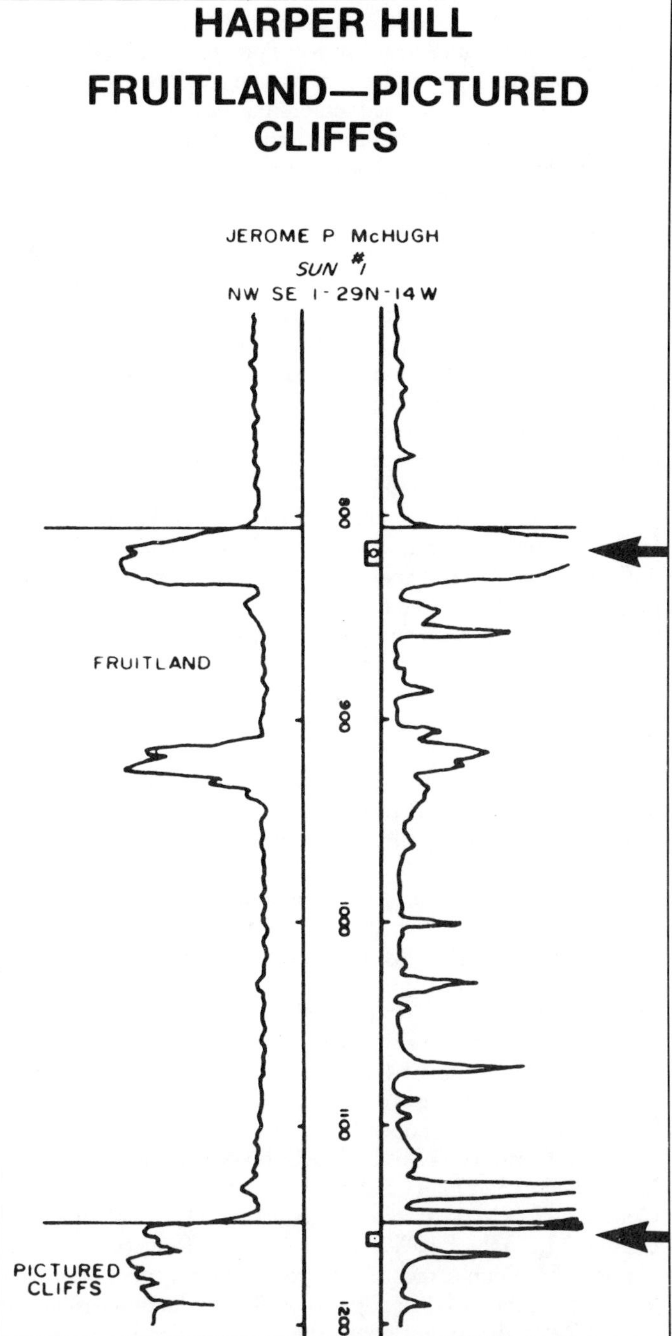

Figure 11—Representative borehole geophysical log for Fruitland Formation gas field. Selected from Fassett et al, 1978. Used with the permission of the Four Corners Geological Society.

Following are brief descriptions of three wells that can be considered representative of San Juan Basin Fruitland Formation coalbed methane production. The first, Phillips

Petroleum Well 6-17, San Juan 32-7 unit, may be considered the grandfather well for San Juan Basin coalbed methane production. It has a 29-year history of production and through 1981 was still a major producer of natural gas. The second and third wells can be considered representative of the new wells being drilled with the benefit of relatively recent knowledge of coalbed methane. These wells are the Amoco Cahn No. 1 and the Dugan Knauff No. 1.

Phillips Petroleum Well 6-17, San Juan 32-7 Unit—This well, completed by Phillips Petroleum Company on August 24, 1953, is the discovery well for the Los Pinos Fruitland, South, gas field. The well, in the northeast corner of San Juan County, New Mexico, is about 6 mi south of the Colorado-New Mexico State Line (NE 1/4, Section 17, T31N, R7W). This hole, drilled with mud and cased to a depth of 3,054 ft, is completed open-hole from 3,054 to 3,240 ft. Following cementing, the hole was drilled with air from a depth of 3,055 ft to TD. The well was completed naturally, with no stimulation. A total of 1.1 Bcf of gas had been produced through 1981. Figure 12, the annual production curve for the well, shows the marked difference that can be expected of gas production from a coalbed methane well, contrasted to a conventional gas well. In a gas well producing from a conventional reservoir, the normal production curve shows a steady decline of gas production over a period of several years. In contrast, in the Phillips Petroleum # 6-17 well, production is shown to have increased steadily from an initial 27.7 MMcf in 1953 to a maximum 57.8 MMcf in 1974. Since 1974, there has been a tendency for gradual decrease, such that 1981 production was 48.6 MMcf. Over this long period (essentially three decades), there has been very little drop in pressure within the well. The initial shut-in pressure in 1953 was 1,504 psi. During a 43-day shut-in test in 1977, shut-in pressure had built back up to 1,472 psi and was still rising at the end of the best period. Figure 13 is a neutron radioactivity log of the open-hole portion of the Phillips Petroleum Well 6-17. As shown on this log, at least five coalbeds are present. Two beds (3 and 5 ft thick) occur in the upper Fruitland coal zone, and three beds (with thicknesses of 4, 8, and 3 ft) are present in the lower coal zone. This comprises a total coal thickness of at least 23 ft in the 186-ft section of open hole.

Amoco Cahn No. 1 Well—The Cahn No. 1 is in the Mt. Nebo Field in San Juan County, 4 mi south of the Colorado-New Mexico state line (1,030′ FNL, 1,600′ FWL, Section 33, T32N, R10W). The following information was obtained from public records or interpreted from geophysical logs of the Cahn No. 1 or adjacent wells. Production casing was set at 2,795-ft depth, just above the basal Fruitland coalbed as correlated from nearby wells. Total hole depth is 2,812 ft, leaving a 17-ft open-hole section below the casing. In the Cahn Gas Com. No. 2 well, 900 ft to the southwest, a correlated 24-ft coalbed was encountered in the interval 2,777 to 2,801, and a 4-ft bed at 2,807 to 2,811.

A 24-hr production test in May 1977 produced 350 Mcf of gas and 239 barrels of water, with a flowing tube pressure of 10 psi on a 0.375-inch choke. Shut-in tubing pressure was 335 psi, and shut-in casing pressure was 483 psi. All available records indicate that no stimulation was performed on the well.

Gas production started in July 1979, and the records shown are through December 1981. During this period, a total of 444 MMcf of gas were produced. Figure 14 shows gas production steadily climbing for the 2.5-year period in which production is here reported; whereas concurrent water production has been steadily declining. Production for the Cahn No. 1 well has required special treatment because the producing reservoir (the basal coalbed in the Fruitland Formation) contains substantial amounts of water, as it does in much of the basin. As indicated in this graph, Amoco appears to have solved the problem of producing gas from wet coalbeds.

Dugan Knauff No. 1 Well—The Knauff No. 1 well is in the Kutz Fruitland gas field, near the center of San Juan County, between the Aztec field on the north and the WAW, Gallegos, and Gallegos South fields to the southwest (1,015′ FNL, 1,650′ FWL, Sec. 31, T28N, R10W). This well is typical of a Fruitland Formation completion employing the more conventional techniques used by Dugan Production Corporation. The well is drilled to TD, cased, and cemented; then a zone is selectively perforated in either a coalbed or an adjacent sandstone; and finally a small frac is employed. In this case, the zone perforated was a single coalbed. From records present at the New Mexico Oil and Gas Commission at Aztec, this well was evidently a nonproducing, preexisting well obtained by Dugan Production Corporation for an attempted Fruitland completion. A cement plug was set at 1,560 ft, and the well was perforated with two shots per ft in the interval 1,515 to 1,521 ft. The well then ejected small amounts of water, a swab was run, and the well kicked off at an estimated 100 Mcf. A frac was performed using 15,000 pounds of 10–20 sand, with 439 barrels of water. A spearhead frac was employed using 100 gallons of 15-% KC1. Formation breakdown occurred at 2,200 psi. The well was shut in at a formation pressure of 1,800 psi. The well was put on-line, and gas production commenced June 1976. Figure 15 shows monthly production for the well from June 1976 through April 1981. Cumulative production as of the end of 1980 was 197 MMcf of gas. The well has been producing naturally against pipeline pressure, without any pumping. There has been no water produced from the well. Production from the well is lower in late summer months than in winter, because of lower pipeline demand. Production from the well has been gradually increasing for the last 4 years. Monthly production has been averaging approximately 3.2 MMcf of gas; daily production, approximately 200 Mcf. Also shown in Figure 15 are the logs of the well, indicating that the location of the perforated zone is directly in the coalbeds in the upper Fruitland coal zone.

Gas Content of Fruitland Coals

Federal government-sponsored R&D in San Juan Basin on the MRCP, discussed here, includes work performed for DOE by TRW, the University of New Mexico, and the USGS and CGS.

Well locations in San Juan Basin where core or drill-chip samples were collected for gas content determinations are shown in Figure 16. Samples were collected from 28 wells in the basin by MRCP.

MRCP Test Site—Under the MRCP, samples were collected by TRW from six wells. Locations of these wells are shown in Figure 16. Four wells are in the west-central portion of San Juan Basin; one is in Rio Arriba County, north of the town of Cuba; and one is in the northeastern corner of San Juan County, just south of the Colorado-New Mexico state line. Data for these six wells are summarized in Table 5, which contains coal thickness data and a summary of the gas desorption measurements made for each well, including average gas content for all coal samples from each well. Also included is the largest value for gas

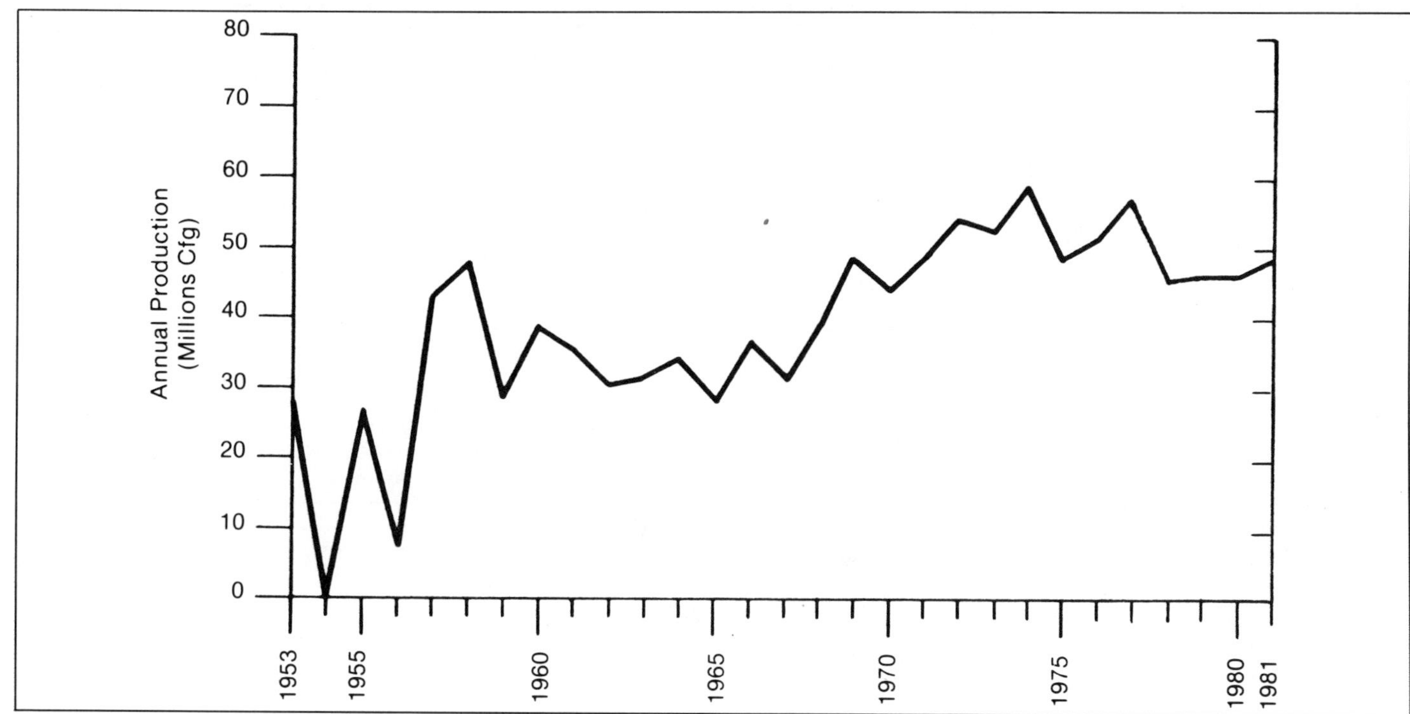

Figure 12—Annual production curve Phillips Petroleum Well 6-17, San Juan 32-7 Unit, South Los Pinos Fruitland gas field, San Juan County, New Mexico. Modified from K.C. Bowman, *in* Fassett et al, 1978. Used with the permission of the Four Corners Geological Society.

contained in any sample from each well. These values (both the average and largest) are expressed in metric and English systems.

Personnel from the Department of Chemical and Nuclear Engineering and from the Bureau of Engineering Research of the University of New Mexico at Albuquerque collected and desorbed 31 drill-cutting samples from 14 wells in the New Mexico portion of the San Juan Basin during the period December 1979 through December 1980.

Locations of these 14 wells are shown on the map of Figure 16. Of the above 14 wells, Numbers 1 to 3 are Southern Union Exploration wells in southeastern San Juan County; Numbers 4 to 8 are Blackwood and Nichols wells along the San Juan River; Numbers 9 and 10 are Northwest Pipeline wells in north-central Rio Arriba County; and Numbers 11 to 14 are Dome Petroleum wells in western Sandoval County.

Average depths of coalbeds sampled for each of the four areas, from south to north, are: 1,150 ft for the Dome Petroleum area in south-central San Juan Basin; 1,300 ft for the Southern Union Exploration area in the southwestern portion of the basin; and 3,070 ft for the Blackwood and Nichols and 3,110 ft for the Northwest Pipeline areas in the north-central and northwestern New Mexico portions of the basin, respectively (see Table 6).

Gas Content—Table 6 presents gas contents determined by desorption of cuttings samples collected during the drilling of the above wells. These reported gas contents have been adjusted in a manner described in detail by Choate et al (1982) to permit use in comparison with other existing core and cuttings desorption data. The adjustment compensates for water and noncoal contaminants included with the cuttings samples.

Gas desorbed from the samples and actually measured while in the desorption canisters is represented in Column 6. Gas lost by the coal cuttings during their travel uphole, and while on the shale shaker at the surface before being sealed in the canisters, is represented in Column 5. The "lost gas" calculations used by Williams and Smith (1981) take into account the nonlinear desorption rates encountered by the coal cuttings during their uphole travel to the surface. Sums of the measured desorbed gas and "lost" gas values are given in Columns 7 and 8, in metric and English units; i.e., cubic centimeters per gram (cc/gm) and cf/ton. The combined lost and desorbed gas values are for the total sample, including the diluting shale cuttings along with the free water in the canisters and the water attached to all cutting surfaces. The gas content values in Columns 9 and 10 represent the authors' best estimate of the gas contained in the coal in-place and are obtained by dividing Columns 7 and 8, respectively, by the coal and in situ water percentage of the total sample. These values are probably close to the total gas-in-place. As expected, cutting samples desorb most of their contained gas while in the canisters, and little residual gas remains. To date, even for very gassy coals, the maximum residual gas in any sample has been only a few tenths of a cc/gm.

Average gas content values are given by well and by the four operator properties, both as measured directly and after adjustment for water and shale dilution. Average adjusted values for the 14 wells range from 73 to 526 cf/ton; highest adjusted sample values in a well range, by well, from 86 to 808 cf/ton (well no. 1 is excluded because the only sample was desorbed in a leaking canister).

Gas Composition—Analyses of gas samples collected from eight of the wells are given in Table 7. In the hydrocarbon fraction of the samples, average methane content (C_1) is 88.1%;

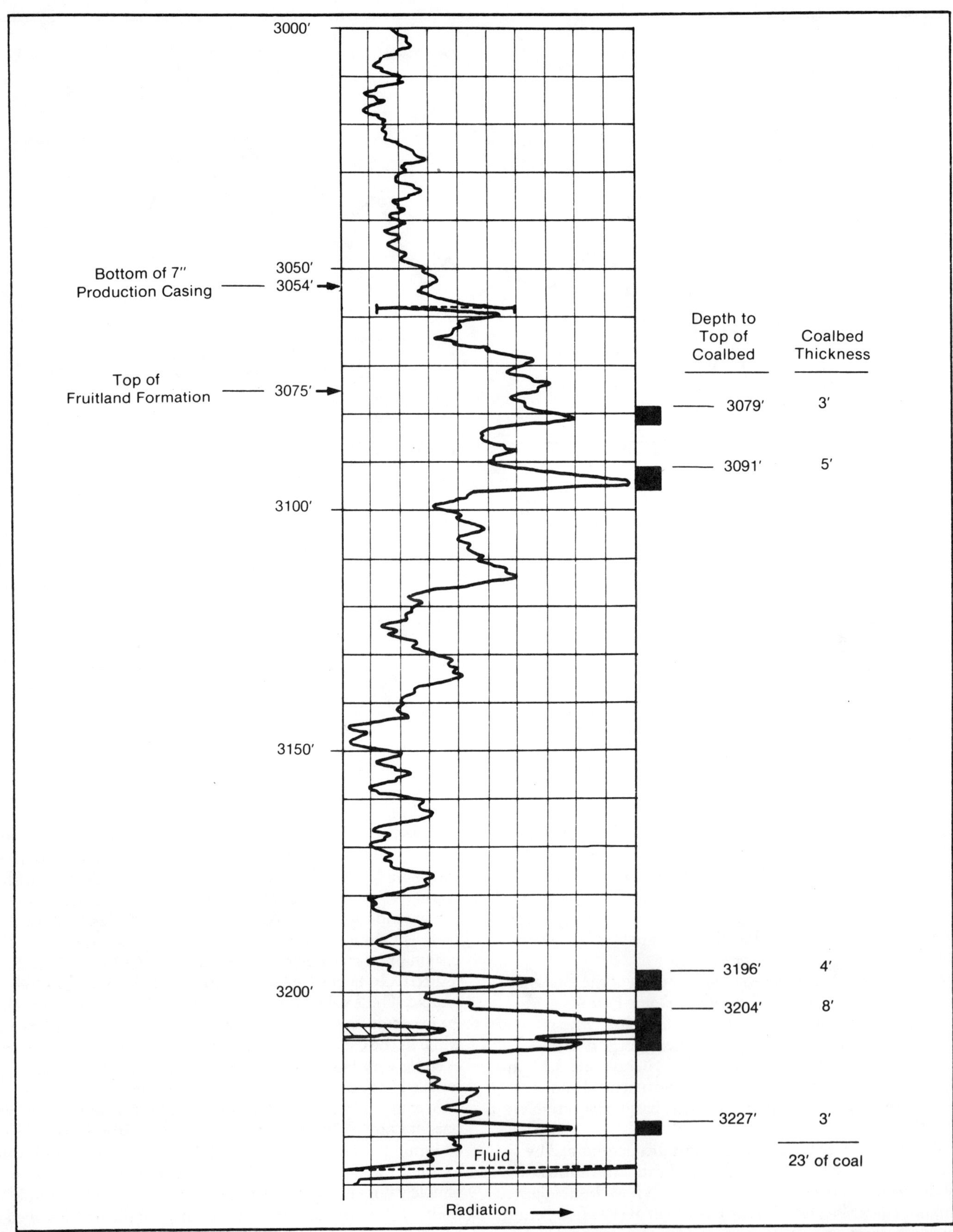

Figure 13—Neutron radioactivity log of open-hole portion of Phillips Petroleum Well 6-17, San Juan 32-7 Unit. NE 1/4, NE 1/4, Sec. 17, T31N, R7W.

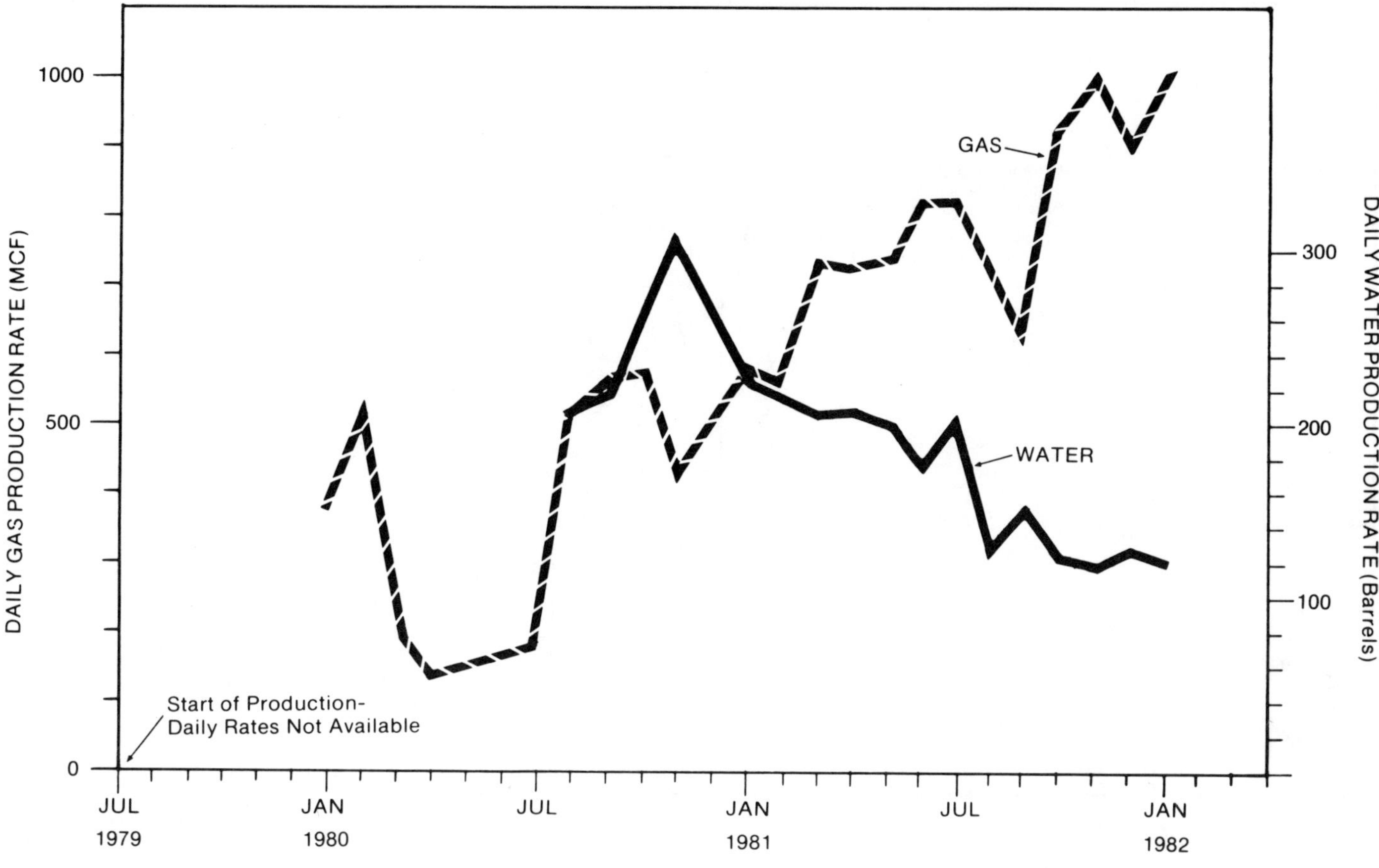

Figure 14—Daily gas and water production rates, Amoco Well Cahn No. 1.

C_2 is 1.6%; and C_3 and higher hydrocarbons are present in only minor-to-trace amounts. In only one sample (well no. 2) is C_2 present in amounts significantly greater than 1%—at 8.55%. In this sample, the ratio of C_1 to total hydrocarbons is 0.90; in the other seven samples, $C_1/C_{(1-4)}$ ranges from 0.987 to 0.996.

In the nonhydrocarbon fraction, N_2 and CO_2 are present in significant amounts. After allowing for minor air contamination, N_2 ranges from approximately 1 to 13% and averages approximately 5%; CO_2 averages 3.2%. This gas composition is relatively typical of gas derived from coalbeds in western sedimentary basins; i.e., very high in C_1 content compared with higher hydrocarbons and containing several percent each of N_2 and CO_2.

U.S. Geological Survey—Colorado Geological Survey–DOE Test Sites—This section includes data for eight wells: The USGS collected samples from six (identified as USGS-1 through USGS-6), and the CGS collected samples from two (designated CGS-1 and CGS-2). Locations are shown on the map of Figure 16. Data for these eight wells are summarized in Table 8. Data in this table include: well site number; sample number as listed in stratigraphic sequence with increasing depth for each well; USBM sample code number; location, as available from the collecting agency; well operator; well name; ground elevation; sample depth; sampling date; and coalbed depth. Also included are coalbed identification, where available; sample weight; and gas content, including lost, desorbed, residual, and total gas in both metric and English units. Of the six USGS wells, well no. 1 was an Energy Reserves Company well; wells 2 through 5 were all drilled by the Coal Branch of the USGS, and well no. 6 was drilled by the American Public Gas Association. The samples collected by the CGS are from the Menefee Formation—the only samples covered in this report that are not from the Fruitland Formation. Locations are given only as "near Durango"; because the rock strata along the Hogback Monocline in the Durango area are relatively steeply dipping, commonly ranging from 25 to 35°, and because the sample depths were very shallow (295 and 310 ft), the samples must have been very close to the Menefee outcrop line.

Because these eight wells were all drilled near the margins of San Juan Basin, at relatively shallow depths, gas values for each well are relatively low, as indicated by the average value for total gas content. Average total gas values in cf/ton for each well are: 129 for USGS-1; 102 for USGS-2; 63 for USGS-3; 13 for USGS-4; 3 for USGS-5; 42 for USGS-6; 5 for CGS-1; and 10 for CGS-2.

Discussion of Methane Resources

Discussion of San Juan Basin Coalbed Methane Province

Tectonic Overview—During this study of most of the intermontane basins containing major coal deposits in the west,

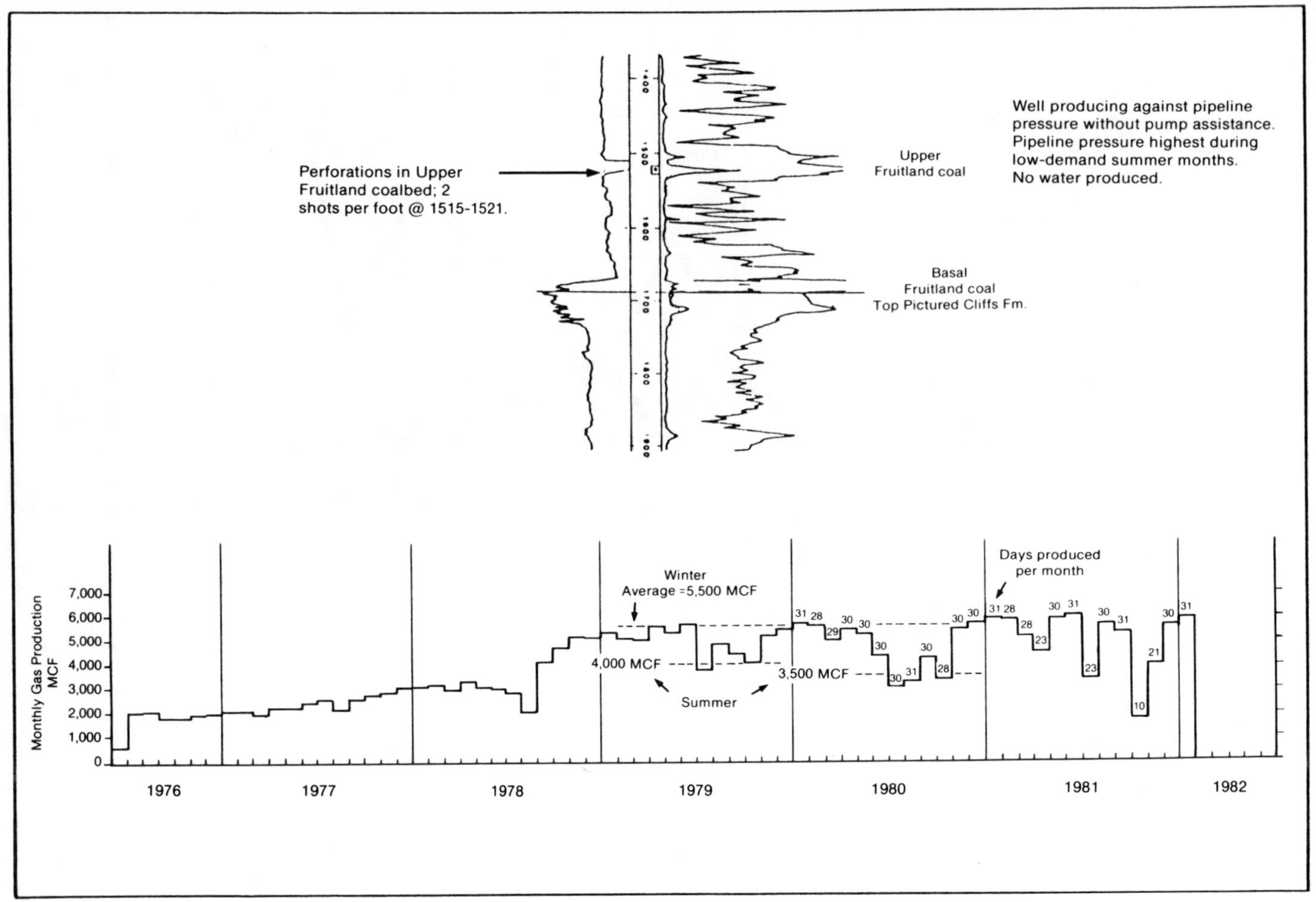

Figure 15—Dugan Well Knauff No. 1, Kutz-Fruitland gas field showing monthly gas production and geophysical logs showing location of perforations in upper Fruitland coalbed.

it has become increasingly clear that methane associated with these coals is not uniformly distributed throughout those basins but commonly has high concentrations—on both regional and local scales—not associated solely with depth of burial. During this study, it has developed that high methane concentration in coalbeds in the west can be most closely associated with the temperature histories of those coalbeds. The higher the temperature to which a coal has been subjected and the longer at that temperature, the greater the coal's rank and the larger the volume of methane produced. In western coal deposits, higher heat regimes have existed not only in association with greater burial depths in basin centers but also wherever those coals have been affected by Cenozoic igneous activity.

In study of western Washington coal deposits (Choate and Johnson, 1980), a coalbed methane province was defined in which methane concentration and coal rank showed remarkably uniform association with distance from the axis of igneous intrusion and associated high heat flow, related to the orogenic development of the Cascade Range. In study of the Piceance Basin of northwestern Colorado (Choate et al, 1981), coalbed methane target areas were defined in the southeastern portion of the area where extremely high methane concentrations were shown to be in proximity to intrusive igneous plutons of intermediate to acidic compositions. In development of this relationship of igneous heating and coalbed methane production in Piceance Basin, it was suggested that the center of such heating could be the San Juan Mountains, and a preliminary map was drawn showing the central relationship of the San Juan Mountains between the southeastern Piceance and northeastern San Juan Basin areas (Choate et al, Fig. 5-22, 1981).

In further development of this theme of igneous activity and high heat flow related to coalbed methane production, Choate and Rightmire (1982) discussed the central relationship of the San Juan Mountain volcanic complex to not only the Piceance and San Juan Basins, but also the Raton Basin, and revised the preliminary heat flow (Fig. 17). As depicted in this figure, the 3.0-HFU contour line—encompassing the San Juan Mountains

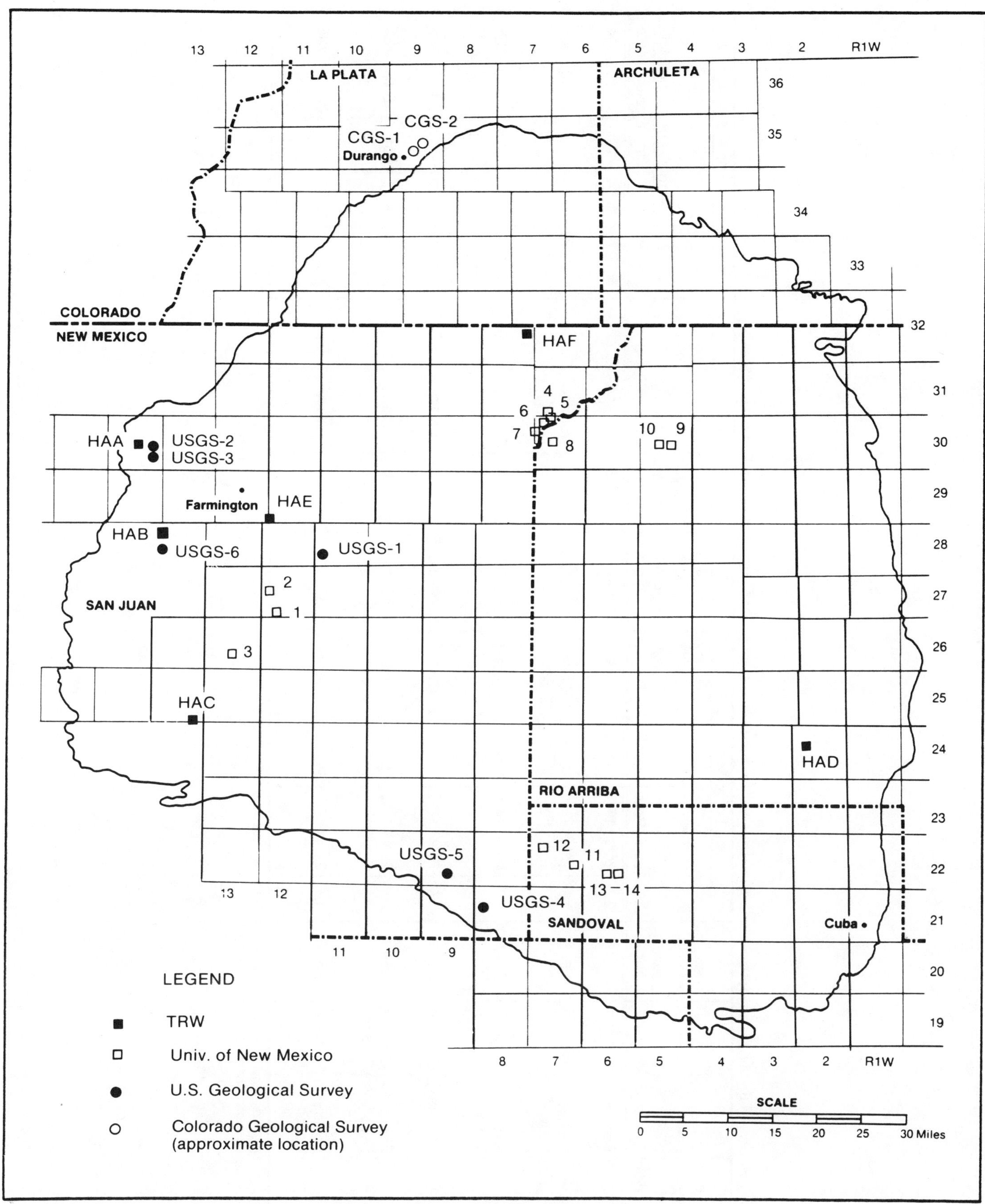

Figure 16—Location map showing drilling sites for coalbed desorption samples from San Juan Basin.

Table 5—Summary of TRW/DOE participation coalbed methane wells in San Juan and Rio Arriba Counties, San Juan Basin, New Mexico.

TRW/DOE Well Designation	Well Operator	Well Name	Surface Elevation (msl)	Well Location				Sample Description							Sample Gas Content			
				Sec.	Twp.	Rng.	County	Sample Type	Depth Interval Sampled (ft)	Total Coal Thickness (ft)	Formation Sampled	Number of Samples	Date Sampled	Average Value		Highest Value	
														cc/gm	cf/ton	cc/gm	cf/ton
HAA	Western Coal	P-70	5,247	22	30N	15W	San Juan	Core	370–414	12.2	Fruitland	2	6/17/78	0.84	27	1.39	44
HAB	Navajo Nation/ USGS	K-3	5,645	18	28N	14W	San Juan	Cuttings	815–835	~18	Fruitland	1	4/29/80	1.50	48	1.50	48
HAC	Navajo Nation/ USGS	A-5	6,070	35	25N	14W	San Juan	Core	684–765	44.6	Fruitland	5	10/21/80	0.72	22	0.81	26
HAD	Jerome P. McHugh	Di's Delight	7,096	17	24N	2W	Rio Arriba	Core	3,040–3,086	46.5	Fruitland	8	11/7/80	0.96	31	3.94	126
HAE	Energy Reserves	Gallegos Canyon #322	5,543	31	29N	12W	San Juan	Core	1,350–1,420	24.6	Fruitland	8	2/18/81	4.71	151	5.13	164
HAF	Southland Royalty	Reese Mesa No. 9	6,667	13	32N	8W	San Juan	Cuttings	3,338–3,368	28	Fruitland	2	4/19/81	18.34	587[2]	22.97	735[2]

NOTES: [1]Gas content determination includes lost gas, desorbed gas, and residual gas.
[2]Values corrected for noncoal contamination.

Table 6—Gas content by cuttings desorption.

Well No.	Sample No.	Location (Sec., Tn., Rn.)	Sampling Interval (ft)	Gas Content: Lost (cc/gm)	Desorbed (cc/gm)	Lost and Desorbed Combined: For Coal, Shale, and Total Water (cc/gm)	For Coal, Shale, and Total Water (cf/ton)	For Coal and In Situ Water Only (cc/gm)	For Coal and In Situ Water Only (cf/ton)	Well Name
1	2	3	4	5	6	7	8	9	10	
Southern Union Exploration										
1	1	32-27-12	1,200–1,210	0.19	0.71	0.90	29	1.69[3]	54[3]	1 Susco-Biddoni
2	1	32-27-12	1,140–1,150	0.38	1.60	1.98	63	3.02	97	1-18 SX Federal
3	1	32-26-13	1,370–1,390	0.61	0.85	1.46	47	2.25	72	1-21 SX Federal
	2		1,370–1,390	0.73	1.02	1.75	56	2.70	86	
	3		1,390–1,400	0.54	0.76	1.30	42	2.00	64	
	4		1,390–1,400	0.47	0.66	1.13	36	1.74	56	
	5		1,390–1,400	0.69	0.97	1.66	53	2.56	82	
	6		1,390–1,400	0.64	0.90	1.54	49	2.37	76	
		Well Averages					47		73	
		Three-Well Averages					46		75	
Blackwood and Nichols										
4	1	32-31-7	3,230–3,250	2.11	2.07	4.18	134	6.72	215	47-1 Northeast Blanco Unit
	2		3,230–3,250	1.95	1.93	3.88	124	6.34	203	
		Well Averages					129		209	
5	1	32-31-7	3,042	0.41	1.379	1.79	57	3.03	97	48-A Northeast Blanco Unit
6	1	5-30-7	2,886	0.47	1.10	1.57	50	2.55	82	45-A Northeast Blanco Unit
	2		3,006–3,015	0.76	1.49	2.25	72	4.18	134	
	3		3,006–3,015	0.53	1.12	1.65	53	3.02	97	
		Well Averages					58		104	
7	1	7-30-7	2,920–2,930	0.55	1.47	2.02	65	5.11	164	31-A Northeast Blanco Unit
	2		2,970–2,979	0.52	1.33	1.85	59	3.66	117	
	3		3,001–3,007	0.70	1.75	2.45	78	4.14	132	
		Well Averages					67		138	
8	1	17-30-7	3,053–3,073	0.77	1.95	2.72	87	4.34	139	24-R Northeast Blanco Unit
	2		3,053–3,073	0.82	2.08	2.90	93	4.13	132	
	3		3,096–3,105	0.39	1.07	1.46	47	3.15	101	
	4		3,125–3,134 3,137–3,143	1.45	1.78	3.23	103	5.83	187	
		Well Averages					83		140	
		Five-Well Averages					79		138	
Northwest Pipeline										
9	1	22-30-5	3,106–3,112	0.21	0.57	0.78	25	1.39	45	71 San Juan 30-5 Unit
	2		3,161–3,183	1.33	1.89	3.22	103	7.19	230	
		Well Averages					64		138	
10	1	21-30-5	3,038–3,040	0	0			Not a Coalbed		
	2		3,119–3,141	1.25	1.22	2.47	79	3.03	97	51 San Juan 30-5 Unit
	3		3,119–3,141	2.51	2.46	4.97	159	6.18	198	
		Well Averages					119		148	
		Two-Well Averages					92		143	

continued

Table 6—(Continued)

Well No.	Sample No.	Location (Sec., Tn., Rn.)	Sampling Interval (ft)	Gas Content						Well Name
				Lost (cc/gm)	Desorbed (cc/gm)	Lost and Desorbed Combined				
						For Coal, Shale, and Total Water		For Coal and In Situ Water Only		
						(cc/gm)	(cf/ton)	(cc/gm)	(cf/ton)	
1	2	3	4	5	6	7	8	9	10	
Dome Petroleum										
11	1	23-22-7	890–904	0.175	1.149	1.324	42	6.05	193	Dome Tesoro 23-1
12	1	?-22-7	1,401–1,408	1.670	4.999	6.669	213	9.14	292	?
13	1	28-22-6	1,305–1,311	0.99	2.63	3.62	116	7.61	243	Dome Navaho 28-22-6 No. 4
	2		1,355–1,375	1.10	3.57	4.67	149	25.24	808	
		Well Averages					133		526	
14	1	27-22-6	1,344–1,350	0.22	0.71	0.93	30	2.30	74	Dome Navaho 27-22-6 No. 1
		Four-Well Averages					105		271	
		14-Well Averages					81		163	
		All Samples Averaged					80		152	

[1]Averaged value of all samples, see end of table.
[2]Composite sample for the well.
[3]Gas values are of too low; leaking valve.
[4]Average value of other samples this well.

and the adjacent portion of the Rio Grande Rift Valley to the east—is central to all three basins. The 2.0-HFU contour lines encompass among them portions of southeastern Piceance Basin, northwestern and southwestern San Juan Basin, and the smaller Raton Basin. Of particular interest, all portions of basin areas within the 2.0-HFU contour lines also contain igneous rocks of either intrusive or extrusive character (Choate and Rightmire, Figs. 2, 3, 1982).

Because of these relationships, increases in coal rank, with associated production of large volumes of methane in western intermontane basin areas, were attributed by Choate and Rightmire (1982) principally to increases in temperature associated with:

- Increased depth of burial
- Regional heating—on a scale of tens of miles—caused by emplacement of major igneous extrusive or intrusive bodies of at least batholith proportions, accompanied by long-term, high-thermal gradients
- Local thermal anomalies—on a scale of up to a few miles—caused by intrusion of small igneous bodies, mostly of intermediate to acidic composition, comprising complexes of stocks, sills, lacoliths, volcanic necks, and dike swarms. Also included are rift zones of known high heat flow.

Fruitland Formation Methane Resource

All existing data indicate that coal rank and gas content of Fruitland coalbeds in San Juan Basin increase from southwest to northeast. Part of this increase in coal rank and gas content can be explained by accompanying increase in depth of burial. However, the gas content in coals of the northern part of the basin seems substantially in excess of that which should be present for the burial depths to which they have been subjected. Some drill-cutting samples desorbed from the northern portion of the basin have contained gas in amounts greater than 500 cf/ton. Measured gas content of coal samples from the northern San Juan Basin ranges from two to eight times that in samples from similar coals found under identical conditions of depositional environment, age, depth of burial, and bed thickness in other coal-bearing basins in the intermontane west. In the following paragraphs, the relationship of coalbed gas content will be stressed relative to probable heat sources, including depth of burial as well as proximity to post-depositional igneous activity and geothermal areas with anomalously high heat flow.

The dominant northwest-southeast structural trend and general shape of the inner portion of the San Juan Basin are shown by the structural contour map on top of the Huerfanito Bentonite Bed (Fig. 3). This volcanic ash deposit in the Lewis Shale represents a time surface of deposition and grades in depth below the top of the Pictured Cliffs Sandstone—and base of the Fruitland coalbeds—from 150 ft in the southwest to 1,250 ft in the northeast portion of the basin (Fassett and Hinds, 1971). As shown on this map, the structural center of the present basin is in the northwest corner of Rio Arriba County, just south of the New Mexico-Colorado state line. Present depth of burial of Fruitland coals is shown in Figure 7, with maximum depth somewhat greater than 4,000 ft.

Table 7—Gas-sample analyses from University of New Mexico test wells.

Well Number (Ref Table 6)	Gas Sample Depth Interval	Gas Composition (in Percent)								Totals	Ratios			
		CH_4 (C_1)	C_2H_6 (C_2)	C_3H_8 (C_3)	i–C_4H_{10} (C_4)	n–C_4H_{10} (C_4)	N_2	O_2	CO_2		C_2/C_1	C_3/C_1	$C_4(i+n)/C_1$	$\frac{C_1}{\text{Total } C_{(1-4)}}$
2	1,140–1,150	81.27	8.55	0.18	<0.01	<0.01	6.5	6.5	0.40	96.9	0.105	0.00222	<0.0001	.903
3	1,370–1,390	89.16	1.07	0.02	<0.01	<0.01				90.23	0.0120	0.00022	<0.0001	.988
4	3,230–3,250	94.076	0.457	0.033	0.011	0.007	2.287	0.171	2.99	100.03	0.00486	0.00035	0.00019	.995
6	3,006–3,015	93.509	0.383	0.008	0.002	0.001	4.030	0.331	1.736	100.0	0.00410	0.00009	0.00003	.996
7	2,970–2,979	90.865	1.097	0.113	0.015	0.010	1.419	0.0	6.481	100.0	0.01207	0.00124	0.00028	.987
8	3,053–3,073	84.845	0.583	0.060	0.008	0.005	6.992	0.789	6.718	100.0	0.00687	0.00071	0.00015	.992
9	3,161–3,183	78.108	0.259	0.020	0.004	0.002	16.944	0.935	3.727	100.0	0.00332	0.00026	0.00008	.996
10	3,119–3,141	93.062	0.384	0.071	0.001	0.006	6.030	0.0	0.445	100.0	0.00413	0.00075	0.00008	.995
Averages (for available data)		88.1	1.60	0.06	0.007	0.005	6.28	0.37	3.21	99.63				

Table 8—Summary of gas desorption measurements for drill-hole samples collected by the USGS and CGS.

Site No.	Sample No.	USBM Sample Code No.	Location		Operator	Well Name	Grnd. Elevation- (ft)	Sample Depth (ft)	Sampling Date	Coalbed Thickness (ft)	Coalbed	Sample Weight (gms)	Gas Content				
			Description	County									Lost (cc)	Desorbed (cc)	Residual (cc)	Total cc/gm	Total cf/ton
USGS-1	1	206	12 Miles Southeast of Farmington	San Juan	Energy Reserves		5,765	1,475	12-16-76	11	Lower Fruitland	808	368	3,382	694	4.2	134
USGS-1	2	207	12 Miles Southeast of Farmington	San Juan	Energy Reserves		5,765	1,485	12-16-76	11	Lower Fruitland	964	208	1,849	1,653	3.8	123
															Average		129
USGS-2	1	498	SW 1/4, Sec. 23, T30N, R15W	San Juan	USGS Coal Branch	23-4, 1977	5,295	464–465	12-6-77	5	Upper Fruitland	404	40	1,525	0	3.9	124
USGS-2	2	499	SW 1/4, Sec. 23, T30N, R15W	San Juan	USGS Coal Branch	23-4, 1977	5,295	586–587	12-6-77	12	Lower Fruitland	734	47	1,755	0	2.5	79
USGS-2	3	354	SW 1/4, Sec. 23, T30N, R15W	San Juan	USGS Coal Branch	23-4, 1977	5,295	589	3-24-78		Lower Fruitland		Data Lost Leaking Canister		0.3		
															Average		102
USGS-3	1	496	NE 1/4, Sec. 25, T30N, R15W	San Juan	USGS Coal Branch	25-2, 1977	5,380	642	12-2-77	7	Upper Fruitland	1,022	62	2,068	0	2.0	65
USGS-3	2	497	NE 1/4, Sec. 25, T30N, R15W	San Juan	USGS Coal Branch	25-2, 1977	5,380	736	12-4-77	23	Lower Fruitland	908	58	1,679	0	1.9	61
															Average		63
USGS-4	1	674	SW 1/4, NE 1/4, Sec. 7, T21N, R8W	San Juan	USGS Coal Branch	No. 6, 1978	6,635	295	6-26-78		Upper Fruitland		130	275	0	0.5	16
USGS-4	2	675	SW 1/4, NE 1/4, Sec. 7, T21N, R8W	San Juan	USGS Coal Branch	No. 6, 1978	6,635	318	6-26-78		Upper Fruitland		95	237	0	0.3	10
															Average		13
USGS-5	1	676	NE 1/4, SW 1/4, Sec. 27, T22N, R9W	San Juan	USGS Coal Branch	No. 8, 1978	6,500	280	7-5-78		Upper Fruitland		0	32	0	0.1	3
USGS-6	1		Sec. 18, T28N, R14W	San Juan	Amer. Public Gas Assoc.			769			Upper Fruitland					2.4	77
USGS-6	2		Sec. 18, T28N, R14W	San Juan	Amer. Public Gas Assoc.			?			Upper Fruitland					0.4	13
USGS-6	3		Sec. 18, T28N, R14W	San Juan	Amer. Public Gas Assoc.			793			Upper Fruitland					0.4	13
USGS-6	4		Sec. 18, T28N, R14W	San Juan	Amer. Public Gas Assoc.			843			Lower Fruitland					2.1	67
USGS-6	5		Sec. 18, T28N, R14W	San Juan	Amer. Public Gas Assoc.			?			Lower Fruitland					1.8	58
USGS-6	6		Sec. 18, T28N, R14W	San Juan	Amer. Public Gas Assoc.			?			Lower Fruitland					1.5	48

continued

Table 8—(Continued)

Site No.	Sample No.	USBM Sample Code No.	Location		Operator	Well Name	Grnd. Elevation- (ft)	Sample Depth (ft)	Sampling Date	Coalbed Thickness (ft)	Coalbed	Sample Weight (gms)	Gas Content				
			Description	County									Lost (cc)	Desorbed (cc)	Residual (cc)	Total	
																cc/gm	cf/ton
USGS-6	7		Sec. 18, T28N, R14W	San Juan	Amer. Public Gas Assoc.			849			Lower Fruitland					0.3	10
USGS-6	8		Sec. 18, T28N, R14W	San Juan	Amer. Public Gas Assoc.			?			Lower Fruitland					0.6	19
USGS-6	9		Sec. 18, T28N, R14W	San Juan	Amer. Public Gas Assoc.			853			Lower Fruitland					2.3	74
USGS-6	10		Sec. 18, T28N, R14W	San Juan	Amer. Public Gas Assoc.			882			Coalbed in Upper Pictured Cliffs					1.5	48
															Average		42
CGS-1	1	160	Durango Area, Colorado	La Plata			7,800+	295	2-3-76	9.0+	Menefee Formation	1,336	40	105	80	0.17	5
CGS-2	1	161	Durango Area, Colorado	La Plata			7,520	311	2-4-76	7.5	Menefee Formation	1,318	145	91	185	0.32	10

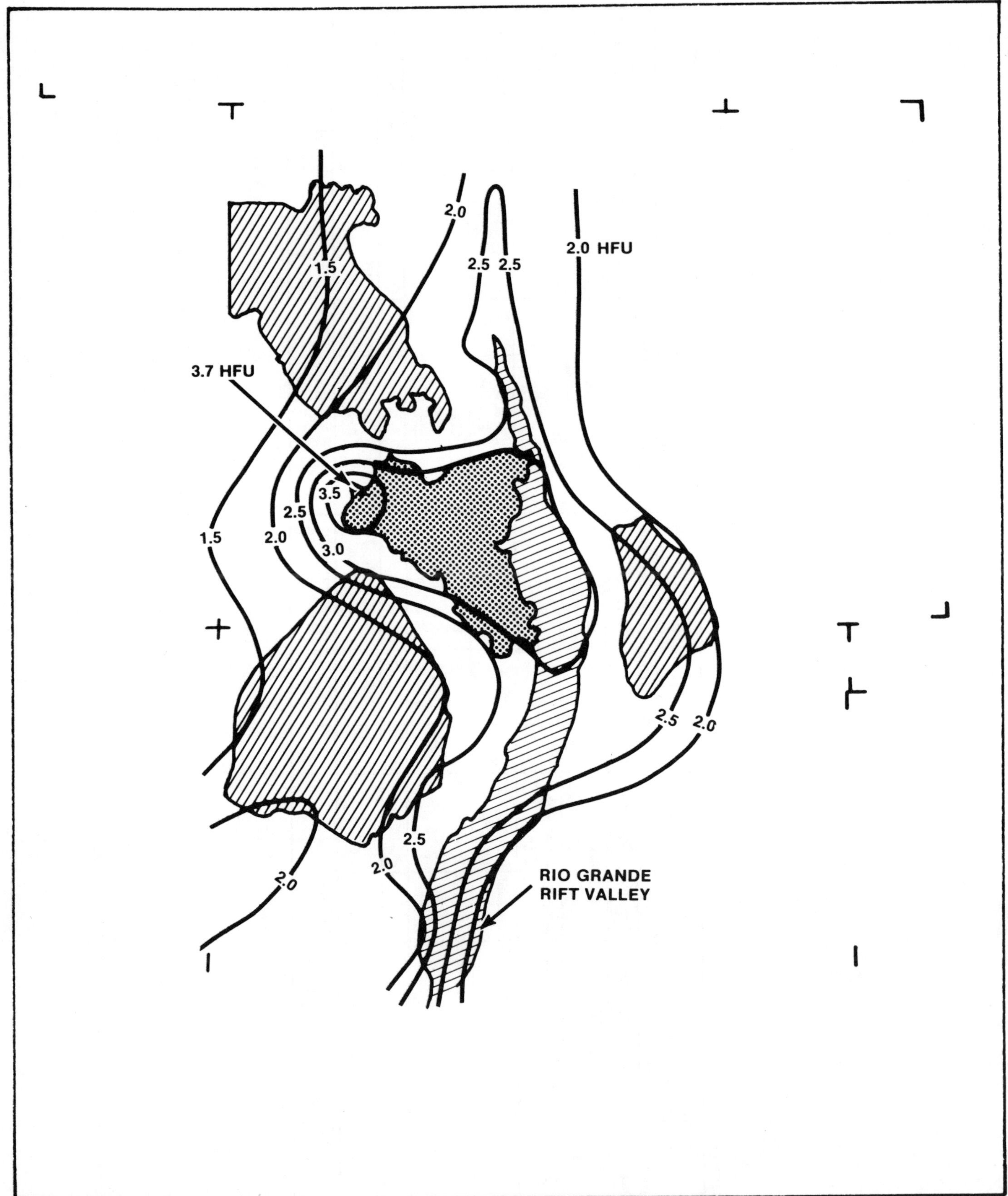

Figure 17—Coal-bearing basins and San Juan Mountains relative to the Rio Grande Rift Valley and generalized heat-flow lines. Choate and Rightmire, 1982.

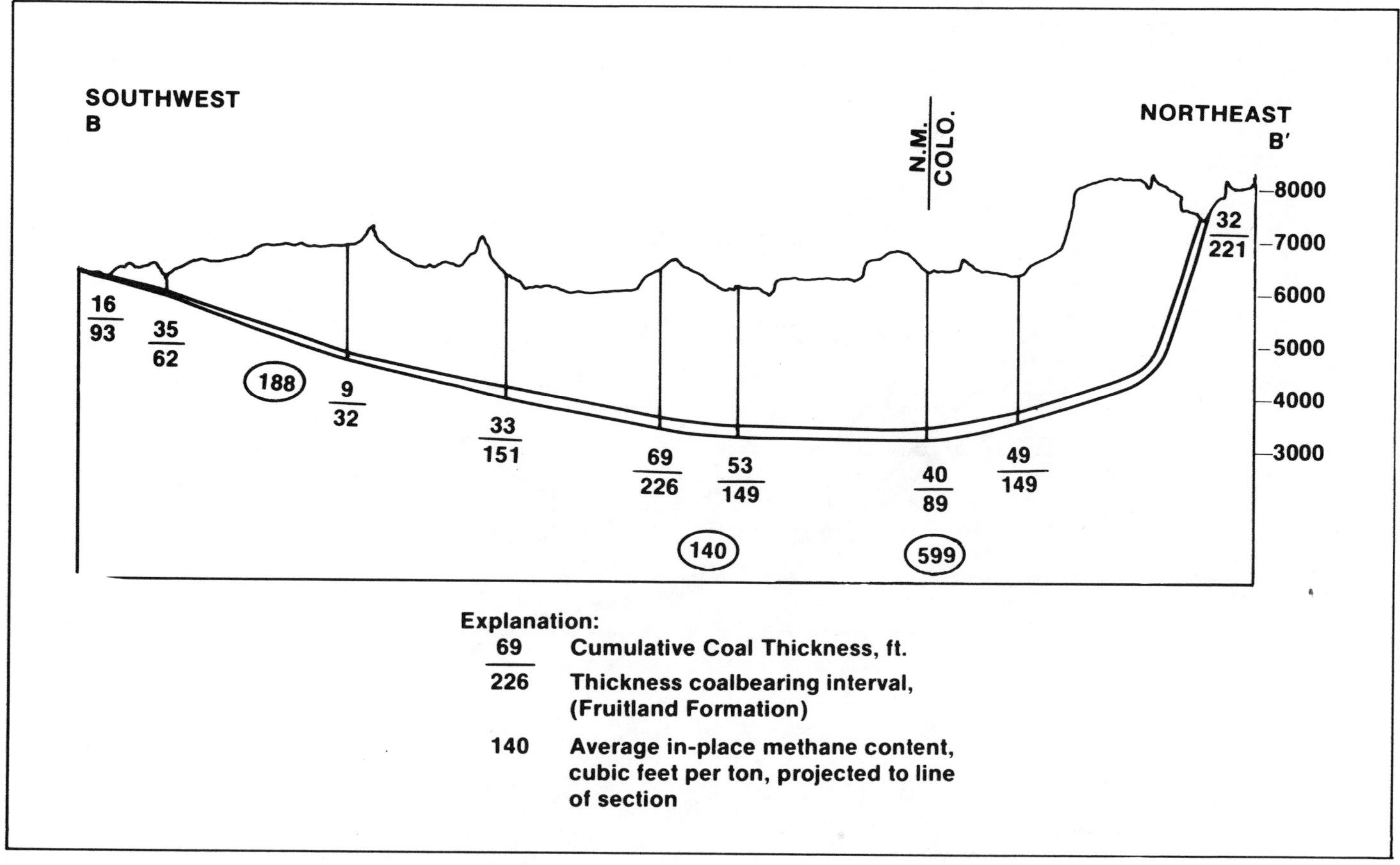

Figure 18—Southwest-northeast cross section B-B′; San Juan Basin, showing cumulative coal thickness, coal-bearing interval thickness, and average methane content.

The area within the central basin, encircled by the Hogback Monocline and underlain by Fruitland coalbeds, comprises approximately 6,250 sq mi and contains 201 billion tons of coal in beds greater than 2 ft thick (Fassett and Hinds, 1971). The thickest individual bed is approximately 40 ft, with maximum total coal thickness somewhat greater than 70 ft (Fig. 8). The axis of maximum coalbed deposition (as indicated in Fig. 8) is somewhat south of the basin's structural axis (as shown in Fig. 3). This isopach map (Fig. 8) clearly indicates the northwest-southeast trend of the ancient beaches behind which the Fruitland coal swamps developed. A southwest-northeast section across the basin, perpendicular to the old beach strandlines (Fig. 18; located in Fig. 8), also shows the increase in elevation in the northern portion of the basin, as the San Juan Mountains are approached from the south.

In Figure 19, the axes shown on the previous six figures, when plotted on one map, indicate that some heat source (other than that associated with burial depth) must be contributing to the high coal ranks and methane concentrations in the northern portion of San Juan Basin. Most of the axes shown have relatively comparable northwest-southeast trends. The depositional axes as represented by the axes of maximum total coal thickness and maximum thickness of an individual coalbed are nearly coincident with each other and are farthest to the southwest. Next to the northeast, by approximately 6 mi, is the axis of maximum overburden; and further to the northeast is the basin's somewhat curvilinear synclinal axis. Of substantial significance, however, is the location of the axes of maximum Btu values and fixed-carbon percentages. Both are shifted substantially northeast of any other axes representing depositional and structural controls. These two axes approximately parallel the heat-flow lines, swinging around the southwestern side of the San Juan Mountains, as shown in Figure 17. It is stressed that only a few data points are available on subsurface coal samples in the Colorado portion of the basin; therefore, the axes shown for Btu and fixed carbon are as yet only approximately located.

Recent unpublished vitrinite reflectance data of the USGS also indicate that the highest rank coals are in the very northern portion of the basin (Rice, 1982, personal communication). Approximate locations are shown in Figure 19 for the two samples with the highest reflectance values of any collected to date in the basin; a value of 1.45 (medium-volatile bituminous) for a Fruitland coal and 3.0 (anthracite) for a Dakota coal. Both samples—in the very northwestern corner of the basin, just southeast of Durango—are even several miles north of the axes shown for maximum Btu and fixed carbon in Figure 19.

The areas in the central portion of the basin, south of the

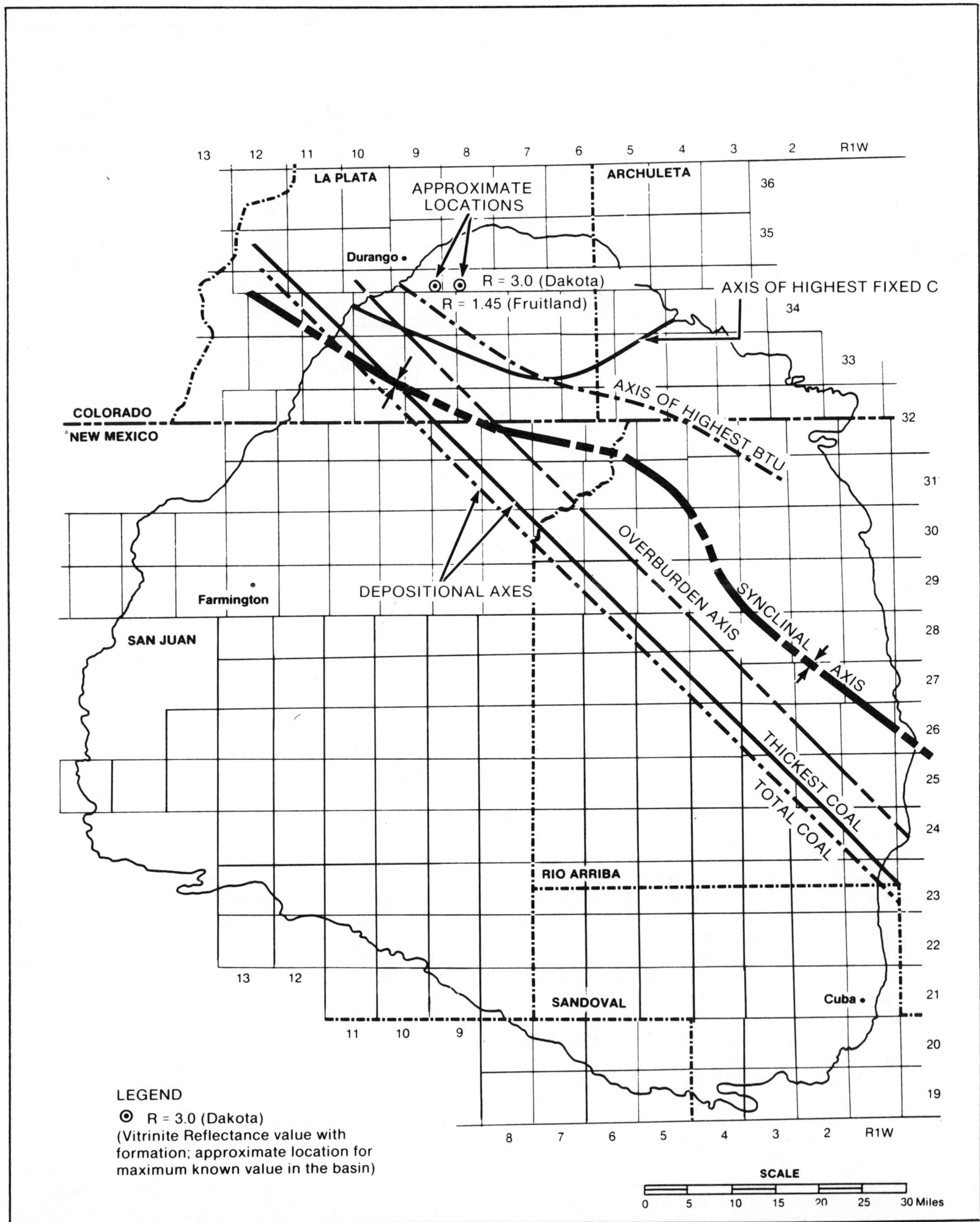

Figure 19—Composite map showing shift from the southwest to the northeast of axes of coal deposition, to maximum overburden, to basin synclinal axis, and then to axes indicating highest coal rank.

Colorado line, are well known to the drilling community as being extremely dangerous because of gas blowouts, with a number of rigs having burned while drilling through overpressured Fruitland Formation coalbeds.

Data on oil and gas well bottom-hole temperature are available for the shallow, southern half of the basin, where temperatures and heat flow are lowest. The wells in San Juan County, where temperature gradients exceed 50°C/km, have a southwest-northeast trend and are mostly coincident with areas producing gas from Fruitland Formation coal zones; i.e., the WAW, Gallegos, Gallegos South, Kutz, and Aztec fields. A second, parallel southwest-northeast–trending zone of similar temperature gradients occurs in northwest Sandoval and southwest Rio Arriba counties. This zone is coincident with an area of potentially high gas concentration as determined from coal sample gas-desorption measurements (as discussed in the following paragraph).

Average coal sample gas desorption data collected to date from the 28 wells (Fig. 16) under the MRCP are summarized in Figure 20. Figure 20 shows the average value for gas content in cf/ton of coal for all coal samples for each well; roof and floor rock samples are excluded. Values for all drill cutting samples have been adjusted to account for water and noncoal contamination. Tables 5, 6, and 7 show all values for gas content in any coal sample from each well. It is important to note maximum gas content values for two reasons: They represent a possible potential for gas concentration in an individual coalbed if the well is selectively completed; and the average value represents a potentially lower limit for the gas present in the beds sampled; i.e., it is representative of the gas contained in the sample while in the desorption canister and an allowance for gas "lost" during sample retrieval under normal drilling conditions.

The contour lines for 50 and 100 cf/ton are relatively close and include most of the basin. The 150-cf/ton average value contour line of Figure 20 was based on the average desorption values given as influenced by geologic relationships discussed earlier. The desorption data to date, as shown in Figure 20, indicate that Fruitland Formation coalbeds, away from the margins of the inner basin, should—almost everywhere—contain gas in concentrations greater than 100 cf/ton. Higher concentrations can be expected in a cresent-shaped arc along the eastern side of the basin, with possibly one small "eye" in the south where gas exceeds 500 cf/ton, and a much larger eye in the north, with gas concentrations also greater than 500 cf/ton.

Estimated Methane Resource

A township-by-township evaluation of gas concentration and total gas content in Fruitland Formation coalbeds was performed for the basin. A map is shown in Figure 21, indicating gas concentration contours in Bcf of gas per sq mi of area contained in Fruitland coalbeds greater than 2 ft thick. Gas concentration contours are shown for 1, 2, and 5 Bcf of gas per sq mi. These data were used in determining total gas contained solely within Fruitland Formation coalbeds greater than 2 ft thick; this value is 31 Tcf.

A target area of high Fruitland coalbed methane potential in San Juan Basin is defined here, comprising 1,900 sq mi, lying within the contour line representing 5 Bcf of gas per sq mi on Figure 21. Though this area seems large as a target, it must be remembered that the entire area is underlain by the coal-bearing Fruitland Formation, which acts as a blanket feeder for methane production; that the depths are relatively shallow with all of the target beds shallower than 4,500 ft and most shallower than 3,500 ft; and that, on the average, gas-in-place within the coalbeds themselves ranges from a minimum of 8 MMcf to over 65 MMcf of gas per acre.

CONCLUSIONS

Understanding coalbed methane production in San Juan Basin may well be the key to understanding coalbed methane production throughout the western United States. The basin exhibits all of the geologic criteria suggesting high coalbed methane production potential and contains a number of wells producing gas from the coalbeds and adjacent reservoir rock.

Methane from coalbeds has been produced in this basin since at least the mid-1950s. Wells completed in the coal-bearing Fruitland Formation during that time (particularly the Phillips, San Juan Unit 32-7 # 6-17) have individually produced up to 1 Bcf and are still producing gas from the coalbeds or from intervals containing significant coal resources.

Amoco Production has recently completed the Cahn # 1 well which, in March 1982, produced approximately 1 MMcfd from a 17 ft thick coalbed at about 2,800 ft.

Coal samples collected in conjunction with the MRCP program have yielded gas contents ranging from about 10 to more than 700 cf/ton. The higher gas contents exceed those expected for the time-depth of burial conditions of the areas from which they were collected. The higher gas content samples from the northern part of the basin are almost certainly influenced not only by their depth of burial but also by an abnormally high geothermal gradient induced by the San Juan Mountain volcanic complex and the Rio Grande Rift.

High gas contents and potential productivity are related to a number of geologic parameters, including:

- Depth of burial
- Increase in rank
- Coal thickness and distribution.

Increase in rank is due collectively to depth of burial and regional and local heating. All of these parameters influence generation of methane from coals in northern San Juan Basin.

Cumulative coal thicknesses in excess of 70 ft have been measured in drill holes along the depositional axis in the Fruitland Formation. This area also partly coincides with the higher rank, medium-volatile bituminous Fruitland coals formed under the influence of the San Juan Mountains' heat source. The combination of cumulative coal thickness, depositional trend, and high rank indicates that the areas nearest the anomalous heat sources may be the best potential target areas.

A detailed evaluation of gas concentration and total gas content within Fruitland Formation coalbeds greater than 2 ft thick indicates the presence of more than 31 Tcf in the coalbeds of that formation alone. A 1,900-sq-mi area defined by gas contents in excess of 5 Bcf per sq mi appears to have the highest potential for methane from coalbeds production. It must be noted that this target area has been defined exclusively on the

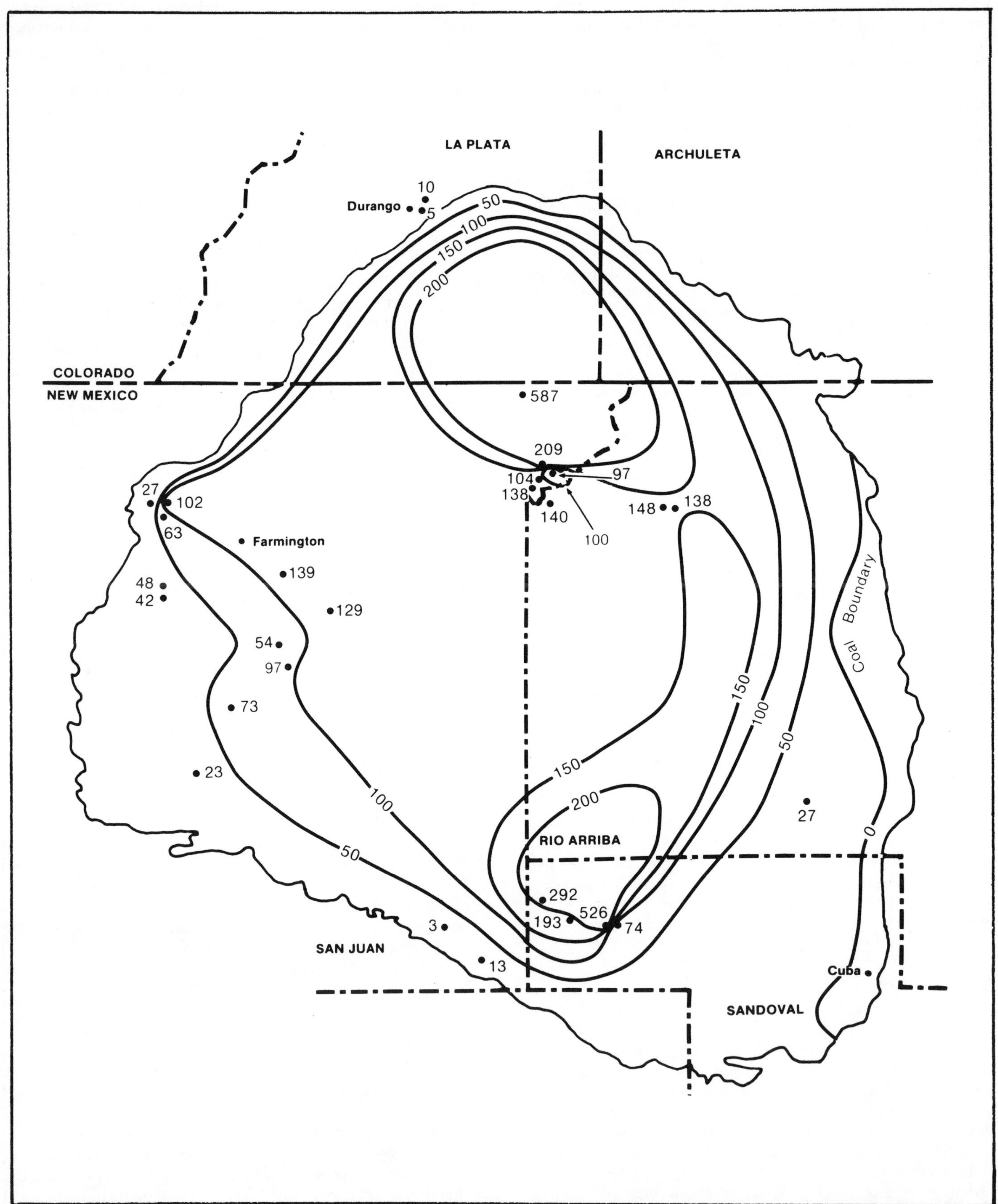

Figure 20—Contour map of averaged values per well of total gas in Fruitland Formation coalbeds; values in cf/ton. Contours drawn to reflect stratigraphy, structure, and coal rank illustrated in previous figures.

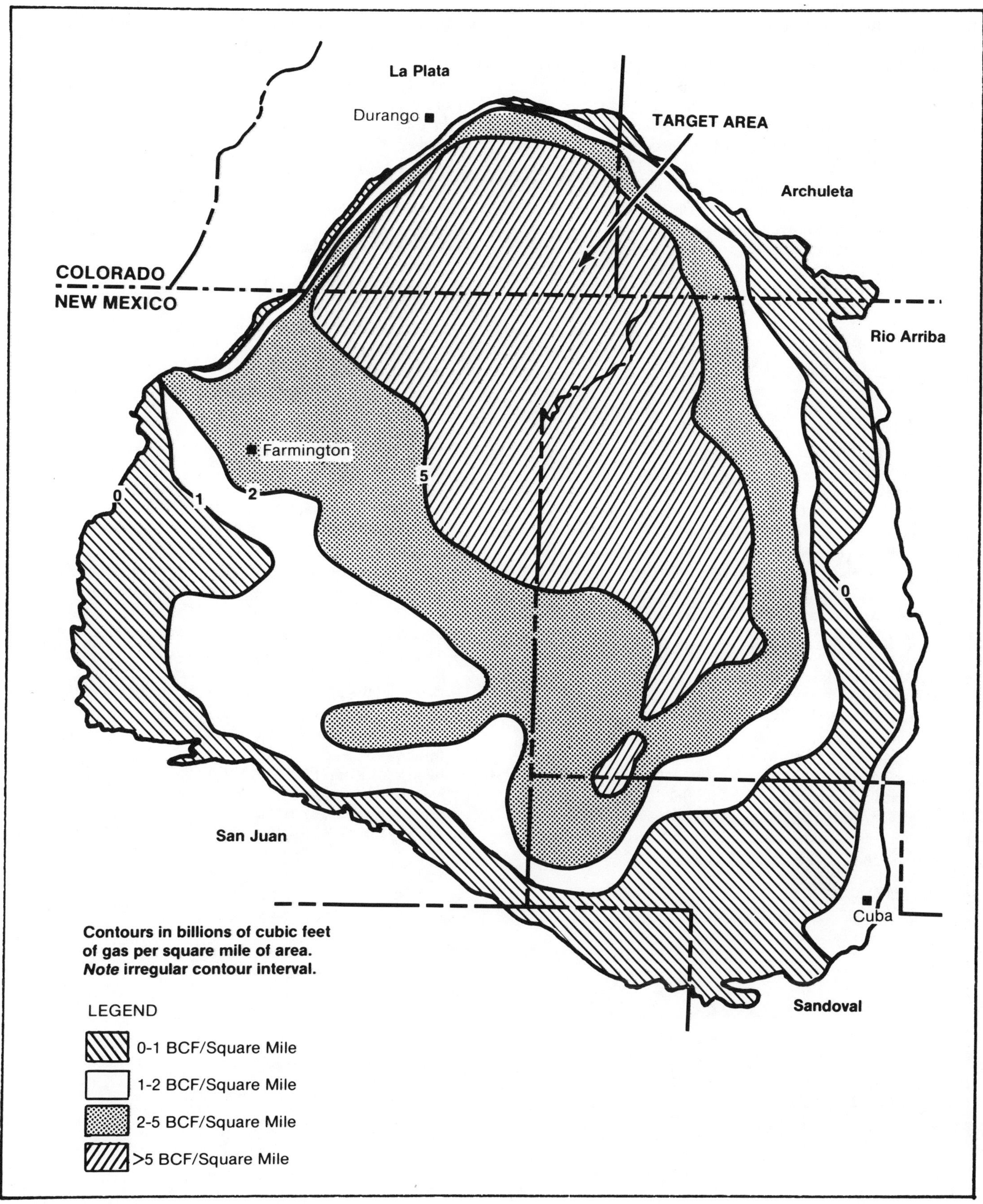

Figure 21—San Juan Basin coalbed methane target area, contoured in Bcf per sq mi.

basis of the Fruitland Formation and does not take into account the methane from coalbeds known to be present in the Mesaverde, Menefee and Crevasse Canyon Formations and the Gallup and Dakota Sandstones. While a preliminary estimate of the Menefee Formation coal resource has been attempted, insufficient coal rank and gas content data are currently available to assess the potential coalbed methane resource associated with this formation. With the exception of two near-outcrop Menefee samples collected near Durango, no gas content data are available for other than the Fruitland Formation. No basinwide coal resource estimates have been made for other coal-bearing formations; therefore, no coalbed methane resource estimates are possible. It is suggested, however, that this additional resource will be small relative to that shown in the Fruitland Formation.

REFERENCES CITED

Choate, R., and C. A. Johnson, 1980, Geologic overview, coal and coalbed methane resources of the western Washington coal region: TRW Energy Engineering Division, Lakewood, Colorado, for U.S. Department of Energy, Morgantown Energy Technology Center, Morgantown, West Virginia.

——, D. Jurich, and G. J. Saulnier, Jr., 1981, Geologic overview, coal deposits, and the potential for methane recovery from coalbeds, Piceance Basin, Colorado: TRW Energy Engineering Division, McLean, Virginia, and Lakewood, Colorado, for U.S. Department of Energy, Morgantown Energy Technology Center, Morgantown, West Virginia.

——, J. Lent, and C. T. Rightmire, 1982, San Juan Basin Report—Upper Cretaceous geology, coal and the potential for methane recovery from coalbeds in Colorado and New Mexico (revision 1): TRW Energy Engineering Division, McLean, Virginia, and Lakewood, Colorado, for U.S. Department of Energy, Morgantown Energy Technology Center, Morgantown, West Virginia.

——, and C. T. Rightmire, 1982, Influence of the San Juan Mountain geothermal anomaly and other Tertiary igneous events on the coalbed methane potential in the Piceance, San Juan, and Raton Basins, Colorado and New Mexico; Proceedings of the Unconventional Gas Recovery Symposium, May 16–18, 1982, Pittsburgh, Pennsylvania: Society of Petroleum Engineers/U.S. Department of Energy, 10805, p. 151–164.

Fassett, J. E., and J. S. Hinds, 1971, Geology and fuel resources of the Fruitland Formation and Kirtland Shale of the San Juan Basin, New Mexico and Colorado: U.S. Geological Survey Professional Paper 676, 76 p.

——, N. D. Thomaidis, M. L. Mathews, and R. A. Ullrich, (eds.), 1978, Oil and gas fields of the Four Corners area, v. I and II: Durango, Colorado, Four Corners Geological Society, 727 p.

Hale, W. E., L. J. Reiland, and J. P. Beverage, 1965, Characteristics of the water supply in New Mexico: New Mexico State Engineer, Technical Report 31.

Kelley, V. C., 1951, Tectonics of the San Juan Basin: Second Field Conference Guidebook, New Mexico Geological Society, p. 124–131.

Kelso, B. S., S. M. Goolsby, and C. M. Tremain, 1980, Deep coal bed methane potential of the San Juan River coal region, southwestern Colorado: Colorado Geological Survey Open-File Report 80-2, 56 p.

Molenaar, C. M., 1977, Stratigraphy and depositional history of Upper Cretaceous rocks of the San Juan Basin area, New Mexico and Colorado, with a note on economic resources, *in* San Juan Basin, v. III: New Mexico Geological Society, p. 159–166.

Peterson, J. A., A. J. Loleit, C. W. Spencer, and R. A. Ullrich, 1965, Sedimentary history and economic geology of San Juan Basin: Bulletin of the American Association of Petroleum Geologists, v. 49, n. 11, p. 2076–2119.

Shomaker, J. W., and M.R. Whyte, 1977, Geologic appraisal of deep coals, San Juan Basin, New Mexico: New Mexico Bureau of Mines and Mineral Resources Circular 155.

——, E. C. Beaumont, and F. E. Kottlowski, 1971, Strippable low-sulfur coal resources of the San Juan Basin in New Mexico and Colorado: New Mexico Bureau of Mines and Mineral Resources Memoir 25.

Williams, F. L., and D. Smith, 1981, Methane recovery from New Mexico coals: University of New Mexico Bureau of Engineering Research, Final Report No. NE-86(81) TRW-915-1, January, 1981, TRW contract H16217JJ0S, 229 p.

Geologic Overview, Coal Deposits, and Potential for Methane Recovery from Coalbeds, Piceance Basin — Colorado

R. Choate
D. Jurich
G. J. Saulnier, Jr.

Piceance Basin in northwestern Colorado has one of the greatest potentials for near-term production of coalbed methane of any basin in the western United States. Coalbeds of the Late Cretaceous Mesaverde Group underlie 6,570 sq mi of the total basin area of 6,680 sq mi. Preliminary estimates based on interpretation of geophysical logs of oil and gas wells indicate that the average value for total coal thickness throughout the basin is approximately 50 ft and that the total coal resource to depths exceeding 10,000 ft is 380 billion tons. Methane contained within the coalbeds comprising this resource is estimated to average 60 trillion cubic feet (Tcf). Lower and upper limits for coalbed methane content in the basin are estimated at 30 and 110 Tcf, respectively. The area with the highest potential for coalbed methane production is found in southeastern Piceance Basin.

INTRODUCTION

The Piceance Basin is in northwestern Colorado in the extreme northeast corner of the Colorado Plateau physiographic province. The basin is bounded by the Axial Basin Uplift and Uinta Mountains to the north; Grand Hogback, White River Uplift, and Sawatch Range to the east; Gunnison Uplift to the south; and Douglas Creek Arch and Uncompahgre Uplift to the west. The basin includes parts of Moffat, Rio Blanco, Garfield, Mesa, Delta, Pitkin, and Gunnison counties and has an areal extent of roughly 7,225 sq mi (Collins, 1976).

The Piceance Basin is an elongate, northwest-trending basin strongly asymmetrical to the northeast (Fig. 1). Although it has a well-defined structure in the subsurface, the basin form is not as evident from the surface. Douglas Creek Arch separates Piceance Basin from the larger Uinta Basin of Utah (Murray and Haun, 1974). Igneous intrusive and extrusive activity in the Somerset-Crested Butte area has modified the structure and surface expression in the southeastern part of the basin.

The surface expression of the basin area is outlined by the upturned resistant sandstones of the Mesaverde Group, which form a distinctive topographic change from the underlying, relatively soft Mancos Shale. The basin measures approximately 150 mi along its axis. Its width diminishes northwest to southeast, from 90 to 47 mi wide from Grand Junction to the Grand Hogback, to less than 20 mi wide at the southeast part. The average topographic relief in the basin is about 7,500 ft, from less than 4,500 ft along the Colorado River to just under 12,000 ft in the West Elk Mountains. Most of the land surface lies 5,000 to 8,000 ft above sea level.

GEOLOGY

Stratigraphy

The coal-bearing formations of Piceance Basin are primarily continental and mixed continental-marine sediments of Cretaceous age. At the end of the Jurassic, highlands emerged to the west of Piceance Basin, and low-lying marshy environments formed that became highly developed in the Cretaceous. The Morrison Formation—a fossiliferous multicolored sequence of shale, mudstone, and sandstone—was deposited at the end of the Jurassic. The Morrison was eroded in the Early to Middle Cretaceous, when the developing Rocky Mountain geosyncline signaled a major change in depositional environments.

Throughout Early Cretaceous, a continental environment prevailed. Shales, sandstones, and some thin coals were deposited. Deposition of reworked older sediments and organic material continued until the mid-Cretaceous, when a seaway representing the Rocky Mountain Geosyncline advanced through the Piceance Basin area from the north. Thousands of

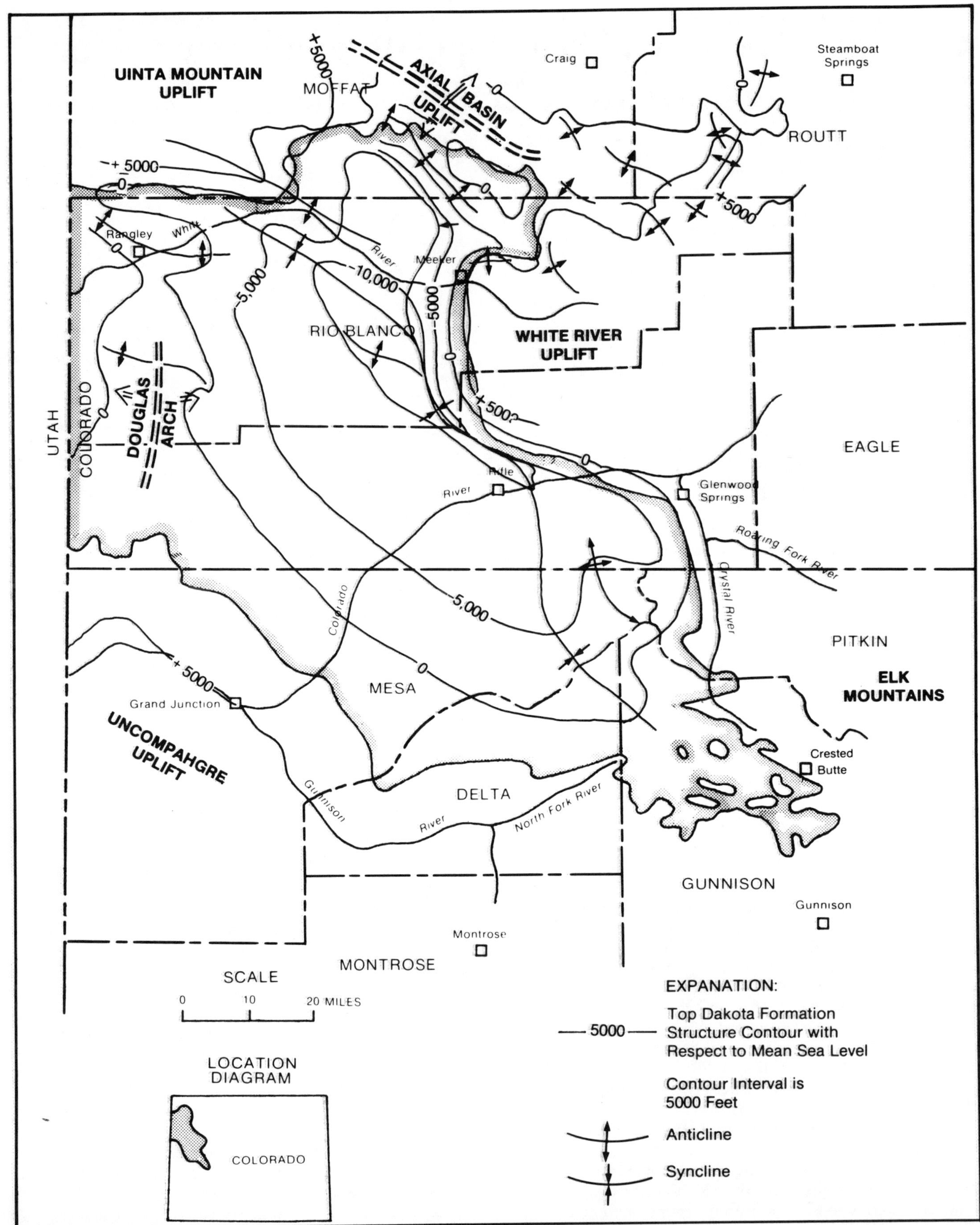

Figure 1—Major structural elements, Piceance Basin and contiguous areas, northwest Colorado. Structural data from Dunn, 1974.

feet of marine and some nonmarine sediments accumulated during the Middle and Late Cretaceous. The total sequence of all Cretaceous formations forms the thickest time period sequence in northwest Colorado (Murray and Haun, 1974).

The Wasatch uplift caused the shoreline to regress southeastward in the Late Cretaceous. With the retreat of the sea, a vast tidal and lowland marsh environment developed, and several thousand feet of clastic and organic sediment were deposited. Coals, sandstones, and some marine and nonmarine shales testify to the fluctuating nature of the environment as the sea level transgressed and regressed. Toward the end of the Cretaceous, continental conditions were dominant in Piceance Basin. Sulfur content is used to determine the marine from nonmarine coals, because high sulfur values are associated with salt marsh and brackish tidal marsh deposits. The majority of coals in the Mesaverde Group (those above the Rollins Sandstone) are considered to be fresh water in origin (Collins, 1976).

The two most plausible explanations for the sandstone/coal/shale associations of the Mesaverde Group are the deltaic complex argument (Collins, 1976) and the barrier beach-coastal plain argument (Young, 1966). Collins (1976) presents strong evidence to support the model that the Mesaverde coals were deposited in a widespread prograding deltaic complex at the western edge of the epicontinental sea in the Rocky Mountain geosyncline. Sea level changes would occur in response to variations in sedimentation rates, compaction, and deltaic growth faulting. Cyclic sedimentation is presumed to result from changes in sea level and the delta distributary system and the development of barrier beaches.

The formation terminology of the Mesaverde Group is complicated, and the principal stratigraphic units possess different names in different locations. The term Mesaverde Group is the most widely accepted and includes all the major coal-bearing units of Late Cretaceous age. Many formation names are restricted to specific coal fields, and no unified nomenclature exists for the entire basin. Figure 2 summarizes some of the terminology used in the basin. The following section summarizes the lithologies of the Cretaceous formations.

Dakota Sandstone (100 to 200 ft)

The Dakota Sandstone contains yellowish brown and gray well-sorted, quartzitic and conglomeratic sandstones with interbedded carbonaceous shales. Lenses of thin coal may be locally present in the upper unit. The sandstones weather rust brown and form ridges.

Mancos Shale (4,000 to 6,000 ft)

The Mancos is a medium- to dark-gray to black fossiliferous marine shale with local interbeds of siltstone, sandstone, and limestone. The shales may be calcareous. The upper Mancos interfingers with the overlying Mesaverde Group.

Mesaverde Group (2,300 to 6,500 ft)

The Mesaverde Group includes the Blackhawk, Prince River, Hunter Canyon, Mount Garfield, Williams Fork, and Iles Formations. The central portion of the group contains the major coalbeds of northwestern Colorado. The sequence contains continental and marine rocks consisting of siltstones, sandstones, shales, and coalbeds. Many thick sandstone members are ridge formers including the Castlegate, Cozette, Fox Canyon, Meeker, Morapos, Sego, Tow Creek, Trout Creek-Rollins, and Twenty Mile Sandstones. The upper unit is shaly and capped by sandstones and the Ohio Creek Conglomerate (Johnson and May, 1978). Major coalbeds are discontinuous and have not been correlated across the basin.

At the end of the Cretaceous, the Laramide orogeny brought a disruption of the depositional conditions that produced the Mesaverde coals. The shoreline regressed to the southeast, the Colorado Plateau began to break up the epicontinental sea into a series of small basins, and continental deposits of the Wasatch and Fort Union Formations began to accumulate. While deep basins developed in the central Rockies in early Cenozoic time, some coal-producing environments persisted during Wasatch depositional as evidenced by thin coals encountered in deep drill holes in Piceance Creek Basin (Murray, 1974).

Miocene through Pliocene time in Piceance Basin is represented by extrusive and intrusive igneous rocks. Some of the extrusive sequences also contain sands, gravels, and water-laid tuffs associated with volcanic activity. Because the authors believe igneous activity had a strong influence on the gas content of Piceance Basin coals, a more detailed account of igneous activity is presented. This material, unless otherwise noted, originates from Collins (1976).

A large number of laccoliths occur in the southeast portion of Piceance Basin. These intrusives cut and alter coal-bearing rocks. The laccoliths, when exposed at the surface, range in composition from granodiorite to quartz monzonite. Although the exact ages of the laccoliths are not known, they are observed to cut rocks of Eocene age. This series of intrusives caused local metamorphism of sedimentary rocks to hornfels and marble. These intrusive bodies are now seen in conspicuous topographic features like Mt. Sopris, Chair Mountain, Snowmass Peak, Capital Peak, and Haystack Mountain (Kent and Arndt, 1980b). Sills of similar composition occur in the Coal Basin and Marble areas.

Basaltic to dacitic vertical dikes from 1 to 50 ft thick, striking from east-west to N 30° W, are found in the Coal Basin area. Radial dikes occur around West Elk Mountain and the Purple Mountain area. The dikes locally affect the strata they intrude and are affected by them, as well. In the Coal Basin area, the dikes are aphanitic below the Rollins Sandstone and porphyritic above it. Coal seams also have been altered by the dikes.

Extrusive basalt flows and some volcanic tuffs and volcano-sedimentary rocks also occur in the basin. The flows are olivine basalts that cap Grand and Battlement Mesas and are found on Sunlight Peak and the West Elk Mountains. These flows are Miocene-Pliocene in age (Murray and Haun, 1974) and do not appear to have affected the rocks or coals in the area.

Structure

Piceance Basin is a northwest-southeast–trending feature, asymmetric to the northeast with a slightly sinuous axis. The lowest point in the basin is more than 13,000 ft deep in the Piceance Creek Basin, northwest of Rifle. The White River Uplift, which causes the S-shape of the basin axis, exposes

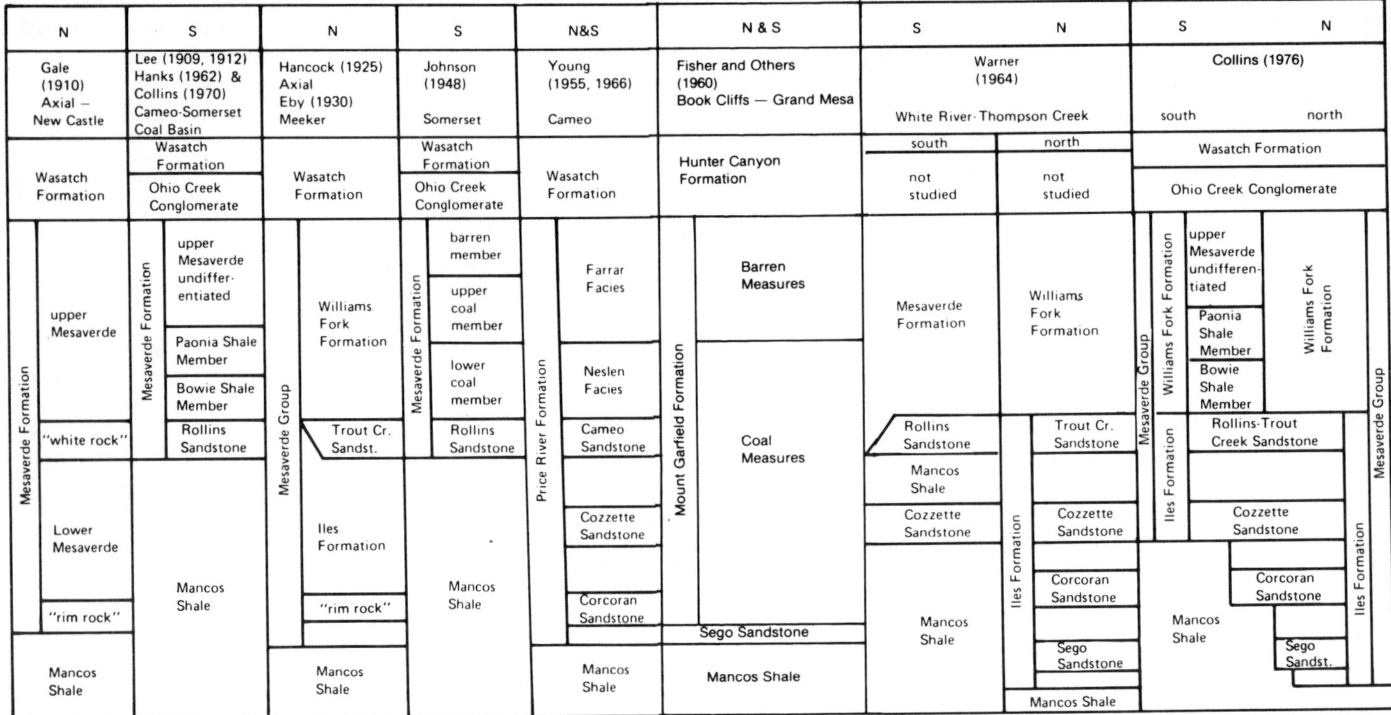

Figure 2—Nomenclature used by various authors for Upper Cretaceous and Early Tertiary rocks in the Piceance Basin. Collins, 1976, reprinted from Colorado School of Mines Quarterly, v. 71, n. 1, by permission of the Colorado School of Mines Press, copyright © 1976; and from Schwochow, 1978, by permission of the Colorado Geological Survey.

Precambrian rocks above 10,000 ft, and the Uncompahgre Uplift contains Precambrian rocks above 8,000 ft, yielding a total structural relief between 26,000 and 28,000 ft. The basin is separated from the structurally similar Uinta Basin of Utah by the broad Douglas Creek Arch.

The Piceance Basin area was a broad platform through the early Paleozoic until the ancestral Eagle Basin was formed during the Pennsylvanian (Quigley, 1965). The rapid growth of the Uncompahgre Uplift and the Ancestral Front Range created a deep basin receiving a complex series of deltaic, fan, and evaporite deposits throughout the Jurassic. The basin was reactivated during the Late Cretaceous, when it was altered by the rising of the White River Uplift. Remnants of the Eagle Basin are found in a few small basins and the larger Piceance Basin. At this time, Piceance Basin was intermontane, receiving a large quantity of continental and lacustrine sediment. During early Pliocene, basalt flows covered what may have been a broad alluvial plain (Murray and Haun, 1974). This period of activity was followed by regional uplift that continues today and that has caused over 5,000 ft of downcutting since the Pliocene (Murray and Haun, 1974). Tertiary sediments are now being exhumed, but the process is slow because of resistant beds in the Green River Formation and basalt caps on the high mesas.

A cross section across the southeastern portion of the basin is shown in Figure 3, and an isopach map of overburden thickness above the Rollins is given in Figure 4.

Piceance Basin is not highly deformed. Faulting and folding occur mainly in the southeast portions, near intrusive activity along bends in the trend of the Grand Hogback and in the Piceance Creek Basin-Sulfur Creek area. High-angle normal faulting and shallow anticlines and synclines are the major expressions of deformation.

The major fold structure in Piceance Basin is the Grand Hogback Monocline, which extends along the eastern boundary of the basin. The flexure probably resulted from the formation of the White River Uplift. Dips along the monocline are variable but generally range between 30 and 70°, with a few instances of vertical to overturned bedding. The Grand Hogback forms a prominent topographic feature, the result of differential resistance to erosion of the sandstones and shales of the Mesaverde Group.

Shallow minor folds occur mainly in the northern and southern portions of the basin. The folds in the southern part of Piceance Basin result from the intrusion of laccoliths and stocks. Coal Basin and Divide Creek Anticlines are examples of this style of folding.

Faulting in Piceance Basin consists primarily of high-angle normal faults with displacements of tens to hundreds of feet. Most faulting is tensional, with fault planes oriented mainly northwest-southeast. Minor faulting is found along bends in the trend of the Grand Hogback associated with igneous intrusions in southeastern Piceance Basin. In a few instances, high-angle reverse faults and thrust faults are present. Collins (1976) points out that severely sheared coal is present in many deformed areas and that the coalbeds may have served as slippage surfaces during periods of folding. These bedding plane faults are difficult to detect, and displacements usually cannot be measured. The Elk Mountain thrust fault predates the laccolithic activity and dike swarms and may have resulted from earlier igneous activity (Godwin and Gaskill, 1964).

Gaskill et al (1973) identify a volcanic center north of

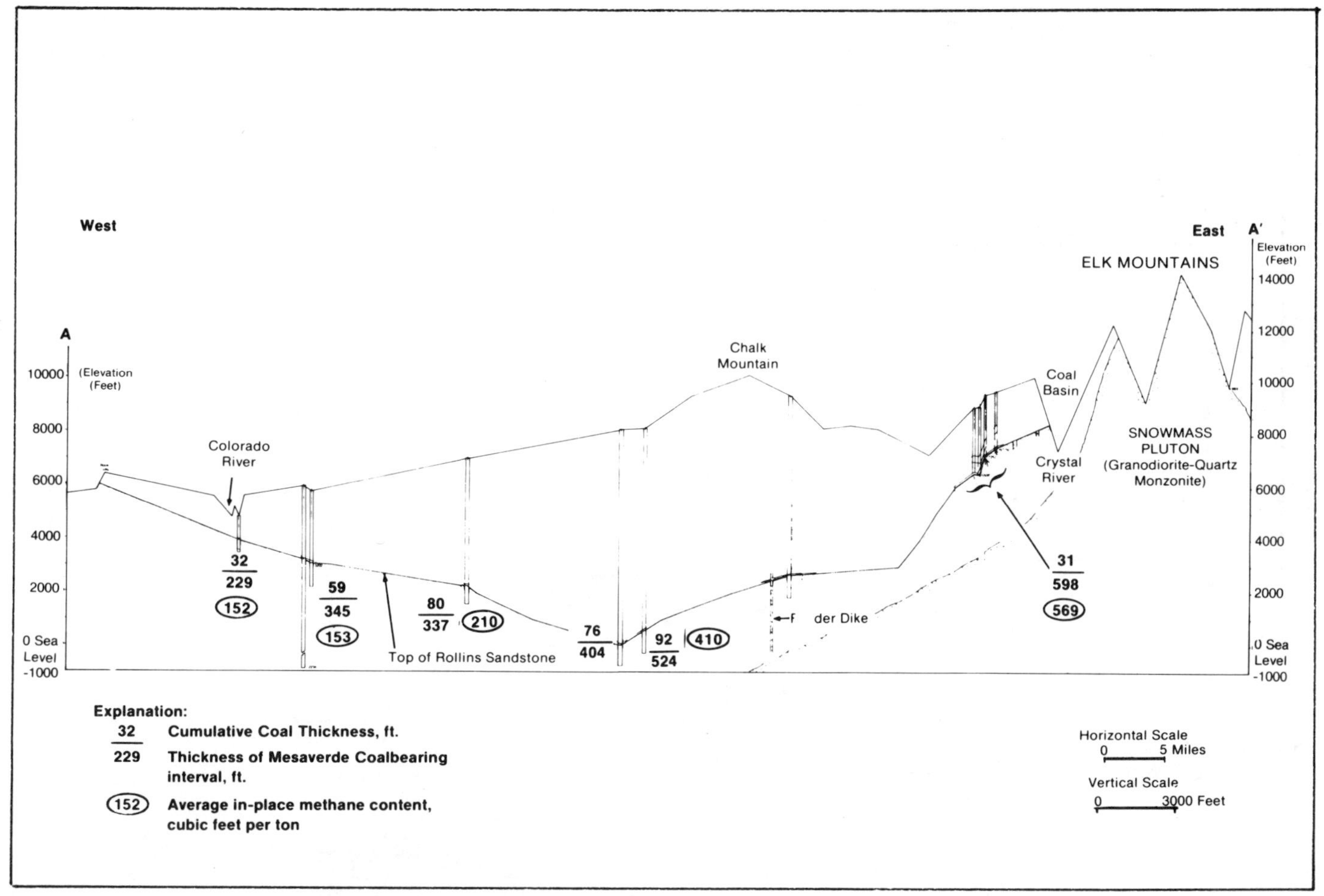

Figure 3—West-east cross section A-A′, Piceance Basin, showing cumulative coal thickness, coal-bearing interval thickness, and average methane content.

Gunnison in the West Elk Mountains. A map of the area by Godwin and Gaskill (1964) (Fig. 5) shows a complex series of dikes, sills, and laccoliths that cause an arching of horizontal strata. Additional thicknesses of volcanics are believed responsible for the Elk Mountain thrust fault. Treasure Mountain Dome, which probably represents more recent plutonic activity (12.5 million years), further disturbed sedimentary strata in the area, especially around Marble. The more silicic composition of this intrusion indicates that possible explosive venting may have taken place in this area (Lipman et al, 1969). Upper Cretaceous rocks intruded by the epizonal intrusives of the West Elk Volcanic center are folded and faulted as a result of this igneous activity.

Gravity measurements indicate that large-scale batholiths lie to the north and south of the West Elk Mountains and are responsible for the more silicic extrusives in the San Juan Mountains and associated stocks to the northeast of the Elk Mountains. The shallow, more mafic intrusives of the West Elk Mountains probably represent a deeper, less differentiated source, although they are probably related to the large-scale magma zone responsible for most of the Early to Middle Tertiary volcanism of central Colorado (Steven, 1975).

COAL RESOURCES

Regional Setting

The depositional environment in Piceance Basin in the Late Cretaceous was an open seaway with a delta and deltaic plain complex prograding to the southeast. The shifting strandline resulted in a complex series of interfingering marine and nonmarine sediments that in turn resulted in persistent swamps where peat accumulated.

The major coalbeds were deposited behind the beach sands as the Cretaceous Sea retreated. Several less significant and highly erratic coalbeds were deposited in advance of the major back-beach coalbeds, when fluctuations in sea level caused the shoreline to advance temporarily up slope. In some areas, acidic waters leached from the coal swamps have bleached the sandstone white. These resistant white sandstone beds stand out as prominent caprocks and benches and indicate the presence of overlying coalbeds. The thickest coalbeds in the basin are located above the Rollins-Trout Creek or its equivalent sandstone, which marks the top of the Iles Formation of the Mesaverde Group (Fig. 2). Similar coals occur above the Cozzette, Corcoran, and Sego sandstones.

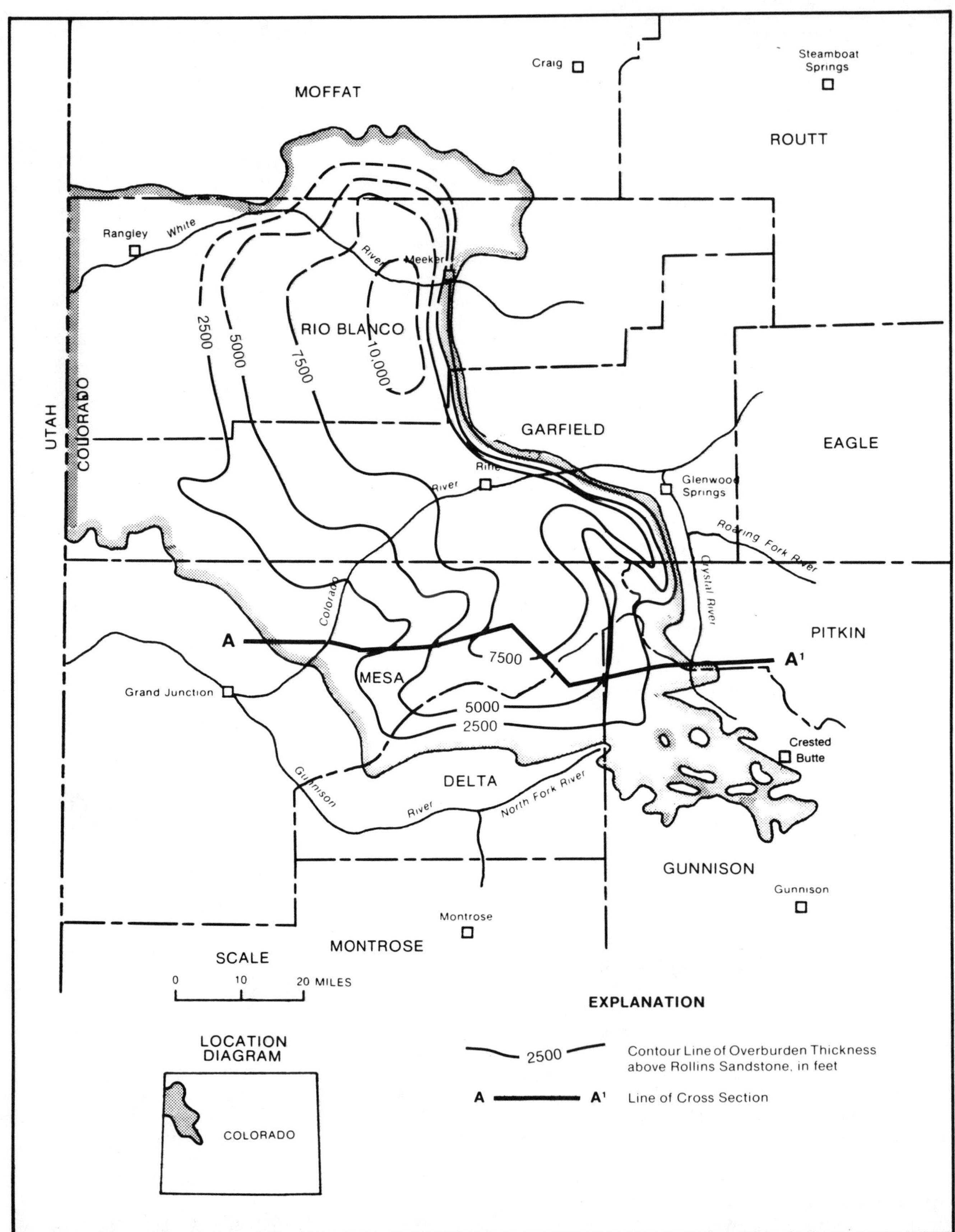

Figure 4—Simplified overburden thickness map.

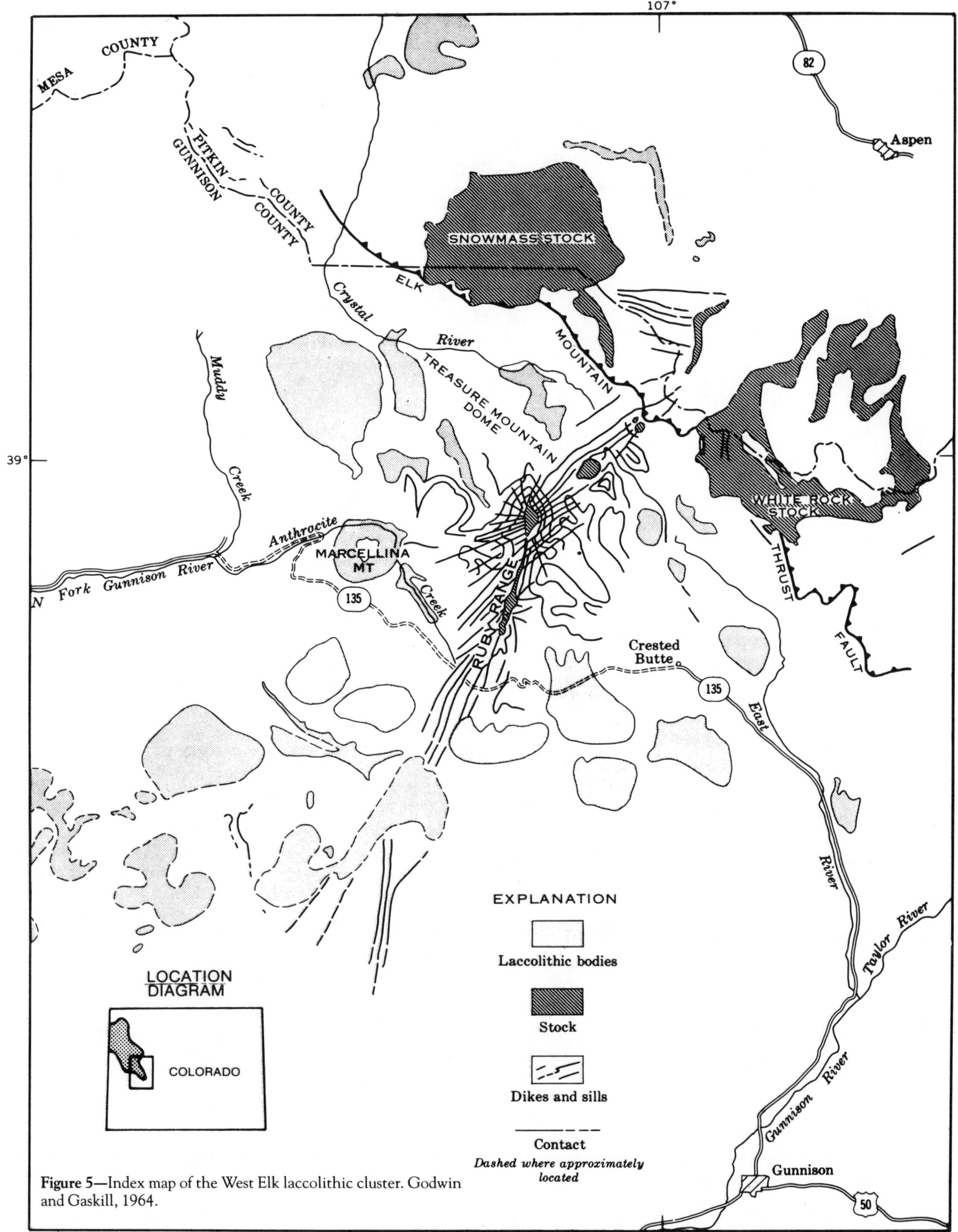

Figure 5—Index map of the West Elk laccolithic cluster. Godwin and Gaskill, 1964.

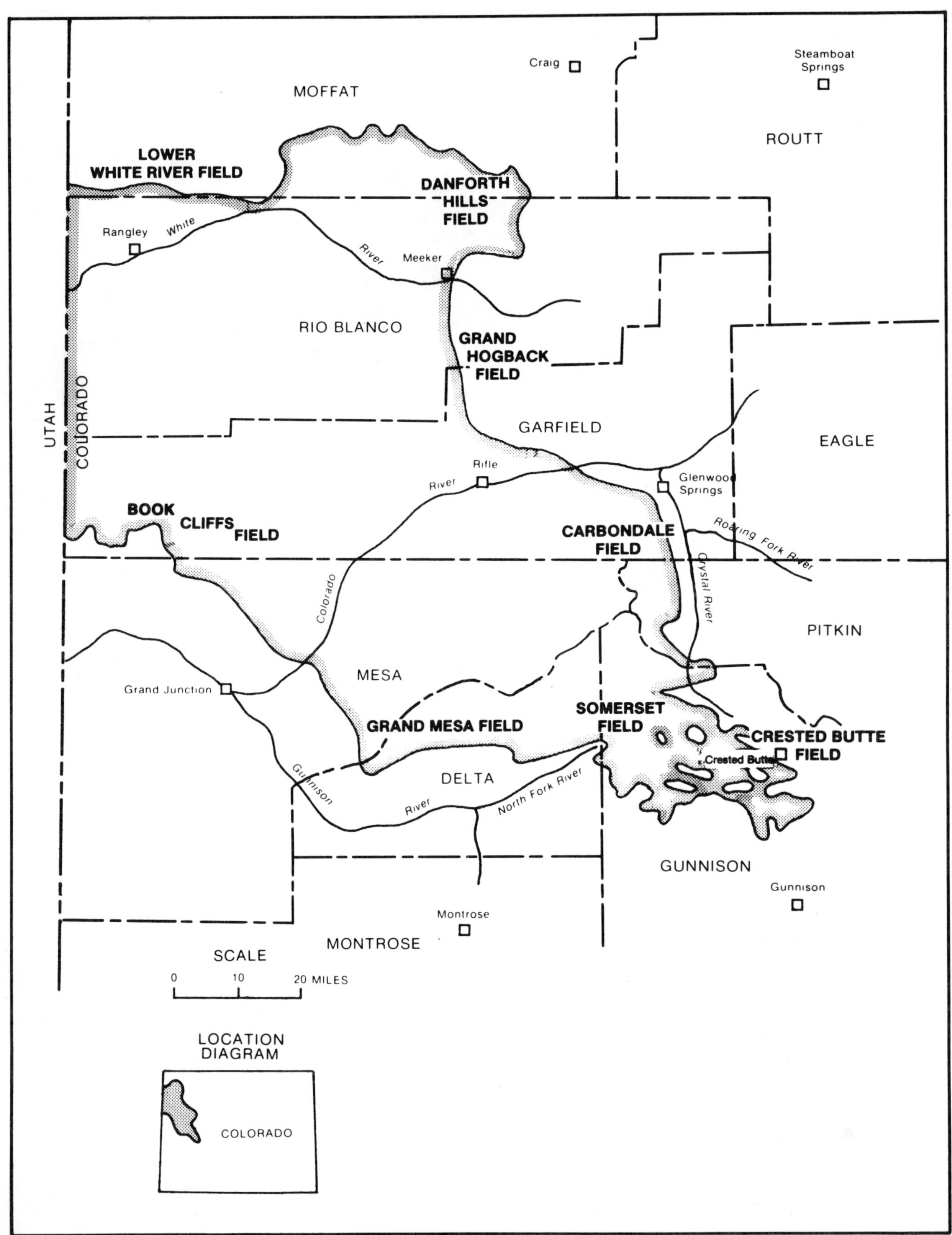

Figure 6—Coal fields of Piceance Basin.

The Iles and Williams Fork Formations, which comprise the Mesaverde Group in the northern and eastern portions of the basin, are the principal coal-bearing formations in Piceance Basin. In the southwest part of the basin, the coal-bearing Mount Garrison Formation is considered the lithologic equivalent of the coal-bearing part of the Mesaverde Group. The periphery of the basin, defined by a continuous outcrop of coal-bearing rocks, has been divided into eight coal fields. Clockwise from the northwest corner of the basin near Rangely, these fields are the Lower White River, Danforth Hills, Grand Hogback, Carbondale, Crested Butte, Somerset, Grand Mesa, and Book Cliffs (Fig. 6).

Coals range in rank from high-volatile C to medium-volatile bituminous over most of the basin, with a general increase from north to south. In the southeastern portion of the basin, abnormally high heat flow—resulting from Tertiary igneous intrusives and related movement—has metamorphosed the coal to semianthracite and anthracite. In some locations, the coal has been altered to graphite.

Hornbaker et al (1976) estimated the coal resource in the basin to a depth of 6,000 ft to be approximately 55 billion tons. Within this resource, significant coking coal reserves totaling 0.45 billion ton have been identified at depths less than 1,000 ft in southeastern Piceance Basin. Collins (1977) estimated the coal resource for beds over 42 inches thick for the entire basin to be approximately 124 billion tons. In this study, the authors—using geophysical logs and in some cases logs covering only a portion of the entire coal section—estimated the coal resource of the basin to be 382 billion tons. Over 75% of the coal-bearing rocks lie at depths greater than 3,000 ft in the interior of the basin. Coal mining has been concentrated on the basin periphery where the coalbeds crop out and are readily accessible. As of 1978, Piceance Basin had produced 87 million tons of coal—14% of the Colorado state total.

Lower White River Coal Field

The Lower White River coal field is along the White River in Moffat and Rio Blanco counties (Fig. 6). The field includes part of the continuous outcrop of the Mesaverde rocks that extends from the Danforth Hills coal field south and west along the White River to the Utah state line. The northeastern tip of the coal field is characterized by strata that dip from 40 to 70° to the south. South of this area, the coals strike north-south and dip from 11 to 16° to the east. The coal-bearing rocks form a narrow hogback on the northern side of the White River in the central portion of the field. Here, the beds have an average dip of 45° to the south. The western portion of the field is a broad shallow anticline where the coal-bearing rocks have been eroded from the center to expose the Mancos Shale.

Original in-place coal resources for the 930-sq-mi area to a depth of 6,000 ft have been estimated to be 11.8 billion short tons (Hornbaker et al, 1976). Coals in this field average high-volatile C bituminous in rank and range in thickness from 8 to 12 ft. Although the coals in this field probably are very similar in occurrence to those in the Danforth Hills coal field, no attempt has been made to correlate the coals of the two fields.

Danforth Hills Coal Field

The Danforth Hills coal field extends southward from the town of Axial to Meeker and is congruent with the mountainous area of the same name (Fig. 6). The Danforth Hills coal field lies in Rio Blanco and Moffat counties and is bounded by Cretaceous coal-bearing rocks that crop out north of the White River, south of the Axial basin, west of the White River Plateau, and east of Strawberry Creek and Citadel Plateau.

Original in-place coal resources, estimated to a depth of 6,000 ft in the approximately 400 sq mi of the coal field, total more than 10.5 billion short tons (Hornbaker et al, 1976). Most of the coal in the field is high-volatile C bituminous in rank.

All minable coalbeds are found in the Iles and William Fork Formations of the Mesaverde Group. Coalbed dips range from 3 to 80°, depending on local structure.

The lower part of the Iles Formation contains coalbeds that are thin and discontinuous, and are generally regarded as unminable. Significant coals occur near the top of the Iles Formation, below the Trout Creek Sandstone, within the Black Diamond coal group (Hancock, 1925) and consist of four to six coalbeds ranging from slightly less than 2 to 14 ft thick. These beds are difficult to correlate outside of the Meeker area. Black Diamond coals are exposed along the eastern and northern margins of the coal field and in the canyons of the Coal Butte area north of Meeker.

Hancock and Eby (1930) divided the Williams Fork Formation into three coal groups. From bottom to top, these are the Fairfield, Goff, and Lion Canyon. Collins (1976) further divided the Fairfield group of Hancock and Eby (1930) into three subgroups: Fairfield, South Canyon, and Coal Ridge.

The Fairfield coal group of the Williams Fork Formation contains six correlated coal zones, including the Major, Bloomfield, Fairfield, Fairfield No. 2, Wesson, and Agency. These zones have been identified primarily in the Meeker area, where up to 14 beds have been correlated. The beds range in thickness from less than 1 to 30 ft.

Coals from the Goff and Lion Canyon groups present in the Meeker area are exposed primarily along the Hogback and the Strawberry Creek escarpment. Hancock and Eby (1930) recorded 16 coals in the Goff group and 30 in the Lion Canyon group that range in thickness from less than 1 to 18 ft.

Grand Hogback Coal Field

The Grand Hogback coal field lies along the eastern margin of Piceance Basin in Garfield and Rio Blanco counties. The field is defined by that portion of the Grand Hogback that lies between the White River and Glenwood Springs (Fig. 6). Coal-bearing strata that crop out in the Grand Hogback coal field generally are continuous with the Danforth Hills coal field to the north and the Carbondale coal field to the south. Original in-place coal resources in the Grand Hogback coal field have been estimated at more than 3.1 billion short tons (Hornbaker et al, 1976). Coal rank ranges from high-volatile C to high-volatile A bituminous.

All the coals occur in the upper 200 to 300 ft of the Iles and the overlying Williams Fork Formation. The Black Diamond coal zone at the top of the Iles Formation, near the town of

Meeker, contains coals of minable thicknesses. These coals pinch out to the south and do not occur south of the town of New Castle (Collins, 1976).

The important coals of the Grand Hogback coal field are found in three principal zones in the Williams Fork Formation known as the Fairfield, South Canyon, and Coal Ridge coal groups. The lowermost coal zone, the Fairfield, is the most persistent, easily correlated, and economically important zone in this group. It progressively thickens to the north, reaching a maximum interval thickness at Rifle Gap, where nine coalbeds occur throughout the 470-ft interval between the Trout Creek (Rollins) Sandstone and the first major sandstone unit in the Williams Fork (Collins, 1976). The South Canyon coal group consists of one or two major beds: the Allen and the overlying Anderson. These coals average high-volatile C bituminous in rank. The Coal Ridge group coals are the most erratic of the three major groups and do not crop out at the exposed sections. The Upper Williams Fork Formation contains several thin, lenticular coalbeds. These coals, known as the Keystone group, occur in minable thicknesses in the northern and southern parts of the field.

Carbondale Coal Field

The Carbondale coal field is along the eastern margin of Piceance Basin in Garfield and Pitkin counties (Fig. 6). The field extends from the Glenwood Springs area, where it merges with the Grand Hogback coal field, to the Coal Basin area north of Chair Mountain, where the West Elk Mountain intrusives separate it from the Somerset and Crested Butte coal fields.

Original in-place coal resource has been estimated at more than 5.21 billion short tons to a depth of 6,000 ft in the 165-sq-mi field. Coals of the Carbondale coal field increase in rank from high-volatile C bituminous in the northern part to medium-volatile bituminous in the southern part. Heat from Tertiary intrusive igneous rocks of the Elk and West Elk Mountains causes anomalously high coal ranks in the southern part of the coal field. In some areas, portions of coalbeds have been altered to anthracite and low- to medium-grade graphite.

All minable coal is found in the Upper Cretaceous Williams Fork Formation of the Mesaverde Group. In this area, the Williams Fork Formation consists of three members, which from oldest to youngest are the Bowie Shale, the Paonia Shale, and an undifferentiated member. Three coal groups occur in the Bowie Shale and Paonia Shale Members, and the undifferentiated member is essentially barren.

The most important coal group, the Fairfield, directly overlies the Rollins Sandstone of the underlying Iles Formation. Coals in the southern part of the coal field range up to 35 ft in individual seam thickness, dip from 12 to 16° to the west, and are reported to be very gassy (Gale, 1907). In the central part of the coal field, the Fairfield group coalbeds consist of two to four beds, average 6 ft in thickness and are high-volatile. A bituminous in rank. In the northern portion of the coal field, four beds are present. From bottom to top, these are: the "A," or Black Diamond; the "B"; the "C," or Wheeler; and the "D," or Pocahontas. The "A" and "D" coalbeds were reported by Gale (1907) to be very gassy and dip from 42 to 44° to the west in this area.

The South Canyon coal group (Collins, 1976) occurs directly above the first persistent sandstone in the Bowie Shale Member. Coalbeds in the South Canyon group are less persistent than those in the Fairfield group. In the Coal Basin area of the Carbondale coal field, only one bed—the Dutch Creek—is of minable thickness and consists of 3 to 20 ft of medium-volatile bituminous coal.

The third and uppermost of the coal groups, the Coal Ridge, occurs in the basal 400 ft of the Paonia Shale Member of the Williams Fork Formation. Coals of this group have not been extensively investigated except in the fields south of Coal Basin and have been mined only at North Thompson Creek and Spring Gulch (Collins, 1976). Coal Ridge group coals are the most erratic of the three groups. In the Coal Basin–Chair Mountain–Marble area, the group consists of up to nine coalbeds ranging from a few inches to 7 ft in thickness. The thickest coalbed occurs directly above the upper sandstone. Coalbeds mined in this area include the Placita and Sunshine and have an average rank of high-volatile A bituminous. In the North Thompson Creek area there are up to 10 coalbeds in the Coal Ridge group. The thickest of these is 9 ft thick and occurs 65 ft above the upper sandstone. The Sunshine is the only named coalbed in this area.

Crested Butte Coal Field

The Crested Butte coal field lies southeast of the Somerset coal field and south of the Carbondale coal field in Gunnison County (Fig. 6). The coal-bearing Mesaverde rocks of this part of Piceance Basin have been extensively affected by the intrusive West Elk Mountain complex. As in the eastern part of the Somerset coal field, the coals have been folded, faulted, and intruded. The Crested Butte coal field is very mountainous, with much of the field lying at elevations above 10,000 ft.

Original in-place coal resources, estimated to a depth of 6,000 ft in the 240-sq-mi area of the Crested Butte coal field, are more than 1.56 billion short tons (Hornbaker et al, 1976). Coal quality in this field ranges from high-volatile C bituminous in the northern and western parts to semianthracite and anthracite in the southeastern part of the field. The field contains important resources of premium-grade high-volatile A and B bituminous coking coal.

The Crested Butte coal field is divided into three districts: the Floresta, Mount Carbon, and Crested Butte.

The Crested Butte district lies in the Slate River valley where some coalbeds crop out near the level of the valley floor but more commonly crop out high in the canyon walls at altitudes exceeding 11,500 ft. Several coalbeds have been identified in the Crested Butte district, and major coals are numbered beginning with the lowermost coal that occurs immediately above the Rollins Sandstone. The No. 1 coal attains a maximum thickness of 6.7 ft in the southeastern part of the field. Coal No. 2 is persistent throughout the field and ranges in thickness from 5 to 6.7 ft. Coal No. 3, also found everywhere in the field, is the most important bed of the Crested Butte district. The No. 3 bed differs from place to place, in thickness as a result of the emplacement of intrusive rocks and associated faulting and folding (Lee, 1912). In several areas, the coal has been crushed; and in the Crested Butte mine, the No. 3 coal ranges from a few inches to 22 ft thick.

The Floresta district lies on the northern flank of the Anthracite Range, an igneous intrusive thrust up through the coal-bearing rocks and displacing all the coalbeds. All the coal occurs in the Paonia Shale Member that dips from 18 to 24° to the north. The Ruby-Anthracite mine produced from the main bed, which averaged 4 ft thick in the mine (Lee, 1912). The coal has been altered to anthracite and is cut by several dikes of igneous rocks.

The Mount Carbon district lies south of the Crested Butte district, separated from it by Mount Carbon, Mount Axtell, and Mount Wheatstone, all erosional remnants of Tertiary lacolithic intrusions. Two coalbeds of significance occur in the Paonia Shale Member of the Williams Fork Formation, in the Mount Carbon district. The lowermost, or Coal No. 1, occurs just above a massive sandstone bed and attains a maximum thickness of 11 ft in the southernmost part of the district. In some areas, heat from the igneous intrusives has metamorphosed the coal from bituminous to anthracitic rank. At the Robinson mine, on the flank of Mount Wheatstone, 4 ft of coal has been metamorphosed to anthracite by the heat of the Mount Wheatstone laccolith.

Somerset Coal Field

The Somerset coal field, along the southeastern edge of Piceance Basin in Delta and Gunnison counties (Fig. 6), is part of a continuous outcrop of Mesaverde coal-bearing strata that stretches more than 100 mi from the Book Cliffs coal field to the northwest, to the Crested Butte coal field to the southeast. The coals occur in the Bowie and Paonia Shale Members of the Williams Fork Formation and are more numerous and better exposed than in the Grand Mesa coal field to the west. Almost all mining and most of the prospecting in this field have been done in the Bowie Shale Member in which there are at least seven thick coalbeds having an aggregate thickness of 38 to 43 ft (Lee, 1912). As in the other coal fields, the coals differ greatly in thickness within short distances and commonly split into as many as six benches, or coalesce into a single bed as thick as 42 ft. In many places, the coals are not well exposed or are burned. Even at the best-exposed section, ash and clinker at several horizons indicate that not all of the coalbeds are exposed.

Original in-place coal resources for beds less than 6,000 ft deep in the 320-sq-mi Somerset field are estimated to be more than 8.04 billion short tons (Hornbaker et al, 1976). Coals range in rank from high-volatile C to high-volatile B bituminous, with coal in the eastern part of the field generally being of coking quality. Coal-bearing strata in the eastern part of the Somerset field are intruded by plutons of the Oligocene West Elk Mountains complex (Murray et al, 1977).

Grand Mesa Coal Field

The Grand Mesa coal field stretches for approximately 50 mi along the southern and southwestern border of Piceance Basin (Fig. 6). It is separated from the Book Cliffs coal field to the north by the Colorado River and is bounded on the south by the Somerset coal field. The Grand Mesa coal field is on the south flank of Grand Mesa, primarily in Delta County, with the northwest portion of the field in Mesa County. The Mesaverde coals in the Grand Mesa coal field occur above the Rollins Sandstones in the Bowie and Paonia Shale Members of the Williams Fork Formation (Murray et al, 1977) (Fig. 7).

Coals in the Grand Mesa coal field range from high-volatile C bituminous to subbituminous A in rank. Original in-place resources estimated to a depth of 6,000 ft in the 530-sq-mi field exceed 8.6 billion short tons (Hornbaker et al, 1976).

The present-day Grand Mesa coal field is divided into two districts: the Palisades, which includes the west slope of Grand Mesa between the Colorado River and the southwestern point of the mesa, and the Rollins district, which extends from the southwestern point of the Grand Mesa eastward to Paonia.

Lee (1912) divided the coal-bearing sequence in the Palisades district into the Bowie and the Paonia Shale Members. The underlying Bowie contains only one significant coalbed in the Palisades district: a 3.3-ft bed in the Stokes Mine. Lee correlated the Bowie coal with the Palisades coal that occurs north of the Colorado River where it is 3.3 ft thick at the Palisades Mine. Lee identified several coalbeds in the Paonia Shale Member, only one of which has been developed. It occurs at the base of the Paonia and ranges in thickness from 4 to 7 ft. Lee correlated this bed with the Cameo north of the Colorado River. Several other coalbeds are found at higher horizons in the Paonia.

All coals in the Rollins district occur in the Paonia Shale Member. A thick coal usually occurs near the base of the member, and other coals occur in the higher horizons and are more numerous than those in the Palisades district. The higher coals increase in number and thickness to the east. The basal coal ranges from 7 to 16 ft in thickness.

Book Cliffs Coal Field

The Book Cliffs coal field stretches for approximately 50 mi between the Colorado River and the Utah state line on the southern rim of Piceance Basin (Fig. 7). Named for the conspicuous Book Cliffs escarpment lying on the north side of Grand Valley, this coal field occupies approximately 800 sq mi in Garfield and Mesa counties. Coals occur principally in the Upper Cretaceous Mount Garfield Formation of the Mesaverde Group. The coalbeds are mostly flat-lying but are tilted and faulted in places.

Original in-place coal resources for the coal field estimated to a depth of 6,000 ft total more than 7.2 billion short tons (Hornbaker et al, 1976). The bulk of the coal in the field is high-volatile C bituminous in rank, with some high-volatile bituminous.

Erdmann (1934) published the first complete description of coal resources for the Book Cliffs coal field. Much of the stratigraphic correlation work done by Erdmann has been restructured by Young (1955), which does not affect Erdmann's coal descriptions but redefines the stratigraphic units in which the coals occur.

Erdmann divided the coals into four zones. These include, in ascending order, the Anchor, Palisade, Cameo, and Carbonera coal zones. The lowest, the Anchor, occurs in the Anchor Mine tongue of the Mancos Shale, which splits the Sego Sandstone into upper and lower benches. The Anchor coal zone has a maximum thickness of 9 ft but is commonly much thinner and contains a high percentage of bony material. The coal has an average rank of high-volatile C bituminous.

The Palisade coal zone occurs immediately above the Corcoran Sandstone (Young, 1955). Coalbeds of the Palisade zone are the most persistent in the Book Cliffs coal field and can be traced from the eastern extent of the field into Utah. The

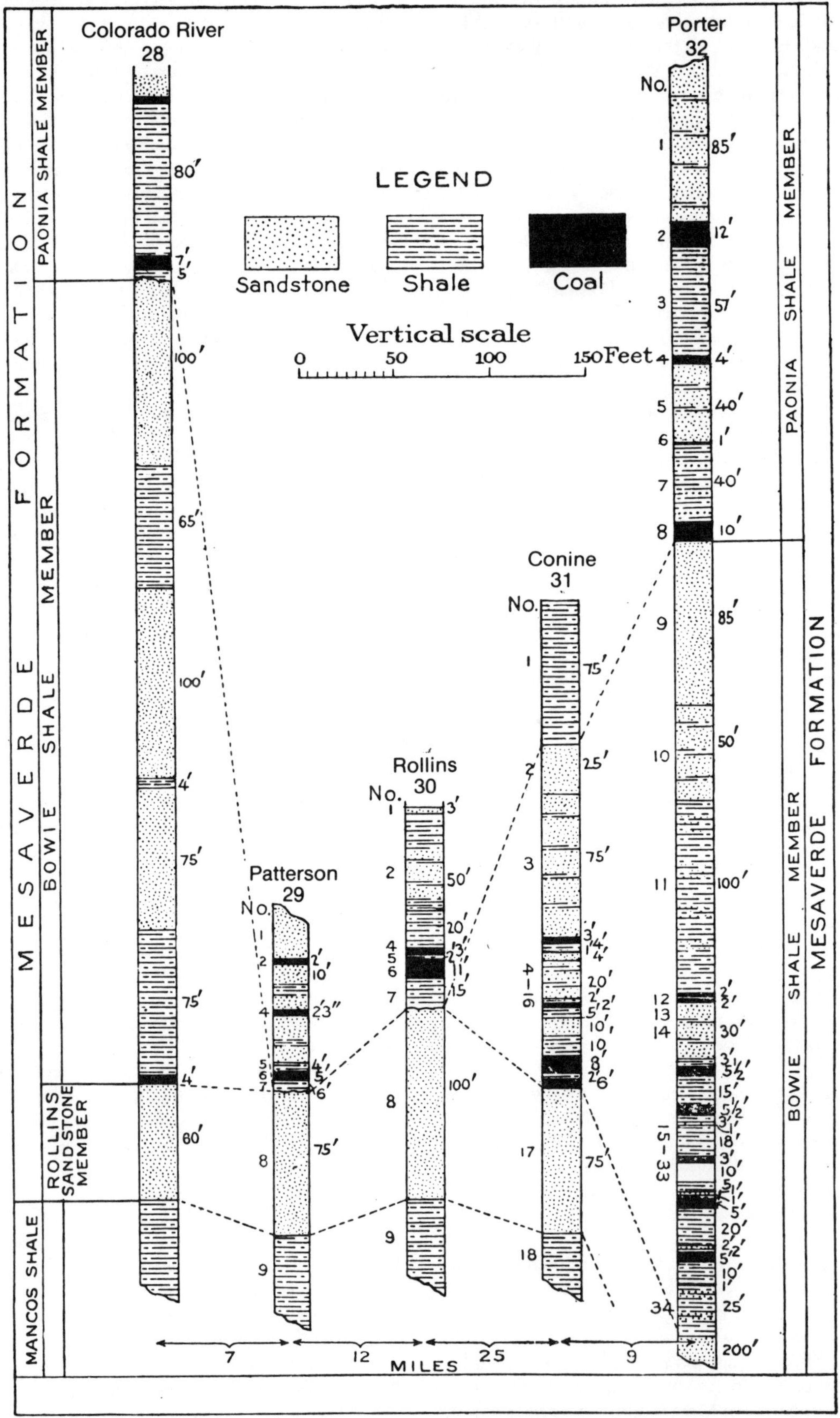

Figure 7—Detailed columnar sections of the coal-bearing part of the Mesaverde Formation in the Grand Mesa and Somerset fields as identified by Lee (1912).

coal zone ranges in thickness from 30 to 60 ft and contains at least seven coalbeds. Only three of these beds are represented at any location (Erdmann, 1934).

The interval between the bases of the Palisade and the Cameo coal zones ranges in thickness from 200 ft in the western part of the field to 450 ft in the eastern part. The Cameo zone can be traced throughout the entire field and occurs directly above the Rollins Sandstone in the eastern part of the field. West of the Book Cliffs mine, the Rollins Sandstone is replaced by several thin sandstone beds separated from the base of the Cameo coal by carbonaceous shale and mudstone. The Cameo coals are thickest in the central portion of the field, often characterized by two beds separated by a thick sandstone unit. Here, the beds have an aggregate thickness from 8 to 23 ft.

The Carbonera coal zone occurs 60 ft above the Cameo zone and is characterized by two and sometimes three coalbeds. The coals have been heavily developed in the western portion of the field. As with the Cameo coals, the Carbonera coals have been burned and are difficult to trace in some parts of the field. The Carbonera zone consists of a series of detached coalbeds whose correlations are based principally on the character and stratigraphic position of the coals. In the western part of the field, the Carbonera coalbeds thicken dramatically in T7S, R102W and thin again to the west. North of the abandoned town of Carbonera, two coals from this zone crop out, with thicknesses of 13 and 7 ft.

POTENTIAL METHANE RESOURCE

Data on Gassy Underground Mines

Data compiled by the U.S. Bureau of Mines clearly indicate that Piceance Basin—particularly the area southeast of a line connecting Rio Blanco and Mack—is the gassiest coal region in the western United States (Murray et al, 1977). Mine emission data from the Mine Safety and Health Administration show that methane released from individual active coal mines in the southeastern Piceance Basin ranges from 18,000 to 2,235,000 cubic feet per day (cf/day). The Coal Basin area mines emit the largest volumes, with four mines averaging over 1 million cf/day with high gas contents per ton of coal.

The gassy nature of Piceance Basin coals has led to numerous safety and regulatory problems. Since the late 19th century, the area around Newcastle (Grand Hogback field) has been the site of many mine fires and explosions. On December 16, 1913, a gas explosion killed 37 miners at the Vulcan Mine. Further south at Sopris, Colorado, 17 miners died in an explosion on March 24, 1922. From 1883 to 1932 in Pitkin, Gunnison, and Garfield counties, 21 gas fires or explosion accidents resulted in 176 deaths (Humphrey, 1960; Kintz and Denny, 1933).

Methane Emission Data from Underground Mine Measurements

The Mid-Continent Coal and Coke Company operates five underground coal mines in the Coal Basin district of the Carbondale coal field. In the Bear Creek, Coal Basin, and Dutch Creek # 2 mines, coal is cut by continuous miners, while the L.S. Wood and the Dutch Creek # 1 mines utilize longwall mining systems. Total production for the five mines is about 1 million tons per year, and all the coal is marginal- to premium-grade medium-volatile bituminous in rank.

The average methane emission for the five mines in 1977 (first quarter, 1977; MESA) totaled 8,226,000 cf/day. In the two longwall operations, the beds dip approximately 13° into the basin away from the outcrop, and the coalbeds are mined along strike for approximately 3,600 ft. Methane drainage holes are drilled into the roof as mining advances (i.e., after the coal is mined). The drainage holes are drilled up into the micro-fracture system of the caving roof at an angle of approximately 60° to the horizontal. Drilled on 75-ft centers for the total advancement, the holes are cased and cemented with 60 ft of 2-inch metal pipe and connected to a 4-inch collection pipe. The collection system is entirely closed to the atmosphere and the internal pressure maintained at a vacuum of approximately 6 inches of mercury. The holes have a total length of 160 ft; the last 100 ft are 1-7/8 inches in diameter and are not cased.

The two drainage systems produce approximately 3.02 million cf/day from a total of about 110 holes. In the L.S. Wood mine, 30 holes have been drilled, but only 15 are producing gas at any one time. Of the 80 holes drilled in the Dutch Creek # 1 mine, only about one-third are in operation and flowing gas at any one time. Individual holes are shut off when atmospheric air begins leaking into the closed system and it starts producing less than 30% methane. Some of the holes, however, produce 100% methane. The mine operators try to maintain an average of 70% methane by opening and closing the various drainage holes.

Surface Drill Hole Data from Other Projects

U.S. Geological Survey

USGS drill hole No. CA-77-2 (Sec 13, T10S, R98W) lies on Plateau Creek at an elevation of 4,830 ft, just upstream from where it flows into the Colorado River. The site is in the Palisade District of the Grand Mesa coal field approximately 4.6 mi upstream from the Cameo Mine. This hole marks the western extent of the cross section for the southeastern Piceance Basin (Fig. 3).

The Colorado Geological Survey collected coal core samples for desorption analysis from the Cameo zone just above the Rollins Sandstone and from the Palisade zone immediately above a lower sandstone, probably the Corcoran. The Cameo zone occurs at an interval from 800 to 875 ft deep and contains five coalbeds with a cumulative thickness of 17 ft (Fig. 8). One sample taken from the highest coal of this zone had a gas content of 80 cubic feet per ton (cf/ton). The Palisade zone contains two beds—1.0 and 3.5 ft thick—within a 20-ft interval. A second sample taken from the thicker coalbed for desorption analysis had a gas content of 223 cf/ton.

Exxon Vega No. 2

The Vega No. 2 well (Sec 34, T9S, R93W, Mesa County) is an important well in understanding the coalbed methane regime within Piceance Basin. Vega No. 2 is the first known attempt at a coalbed methane completion in the basin. The well is near the synclinal axis of the basin and has been completed in the Cameo coal zone, immediately above the Rollins Sandstone, at 8,000 ft. A careful evaluation of all completion and stimulation operations employed in this well, along with daily production

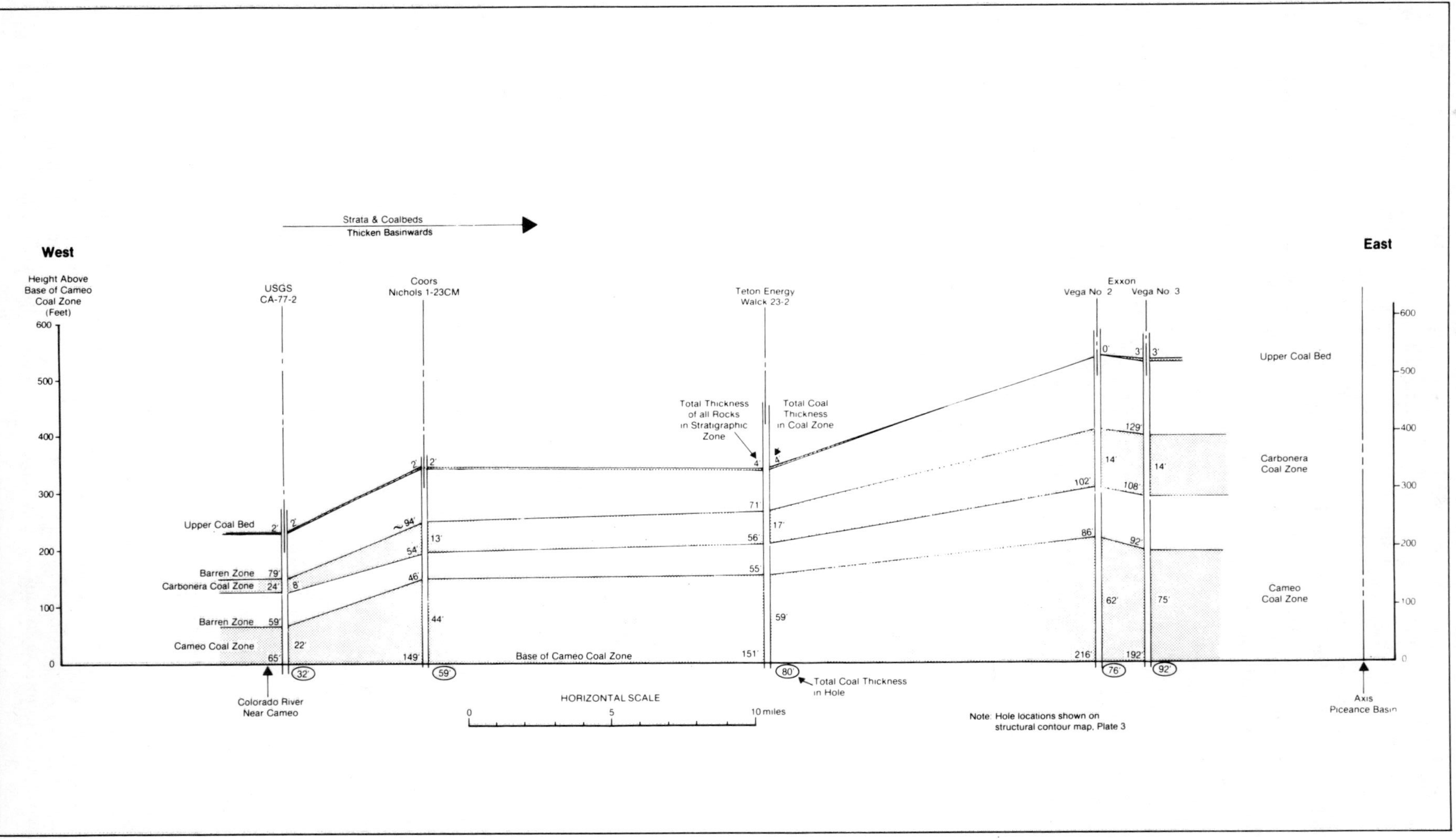

Figure 8—East-west section from the Colorado River to Piceance Basin synclinal axis showing correlation of coal zones in five wells, with the Cameo and Carbonera coal zones of the Book Cliffs coal field.

records involving both gas and water flow, provides valuable information on how to produce Piceance Basin coalbed methane. Following is a detailed development history of this well compiled from sources that include Exxon geologists and engineers, official state agency files, log service companies, and others.

Vega No. 2 well, near the center of the east-west cross section of Figure 3, is about 12 mi east of the Teton Energy well, Walck 23-2, one of the major wells for which coal-core desorption tests have been performed in the MRCP. Vega No. 2 bottomed at 8,888 ft in the Corcoran Sandstone. Planned potential gas targets were the Corcoran and the overlying Cozzette and Rollins Sandstones.

On completion of drilling, geophysical logs were run, including caliper, spontaneous potential (SP), resistivity (dual-induction), gamma ray, and formation density. Exxon attempted unsuccessfully to complete in the Cozzette, with perforations between 8,615 and 8,620 ft. Subsequently, a completion was attempted in the Cameo coalbeds above the Rollins Sandstone.

Coalbeds at Vega No. 2 comprise two coal zones:

1. The upper Carbonera coal zone, at a depth interval of 7,648 to 7,750 ft, consisting of five coalbeds ranging in thickness from 1 to 4 ft, and totaling 14 ft of coal in a 102-ft section.
2. The lower Cameo coal zone, at a depth interval of 7,836 to 8,052 ft, consisting of five coalbeds ranging in thickness from 4 to 25 ft, and totaling 62 ft of coal in a 216-ft section.

Coal and coalbed methane resources calculated for the Vega No. 2 area, per sq-mi section of land, using a total coal thickness of 76 ft, a tonnage factor of 1.152 million tons per ft of coal per sq mi of land, and the average core sample methane desorption value of 410 cf/ton as determined for Vega No. 3, are: 87.6 million tons of coal and 35.9 billion cubic feet (Bcf) of gas.

To facilitate completion, a cement plug was placed between 8,140 and 8,600 ft, and the casing was perforated in the following zones:

7,847-7,851	1 shot/1.5 ft	3 shots
7,863-7,879	1 shot/1.5 ft	9 shots
7,985-8,007	1 shot/2 ft	12 shots
8,046-8,062	1 shot/2 ft	4 shots

All perforations occur within the boundaries of the four major coalbeds, which are very clearly indicated on the gamma ray, density, and dual-induction logs.

The formation was broken down with 9,200 gallons of water containing 2% potassium chloride (KCl) over the four zones. Next, a massive frac was begun, requiring 143,200 gallons of Versa gel 2500, with initial proppant consisting of 50,000 pounds of 100-mesh sand, to be followed with 222,400 pounds of 20–40 mesh sand.

All of the 10-mesh sand was pumped using 61,760 gallons of Versa gel. After pumping in 120,960 pounds of 20–40 mesh sand with 30,240 gallons of Versa gel, however, the frac system sanded up and pumping the remaining Versa gel and sand required in the original design ceased.

From this point on, evaluation of the Vega No. 2 well provides information on the intricate relationship of coalbed methane gas production to water present in a coalbed. Even the limited data available on U.S. coalbed methane indicate that it is generally necessary to keep formation water content to very low levels if coalbed methane is to desorb and flow. In the Vega No. 2 well, the rate of water production steadily decreased during a 30-day cleanup period. During the last 6 days of this 30-day period, water production was, consecutively, 75, 48, 36, 36, 29, and 12 barrels per day. Nine days later, the well produced for a 24-hour test; gas flow of 440 thousand cubic feet (Mcf) was analyzed to have a heating value of 1,072 Btu per cu ft.

After this test, the well was shut-in awaiting connection to the Rocky Mountain Natural Gas Co. pipeline; production was initiated on February 4, 1981. The well flowed naturally against pipeline pressure of 550 pounds per square inch (psi). The first day's production exceeded 1 million cubic feet (MMcf), and included a small amount of water. On the second day, flow rate dropped to as low as 180 Mcf per day and then recovered to produce a total of 520 Mcf for the 24-hr period, along with 60 to 100 barrels of water. During the following three days, gas production was 444, 432, and 432 Mcf, and water production for each day was 175 barrels. These data indicate the fracture zone had filled with water during the long shut in, which retarded gas flow. However, data on water production during the final cleanup stages indicate that after a relatively short period of production, the accumulated water in the fracture system should be removed. Because of the tight nature and low permeability of Mesaverde strata, the steady-state future flow of water into the system should be quite low. At that point, gas production from the well should recover.

Exxon Well Vega No. 3

Vega No. 3 well (Sec 10, T10S, R93W, Mesa County), 1.9 mi southeast of Vega No. 2, is one of the wells included in the east-west cross section of Figure 3. This well is important, because coring was performed in the Cameo coal zone above the Rollins Sandstone, and methane desorption measurements were made on the coal core samples. Desorption data from Vega No. 3, in conjunction with the detailed well history and coalbed methane production records available on Vega No. 2, provide the most complete composite picture for coalbed methane potential available for any area in the Piceance Basin at the time of the preparation of this report.

Coalbeds in the Vega No. 3 well, above the Rollins Sandstone, comprise the following sequence:

1. An upper coalbed, 3 ft thick, at a depth of 7,128 ft
2. A middle coal zone, at a depth interval of 7,131 to 7,260 ft, consisting of four coalbeds ranging in thickness from 1 to 6 ft and totaling 14 ft of coal in a 108-ft section
3. A lower coal zone, at a depth interval of 7,460 to 7,652 ft, consisting of seven coalbeds ranging in thickness from 2 to 27 ft and totaling 75 ft of coal in a 192-ft section

Within the Cameo and Carbonera coal zones at Vega No. 3, 11 coalbeds ranging in thickness from 1 to 27 ft and totaling 89 ft of coal were encountered within a stratigraphic distance of 392 ft.

Coalbed sequence and spacing in the Vega No. 3 well, as measured on geophysical logs, correlate closely with the Vega No. 2 and with the coal-zone lithologies to the west: in the Teton Energy well Walck 23-2, south of Colbran; the Adolph

Coors wells Nichols 1-24 and 1-23 CM, west of Mesa; and the USGS well CA-77-2, near the Cameo Mine on the Colorado River (Figs. 3, 8). In all of these wells, coalbeds occur principally in two coal zones that extend uninterrupted from the Colorado River at Cameo, eastward to the Piceance Basin synclinal axis near the Vega wells. The lower of these two coal zones is almost certainly the Cameo coal zone of the Book Cliffs and Grand Mesa coal fields; and the upper, the Carbonera coal zone. The relative coalbed stratigraphic relationships and coal thicknesses in these wells are summarized in Figure 8. Also indicated on this illustration is the thickening of the coal zones toward the basin center, along with the barren zones between the coal zones. Thickness of individual coalbeds and total coal thickness also increase toward the basin axis.

An isolated coalbed ranging in thickness from 2 to 4 ft commonly occurs from 70 to 130 ft above the top of Carbonera coal zone. This bed has been extremely useful as a guide during MRCP drilling operations. During drilling, penetration of this bed can be seen on the geolograph log and confirmed by drill cuttings, and permitted core points to be chosen with accuracies of 1 to 3 ft during MRCP operations.

Methane desorption values for two coal-core samples from the Vega No. 3 have been reported by the Colorado Geological Survey (1980). Both core samples came from the same bed at a recorded depth of 7,598 ft. This depth correlates with a 13.5 ft thick bed, just above the 27 ft thick basal bed of the Cameo zone. Total gas reported for the two samples is 381 and 438 cf/ton, with an average for the two samples of 410 cf/ton.

Coal and coalbed methane resources calculated for the Vega No. 3 area, per 1-sq-mi section of land, using a total coal thickness of 92 ft, are: $(92 \times 1.152) = 106.0$ million tons of coal, and 43.5 billion cfg/sq mi.

U.S. Steel Exploration Core Holes

U.S. Steel has leased approximately 10 sq mi of federal land in T10S, R89W, and R90W in the Coal Basin area of the Carbondale coal field (Fig. 9). To retain the lease, it must develop the coal during the 1980s. To determine ventilation requirements for an underground mine design, U.S. Steel drilled four exploration holes and cored and sampled four coalbeds for desorption analysis. These samples from the Placita, "V," "B," and "A" beds were found to have gas contents that range from 107 to 1,041 cf/ton.

Instead of crushing the samples after desorption to release the residual gas for direct measurement, a USBM curve was used to estimate this component of the total gas. The USBM has determined that the curve does not consistently predict residual gas content accurately and its use has been discontinued by the bureau (P. Diamond, 1981, personal communication). The total gas content values reported by U.S. Steel are therefore considered as approximate values by these authors.

A summary of the samples collected at each of the holes, the desorption results, and a methane resource estimate for the test area are presented in Table 1.

The weighted average gas content for all of the samples was calculated to be 569 cf/ton. Using bed thicknesses measured in the core holes and a tonnage multiplier (1,800 tons per acre-ft of bituminous coal or 1.152 million tons per section-ft) shows the coal resource to be 43.1 million tons per section of land. The total methane resource in the four sampled coalbeds is estimated to be 24.5 Bcf per sq mi.

Union Oil Well Overland No. 1

Overland No. 1 well (Sec 13, T11S, R92W, Delta County) is approximately halfway between the Exxon Vega wells in the center of the basin to the west, and the U.S. Steel exploration holes at Coal Basin. The hole was spudded on September 7, 1957, completed October 10, 1957, and drilled to a depth of 7,600 ft from a surface elevation of 9,421 ft. Depth to the Rollins Sandstone is 6,757 ft.

In this well, core samples were recovered from a thick coalbed just above the Rollins Sandstone that had been intruded by a sill of acidic igneous rock. In the core-log description, coal is reported occurring in core Nos. 29 and 30 at a depth interval between 6,724 and 6,741 ft. Although no documentation of the intruded igneous sill was made in the written report, suspicions were evidently aroused, and the events were reported to the USGS. Averitt (USGS Coal Branch office in Denver) gives a brief description of this rare core sample as follows (Averitt, 1975, p. 74):

> "...in a well drilled in Sec. 13, T. 11 S., R. 92 W., in which the coal is 6,723 feet below the surface...the drill penetrated 14 feet of natural coke, underlain by an estimated 12–14 feet of quartz latite, which is under-lain by 12 feet of coal."

This intrusion of a coalbed by igneous magma is comparable to similar occurrences in the highly gassy coalbeds directly to the east in Coal Basin and to the southeast in the Crested Butte field. In these two areas, the intruding dikes and sills are commonly in proximity to larger igneous bodies—stocks and plutons—of comparable acidic compositions. The sill that cuts the coalbed in Overland No. 1 well probably formed from a feeder dike originating from an igneous pluton at greater depth as indicated in the cross section in Figure 4. The complex of stocks and plutons at Coal Basin, Crested Butte, and probably below the Overland No. 1 well may represent cupolas and roof pendents of a much larger, deeper, igneous body of batholithic proportions. If so, the potential for regional coalbed methane being generated as heat was released by cooling magmatic bodies lying below the basin's coalbeds, is extremely high in the entire southeastern portion of Piceance Basin.

Hawk's Nest Mine

The Hawk's Nest Mine property is located in Sections 1, 2, 3, 10, 11, and 12, T13S, R90W in the Somerset coal field. The two operating mines, Hawk's Nest East (No. 2) and Hawk's Nest West (No. 3), both produce from the "E" coalbed. This bed averages 9.0 ft thick and dips from 2 to 3° to the northwest. The coal is premium-grade, high-volatile B bituminous and is shipped to U.S. Steel in Provo, UT.

Colorado Geological Survey personnel collected coal samples for desorption analysis from four exploration drill holes on the Hawk's Nest property. Lithologic logs, coalbed thicknesses, and gas desorption results were collected for a total of 25 samples from the "A" to "D" and Wild coalbeds. Gas contents ranged from 80 to 245 cf/ton. The weighted average gas content of beds "A" through the Wild, 157 cf/ton, was used to estimate the methane resource of the "E" and higher coalbeds.

The coal resource of all beds encountered in the four drill holes was calculated to determine the amount of coalbed methane contained in the study area. This resource,

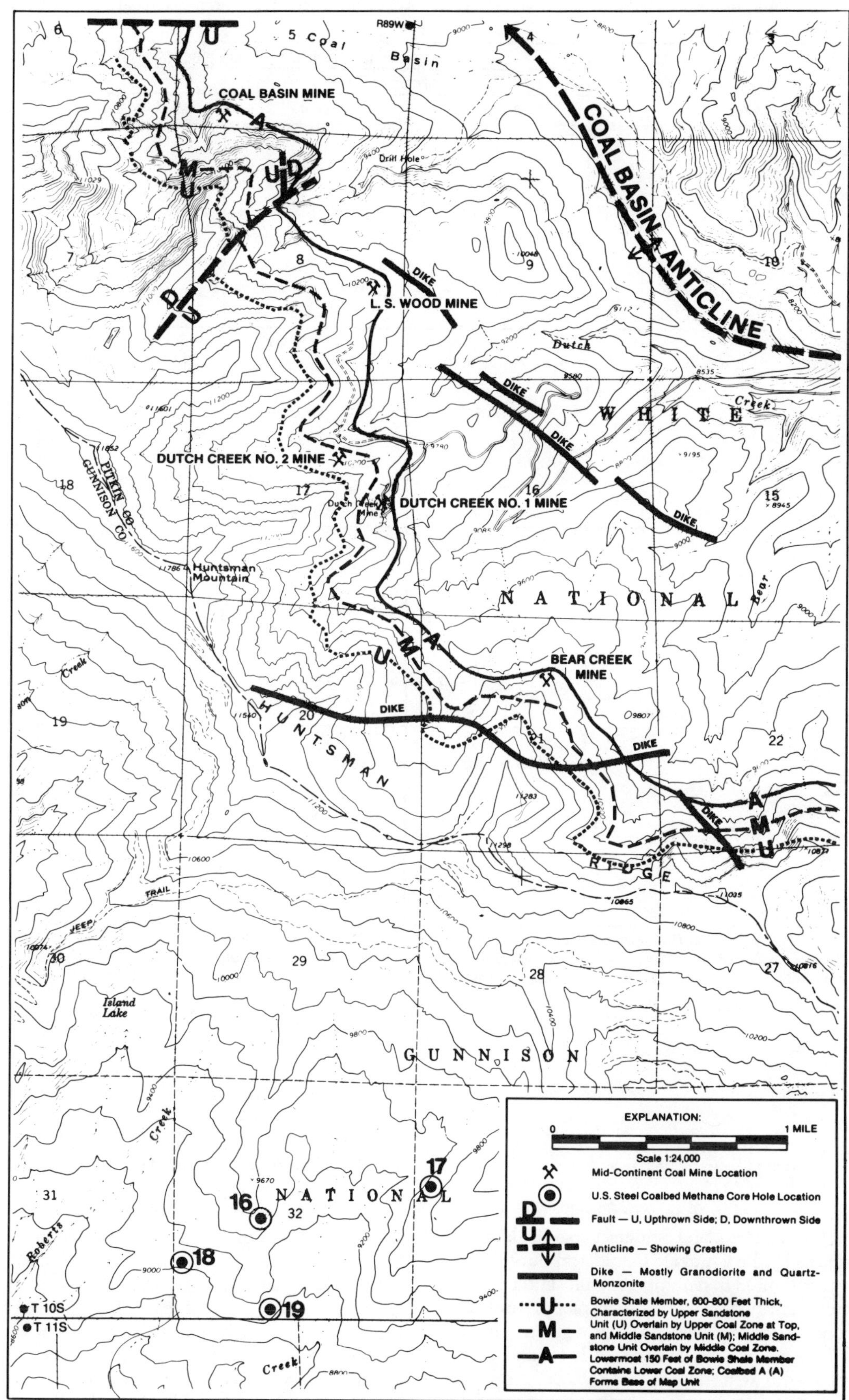

Figure 9—Geologic and topographic map of the coal basin area showing locations of the Mid-Continent Mines and the U.S. Steel coalbed methane drilling sites. Modified from Kent and Arndt, 1980b.

Table 1—Core sample gas desorption results, coal resource, and methane resource for U.S. Steel's property, Coal Basin, Carbondale coal field, Gunnison and Pitkin Counties, Colorado.

Coal Bed	Sample Parameters		Hole Number and Desorption Results			
			Summer 1979		Summer 1980	
			No. 16	No. 17	No. 18	No. 19
Placita	Coring Interval (ft)		1,570.2–1,575.7	1,219.7–1,226.4	Data Not Available	Data Not Available
	Coalbed Thickness (ft)		5.5	6.7		
	Number of Samples		4	3		
	Gas Content (cf/ton)	Range	720.3–812.0	107.4–620.8		
		Average	765.3	435.9		
V	Coring Interval (ft)		1,847.5–1,859.6	1,506.9–1,518.5	2,018–2,032.6	2,093.6–2,107.1
	Coalbed Thickness (ft)		12.1	12.0	14.1	14.6
	Number of Samples		8	4	4	5
	Gas Content (cf/ton)	Range	641.4–760.2	409.7–708.8	642–851	411–1,041
		Average	716.4	630.4	759	527
B	Coring Interval (ft)		2,251.4–2,260.0	1,964.9–1,971.9	2,467.2–2,475.1	2,530.1–2,539.7
	Coalbed Thickness (ft)		8.6	7.0	7.9	9.6
	Number of Samples		2	1	3	3
	Gas Content (cf/ton)	Range	475.0&485.9	249.5	395–740	537–658
		Average	480.5	249.5	537	628
A	Coring Interval (ft)		2,269.6–2,277.7	No Samples Collected	2,480.9–2,488.2	2,546.6–2,554.3
	Coalbed Thickness (ft)		8.1		7.5	7.7
	Number of Samples		2		2	3
	Gas Content (cf/ton)	Range	490.9&676.3		520-568	285–486
		Average	520.9		553	329

Note: Tonnage calculations assume 1,800 tons per acre-ft for bituminous coal; ie, 1.152 million tons per acre-ft per 640-acre, section of land. Tonnages are probably higher than calculated because coal rank is medium-volatile B bituminous. Weighted average gas content for all four beds = 569 cf/ton.

Totals

approximately 85.8 million tons of coal per sq mi, contains 13.5 Bcf of methane.

Colorado Geological Survey

Methane desorption test results from core samples taken in the Danforth Hills, Grand Mesa, and Somerset coal fields are from a report presented by the Colorado Geological Survey. Data in that report are limited to desorption results and the locations of test sites identified on a small-scale map. Detailed locations and drill hole surface elevations are not given. The desorption results show average gas contents of 3, 15, and 45 cf/ton, respectively. Along with those from TRW well tests, these results indicate low gas contents in shallow coalbeds of the northern coal fields and generally in coals covered by less than 1,000 ft of overburden.

Rio Colorado Well No. 1 Cactus Valley 23-6-9

No. 1 Cactus Valley (NW/4, NW/4, Sec 23, T6S, R90W, Garfield County) was spudded October 24, 1979, by Rio Colorado Oil and Gas Company to explore the coalbed methane potential of the thick sequence of coalbeds in South Canyon, south of the Colorado River, about 5 mi west of Glenwood Springs and 0.8 mi south of the South Canyon mine in Garfield County. A cross-sectional sketch at South Canyon from Gale (1910) shows at least nine coalbeds ranging in thickness from 4 to 30 ft and a total of 96 ft of coal in a 640-ft stratigraphic section. The South Canyon coalbeds are in the Bowie Shale Member, at the base of the Williams Fork Formation of the Mesaverde Group (Collins, 1976). The coalbeds crop out on the northeast slope of Coal Ridge, which forms part of the Grand Hogback. According to Gale (1910), the South Canyon mine produced from the Wheeler bed. The attitude of this coalbed is N 60° W, dipping 52° SW. Ground elevation at the South Canyon mine adit is about 6,400 ft, and 6,681 ft at the No. 1 Cactus Valley well. Horizontal distance between the well and the mine adit as projected in the downdip direction is approximately 2,700 ft. Gas Commission files show that coring was planned for the intervals of 3,250 to 3,450 ft and 3,700 to 4,000 ft.

The projection of the thicker coalbeds from the surface, in the downdip direction, intercepts the planned coring intervals. After completion, published reports indicated Rio Colorado planned to conduct production testing for gas in the most promising coalbeds identified during drilling and coring.

During coring, coal samples were collected by USBM personnel. Release of the desorption test results by the USBM has not yet been approved by Rio Colorado nor have details of drillstem tests, fracing, or flow tests been released by Rio Colorado. A scout report by the Petroleum Information Service provides some additional data: casing was set to a total depth of 4,393 ft; perforated in the intervals of 2,440 to 2,550, 2,740 to 3,320, and 3,868 to 4,130 ft; a foam frac was performed; and the well was undergoing pressure testing.

The heating values of four as-received coal samples from surface and mine outcrops collected at South Canyon (Collins, 1976) range from 11,643 to 13,127 Btu. These values indicate coal rank ranging from high-volatile C to high-volatile A bituminous. However, directly downdip from No. 1 Cactus Valley along the Grand Hogback Monocline, about 5 mi to the southwest, these coalbeds occur below 9,000 ft. Here, Freeman (1979) has measured vitrinite reflectance values of 2.1 from drilling samples and has mapped the coal as semianthracite in rank. Gale (1910, p. 131) notes that in the underground workings at South Canyon, "there is a large volume of gas in the coal, and Wolf safety lamps are used." No. 1 Cactus Valley, located only 11 mi north of the extremely gassy Coal Basin mines, should be favorable for coalbed methane production. However, because of the steep dip of the coalbeds on the Grand Hogback monocline, methane may have had an unusually favorable path for updip migration and could have been lost. If coalbed methane production proves unfavorable at the No. 1 Cactus Valley, another attempt might be made at the center of the narrow syncline, 5 mi to the southwest.

Chevron Well Skonberg No. 1

The bottom 1,850 ft of the Skonberg No. 1 (Sec. 9, T7S, R93W, Garfield County) well was completed open-hole in coal-rich strata of the Williams Fork Formation above the Rollins Sandstone. In the 1,584-ft open-hole interval from 7,054 to 8,614 ft, six coal zones occur that contain coalbeds varying in thickness from 1 to 14 ft with an aggregate thickness of more than 100 ft of coal. Coal strata are encountered in 10% of the open-hole interval.

The Skonberg No. 1 well is on the northwestern nose of the Divide Creek Anticline (Fig. 3). At this location, the coal sequence in this hole should fall in the low-volatile bituminous rank, according to Freeman's map of coal rank based on drill cuttings and vitrinite reflectance measurements (Freeman, 1979). To date, only limited data are available on this well. Johnson et al (1979) present a cross section indicating that this is a producing gas well. Geophysical logging performed on March 1 and March 15, 1966, indicated Skonberg No. 1 was the discovery well for the Mam Creek gas field.

The 91 MMcf produced from Skonberg No. 1 must be derived from the coalbeds and the interbedded sandstones. Although there is no way to divide this production, between coalbeds and tight gas sands, all the gas probably was derived from the coalbeds, regardless of the producing zone.

Amoco Production Company

Amoco Production Company has applied to drill three wells and has requested a permit to degasify the coalbeds above the Rollins Sandstone. The Green Mountain Unit No. 1 (Sec. 20, T12S, R92W) has a projected depth of 3,000 ft. The Bowie Unit No. 1 (SE/4, NE/4, Sec 15, T12S, R91W) was spudded August 23, 1982, and drilled to a depth of 3,391 ft. Planned location of the Somerset Unit No. 1 is in the SE/4, NW/4, Sec 9, T12S, R90W.

All three wells are located on U.S. government land, on which Amoco holds the oil and gas leases but not the coal rights. Since some of the target coals may be less than 3,000 ft deep (potentially minable), a legal decision was sought from the solicitor's office at the Department of the Interior to determine ownership of the coalbed methane. A favorable opinion was recently forwarded to the USGS Conservation Division, which issues the drilling permits, and AMOCO is now proceeding with the drilling program.

The planned Amoco wells lie in a very favorable location within the southeastern Piceance Basin, southeast of the Exxon Vega properties, southwest of the U.S. Steel Coal Basin exploration area with its very high coalbed methane indications, and north of the gassy Somerset coal field. As shown on the depth of overburden map (Fig. 4), depth to the Rollins Sandstone and the base of the coalbeds at the Amoco sites should range from approximately 3,300 to 4,100 ft.

Aries Resources Wells Leon Lake Nos. 1 and 2

Aries Resources, Inc., drilled two exploratory coalbed methane wells in 1981: Leon Lake No. 1 (SW/4, Sec. 12, T12S, R94W) and No. 2 (NE/4, Sec. 14, T12S, R94W), in northern Delta County, west of the Amoco wells described above. These were planned as exploration wells for coalbed methane in the coal zone immediately above the Rollins Sandstone. Leon Lake No. 1, spudded August 8, 1981, and drilled to a depth of 4,650 ft, was perforated in four zones between 4,550 and 4,397 ft. The well is now shut in and no production flow test data are available.

Methane Recovery from Coalbeds Project Data

TRW and DOE have cooperated with oil and gas companies to drill sites in three areas of the Piceance Basin: the Rangely area, the Douglas Creek area of Rio Blanco County in the northwestern Piceance Basin, and the Plateau Creek area of Mesa County in the southeastern Piceance Basin.

During the early stages of the MRCP, the cooperative testing program was essentially unknown to the oil and gas industry and it was difficult to arrange testing operations with potential cooperators. Some of the earliest cooperators in the program were Western Fuels, Fuelco, and Twin Arrow Drilling, all with operations in the northwestern portion of Piceance Basin. Later in the program, two coal characterization tests that involved coring, sample desorption, and a drill stem test, and one production well test involving perforating, fracing, and flow testing well tests were performed in the Plateau Creek area of Mesa County with Adolph Coors and Teton Energy Companies.

Teton Energy Well Walck 23-2

Teton Energy well Walck 23-2 (NW/4, NW/4, Sec 23, T10S, R95W) was spudded March 10, 1980, at a ground elevation of 7,017 ft in the Wasatch Formation and bottomed at the base of the Mesaverde Group. The gas targets of Teton Energy were, in descending order, the Rollins, Cozzette, and Corcoran Sandstones. The MRCP objective was to determine the gas content of the coalbeds immediately above the Rollins Sandstone.

Coalbeds encountered in lower Mesaverde strata in the Walck 23-2 well comprised the following zones, as measured from the geophysical logs that are 6.5 ft deeper than on-site measurements:

1. An upper coalbed, 4 ft thick, at a depth of 4,483 ft
2. A middle coal zone, at a depth interval of 4,557 to 4,613 ft consisting of five coalbeds ranging in thickness from 1 to 4 ft, and totaling 17 ft of coal in a 56-ft section
3. A lower coal zone, at a depth interval of 4,667 to 4,814 ft, consisting of eight coalbeds ranging in thickness from 1 to 14 ft, and totaling 59 ft of coal in a 147-ft section

Within a 311-ft stratigraphic interval, 14 coalbeds were encountered, ranging in thickness from 1 to 14 ft and totaling 80 ft of coal. The lower coal zone in the well correlates with the Cameo coal zone and the middle coal zone with the Carbonera coal zone of the Book Cliffs and Grand Mesa coal fields.

Coring and sampling were restricted to the lower coal zone. Three intervals were cored from 4,682 to 4,696; 4,726 to 4,759; and 4,798 to 4,819 ft (these are on-site measurements and are shallower than the geophysical logs show). Within the cored intervals, the strata consisted of 41.5 ft of coal, 5.0 ft of carbonaceous shale, and 21.5 ft of sandstone. Eleven feet of the Rollins Sandstone immediately underlying the basal coalbed was deliberately cored to obtain data. A 3.5-inch-diameter core was recovered using a 60-ft core barrel. Core recovery in the shale and sandstone was essentially 100%; recovery in coal was approximately 25%.

A total of 16 methane desorption samples was collected: 13 from cores and 3 from drill cuttings. Results of sample desorption testing are summarized in Table 2.

Three samples considered most representative of methane content for individual coalbeds were collected from the basal 14 ft thick bed with an average gas content of 305 cf/ton, and three samples collected from the next higher, 12 ft thick bed had an average gas content of 231 cf/ton.

Coal and methane resources per sq mi for the area adjacent to the Walck 23-2 well are summarized in Table 3. These estimates are based on the coalbed thicknesses measured from the density log and the gas desorption results for eight core samples representing pure coal from within three coalbeds (Fig. 10). The total coal resource is 92 million tons per sq mi of area, and the total methane resource is 20.5 Bcf per sq mi. Gas concentration based on a weighted average is 223 cf/ton and an arithmetic average is 220 cf/ton.

Compositions of gas samples collected during desorption from seven of the samples are given in Table 4. These analyses indicate that the average composition is 77% methane, 11% ethane, 2% propane, and less than 0.3% total of all other hydrocarbons. Nonhydrocarbons average 8% carbon dioxide and 2% nitrogen.

Three samples of roof and floor rock, labeled R-1, R-2, and R-3, were taken for physical property analysis. Sample R-1, taken 2 ft below the 8.0-ft coalbed in core No. 2, is a 4.5 ft thick, carbonaceous shale. Sample R-2, collected 1.1 ft above the 12.2-ft coalbed in core No. 2, is a medium-grained sandstone. Sample R-3 was taken from the Rollins Sandstone 3.5 ft below the 13.5-ft coalbed of core No. 3.

Test results are summarized below.

Sample Number	Triaxial Comp. Strength (psi)	Uniaxial Comp. Strength (psi)	Permeability (millidarcies)	Bulk Density (gm/cc)
R-1	—	9,170	0.12	2.30
R-2	20,400	—	0.03	2.50
R-3	—	—	0.44	2.52

Adolph Coors Well, Nichols 1-24

The Adolph Coors Nichols 1-24 (NW 1/4, NW 1/4, Sec 23, T10S, R97W) was initially drilled in 1978 to a depth of 3,600 ft. During the months of April, May, and June 1980, the well was plugged back to 2,720 ft and the casing perforated over an 8-ft interval (2,625 to 2,633 ft) in the Cameo coal zone and the well tested for coalbed methane production. No attempt was made to stimulate the basal coalbed immediately above the Rollins Sandstone because of potential water inflow. In this well, 11 coalbeds having an aggregate thickness of 75 ft occur within a 360-ft interval. The coalbeds range in thickness from 2 to 13 ft. Geophysical logs run on this hole include compensated density, neutron, induction electric, gamma ray, and SP.

The well was stimulated using 52,500 gallons of 75% quality nitrogen foam and 49,000 pounds of sand proppant. Production testing, after post-frac well cleanup and swabbing over a 15-day interval following the treatment, resulted in no sustained methane production. A conventional step-rate injectivity test was performed to evaluate the stimulation treatment. The injection test results show that a previously fractured formation accepted the injected fluid. The results of this testing are presented in detail in a MRCP well test report on the subject well (TRW, 1981).

Coal samples were collected from the nearby Adolph Coors Nichols 1-23 CM well to confirm the presence of producible quantities of gas here.

Gas production from this well was unsuccessful, probably because of the completion casing and cementing program used in drilling and completing in the original deeper target in 1978.

Adolph Coors Well, Nichols 1-23 CM

The Adolph Coors Nichols 1-23 CM (Sec. 23, T10S, R97W) was spudded September 30, 1980, at a surface elevation of 5,966 ft. The target depths of this well were 7,700 and 8,000 ft in the Dakota Sandstone and the Jurassic Salt Wash Sandstone

Table 2—Summary of coalbed methane sampling results to date for Teton energy well, Walck 23-2, southeastern Piceance Basin, Colorado.

Sample Number	Core Number	Coring Interval	Coalbed or Rock Unit Name	Sample Lithology	Sample Weight (gm)	Total Gas cc/gm	Total Gas cf/ton	Gas Composition Sample
2	1	<4,680	Mesaverde Gr.	Noncarb. Shale	1,046	.06	2	
1	1	4,682–4,696	"Cameo Zone"	Coal and Shale	2,255	1.48	48	X
8	2	4,726.0–4,726.8	"Cameo Coals"	Coal, dull	3,804	1.08	35	
6	2	4,730.5–4,731.3	"Cameo Coals"	Coal	3,003	3.68	118	X
7	2	4,734.9–4,735.7	"Cameo Zone"	Carb. Shale	3,864	0.31	10	
3	2[1]	4,752–4,753	"Cameo Coals"	Coal, cuttings	486	4.48	144	
10	2	4,753.0–4,753.8	"Cameo Coals"	Coal	2,313	6.09	195	
4	2[1]	4,755–4,757	"Cameo Coals"	Coal, cuttings	2,278	4.85	155	X
9	2	4,755.1–4,755.7 and 4,756.5–4,756.7	"Cameo Coals"	Coal	1,486	9.85	316	
5	2	4,755.7–4,756.5	"Cameo Coals"	Coal	2,009	5.72	183	X
14	3	4,798.3–4,799.1	"Cameo Coals"	Coal	1,621	9.85	315	
13	3	4,801.0–4,801.8	"Cameo Coals"	Coal	1,733	8.93	286	
12	3	4,804.0–4,804.8	"Cameo Coals"	Coal	1,829	9.76	313	X
11	3[1]	4,807–4,808	"Cameo Coals"	Coal, cuttings	397	5.58	179	X
15	3	4,808.0–4,808.3	"Cameo Coals"	Coal	631	5.95	190	
16	3	4,816.0–4,816.8	Rollins SS	Sandstone	2,805	2.83	91	X

1. Drill cuttings off the shale shaker while coring.
2. Sample depths given above are as measured on site at the time of coring. Top of the Rollins Sandstone, used as a marker horizon, is 6.5 ft deeper on the geophysical logs.

Table 3—Coal and methane resource per section of land at Teton energy well Walck 23-2.

Bed Thickness (ft)	Sample Desorption Value: Individual Sample Value (cf/ton)	Sample Desorption Value: Average (cf/ton)	Coal Resource* per Section of Land (millions of tons)	Methane Resource per Section of Land (Bcf)	Coal Resource per Section of Land (millions of tons)	Methane Resource per Section of Land (Bcf)
(beds not sampled) 46					52.99	11.82 (223 × 52.99)
8.0	35,113	74	9.22	0.68		
12.2	195,316,183	231	14.05	3.25		
13.5	315,286,313	305	15.55	4.74		
33.7	220 (average)	Totals	38.8	8.67	53.0	11.82

*Tonnage factor of 1.152 million tons of bituminous coal per foot of coal per sq mi of area.

Arithmetic average = 220 cf/ton
Weighted average = 223 cf/ton

Total coal resource, all beds = 92 million tons per section of land.
Total methane resource, all beds = 19 Bcf gas per section of land.

Member of the Morrison Formation, respectively. The MRCP test involved the coalbeds immediately above the Rollins Sandstone. This test site was selected by TRW for desorption drill stem testing to determine why negligible methane production had occurred at the Adolph Coors Nichols 1-24 well test site approximately 0.5 mi to the east.

The coalbeds occur in two zones:

1. Upper zone (Carbonera coals) is 54 ft thick, containing four beds—from top to bottom: 3.8, 1.2, 2.0, and 6.0 ft thick—a total of 13 ft of coal.
2. Lower zone (Cameo coals) is 149 ft thick, separated from the upper zone by 46 ft of barren rock. This zone contains nine coalbeds that range from 0.8 to 11.4 ft thick with a cumulative thickness of 44 ft. The lowest coalbed occurs directly on top of the Rollins Sandstone.

A total of 57 ft of coal in 13 beds was encountered in a 249-ft interval occurring immediately above the Rollins Sandstone.

The Cameo coal zone was cored and samples collected from

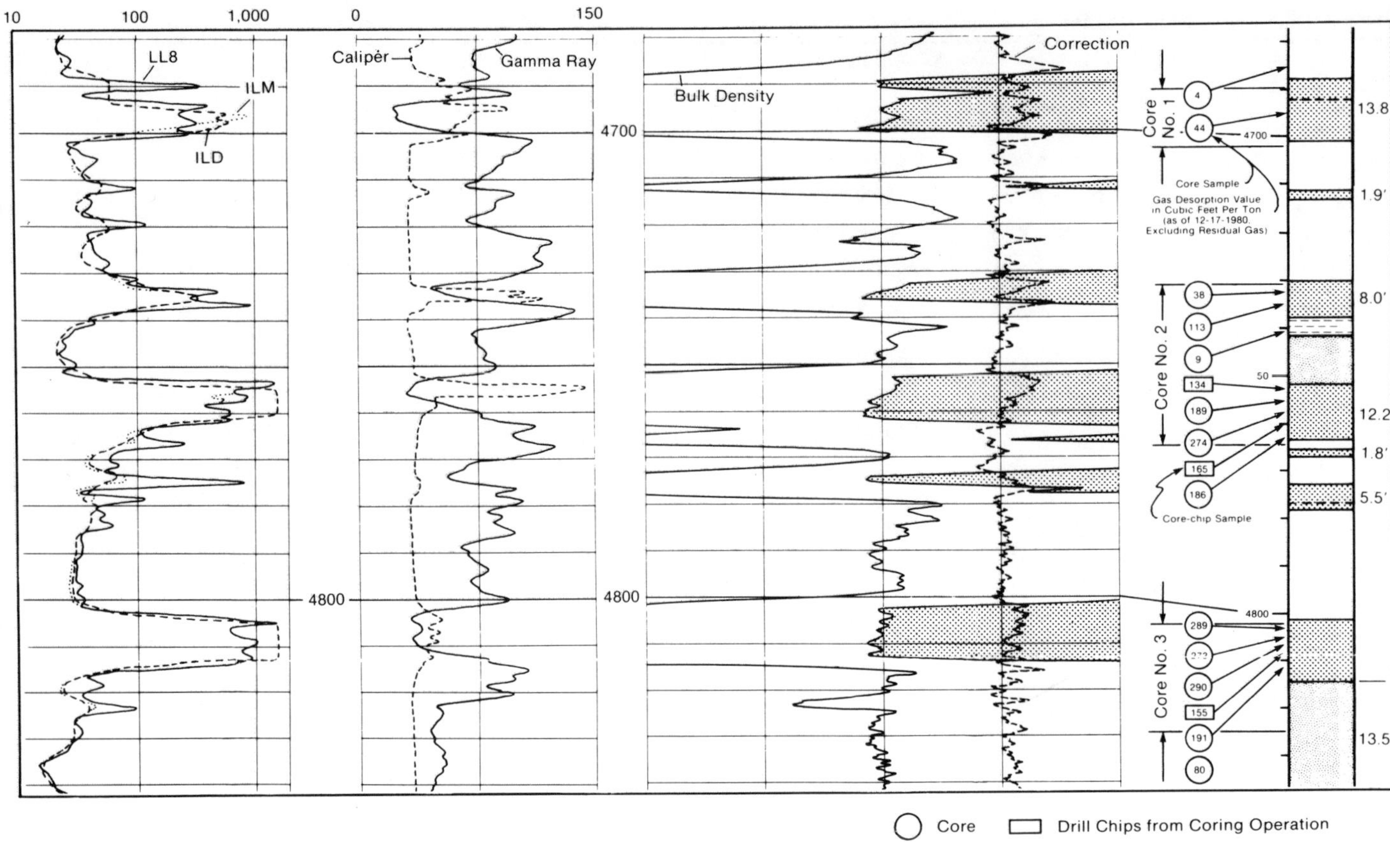

Figure 10—Teton Energy Walck well, logs and gas content data.

Table 4—Summary of gas compositions for Teton energy well, Walck 23-2, southeastern Piceance Basin, Colorado. Data recalculated on an air-free basis and normalized to 100%.

Component	Symbol	Sample Number and Percent							
		1	4	5	6	11	12	15	Average
Methane	C_1	66.94	73.84	80.18	80.18	78.28	85.10	73.23	76.82
Ethane	C_2	7.73	14.64	7.86	7.99	15.06	8.04	16.59	11.13
Propane	C_3	2.12	4.77	0.76	0.74	0.88	0.74	1.61	1.66
I-Butane	C_4	0.18	0.39	0.13	0.15	—	0.10	0.16	0.16
N-Butane		0.12	0.32	0.10	0.11	—	0.09	0.24	0.14
Higher Hydrocarbon	C_{5+}	—	0.06	—	—	—	—	—	0.01
Nitrogen	N_2	13.77	0.34	0.10	0.12	0	0.10	2.49	2.42
Carbon Dioxide	CO_2	9.14	5.64	10.87	10.71	5.78	5.83	5.68	7.66
Totals		100.00	100.00	100.00	100.00	100.00	100.00	100.00	100.00

Analyses performed by Spectron Labs, Denver, Colorado.

depths ranging from 2,631 to 2,770 ft. A total of 11 desorption samples was collected—nine from coalbeds and two from roof rock. Desorption measurements yielded gas content values that range from 113 to 329 cf/ton for the coal samples, and 3 and 34 cf/ton for the two roof rock samples. A summary of the desorption results is given in Table 5.

Coal and methane resources per sq mi for the area adjacent to the Nichols 1-23 CM well are calculated to be 27.7×10^6 tons and 14.4 Bcf, respectively. These estimates are based on the coalbed thicknesses measured from the density log and the gas desorption results for eight core samples that represent pure coal from within three coalbeds. The total coal resource is 66.2 million tons per sq mi of area. Desorbed gas concentration based on a weighted average is 217 cf/ton and on an arithmetic average is 222 cf/ton.

Two drill stem tests were conducted on the lowermost coals

Table 5—Summary of coalbed methane sampling results to date for Adolph Coors Well, Nichols 1-23 CM, Mesa County, southeastern Piceance Basin, Colorado.

Sample Number	Core Number	Canister Number	Coring Interval	Coalbed or Rock Unit Name	Sample Lithology	Gas Content (cf/ton)
1	—	57	2,631.0–2,633.0	"Cameo Coals"	Coal, chips	131
2	1	16	2,670.7–2,671.4	Mesaverde Gr.	Sandstone	3
3	1	15	2,672.5–2,673.3	"Cameo Coals"	Coal	114
4	2	152	2,714.0–2,714.9	"Cameo Coals"	Coal	214
5	2	146	2,721.0–2,721.7	"Cameo Coals"	Coal	236
6	2	38	2,729.2–2,729.7	"Cameo Coals"	Coal	113
7	2	158	2,730.0–2,730.8	"Cameo Coals"	Coal	247
8	3	102	2,750.9–2,751.9	"Cameo Coals"	Coal	329
9	3	5	2,765.4–2,766.0	Mesaverde Gr.	Sandstone	34
10	3	151	2,766.4–2,767.1	"Cameo Coals"	Coal	252
11	3	115	2,768.2–2,769.1	"Cameo Coals"	Coal	270

1. Sample number 1 comprises chips collected off the shale shaker; all others are core samples.
2. Sample depths given above are as measured on site at the time of coring. Top of the Rollins Sandstone, used as a marker horizion, is 6.6 ft deeper on the geophysical logs. Coolbed tops range from 3.6 to 5.2 ft deeper on the geophysical logs.

immediately above the Rollins Sandstone. The first test was performed on the lower of the two beds. A very small show of gas and 40 ft of borehole fluid were recovered. During the second test on the higher coalbed, 2 ft of borehole fluid was recovered and no show of gas was recorded. A summary of the formation pressures in pounds per square inch gauge (psig) is as follows:

	DST No. 1 (2,754.0 ft to 2,769.5 ft)		DST No. 2 (2,708 ft to 2,723 ft)	
psig	Inside	Outside	Inside	Outside
Initial Flow	13.5	26.6	8.1	13.3
Initial Shut-In	120.7	132.7	13.5	26.6
Final Flow	26.9	26.6	8.1	13.3
Final Shut-In	93.9	106.2	13.5	26.6
Initial Hyd.	1,240.1	1,265.8	1,187.1	1,239.5
Final Hyd.	1,226.8	1,239.5	1,173.8	1,160.6

Samples of the shale roof rock were analyzed for permeability and compressive strength. The tests indicate low permeability, from less than 0.01 to 0.07 millidarcy, and an average compressive strength of 17,500 psi.

Dual induction, compensated neutron, formation density, gamma, and SP logs were run after reaching total depth.

Douglas Creek Area Wells

TRW cooperated on four exploration wells in the Douglas Creek area, two with Fuelco, and two with Twin Arrow Drilling Company. All four wells were spudded in the upper part of the Mesaverde Group. Results of desorption tests for samples from these wells yielded less than 100 cf/ton methane, with few exceptions. The following is a brief summary of each well:

Fuelco Well, Cathedral 0-28-3-101-S—The Cathedral 0-28-3-101-S well (Sec. 28, T3S, R101W) penetrated five coalbeds that ranged from 4 to 12 ft thick occurring within a 220-ft interval from 1,380 to 1,600 ft. Two coal samples were taken from the two deepest coalbeds for desorption analysis and indicate gas contents of 18 and 81 cf/ton.

Twin Arrow Well, C&K # 1-13—The C&K well No. 1-13 (SW/4, NW/4, Sec. 13, T3S, R101W) was being abandoned by Twin Arrow when TRW isolated and performed flow testing on the coalbeds. If flow was observed, production testing was contemplated. Geophysical logs available for this hole include induction-electric log, compensated density log, and a cement bond log.

Five zones were tested: 573 to 581; 627 to 633; 661 to 665; 726 to 736; and 801 to 810 ft. A bridge plug was set at 1,050 ft and the hole was pressure and flow tested. Pressure and flow were negligible. Casing was cemented from 1,050 to 320 ft; then perforated in the test zones using a one-shot-per-ft spacing. The coal zones were then treated with 500 gallons of 7.5% HF. The hole was swabbed dry and pressure-tested. The shut-in pressure was zero and no flow was observed.

Twin Arrow Well, C&K # 4-14—The C&K well No. 4-14 (Sec. 14, T3S, R101W) was spudded on November 25, 1978, and drilled to a depth of 925 ft before it was abandoned because of a stuck tool. TRW collected 150 ft of core from 636 to 828 ft and recovered 78% of the section. After cementing the abandoned hole, the rig was moved approximately 30 ft and a twin hole was drilled and logged. Ten coalbeds occurring within a 520-ft interval were identified on the density log. Nine samples, from 685 to 986 ft, were taken for desorption analysis. The samples had gas contents that ranged from 86 to 243 cf/ton (averaging 127 cf/ton) for the coal, and from 24 to 214 cf/ton (averaging 83 cf/ton) for the carbonaceous shale and coal samples. All of the coal sampled for desorption was high-volatile B bituminous.

Fuelco Well, D-26-3-101-S—TRW collected 96 ft of core at well D-26-3-101-S (Sec. 26, T3S, R101W). Six coalbeds were intercepted during coring operations, two of which were sampled for desorption analysis. Three samples, two from an 8.7-ft bed at a depth of 1,209 to 1,218 ft, and one from a 2.0-ft bed at a depth of 1,223 to 1,225 ft, had gas contents that ranged from 16 to 26 cf/ton.

Rangely Area Wells

TRW participated with Western Fuels on four wells located on the northeast flank of the Rangely Anticline in the Lower White River coal field. All of the wells were spudded in the Mesaverde Group and drilled to the Rollins Sandstone at depths ranging from 800 to 1,400 ft. Desorption sampling results average 50 cf/ton, with a range from 3 to 90 cf/ton. Borehole geophysical logs run in these wells include electric, gamma-ray, gamma-gamma, and caliper logs.

Discussion of Methane Resources

A unique set of circumstances involving Piceance Basin create a testing ground for ideas about the origin and occurrence of coalbed methane. The combination of aggregate coal thickness, variable depth of overburden, and local and regional igneous activity gives Piceance Basin a broad range in coalbed methane physical properties and characteristics. Variability in gas content per unit volume of coal and detailed stratigraphic data make cross sections and areal plots particularly useful in identifying target areas (Figs. 3, 4, 5).

Early work on the vertical variability of coal rank by Erdmann (1934) illustrates how coal rank increases with depth in the Book Cliffs coal field. He pointed out that with increased depth, the heating value of the coals and the degree of carbonization increase, while the percentage of volatile matter decreases. These trends are indicative of the potential for increased methane content. In addition, the coal zones and aggregate coal thickness increase from west to east as shown on Figure 8, and this trend also correlates with increased methane generated.

Reviewing the cross sections, well histories, and maps showing gas content of coals and gassy mines suggests that gas contents of coalbeds increase with proximity to the area of igneous intrusive activity. Kent and Arndt (1980a, p. 43) point out that,

> "....local advances in rank of coal in the Thompson Creek mining area and to the south resulted from accelerated thermal metamorphism caused by very steep geothermal gradients during and following localized emplacements of magma masses in Oligocene time."

From data in this report, it appears that such regional thermal metamorphism was at least partly responsible for creating the large measured methane resource. Steven (1975) postulated two Laramide Tertiary batholiths to the east and south of the basin. These batholiths could have been the source of the laccoliths, sills, and silicic volcanics observed in the region and would cause the elevated heat flow values, providing additional support for the concept of regional thermal metamorphism. As the regional metamorphism is enhanced by local metamorphism owing to epizonal intrusive activity, coal rank increases still more, as does gas content. When the anthracite and meta-anthracite rank is reached in local baked zones along sills and dikes, both volatile matter and gas content in the coalbed may drop off. At present, the medium- to low-volatile bituminous ranks appear to be the most gassy in southeastern Piceance Basin. Figure 11 from Freeman (1979) shows this relationship, especially for the Coal Basin-Thompson Creek areas, and illustrates the correspondence of the semianthracite range and the thickest sections on the depth of overburden map.

Dolly and Meissner (1977) discuss the high gas content of mines in the Raton Basin. Their review of coal rank, gas content, and occurrence of igneous rocks reveals associations similar to those found in Piceance Basin. They note:

> "In addition to the 'average' coal ranks reported in the various mining areas of the basin and those predicted regionally at depth, significant volumes of coal have been locally metamorphosed to considerably higher rank in 'natural coke' zones associated with widespread igneous dike and sill intrusion prevalent in all parts of the basin" (p. 262).

In Raton Basin, Dolly and Meissner (1977) found that coal mines in higher rank coals were very gassy and thought this provided evidence of an association of high methane gas content and levels of organic metamorphism.

The relationship between the depth of overburden and depth of burial to changes in coal rank—and, hence, gas content—is difficult to quantify because the total thickness of Eocene deposits is unknown. The age of uplift is considered to be Pliocene, however (Larson et al, 1975), and the coalbeds lay at great depth during most of the Tertiary. Coalification and rank increase owing to depth of burial seem quite likely, especially in light of the previously mentioned data collected by Erdmann (1934).

Whether methane production potential is high in any prospect area depends on several factors besides coal rank. If an area is heavily fractured and the coals are deep, production could be high, as in Coal Basin and Thompson Creek. Shallower, fractured coals are less likely to be gassy, because the gas may have escaped—particularly if the coals lie above the water table. Data collected for this report indicate that in the southeastern part of the basin, productive depths begin at depths over 2,000 ft, except where adjacent to igneous intrusions.

Sandstone beds, often associated with the coals, may act as reservoirs for generated methane. Many of the productive sands, such as at the Divide Creek gas field, may be producing coalbed methane (Murray et al, 1977). Coals naturally or artificially fractured may reverse the process and produce gas and water until a dewatered system would yield only gas. If the sandstones were highly permeable, however, they would be the best target.

This analysis has led to the following proposed guidelines for future targets in Piceance Basin. The southeastern Piceance Basin has a high coalbed methane content where: 1) overburden thickness is greater than 2,000 ft; 2) heat flow is greater than 1.5 to 2 heat flow units; 3) epizonal igneous intrusives are nearby; 4) faulting is well developed; 5) aggregate coal thickness is large; and 6) fold axes, both synclinal and anticlinal, occur. Prime target areas are those where two or more of these guidelines apply.

Estimated Resource Volume

Coal resources of Piceance Basin were estimated on a field-by-field basis by Hornbaker et al (1976) to a depth of 6,000 ft, as summarized in the columns that follow.

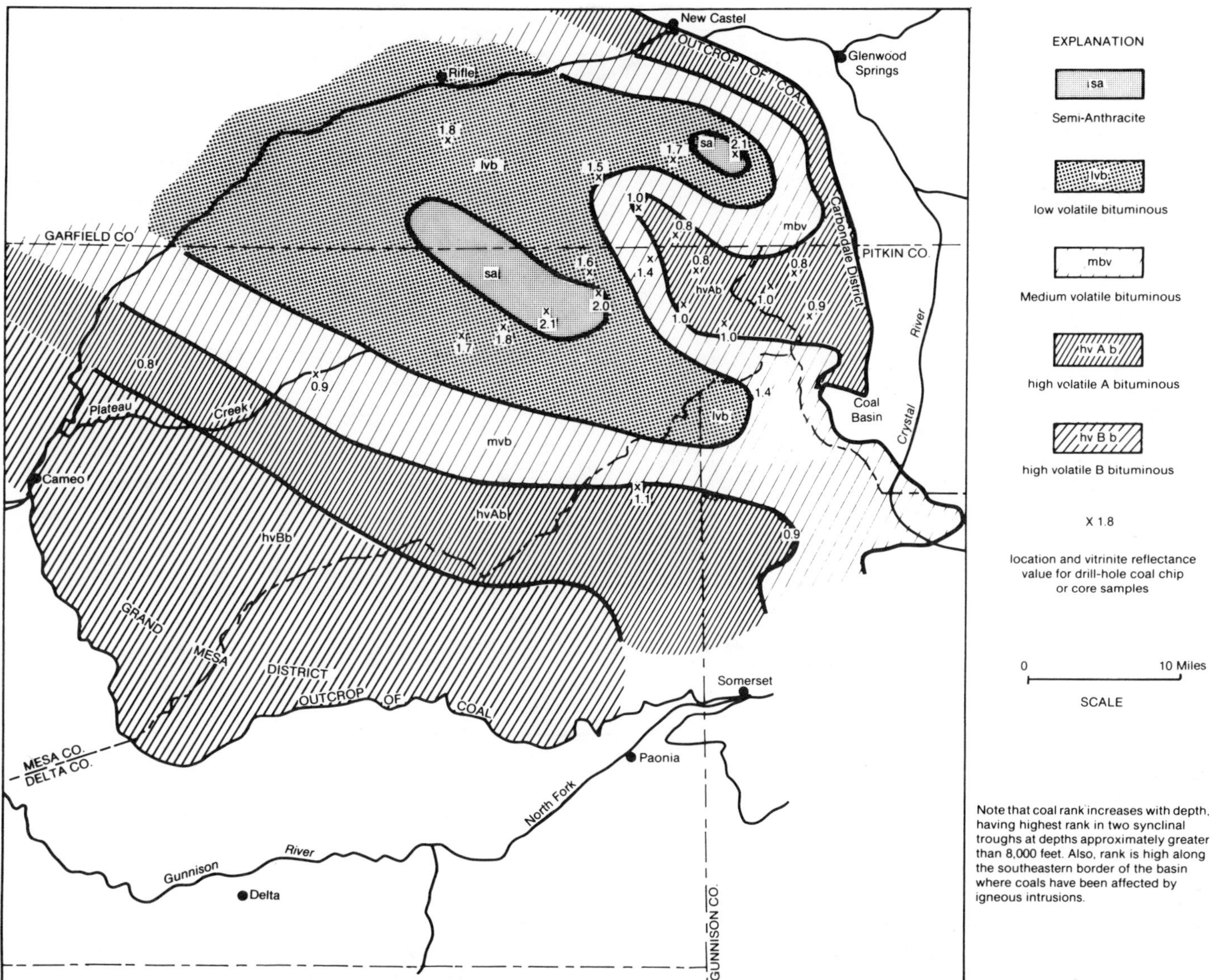

Figure 11—Map of coal rank boundaries in southeastern Piceance Basin as determined from vitrinite reflectance measurements from drill-hole chips and core samples. From Freeman, 1979.

	Size of Area Considered* (sq mi)	Coal Resources* (billions of tons)	Total Thickness of Coal Represented** (ft)
Lower White River	930	11.8	11.0
Danforth Hills	400	10.5	22.8
Grand Hogback	160	3.0	16.3
Carbondale	165	5.2	27.4
Crested Butte	240	1.6	5.8
Somerset	320	8.0	21.7
Grand Mesa	160	8.6	16.3
Book Cliffs	800	7.2	7.8
	3,545	55.9 Total	13.7 Average

*Data from Hornbaker et al (1976).

**Total thickness determined by using a tonnage factor of 1.152 million tons of bituminous coal per sq mi of area per ft of coal.

Total area of Piceance Basin underlain by Mesaverde coalbeds measured by planimeter is 6,570 sq mi. The area of the basin used by Hornbaker et al, where coals were less than 6,000 ft deep, was 3,545 sq mi, more than half the basin's total area. Hornbaker et al calculated the total coal resource for depths less than 6,000 ft for this area to be 55.9 billion tons. In the final column above, the total coal thickness for each field ranged from 5.8 to 27.4 ft and averaged 13.7 ft of coal for the entire 3,545-sq-mi area.

It is the authors' opinion that this average value of 13.7 ft for the total thickness of coal underlying half of Piceance Basin substantially underestimates the amount of coal actually present. In our approach, a preliminary map of total coal thickness in the Piceance Basin was constructed, based principally on borehole geophysical logs. This map incorporates data from this study, Fender and Murray (1978), recent USGS

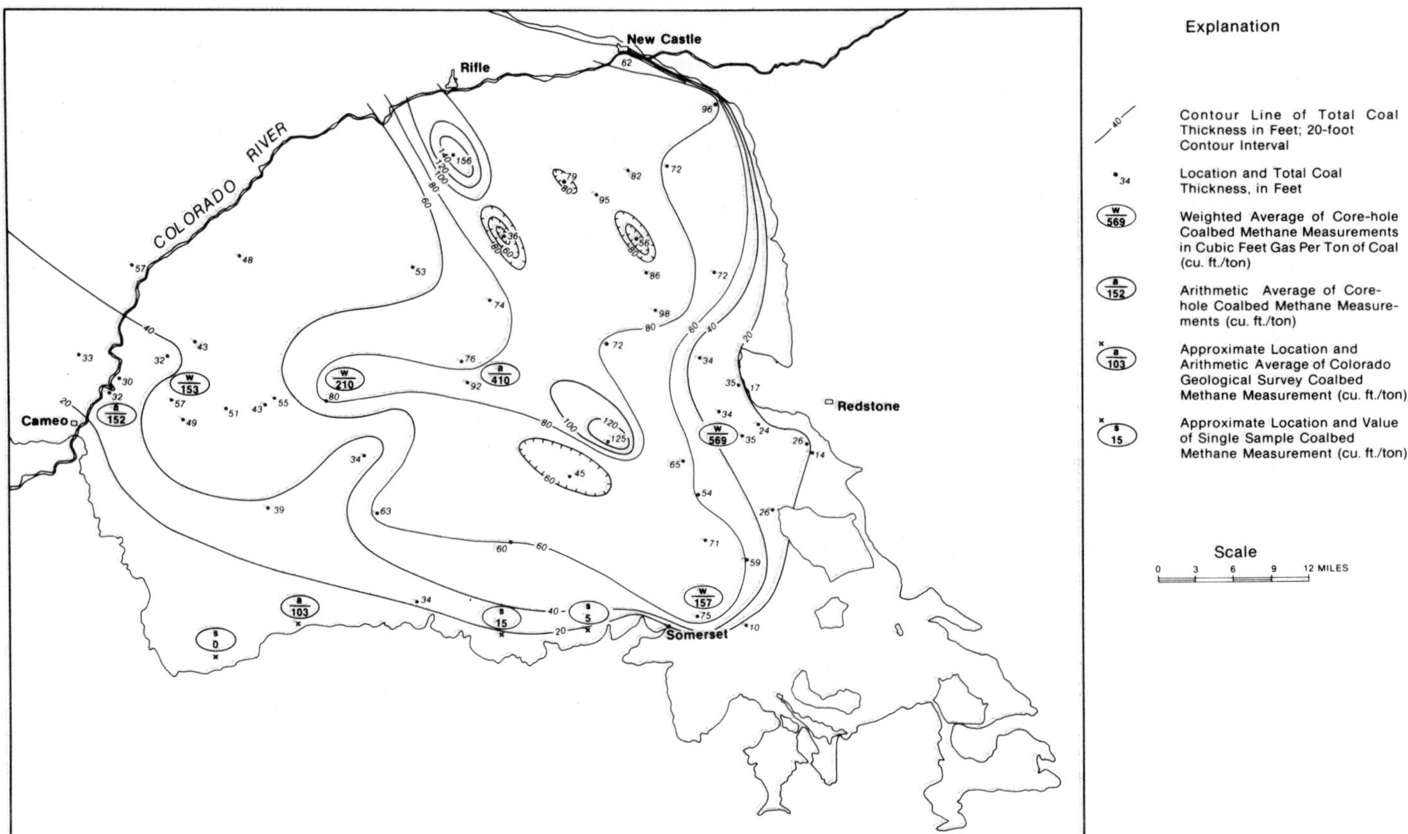

Figure 12—Isopach map of total thickness and gas content values of Mesaverde coals south of the Colorado River.

drilling programs, and some selected data from underground mine measurements reported principally in old USGS Bulletins. Data points are concentrated south of the Colorado River, where data were sufficient to construct a preliminary isopach map of total thickness of Mesaverde coals to a 20-ft contour interval (Fig. 12). North of the Colorado River, data points are insufficient for this type of mapping, but enough points are present to sketch an approximate location for the 40-ft contour line.

In estimating coalbed thickness distribution for the area south of the Colorado River, the areas between each 20-ft contour line were measured with a planimeter as shown below:

Contour Interval (ft)	Area* (sq mi)	Total Coal** (billions of tons)
0-20	536	6.2
20-40	300	10.4
40-60	608	35.1
60-80	521	42.0
80-100	413	42.8
100-120	35	4.5
120-140	14	2.0
140-160	5	0.8
Totals	2,432	144.0

*Values include separate calculations for average thickness of coal for high and low points on the coal isopach map.

**Tonnage factor is 1.152 million tons of coal per ft of coal per sq mi of area.

As shown, the basin underlain by Mesaverde coals south of the Colorado River is 2,430 sq mi; the total coal resource present is 144 billion tons; and the average thickness is 51 ft.

For the 4,140-sq-mi area north of the Colorado River underlain by Mesaverde coal, values of 40 ft for the lower limit, 60 ft for the upper limit, and 50 ft for the average value will be used. Thus, for this area, Mesaverde coal present is estimated to be:

$$4{,}140 \text{ sq mi} \times 50 \text{ ft} \times 1.152 \times 10^6 \text{ tons} = 238 \text{ billion tons.}$$

The methane gas resource for the area south of the Colorado River will be estimated using the following values: 150 cf/ton as a lower limit; 450 cf/ton as an upper limit; and 250 cf/ton as an average value. For the area north of the Colorado River, with its demonstrated lower methane concentrations, 25 cf/ton is the lower limit; 200 cf/ton is the upper limit; and 100 cf/ton is the average that will be used.

Using these values, the estimated methane resource for Piceance Basin areas south and north of the Colorado River, separately, is shown in Table 6. The estimated coalbed methane resource in the Piceance Basin, south of the Colorado River, is 36 Tcf; and north of the river, 24 Tcf. For the entire basin, the methane resource is estimated at 60 Tcf, with an estimated lower limit of 28 and an upper limit of 112 Tcf. These values have been rounded to 60, 30, and 110 Tcf, respectively.

Table 6—Methane resource estimates for Piceance Basin areas south and north of the Colorado River.

Area	Coal Resource (billions of tons)	Methane Concentration (cf/ton)			Methane Resources (Tcf)		
		Lower Limit	Upper Limit	Estimated Average	Lower Limit	Upper Limit	Estimated Average
South of Colorado River	144	150	450	250	21.6	64.8	36.0
North of Colorado River	238	25	200	100	5.9	47.6	23.8
Totals	382				28	112	60
Rounded Values	380				30	110	60

CONCLUSIONS

Piceance Basin, one of the largest coal basins in Colorado with an estimated 380 billion tons of coal in place, also contains some of the gassiest mines in the western United States. Individual active mines in the basin emit as much as 2.2 Mcf of methane per day. The gassiest coals occur in the southeast portion of the basin at depths exceeding 2,000 ft. Limited exploration in the basin has encountered coalbeds where gas concentrations exceed 750 cf/ton. A well completed near the basin center had an initial production exceeding 1 Mmcf of gas in a 24-hr period. The potential depths from which gas may be economically recovered may exceed 10,000 ft.

The high gas content of southeastern Piceance Basin coals probably is caused by combination of the large overburden thickness (which may exceed 10,000 ft), and the high heat flows present in that part of the basin. Examination of heat flow maps indicates the highest heat flow values are coincident with some of the gassiest mines. Coal rank also increases with geothermal heating, and high-quality coking coal mines are gassy. A cross section constructed from available data shows that methane content of coalbeds increases with depth and proximity to Tertiary intrusives.

Gas contents of beds in the northern basin do not appear to be high, but the deepest coalbeds lie at depths of 10,000 ft and have not been tested for their methane content. Many sandstone gas fields and structural and stratigraphic traps occur in the Piceance Basin, and a large percentage of these are in the coal-bearing Mesaverde Group. It is probable that many of the fields derived much of their gas from coalbeds.

Productive gas-bearing coal sequences may produce water and gas. Although the Mesaverde Group is not a highly productive aquifer, the producing zones are probably under both gas and artesian pressure. With continued production, water yields will probably be reduced significantly, which could yield to more cost-effective wells and better production from the coalbeds as they are dewatered. Since deep ground water in Piceance Basin is sometimes of poor quality, production of water-bearing gassy zones may lead to waste-water disposal problems.

The following criteria are suggested for delineating target areas for future exploratory drilling in southeastern Piceance Basin:

- Areas where overburden thickness is greater than 2,000 ft
- Areas of thick coals, particularly near the basin center
- Areas of high heat flow, especially where greater than 1.5 to 2 heat flow units
- Areas where epizonal intrusive igneous activity has caused folding and arching of coal-bearing strata, regardless of depth
- Areas along major synclines and anticlines
- Areas where igneous activity has faulted and fractured coal-bearing sequences
- Areas downdip from gassy mines in the direction of higher heat flow values
- Coincidence of any two of the above criteria would constitute a "prime" target area.

Figure 13 shows the authors' estimate of the principal target area in southeastern Piceance Basin. Cumulative coal thickness (Fig. 8), depth of overburden (Fig. 4), coal rank (Fig. 11), and coalbed methane content (Fig. 12) were used to arrive at this area, defined as having the highest coalbed methane production potential in the basin.

ACKNOWLEDGMENTS

We wish to thank Adolph Coors Company, Teton Energy, Twin Arrow, and Western Fuels for their cooperation with DOE-TRW during their drilling operations by providing the opportunity to core and test coalbeds within Piceance Basin to depths of 5,000 ft.

REFERENCES CITED

Averitt, P., 1975, Coal resources of the United States, January 1, 1974: U.S. Geological Survey Bulletin 1412, 131 p.

Collins, B. A., 1976, Coal deposits of the Carbondale, Grand Hogback, and southern Danforth Hills coal fields, eastern Piceance Basin, Colorado: Quarterly of the Colorado School of Mines, v. 71, n. 1, 138 p.

———, 1977, Geology of the coal basin area, Pitkin County, Colorado, *in* H. K. Veal, ed., Exploration frontiers of the central southern Rockies, symposium: Rocky Mountain Association of Geologists, p. 363–377.

Colorado Geological Survey, 1980, Evaluation of the methane potential of unmined-unminable coalbeds in Colorado, review of coal methane resource for western basins: U.S. Department of Energy, Morgantown Energy Technology Center, Morgantown, West Virginia, 7 p.

Dolly, E. D., and F. F. Meissner, 1977, Geology and gas exploration potential, Upper Cretaceous and Lower Tertiary strata, northern Raton Basin, Colorado, *in* H. K. Veal, ed., Exploration frontiers of the central and southern Rockies, symposium: Rocky Mountain Association of Geologists, p. 247-270.

Dunn, H. L., 1974, Geology of petroleum in the Piceance

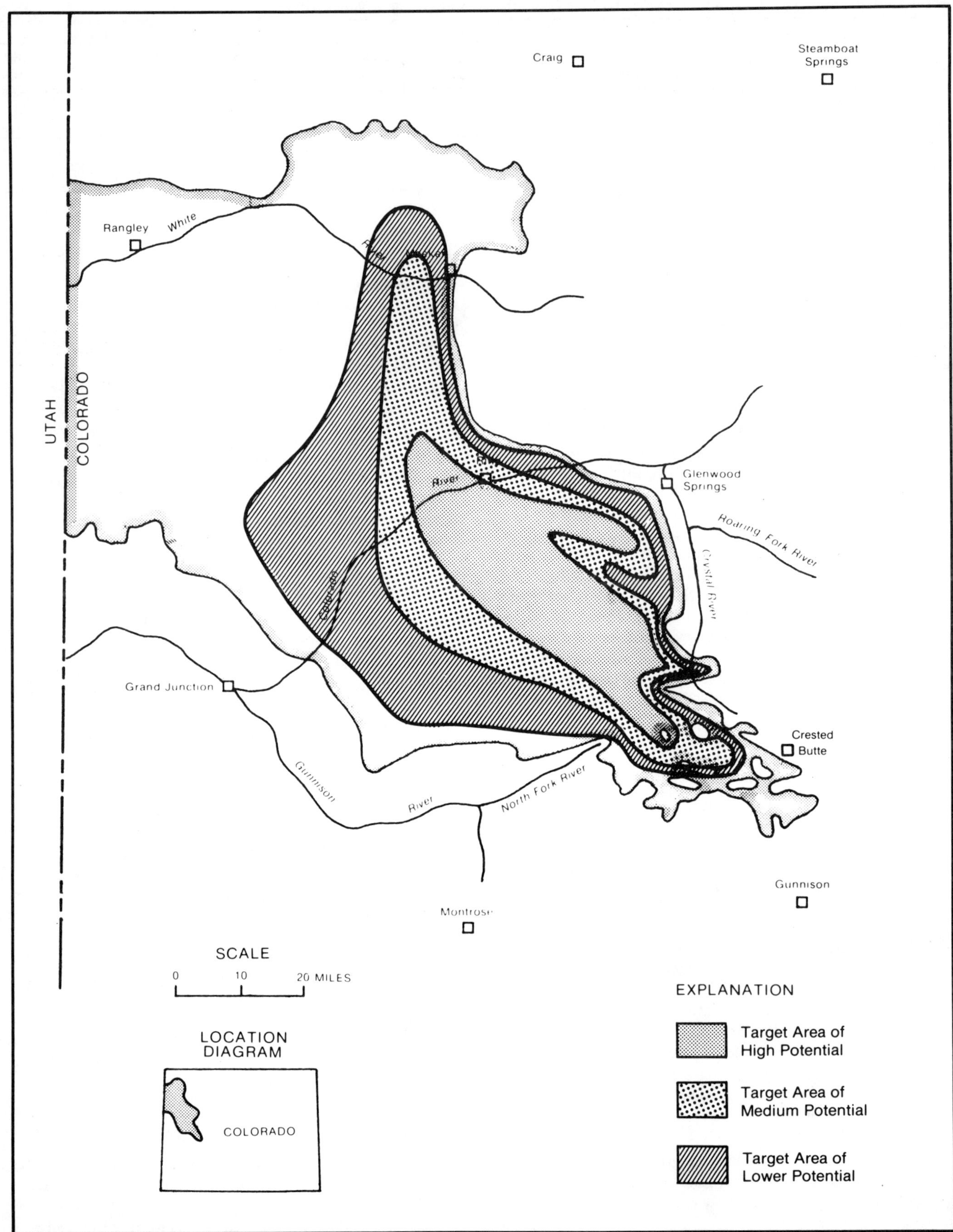

Figure 13—Piceance Basin coalbed methane target area.

Creek Basin, northwestern Colorado, *in* D. K. Murray, ed., Guidebook to the energy resources of the Piceance Creek Basin, Colorado: Rocky Mountain Association of Geologists 25th Annual Field Conference, p. 217–223.

Erdmann, C. E., 1934, The Book Cliffs coal field in Garfield and Mesa counties, Colorado: U.S. Geological Survey Bulletin 851, 150 p.

Fender, H. B., and D. K. Murray, 1978, Data accumulation on the methane potential of coalbeds of Colorado: Colorado Geological Survey Open-File Report 78-2, 25 p.

Freeman, V. L., 1979, Preliminary report on rank of deep coals in part of the southern Piceance Creek Basin, Colorado: U.S. Geological Survey Open-File Report 79-725, 10 p.

Gale, H. S., 1907, Coal fields of the Danforth Hills and Grand Hogback, in northwestern Colorado: U.S. Geological Survey Bulletin 316-E, p. 264–301.

——, 1910, Coal fields of northwestern Colorado and northeastern Utah: U.S. Geological Survey Bulletin 415, 265 p.

Gaskill, D. L., B. L. Bartlson, and E. E. Larson, 1973, West Elk volcanic center, Gunnison County, Colorado—A preliminary report (abs.): Geological Society of America Rocky Mountain Section Annual Meeting, v. 5, n. 6, 481 p.

Godwin, L. H., and D. L. Gaskill, 1964, Post-Paleocene West Elk laccolithic cluster, west-central Colorado, *in* Geological survey research 1964: U.S. Geological Survey Professional Paper 501-C, p. C66–C68.

Hancock, E. T., 1925, Geology and coal resources of the Axial and Monument Butte quadrangles, Moffat County, Colorado: U.S. Geological Survey Bulletin 757, 134 p.

—— and J. B. Eby, 1930, Geology and coal resources of the Meeker quadrangle, Moffat and Rio Blanco counties, Colorado: U. S. Geological Survey Bulletin 812-C, p. 191-242.

Hornbaker, A. L., R. D. Holt, and D. K. Murray, 1976, 1975 Summary of coal resources in Colorado: Colorado Geological Survey Special Publication 9.

Humphrey, H. B., 1960, Historical summary of coal-mines in the United States, 1810–1958: U.S. Bureau of Mines Bulletin 586, 280 p.

Johnson, R. C., and O. May, 1978, Preliminary stratigraphic studies of the upper part of the Mesaverde Group, the Wasatch Formation, and the lower part of the Green River Formation, DeBeque area, Colorado, including environments of deposition and investigations of Palynomorph assemblages: U.S. Geological Survey Miscellaneous Field Studies Map MF-1050, 2 sheets.

——, M. P. Grancia, and N. C. Dessenberger, 1979, Cross section B-B′ of Upper Cretaceous and Lower Tertiary rocks, southern Piceance Creek Basin, Colorado: U.S. Geological Survey Miscellaneous Field Studies Map MF-1130 B.

Kent, B. H. and H. H. Arndt, 1980a, Geology of the Thompson Creek coal mining area, Pitkin County, Colorado, as related to subsurface hydraulic mining potential: U.S. Geological Survey Open-File Report 80-507, 81 p.

——, 1980b, Geology of the Carbondale coal mining area, Garfield and Pitkin counties, Colorado, as related to subsurface hydraulic mining potential: U.S. Geological Survey Open-File Report 80-709, 93 p.

Kintz, B. M., and E. H. Denny, 1933, Explosions in Colorado coal mines, 1883 to 1932: U.S. Bureau of Mines Information Circular 6753, 20 p.

Larson, E. E., M. Ozima, and W. C. Bradley, 1975, Late Cenozoic basic volcanism in northwestern Colorado and its implications on the origins of the Colorado River System, *in* B. F. Collins, ed., Cenozoic history of the southern Rocky Mountains: Geological Society of America Memoir 144, p. 155–178.

Lee, W. T., 1912, Coal fields of Grand Mesa and the West Elk Mountains, Colorado: U.S. Geological Survey Bulletin 510, 237 p.

Lipman, P. W., F. E. Mutschler, B. Bryant, and T. A. Steven, 1969, Similarity of Cenozoic igneous activity in the San Juan and Elk Mountains, Colorado, and its regional significance: U.S. Geological Survey Professional Paper 650-D, p. D33–D42.

Murray, D. K., (ed.,), 1974, Guidebook to the energy resources of the Piceance Creek Basin, Colorado: Rocky Mountain Association of Geologists 25th Annual Field Conference, 302 p.

—— and J. D. Haun, 1974, Introduction to the geology of the Piceance Creek Basin and vicinity, northwestern Colorado, *in* D. K. Murray, ed., Guidebook to the energy resources of the Piceance Creek Basin, Colorado: Rocky Mountain Association of Geologists 25th Annual Field Conference, p. 379-405.

——, H. B. Fender, and D. C. Jones, 1977, Coal and methane gas in the southeastern part of the Piceance Creek Basin, *in* A. K. Veal, ed., Exploration frontiers of the central and southern Rockies, symposium: Rocky Mountain Association of Geologists, p. 379-405.

Quigley, D. M., 1965, Geologic history of Piceance Creek-Eagle Basins: Bulletin of the American Association of Petroleum Geologists, v. 49, n. 11, p. 1974–1996.

Schwochow, S. D., 1978, Mineral resources survey of Mesa County—a model study: Colorado Geological Survey Resource Series 2, 110 p.

Steven, T. A., 1975, Middle Tertiary volcanic field in the southern Rocky Mountains, *in* B. F. Collins, ed., Cenozoic history of the southern Rocky Mountains: Geological Society of America Memoir 144, p. 75–94.

TRW Energy Engineering Division, 1981, Well test report: Adolph Coors, Nichols 1-24, Mesa County, Colorado: U.S. Department of Energy, Morgantown Energy Technology Center, Morgantown, West Virginia, February, 1981, contract No. DE-AC21-78MC09089.

Young, R. G., 1955, Sedimentary facies and intertonguing in the Upper Cretaceous of the Book Cliffs, Utah-Colorado: Geological Society of America Bulletin, v. 55, p. 177–202.

——, 1966, Stratigraphy of coal-bearing rocks of the Book Cliffs, Utah-Colorado, in central Utah coals: Utah Geological and Mineral Survey Bulletin 80, p. 7–20.

Geologic Overview, Coal Deposits, and Potential for Methane Recovery from Coalbeds of the Uinta Basin—Utah and Colorado

M. A. Adams
J. N. Kirr

The Uinta Basin is an east-west asymmetrical syncline located in Utah and Colorado over approximately 14,450 sq mi. For the purpose of this report, the Uinta Basin includes the Wasatch Plateau, a southwest-trending appendage from the Uinta Basin. The in-place coalbed methane resource is estimated to be at least 0.8 and possibly as much as 4.6 Tcf. This resource was evaluated based on desorption data from the Utah Geological and Mineral Survey and Mountain Fuel Resources, Inc.

Major coal-bearing strata of the Uinta Basin and Wasatch Plateau are contained in the Upper Cretaceous Mancos Shale and Mesaverde Group. The area contains seven coal fields or regions located along the perimeter of the basinal area. The Wasatch Plateau and Book Cliffs coal fields are the most extensively developed of the seven, and their major coal deposits are found in the Mesaverde Group Blackhawk and Price River Formations. Little is known about the coal resource at depth in the basin.

INTRODUCTION

The Uinta Basin is one of six major physiographic sections comprising the Colorado Plateau province. Located in the northernmost part of the province, the basin covers 14,450 sq mi in the northeastern corner of Utah and a small part of northwest Colorado (Fig. 1). It slightly overlaps the Piceance Basin in Colorado. The Uinta Coal Basin includes the western part of the Uinta Basin Section and the Wasatch Plateau of the High Plateau Section of the Colorado Plateau province. The plateau contains gently rolling to flat-lying strata, with rugged terrain dissected by canyons.

The basin is an east-west asymmetrical syncline lying immediately south of the Uinta Mountains. To the west lie the Wasatch Mountains and the majority of the High Plateau section. The Uncompahgre Plateau is southeast of the basin. The basin's southern terminus is a double escarpment called the Tavaputs Plateau, also referred to as the Book Cliffs and Roan Cliffs. This escarpment descends 3,000 ft to the Canyon Lands section of the Colorado Plateau province (Thornbury, 1965).

The Uinta is a basin both structurally and topographically, with an average elevation of over 5,000 ft. Near its southern margin (in the Tavaputs Plateau), elevations exceed 10,000 ft (Thornbury, 1965). Much of the area has rugged topography as a result of erosional dissection, although numerous broad benches believed to be pediments are located near the southern base of the Uinta Mountains. In the central portion of the basin, badland topography has developed extensively.

Compared with other areas of the Colorado Plateau Province, the Uinta Basin is not structurally complex (Fig. 2). It is a broad, east-west–trending syncline, asymmetrical on the north side. The area has a broad, gentle south flank with comparatively shallow dips. On the northern edge of the basin, a sharp increase in dip occurs where the strata turn upward against the south flank of the Uinta Mountains (Fig. 3). Faulting within the basin is limited primarily to the northern boundary, where the Uinta Mountains uplift caused high-angle normal faults with displacement generally less than 250 ft (Doelling and Graham, 1972).

The Wasatch Plateau, a southwest-trending appendage from the Uinta Basin, is structurally more complex. The plateau, which belongs to the High Plateau section of the Colorado Plateau Province, is located between the San Rafael Swell on the east and the Great Basin on the west and carries characteristics of both areas. Strata of the eastern area dip gently westward as an extension of the west flank of the San Rafael Swell (Spieker, 1931). The western part of the area is a monoclinal fold more characteristic of the Great Basin (Childs, 1950). Faulting, common in the Great Basin, is the major structural disturbance in the plateau. Normal faults, which resulted from post-Cretaceous tensional stresses, are found in four major north-south–trending fault zones: Musinia, Joe's Valley, Pleasant Valley, and North Gordon. These four fault zones have produced prominent ridges and valleys in the plateau and have altered the paths of many minor streams (Doelling, 1972).

STRATIGRAPHY

Geologic formations in the Uinta Basin range in age from Late Precambrian to Quaternary (Table 1). Exposed strata are primarily quartzite, limestone, shale, and conglomerate. Coal-bearing zones are contained in Mesozoic strata that—along

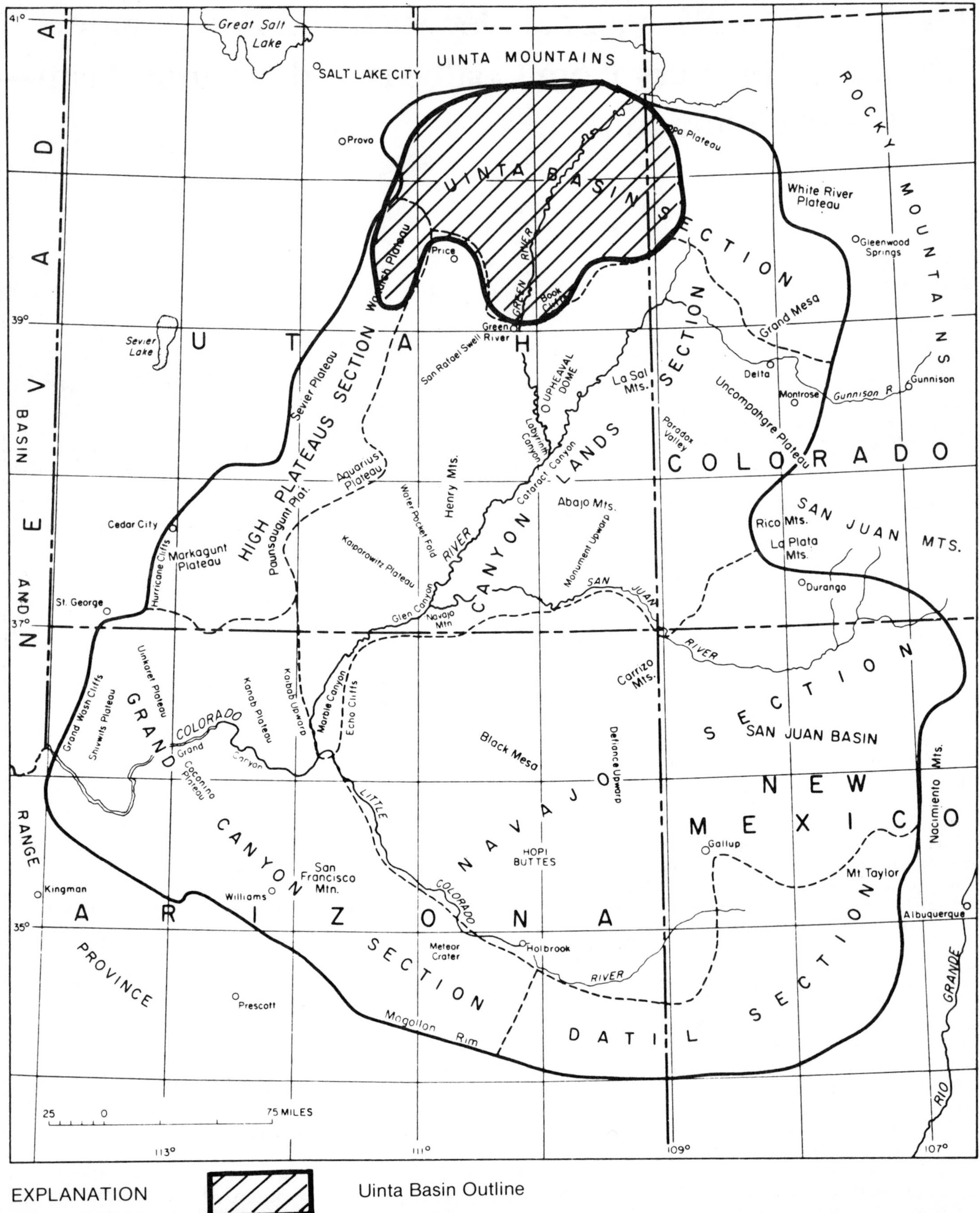

Figure 1—Index map of the Colorado Plateau Province with sectional boundaries, showing the location of the Unita Basin report area. After Hunt, 1956, from Thornbury. Copyright © 1965. Reprinted by permission of John Wiley and Sons, Inc.

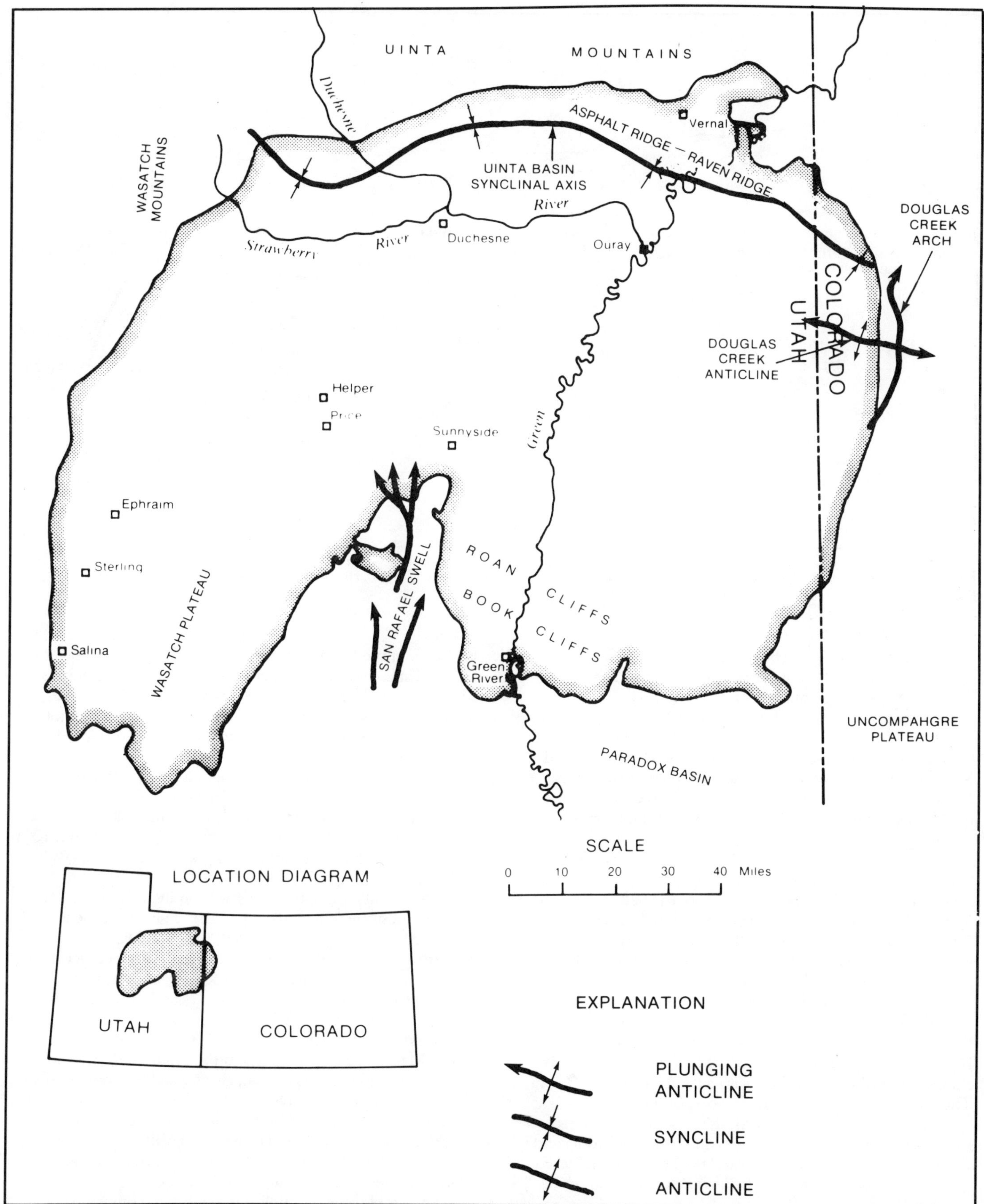

Figure 2—Prominent structural features within and surrounding the Uinta Basin. From Osmond, 1964. Used with the permission of the Utah Geological Association.

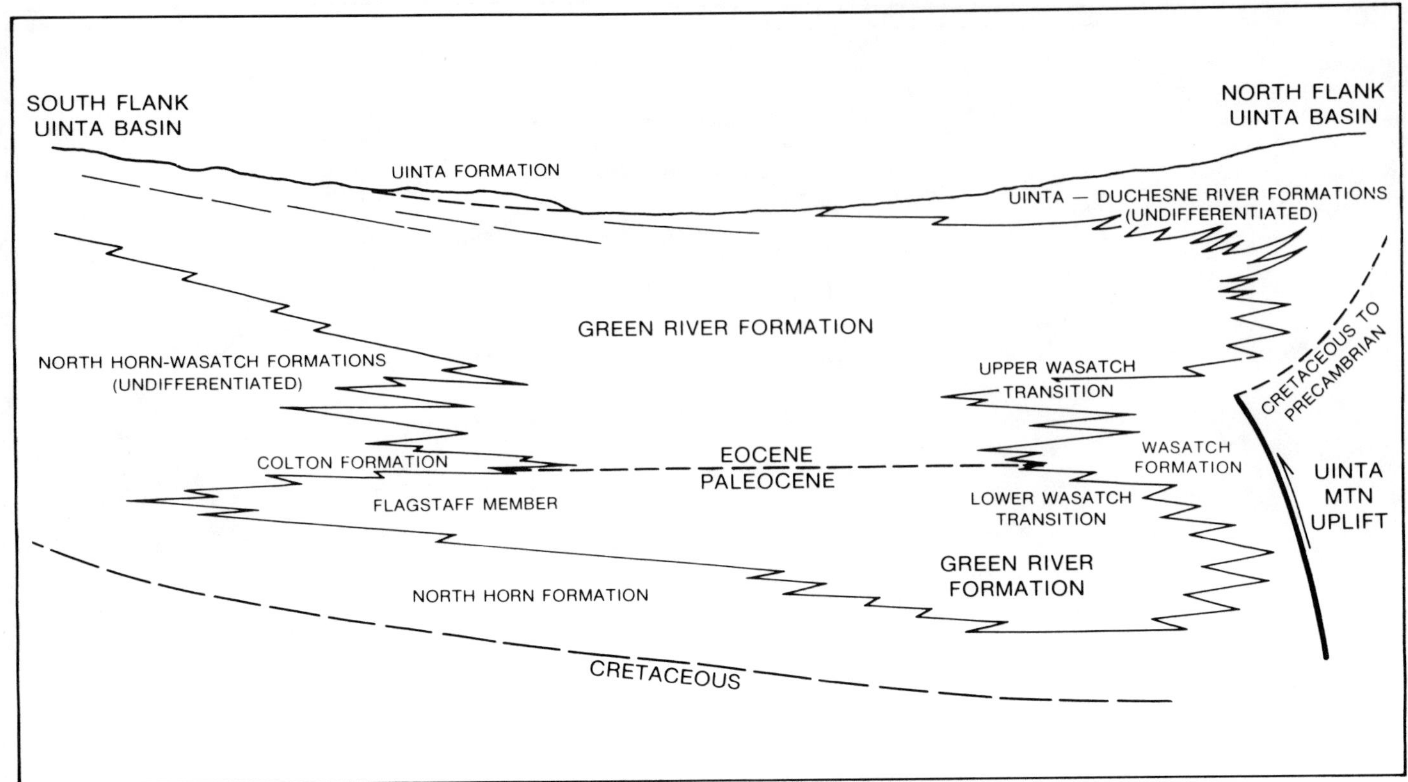

Figure 3—Diagrammatic cross-section north to south across the western part of the Uinta Basin showing relationships among formations of Paleocene to Oligocene age. From Hood and Fields, 1977.

with Cenozoic strata—will be discussed.

The Mesozoic strata range from approximately 5,000 to 20,000 ft thick and include all strata of the Triassic, Jurassic, and Cretaceous Systems. Triassic strata include, in ascending order, the Moenkopi Formation, Shinarump Conglomerate, Chinle Formation, and Navajo Sandstone. The Chinle and Moenkopi Formations are the principal Triassic Age units. The Moenkopi Formation consists of brown-to-green silty shale, and the Chinle Formation consists of red-brown and green silty shale.

The Jurassic strata include the Carmel Formation, Entrada Sandstone, Curtis Formation, and Morrison Formation. The Morrison Formation is considered one of the best marker beds of Jurassic Age strata, and it includes varicolored shales interbedded with sandstones of fluvial origin. This formation contains dinosaur fossils found in Dinosaur National Monument near Jensen, Utah (Walton, 1944).

Cretaceous strata within the Uinta Basin consist of, in ascending order, the Cedar Mountain Formation, Dakota Sandstone, Mancos Shale, Mesaverde Group, and part of the North Horn Formation. The Cedar Mountain Formation is composed of varicolored shale with a basal conglomerate unit. The Dakota Sandstone defines the base of the Upper Cretaceous over the Uinta Basin. In the western portion of the basin, it is a coarse-grained, cross-bedded, tan sandstone; and to the east, the sandstone coarsens and thickens. It is known to split into numerous sandstone beds separated by sandy shale (Walton, 1944).

The valley forming Mancos Shale conformably overlies the Dakota Sandstone. The shale intertongues with the upper boundary of the Mesaverde Group sandstone. The Mesaverde Group, exposed in the northern Uinta Basin in the Tabby Mountain and Vernal areas, is divided into two distinctive sections: the lower (of marine sandstone); and the upper (of brackish-water sandstones, sandy and carbonaceous shales, and coal) (Walton, 1944).

The Tabby Mountain area exposes a representative section of the Mesaverde Group for the western part of the basin. The lower part is more than 850 ft thick and comprised of massive sandstone with fossils. The upper section is greater than 2,100 ft thick and consists of sandstone, sandy shales, and coal (Walton, 1944).

In the Vernal region, the lower section includes a near-shore marine sandstone that thins to the east. The sandstone is composed of two members: Asphalt Ridge and Rim Rock Sandstones, separated by a shale parting. The upper section of the Mesaverde Group includes brackish-water sandstone interbedded with brightly colored shales and thin coalbeds. This section is referred to as the Williams Fork Formation and is poorly exposed throughout the region (Walton, 1944).

The Mesaverde Group, found in the Wasatch Plateau and Book Cliffs coal field to the east, is comparable lithologically to the aforementioned Mesaverde Group (Fig. 4). The group in the plateau includes the basal Star Point Sandstone and the Blackhawk and Price River Formations (Walton, 1944).

The Star Point Sandstone intertongues with the underlying Mancos Shale. The unit ranges in thickness from 200 to 450 ft and consists of three prominent ridge-forming sandstone beds, separated by shale or thin sandstone partings (Spieker, 1931). The sandstone thins to the east over the plateau and adjacent

Table 1—Geologic units that outcrop in the northern Uinta Basin. From Hood and Fields, 1977.

<table>
<tr><th colspan="2" rowspan="2">Geologic Time</th><th colspan="2">Geologic Unit</th></tr>
<tr><th>Western Part of Basin</th><th>Eastern Part of Basin</th></tr>
<tr><td rowspan="8">Cenozoic</td><td rowspan="2">Quaternary</td><td colspan="2">Alluvium, gravel surfaces, talus deposits, dune sand and other windblown deposits</td></tr>
<tr><td colspan="2">a/ Glacial deposits, alluvium of Pleistocene age, and terrace deposits</td></tr>
<tr><td rowspan="6">Tertiary</td><td colspan="2">b/ Browns Park Formation</td></tr>
<tr><td>Extrusive igneous rocks</td><td></td></tr>
<tr><td colspan="2">a/ Duchesne River Formation</td></tr>
<tr><td colspan="2">a/ Uinta Formation</td></tr>
<tr><td colspan="2">Green River Formation</td></tr>
<tr><td colspan="2">Wasatch Formation</td></tr>
<tr><td rowspan="11">Mesozoic</td><td rowspan="4">Cretaceous</td><td>a/ Currant Creek Formation</td><td></td></tr>
<tr><td colspan="2">b/ Mesaverde Group</td></tr>
<tr><td colspan="2">Mancos Shale (including b/ Frontier Sandstone Member)</td></tr>
<tr><td colspan="2">Dakota Sandstone and Cedar Mountain Formation, undivided</td></tr>
<tr><td rowspan="4">Jurassic</td><td colspan="2">Morrison Formation</td></tr>
<tr><td colspan="2">b/ Curtis Formation</td></tr>
<tr><td colspan="2">b/ Entrada Sandstone</td></tr>
<tr><td>Twin Creek Limestone</td><td>Carmel Formation</td></tr>
<tr><td>Jurassic and Triassic</td><td colspan="2">a/ Glen Canyon Sandstone, Nugget Sandstone</td></tr>
<tr><td rowspan="2">Triassic</td><td colspan="2">Chinle Formation (including b/ Gartra Member)</td></tr>
<tr><td>Mahogany Formation
Thaynes Formation (or Group)
Woodside Formation</td><td>Moenkopi Formation</td></tr>
<tr><td rowspan="6">Paleozoic</td><td>Permian</td><td colspan="2">Park City Formation (or Group)</td></tr>
<tr><td>Permian and Pennsylvanian</td><td colspan="2">a/ Weber Quartzite</td></tr>
<tr><td>Pennsylvanian</td><td colspan="2">b/ Morgan Formation</td></tr>
<tr><td>Pennsylvanian and Mississippian</td><td colspan="2">Manning Canyon Formation</td></tr>
<tr><td>a/ Mississippian</td><td>Upper Mississippian Rocks, undivided
Lower Mississippian Rocks, undivided</td><td>Mississippian Rocks, undivided</td></tr>
<tr><td>Cambrian</td><td>Tintic Quartzite</td><td>Lodore Formation</td></tr>
<tr><td rowspan="3"></td><td rowspan="3">Precambrian</td><td>Red Pine Shale of Uinta Mountain Group</td><td></td></tr>
<tr><td colspan="2">Unnamed quartzite unit of Uinta Mountain Group</td></tr>
<tr><td>Lower Part of the Uinta Mountain Group, undivided</td><td></td></tr>
</table>

Book Cliffs area, illustrating the retreating phase of the Mancos sea (Spieker and Reeside, 1925).

A sharp, conformable boundary separates the underlying Star Point Sandstone from the overlying Blackhawk Formation. The member is a slope-forming unit, sandwiched between two prominent ridge formers. The formation ranges from 700 to 1,000 ft thick and consists of sandstone, shale, and coal (Spieker, 1931).

The Blackhawk Formation in the Wasatch Plateau and the Book Cliffs is the major coal-bearing unit of the Mesaverde Group. Thick coalbeds are found in the lower half of the formation with thin coalbeds present through the total unit. The Hiawatha coalbed is the formation's basal bed and clearly separates the Blackhawk Formation from the Star Point Sandstone (Spieker, 1931).

The overlying Price River Formation is distinctive from the Blackhawk Formation because it is generally coarser grained and contains conglomerates (Spieker, 1931). In addition to the conglomerate, the Price River Formation also consists of sandstone, conglomerate, and a small amount of shale.

The Cretaceous and Tertiary age North Horn Formation consists of varicolored shales with interbeds of sandstone. The formation also contains small amounts of conglomerate, coal, and limestone. The type locality for this formation is the North Horn Mountain, where the formation is 1,650 ft thick. To the east, this formation thins to less than 400 ft thick (Williams, 1950).

The Eocene Series of the Cenozoic Tertiary System includes the Wasatch, Green River, and Uinta Formations. The Wasatch Formation—also known as the Colton Formation—has an exposure in the extreme eastern part of the Uinta Basin, where it forms Raven Ridge. The formation, composed of varicolored

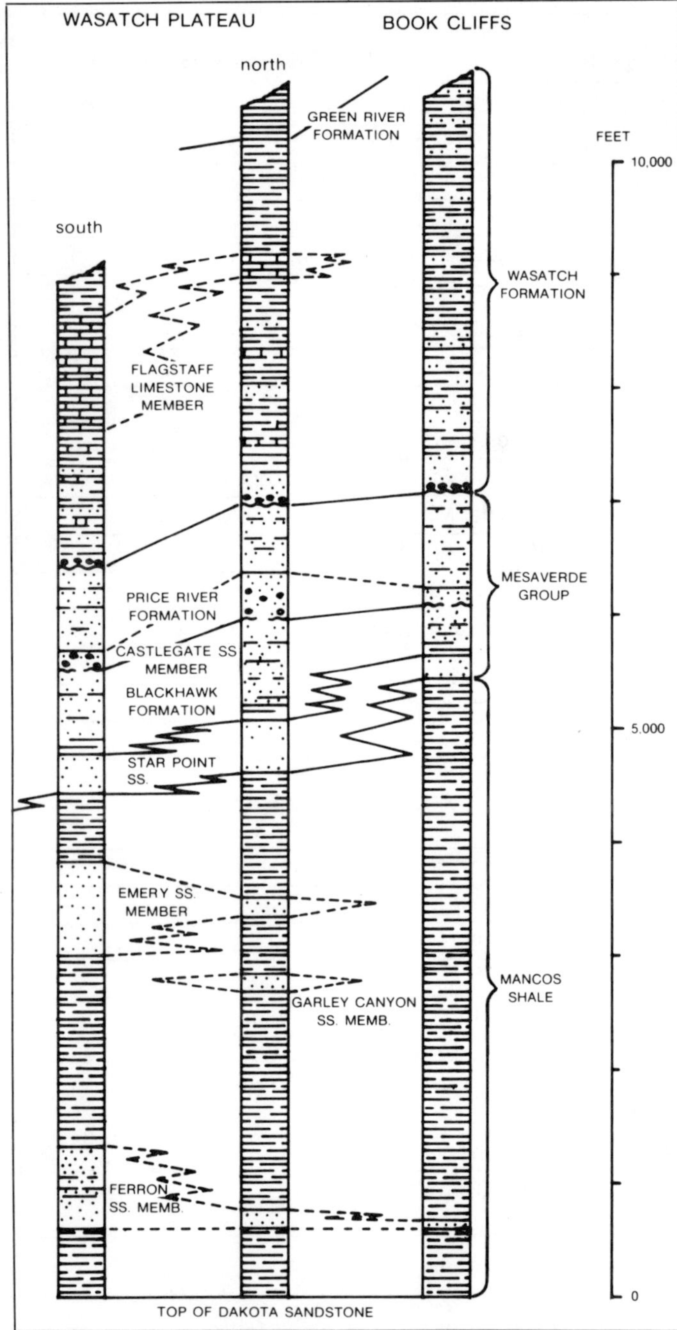

Figure 4—Columnar sections of the Cretaceous and Tertiary formations of the Wasatch Plateau and Book Cliffs. Spieker and Reeside, 1925.

shales with thin bedded sandstone, is 850 ft thick at Raven Ridge and thickens to greater than 3,000 ft to the west. In many areas, it is difficult to distinguish the Wasatch Formation from the underlying North Horn and Tuscher Formations.

In the western part of the basin, the Wasatch Formation is completely overlapped by the Green River Formation; while to the far southwest, the Wasatch Formation of the Wasatch Plateau is devoid of any definable overlying Green River Formation (Spieker and Reeside, 1925). Outcrops of the overlying Green River Formation occur in the southern part of the Uinta Basin. This formation includes gray-green to white clay shale and hard blocky sandstones a few feet thick, in addition to oil shales and bituminous sandstone in the middle of the section. The formation's thickness ranges from 1,800 ft at the Duchesne County line to greater than 5,000 ft near the town of Duchesne, Utah. In the south and middle portions of the basin, Green River strata are conformably overlain by the Uinta Formation, whereas the north side of the basin is marked by an unconformity. In this area the Duchesne River Formation directly overlies the Green River Formation.

The Uinta Formation outcrops in the central area of the Uinta Basin. The formation is conformable, with the overlying Duchesne River Formation to the north and the underlying Green River Formation to the south. The unit is composed of variegated shales and sandstone beds that vary from less than 1 to greater than 50 ft thick. Total formation thickness ranges from 700 to 5,400 ft, depending on the boundary placement above and below the unit.

It is questionable if the Duchesne River Formation was deposited within the Eocene or Oligocene depositional cycle. The formation is composed of reddish-orange shales, buff sandstone with shale interbeds, and conglomerate. The formation ranges from 1,300 to 3,000 ft thick and is subdivided into four lithostratigraphic units. In ascending order, they are: the Brennan Basin Member, Dry Gulch Creek Member, Lapoint Member, and Starr Flat Member (Anderson and Picard, 1972).

The Bishop Conglomerate is the main member of the Miocene Series. It lies with an angular unconformity on rocks varying in age from Precambrian to Oligocene. The conglomerate consists of boulders 1 to 6 ft in diameter, intermixed with gravel and sand, and it ranges up to a maximum of 500 ft thick.

Quaternary

Pleistocene glacial drift covers the north edge of the Uinta Basin, where river valleys emerge from the mountains. Alluvium is evident along most of the larger stream channels.

COAL RESOURCES BY AREA

Coals of the Uinta Basin and Wasatch Plateau, which have been studied and analyzed, are contained within seven coal fields or regions located around the Uinta Basin's perimeter and over the majority of the Wasatch Plateau (Fig. 5). The Uinta Basin includes a small part of northwest Colorado, west of the Douglas Arch; however, because of that area's limited coal content, it will not be discussed in this report.

Counterclockwise from the northeast corner of the basin, the coal areas are the Vernal coal field, Tabby Mountain coal field, Sevier-Sanpete region, Wasatch Plateau coal field, Emery coal field, Book Cliffs coal field, and Sego coal field. Important coals within this region are of Cretaceous age (Table 2).

Environment of Deposition—Coal

During the Cretaceous period, marine water covered the Uinta Basin from its east edge to the western boundary of the Basin and Range province. From Early to Late Cretaceous, subsidence and deposition occurred in the Uinta Basin until the tectonic activity of the Laramide orogeny pushed the shoreline eastward. Transgression and regression of the seas and the associated deltaic environment created conditions favorable to

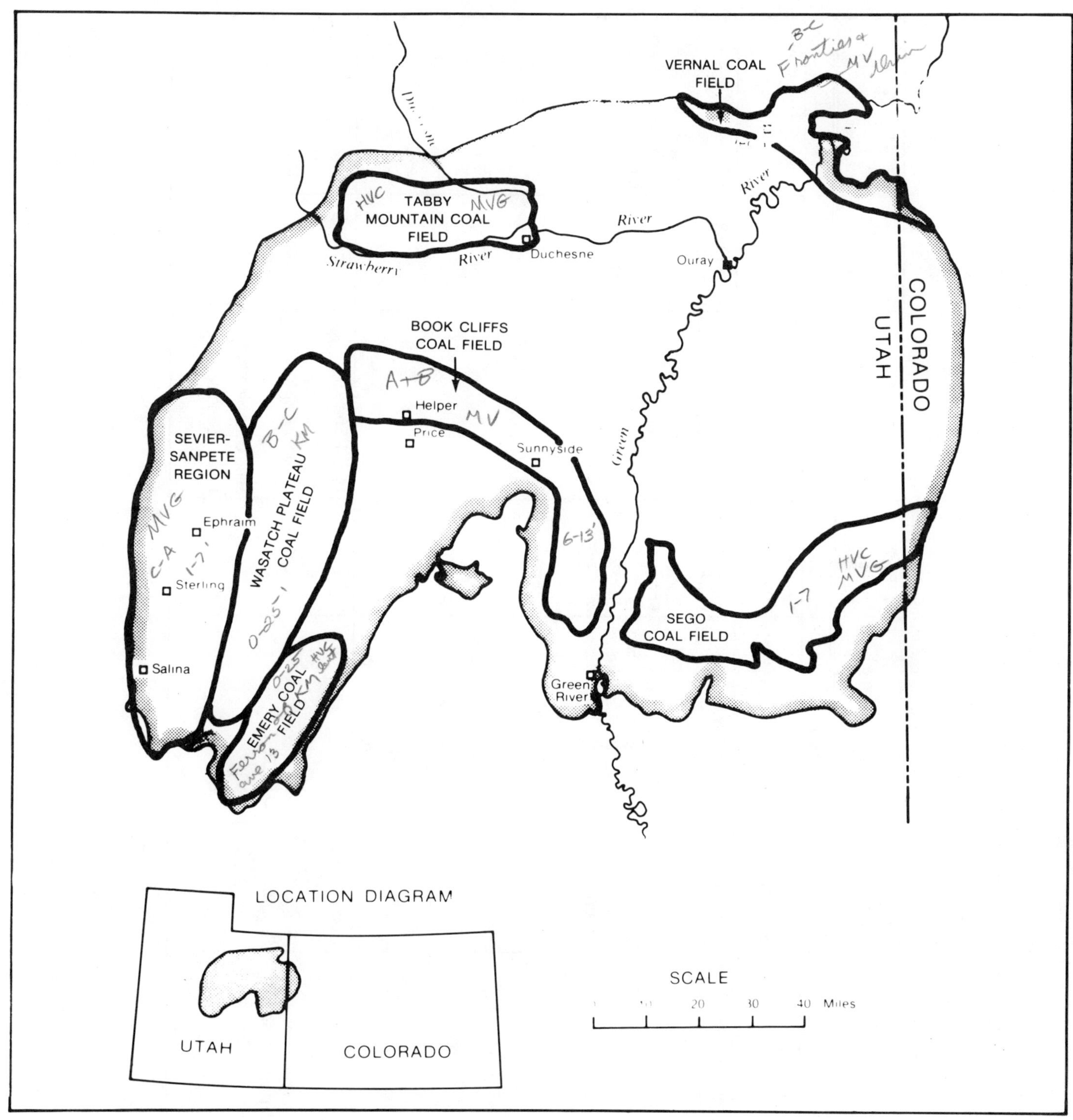

Figure 5—Map showing the location of the Uinta Basin coal fields and region.

the growth and accumulation of large quantities of vegetation that formed coalbeds up to 25 ft thick. Coals formed during the transgressive phase of the seas are the Ferron and Frontier coals, located in the Tabby Mountain and Vernal coal fields. Coals formed during the regressive cycle of the seas are contained in the Blackhawk Formation of the Wasatch Plateau coal field, the Book Cliffs coal field, and the Neslen Member coal of the Sego coal field (Doelling, 1972).

Vernal Coal Field

The Vernal coal field, located in northeast Utah, lies principally in Uintah County and covers the northeast corner of the basin. It is 40 mi long east to west; 24 mi wide north to south. The major coalbeds are found in two Cretaceous age formations, the Frontier Sandstone and the Mesaverde Formation. Coalbeds within the two formations generally are thin, split, and steeply dipping (Doelling and Graham, 1972).

Table 2—Divisions of the major Cretaceous coal-bearing units within the Uinta Basin coal fields and region. After Doelling, 1972.

Period	Geologic Unit	Vernal Coal Field	Tabby Mountain Coal Field	Sevier-Sanpete Region	Wasatch Plateau Coal Field	Emery Coal Field	Book Cliffs Coal Field	Sego Coal Field
Tertiary		Wasatch Formation		Flagstaff Limestone				
Cretaceous			Currant Creek Formation	North Horn Formation* (Wales) Coal Field)	North Horn Formation	North Horn Formation	North Horn Formation	Tuscher Formation
	Mesaverde Group	Williams Fork Formation* Rim Rock Sandstone* Asphalt Ridge Formation	Upper Mesaverde Group* Lower Mesaverde Group	Price River Formation Blackhawk Formation* (Mount Pleasant and Salina Canyon Coal Fields) South Flat Formation Six Mile Canyon Formation* (Sterling Coal Field) Funk Valley Formation Allen Valley Shale Sanpete Formation	Price River Formation Blackhawk Formation* Star Point Sandstone	Price River Formation Blackhawk Formation Star Point Sandstone	Price River Formation Blackhawk Formation* Star Point Sandstone	Price River Formation* (Neslen Member)
	Mancos Shale	Frontier Sandstone* (Upper Member)	Upper Shale Member Frontier Sandstone* Mowry Shale	Mancos Shale	Mancos Shale	Masuk Shale Emery Sandstone Blue Gate Member Ferron Sandstone* Tununk Shale Member	Mancos Shale	Mancos Shale

*Major coal formation.

Table 3—Average ultimate analyses (as-received) of coalbeds and zones within the Uinta Basin coal fields. From Doelling, 1972; and Doelling and Graham, 1972.

Source	Coal Field	Coalbed/Zone	Sulfur Wt. %	Hydrogen Wt. %	Carbon Wt. %	Nitrogen Wt. %	Oxygen Wt. %	Number of Tests
1	Vernal West	Frontier Zone	1.7	5.5	63.7	1.1	18.6	6
2	Wasatch Plateau Central	Hiawatha Bed	0.6	6.0	71.6	1.4	13.8	5
		Bear Canyon Bed	0.5	6.3	74.2	1.4	13.3	5
2	Emery North	Upper Coal Zone No. 1 Bed	1.0	5.1	70.7	1.3	11.8	3
2	Book Cliffs West	Castlegate A Bed	0.5	5.7	71.8	1.4	14.5	1
		Castlegate B Bed	0.7	5.6	69.6	1.4	13.8	2
		Royal Blue Bed	0.7	5.6	72.3	1.3	15.1	1
		Castlegate C Bed	0.5	5.8	73.1	1.4	14.5	1
		Castlegate D Bed	0.4	5.7	74.0	1.4	11.2	1
		Kenilworth Bed	0.6	5.9	71.9	1.4	13.1	1
	East	Lower Sunnyside Bed	1.0	5.8	73.8	1.5	13.1	2
		Sunnyside Bed	1.1	5.7	73.4	1.6	12.6	2
1	Sego West	Palisade Bed	0.7	5.2	63.9	1.5	16.3	2
		Chesterfield Bed	0.8	5.3	67.2	1.5	4.9	3
		Ballard Zone	0.7	5.3	62.6	1.4	15.5	2

Source: 1. Doelling and Graham, 1972.
2. Doelling, 1972.

In this field, the Frontier Sandstone ranges from 150 to 300 ft thick and is composed of clay shale, silty and sandy shale, carbonaceous shale, sandy siltstone, sandstone, and coal. The sandstone is divided into lower and upper members. The lower member ranges from 75 to 203 ft thick and consists of clay shale, silty to sandy shale, and shaly sandstone. The upper member, the major coal-bearing unit, is divided into three units: a lower, thick, massive sandstone; a 10 to 50 ft thick coal zone; and a resistant sandstone caprock. The coal generally is split and is ranked as high-volatile B or C bituminous with a medium sulfur content (Doelling and Graham, 1972). Tables 3 and 4 show a comparison of ultimate and proximate analyses of coalbeds and zones within the seven coal fields of the Uinta Basin.

The Mesaverde Formation is divided into four units: the Asphalt Ridge Member, the Rim Rock Member, and the third and forth units. The coals within this formation generally are subbituminous C rank. The Asphalt Ridge Sandstone is the basal unit of the formation and is approximately 110 ft thick along Asphalt Ridge. The member is composed of yellow to white, fine- to medium-grained, subangular sandstone and siltstone. The Rim Rock Sandstone is composed of fine- to coarse-grained sandstone interbedded with shale, coal, and carbonaceous shale and ranges from 150 to 530 ft thick. The coal within this unit is found east of the Green River and consists of a few lenticular, thin coalbeds (Doelling and Graham, 1972).

The third and fourth units also are known as the Upper Mesaverde Williams Fork Formation. The third unit is a medium-resistant interbedded sandstone and shale, while the fourth unit is composed of soft sandstone and shale. Within the third and fourth units are numerous lignitic coalbeds and a few bituminous coals. The coal has a low sulfur content and improves in quality to the east.

Original resources in the field for the Frontier Sandstone and Mesaverde Formation have been estimated to be 164.3 and 12.9 million short tons, respectively (Doelling and Graham, 1972).

Tabby Mountain Coal Field

The Tabby Mountain coal field (see Fig. 5) is located on the southwest flank of the Uinta Mountains. Extending 35 mi east to west and varying from 1 to 3 mi north to south, the field covers an estimated 70 sq mi of Wasatch and Duchesne counties in Utah (Doelling and Graham, 1972).

Coals in this field are contained in two steeply dipping Upper Cretaceous formations: the Frontier Sandstone and the Mesaverde Formation. These two formations have multiple coal zones, and each contains thick coalbeds. The Frontier Sandstone is a gray to yellow, fine- to coarse-grained, thin to thickly bedded sandstone with subbituminous C rank coal. In one section of the sandstone, along Red Creek in Sections 16 and 21, T.15S., R.9W., 27 ft of coal was measured in four coalbeds (Doelling and Graham, 1972). The upper section of the Mesaverde Formation contains the greatest number of coalbeds of the two formations. The formation ranges from 550 to 4,000 ft thick within this coal field and contains coal ranked as high-volatile C bituminous. In one section of the Mesaverde Formation, along Red Creek in Sections 22, 23, and 26, T.15S., R.9W., over 29 ft of coal was measured (Doelling and Graham, 1972).

Total resources in the coal field are estimated to be 1.8 billion short tons, with approximately 231 million short tons found in coalbeds greater than 4 ft thick and under less than 3,000 ft of cover. Of the 231 million tons, 168 are located in the Mesaverde Formation; the Frontier Sandstone contains the remaining 63 million tons. Total past production in the field is

Table 4—Selected proximate analyses (as-received) of coal within the Uinta Basin coal fields and region. From Doelling, 1972; and Doelling and Graham, 1972.

Source	Coal Field or Region	Coal	No. of Samples	Moisture (%)	Volatile Matter (%)	Fixed Carbon (%)	Ash (%)	Sulfur (%)	Btu/lb
1, 4	Vernal	Frontier Formation Average	—	9.1	39.4	48.5	12.5	1.6	11,509
		Mesaverde Formation Average	—	12.8	42.7	39.9	17.5	0.8	8,950
4	Tabby Mountain	Frontier Sandstone Average	—	15.4	34.3	36.3	13.0	0.8	8,450
		Winchester Bed	3	19.4	33.5	36.9	10.2	0.7	8,110
1		Mesaverde Lower Zone	1	19.2	37.0	37.6	6.2	0.7	8,723
		Mesaverde Formation Average	—	13.6	38.8	41.5	6.7	1.1	10,285
		Fraughton Bed	2	15.4	39.2	39.4	7.1	0.9	10,321
2	Wasatch Plateau	Field Average	—	6.1	43.4	47.4	6.7	0.6	12,589
	South	Ferron	1	11.9	36.4	42.7	9.0	0.5	—
		Hiawatha Bed	2	6.6	41.0	47.0	5.7	0.7	12,345
		Upper Hiawatha Bed	1	12.9	—	—	5.5	0.5	—
		Muddy No. 1 Bed	1	8.4	39.1	45.2	7.3	0.5	11,922
		Ivie Bed	2	13.4	36.2	43.8	6.7	0.6	10,570
	Central	Hiawatha Bed	4	5.1	45.0	45.2	4.9	0.6	13,243
		Blind Canyon Bed	5	4.8	41.1	45.6	9.1	0.6	12,650
		Bear Canyon Bed	6	6.0	43.8	45.5	4.6	0.5	13,410
2	North	Hiawatha Bed	2	7.3	37.5	49.1	4.9	0.7	11,696
		Hiawatha Lower Split	2	9.6	38.1	45.3	3.9	0.5	10,550
		Wattis Bed	1	7.3	41.0	47.7	4.0	0.9	12,530
		Castlegate "A" Bed	5	7.6	40.6	45.7	5.0	0.5	12,500
		Bob Wright Bed	1	2.8	46.0	46.9	4.3	0.5	13,070
2, 4	Sevier-Sanpete	Mount Pleasant Field	7	3.4	42.9	45.6	8.1	0.8	12,890
	Region	Salina Canyon Field	12	7.6	42.2	42.2	9.8	0.5	11,367
		Sterling Field	1	8.2	42.6	43.2	6.1	0.9	11,767
		Wales Field	9	5.0	34.2	46.4	14.8	4.3	10,119
2	Emery	Field Average	(#)	(47)	(46)	(46)	(47)	(46)	(44)
				7.4	37.7	44.8	8.5	1.0	11,450
	North	Lower Coal Zone, "C" Bed	3	8.1	35.6	41.4	14.9	0.9	10,245
		Upper Coal Zone, "I" Bed	4	4.8	40.6	46.9	7.8	1.9	12,498
	South	Ferron Bed	1	23.6	32.6	33.3	10.6	2.9	7,823
		Lower Coal Zone, "A" Bed	3	13.0	33.4	41.4	12.2	0.4	8,801
		"M" Bed	1	10.2	36.2	37.6	16.0	0.9	9,526
		Upper Zone	2	12.4	36.5	41.5	9.6	1.7	10,232
		Ivie Bed	3	12.9	36.3	43.4	7.4	0.6	10,570
3	Book Cliffs	Field Average	—	4.8	—	49.1	6.7	0.9	12,760
		Castlegate Area Average	—	4.3	42.6	46.4	6.6	0.5	12,825
		Soldier Canyon Area Average	—	4.8	38.6	49.3	7.0	0.5	12,531
		Sunnyside Area Average	—	5.0	38.2	50.5	6.4	1.1	12,648
		Woodside Area Average	—	5.5	37.5	50.1	6.7	0.7	12,664
2	East	Lower Sunnyside Bed	4	5.9	36.2	51.7	6.4	0.8	12,420
		Sunnyside Bed	2	4.7	38.6	51.3	5.5	1.1	13,384
2	Book Cliffs West	Castlegate "A" Bed	6	5.4	41.1	48.0	5.5	0.5	12,827
		Castlegate "B" Bed	4	4.8	41.9	45.6	7.9	0.5	12,650
		Castlegate "C" Bed	3	3.9	41.7	49.4	4.8	0.5	13,118
		Castlegate "D" Bed	2	4.6	45.2	43.1	7.5	0.4	12,730
		Gilson Bed	2	6.0	39.7	50.0	4.4	0.5	13,085
		Sunnyside Bed	2	4.7	—	—	4.5	0.5	—
		Beckwith Zone	1	4.8	33.6	50.2	11.4	1.2	—
		Rock Canyon Bed	4	5.2	39.1	49.4	6.5	1.1	12,600
		Kenilworth Bed	4	4.5	41.5	45.9	8.2	0.4	12,553
		Liberty Bed	2	5.2	43.1	45.0	6.6	0.9	12,703
		Aberdeen Bed	2	4.5	40.9	48.1	6.8	0.5	12,642
		Royal Blue Bed	2	4.2	42.6	48.5	4.7	0.7	12,940
		Royal No. 2 Bed	2	3.0	43.0	47.8	6.3	0.5	13,136
1, 3	Sego	Field Average	(#)	(27)	(23)	(23)	(27)	(27)	(18)
				9.1	34.7	46.8	11.1	0.6	10,940
1	East	Carbonera Zone	1	10.0	42.2	41.2	6.6	0.6	10,270

continued

Table 4—(Continued)

Source	Coal Field or Region	Coal	No. of Samples	Moisture (%)	Volatile Matter (%)	Fixed Carbon (%)	Ash (%)	Sulfur (%)	Btu/lb
1	West	Palisade Zone	(#)	(6)	(5)	(5)	(6)	(6)	(4)
				7.7	35.9	44.6	10.9	0.6	11,610
		Ballard Zone	4	10.9	32.2	42.7	14.2	0.6	10,230
		Chesterfield Zone	(#)	(15)	(12)	(12)	(15)	(15)	(8)
				8.4	34.7	46.5	10.9	0.6	11,247

Source:
1. Doelling and Graham, 1972.
2. Doelling, 1972.
3. Doelling, Smith, and Davis, 1979.
4. Keystone Manual, 1979.

estimated as only a few thousand tons (Doelling and Graham, 1972).

Sevier-Sanpete Region

The Sevier-Sanpete region consists of four separate coal areas (see Fig. 5) in Sevier and Sanpete counties in central Utah, covering approximately 77 sq mi. These fields are the Mount Pleasant, Sterling, Wales, and Salina Canyon.

The Mount Pleasant coal field is located in the northeast corner of the region. Major coalbeds occur in the Upper Cretaceous lower Blackhawk Formation with minor, thin coalbeds in the North Horn, Price River, and upper Blackhawk Formations. The lower Blackhawk Formation contains six coalbeds referred to as A to F in descending order. These coalbeds possibly are continuous with the Wasatch Plateau coals. The thickest coalbed is the top A bed, which averages 5.7 ft thick and occurs approximately 955 ft from surface level. The six coalbeds are high-volatile C to A bituminous in rank (Doelling, 1972). Coal resources in the Blackhawk Formation for this field are estimated to be 249 million tons (Doelling, 1972).

The Sterling coal field is located on the west edge of the Wasatch Plateau in the central part of the region. The Cretaceous Six Mile Canyon Formation is the major coal-bearing zone and contains six coalbeds or zones that range from 1 to 4 ft thick. The coalbeds or zones are separated by sandstone intervals that range from 3 to 75 ft thick. The coal primarily dips 15° to 30° southeast, with a limited area in the north end of the field dipping 20° to 50° east (Doelling, 1972). The split, lenticular coalbeds have two principal joint patterns: one oriented north to south, and the other east to west. One analysis has been completed for the B coalbed, ranking it as high-volatile C bituminous (Doelling, 1972). Minor coalbeds are located in the Blackhawk Formation and Castlegate Sandstone in the Sterling coal field. The Blackhawk Formation contains an estimated 2 million tons, with resource estimates for most coal within the field lacking because of limited studies of the area.

The Wales coal field is located on the east flank of the Gunnison Plateau in the northern half of Sanpete County. Major coalbeds are in the middle unit of the Cretaceous North Horn Formation, with minor coals found in the Colton and South Flat Formations. Coalbeds within this field are steeply dipping, lenticular, and badly split by bone and shale. The coals are high in ash and sulfur content and generally are bituminous in rank. The coalbeds dip 10° to 16° west and have been known to dip as steeply as 64° west. The last mining operation within the Wales coal field occurred in 1955; the total amount of coal produced was 175,000 tons (Doelling, 1972). Resource estimates are not available for this field.

The Salina Canyon coal field, in the southern half of the Sevier-Sanpete region, contains three important coal zones in the Cretaceous upper Blackhawk Formation. The two uppermost zones contain thin beds, whereas the lower zone contains thick coalbeds that range up to 6.5 ft thick. These coals are located approximately 110 ft above the base of the Blackhawk Formation. As of 1955, 430,000 tons of coal had been mined out of the field, with a total resource estimate for the Blackhawk Formation being approximately 86.4 million tons (Doelling, 1972).

Wasatch Plateau Coal Field

The Wasatch Plateau coal field (see Fig. 5) covers approximately 1,100 sq mi of Emery, Carbon, Sevier, and Sanpete counties. The coalbeds are found in the Upper Cretaceous Blackhawk Formation and dip gently in a west-northwesterly direction. Numerous fault zones cut through the coal field parallel to the north-south axis of the field, displacing the coals. On the field's east boundary, coal crops out on cliff faces; while to the west, the Wasatch monocline and faulting cause the coal to drop under 2,500 ft of cover (Doelling, Smith, and Davis, 1979). The northeast portion of the field is separated from the Book Cliffs coal field by the North Gordon Fault Zone. Along the south boundary of the coal field, the coal is almost entirely under volcanics.

The Blackhawk Formation is 700 to 1,500 ft thick and consists of sandstone, shale, carbonaceous shale, and coal. The coalbeds are located in the lower 300 ft of the formation. Twenty-two coalbeds greater than 4 ft thick have been identified, with the most important being the Hiawatha, Castlegate "A," Blind Canyon, and Wattis coalbeds (Doelling, 1972).

The coalbeds' rank decreases toward the southern end of the field. To the north, they are ranked as high-volatile B bituminous; while coalbeds in the southern end of the local field are high-volatile C bituminous, although there are local areas of high-volatile A bituminous coal (Doelling, 1972).

Resources of the Wasatch Plateau coal field are 6,230 million tons, with potential reserves (comprised of coalbeds greater than 4 ft thick and under less than 3,000 ft of cover) estimated to be 3,888 million tons (Doelling, 1972).

Emery Coal Field

The Emery coal field (see Fig. 5), which covers approximately 210 sq mi, lies along the Emery-Sevier county line directly east of the lower third of the Wasatch Plateau coal field (Doelling, 1972).

The important coalbeds are found in the upper portion of the Ferron Sandstone Member of the Mancos Shale. The coal crops out in the southeast margin of the field below the sandstone cliffs, where the beds dip 3° to 5° westward. The coal thins to the north and disappears in outcrops. To the west and northwest, coal is buried under overlying formations, and to the south, beneath volcanics. Thirteen coalbeds have been defined in four zones, with total coal zone thickness ranging from 400 to 500 ft. The coalbeds are labeled A through M, in ascending order, with the most important coalbeds being the A and C beds in the lower coal zone and the I bed in the upper coal zone. Coalbeds within the Ferron Sandstone Member are high-volatile C bituminous and are generally lenticular and discontinuous (Doelling, 1972). The 13 coalbeds have an average thickness of 13 ft, with six coalbeds greater than 4 ft thick. In some instances, two coalbeds will coalesce to form a coal body up to 25 ft thick (Doelling, Smith, and Davis, 1979).

In the Emery coal field, original coal resources were estimated to be greater than 1.4 billion tons in beds greater than 4 ft thick and under less than 3,000 ft of cover. Seventy-five percent of the resources are estimated to be under less than 1,000 ft of cover (Doelling, 1972).

Book Cliffs Coal Field

The Book Cliffs coal field (see Fig. 5) extends from the North Gordon fault zone, 70 mi southeast along the Book Cliffs to the Green River. The coal field averages 4 mi wide and covers an estimated 280 sq mi. The field is a northward-dipping monocline with coal exposed along cliffs at the south boundary of the Uinta Basin. The cliff-front is irregular and deeply cut by perpendicular drainages (Doelling, 1972).

Within the field, major coal deposits are found in the Upper Cretaceous Blackhawk Formation, ranging in thickness from 400 to 1,300 ft. The formation was deposited during the sea's eastward regression in the Cretaceous period, causing coalbeds to the west to be slightly older than those in the east. Individual coalbeds dip an average of 3° to 7° northeast and reach a maximum of 25 ft thick with average coalbed thickness ranging from 6 to 13 ft (Fig. 6). The coal has a low sulfur content and a high-volatile B bituminous rank.

Along the western margin of the coal field, partings of carbonaceous mudstone and/or bone shale divide some of the coalbeds into splits. To the east, a few coalbeds contain partings of carbonaceous mudstone or sandstone tongues. Volcanic ash partings have been discovered in Upper Cretaceous coals, which are older than the Upper Cretaceous Blackhawk Formation.

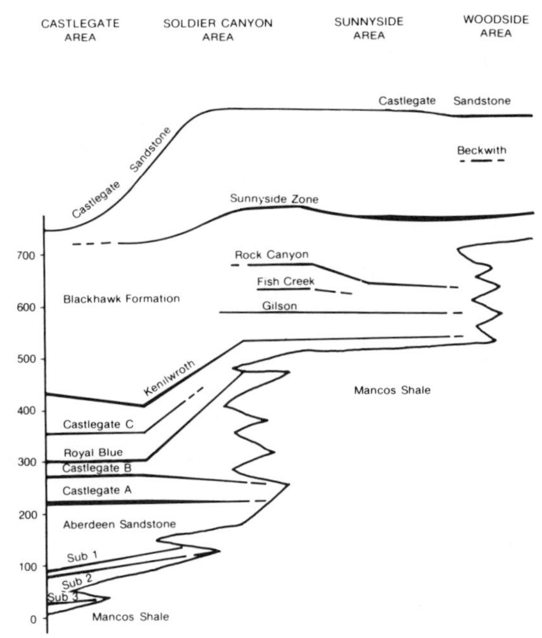

Figure 6—Correlation of coalbeds of the Book Cliffs coal field. Doelling, Smith, and Davis, 1979.

Coal within the Book Cliffs coal field has well-developed cleat and contains cone-shaped masses of slickensided rock referred to as "bells," "pots," "camel-backs," or "tortoises" (Young, 1976).

The coal field is subdivided into four areas: Castlegate, Soldier Canyon, Sunnyside, and Woodside. Important coalbeds and zones within the Book Cliffs coal field are: the Spring Canyon, Castlegate, and Kenilworth in the Castlegate area; the Gilson and Rock Canyon in the Soldier Canyon area; the Lower Sunnyside in the Sunnyside area; and the Beckwith zone in the Woodside area.

The western half of the Book Cliffs coal field has seven widely traceable coalbeds in the Blackhawk and Price River Formations. Each of the seven beds is greater than 14 ft thick, and all occur in the lower 500 ft of the Blackhawk Formation. The coalbeds are, in ascending order: the Aberdeen (Castlegate "A"), Kenilworth (Castlegate "D"), Gilson, Fish Creek, Rock Canyon, Lower Sunnyside, and Upper Sunnyside. In general, the coals are highly attrital, with a few vitrain bands with the majority of the coal showing closely spaced cleats (Young, 1976).

The Castlegate "A" coalbed, also referred to as the Aberdeen bed, is exposed near Castlegate, Utah, and can be traced for 18 mi from Spring Canyon to Coal Creek Canyon. The maximum known thickness for this bed is 19.8 ft, at the Kenilworth mine. There is an interval of approximately 200 ft between the Castlegate "A" bed and the Castlegate "D" bed (Young, 1976).

The Castlegate "D" or Kenilworth coalbed is a widespread lenticular coal with a maximum recorded thickness of 18.8 ft at

Kenilworth, Utah, although the bed averages 6.7 ft. At the coalbed's thickest point, it is assumed that the Castlegate "D" bed has coalesced with the Gilson and Fish Creek coalbeds. The interval between the Castlegate "D" and Gilson coalbeds is known to range up to 47 ft (Young, 1976).

The Gilson coalbed reaches a maximum thickness of 11 ft in areas where it has not coalesced with the Castlegate "D" bed. The Gilson bed is usually one bench, but local shale partings split the coalbed into two or three benches. Between the Gilson and Fish Creek coalbeds, there is an interval of sandstone and shale ranging up to 60 ft thick.

The Fish Creek coalbed is not well recorded. Where the coalbed separates from the Castlegate "D" coalbed, it is known to have a maximum thickness of 5 ft. The Fish Creek and Rock Canyon coalbeds are separated by a 10- to 35-ft interval of sandstone, shale, and sandy shale. The Rock Canyon coalbed has a maximum recorded thickness of 7.8 ft near Soldier Creek, with the bed thinning to the east. The Rock Canyon and Lower Sunnyside coalbeds are separated by 60 to 160 ft of sandstone and shale (Young, 1976).

The Sunnyside coalbed is the most widespread and uniformly thick coal in the western Book Cliffs coal field. The coal can be traced southeast from Kenilworth to Woodside, Utah. North of Whitmore Canyon, the coal splits into three beds referred to as the lower, middle, and upper. The lower bed thins rapidly and pinches out, but where it does exist in quantity, it is an excellent coking coal. The middle bed, also referred to as the lower Sunnyside bed, is found directly on a prominent sandstone bed. The upper coalbed, or upper Sunnyside, is lenticular and has reached a maximum thickness of 22 ft. The coal in this bed is also an excellent coking coal (Young, 1976).

The resource estimate for the Book Cliffs field is approximately 3,700 million short tons, which includes coalbeds greater than 4 ft thick and under less than 3,000 ft of cover. As of 1972, approximately 209 million short tons of coal had been removed from the coal field (Doelling, 1972).

Sego Coal Field

The Sego coal field (see Fig. 5) is located in east central Utah along the Book Cliffs, between the Green River and Colorado state line. The coal field is 65 mi wide and encompasses approximately 390 sq mi (Doelling and Graham, 1972).

The Upper Cretaceous Neslen Member of the Price River Formation contains four major coal zones in the field. The Neslen Member consists of quartzose sandstone, subgraywackes, siltstone, shale, and coal. The coal, found in beds that range from 1 to 7 ft thick, is high-volatile C bituminous in rank (Doelling and Graham, 1972). The coalbeds dip gently to moderately northward into the Uinta Basin.

The four coal zones, in ascending order, are the Palisade, Ballard, Chesterfield, and Carbonera. The Palisade coal zone is 40 to 50 ft thick, and the Ballard coal zone is approximately 10 ft thick. The Ballard and Chesterfield coal zones are separated by a thick, cliff-forming sandstone. The Chesterfield zone is 50 ft thick, and is the only coal zone to produce significant quantities of coal in the Thompson and Sego Canyon areas. The Carbonera coal zone ranges from 100 to 150 ft thick, and outside of the developed area the coal thickens locally (Doelling and Graham, 1972).

Estimated original coal resources in the Sego coal field are 294 million tons of coal in beds greater than 4 ft thick. The Carbonera coal zone contains 9 million tons of resources; the Chesterfield coal zone, 116 million tons; Ballard coal zone, 93 million tons; and the Palisade coal zone, 76 million tons. An estimated 2.7 million tons of coal have been mined from the Sego field, the majority removed from the Thompson and Sego Canyon areas (Doelling and Graham, 1972).

POTENTIAL METHANE RESOURCE

Methane Data in Existing Literature

The Utah Geological and Mineral Survey (UGMS) worked under a grant from the U.S. Bureau of Mines (USBM) and, later, the Department of Energy (DOE), to collect data on the methane content of the various coals in Utah. As a part of these studies, the UGMS, from 1975 to 1979, collected 164 core samples from 7 of the more than 20 coal fields listed for Utah. The core samples collected were mainly from coals, although nine samples were taken from rocks adjacent to the coalbeds. Sampling was confined to those areas being actively mined and for which significant data are available. Data collected by UGMS during the 1975 to 1979 project period were tabulated and reported as a part of its Special Study Series in August 1979 (Doelling, Smith, and Davis, 1979).

Deeper mining and the higher methane content generally associated with deeper coalbeds prompted USBM to begin collecting data on the measured methane emission rates of U.S. coal mines with an emission of at least 100,000 cubic feet/day (cfd), and from mines in coalbeds with total mine emissions in excess of 1 million cfd. Data from this source relevant to the Uinta Basin are shown in Table 5. While it cannot be utilized directly to determine the in-place gas content of the coals, this information is useful to delineate those areas having higher gas contents. In 1975, five mines were recorded as having emission rates of 100,000 cfd or greater. Four are located in the Book Cliffs coal field and one in the Emery coal field (Irani et al, 1977).

USBM also is conducting research to determine the effectiveness of long holes in degasifying an area. After a section of Kaiser Steel Company's Sunnyside No. 1 mine in the Book Cliffs coal field was closed to mining because of excessive methane emissions, USBM drilled two horizontal holes from outside entries to degasify the virgin coal in advance of mining operations (Perry et al, 1978). Gas flow from the holes was monitored for a 9-month period following completion of the holes. This procedure decreased face emissions by about 40% in that section of the mine.

In 1978, Mountain Fuel Resources, Inc., and DOE initiated a demonstration project for methane recovery from deep coals in the Book Cliffs coal field. As part of this study, Mountain Fuel drilled three vertical wells into coalbeds to demonstrate drilling and fracturing techniques for methane recovery from coals. The first two wells, WP-1 and WP-2, were drilled in late 1979, and the third well, WU-1, was completed late in 1981. Indications from these wells are that the recovery of gas in sufficient quantities for commercialization may be feasible from this coal field (Mountain Fuel Resources, Inc., 1978, and Mountain Fuel Supply Co., 1980).

Methane Data for Uinta Basin Coal Fields

As noted, UGMS has collected coal samples and determined their methane content by desorption. These samples were collected from seven coal fields, four of which are part of this study. Fields for which data are available are the Wasatch,

Table 5—Average daily methane emission from mines with emission rate of 100,000 cfd or greater. From Irani et al, 1977.

Coal Field	Mine	Location	Coalbed	1971 MMcfd	1973 MMcfd	1975 MMcfd
Book Cliffs	Sunnyside No. 1 Kaiser Steel Corp.	Carbon County, Utah	Upper Sunnyside Lower Sunnyside	1.2	1.9	1.4
	Kennilworth North American Coal Corp.	Carbon County, Utah	Upper Sunnyside Lower Sunnyside	0.9	0.7	—
	Sunnyside No. 3 Kaiser Steel	Carbon County, Utah	Upper Sunnyside Lower Sunnyside	0.3	0.3	0.4
	Soldier Canyon Premium Coal Co.	Carbon County, Utah	Upper Sunnyside Lower Sunnyside	0.2	0.2	0.2
	Carbon No. 2 Carbon Fuel Co.	Carbon County, Utah	Upper Sunnyside Lower Sunnyside	0.2	0.1	—
	Braztah No. 3 Braztah Corp.	Carbon County, Utah	Castlegate B	—	—	0.2
Emery	Emery Consolidation Coal Co.	Emery County, Utah	Ferron J	—	—	0.1

Emery, Book Cliffs, and Sego coal fields (see Fig. 5 for locations).

The Wasatch Plateau coal field contains several coalbeds, eight of which are presently being mined in 14 active coal mines (Keystone, 1979). UGMS has collected 18 coal samples from eight coalbeds in this field (Doelling et al, 1979). Results of this analysis indicate that coals of the Wasatch field have a low methane gas content, with the desorbed gas values ranging from zero in the Upper O'Connor to a high value of 53 cubic feet/ton (cf/ton) in the Hiawatha bed. The high value on the Hiawatha bed was obtained from a sample taken from in-mine drilling. On the average, the Hiawatha bed had the highest desorbed methane content at 20 cf/ton; the lowest average gas content was in the Upper O'Connor, which desorbed 0.5 cf/ton. Average gas content for all samples taken in the field was 11 cf/ton.

The Emery coal field has several mines presently in operation, and, as of 1972, the principal beds being mined were the Ferron and Ivie coalbeds (Doelling, 1972). Samples were collected by UGMS from several sites in the Upper Ferron, Ferron, Lower Ferron, and Ferron "A" beds. Values of the total desorbed gas range from zero to a high of 18 cf/ton. The average methane content for all the samples from this field is 5 cf/ton.

Of all coal-bearing areas within the Uinta Basin, none has been explored or mined more than the Book Cliffs coal field, with 25 mining operations reported in eight coalbeds (Doelling, 1972). Some of these mining operations also have the basin's highest methane emission rate as measured by USBM.

USBM used the closed section in the Sunnyside No. 1 for a research project to determine the effectiveness of long holes in degasifying in advance of mining operations (Perry, Aul, and Cervik, 1978). Two holes, measuring 430 and 450 ft, respectively, were drilled horizontally into the coalbed. The initial gas flow from these holes was 160,000 and 127,000 cfd, respectively. Total gas production 16 days after completion declined to slightly over 144,000 cfd. At the end of a 9-month period, the two holes had produced over 35 MMcf of commercial-quality gas (Perry, Aul, and Cervik, 1978).

UGMS also collected 76 other coal samples from 12 coalbeds or zones in the Book Cliffs field. Coals of the Book Cliffs field, as expected from the emission rates, show a higher total gas content compared with other coal-bearing regions of the Uinta Basin. UGMS data show the gas concentrations as ranging from a low of 3 cf/ton in the Castlegate A to a high of 353 cf/ton in the Castlegate D coalbed. The lowest average coalbed methane content was found in the Gilson, with a total gas value of 36 cf/ton, and the highest average total gas value occurred in Subseam 1, below the Aberdeen Sandstone, with a value of 273 cf/ton. Average value for all samples taken from the Book Cliffs field is 100 cf/ton (Doelling, Smith, and Davis, 1979).

An interesting fact noted by UGMS was that coalbeds north of the Carbon-Emery County line are moderately gassy (approximately 32 to 160 cf/ton) to gassy (greater than 160 cf/ton), whereas coals south of the county line contain much less gas, although there is no difference in coal quality (Doelling, Smith, and Davis, 1979). No explanation for this phenomenon is apparent, but Doelling, Smith, and Davis (1979) note the coincidence of high magnetic readings in areas of gassy coal and suggest that materials in the basement may have been responsible for increasing temperature or pressure, which would favor gas formation or the formation of coking coal.

Gas content of coalbeds in the Book Cliffs field prompted Mountain Fuel Resources, Inc., to undertake a project to investigate the commercial recovery of methane from deep coalbeds. Working in cooperation with DOE, Mountain Fuel selected three sites to drill vertical wells. The first two wells were drilled in late 1979 about 23 mi northeast of Price, Utah, in Carbon County. These two wells were located approximately 1,800 ft apart, and were drilled to depths of 3,000 and 3,177 ft. One well was a cased-hole completion in the 6 ft thick Sunnyside coal. The other well was an open-hole completion across the 13 ft thick Gilson coal (Allred and Coates, 1982).

Well WU-1, located approximately 12 mi north of Price, Utah, in Carbon County, was drilled to a total depth of 4,110 ft. Coals were encountered between the depths of 3,494 and 4,019 ft. The wells were completed in 10 zones, with the coalbeds ranging from 2 to 6 ft thick (Allred and Coates, 1982).

The coal samples collected for desorption purposes from the three wells included 14 conventional core samples from Wells WP-1 and WP-2. Four coal chip samples collected from Well WU-1 had significantly lower gas contents than those measured from the conventional core samples. Desorbed values ranged from a low of 153 cf/ton to a high of 443 cf/ton (Allred and

Coates, 1982).

The eastern extension of the Book Cliffs area is known as the Sego coal field. Like the Book Cliffs field, it is also a narrow, elongated field with outcrops along the southern boundary of the basin. The Green River is the boundary separating the Book Cliffs from the Sego field. Doelling (1972) noted that mining activity was being carried out in the Palisade, Chesterfield, and Ballard coal zones.

UGMS collected 26 samples and desorbed them to determine their gas content. In general, all the coal zones showed a low gas content. Average values of total gas in cf/ton for the four major coal zones are: Carbonera, 18; Chesterfield, 21; Ballard, 8; and Palisade, 7.

In the Vernal coal field, Tabby Mountain coal field, and Sevier-Sanpete region, testing for availability of methane gas has not been attempted because of the coalbeds' poor quality and because of the availability of higher quality and thicker coalbeds within the basin. Small-scale coal production has occurred in all three fields.

The basinal area not located within the seven coal fields or regions has not been analyzed for either its coal resource content or its coalbed methane content.

Estimated Resource Volume

Presently available information is inadequate to estimate accurately the methane content of Uinta Basin coals. The exact prediction of total in-place methane resource for the basin would require more extensive coal resource evaluation and desorption analyses. Moreover, the extent of the coals in the central part of the basin has not yet been ascertained, and, if present, such coals would represent a substantial resource (UGMS, 1980, personal communication). UGMS is continuing an expanded program of methane desorption that will help delineate the resource more accurately in the future. However, based on the information presented in the preceding sections, it is possible to estimate the methane resource in those areas for which data are available.

Several methods can be used to determine an area's total methane content. If desorption data and the area's total coal reserves are known, the computations are quite simple. Data available for the Uinta Basin are fairly complete but do not allow complete assessment of each field on a bed-by-bed basis. To derive a sense of uniformity in calculations, it was decided to take the low values for each bed having desorption data and from this determine the average low methane content for each field for which data were available. The same procedure was applied to determine the average high methane content for each field. To determine the average methane content of a field, all the desorption data were used. These figures then were multiplied by known or estimated resources for the field to determine the potential methane resource.

The Wasatch Plateau coal field is estimated to contain 6,230 million tons of coal in beds greater than 4 ft thick and under less than 3,000 ft of overburden. The average low methane content for the Wasatch area is 5 cf/ton and the average high value, 11.9 cf/ton. Using these values gives an estimated low methane resource of 29.9 Bcf and a high resource value of 74.1 Bcf of gas. The average gas content is 11 cf/ton for an estimated field total of 68.5 Bcf.

The Emery coal field is not as extensive, and the gas content is somewhat lower. The average low methane content is 2 cf/ton, and the high value is 10 cf/ton. Coal resources for this field are estimated to be 1.4 billion tons, which would make the estimated low methane resource 2.7 Bcf of gas, and the high value, 13.9 Bcf. The average methane content is 5 cf/ton, which makes an average reserve of 6.72 Bcf.

The high gas content of the Book Cliffs coal field makes it the largest and best potential methane reservoir. The average low and high methane values by desorption are 52 and 192 cf/ton, respectively. Estimated coal resources for the Book Cliffs have been placed at 3.7 billion tons, which would produce average low and high methane contents of 193.1 and 708.9 Bcf, respectively. Overall average methane content is 100 cf/ton, or 368.2 Bcf of gas. Data that were gathered in the Whitmore Park area had substantially higher gas values. The average of their samples was 328 cf/ton. If that is taken as an average field value, the overall average total gas content for the area is 1,211.8 Bcf.

The Sego field does not have the gas content of the Book Cliff area, but the average low desorbed gas content (3 cf/ton) and the average high desorbed content (44 cf/ton) are better than those of the Emery and Wasatch coal fields. Total resources for the Sego field are 294 million tons, which would provide low and high methane resources of 735 million and 12.8 billion cu ft of gas, respectively. The average value of desorbed gas is 11 cf/ton, which makes the field average 3.2 Bcf.

Detailed gas contents or data regarding average gas in the Vernal, Tabby Mountains, and Sevier-Sanpete coal fields are not available. The total coal resources of these three areas are estimated to be 2,314.6 million tons of coal. It appears that the coals in these areas contain some quantity of gas. The lack of information seems to indicate that the content is probably low, so that any estimate for those areas should be low. The lowest gas values for any field were found in the Emery coal field. If the values of desorbed gas for that field are applied to the areas for which no gas data are available, it is possible to make at least a general statement about gases in those regions. Using this rationale, the low and high gas contents are 4.4 and 22.6 Bcf, respectively. The average value would be 11.3 Bcf of gas.

Using the resource figures generated by this method, total estimated gas resources for the seven coal areas of the Uinta Basin can be calculated. UGMS data would estimate the low and high values to be 230.8 and 832.3 Bcf of gas, respectively. An average value would be 457.9 Bcf. If the higher average value of 328 cf/ton from the Mountain Fuel project is used for the seven coal areas having an estimated 13.9 billion tons of reserves, the total average gas content would be 4,552 Bcf of gas.

CONCLUSIONS

The Uinta Basin of Utah and Colorado contains a significant amount of methane gas in the Upper Cretaceous coal-bearing units. Information regarding the potential methane resource is available for the Wasatch Plateau, Emery, Book Cliffs, and Sego coal fields, all of which are located in the southern half of the basin. Specific gas content information is lacking for the Vernal, Tabby Mountain, and Sevier-Sanpete coal areas; and for the deeper parts of the basin, both coal and gas information are lacking. Gas contents for such areas are speculative.

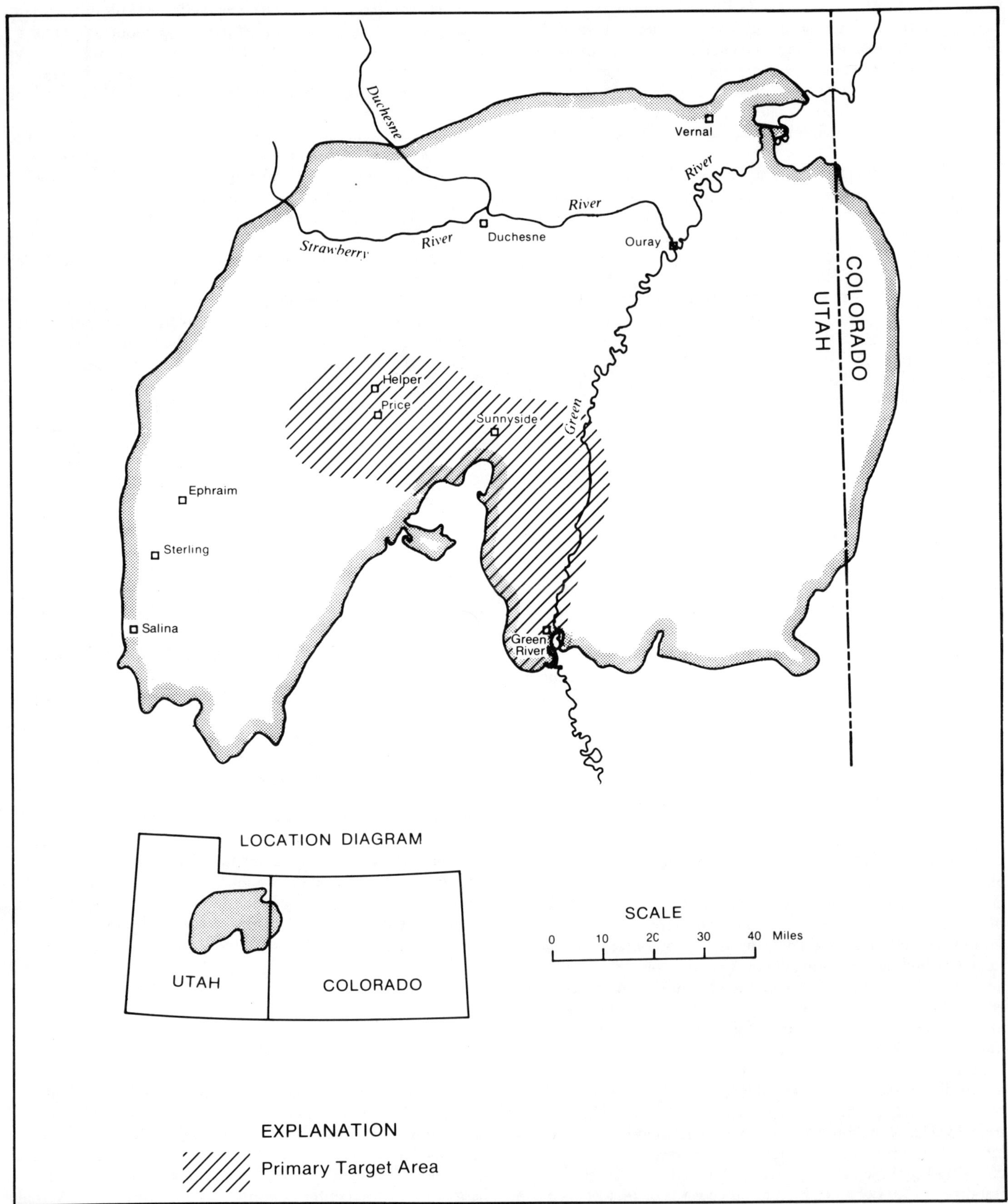

Figure 7—Primary methane target area of the Uinta Basin.

The primary methane target area shown in Figure 7 covers approximately 2,070 sq mi and includes the Book Cliffs coal field and additional areas in parts of Carbon, Grand, and Emery counties in Utah. The primary area was selected because:

- Of the four coal fields where coal samples were tested for their gas content by the UGMS, the data show highest gas yields from the Book Cliffs coal field, north of the Carbon-Emery County line.
- Two wells drilled by the Mountain Fuel Resources, Inc., in the Book Cliffs field, showed high gas contents ranging from 187 to 443 cf/ton per coalbed.
- Numerous mines within the target area provide emission data reflecting large amounts of methane gas being released from the Sunnyside coalbeds.

Information for the Wasatch Plateau, Emery, and Sego coal fields provides only limited gas content data. Data from UGMS show significantly lower gas content than samples taken in the target area. Additional testing of the methane gas content should be undertaken in these areas, helping to delineate these areas further with respect to development of their methane resource potential.

Because of the absence of methane gas data for the Vernal, Tabby Mountain, and Sevier-Sanpete coal areas, and the deeper portions of the basin, a testing program should be developed to explore their methane potential.

REFERENCES CITED

Allred, L. D., and R. L. Coates, 1982, Methane recovery from deep coalbeds in the Book Cliffs: SPE/DOE Unconventional Gas Symposium, May 16–18, 1982, Pittsburgh, Pennsylvania, Paper 10819, p. 245–252.

Anderson, D. W., and M. D. Picard, 1972, Stratigraphy of the Duchesne River Formation (Eocene-Oligocene?), northern Uinta Basin, northeastern Utah: Utah Geological and Mineralogical Survey Bulletin 97, 29 p.

Childs, O. E., 1950, Geologic history of the Uinta Basin, *in* Guidebook to the geology of Utah, Number 5: International Association of Petroleum Geologists, p. 49–59.

Doelling, H. H., 1972, Central Utah coal fields, Sevier-Sanpete, Wasatch Plateau, Book Cliffs, and Emery, Monograph Series No. 3: Utah Geological and Mineralogical Survey, 570 p.

———, and R. L. Graham, 1972, Eastern and northern Utah coal fields: Vernal, Henry Mountains, Sego, La Sal-San Juan, Tabby Mountain, Coalville, Henry's Fork, Goose Creek, and Lost Creek, Monograph Series No. 2: Utah Geological and Mineralogical Survey, 409 p.

———, A. D. Smith, and F. D. Davis, 1979, Methane content of Utah coals: Utah Geological and Mineralogical Survey, Special Studies 49, p. 1– 43.

Hood, J. W., and F. K. Fields, 1977, Water resources of the Northern Uinta basin area; Utah and Colorado with special emphasis on groundwater supply: U.S. Geological Survey, Open–File Report 77–730.

Hunt, C. B., 1956, Cenozoic geology of the Colorado Plateau: U.S. Geological Survey Professional Paper 279, 87 p.

Irani, M. C., et al, 1977, Methane emission from U.S. coal mines in 1975, a survey: U.S. Bureau of Mines Information Circular 8733, 55 p.

Keystone, 1979, Keystone coal industry manual: New York, Mining Informational Services, McGraw-Hill Mining Pub., p. 560–570.

Mountain Fuel Resources, Inc., 1978, Demonstration project for methane recovery from unminable coalbeds, v. 1, Technical proposal submitted to DOE in response to RFP No. EY-78-R-21-8217, p. 201–252.

Mountain Fuel Supply Company, 1980, Demonstration project for methane recovery from unminable coalbeds, Progress Report No. 11—1 January to 31 January, 1980: Prepared for the U.S. Department of Energy, Contract No. DE-ACZI-79MC 10734.

Osmond, J. C., 1964, Tectonic history of the Uinta Basin, Utah, *in* Guidebook to the geology and mineral resources of the Uinta Basin: Intermountain Association of Petroleum Geologists, p. 47–58.

Perry, J. H., G. N. Aul, and J. Cervik, 1978, Methane drainage study in the Sunnyside coal bed, Utah: U.S. Bureau of Mines Report of Investigations 8323, 17 p.

Spieker, E. M., 1931, The Wasatch Plateau coal field, Utah: U.S. Geological Survey Bulletin 819, 210 p.

———, and J. B. Reeside, Jr., 1925, Cretaceous and Tertiary formations of the Wasatch Plateau, Utah: U.S. Geological Survey Bulletin, v. 36, p. 435–454.

Thornbury, W. D., 1965, Regional geomorphology of the United States: New York, John Wiley and Sons, Inc., p. 405–441.

Walton, P. T., 1944, Geology of the Cretaceous of the Uinta Basin, Utah: Geological Society of America Bulletin, v. 55, p. 91–130.

Williams, M. D., 1950, Tertiary stratigraphy of the Uinta Basin, *in* Guidebook to the Geology of Utah, Number 5: International Association of Petroleum Geologists, p. 101–114.

Young, R. G., 1976, Genesis of western Book Cliffs coal: Brigham Young University Geology Studies, v. 22, pt. 3.

Geologic Overview, Coal, and Coalbed Methane Resources of the Greater Green River Coal Region — Wyoming and Colorado

J. P. McCord

The Greater Green River coal region includes an area of approximately 21,000 sq mi in southwestern Wyoming and northwestern Colorado. It includes the Hams Fork Coal Region in the western Wyoming Overthrust Belt, and the Green River, Great Divide, Washakie, and Sand Wash structural basins. Coal is found in Upper Cretaceous and Tertiary formations that crop out on the flanks of the contiguous uplifts and dip into the basinal areas. A conservative estimate of the coal resource within this region, to a depth of 3,000 ft (6,000 ft in the Yampa/Sand Wash Basin), is approximately 82 billion short tons.

The presence of methane in coal within the Greater Green River coal region is indicated by both indirect evidence and measured gas content of coal samples. Indirect evidence includes:

- Coal mine methane emissions
- Blowouts and mud-log gas kicks experienced while drilling through coalbeds
- Dry gas fields with production from coal-bearing intervals.

Gas content measurements of coal samples are available for several locations within the Greater Green River coal region. This limited data base indicates a methane content ranging from less than 10 cubic feet/ton (cf/ton) of coal to more than 500 cf/ton. Based on these data and the estimated magnitude of the coal resource, it is estimated that the Greater Green River coal region contains between 0.2 and 30 trillion cubic feet (Tcf) of coalbed methane.

INTRODUCTION

The Greater Green River coal region occupies approximately 21,000 sq mi in southwestern Wyoming and northwestern Colorado (Fig. 1). It is bordered on the west by the western Wyoming thrust belt, on the north and northeast by the Gros Ventre and Wind River mountain ranges, on the east by the Rawlins and Sierra Madre Uplifts, and on the south by the Uinta Mountains and the Axial Basin anticline. In Wyoming, nearly all of Sweetwater County and significant portions of Uinta, Lincoln, Sublette, and Carbon counties are located in the Greater Green River coal region. Colorado counties within the region include portions of Moffat, Routt, and Rio Blanco.

The climate in the Greater Green River coal region is semiarid, with temperature extremes of above 100°F in the summer months and below −35°F in the winter. Annual precipitation varies from a low of 5 inches in the Red Desert of the Great Divide approximately 17 inches at Kendall, Wyoming, near the headwaters of the Green River (National Oceanic and Atmospheric Administration, 1974).

The economic base of this region is mineral resources, agriculture, transportation, and tourism. The principal mineral resources include natural gas, oil, coal, oil shale, uranium, and trona. The presence of a vital and growing oil and gas and coal industry in the region—with existing support industries and transportation systems—should contribute to the potential commercialization of the coalbed methane resource. Agricultural activities in the region include cattle and sheep raising and the growing of native hay along the major stream valleys.

STRUCTURE

The Greater Green River coal region includes five structural units:

- Western Wyoming Thrust Belt
- Green River Basin
- Great Divide Basin
- Washakie Basin
- Rock Springs Uplift.

Western Wyoming Thrust Belt

Within the western Wyoming thrust belt, the major structural features across the Hams Fork coal region include the Crawford Mountain thrust fault, the Absaroka thrust fault, the Lazeart

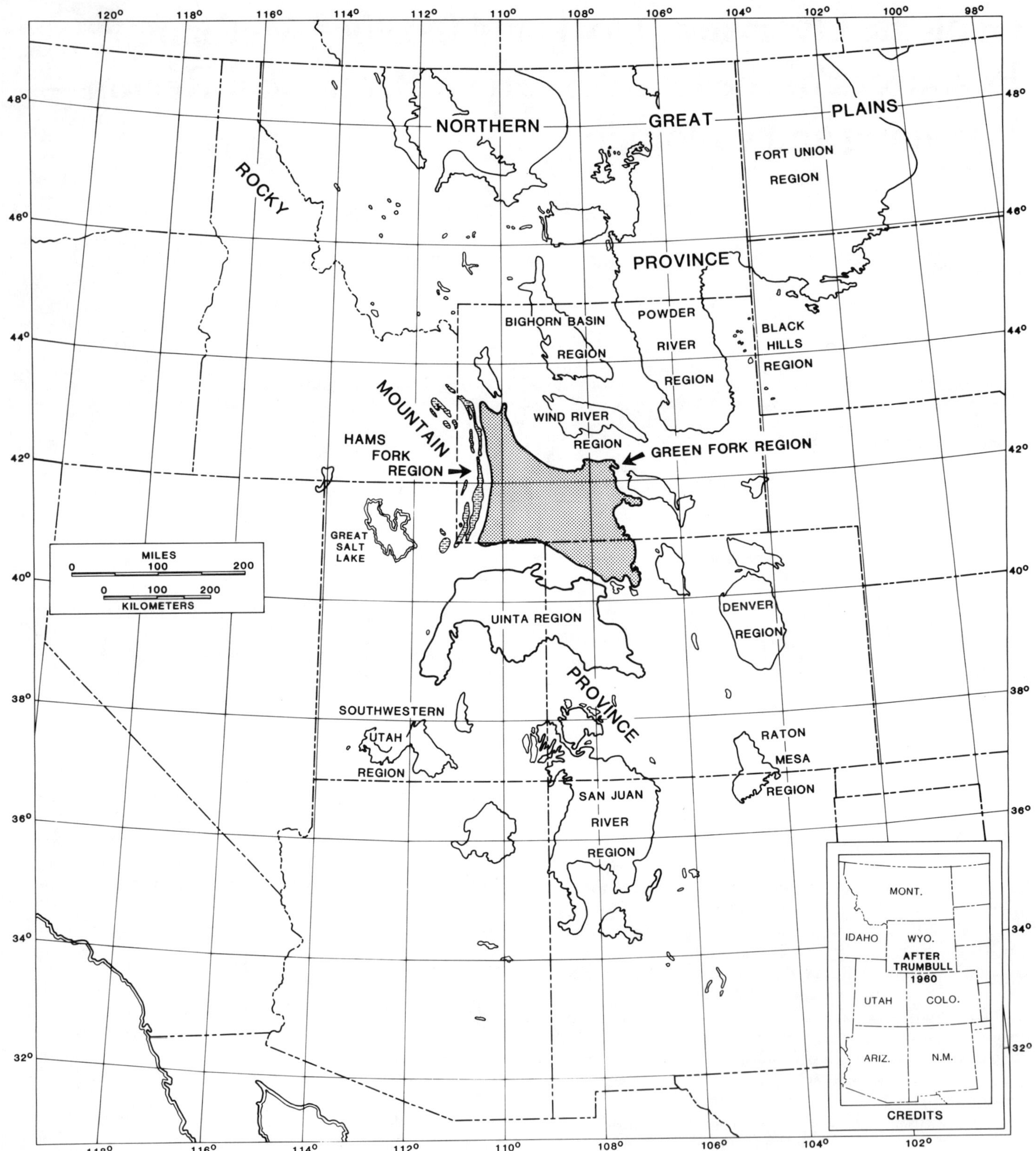

Figure 1—Regional setting of the Greater Green River coal region. (RMAG). Used with the permission of the Rocky Mountain Association of Geologists.

Syncline, and the Darby-Hogback-Prospect thrust fault group. In addition, there are several concealed thrust faults underlying the Hams Fork Plateau; and steeply dipping faults—both normal and reverse—are common in the western part of the region.

The southernmost mapping of the Crawford Mountain thrust fault is located approximately 3 mi east of Woodruff, Utah. From this point, the thrust fault has been traced north, passing just west of Sage, Wyoming, to a point approximately 15 mi north of Cokeville, Wyoming. Stratigraphic throw along the Crawford Mountain thrust faults varies from 22,000 ft on the south to 18,000 ft on the north. The fault dips between 30 and 70° to the west, averaging about 40°. Lateral displacement along the fault has been estimated at a minimum of 5 mi and may be considerably more. Movement along the Crawford Mountain thrust fault has been placed between late Early Cretaceous and early Eocene and is associated with the Sevier orogenic period. The fault cuts the late Early Cretaceous age Sage Junction Formation and is unconformably overlain by the Eocene Wasatch Formation (Rubey et al, 1975).

The Absaroka thrust fault can be traced a distance of over 200 mi. This north-south–striking thrust fault approximately bisects the western Wyoming overthrust belt passing about 10 mi east of Evanston, Wyoming, and about 7 mi west of Kemmerer, Wyoming. Stratigraphic separation along the Absaroka thrust fault places Paleozoic strata ranging in age from Cambrian to Devonian in fault contact with strata of Late Cretaceous age. Stratigraphic throw ranges from about 10,000 to 15,000 ft. The fault dips between 35 and 70° to the west, but the dip probably decreases considerably to the west with depth. Lateral displacement is conservatively estimated at between 10 and 15 mi (Rubey, 1958; Rubey and Hubbert, 1959). Movement along the Absaroka thrust fault occurred during Upper Cretaceous time and is the last event attributed to the Sevier orogenic period.

The Lazeart syncline closely parallels the Absaroka thrust fault and is approximately 2 mi east of it. This syncline, first recognized and named by Veatch (1907), extends over 80 mi in a north-south direction. South of Kemmerer, Wyoming, portions of the syncline are concealed under Tertiary cover. The syncline is asymmetric. The eastern limb dips about 30° to the west, while the beds on the western limb are vertical to approximately 35° past vertical (55° to the west). Seismic surveys have shown that the Lazeart syncline is a near-surface drag-type fold. At depth, the strata are not involved in the folding (Cook, 1977).

The Hogback-Darby-Prospect thrust fault group defines the eastern edge of the western Wyoming thrust belt and extends from the northern flank of the Uinta Mountains north to a point about 5 mi south of Victor, Idaho. The trace of this group of thrust faults is concave to the west. Stratigraphic separation along the Darby thrust fault places Cambrian to Devonian Paleozoic strata in fault contact with strata of Late Cretaceous age. Along the Hogback and Prospect thrust faults, stratigraphic separation places Cambrian and Devonian Paleozoic strata in fault contact with Paleocene and Eocene Tertiary strata. The fault surfaces on this group of thrust faults dip gently to the west at angles between 15 and 20° (Blackstone, 1977).

The Green River Basin

The Green River Basin is a broad synclinal basin with an area of approximately 10,000 sq mi. The principal synclinal axis trends about N 30° E and lies approximately 20 mi west of the axis of the Rock Springs anticline. Except along the margins of the basin, the beds are essentially horizontal or dip at very low angles, generally less than 1.5° (Bradley, 1964).

To the north and northeast, the Green River structural basin butts up against the Wind River Mountains. Along this front, the Continental fault and Pinedale anticline (trending approximately N 30° W) are associated with a thrust fault and deep syncline (Berg, 1961). Sedimentary rocks attain a thickness of approximately 30,000 ft in the trough of this deep syncline (Krueger, 1960). Dips in this area are to the south and range from 0.5 to 5°. To the east, the basin is bounded by the Rock Springs Uplift. Here, the beds dip 3 to 12° to the west. On the west, the basin is bounded by the western Wyoming thrust belt. Here, beds dip between 2 and 8° to the east.

The southern margin of the Green River structural basin is more complex. The basin butts up against the northern flank of the Uinta Mountains. Two faults contribute to the complexity along the basin margin. The Henrys Fork fault extends from the Green River approximately 2 mi directly east of Linwood, Utah, westward about 20 mi to Phil Pico Mountain. This fault is a high-angle, reverse fault except on its west end, where it becomes a normal fault. The beds on the north, or down-thrown, side are steeply upturned or overturned in the drag zone. The Uinta fault, which extends from Diamond Peak in the extreme northwestern part of Colorado westward into Utah at least as far as Burnt Fork (Bradley, 1964), is a vertical fault with the northern side being down-thrown. Seismic data in this area suggest that one of the deeper parts of the basin can be found on the northern side of this fault (Fidlar, 1950).

A few folded structures are known to exist in the basin, and the most important is the Moxa Arch. This feature is a broad, gently folded basement uplift, approximately 120 mi long and extending from the north flank of the Uinta Mountains to the Big Piney-LaBarge Platform. The major folding of the arch occurred during middle Upper Cretaceous. Subsurface data indicate that erosion removed the basal Mesaverde (Rock Springs and Blair Formation equivalents) from the crest. The arch was then buried by middle Mesaverde rocks of the Ericson Formation that were deposited on the unconformity. Younger Cretaceous and Tertiary sediments do not reflect this arch (Wach, 1977).

The Rock Springs Uplift

The Rock Springs Uplift is located in the central part of Sweetwater County, Wyoming, and is an anticlinal structure of Laramide age. The axis of the anticline is generally north-south and plunges in both directions: approximately 1.5 to 3° to the north and 4° to the south. The topographic expression of this doubly plunging anticline is that of a crude ellipse elongated in the north-south direction.

The anticline is asymmetrical, with the western limb dipping between 10 and 15° to the west and the eastern limb dipping between 5 and 8° to the east. Many normal faults with throws generally less than 100 ft cut the uplift and trend generally east-west. Some of these faults extend for up to 15 mi (Welder and McGreevy, 1966). According to Love (1961), Precambrian rocks at the apex of the Rock Springs Uplift are approximately 17,000 ft above the lowest Precambrian rocks in the adjacent Green River and Washakie Basins.

Great Divide Basin

The Great Divide Basin, variously called the Great Divide, Red Desert, and Shoshone Basin, is a large topographic and structural basin with interior drainage. It is bounded on the west by the Rock Springs Uplift, on the north by the Wind River Mountains and the Bison Basin, on the east by the Rawlins Uplift, and on the south by the Wamsutter Arch.

The Great Divide Basin is generally a simple synclinal basin modified by broad shallow folds and widespread small-scale faults. The synclinal axis trends generally north-south in the southeast part of the basin and curves around to approximately N 60° W in the northwestern part of the basin. In the west and southwest parts of the basin, the strata dip from 2 to 3° toward the east and northeast. In the east, the strata dip up to 20° west on the western flank of the Rawlins Uplift. The structure of the northern part of the basin is complicated by normal faulting superimposed on a thrust fault associated with the Wind River and Green Mountains.

Washakie Basin

The Washakie Basin is a shallow syncline that covers an area of about 3,000 sq mi in southcentral Wyoming and northwestern Colorado. This area includes parts of Sweetwater and Carbon counties, Wyoming, and parts of Moffat and Routt counties, Colorado. The Colorado portion of the basin is known as the Sand Wash Basin.

The Washakie Basin is bordered on the west by the Rock Springs Uplift, on the north by the Great Divide Basin, on the east by the Sierra Madre Uplift, and on the south by the Sand Wash Basin. It is separated from the Great Divide Basin along the Wamsutter Arch. Along the north and east sides of the basin, the dip of the beds ranges from 3 to 5° toward the center of the basin. Along the west and southwest sides, the dip ranges from 8 to 12° toward the basin center. Away from the edges of the basin, the strata flatten out and are essentially horizontal in the central part of the basin. This essentially uniform synclinal structure is broken by a west-trending uplift area along the Wyoming-Colorado state line. This area is referred to as the Cherokee Ridge Arch and separates the primary Washakie Basin from its subsidiary Sand Wash Basin. Cherokee Ridge is a gentle westward-plunging anticlinal structure with many small subsidiary folds and tensional faults. The total structural rise of the basement rocks in the lowest point on the ridge has been estimated to be less than 1,500 ft above the lowest point in the Sand Wash Basin to the south, and between 1,500 and 2,000 ft above the lowest point in the Washakie Basin to the north (Haun, 1962).

Sand Wash Basin

The Sand Wash Basin is a southeasterly trending synclinal prong of the Washakie Basin. It extends from the southeastern flank of the Rock Springs Uplift in a southeast direction toward Craig, Colorado. It is bounded on the north by the Cherokee Ridge and on the east by the Sierra Madre-Park Range. To the southeast, the basin is bounded by a series of folds associated with the Axial fold belt. To the southwest, the basin is bounded by the Uinta Mountains Uplift. In the deepest part of the Sand Wash Basin, the top of Precambrian rocks is estimated to be 17,500 ft below sea level. Cambrian rocks are exposed to the southeast in the Uinta Mountains at an elevation of 9,000 ft. This results in a structural relief of 26,500 ft in a horizontal distance of 30 mi. In the Park Range, east of the basin, the highest Precambrian rocks are at 11,940 ft at the summit of Mt. Ethel. This indicates a structural relief of over 29,000 ft in this area (Haun, 1962).

STRATIGRAPHY

Sedimentary rocks within the Greater Green River coal region range in age from Lower Cambrian to Quaternary and can be divided into two broad groups based on depositional environments. The older group is a Paleozoic/Mesozoic sequence of rocks predominantly of marine origin. The younger group is a thinner sequence of Cretaceous/Cenozoic rocks of predominantly continental origin. Together, these groups range to over 40,000 ft in sedimentary thickness. Figure 2 summarizes the stratigraphic units in the Greater Green River coal region.

The coal-bearing formations in this region are found in the Cretaceous/Cenozoic sequence. In the following discussion, summarized from McGookey et al, 1972, the Cretaceous/Cenozoic stratigraphic units will be discussed in relation to the evolving geologic environments of deposition.

Cretaceous

Lower Cretaceous sedimentary rock units are predominantly fluvial conglomeratic sandstones and varicolored shales. The distribution, thickness, and lithologic characteristics of the basal part of these units record a period of orogeny to the west (western Utah and Idaho) that contributed conglomerates and coarse clastics eastward into a slowly subsiding asymmetrical basin. The trough of the basin was located in the vicinity of the present-day Idaho-Wyoming border. Sediments of the Gannett Group rest conformably on the Upper Jurassic Morrison Formation. East of the trough is evidence of Early Cretaceous lakes. These fresh-water lakes appear to have been very large and contributed sedimentary sequences characterized by fresh-water limestone. The make-up of the Gannett Group reflects both the orogenic and tectonic activity of the time and the development of the large fresh-water lakes to the east. The basal unit, the Ephraim Conglomerate, ranges in thickness up to 1,180 ft and is overlain by a sequence of fresh-water limestone (Peterson Limestone), another conglomerate/sandstone unit (Bechler Formation), a second limestone unit (Draney Limestone), and finally a red calcareous mudstone (Smoot Formation). The limestone and mudstone units range from 250 to 370 ft in thickness, and the conglomeratic Bechler Formation ranges up to 1,700 ft in thickness.

At the end of Lower Cretaceous time, the Rocky Mountain region was subjected to the first Cretaceous marine transgression. Transgression was from the north and formed a narrow north-south seaway from the Gulf of Mexico to the Arctic Ocean. The following sedimentary sequences in the region are generally correlative and transgressive in nature: Bear River Sandstone-Aspen Shale; Dakota Sandstone-Mowry Shale; Fall River Sandstone (member of Cloverly Group)-Thermopolis Shale; and Muddy Sandstone-Mowry

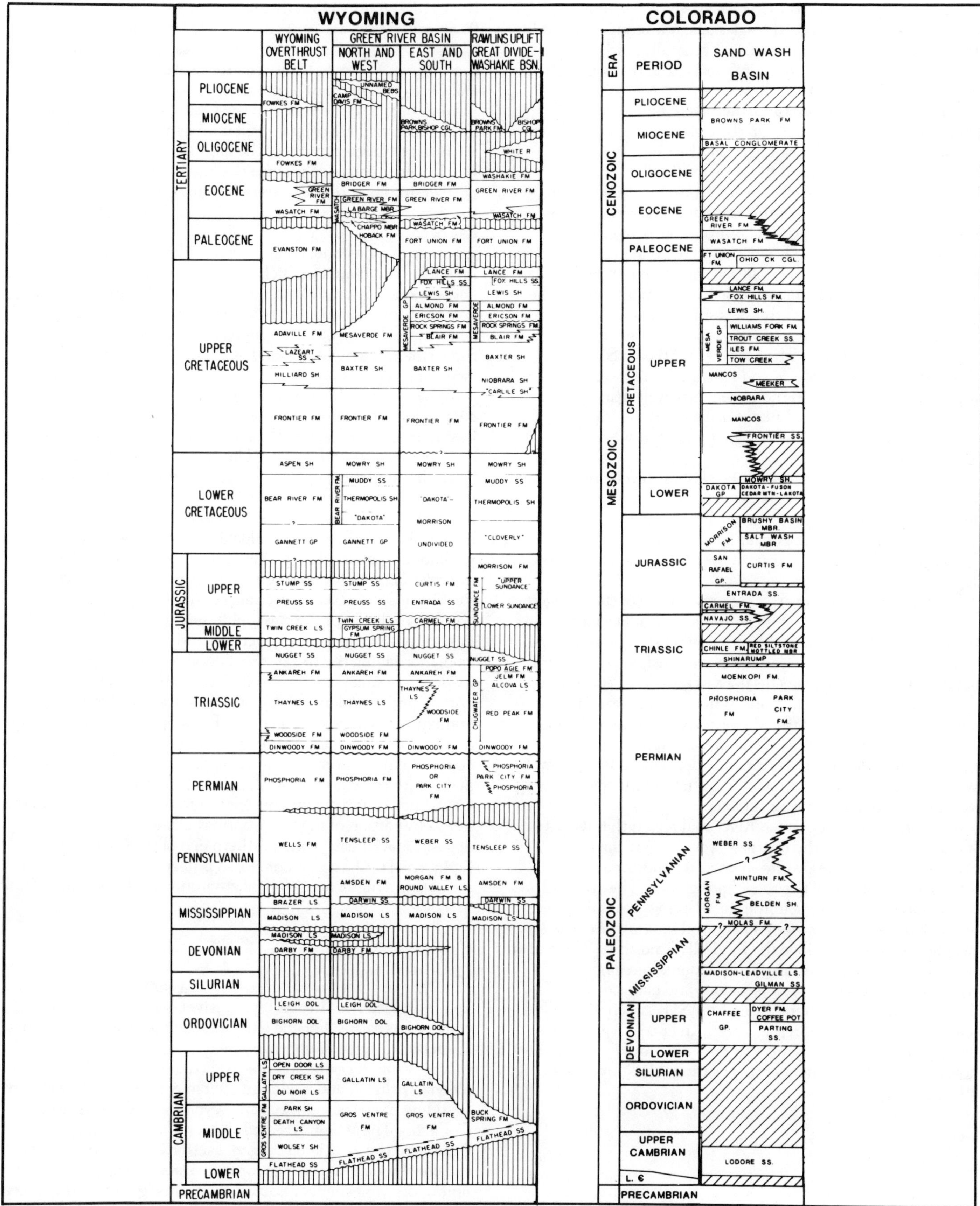

Figure 2—Generalized stratigraphic column of the Greater Green River coal region.

Shale. Generally, the Lower Cretaceous sandstone units are characterized by interbedded black shale, mudstone, coal, and fresh-water limestone. The shales are generally hard, gray, and fissile and contain fish scales, silty streaks, and siliceous beds (McGookey et al, 1972).

During Late Cretaceous time, there were four major transgressive-regressive cycles in the region. These cycles have been referred to, oldest to youngest, as the Greenhorn, the Niobrara, the Claggett, and the Bearpaw (McGookey et al, 1972). The Greenhorn cycle of early Upper Cretaceous time resulted in the deposition of the Frontier Formation and the Mancos and Baxter Shales. The primary source of clastic sediments for the Frontier Formation were uplifted areas in Idaho and northern Utah. Rapid accumulation of Frontier sediments resulted in formation of a deltaic environment. The Frontier Formation reflects this depositional environment and consists of a thick sequence of sandstone, siltstone, shaley claystone, and coal beds. The Mancos and Baxter Shales (as well as the equivalent Steele and Hilliard Shales) were deposited primarily late in the Greenhorn cycle, through the Niobrara cycle. Great thicknesses of dark gray marine shale were deposited in a foreland trough in the southwestern part of Wyoming.

During the regressive phase of the Niobrara cycle in southwestern Wyoming, coals were deposited in swamps that developed landward from the prograding shallow water and littoral sand barriers. The Lazeart Sandstone and Adaville Formation and some basal units of the Mesaverde Group were deposited during this regressive stage.

Tectonic activity during the regressive phase of the Claggett cycle—including vertical uplift, batholithic emplacement, and volcanic activity in western Montana—and the culmination of thrusting in the Sevier orogenic belt provided for an eastward flood of clastic sediments. This flood of sediments contributed to the Mesaverde Group Iles and William Fork Formations in northwest Colorado and the Mesaverde Group Rock Springs Formation in southwestern Wyoming, and represents a transition from marine to continental deposition. In southern Wyoming and northwestern Colorado, important coal deposits were deposited in this transition zone.

Rocks in the Greater Green River coal region that record the transgressive phase of the Bearpaw cycle include the Lewis Shale and the Mesaverde Group Ericson Sandstone and Almond Formations of the Rock Springs area. The Almond Formation is coal-bearing in the Rock Springs coal field. Progradational shallow water and littoral sandstones of the Bearpaw regression include the Fox Hills Sandstone. The end of the Cretaceous period in the Rocky Mountain region is characterized by a pattern of piedmont-and-plains fluvial deposition in intermontane basins formed between vertically uplifted ranges. In the Greater Green River coal region, these fluvial deposits are represented by the Lance and Evanston Formations and are generally coal bearing (McGookey et al, 1972).

Cenozoic

Paleocene

During Late Cretaceous time and continuing episodically until Eocene time, a period of mountain building called the Laramide orogeny was the dominant geologic force in the Rocky Mountain region. During this period, the Greater Green River coal region was generally a single large intermontane basin surrounded by areas of uplift. These uplift areas included the western Wyoming thrust belt to the west, the Wind River and Granite mountains to the north, the Sierra Madre and Rawlins uplifts to the east, and the Uinta Mountains to the south. Each of these areas contributed significant quantities of sediment to different parts of the basinal area.

Paleocene rock units include the Hoback Formation in the northeastern part of the western Wyoming thrust belt and the Fort Union Formation—the dominant and most extensive Paleocene unit in the region. The Hoback Formation is a sequence of sandstone, conglomerate, siltstone, and shaley limestone and ranges in thickness from 8,000 to 16,000 ft. The Fort Union Formation is a thick sequence of sandstone, shale, and coalbeds and ranges in thickness from 2,000 to 6,000 ft. In general, the Fort Union is composed of sediments washed out of the surrounding highlands and associated fluvial, paludal, and lacustrine deposits.

In the south and southeastern part of the Greater Green River coal region, in what is now the Washakie and Sand Wash Basins, massive amounts of clastic sediments were washed out of the Sierra Madre and Rawlins Uplift area to the east. Short rivers draining these uplifts coalesced in the basins into a larger river system moving northward into what is now the Great Divide Basin. This drainage system gave rise to a mixture of high- to low-energy fluvial, paludal, lacustrine, and piedmont-type sandstones, shales, lignites, and coals. Sediments from the Granite Mountains and Sweetwater Uplift north of the Great Divide Basin were carried south into the northern portions of the Rock Springs Uplift area and the Great Divide Basin. These areas were apparently slowly subsiding and had low gradients, giving rise to extensive flood plains and swamps. Evidence for these conditions includes extensive channel systems and abundant thick coalbeds in the Fort Union Formation at the north end of the Rock Springs Uplift.

At the south end of the Green River Basin proper, depositional environments were mixed and included flood plains with associated alluvial, paludal, and lacustrine deposits. Coals and lignites are present but not to the extent found in the northern portion of the Rock Springs Uplift area. In the depositional trough at the northern end of the Green River Basin, the Fort Union is composed primarily of sandstone, indicating a high energy fluvial environment (Robinson, 1972).

Eocene

Eocene sediments were deposited as a continuous unit before uplifts divided the Greater Green River coal region into its component basins. Eocene units include the Wasatch, Green River, and Bridger Formations.

The Wasatch Formation is a red-bed unit of mixed fluvial, alluvial, paludal, and piedmont character that was deposited in subsiding basins of southwestern Wyoming and northwestern Colorado. The Wasatch records both the subsidence of these basins and the associated orogenic activity. This formation is present throughout the component basins as outcrop or as the unit underlying younger Eocene sediments (generally the Green River Formation).

The Green River Formation is a lacustrine unit that extensively intertongues with the underlying Wasatch

Formation. According to Bradley (1964), the Green River Formation has the form of a large lens of fine-grained, calcareous sedimentary rock that weathers white, gray, and pale green and represents deposition in Eocene Lake Gosiute that at times covered almost the entire area of the Greater Green River coal region. This lens-like unit is underlain and overlain by a thick body of fluvial sediments. The underlying unit is the Wasatch Formation. The overlying unit is the Bridger Formation. The episodic growth and shrinkage of Eocene Lake Gosiute resulted in the complex intertonguing of the Green River Formation with both the Wasatch and Bridger Formations; and locally around the margins, the Bridger rests directly on the Wasatch.

The Wasatch, in addition to its main body, includes at least four tongues: the Red Desert, the Niland, the Cathedral Bluffs, and the New Fork. The Green River Formation includes three members and three tongues. The members include the Tipton Shale, the Wilkens Peak, and the Laney Shale. The tongues include the Luman, the Tipton, and the Fontenelle.

The Bridger Formation is the youngest Eocene formation in southwestern Wyoming. It spreads over nearly all of the central parts of the Green River and Washakie Basins, into the southwest part of the Sand Wash Basin adjacent to the Uinta thrust. This formation marks a return to the mixed fluvial environments characteristic of the Wasatch Formation. However, the Bridger does not display the red coloration typical of the Wasatch. Two distinctive characteristics of the Bridger Formation are the presence of extensive volcanic ash beds, which comprise up to 20% of the formation in some areas, and the tendency for Bridger outcrops to erode to a badland-type topography.

COAL RESOURCES

As defined by the U.S. Geological Survey (USGS), coal fields in the Greater Green River coal region are part of the Rocky Mountain coal province. The region has been broken down into two major subregions: the Hams Fork and the Green River coal areas. These major areas are further subdivided into 11 coal fields. Six of these coal fields are in the Green River coal region, and include the Rock Springs, Great Divide Basin, Little Snake River, LaBarge Ridge, and Henrys Fork in Wyoming, and the Yampa coal field in Colorado (Fig. 3). Four coal fields are in the Hams Fork coal region and include the Evanston, Kemmerer, Greys River, and McDougal coal fields (Fig. 4). Figure 5 summarizes the major coal-bearing formations in the Greater Green River coal region. Table 1 presents the general characteristics of coal from the area of the Greater Green River coal region (Glass, 1978).

Hams Fork Coal Region

The Hams Fork coal region is located in the western Wyoming thrust belt and extends from extreme southwestern Uinta County north through Lincoln County into southern Teton County. It is the fifth largest coal-producing area in Wyoming (Glass, 1977). Coal-bearing rocks occur in the Lower Cretaceous Bear River Formation, the Upper Cretaceous Frontier, Blind Bull, and Adaville Formations, and the Paleocene Evanston Formation. The Frontier and Adaville Formations are the most important, with the Bear River, Blind Bull, and Evanston Formations being minable only locally. All together, the coal-bearing rocks are found in over 9,000 ft of the estimated 20,000 ft of post-Jurassic strata. Only the Frontier and Adaville Formations will be discussed below. This discussion is summarized from Glass (1977).

The Upper Cretaceous Frontier Formation coals are generally less than 6 ft thick, but the main Kemmerer coalbed is up to 20 ft thick. The important coals in the Frontier in this region include the Lower Carter, the Spring Valley, the Willow Creek, and the Kemmerer coal groups.

The Lower Carter coal group is the lowest zone of coals in the Frontier Formation. The only named coal in this group is the League of Nations coalbed, prospected southeast of Kemmerer in the Kemmerer coal field. The League of Nations coalbed is usually split into two or three benches, separated by thick shaley partings. It is high in ash and, at its thickest, varies from 2 to 5 ft. There are no published analyses of coal quality.

The Spring Valley coal group contains three persistent coalbeds ranging up to 10 ft in thickness. Outcrops of the Spring Valley coal group have been traced for more than 50 mi in the southern half of the Kemmerer field. However, outcrops of this group have not been recognized in the northern half of the field (Hunter, 1950). Townsend (1960) reported the Spring Valley coals as high-volatile C bituminous.

The Willow Creek coal group is best developed in the Kemmerer field, where outcrops have been mapped for over 50 mi. Most of the past mining activity has been in the vicinity of Kemmerer, extending south to the southern Lincoln County boundary. The thickness of coals mined from this group has ranged from 3 to 11.2 ft, with the Willow Creek No. 5 bed being thickest. Presently, there is no mining activity on these beds. According to Schroeder (1976, 1977), the Willow Creek coal group is not recognized in the southern half of the Kemmerer field. The U.S. Bureau of Mines (USBM) examined the coking properties of coal from the Willow Creek group and found some of these coals to produce a weak coke (Toenges et al, 1945). However, it was concluded that Willow Creek coals were not suitable for metallurgical-grade coke without blending. Willow Creek coals have been reported as high-volatile A to high-volatile B bituminous in rank.

The Kemmerer coal group—the uppermost coal group in the Frontier Formation—usually consists of four coals. From oldest to youngest, these are the "B" seam, the "A" seam, the main Kemmerer, and the Radio. Outcrops of these coals have been mapped for more than 60 mi in the Kemmerer coal field. Minable coals in this group range up to 20 ft in thickness but average 5 to 10 ft. Rank of coals from the Kemmerer group ranges from high-volatile B bituminous to high-volatile C bituminous.

The Adaville Formation contains up to 32 subbituminous coalbeds, many of which are the thickest Upper Cretaceous coals in Wyoming (Glass, 1976). Adaville Formation coals crop out on the western side of the Kemmerer coal field in three distinct basins over a combined distance of more than 60 mi. Dips in these basins range from about 17 to 45° to the west. The stratigraphic character of individual coal seams is highly variable, making correlation of individual beds very difficult.

Coalbeds in the Adaville Formation are more numerous and thickest in the central basin located just west of Kemmerer. Here, there are numerous beds exceeding 10 ft in thickness and up to seven beds with thicknesses between 4 and 8 ft. The thickest bed is the Adaville No. 1, attaining a thickness of up to

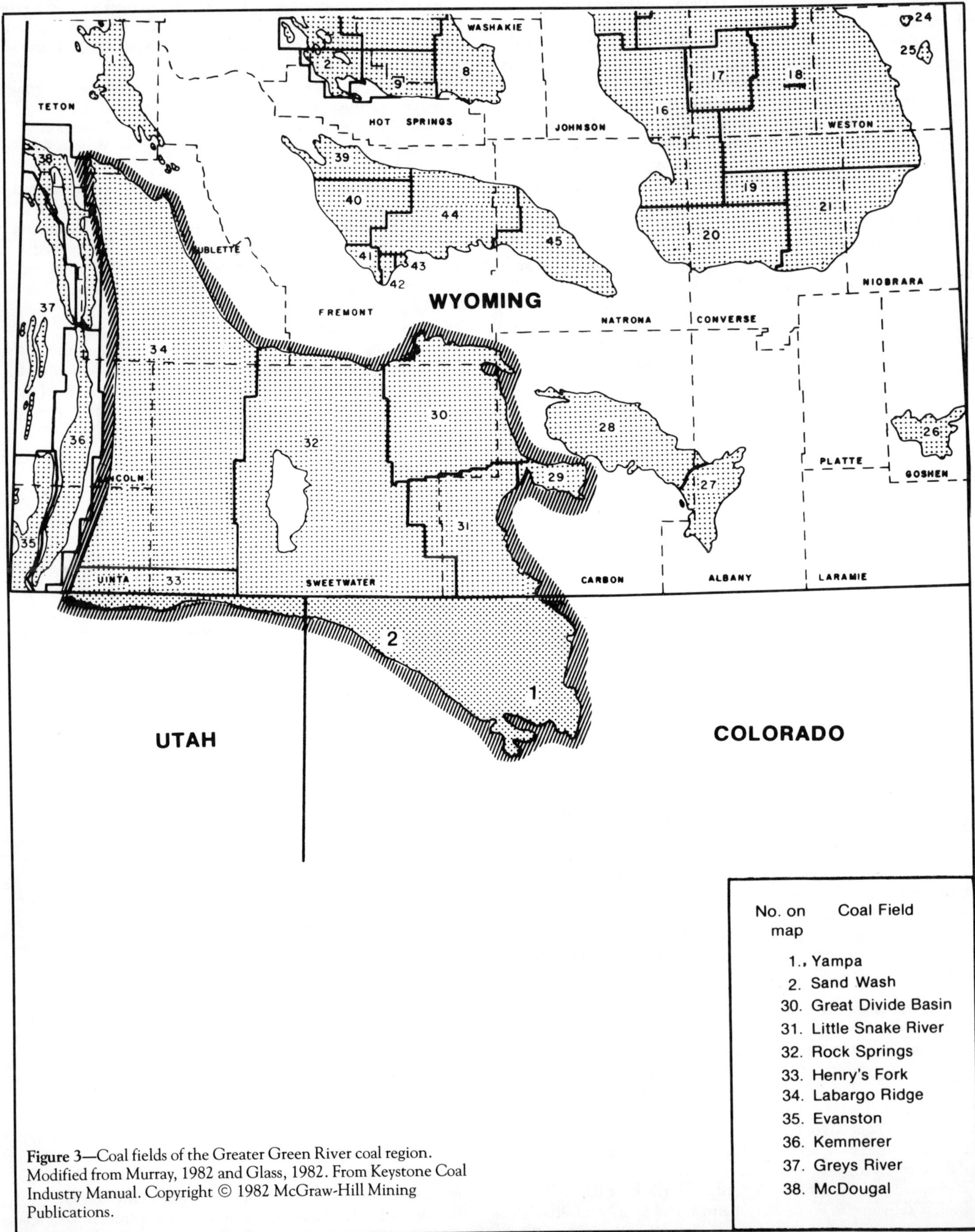

Figure 3—Coal fields of the Greater Green River coal region. Modified from Murray, 1982 and Glass, 1982. From Keystone Coal Industry Manual. Copyright © 1982 McGraw-Hill Mining Publications.

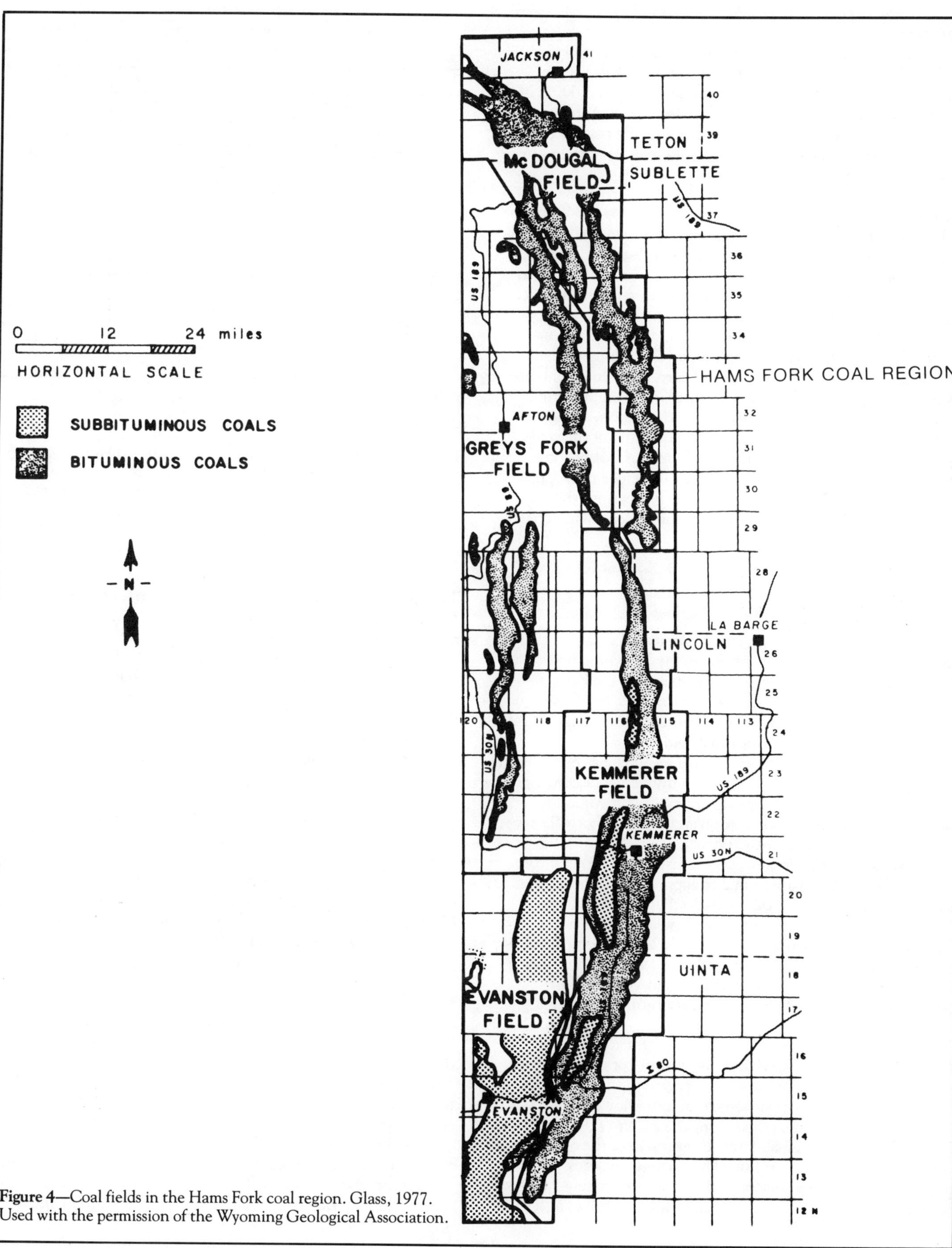

Figure 4—Coal fields in the Hams Fork coal region. Glass, 1977. Used with the permission of the Wyoming Geological Association.

118 ft. Rank of coals from the Adaville Formation ranges from subbituminous A to C.

The most recent formal coal resource estimate for the Hams Fork coal region was made by Berryhill et al, in 1950. This estimate of 4.87 billion tons of coal under less than 3,000 ft of overburden today appears conservative in view of more recent strippable coal reserve estimates on the Adaville coals. Recent estimates place the strippable reserves alone at more than one billion tons (Glass, 1977).

Green River Coal Region

The Green River coal region occupies approximately 18,500 sq mi in southwestern Wyoming and northwestern Colorado. The Wyoming portion of the region (approximately 15,000 sq mi) is the largest coal-bearing area in that state. There are six formally named coal fields in the region: the Rock Springs, Great Divide Basin, Little Snake River, Labarge Ridge, and Henrys Fork coal fields in Wyoming, and the Yampa field in northwestern Colorado. For the most part, these coal fields are located on the flanks of uplift areas that border on or intrude into the three major structural basins of the region. Almost all information concerning the quality and quantity of the coal resource comes from data on these coal fields. Coal-bearing formations in the Green River coal region include the Upper Cretaceous Mesaverde Group and Lance Formation, the Paleocene Fort Union Formation, and the Eocene Wasatch Formation.

	GREEN RIVER COAL REGION OF WYOMING	GREEN RIVER COAL REGION OF COLORADO	HAMSFORK
EOCENE	WASATCH	WASATCH	WASATCH
PALEOCENE	FORT UNION	FORT UNION	EVANSTON
UPPER CRETACEOUS	LANCE	LANCE	ADAVILLE
	LEWIS	LEWIS SHALE	
	MESAVERDE GROUP: ALMOND	MESAVERDE GROUP: WILLIAMS FORK	
	ERICSON		
	ROCK SPRINGS	ILES	
	BLAIR		
			HILLIARD
			FRONTIER
LOWER CRETACEOUS			ASPEN
			BEAR RIVER

NON-COAL-BEARING ROCKS

Figure 5—Generalized stratigraphic column of coal-bearing rocks in the Greater Green River coal region. Modified from Glass, 1978. Used with the permission of the Geological Survey of Wyoming.

The Rock Springs Coal Field

The Rock Springs coal field is located in central Sweetwater County, Wyoming, and covers approximately 3,000 sq mi. This coal field is the most important one in the Wyoming portion of the Green River coal region and is centered on the Rock Springs Uplift. Table 2 summarizes the stratigraphy in this area. Table 3 summarizes the coal quality of the different coal-bearing formations in the Rock Springs coal field.

Historically, the most important coal resource has been found in the Mesaverde Group. This group is divided, in ascending order, into the Blair, Rock Springs, Ericson, and Almond Formations. Only the Rock Springs and Almond Formations contain commercial quantities of coal.

The Rock Springs Formation contains at least 12 coalbeds, thicknesses of which vary from 2 to approximately 14 ft. In addition, there are numerous beds less than 2 ft thick. Generally, the beds are persistent and average about 6 ft in thickness. Total thickness of coal in beds 2 or more ft thick has been estimated to be 80 to 90 ft (Schultz, 1910). Coals in the Rock Springs Formation are generally high-volatile C bituminous in rank and have a sulfur content less than 1%.

The Rock Springs Formation is overlain by the Ericson Formation, which is barren of coal. This formation is 800 to 1,000 ft in thickness. Overlying the Ericson Formation is the Almond Formation. This unit, the uppermost formation of the Mesaverde Group, contains some coal in the lower part. The Almond Formation is exposed along the eastern side of the Baxter Basin. However, all but the lower part of the formation is covered along the western side. The average thickness of individual coalbeds in this formation is approximately 5 ft, with the maximum thickness approaching 9 ft (Berryhill et al, 1950). Total thickness of beds 2 or more ft thick ranges between 15 and 30 ft (Schultz, 1910). Coal in the Almond Formation ranges in rank from subbituminous C to subbituminous B.

In the upper part of the Lance Formation is the coal-bearing sequence locally known as the Black Butte coal group. This group includes five minable coalbeds that average 6.5 ft in thickness (Berryhill et al, 1950). The thickest bed is 9.5 ft and is separated by two shale partings that total about 5 inches. The five minable coals within the Black Butte coal group are designated (from oldest to youngest): the Hall, Maxwell, Black Butte, Gibraltar, and Overland. As a group, on an as-received basis, these coals average 20.8% moisture, 5.5% ash, 0.77% sulfur, and 9,780 Btu/lb (subbituminous B).

Above the Fort Union basal conglomerate is the coal-bearing sequence locally known as the Black Rock coal group. In the lower part of this group, some beds attain a thickness of 26 ft. Measured sections of the thicker seams in this coal group average 6.0 ft (Berryhill et al, 1950). Coal of the Black Rock group ranges in rank from subbituminous C to subbituminous B.

Recent mapping by Roehler (1976a, 1976b, 1977) has shown three persistent coalbeds within the Black Rock group. In order of oldest to youngest, these beds are the Little Valley, Hail, and Big Burn. The Big Burn bed ranges up to 9 ft in thickness, and the Little Valley bed ranges up to 15 ft in thickness.

There are several unnamed and one named Wasatch coalbeds in the Black Buttes area. The named bed is the Nuttal and is a subbituminous coal ranging in thickness up to 8 ft. The unnamed coals are also subbituminous in rank and average 6 to 8 ft in thickness. Though these are presently not mined, there is a possibility that mining will begin in the next several years.

According to Berryhill et al (1950), the Rock Springs coal

Table 1—General characteristics of coal in the Greater Green River coal region.

Geographic Area	Rank[1]	Bed Thickness (ft)			Moisture (%) (AR)[2]		Ash (%) (AR)[2]		Sulfur (%)[2]		Heating Value (Btu/lb) (AR)[2]	
		Maximum	Range[1]	Average[1]	Range	Average	Range	Average	Range	Average	Range	Average
Southern Wyoming	Bituminous/ Subbituminous	50	4–35	20[1]	10.2–20.5	12.4	4.2–12.2	7.1	.4–.9	.52	9,270–11,100	10,500
Western Wyoming	Bituminous/ Subbituminous	110	5–90	16.5	20.4–20.9	20.8	3.0–4.8	4.5	.6–.7	.6	7,500–10,200	9,600
Northwest Colorado	Bituminous/ Subbituminous	42	6–13	9.0	6.3–23.0	16.1	3.9–13.8	7.2	.2–2.7	.8	8,250–12,560	10,330

[1]Beds currently mined
[2](AR) = as received

Table 2—Cretaceous-Tertiary stratigraphic units in the Rock Springs coal field. Yourston, 1955. Used with the permission of the Wyoming Geological Association.

Age	Group	Formation	Thickness*	Character
Eocene		Wasatch	1,000–2,500	Variegated claystone and shale, carbonaceous shale and coal, drab sandstone and conglomerate
Paleocene		Fort Union	6,000–9,400	Sandstone, shale, and conglomerate; many beds of coal known locally as the Black Rock coal group occur near the base of the formation
		Lance	1,500	Sandstone, clay, and coal known locally as the Black Butte coal group occurring in the upper part of the formation; massive white to yellow sandstone base
		Lewis Shale	750	Gray, soft, marine shale, with many gray and brown (600+) lenticular sandstone beds; many concretions
Upper Cretaceous	Mesaverde Group	Almond	700–950 (300–500)	White and brown soft sandstone, gray sandy shale, coal and carbonaceous shale occurring in the lower part of the formation
		Ericson	800–1,000 (300–500)	White and gray massive sandstone; lenticular conglomerate beds in the upper part
		Rock Springs	600–1,400 (1,350)	White to brown sandstone, interbedded shale and clay; numerous coal beds with an aggregate thickness of approximately 90 ft
		Blair	1,000–1,200 (600+)	Massive yellow and brown sandstone at top (Golden Wall) underlain by shale, sandy shale, and tan sandstone
Lower Cretaceous		Baxter Shale	1,000+	Gray to black soft shale and shaly sandstone
		Unexposed Shale	1,800±	
		Frontier	500	Gray to tan sandstone interbedded with yellow to dark gray marine shale, carbonaceous shale, and coal with some white tuff and bentonite beds
		Aspen Shale	200	Black and gray shale, weathers silvery gray with thin bentonite beds

*Thicknesses shown are from Schultz (1944), except those in parentheses, which are from Dobbin (1944).

Table 3—Summary of coal quality in the Rock Springs coal field, on an as-received basis. Modified from Glass, 1978; Root et al, 1973. Used with the permission of the Geological Survey of Wyoming.

		Quality					
Formation	Coalbed	Moisture (%)	Volatile Matter (%)	Fixed Carbon (%)	Ash (%)	Sulfur (%)	Heating Value (Btu/lb)
Rock Springs	3	16.7	35.7	45.4	2.2	.9	10,520
	1	8.5–17.6	34.5–36.5	43.9–50.4	4.0–6.0	.8–1.0	10,480–11,830
	7	5.0–16.5	33.8–40.3	44.3–52.6	2.4–6.4	.6–1.1	10,640–13,110
	11	6.6–8.5	38.4–39.7	47.7–48.5	4.6–6.7	.7	12,379–12,572
Almond	Average	16.4	31.0	47.7	5.0	.6	9,727
Lance	Average	20.8	—	—	5.5	.77	9,780
Fort Union	Deadman	20.5	29.1	40.7	9.7	.47	9,350
Wasatch	Average	11.3–19.6	33.0	40.4	7.2–8.1	.74–1.5	8,770–9,610

field contained an estimated 12,726.68 million short tons of original reserves, of which 9,878.04 million short tons were bituminous coals and 2,848.64 million short tons were subbituminous coals. Recent work by Roehler (1976a, 1976b) had increased the subbituminous coal resource estimate by more than 1 billion short tons.

The Great Divide Basin Coal Field

The Great Divide Basin coal field, as discussed here, includes approximately 2,500 sq mi in the northeastern part of the Green River coal region. This field is largely in Sweetwater County and extends northward into Fremont and westward into Carbon counties. It is bounded on the north by the Green Mountains and the Wind River Uplift, on the east by the Rawlins Uplift, on the south by the Wamsutter Arch, and on the west by the Rock Springs Uplift. Minable coals in the basin are found in the Late Cretaceous Mesaverde Group and Lance Formation and in the Tertiary Fort Union and Wasatch Formations.

Three coal zones are found in the Mesaverde Group. The lower zone contains four to six discontinuous, thin, relatively dirty beds of coal. The middle zone locally attains minable thicknesses south of T24N. In Section 22, T21N, R88W, a coal from this zone was mined in the past and ranged in thickness up to approximately 8.5 ft. The upper coal zone contains at least four thin coal beds, all of poor quality. Analyses of these Mesaverde coals are unavailable. However, Mesaverde coals nearby are of high-volatile C bituminous rank (Berryhill et al, 1950).

Lance Formation coalbeds are numerous and occur throughout the formation. Both the number of beds and bed thicknesses increase to the south. The coalbeds range in thickness from 10 inches to 12 ft and average about 6 ft. On an as-received basis, analysis of one weathered sample shows 4.1% ash, 0.3% sulfur, and a heating value of 9,023 Btu/lb. Analyses on correlative coals from nearby basins to the east show a somewhat higher heating value. It is probable, therefore, that Lance Formation coals range from subbituminous C to subbituminous A in rank (Berryhill et al, 1950).

The lower 800 to 1,800 ft of the Fort Union Formation is barren of coal. Above this barren unit are two coal-bearing members separated by another barren unit. In the southeastern part of the field, both coal-bearing members contain workable beds. However, the number and thickness of the beds decrease to the north. In the southeastern part of the basin, coalbeds from these members range in thickness up to approximately 21 ft and average about 4.6 ft. In the western part of the basin, they locally reach 30 ft in thickness. No analyses are available for the eastern Fort Union coals, but it is probable that they are subbituminous in rank.

Wasatch coals seem best-developed on the east side of the basin, where a minimum of 10 coalbeds crop out. These range in thickness from 5 to 42 ft, tend to be lenticular, and are reported to be subbituminous in rank (Glass, 1977).

Combined resource estimates from Pipiringos (1961), Masurksy (1962), and Berryhill et al (1950) indicate that the total measured and indicated reserves in the Great Divide Basin are 3,560.96 million short tons. Except for a small amount of bituminous Mesaverde Group coal situated along the eastern margin of this field, all coal is subbituminous in rank.

The Little Snake River Coal Field

The Little Snake River coal field is located on the eastern flank of the Washakie Basin and includes an area of approximately 1,500 sq mi. This area is bounded on the east by the Sierra Madre Uplift, on the north by the Wamsutter Arch, and on the south by the Colorado-Wyoming state line. In this field, the Upper Cretaceous Mesaverde Group and Lance Formation, and the Tertiary Fort Union and Wasatch Formations, contain numerous workable coalbeds.

The Mesaverde Group coal is found in the middle and upper members. Coals in the middle member range in thickness from about 3 to 12 ft. These coalbeds are both more numerous and thicker in the southern part of the coal field and tend to decrease in number and thin to the north. The upper member contains several coalbeds that outcrop along the central eastern flank. These beds range in thickness from 5 to 11 ft. In general, the Mesaverde coals are lenticular. However, where one bed thins or pinches out, another bed is likely to thicken. This results in the total thickness of the Mesaverde coals remaining fairly constant over a wide area. Mesaverde coals in this coal field are generally high-volatile C bituminous in rank and have an average heating value, on an as-received basis, of about 10,500 Btu/lb.

Lance coalbeds are generally found between shales. This stratigraphic position means the coalbeds are usually covered by scree, and the lack of exposures makes correlation very difficult. These coals range from 2 to more than 8 ft in thickness and are generally subbituminous.

According to Sanders (1975), Fort Union Formation coals are found in both the lower part and the uppermost 600 to 700 ft of the formation. The coalbeds in the lower part are generally thin and discontinuous but may range up to 5 ft in thickness. In the uppermost part of the formation, up to five coalbeds crop out in what is known as the Cherokee coal district. This coal district lies along the south flank of the Wamsutter Arch, which separates the Great Divide and Washakie Basins. These coalbeds range in thickness from 5 to 30 ft and are generally subbituminous C to lignite in rank.

The upper Fort Union coals, as defined by Sanders (1975), are considered to be Wasatch Formation coals by Glass (1978). Glass defines two unmined beds in the Cherokee district as the Upper and Lower Cherokee beds, and these correlate to the B and C beds of Sanders. The Upper Cherokee (or B) bed ranges in thickness from 10 to 18 ft and normally has a 1- to 2-ft parting. The Lower Cherokee (or C) bed is generally 40 to 70 ft below the upper bed and ranges from 20 to 32 ft in thickness. In places, these two beds coalesce and form a single bed 30 to 40 ft thick.

According to Berryhill et al (1950), the Little Snake River coal field contained an estimated 1,938.93 million short tons of original reserves, of which 23.38 million short tons were bituminous coals and 1,915.55 million short tons were subbituminous coals.

Labarge Ridge Coal Field

The Labarge Ridge field is the only named field in the western part of the Green River coal region. This field is west of the town of Labarge, in parts of T26-28N R113-114W, in Lincoln and Sublette counties, and is bounded on the west by the western Wyoming thrust belt. The coal-bearing rocks are restricted to two small areas totaling about 25 sq mi. The coalbeds are thought to be from the Upper Cretaceous Adaville Formation. No information is available on these beds in the subsurface.

Between 7 and 10 coalbeds crop out on the flanks of the Dry Piney anticline. These coalbeds range from about 1 to 8 ft in thickness and dip to both the east and west between 20 and 50°. The rank of the coal from this field is subbituminous B.

Berryhill et al (1950) infer that up to 6.97 million short tons of subbituminous coal are available in the Labarge field.

Henrys Fork Coal Field

The Henrys Fork coal field is located on the northern flank of the Uinta Mountains in the vicinity of Manila and Linwood, Utah. Early investigations in this area (Gale, 1910) reported that a few coalbeds had minable thicknesses; however, the area has received little subsequent interest. The small amounts of coal observable apparently dip very steeply to the north, but this may be accentuated by faulting. No information is available on the quantity or quality of coal.

Yampa Coal Field

Cretaceous, Paleocene, and Eocene coal-bearing formations crop out within the Yampa River drainage area of the Sand Wash Basin and constitute the Yampa coal field. This is the only coal field in the Colorado portion of the Green River coal region. Figure 6 is a generalized column section of the sedimentary rocks in the Yampa coal field.

Coalbeds are found throughout the Yampa coal field in the Upper Cretaceous Mesaverde Group and Lance Formation and the Tertiary Fort Union and Wasatch Formations. Mesaverde Group coals range in rank from anthracite to noncoking, high-volatile C bituminous. Anthracite and semianthracite Mesaverde coals are very localized and result from contact metamorphism of bituminous coals by igneous intrusives. Lance, Fort Union, and Wasatch Formation coals are generally subbituminous B or C in rank.

Coals in the Mesaverde Group are found in both the Iles and Williams Fork Formations and are subdivided into three coal zones. Table 4 summarizes the various coalbed nomenclature for these zones (Murray, 1977).

The lower zone coals are found in the upper 400 ft of the Iles Formation. There are five main coalbeds in this zone, ranging in thickness from 2 to 10 ft. However, the lenticular nature of these coalbeds makes it very difficult both to estimate reserves and to assure encountering coals from this zone in the subsurface (Bass et al, 1955). Lower zone coal is generally high-volatile C bituminous in rank, on an as-received basis.

The middle zone coals occur in the interval between the Trout Creek and Twentymile Sandstone marker beds. The Trout Creek Sandstone is the uppermost unit of the Iles Formation and is a fine-grained, massive, white sandstone bed about 100 ft thick. The Twentymile Sandstone is a member of the Williams Fork Formation and is similar in composition and color to the Trout Creek Sandstone. It ranges in thickness from 100 to 200 ft and is separated from the Trout Creek Sandstone by 900 to 1,100 ft of intervening strata. However, most of the thick coalbeds in the middle zone lie in the 400-ft interval immediately above the Trout Creek Sandstone. There are five main coalbeds in the middle zone, ranging in thickness from 3 to more than 20 ft. Middle zone coal is generally high-volatile C bituminous in rank, on an as-received basis.

The upper zone coals are found in the uppermost 650 ft of the Williams Fork Formation above the Twentymile Sandstone marker bed. Nine main coalbeds are found in this zone, ranging in thickness from 3 ft to more than 10 ft. Upper zone coal is generally high-volatile C bituminous in rank, on an as-received basis.

Coalbeds in the Lance Formation are found in the lower one-third of the formation. Along the southeast flank of the Sand Wash Basin, two Lance coalbeds have been mined in the past. One of these is the Lorella bed, which ranges in thickness from 3 to 10 ft. The Lorella coalbed is stratigraphically located about 50 ft above the base of the Lance. The other mined coalbed is the Kimberly, which ranges in thickness from about 8 to 12 ft. The rank of these Lance coals is generally subbituminous B or C (Bass et al, 1955). It is possible that there are at least several more coalbeds in the upper two-thirds of this formation, concealed under Tertiary units.

On the eastern flank of the Sand Wash Basin, three Fort Union coalbeds have been mined in the past. Two of these are called the Campbell beds and are stratigraphically located near the base of the formation. These beds are each about 6 ft in thickness and are separated by a 3-ft parting. The other mined bed is called the Seymour bed and ranges in thickness from about 10 to 17 ft. These coals are generally subbituminous B or C in rank.

The Wasatch Formation contains several coal-bearing zones, as evidenced by outcrops located in the northern part of the Sand Wash Basin. These are probably correlative with the Wasatch coals found in the Little Snake River field. These coals tend to be lenticular, with a thickness ranging from about 6 ft to 20 ft. In some areas, these coals may be strippable. No data are available concerning the rank of Wasatch coals in the Sand Wash Basin. However, they probably have the same range as the Wasatch coals of the Rock Springs and Little Snake River coal fields; that is, subbituminous B and C (Bass et al, 1955). Resource estimates of coal in the Sand Wash Basin do not include Wasatch coals.

According to Murray (1979), the Colorado portion of the Green River coal region (Yampa field and Sand Wash Basin) probably contains over 60 billion short tons of coal resources with up to 6,000 ft of overburden. This number should be considered a very rough and conservative estimate because of the lack of a detailed evaluation of coal resources below minable depths. A major part of the region contains multiple beds of coal in several formations below 3,000 ft (Jones et al, 1978). Of the total resource, an estimated 1.0 billion short tons are strippable (Speltz, 1976).

Summary of the Estimates of the Total Coal Resource in the Greater Green River Coal Region

Table 5 summarizes the estimated total original in-place coal resources of the Greater Green River coal region. The estimated total resource of more than 81 billion short tons should be considered extremely conservative. Very little information is available on the quantity of coal below 3,000 ft. It is probable

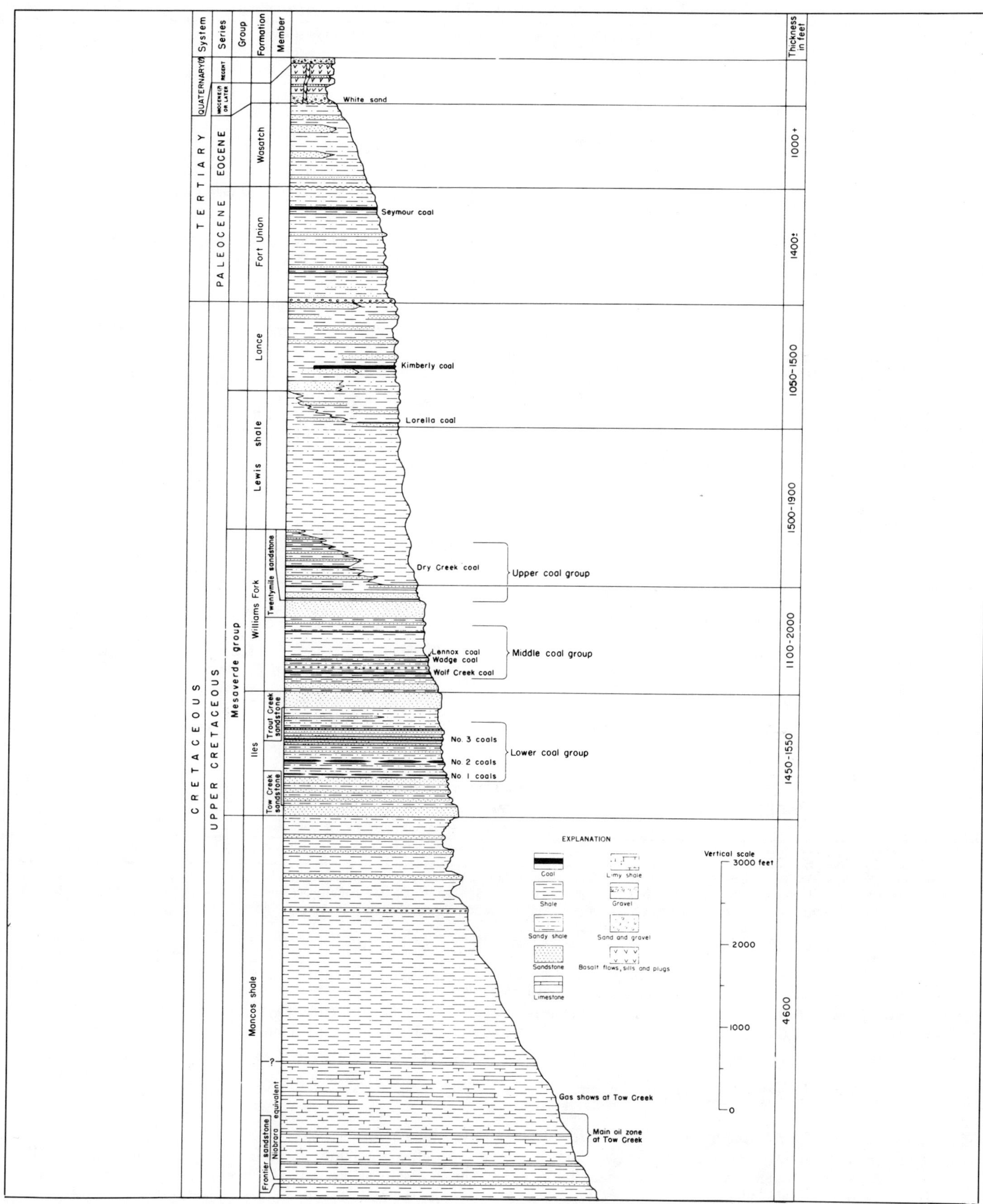

Figure 6—Generalized columnar section of the Upper Cretaceous, Tertiary, and Quaternary Units in the Yampa Coal Field. Bass et al, 1955.

Table 4—Mesaverde Group coalbed nomenclature, Yampa field, Colorado Green River coal region (descending order).

Coal Zone	Letter Designation	Other Names
Upper Zone	S	
	R	
	Q	Dry Creek Bed
	P	
	O	
	N	
	M	Carey Bed
	L	Crawford Bed, Sleepy Cat Bed
	K	
Middle Zone	J	Lennox Bed, Searcy Gulch Bed, Kellogg Bed
	I	Wadge Bed
	H	Wolf Creek Bed
	G	
	F	
Lower Zone	E	
	D	
	C	Webber Bed, Butcher Knife Bed, Rice Mine Bed
	B	Bear River Bed, Sun Mine Bed
	A	No. 2 Bed, Pinnacle Bed
		No. 1 Bed, Brooks Bed, Curtis Bed

Table 5—Summary of estimated total, original, in-place coal resources of the Greater Green River coal region (measured, indicated, and inferred)*. From Berryhill et al, 1950; Roehler, 1976a, 1976b; Masursky, 1962; Pipiringos, 1961; Hornbaker et al, 1976.

		Estimated Coal Resources (million short tons)		
Coal Region	Coal Field	Bituminous	Subbituminous	Total
Greater Green River Basin	Rocks Springs	9,878.04	3,988.0	13,866.04
	Great Divide Basin	—	3,560.96	3,560.96
	Little Snake River	23.38	1,915.55	1,938.93
	Yampa/Sand Wash Basin	38,605.00	19,302.00	57,907.00
	Labarge Ridge	—	6.97	6.97
	Subtotal	48,506.42	28,773.48	77,279.90
Hams Fork	Kemmerer	3,079.92	1,362.41	4,442.33
	Greys Fork	.68	—	.68
	McDougal	117.08	—	117.08
	Evanston	—	314.5	314.45
	Subtotal	3,197.68	1,676.86	4,874.54
Total		51,704.10	30,450.34	82,154.44

*All resource estimates, except for the Yampa/Sand Wash Basin are on coalbeds ranging from 0–3,000 ft in depth. In the Yampa/Sand Wash Basin the resource estimate is on coalbeds ranging from 0–6,000 ft in depth. In both cases, multiple coalbeds from several formations exist below the interval for which the resource is estimated. Therefore all resource estimates for the Greater Green River coal region should be considered very conservative.

that vast quantities of coal in multiple zones of the coal-bearing formations are present in the subsurface of all structural basins to depths greater than 10,000 ft.

POTENTIAL METHANE RESOURCE

The presence of coalbed methane in the Greater Green River coal region is evidenced by limited measurements of coal gas content and by indirect indicators. Gas content measurements of coal samples are available for several locations. These data were gathered as part of the Department of Energy's (DOE) Methane Recovery from Coalbeds Program (MRCP). Indirect indicators of coalbed methane include:

- Coal mine methane emissions
- Gas kicks experienced while drilling through coalbeds
- Gas fields with significant dry gas production from coal-bearing formations.

Coalbed Methane Data

Coalbed methane data were collected as part of DOE's MRCP at several locations in the Greater Green River coal region. Under the MRCP, cooperative agreements were made with several operators drilling conventional oil/gas wells through coal-bearing formations. Under these agreements, coal seams of interest were sampled to determine methane content and, in

Table 6—Summary of locations in the Greater Green River coal region where coalbed methane content data are available.

Location	Formation	Collecting Agency	Number of Samples
Yampa Field	Williams Fork	CGS	6
Yampa Field	Williams Fork	USGS	5
Sand Wash Basin	Williams Fork	DOE/TRW	13
Western Green River Basin	Mesaverde	DOE/TRW	9

Table 7—Summary of coal methane content data for coalbeds sampled by the USGS and the CGS in the Green River coal region.

Location	Formation	Date Sampled	Coalbed Name	Depth of Coalbed (ft)	Coalbed Thickness (ft)	Estimated Gas Content (cf/ton)
Yampa Field	Williams Fork	8/23/76	Wadge	1,283	11.5	0
Yampa Field	Williams Fork	8/26/76	Wadge	1,393	11.0	10.2
Yampa Field	Williams Fork	9/18/76	U. Wolf Cr.	483	1.3	1.2
Yampa Field	Williams Fork	9/19/76	Wadge	335	8.5	7.8
Yampa Field	Williams Field	10/5/76	U. "A", Wolf Cr.	1,104	4.5	3.5
Yampa Field	Williams Fork	10/6/76	Lower "A", "B" Wolf Creek	1,123	9.7	2.6
USGS-IC-H Run 1	Williams Fork	9/9/78	—	176.5–197.1	20.6	3.6
USGS-IC-H Run 19	Williams Fork	9/21/78	—	664–647.6	3.6+	2.9
USGS-IC-H Run 28	Williams Fork	9/23/78	—	720.9–723.9	3.0	0
USGS-IC-H Run 36	Williams Fork	9/25/78	—	766.3–775.3	9.0	16.1
USGS-IC-H Run 41	Williams Fork	9/26/78	—	799.8–807.1	7.3	5.7

some cases, were tested to determine production potential. In addition, the USGS and the Colorado Geological Survey (CGS) incorporated methane content determination in several of their coal resource assessment programs. Table 6 summarizes the locations where coalbed methane content data were obtained.

Yampa Coal Field

In late 1976, the CGS collected coal core samples from several locations in the Yampa coal field for methane content determination. Coal samples were obtained from the Mesaverde Iles and Williams Fork Formations and ranged in depth from less than 500 to approximately 1,400 ft. In late 1978, the USGS collected coal samples from one corehole in the Yampa field for methane content determination. Coal samples were obtained from the Mesaverde Williams Fork Formation and ranged in depth from less than 200 to approximately 800 ft. Table 7 summarizes the data obtained by the CGS and the USGS.

Sand Wash Basin

In late 1978, TRW Energy Systems personnel directed sampling and testing on Energy Reserves Group Robert A. Van Dorn Well # 1 located in Moffat County, Colorado, Section 29, Township 7N, Range 9W. This well was tested as part of the DOE MRCP to provide gas content data on Sand Wash Basin coal. Operations included coring, drill stem tests, and geophysical logging.

The Robert A. Van Dorn Well # 1 was spudded in the Lance Formation and drilled to a TD of 5,026 ft. The hole was bottomed in the Trout Creek sandstone. The top of the Trout Creek sandstone marks the contact between the overlying Williams Fork Formation and the underlying Iles Formation. In this area, these two formations comprise the Mesaverde Group. This well was drilled primarily to test the Fairfield Group of coal seams found in the Williams Fork Formation. Table 8 summarizes the stratigraphic units encountered during drilling.

As can be seen in Table 8, coal seams were encountered in both the Lance and the Williams Fork Formations. Up to eight seams were encountered in the Lance Formation, ranging in thickness from 1 to 6 ft. These seams are shallow and outcrop south of Craig. For this reason, no cores or tests were planned for these coals. However, the literature indicates that they are generally subbituminous B or C.

In the Williams Fork Formation, up to 22 individual coal seams of the Fairfield group were either drilled through or cored. Seam thickness ranged from 2 to 15 ft and cumulative coal thickness was approximately 140 ft. These coal seams are grouped in three distinct zones. The upper zone (3,640 to 4,020 ft) contains 12 seams ranging in thickness from 2 to 6 ft. The middle zone (4,270 to 4,300 ft) contains two seams ranging in thickness from 2 to 6 ft. The lower zone (4,630 to 5,016 ft) contains 8 seams ranging in thickness from 2 to 15 ft. Table 9 summarizes the methane content of conventional core samples obtained from this well.

In addition to obtaining coal samples, two drill stem tests were attempted to estimate the production potential of intervals containing coalbeds. Both tests were accomplished using a Multi-flow Evaluator over a straddle-packed, open-hole interval and included test periods of the following duration:

- Initial flow—5 minutes
- Initial shut-in—20 minutes
- Final flow—60 minutes
- Final shut-in—240 minutes.

Table 8—Stratigraphy of the energy reserves group Robert A. Van Dorn Well # 1.

Formation	Age	Depth Interval (ft)	General Lithology
Lance	Upper Cretaceous	Surface–1,170	Interbedded gray shale; light-tan, fine-grained sandstone; and several thin coal seams
Lewis Shale	Upper Cretaceous	1,170–3,330	Dark gray to bluish, homogenous, marine shale
Williams Fork	Upper Cretaceous	3,330–5,016	Interbedded sandstone, sandy shale, shale, and coal seams
Iles	Upper Cretaceous	5,016–TD (5,026)	Trout Creek Sandstone; massive, fine-grained, white sandstone

Table 9—Summary of depth, bed thickness, rank, and methane content of coal sampled from the Energy Reserves Group, Robert A. Van Dorn # 1 Well, Section 29 T7N R90W, Moffat County, Colorado.

Coalbed Depth Interval	Geologic Formation	Thickness of Coalbed (ft)	Sample Interval (ft)	Rank of Sample (mmmF[1])	Methane Content (cf/ton)
3,652–3,652.6	Williams Fork	.6	3,652–3,652.6	HvBb[2]	254
3,674.7–3,676.7	Williams Fork	1.4	3,675–3,675.8	HvBb	366
3,923–3,924	Williams Fork	1.0	3,923–3,924	HvBb	147
4,654–4,668	Williams Fork	14.0	4,654–4,655	HvBb	332
			4,655–4,656	HvBb	258
			4,656–4,657	HvBb	349
			4,657–4,658	HvBb	309
			4,658–4,659	HvAb[3]	272
			4,659–4,660	HvBb	326
4,704.4–4,706	Williams Fork	1.4	4,704.4–4,705	HvAb	336
			4,705–4,706	HvBb	369

[1]mmmF—moist, mineral-matter free.
[2]HvBb—high-volatile B bituminous.
[3]HvAb—high-volatile A bituminous.

The first test was accomplished on the depth interval of 3,700 to 3,800 ft. Within this interval, there were three coal seams with an aggregate thickness of 18 ft. Based on the data obtained, it was determined that this test probably failed because of the plugging of the tool.

The second test was accomplished on the depth interval of 4,634 to 4,714 ft. Within this interval was one coal seam approximately 14 ft thick. Table 10 summarizes the pressure data from this test.

During this test, there was a slight show of gas to surface at the end of the final flow period (minute 58). Drill pipe recovery included:

- 1,478 ft of gas-cut, water-cut, drilling fluid
- 1,478 ft of gas-cut, mud-cut, water
- 1,477 feet of gas-cut water.

Table 10—Summary of pressure data from DST No. 2, Energy Reserves Group, Robert A. Van Dorn Well # 1.

Description	Pressure (psi)
Initial hydrostatic mud	2,197
Initial flow # 1	118
Initial flow # 2	331
Initial shut-in	2,011
Final flow # 1	313
Final flow # 2	1,964
Final shut-in	2,020
Final hydrostatic	2,197
Bottomhole temperature—138°F	

Green River Basin

In early 1979, TRW Energy Systems personnel directed sampling and testing on the Belco Petroleum S-29-27 well located in Sublette County, Wyoming, Section 28, Township 30N, Range 113W. This well was tested as part of the DOE MRCP to provide gas content data on Green River Basin coal. Operations included conventional and sidewall coring and geophysical logging.

The Belco S-29-27 well was spudded in the Wasatch Formation and drilled to a TD of 3,539 ft—approximately 100 ft into the Mesaverde Formation. This well was drilled primarily to test Paleocene sandstones overlying the Mesaverde. Table 11 summarizes the stratigraphic units encountered during drilling.

In this well, five Mesaverde coal seams with an aggregate thickness of about 20 ft were encountered. Individual seam thickness ranged from 2 to 5 ft. Three seams were conventionally cored and sampled. Table 12 summarizes the methane content of these samples.

Table 11—Stratigraphy of the Belco S-29-27 Well, Sublette County, Wyoming.

Formation	Age	Depth Interval (ft)
Wasatch	Eocene	Surface–718
Paleocene (variously Almy or Fort Union)	Paleocene	718–3,430
	Unconformity	
Mesaverde Formation	Upper Cretaceous	3,430–TD

Indirect Indicators of Coalbed Methane

Occurrence of Methane in Coal Mines

Occurrence of methane in coal mines, if emission rate is measured, allows for an empirical estimate of coal methane content. Studies by the USBM in eastern coal mines indicate that a mine emission rate ranges from six to nine times the methane content of the coals being mined (Kissell et al, 1973).

The Apex # 2(A) Mine in the Yampa coal field, Section 22, T4N, R86W, Routt County, Colorado, has an emission rate of 11,400 cf/day at a mining rate of 60 tons per day (190 cf/ton of coal mined) (Fender and Murray, 1978). The coal being mined is the Pinnacle bed of the Iles Formation, with an overburden depth down to 400 ft. Based on the above empirical relationship, this coal ranges in methane content from approximately 20 to 30 cf/ton.

Methane Encountered in Shallow Drill Holes

Rocky Mountain Energy Corporation (RMEC), the coal resource company of Union Pacific Railroad, has encountered significant flows of gas from numerous shallow wells drilled for coal resource evaluation in the Rock Springs coal field (Cox, 1978, personal communication). Table 13 summarizes the locations of these wells and the depth where the methane was encountered. In addition to these wells, there is an artesian spring in Section 21, T17N, R101W, which bubbles gas.

Gas Fields Producing from Coal-bearing Formations

Review of the Wyoming Geological Association's oil and gas field summaries indicates that numerous dry gas fields throughout the Greater Green River coal region produce from coal-bearing formations. While this is not a direct indication of coalbed methane, it does suggest a possible link. To establish the relationship of coal intervals to production intervals in these fields, detailed study of geophysical logs and other available stratigraphic data is required. Table 14 summarizes the dry gas fields in the Wyoming portion of the Greater Green River coal region that produce from coal-bearing formations.

ESTIMATED RESOURCE VOLUME

The paucity of data on the methane content of coals in the Greater Green River coal region makes an accurate resource estimate very difficult. The specific types of data required to allow reasonable estimates of the methane resource include the methane content of coals from each coal-bearing formation and quantitative data on how these methane contents vary with coal quality and with depth of burial. To date, the available data on the methane content of coals in the Greater Green River coal region are limited to information on Upper Cretaceous Mesaverde Group coals at widely scattered locations. Minimum and maximum values of the methane contents of these samples are used to estimate the potential in-place coalbed methane resource for the bituminous coals of the Greater Green River coal region. The chosen minimum value is 2.9 cf/ton. This methane content was determined by the CGS from a sample taken at center of the southwest quarter of Section 23, T4N R91W, from a depth of approximately 647 ft. The chosen maximum value is 539 cf/ton, determined by MRCP from a sample taken from Well S-29-27, Section 28 T30N, R113W, at a depth of approximately 3,495 ft.

To date, subbituminous coals from the Greater Green River coal region have not been sampled for methane content determination. Therefore, to estimate the potential coalbed methane resource volume in this area, minimum and maximum methane content are assumed. Values of 0 cf/ton and 100 cf/tor are offered to present a conservative range of the potential

Table 12—Summary of depth, bed thickness, and methane content of coal sampled in well S-29-27, Section 28 T30N R113W, Sublette County, Wyoming.

Coal Seam Depth Interval (ft)	Geologic Formation	Thickness of Coalbed (ft)	Sample Interval (ft)	Methane Content (cf/ton)	Coal Rank (as received)
3,478–3,482	Mesaverde	4.0	3,479.1–3,479.6	505	hvAb[1]
			3,479.6–3,480.6	487	hvAb
			3,480.6–3,481.6	436	hvAb
3,492–3,497	Mesaverde	5.0	3,494.8–3,495.8	535	hvBb[2]
			3,495.8–3,496.5	539	hvAb
3,526–3,528	Mesaverde	2.2	3,526.5–3,527.5	463	hvAb
			3,527.5–3,528.2	515	hvAb

[1]hvAb—high-volatile A bituminous.
[2]hvBb—high-volatile B bituminous.

Table 13—Shallow wells that encountered methane in the Rock Springs coal field.

Well Location	Area	Depth to Methane (ft)	Formation	Comments
Sec. 5 T21NR103W (SENW)	Long Canyon	300–320	Rock Springs	Hydrologic observation well
Sec. 5 T21NR103W (NESE)	Long Canyon	300–320	Rock Springs	—
Sec. 16 T19NR99W	Bitter Creek	240	Wasatch-Ft. Union	—
Sec. 15 T20NR101W (NWNW)	Leucite Hills	240	Almond	Big blow—damaged rig
Sec. 15 T20NR101W (NESW)	Leucite Hills	500	Almond	Big blow—required 500 bags of cement to plug it
Sec. 9 T17NR101W (NENW)	Salt Wells	100	Almond	—
Sec. 9 T17NR101W (SESW)	Salt Wells	220	Almond	Big blow
Sec. 16 T17NR101W (SWSW)	Salt Wells	50–100	Almond	

Table 14—Summary of dry gas fields in the Wyoming portion of the Greater Green River coal region that produce from coal-bearing formations.

Field	Township/Range	Producing Formation(s)
Hiawatha, East	T12N R99-100W	Fort Union
Kinney	T13N R99-100W	Wasatch
Baggs South/Pole Gulch	T12N R92-93W	Wasatch/Fort Union
Little Snake	T12N R94-95W	Wasatch/Fort Union
Smith Ranch	T12N R93W	Wasatch/Fort Union
State Line	T12N R93-95W	Wasatch/Fort Union
Westside Canal	T12N R91-92W	Fort Union/Lance/Mesaverde (Williams Fork)
Chimney Butte	T28N R112-113W	Wasatch/Fort Union (Almy)/Mesaverde
Goat Hill	T31N R113W	Fort Union (Almy)
Craven Creek	T24N R114W	Wasatch/Fort Union
Robbers Gulch	T14N R91-92W	Mesaverde (Almond)
Haystack	T14N R96W	Mesaverde (Almond)
Barrel Springs	T16N R93W	Mesaverde (Almond)
Salazar	T16N R95W	Mesaverde (Almond)
Bitter Creek	T16-17N R99W	Mesaverde (Almond)
Higgins	T17N R98-99W	Mesaverde (Almond)
Sand Butte	T17N R99W	Mesaverde (Almond)
Creston III Unit	T17-18N R91-92W	Mesaverde (Almond)
Wild Rose	T17-19N R93-95W	Mesaverde (Almond)
Delaney Rim Unit	T17-18N R97-98W	Mesaverde (Almond)
Stage Stop	T18N R99W	Mesaverde (Almond)
Tierney, North	T19-20N R94W	Mesaverde (Almond)
Ten Mile Draw	T21N R99W	Mesaverde (Almond)
Roser	T21N R100W	Mesaverde (Almond)
Red Desert	T22N R96W	Mesaverde (Almond)
Dry Piney	T27-28N R114W	Frontier
Labarge East	T26-27N R112W	Mesaverde
Fontenelle II	T25-26N R112W	Frontier
Fabian Ditch	T20N R112W	Frontier
Clay Basin	T3N R23-24E	Frontier

methane resource.

The total coalbed methane resource in the Greater Green River coal region is conservatively estimated to range from 150 billion cubic feet (Bcf) to more than 30 trillion cubic feet (Tcf). Table 15 summarizes the estimated coalbed methane resources in this region for both bituminous and subbituminous coals.

CONCLUSIONS

The Greater Green River coal region contains substantial coal resources (both bituminous and subbituminous) that extend from outcrop (on the rims of the basins) to depths greater than 10,000 ft (in the deepest parts of the basins). Coalbed methane associated with these resources is also potentially very substantial. However, methane characterization data, to date, are limited to only a few widely scattered locations where bituminous coals have been sampled. This small, limited sample is only the first step in the development of the framework required to characterize overall distribution of the coalbed methane resource in the Greater Green River coal region.

In the initial phase of the development of the framework data base, only coal in the upper 5,000 ft should be sampled and tested. The magnitude and distribution of this resource should

Table 15—Estimated coalbed methane resource in the Greater Green River coal region.

Coal Region	Coal Rank	Estimated Coal Resources (millions short tons)	Methane Content of Coal Minimum (cf/ton)	Methane Content of Coal Maximum (cf/ton)	Total Estimated Methane Resource Minimum (cu ft)	Total Estimated Methane Resource Maximum (cu ft)
Green River	Bituminous	48,506.42	2.9	539.0	$.14 \times 10^{12}$	26.1×10^{12}
	Subbituminous	28,773.48	0.0	100.0	0	2.9×10^{12}
Hams Fork	Bituminous	3,197.68	2.9	539.0	$.01 \times 10^{12}$	1.7×10^{12}
	Subbituminous	1,676.86	0.0	100.0	0	$.17 \times 10^{12}$
Total		82,154.44	—	—	$.15 \times 10^{12}$	30.0×10^{12}

The source or rationale for the minimum and maximum methane content for coals in the Greater Green River coal region used in the above table are as follows:

Bituminous Coal – The minimum methane content is taken from data generated by the U.S. Geologic Survey on coals sampled from the Williams Fork Formation in the Yampa Field (location C-IC-H Run 19)

The maximum methane content is taken from data generated by U.S. Department of Energy on coals sampled from the Mesaverde Formation, Sublette County, Wyoming (Well S-29-27, Sample Depth = 3,496 ft).

Subbituminous Coal – To date, there is no methane content data available on subbituminous coal in the Greater Green River coal region. The assigned minimum and maximum methane contents are offered to present a conservative range of the potential methane resource.

be fairly well-understood to this depth, and its economic producibility should be demonstrated before deeper sampling and testing are undertaken. Therefore, the basic criteria for establishing the subareas for the initial sampling and testing of coals in the Green River coal region are simply to identify those areas within the region that contain one or more of the coal-bearing units within 5,000 ft on the surface. Figure 7 delineates these areas, within which smaller initial target areas can be defined based on localized structural controls and previous experience in encountering methane in drill holes. Structurally attractive areas include zones along fold axes, where methane desorbed from coal can collect in-place in the open fractures that develop in the stressed rocks, and at the intersections of two or more linear structural features. In addition, gas fields that produce from coal-bearing formations should be examined in detail.

ACKNOWLEDGMENTS

The author gratefully acknowledges the assistance of those organizations and persons who provided information, data, and advice during the preparation of this report. Among these, special thanks are due to Gary Glass (Wyoming Geological Survey), Carol Tremain (Colorado Geological Survey), USGS Coal Branch, Belco Petroleum, and the Energy Reserves Group.

In addition, the cooperation of the following organizations in permitting the reproduction of various figures, tables, and data contained in the text is greatly appreciated: U.S. Geological Survey, Rocky Mountain Association of Geologists, Wyoming Geological Survey, Colorado Geological Survey, McGraw-Hill Publications, and the Wyoming Geological Association.

REFERENCES CITED

Bass, N.W., J. B. Eby, and M. R. Campbell, 1955, Geology and mineral fuels of parts of Routt and Moffat counties, Colorado: U.S. Geological Survey Bulletin 1027, p. 143–250.

Berg, R. R., 1961, Laramide tectonics of the Wind River Mountains, *in* Symposium on Late Cretaceous rocks of Wyoming: Wyoming Geological Association Guidebook, 16th Annual Field Conference, p. 70–80.

Berryhill, Jr., H. L., D. M. Brown, A. Brown, and D. A. Taylor, 1950, Coal resources of Wyoming: U.S. Geological Survey Circular 81, p. 78.

Blackstone, D. L., 1977, The overthrust belt salient of the Cordilleran Fold Belt western Wyoming—southeastern Idaho—northeastern Utah: Wyoming Geological Association Guidebook, 29th Annual Field Conference, p. 367–384.

Bradley, W. H., 1964, Geology of the Green River Formation and associated Eocene rocks in southwestern Wyoming and adjacent parts of Colorado and Utah: U.S. Geological Survey, Prof. Paper 496–A, 86 p.

Cook, W. R. III, 1977, The structural geology of the Aspen Tunnel Area, Uinta County, Wyoming: Wyoming Geological Association Guidebook, 29th Annual Field Conference, p. 397– 405.

Fender, H. B., and D. K. Murray, 1978, Data accumulation on the methane potential of the coal beds of Colorado: Colorado Geological Survey, Open-file Report 78-2, 25 p.

Fidlar, M. M., 1950, Structural features of the Green River Basin: Wyoming Geological Association Guidebook, 5th Annual Field Conference, p. 86–87.

Gale, H. S., 1910, Coal fields of northwestern Colorado and northeastern Utah: U.S. Geological Survey Bulletin 415, p. 233–239.

Glass, G. B., 1976, Review of Wyoming coal fields: Geological Survey of Wyoming Public Information Circular 4.

——, 1977, Update on the Hams Fork Coal Region: Wyoming Geological Association Guidebook, 29th Annual Field Conference, p. 689–706.

——, 1978, Wyoming coal fields: Geological Survey of Wyoming Public Information Circular 9, 91 p.

——, 1982, Wyoming: description of seams, *in* 1982 Keystone coal industry manual: New York, Mining Informational Services, McGraw-Hill Mining Pub., p. 660–685.

Haun, J. D., 1962, Introduction to the geology of northwest Colorado, *in* Exploration for oil and gas in northwestern

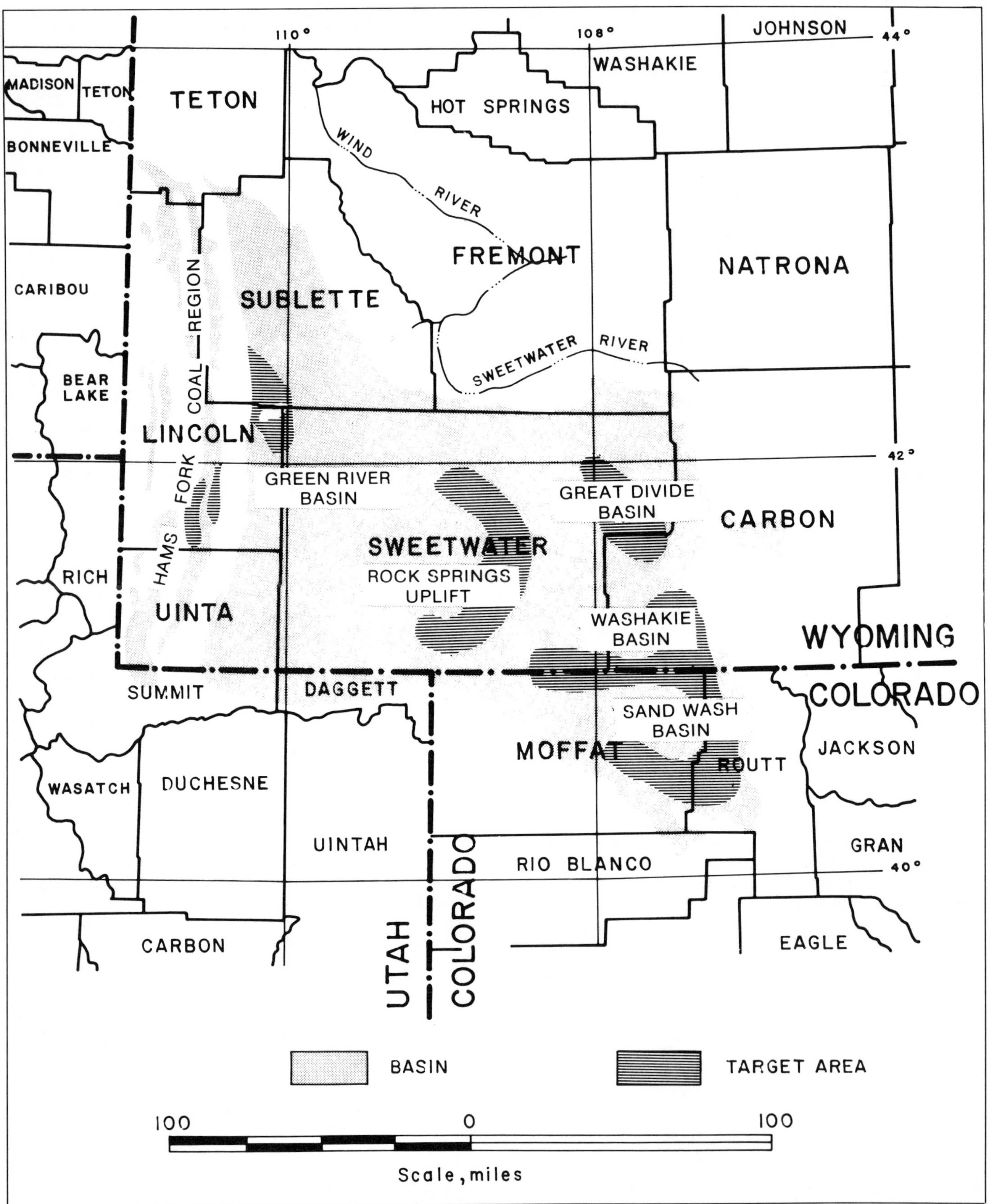

Figure 7—Generalized locations of the primary target areas in the Greater Green River coal region.

Colorado: Rocky Mountain Association of Geologists, p. 7–15.

Hornbaker, A. L., R. D. Holt, and D. K. Murray, 1976, Summary of coal resources of Colorado: Colorado Geological Survey Special Publication No. 9.

Hunter, Jr., W. S., 1950, The Kemmerer coal field in Wyoming: Wyoming Geological Association Guidebook, 5th Annual Field Conference, p. 123–132.

Jones, D. C., et al, 1978, First annual report, evaluation of coking-coal deposits in Colorado: Colorado Geological Survey Open-File Report 78-1, 18 p.

Kissell, F. N., C. M. McCulloch, and C. H. Elder, 1973, The direct method of determining methane content of coalbeds for ventilation design: U.S. Bureau of Mines Report of Investigations 7767.

Krueger, M. L., 1960, Occurrence of natural gas in the western part of Green River Basin: Wyoming Geological Association Guidebook, 15th Annual Field Conference, p. 195–209.

Love, J. D., 1961, Definition of the Green River, Green Divide and Washakie Basins, southwestern Wyoming: Bulletin of the American Association of Petroleum Geologists, v. 45, no. 10, p. 1749–1755.

Masursky, H., 1962, Uranium-bearing coal in the eastern part of the Red Desert Area, Wyoming: U.S. Geological Survey Bulletin 1099-B.

McGookey, D. P., et al, 1972, Cretaceous system, *in* Geologic atlas of the Rocky Mountain Region: Rocky Mountain Association of Petroleum Geologists, p. 190–228.

Murray, D. K., 1977, Colorado: seam description and analyses, *in* 1977 Keystone coal industry manual: New York, Mining Informational Services, McGraw-Hill Mining Pub., p. 575–591.

——, 1979, Colorado: seam description and analyses, *in* 1979 Keystone coal industry manual: New York, Mining Informational Services, McGraw-Hill Mining Pub., p. 458–481.

——,1982, Colorado: description of seams, *in* 1982 Keystone coal industry manual: New York, Mining Informational Services, McGraw-Hill Mining Pub., p. 512–534.

National Oceanic and Atmospheric Administration, 1974, Climate of the States, v. II, Western States: p. 961–975.

Pipiringos, G. N., 1961, Uranium-bearing coal in the central part of the Great Divide Basin: U.S. Geological Survey Bulletin 1099-A.

Robinson, P., 1972, Tertiary history, *in* Geological atlas of the Rocky Mountain Region: Rocky Mountain Association of Geologists, p. 233–256.

Roehler, H. W., 1976a, Geology and mineral resources of the Cooper Ridge NE Quadrangle, Sweetwater County, Wyoming: U.S. Geological Survey Geologic Quadrangle Map GQ-1363.

——, 1976b, Geology and mineral resources of the Sand Butte Rim NW Quadrangle, Sweetwater County, Wyoming: U.S. Geological Survey Open-File, Report 76-495.

——, 1977, Coal resources in the Mud Springs Ranch Quadrangle, Sweetwater County, Wyoming: U.S. Geological Survey Open-File Report 77-77.

Root, F. K., G. B. Glass, and D. W. Lane, 1973, Sweetwater County, Wyoming: Geological Survey of Wyoming County Resource Series, CRS-2.

Rubey, W. W., 1958, Geology of the Bedford Quadrangle, Wyoming: U.S. Geological Survey Map GQ-109.

——, and M. K. Hubbert, 1959, Role of fluid pressure in mechanics of overthrust faulting, pt. II, Overthrust belt in geosynclinal area of western Wyoming in light of fluid-pressure hypothesis: Geological Society of America Bulletin, v. 70, p. 115–206.

——, S. S. Oriel, and J. I. Tracey, 1975, Geology of the Sage and Kemmerer 15 Minute Quadrangles, Lincoln County, Wyoming: U.S. Geological Survey Professional Paper 855.

Sanders, R. B., 1975, Geologic map and coal resources of the Creston Junction Quadrangle, Carbon and Sweetwater counties, Wyoming: U.S. Geological Survey Coal Investigations Map C-73.

Schroeder, M. L., 1976, Preliminary geological map and coal resources of the Ragan Quadrangle, Uinta County, Wyoming: U.S. Geological Survey Open-File Report 76-663.

——, 1977, Preliminary geological map and coal resources of the east half of Guild Hollow Quadrangle, Uinta County, Wyoming: U.S. Geological Survey Open-File Report 77-427.

Schultz, A. R., 1910, The southern part of the Rock Springs coal field, Sweetwater County, Wyoming: U.S. Geological Survey Bulletin 381, p. 214–281.

Smith, J. B., M. F. Ayler, C. C. Knox, and B. C. Pollard, 1972, Strippable coal reserves of Wyoming: U.S. Bureau of Mines Information Circular 8538.

Speltz, C. N., 1976, Strippable coal resources of Colorado—Location, tonnage, and characteristics of coal and overburden: U.S. Bureau of Mines Information Circular 8713, 70 p.

Toenges, A. L., J. D. Davis, L. A. Turnbull, and J. M. Schopf, 1945, Reserves, bed characteristics and coking properties of the Willow Creek coalbed, Kemmerer District, Lincoln County, Wyoming: U.S. Bureau of Mines Technical Paper 673.

Townsend, D. H., 1960, Economic report on the Kemmerer coal field: Wyoming Geological Association Guidebook, 15th Annual Field Conference, p. 251–255.

Veatch, A. C., 1907, Geography and geology of a portion of southwestern Wyoming: U.S. Geological Survey Professional Paper 56.

Wach, P. H., 1977, The Moxa Arch, an overthrust model?: Wyoming Geological Association Guidebook, 29th Annual Field Conference, p. 651–664.

Welder, G. E., and L. J. McGreevy, 1966, Ground-water reconnaissance of the Great Divide and Washakie Basins and some adjacent areas, southwestern Wyoming: U.S. Geological Survey Hydrologic Investigations Atlas, HA-219.

Yourston, R. E., 1955, Rock Springs coal field, *in* Green River Basin: Wyoming Geological Association Guidebook, 10th Annual Field Conference, p. 197–202.

Geologic Overview, Coal, and Coalbed Methane Resources of the Wind River Basin—Wyoming

H. H. Rieke
J. N. Kirr

The Wind River Basin is an east-west–trending intermontane asymmetrical syncline covering an estimated 8,100 sq mi in central Wyoming. The basin contains comparatively minor structural deformation in the form of folding and faulting. Structurally, the surrounding area is more complex, with major thrust fault zones traversing northwest to southeast over the region.

Major coal-bearing strata of the Wind River Basin are contained in the Upper Cretaceous Mesaverde Formation. The area is divided into seven coal fields and regions: Muddy Creek coal field; Pilot Butte coal field; Hudson (Lander) coal field; Beaver Creek coal field; Big Sand Draw coal field; Alkali Butte coal field; and Arminto (Powder River) coal field. Their major coal deposits are found in the Mesaverde and Meeteetse Formations, along the basin perimeter. The Alkali Butte coal field contains three coalbeds greater than 4 ft thick, the most important being the Signor, Beaver, and Shipton coals, with a rank of subbituminous C. The Hudson coal field contains the Signor coalbed or zone, which has been traced some 22 mi from the Alkali Butte coal field to the east. Coal contained within this zone is ranked as subbituminous. The remaining five coal fields have not been extensively developed. Thin and split coalbeds are prevalent in a few of these fields, and in many instances top-quality coals are scattered, making it difficult to justify development. Little is known about the basin's coal resource at depth.

The data collected suggest that a large part of the Wind River Basin may have some potential for coalbed methane production. However, a primary target area of about 1,500 sq mi, located in the northeastern part of the basin (Fremont County and western Natrona County) has been designated as having the highest potential for coalbed methane production. This target was determined by evaluating desorption data obtained from U.S. Geological Survey coal cores from the Wind River Basin Indian Reservation and integrating the data with coalbed geology. On the basis of this information, an estimated range of 5.2 to 2,225 billion cubic feet of gas may be contained in the Wind River Basin. Ranges for the expected inplace methane resource have been established for the coal fields of the Wind River Basin, based on available data.

INTRODUCTION

The Wind River Basin, an asymmetrical intermontane syncline encompassing an area of about 8,100 sq mi in west-central Wyoming (Fig. 1), is named after the northward-flowing Wind River, a tributary of the Big Horn River. The coal-bearing area generally coincides with the topographic and structural basin; however, the Wind River coal basin is limited to that portion underlain by Upper Cretaceous or younger rocks. Glass and Roberts (1978) chose this definition because the Mesaverde Formation is the oldest important coal-bearing unit in the basin.

The Wind River topographic basin extends for approximately 180 mi in a southeasterly to northwesterly direction and is located entirely in Fremont County and the western part of Natrona County, Wyoming (Fig. 1). This teardrop-shaped basin is about 75 mi wide from north to south, bounded along its southwestern side by the granitic Wind River Mountains and on its south side by the Granite Mountains. The basin is bordered on the north by the Owl Creek and Absaroka Mountains and the Casper Arch. It is the third-largest coal-bearing basin in the state, outranked only by the Green River and Powder River Basins. The approximate dimensions of the area underlain by coal are 125 mi long and 45 mi wide. The deepest part of the basin lies in the northeastern section close to the Owl Creek Mountains and the Casper Arch, where sedimentary strata reach a maximum thickness of 33,000 ft. Paleozoic and Mesozoic rocks, with the exception of the Silurian, are present. Three-fourths of the basin's area is covered by Tertiary-age rocks (Paape, 1968) that conceal, for the most part, the important coal-bearing Upper Cretaceous Formations cropping out along the southern boundary of the basin near the town of Lander. An estimated 39,000 cu mi of Cambrian to Eocene sediments fill this basin.

GEOLOGY

Shelton and Ross (1971) catalogued the significant geologic features of the Wind River Basin and compared them with the Bighorn and Powder River Basins. They studied the basin's physiography, sedimentology, structure, igneous activity, and stratigraphic framework through geologic time. These factors are important in delineating the quantity and quality of the basin's coalbed methane resources.

Recognizing the stratigraphic position of marine transgressions is important in correlating and subdividing strata. Sedimentary rocks in the basin fall into two general depositional environments: Paleozoic and Mesozoic sediments, predominantly marine and containing continuous coalbeds in

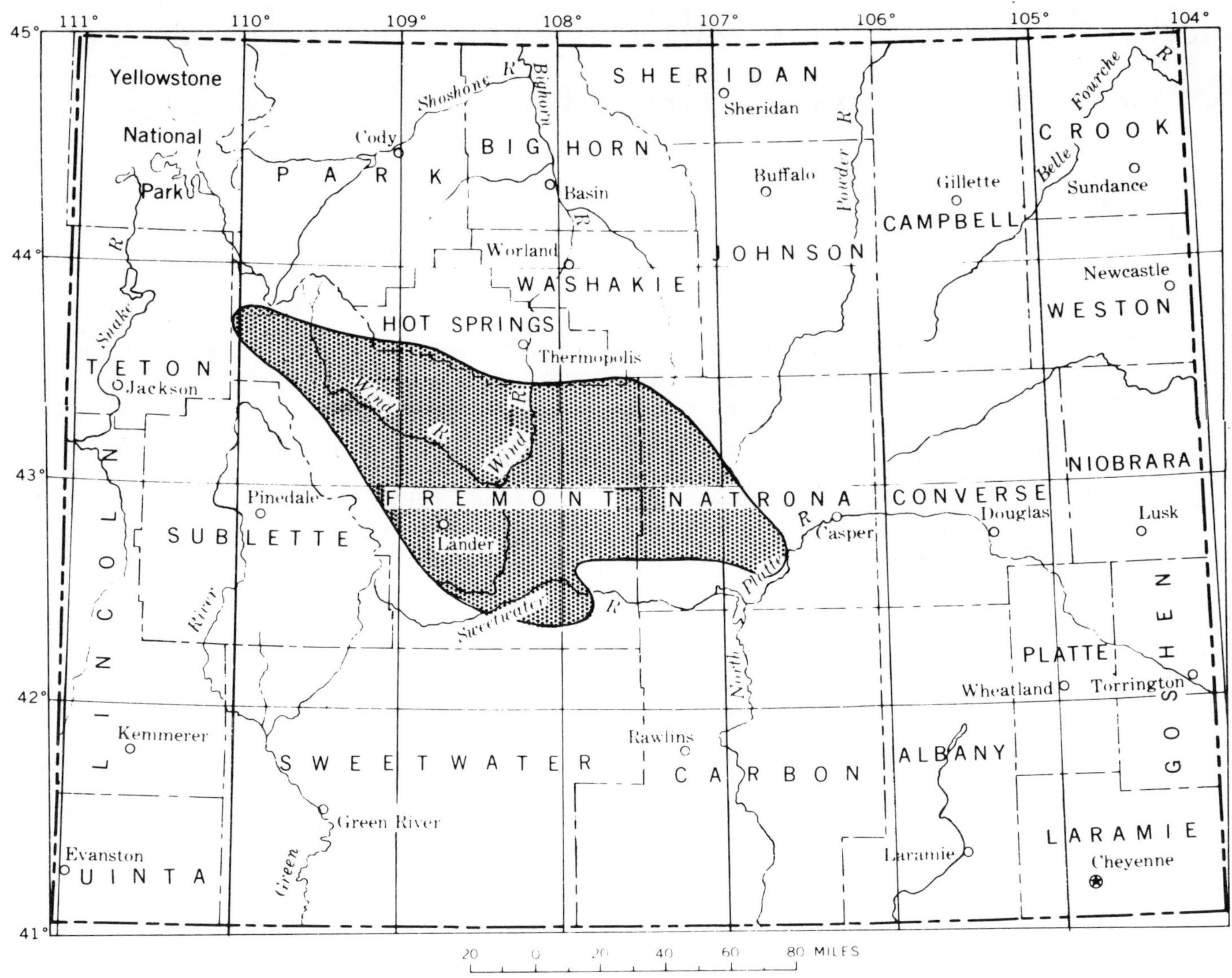

Figure 1—Location of the Wind River structural basin in Wyoming. Keefer and Van Lieu, 1966.

the Upper Cretaceous sequence; and the Cenozoic-to-recent deposits, predominantly continental in origin. The restricted Cenozoic depositional environments are an important factor in the formation of coal and coalbed methane. Numerous detailed stratigraphic studies of the Paleozoic, Mesozoic, and Cenozoic formations have been published (e.g., Keefer and Van Lieu, 1966; Thomas, 1948) and are summarized in Figure 2 in a generalized stratigraphic nomenclature chart for the State of Wyoming and the Denver Basin. Figure 3 is a composite stratigraphic column for the Wind River Basin.

Depositional Environments and Stratigraphy

Igneous and metamorphic rocks of Precambrian age form the core of the mountain ranges and underlie sedimentary rocks within the basin (McGreevy et al, 1969). Smithson and Ebens (1971) reported that the Wind River Range is composed mostly of Precambrian granite gneisses, augen gneisses, and migmatites that are metamorphic rocks typical of old shield areas. The Precambrian rocks of the Owl Creek Mountains cover only about 50 sq mi and consist of granite and granite gneiss. In the Washakie Range, just west of the Indian reservation, the Precambrian rocks consist of gray and pink granite and granite gneiss, cut by thin pegmatite dikes and quartz veins. Schistose rocks are common (Keefer, 1957). In the Immigrant Trail, Dubois, and Big Sandy areas of the basin, Precambrian and Paleozoic rocks have been thrust over the Mesozoic coal-bearing formations.

During the Paleozoic and most of the Mesozoic, the Wind River Basin was part of a stable shelf that lay east of the Cordilleran geosyncline (Keefer, 1965). Sediments representing all geologic periods except the Silurian were deposited across the region during repeated epicontinental marine transgressions. Stratigraphic sequences are thicker and more complete in the western part of the basin than in the east, where they are thin and discontinuous. Some units disappear to the east because of truncation or nondeposition (Fig. 2). The depositional environments were influenced locally by slight fluctuations in sea level and by tectonic movements limited to broad upwarps and downwarps showing little direct relation to the Laramide structural trends that developed later.

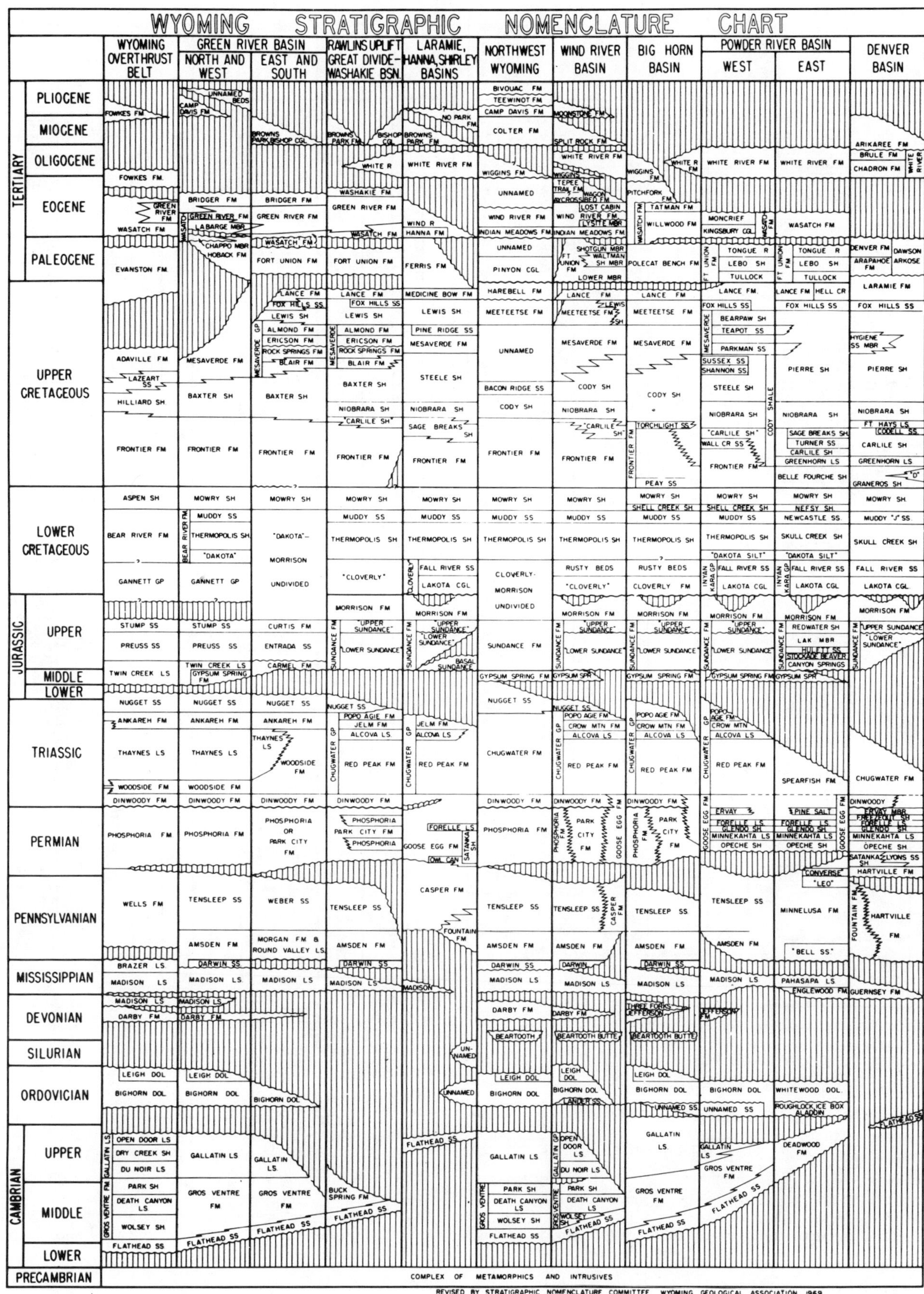

Figure 2—Stratigraphic correlations for Wyoming and the Denver Basin. Used with the permission of the Wyoming Geological Association.

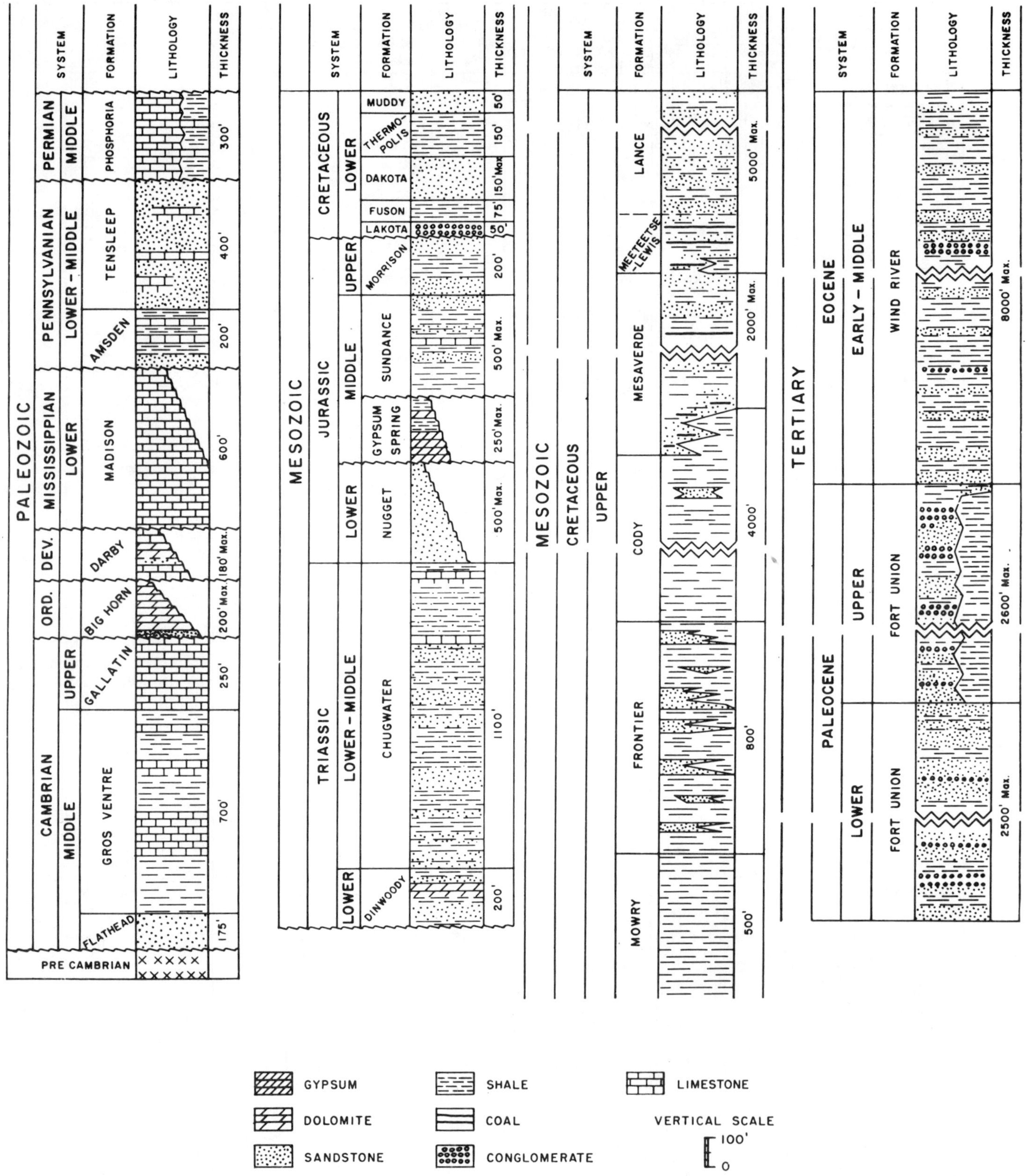

Figure 3—A composite columnar stratigraphic section providing total maximum thickness of sediments in the Wind River Basin. Paape, 1968.

Cambrian rocks unconformably overlie Precambrian rocks and are represented by a marine sandstone, shale, and limestone sequence. The most generally accepted nomenclature for this sequence is, from the base upward: Flathead Sandstone, Gros Ventre Formation, and Gallatin Limestone.

Mississippian time is represented by the Madison Limestone, named for the Madison Range in the Three Forks region of southwestern Montana. Pennsylvanian rocks are represented by two formations: the Amsden and Tensleep Sandstone, representing Early and Middle Pennsylvanian time. No coal is known to occur in the Carboniferous rocks of the Wind River Basin.

McKelvey (1959) suggested that the Permian sequence be divided into four major rock groups: red beds; carbonate; mudstone, chert, and phosphorite; and sandstone (Fig. 2). Respectively, these are assigned to the Goose Egg Formation, Park City Formation, Phosphoria Formation, and Shedhorn Sandstone. Because the Phosphoria fossils are unique and the facies relationships are diverse, detailed correlations within the Phosphoria are difficult. (Lane [1973] discusses this nomenclature with respect to subsurface data.) The name Park City Formation is commonly used because the Permian rocks in the western Wind River Basin are primarily carbonates.

Triassic rocks are underlain by Permian beds throughout west-central Wyoming (Picard, 1978). Marine Permian rocks (Phosphoria Formation) in the western part of the Wind River Basin intertongue eastward and southeastward with red beds (Goose Egg Formation) that are difficult to distinguish from red beds of Triassic age.

Red-bed deposition dominated during the Triassic. In earliest Triassic time, siltstones and carbonates were deposited in a marine environment in the central and western parts of the basin. A westward shift in the red-bed depositional environment resulted in nearly 1,000 ft of red shale, siltstone, and sandstone uniformly deposited over the basin on broad alluvial, deltaic, and littoral plains and mud flats (Keefer, 1965).

The Jurassic of central Wyoming also represents a complex depositional system with diverse sediment types and widespread unconformities (Keefer, 1965). The pattern of unconformities within this rock sequence indicates considerable tectonic activity. Early Jurassic rocks are characterized by continental conditions; whereas during the Middle Jurassic Series, a marine environment persisted. Upper Jurassic sediments are characterized by several hundred feet of sandstone and variegated shale of the Morrison Formation. Because the uppermost Jurassic and lowermost Cretaceous rocks are similar, in many places the position of the Jurassic-Cretaceous boundary is uncertain (Keefer, 1965). Locally, a series of lenticular conglomerates near the middle of the sedimentary sequence may mark the systemic boundary. At the close of the Jurassic, highlands began to form in the geosynclinal area west of Wyoming, and the major sites of deposition were shifted eastward.

During the Early Cretaceous, black marine shales were formed as shallow marine deposits. The oil-producing Muddy Sandstone has characteristics of both marine and nonmarine origins. Keefer (1965) suggests that low islands with fresh-water streams and abundant vegetation existed, and widespread bentonite beds reflect volcanic activity.

Upper Cretaceous rocks reflect a complex history of transgression and regression. Vast amounts of clastic debris were shed into the basin from the west. As the marine environment shifted eastward in early Late Cretaceous time, an extensive regressive sandstone (basal Frontier Formation) was deposited across the Mowry black marine shale. A period of continental sedimentation followed, with swamps, lagoons, and marine embayments dominating the depositional environments.

A major marine transgression followed deposition of the Frontier, as more than 2,000 ft of marine shale were deposited in the Cody sea. Maximum transgression was reached in middle Niobrara time, followed by a long period of regression. Another 2,000 ft of very fine-grained shaly marine sandstone accumulated. The final regressive deposit is the basal sandstone of the Mesaverde Formation (Keefer, 1965).

Much intertonguing takes place between the Cody and Mesaverde Formations in a transition from shallow offshore sedimentation to an environment that deposited 2,000 ft of shaly, sandy marine sediments.

Both the Mesaverde and Meeteetsee Formations are the result of sedimentation in flood-plain, coastal swamp, and lagoonal environments. The land surface sloped gently eastward, and existing highlands in eastern Idaho continued to furnish large amounts of clastic debris and some volcanic ash to eastward-flowing rivers.

Laramide tectonic movements began with the deposition of the Lance Formation (Upper Cretaceous). These movements had significant influence on the sedimentation pattern through the Eocene. Keefer (1965) suggested that the upper surface of the Meeteetsee/Lewis be considered a plane of "zero" deformation. At the beginning of Lance deposition, broad upwarps began to form in the areas now occupied by the Granite Mountains along the basin's southern edge and by the Washakie Range. The major structural features of the basin were defined by latest Cretaceous time (Paape, 1968).

The Wind River Mountains began to rise along the western side of the Wind River Basin at the beginning of the Paleocene. Both the Granite and Washakie Mountains continued to rise, and anticlinal folding took place in some of the marginal areas of the basin.

The basin trough continued to subside during the Paleocene and, during the latter part, it was flooded by a large body of water known as Waltman Lake. Hyne et al (1979) noted that Waltman and Modern Lakes are similar in size, and sediments consist of organic-rich black shales. These shales are thought to be the source rock for much of the oil and gas found in the Lower Tertiary formations of the northeastern part of the basin.

During the early Eocene, the rate of folding and uplift was accelerated. Large-scale reverse faulting took place in the northwestern part of the basin, and alluvial fans formed in front of the rising high lands. The southern Big Horn Mountains and the Casper Arch were also formed along extensive reverse faults as the Wind River Basin was filling with up to 8,000 ft of lower Eocene sediments.

Basin subsidence and the tectonic activity of the Laramide orogeny ended by the middle Eocene. Renewed folding and faulting of existing structural features took place after the Wind River Formation was deposited and small lakes formed in low areas. Large-scale normal faulting was probably related to regional orogenic uplift that occurred in the Late Tertiary. Keefer (1965) reported that in many places, this faulting resulted

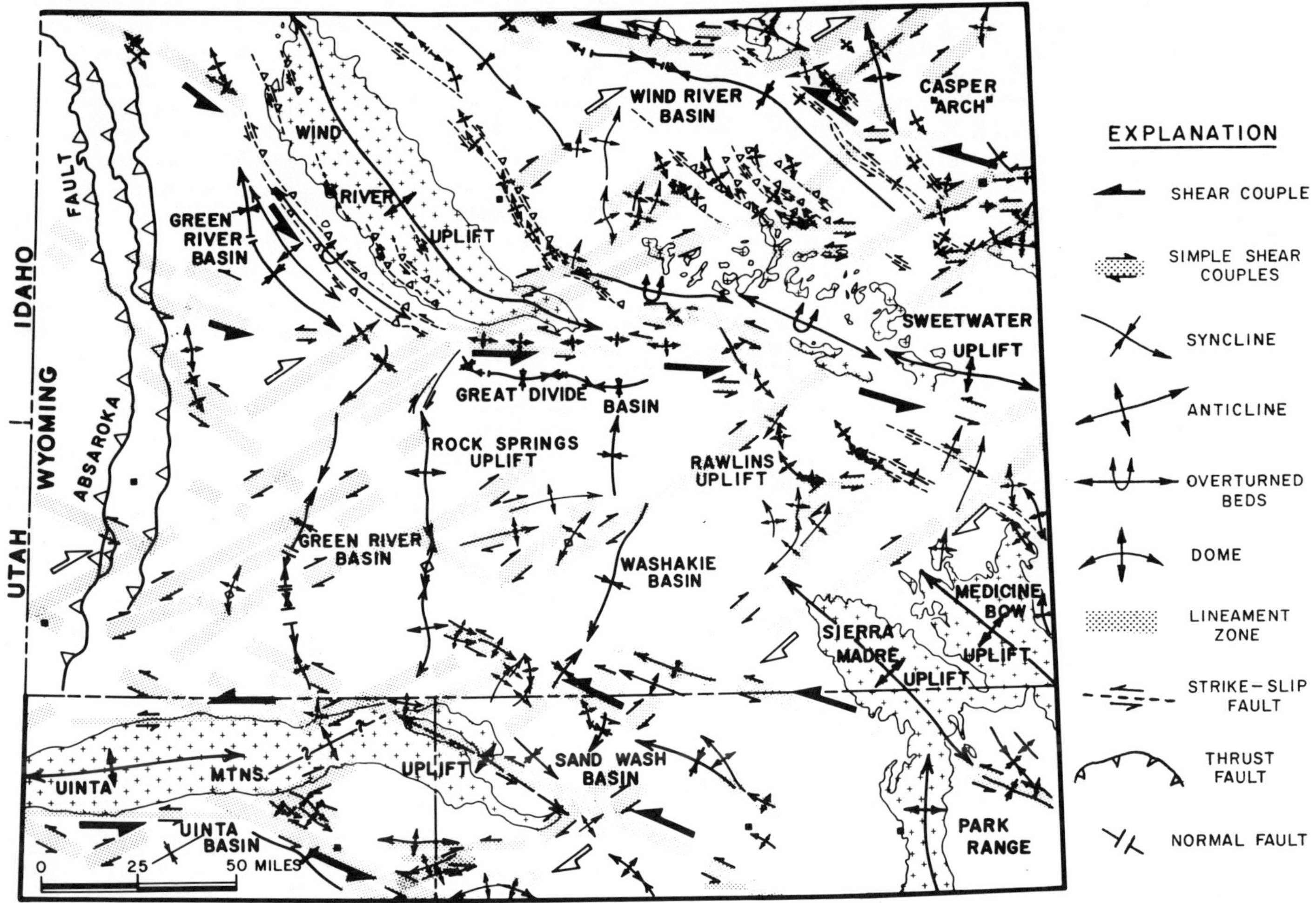

Figure 4—Regional structure and simple shear zones for southwest Wyoming and northern Colorado. Thomas, 1971. Used with the permission of the Wyoming Geological Association.

in the collapse of major Laramide uplifts along former reverse faults. Regional tilting and normal faulting continued into the Miocene.

The sedimentary history of the basin during Middle and Late Tertiary times is one of nearly continuous aggradation. The Oligocene, Miocene, and Pliocene rocks are predominantly volcanic, in sharp contrast to the predominantly nonvolcanic, locally derived clastic material of the lower Eocene.

By the Pliocene, the entire region (except for the highest mountain peaks) was buried. During late Pliocene, this area was uplifted 3,000 to 4,000 ft and a long period of degradation set in. Keefer (1965) noted that this continues today, and the structural and physiographic features present are not very different from those of the early Eocene.

Structure

The Wind River Basin in central Wyoming is typical of the large sedimentary and structural basins that formed in the Rocky Mountain region during Laramide deformation (Fig. 1) (Keefer, 1970). Since it is an integral part of the Rocky Mountain structural complex of folded and faulted sedimentary rocks, the tectonic history of the basin generally will be related to that of the whole system. The results of tectonic activity in the Wind River, Bighorn, and Greater Green River Basins are summarized in Figure 4. Flat-lying lower Eocene strata—part of the thick sequence of basin-fill sediments that accumulated during the major phases of tectonism in the latest Cretaceous and Early Tertiary—occupy the central part of the basin. The basin is completely surrounded by broad belts of folded and faulted Precambrian, Paleozoic, and Mesozoic rocks that form the flanks and cores of the adjacent mountain range and anticlinal complexes. Paape (1968) noted that the basin's shape appears to be a result of the complex compressional forces that caused extensive reverse faulting along all flanks of the basin during the Laramide orogeny. Figure 5 shows major structural features of the Wind River Basin.

The Wind River Basin is markedly asymmetric. The structurally deepest parts are close to the Owl Creek and Bighorn Mountains on the north and the Casper arch on the east. The main trough of the basin lies 3 to 5 mi south of these mountains and appears to intersect the north end of the Casper arch at a right angle. Asymmetry toward the southwest and a pronounced northwest alignment (N 40° W) of individual folds and faults dominate the structural pattern across the entire region (Keefer, 1970). The present Wind River Basin is composed of two distinct synclinal areas: one trending east-west along the northern flank of the basin, which will be referred to as the main northern basin syncline, and the other a secondary

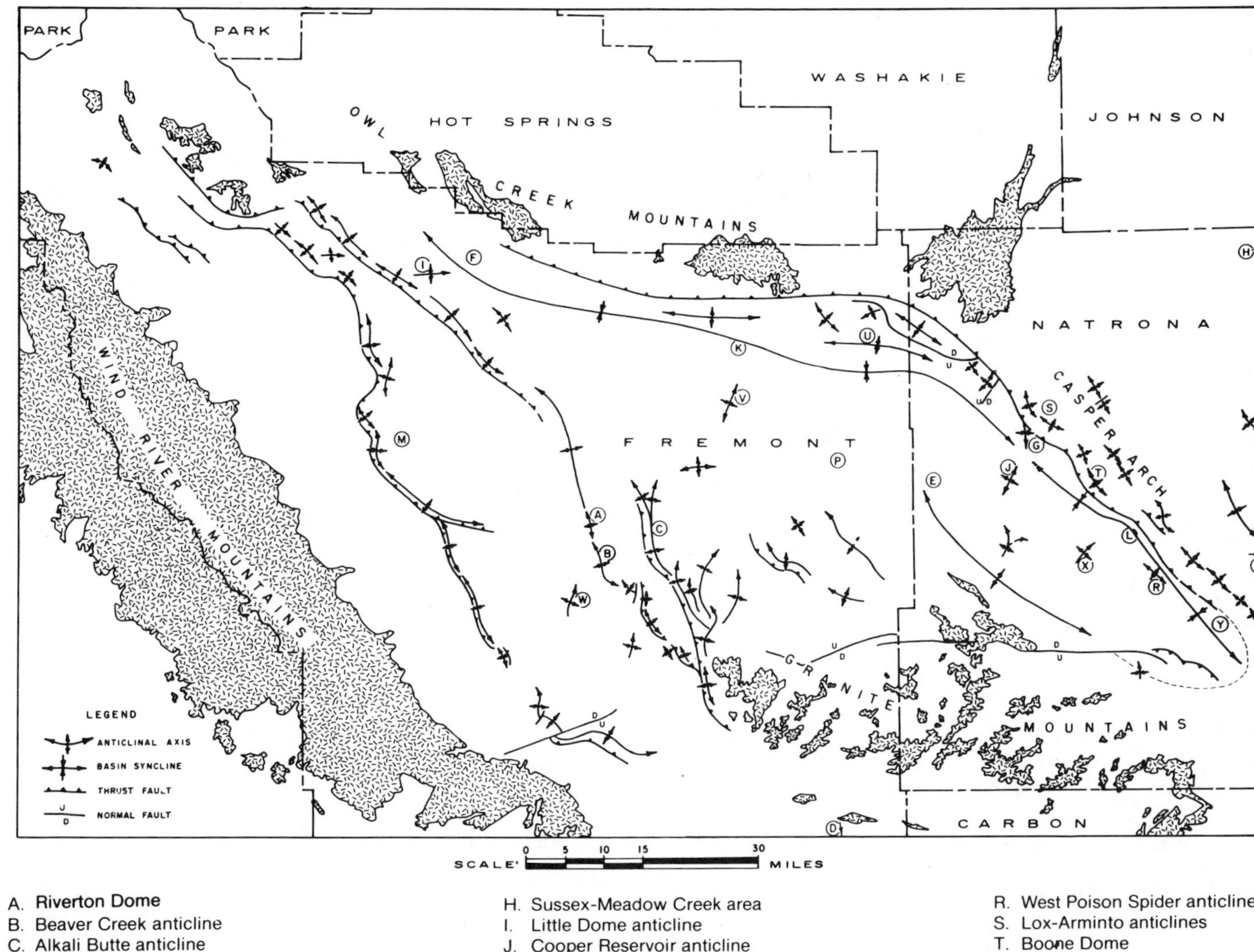

A. Riverton Dome
B. Beaver Creek anticline
C. Alkali Butte anticline
D. Southeast Wind River Mountain — Lost Soldier area
E. Raderville area
F. Shotgun Butte area
G. Waltman area
H. Sussex-Meadow Creek area
I. Little Dome anticline
J. Cooper Reservoir anticline
K. Northern major basin syncline
L. Southeastern minor basin syncline
M. Steamboat Butte-Sage Creek trend
N. Glenrock area
P. Squaw Butte unit area
R. West Poison Spider anticline
S. Lox-Arminto anticlines
T. Boone Dome
U. Lost Cabin anticline
V. Poison Creek anticline
W. Nine Mile anticline
X. Sun Ranch anticline
Y. Poison Spring Creek area

Figure 5—Major structural features of the Wind River Basin. Paape, 1968.

synclinal area extending along the southeastern margin of the basin, which will be referred to as the minor southeastern basin syncline (Paape, 1961).

The structures of the Precambrian granitic and metasedimentary core rocks of the Wind River range have not been extensively studied, but faults, shear zones, dikes, and folds generally have northwest, north, or northeast trends. Keefer (1970) states that the latest movements on northwest-trending structures may be of Laramide age and that east- and northeast-trending structures may have resulted from Precambrian deformations.

Along the east flank of the Wind River Range, Paleozoic and Mesozoic strata form a linear outcrop belt with uniform northeastward (basinward) dips of 12 to 15°. The monoclinal continuity is interrupted by a series of sharply folded northwest-trending anticlines that extend along the west margin of the basin for 90 mi (Fig. 5). The 13 individual folds that occur along this trend have structural closures ranging from 500 to 4,000 ft (Keefer, 1970).

The Owl Creek Mountains along the north margin of the Wind River Basin include a complex group of structures with diverse trends and structural behavior. West of Mexican Pass, the range is partitioned into several horst-like blocks of Precambrian rocks flanked by Paleozoic rocks. Individual blocks, trending northwest, are almost completely surrounded by reverse faults or monoclinal flexures. The adjacent basin margin is also intensely deformed into sharply folded asymmetric anticlines and synclines. East of Mexican Pass, the structure of the range is virtually a single broad anticlinal arch with a gently dipping (12 to 15°) north flank bordering the Bighorn Basin and a steep-to-overturned south flank that overrode the north margin of the Wind River Basin along a continuous reverse fault zone (South Owl Creek Mountains fault).

The plunging southwest end of the Bighorn Mountains forms the northeast margin of the Wind River Basin. Paleozoic and

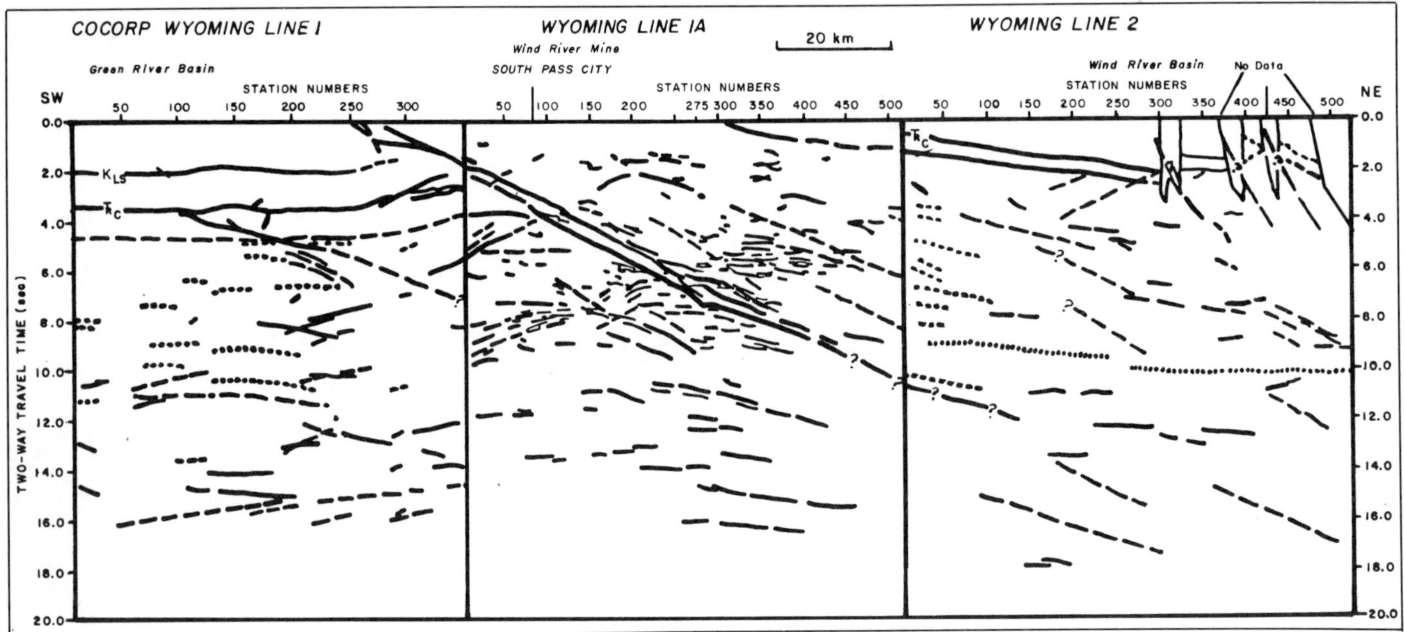

Figure 6—COCORP seismic data in southwestern Wyoming. Profile runs transversely across the Wind River Basin, Wind River Mountains into the eastern part of the Green River Basin. Schilt et al, 1979. Deep reflections beneath the Wind River Basin may correspond to the Moho (11.0 and 15.0 Sec.).

Mesozoic strata, which dip 15 to 25° basinward, have been faulted over the basin margin along the buried South Owl Creek Mountains faults. Structural relief here probably exceeds 30,000 ft in places.

The Casper arch is a major, but not deeply eroded, structural upwarp whose steep-to-overturned west limb coincides with the east margin of the Wind River Basin. A nearly continuous northwest-trending series of subsidiary anticlines is superimposed along the west edge of the major arch.

The Granite Mountains, along the south edge of the Wind River Basin, were uplifted several thousand feet and deeply eroded during Laramide deformation. Then, owing to extensive downfaulting and downfolding in post-early Eocene time, the central Precambrian core collapsed and was buried by Middle and Upper Tertiary sediments. Erosion has now exposed part of the basement complex (Keefer, 1970).

The south margin of the Wind River Basin is interrupted by a series of northwest-plunging anticlines that, from west to east, include Beaver Creek, Sand Draw, Alkali Butte, Conant Creek, Muskrat, Dutton Basin, and the Rattlesnake Hills (Blackstone, 1948) (Fig. 5). Older rocks exposed in these folds extend beneath the unconformable Eocene-age Wind River sediments and structural detail cannot be observed (Brenner, 1972). A somewhat similar situation exists in the northwest part of the basin, where the Dry Mountain-Sheldon trend and the Circle Ridge, Maverick Springs, Little Dome line of folding extends basinward beneath the Wind River Formation. Blackstone (1948) stated that geophysical data do not show these folds continuing across the central part of the basin between the northern and southern margins.

As shown by Figure 5, many of the folds are associated with faults, especially in the northeastern and western parts of the basin. Dominant types of faults include high-angle reverse, normal, and thrust faults. Schilt et al (1979) developed a composite seismic section running transversely across the Wind River Basin, Wind River Mountains, and the Green River Basin (Fig. 6).

The Laramide uplifts have brought Precambrian basement to the surface next to deep sedimentary basins with basement relief on the order of 6.2 mi (10 km). Many of these uplifts are flanked by moderate- to low-angle thrust faults, with crystalline rocks thrust over sediments. Debate on whether these faults steepen and become nearly vertical within the crust, maintain a shallow dip well into the crust, or flatten out at shallow depths without significantly penetrating the crust has not been resolved. A fault steepening to become vertical would imply a vertical principal compressive stress; while a shallow-dipping fault would indicate a horizontal principal compressive stress. If movements were largely a result of gravity sliding, a fault might be expected to flatten without extending very deep into the crust. Schilt et al (1979) discovered a fault plane dipping 30 to 35° that penetrated to a depth of at least 15.5 mi (25 km) (Fig. 6). This fault geometry demonstrates a compressional stress regime at the time of faulting and implies that vertical and horizontal displacements of at least 8 and 13 mi, respectively, occurred in the area (Schilt et al, 1979).

The fault plane extends nearly to the base of the crust (and perhaps deeper), implying that the crust essentially fractured as a rigid plate (Fig. 6). As the fault approaches the base of the crust, it may flatten to parallel the Mohorovicic Discontinuity; or, if it were offset initially, it may have since conformed to a uniform depth (Schilt et al, 1979).

Both the north and south margins of the range are bounded by normal faults, along which Late Tertiary movement caused the mountain block to be dropped down with respect to the adjacent basin. The west end of the range overrode the southwestern arm of the Wind River Basin along the buried Immigrant Trail reverse fault that formed during the Laramide

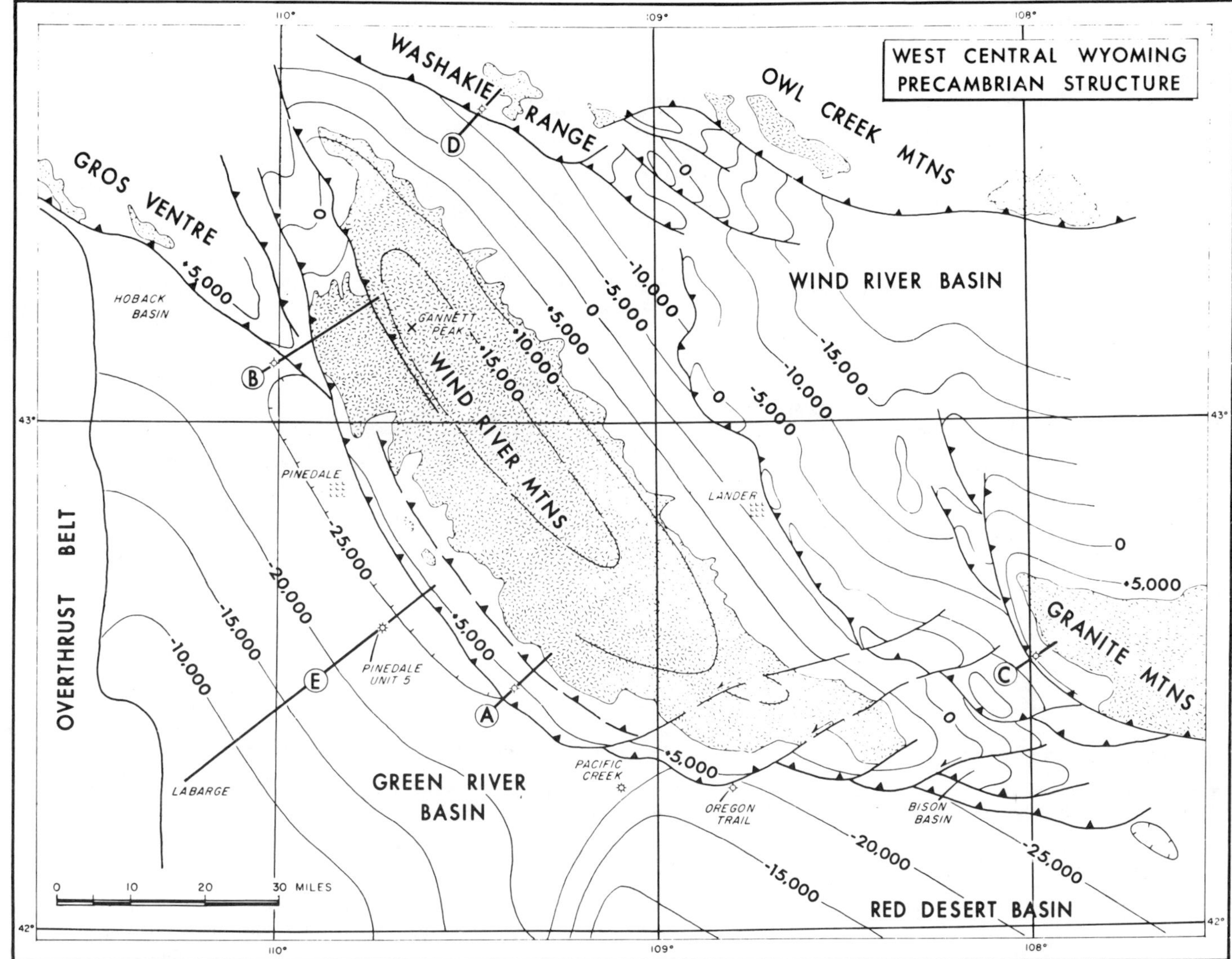

Figure 7—Generalized Precambrian structure of west-central Wyoming showing location of cross sections: (A) Big Sandy Area; (B) New Fork Area; (C) Immigrant Trail Area; (D) Dubois Area; and (E) stratigraphic section. Berg, 1961. Used with the permission of the Wyoming Geological Association.

orogeny. Paleozoic and Mesozoic strata dip 10 to 15° north off the north flank of the Granite Mountains; but along the basin margin, these rocks were folded into several large north- and northwest-trending anticlines that project far into the interior of the basin (Keefer, 1970).

The implications of Precambrian igneous rocks being thrust over coal-bearing rocks would be important in the exploration for coalbed methane (Figs. 7, 8). The characteristics of these thrusts are:

- Folded Precambrian crystalline rocks
- Variable low-angle dip, steepening with depth
- Overturned and deformed sequence of Paleozoic rocks underlying the Precambrian thrust mass
- Cretaceous rocks below the thrust.

Such thrusts imply small movements along a large number of closely spaced joints in the more competent igneous rocks (Berg, 1961).

The northern boundary of the basin is complex and difficult to interpret, because older structures are masked by the Wind River Formation at the foot of the Owl Creek Mountain uplift. The basin margin is overthrust with a maximum horizontal displacement of 6 mi postulated south of the canyon of the Big Horn River where Eocene rocks overlap the fault trace (Brenner, 1972).

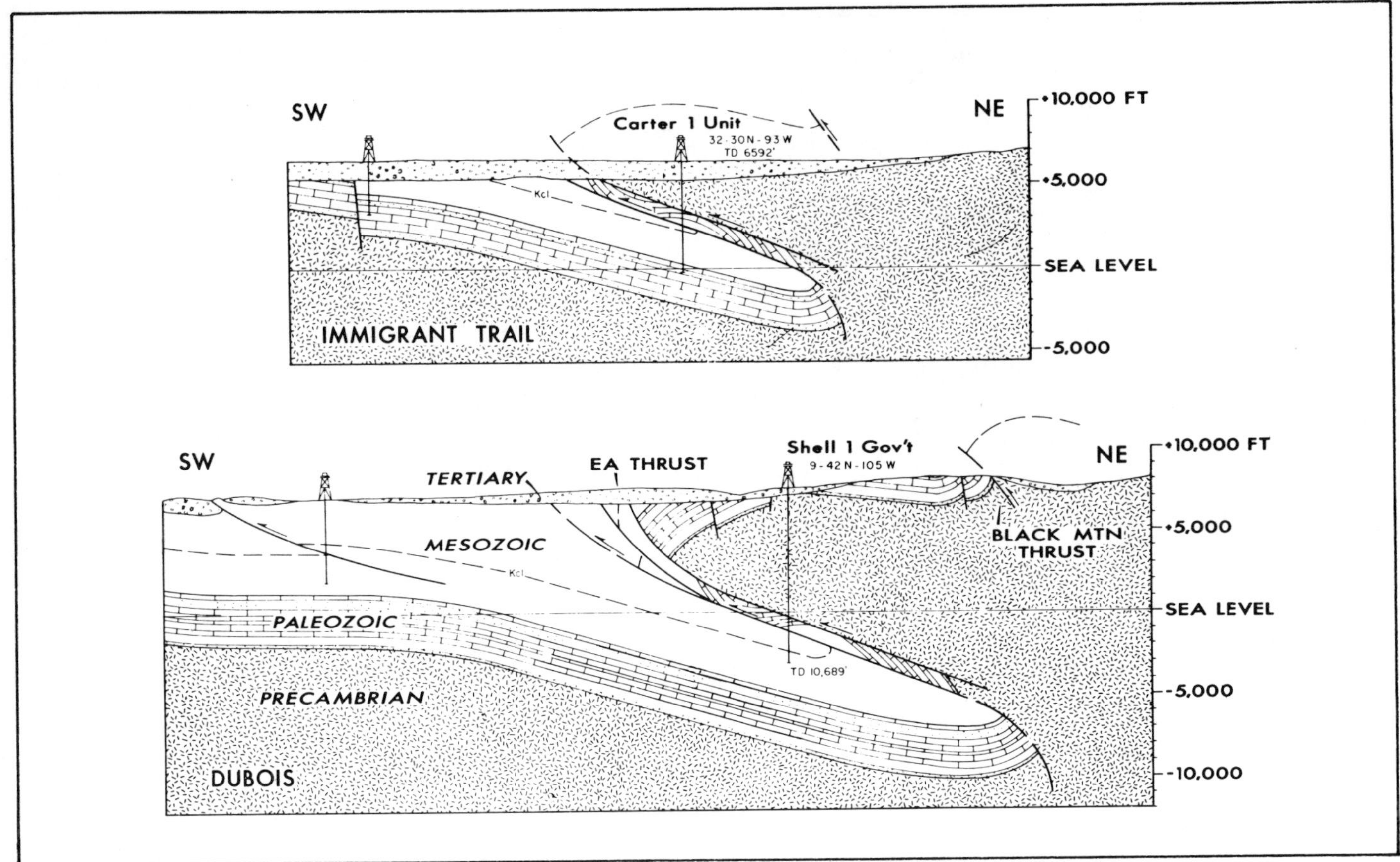

Figure 8—Structure sections of Immigrant and Dubois areas. Surface geology in Dubois section, from Love, 1939. "Kcl" horizon is Lower Cretaceous Cloverly Sandstone. Berg, 1961. Used with the permission of the Wyoming Geological Association.

COAL RESOURCE

The Wind River coal region in west-central Wyoming could contain substantial coal resources that extend from outcrop along the rim of the basin to depths of over 14,000 ft near the synclinal axis in the deep northeastern part of the basin (Fig. 9).

Regional Setting

Glass (1978) reported that Wyoming's coals occur in Cretaceous (66 to 135 million years ago) or Early Tertiary (38 to 66 million years ago) deposits. During both periods, depositional environments and climates were periodically well suited to the development of densely vegetated swamps. Peats that accumulated in these swamps were transformed into the nearly trillion tons of coal that still underlie 41% of the state of Wyoming (Glass, 1978).

Wyoming is designated as a coal-producing district by the U.S. Bureau of Mines (USBM). The state's coal areas are divided into ten major regions, basins, or fields covering approximately 40,000 sq mi. Of the nation's coal resources, 24% lies under less than 6,000 ft of overburden in the state.

The formations that characteristically contain coalbeds are thick, usually ranging from 700 to 7,000 ft. Thicknesses of the various Tertiary formations vary from basin to basin, but the Cretaceous formations exhibit gradual regional thickening or thinning across Wyoming.

The most widespread coal-bearing rocks in Wyoming are Cretaceous and usually crop out as narrow bands of upturned rock around the margins of the larger structural basins and uplifted areas in the state. The Wind River Basin is subdivided into seven individual coal fields (Fig. 10), distributed along the edge of the basin (Woodruff and Winchester, 1912):

- Muddy Creek
- Pilot Butte
- Hudson (Lander)
- Beaver Creek
- Big Sand Draw
- Alkali Butte
- Arminto (Powder River).

All the coal-bearing formations crop out along the flanks of the Wind River Basin. Coalbeds also crop out as irregularly exposed, linear bands in the thrust belt of western Wyoming. Flat-lying, generally barren Tertiary rocks occupy the central portions of most of the coal-bearing areas where they overlie older Cretaceous rocks. Tertiary rocks also commonly exhibit steeper dips as they approach the margins of the coal-bearing basins and regions (Glass, 1978).

The coal-bearing formations contain numerous coal seams separated from one another by as little as a few inches of shale or

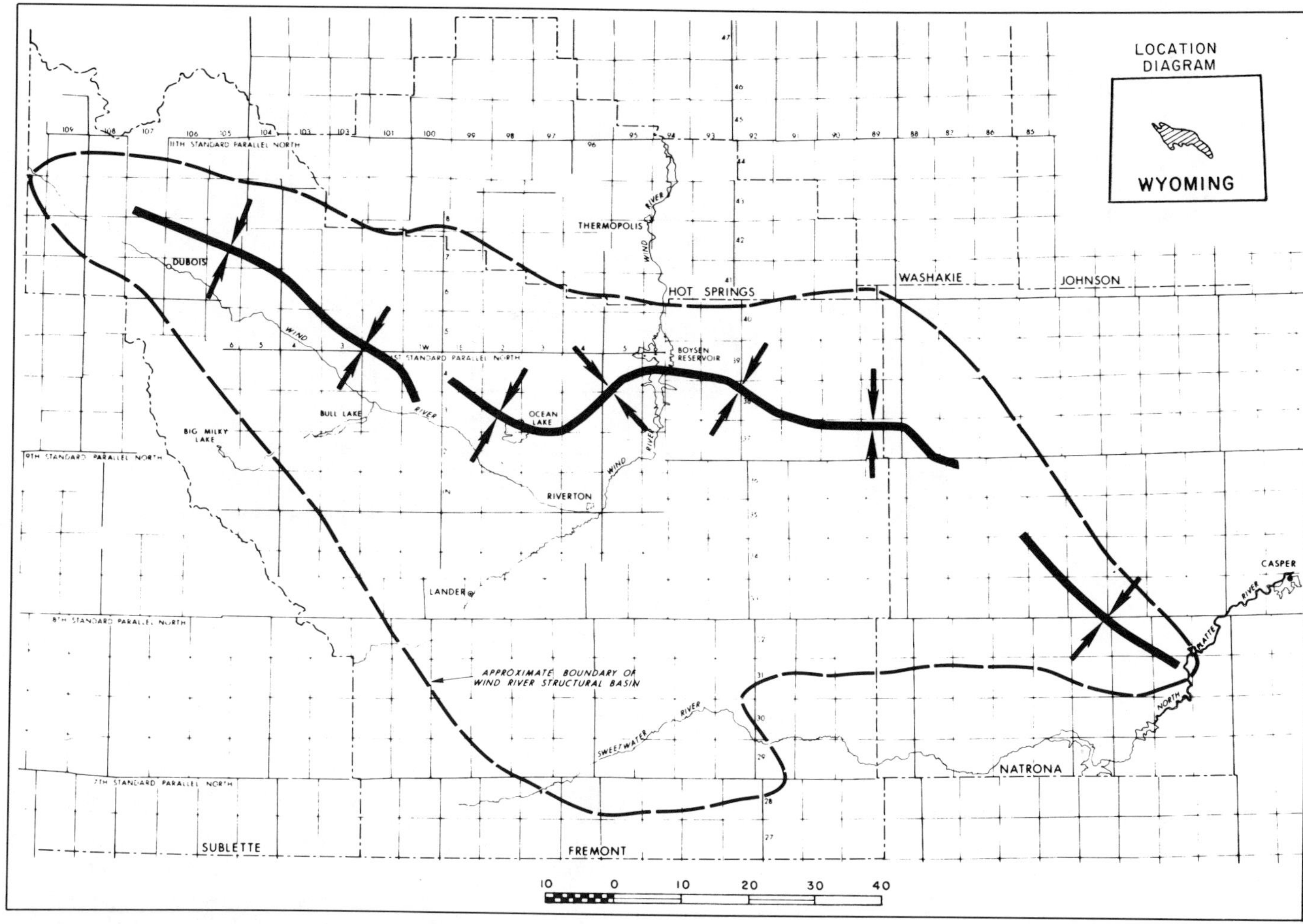

Figure 9—Synclinal axis of the Wind River Basin, Wyoming.

claystone to hundreds of feet of rock that may vary from coarse sandstone or conglomerate to siltstones, claystones, and shales (Fig. 11). Although the Cretaceous coals interspersed in these rocks are generally less than 10 ft thick, a few Cretaceous coals are 30 to 100 ft thick in westernmost Wyoming (Glass, 1978).

Coalbeds of resource thickness (> 14 inches) occur in the Frontier, Mesaverde, Meeteetse, and Lance Formations of Cretaceous age, and in the Fort Union and Wind River Formations of Tertiary age (Windolph et al, 1980).

The Wind River Basin coal resource is large, and most is contained in thick beds lying at great depths (> 11,000 ft) within the basin. Most earlier coal resource calculations were restricted to outcrop areas along strike of the coalbeds and for relatively short distances downdip (Woodruff and Winchester, 1912; Berryhill et al, 1950; Thompson and White, 1952; Averitt, 1975; Glass, 1975). Until recently, little was known of the extent of Wind River coalbeds downdip to the center of the basin. For this reason, early calculations of coal resources grossly understate the amount of coal that certainly exists in the center of the basin.

At the request of the U.S. Bureau of Indian Affairs, the U.S. Geological Survey (USGS) initiated a coal-exploratory drilling program during the fall of 1978 to obtain fresh coal samples, core samples, and electric logs. Mr. John F. Windolph, Jr. (USGS, Reston, Virginia), was the project geologist in charge. Data from these shallow drill holes (< 1,200 ft), together with data from geologic field mapping and numerous measured sections, have been used to evaluate the coal resources of the Wind River Indian Reservation (Fig. 12).

Rank

Wyoming coal ranges in rank from lignite to high-volatile A bituminous. Some occurrences of lignites have been reported from the Wind River Basin (Unfer, 1951), representing a southern extension of the Tertiary lignite deposits of Montana and North Dakota.

Subbituminous coals—all Tertiary or Late Cretaceous in age—occur in all Wyoming coal-bearing areas except the Black Hills Region; and subbituminous coals usually occupy the central parts of the coal basins. Glass and Roberts (1978) state that the rank of the Wind River coals is subbituminous C.

Older coalbeds in any given Wyoming coal field generally are higher in rank than the younger beds. Rank of individual beds

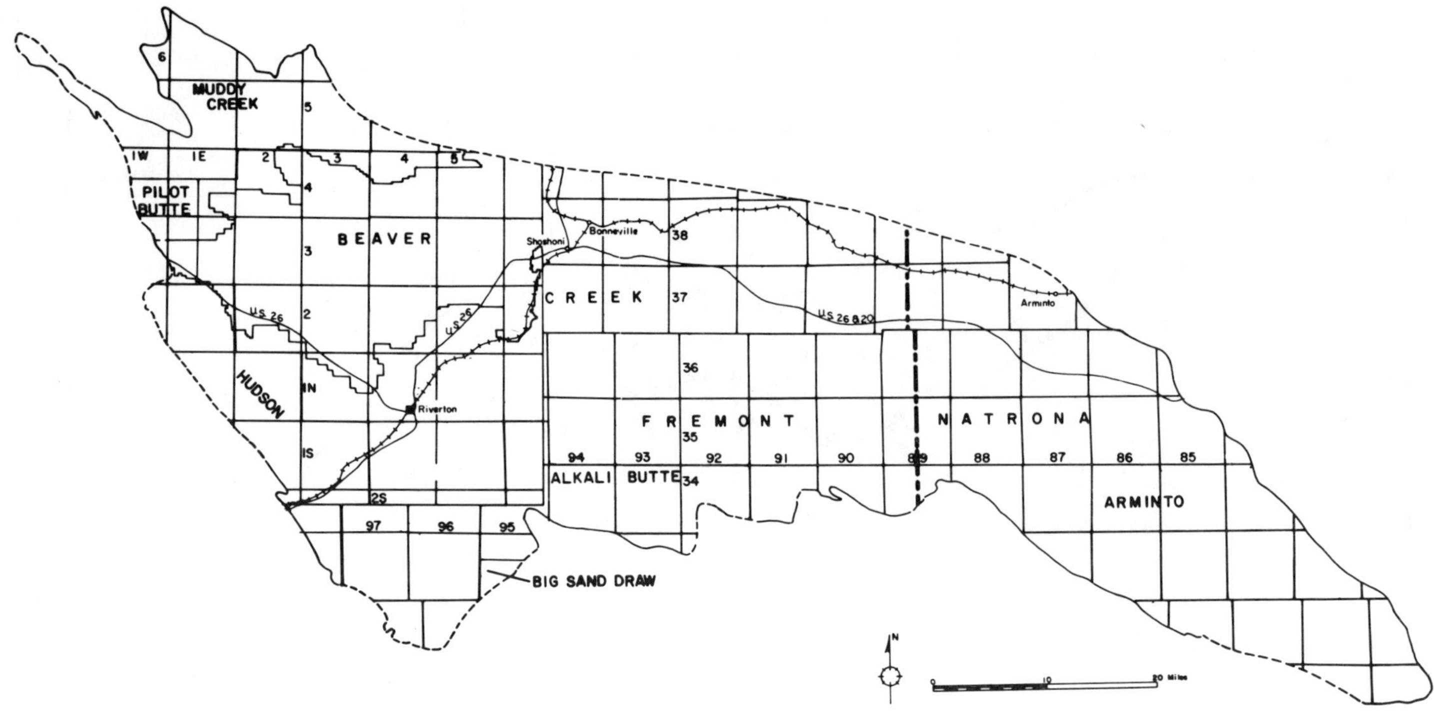

Figure 10—Individual coal fields in the Wind River coal basin. Glass and Roberts, 1978. Used with the permission of the Wyoming Geological Association.

in a field also seems to increase toward the troughs of the structural basins. Both of these variations in rank have been attributed to increases in depth of burial (Unfer, 1951).

Chemical Composition

Moisture, volatile matter, and fixed-carbon contents of Wyoming coals vary widely across the state, in response to variations in rank. The bituminous and higher rank subbituminous Cretaceous and Tertiary coals are very similar, measured on an as-received basis. Moisture contents are less than 15%, volatile matter contents are between 30 and 40%, and fixed-carbon contents are greater than 40%.

Ash and sulfur contents are characteristically low and related more to depositional histories of the seams than to any differences in rank (Glass, 1978). Consequently, variation in ash content, in particular, is quite irregular.

Proximate and ultimate analyses for select Wind River Basin coal fields are given in Table 1.

Coalbed Stratigraphy

Accurate coalbed correlation is essential to the determination of coal resources and reserves, especially those that are particularly gassy. The subbituminous coals of the Wind River Basin have only been correlated within individual coal fields. Coalbeds are shown in geologic columns for each field in the Wind River Basin, and Figure 13 provides a generalized geologic column showing key coalbeds or coal-bearing zones and major beds in the Wind River Basin. Published correlations of individual coals across most of the coal-bearing areas or even across a coal field are rare, and, for this reason, correlation of one coal seam across wide areas is not yet possible. In fact, correlation of some coal-bearing formations from one basin or region to another is speculative. In the authors' opinion, some of the coals can be effectively projected and correlated to those coalbeds in the Bighorn Basin just north of the Wind River Basin. Campbell (1917) pointed out that the Wind River Basin is similar to the Bighorn Basin, except that the Late Tertiary formations more completely conceal the coal-bearing rocks. For this reason, less is known regarding the coalbeds, which have been mined only locally. The coal is subbituminous and compares favorably with coals of the same rank in adjacent Wind River fields.

Although coals are reported in the Frontier, Cody, Lance, Fort Union, and Wind River Formations, the thicker and more important coals are limited to the Upper Cretaceous Mesaverde Formations (Glass, 1978).

Stratigraphic details of the coal-bearing formations in the Wind River coal fields are presented in order from oldest to youngest. This is also the order of importance from the best and most abundant coal to the poorest, least abundant coal amount (Thompson and White, 1952).

Frontier Formation

According to Keefer and Troyer (1964), the Frontier Formation in the Wind River Basin is a resistant sequence of sandstone and shale overlain by the Cody Shale and underlain by the Mowry Shale. It is an important oil- and gas-producing formation in many fields in the Wind River Basin. Thompson et al (1949) have made a detailed surface and subsurface study of the formation in the basin.

The Frontier Formation consists, for the most part, of

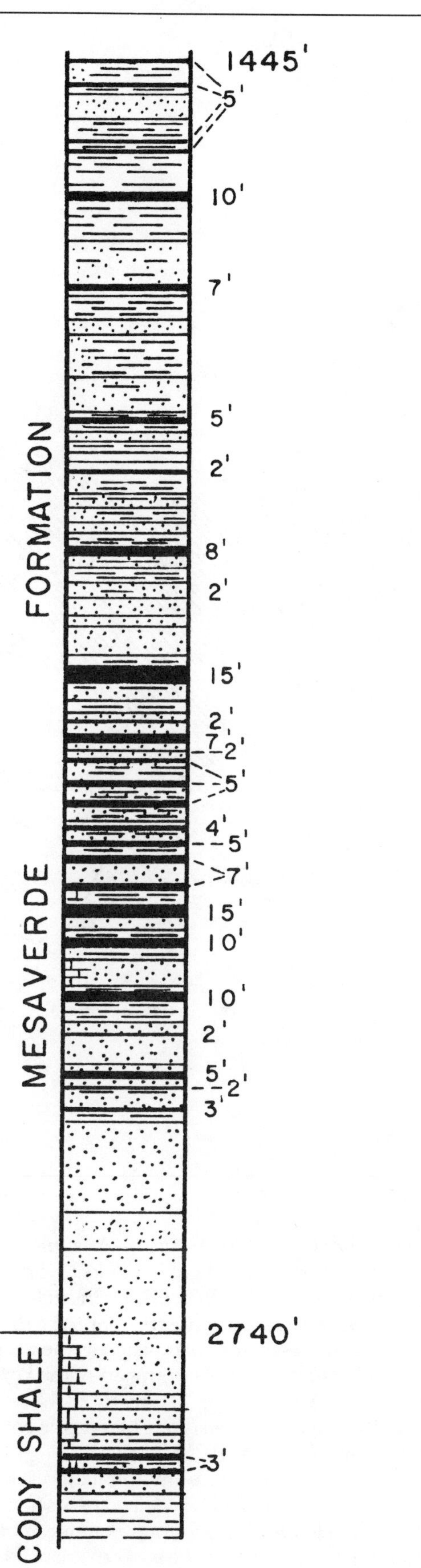

Figure 11—Twenty-nine Mesaverde coalbeds greater than 2 ft thick.

alternating gray to black shale; gray, fine- to medium-grained, massive- to thin-bedded sandstone; and a few thin beds of tuff and bentonite. Chert pebbles as large as 1/4-inch in diameter are present in the upper 30 ft. Abundant marine fossils of early Niobrara age occur in the upper part. The contact between the Frontier Formation and the overlying Cody Shale is generally well marked by a lithologic and topographic change from ledgy sandstone below to soft, weathered shale above (Keefer and Troyer, 1964).

Because of facies changes and the transitional nature of the rock within the contact zones, the formation boundaries are not consistent over long distances. Keefer (1972) reported that this causes the thickness of the Frontier Formation to vary considerably from place to place.

Frontier Coalbed Stratigraphy—Very little is known about the extent of the coals in the Frontier Formation. Windolph (1981, personal communication) noted that the upper Frontier coals were mined on the Wind River Indian Reservation in an area covered by the Blue Hole quadrangle. Two coalbeds are present: one 3 ft thick and the other 2 ft thick, and separated by a 6-ft split. The coal is about 150 ft below the Cody Shale and lies on the top of the First Wall Creek sand.

Mesaverde Formation

The lithology of the Mesaverde Formation of the Wind River Basin bears little resemblance to that of the formation at its type section. The Mesaverde consists of light- to dark-gray sandstone interbedded with shale, siltstone, ironstone, and coal. Tonsteins appear for the first time in the coalbeds (Windolph, 1981, personal communication). Thompson and White (1952) report that the sandstones are more massive, more resistant, and more quartzose than those of the underlying Cody Shale. The units are highly lenticular; most beds can be traced less than 5 mi and several for only a few hundred feet. Perhaps the most persistent bed within the formation is a fine-grained, white to light-gray sandstone averaging 50 ft in thickness, considered to be the basal bed of the formation throughout most of the area (Thompson and White, 1952).

The Mesaverde Formation appears to overlie the Cody Shale conformably and is unconformably overlain by the Lance, Fort Union, or Wind River Formations. The western sections are thinnest where the Fort Union or the Wind River Formations overlap; in the east, the formation is thicker where unconformably overlain by the Lance Formation (Thompson and White, 1952). The maximum angular unconformity observed between rocks of the Mesaverde and Lance Formations is about 5°; between the Mesaverde Formation and rocks of the Fort Union Formation, it is 30°; and between rocks of the Mesaverde and Wind River Formations, about 40° (Thompson and White, 1952).

Mesaverde lithology results from transgressive and regressive events, described in considerable detail by Keefer and Rich

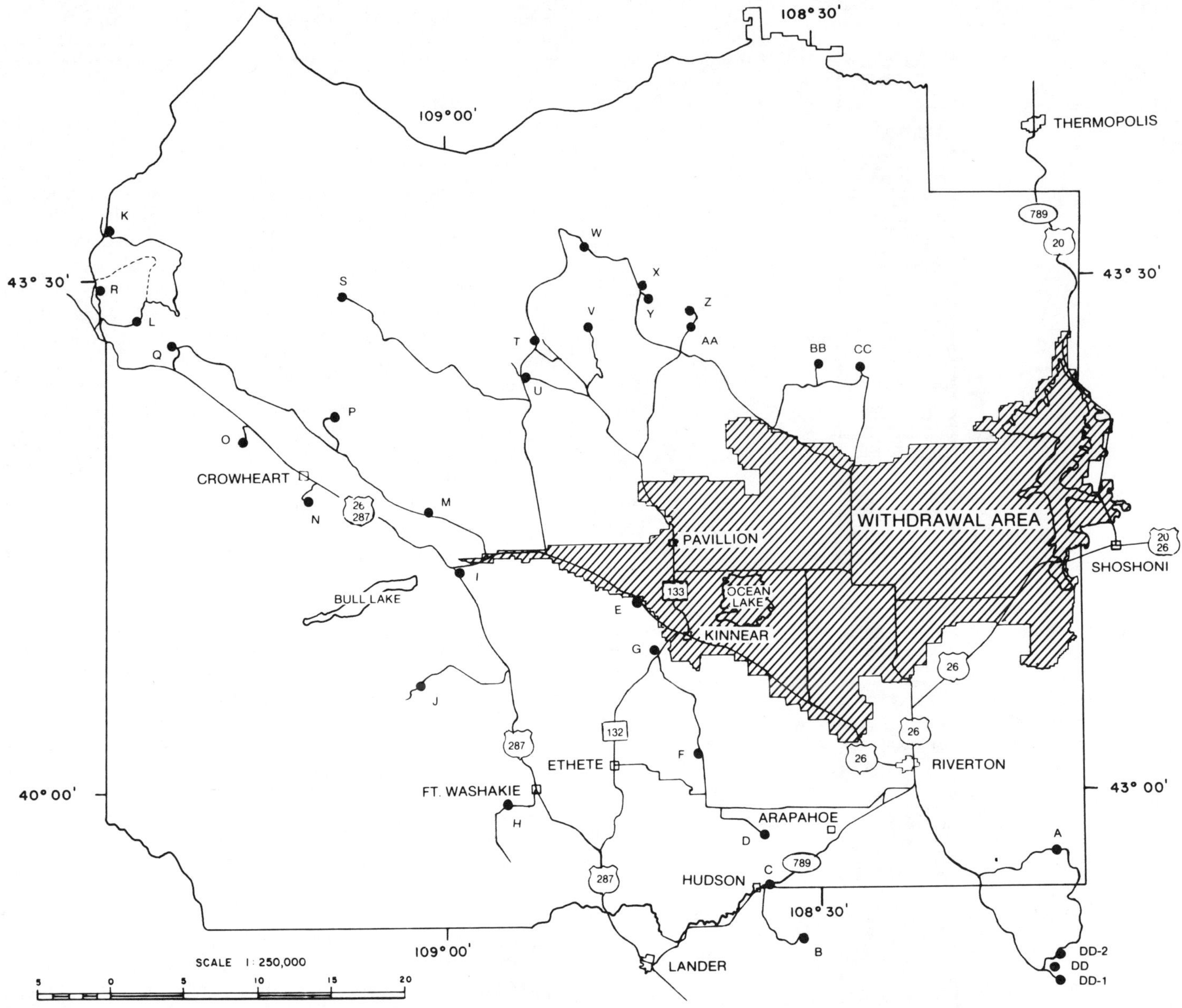

Figure 12—USGS drilling sites on the Wind River Reservation. Windolph et al, 1980.

(1957), Barwin (1961), Weimer (1961), and Rich (1958). Weimer (1961) proposed that a large delta, the Rawlins, was actively deposited in northwest Colorado and south-central Wyoming during Late Cretaceous time.

Deposition of the lower Mesaverde Formation in the western area of the basin indicates a widespread eastward regression of the Late Cretaceous seas, resulting in thick accumulations of sediments in nearshore marine, brackish water, fluvial, and swampy environments containing widespread peat deposits. Minor oscillations of the shoreline caused the interfingering of littoral and offshore marine deposits seen at the Mesaverde-Cody contact (Shapurji, 1978). Figure 14 is a generalized lithofacies map of the nonmarine (swamp, lagoonal, fluvial, and deltaic) and marine facies.

Mesaverde Coalbed Stratigraphy—As a rule, coalbeds in the western portion of the Hams Fork Coal Region are geologically older than those in the Wind River Basin (Fig. 15) (McCord, 1980). Green River coals are thought to be the same age as the Wind River Mesaverde coals, whereas the youngest Mesaverde coals occur in the southwestern part of the Powder River Basin (Fig. 16). The Bighorn Basin directly to the north of the Wind River Basin contains persistent Mesaverde coals in the south and southwest (Glass, 1978).

Where basal nonmarine environments existed in the southwestern basin, a basal coalbed exists just above the Cody Shale. Windolph's USGS group examined outcrops near the top of the Cody Shale in the western area of the basin and reported coarsening-upward grain sizes in the Cody and

Table 1—Summary of Signor and Lander coal analyses from the Alkali Butte and Hudson (Lander) coal fields, Fremont County, Wyoming. Modified from Glass, 1978. Used with the permission of the Geological Survey of Wyoming.

Alkali Butte

As-Received	Single Analysis
Moisture (%)	26.0
Volatile Matter (%)	30.7
Fixed Carbon (%)	38.1
Ash (%)	5.1
Sulfur (%)	0.6
Heat Value (Btu/lb)	8,760

Hudson (Lander)

As-Received Basis	Range (37 analyses)[1]	Average: Proximate (37 analyses)	Average: Ultimate (7 analyses)
Moisture (%)	16.0–24.7	21.2	
Volatile Matter (%)	29.1–38.1	32.9	
Fixed Carbon (%)	33.2–45.0	39.8	
Ash (%)	2.8–11.8	6.1	4.6
Sulfur (%)	0.3–1.3	0.6	0.6
Hydrogen (%)			6.2
Carbon (%)			55.2
Nitrogen (%)			1.2
Oxygen (%)			32.2
Heat Value (Btu/lb)	8,870–10,510	9,480	

[1]These analyses came from the Poposia, Poposia No. 2, Rogers, George, Big Dunne, Dunne Little, Schroeder, Wyoming Central, Hickey, McKinley, Indian, and Mitchell No. 2 deep mines. Samples were a mixture of channel, tipple, and delivered samples. Usually the entire bed thickness was not sampled, and many samples were cleaned or screened before analysis. Analyses were done by the U.S. Bureau of Mines and published in various reports listed in the references.

intertonguing with the basal sand units of the Mesaverde Formation, of which both may indicate regression of the Cretaceous sea. Sedimentary structures characteristic of mouth bars, point bars, and distributary channel fillings were observed in the lower part of the Mesaverde Formation. These sedimentary structures also indicate emergence during regression of the sea. During this emergence, protective barriers and widespread swampy platforms formed; vegetation accumulated on these platforms and is preserved as coal. Thin, discontinuous coalbeds that typically grade into carbonaceous shale formed in marshes near barrier islands. Thin-to-thick, laterally extensive Mesaverde coals that grade laterally into carbonaceous shale formed in marshes on emergent levee, overbank, and bayfill deposits on the landward side of the lagoon. Deposition of these detrital deposits, as well as tidal-creek channel sandstones, caused the coalbeds to split and merge. Figure 17 illustrates similar types of depositional environments that existed in the upper part of the Almond Formation (Upper Mesaverde). Most coalbeds in this sequence are thick, lenticular, irregularly distributed, and probably related to crevasse splays and distributary channels (Windolph et al, 1980). A thicker and more widespread sheet-like coalbed (Signor coal) overlies this zone and may have been deposited under more stable conditions found in the back-bay swamp of a lower delta plain environment.

An upper delta plain environment is suggested by the overlying sedimentary sequence that consists of 1,000 ft of interbedded lenticular sandstone, siltstone, gray shale, carbonaceous shale, thick-to-thin coalbeds that are commonly discontinuous laterally, and thin fresh-water limestone beds.

The uppermost unit of the Mesaverde Formation is a white, coarse- to medium-grained blanket sandstone containing several thin carbonaceous shale beds. It appears to be unconformable upon the underlying sequence and is interpreted as an alluvial plain deposit. The basal shale units of the overlying Meeteetse Formation record a transgression of the Cretaceous sea (Windolph et al, 1980).

Downey Coalbed—Through detailed drawings of measured stratigraphic sections, Windolph (1981, personal communication) has found that the Downey coalbed in the Big Sand Draw coal field correlates with the Signor coalbed in the Alkali Butte coal field. The Downey coalbed was thought to be the lowest commercial coal occurring in the Mesaverde Formation and is typically 28 ft thick where mined in the Big Sand Draw coal field (Thompson and White, 1952).

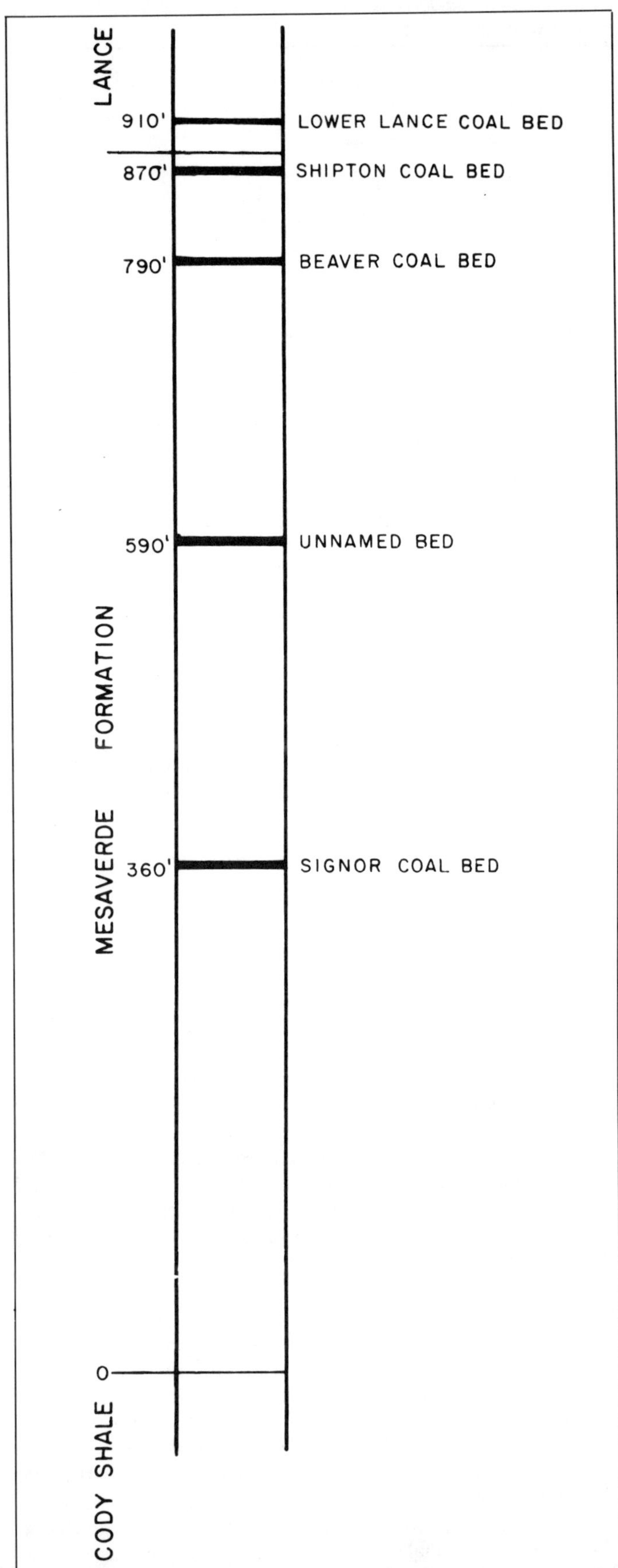

Figure 13—Generalized stratigraphic relationship of the key coalbeds or coal-bearing zones in the Wind River Basin.

Signor Coalbed—The Signor coalbed is probably the most persistent Mesaverde coal yet identified in the southwestern part of the Wind River coal basin (Fig. 18). The Signor bed can be traced for more than 10 mi in the Alkali Butte coal field, and has a maximum reported thickness of 16.5 ft. Although it averages 7 to 10 ft thick over much of its outcrop, it thins in places to less than 2 ft. The coal has been deep mined, and dips at 13 to 31°. Windolph et al (1980) tentatively correlated a coalbed measuring 7 ft thick in the Hudson quadrangle with the Signor bed some 22 mi toward the east in the Alkali Butte coal field (Fig. 14).

The coalbeds delineated in the classical Hudson stratigraphic section are not associated with the Signor coalbed. These coals are stratigraphically lower and are not present in the Alkali Butte area but do appear in the Stanolind Oil and Gas Company well (SE 1/4 SE 1/4, 3, 33N, 96W) (Fig. 10). In the Hudson coal field, the Signor coalbed is cut out by two unconformities (Windolph, 1981, personal communication). Most of the Mesaverde Formation is removed by the Fort Union and overlying deposits; elsewhere, the Wind River Formation cuts out the Fort Union Formation.

Lord et al (1913) published an analysis of the lower 4.2 ft of the 7.9 ft thick coal channel sample of the Signor coalbed from the Alkali Butte coal field (Table 1). Glass (1978) summarized the analyses of the Lander coalbed from the Hudson coal field (Table 1).

Kinnear Coalbed—This thin Mesaverde coal crops out for about 1 mi in the Pilot Butte coal field in the northwestern part of the basin, where it was once deep mined (Glass, 1978). The Kinnear coal has a maximum reported thickness of 2.8 ft, although 2 ft is representative thickness (Woodruff and Winchester, 1912). The bed dips 29 to 30° northeast and is the only coal identified in the Pilot Butte coal field. Because this coalbed lies about 280 ft above the Mesaverde/Cody contact, it could be equivalent to the Signor coalbed. Table 2 provides a single analysis of a 2.8 ft thick channel sample of this coal (Lord et al, 1913).

Unnamed Coalbeds—Between the Signor and Beaver coalbeds are numerous discontinuous beds up to 15 ft thick. These coals are thought to have formed in an upper delta plain environment (Windolph et al, 1980).

Beaver Coalbed—The Beaver coalbed is a subbituminous Mesaverde coal that crops out in the Alkali Butte coal field. A fairly persistent coal, occurring about 430 ft above the Signor bed (Glass, 1978), the coal maintains a thickness of 3 ft for more than 1 mi along its outcrop. There are no analyses of this coalbed, and all calculated resources of this coal lie within the Wind River Indian Reservation.

Shipton Coalbed—The Shipton coalbed is a persistent Mesaverde coal in the Alkali Butte coal field. It overlies the Beaver coal by about 65 ft, and the Signor coalbed by about 495 ft. It crops out for more than 3 mi, has a maximum thickness of 9.5 ft, and averages close to 5 ft thick. It dips 18 to 25° to the north or west and has been deep mined.

A single analysis of a channel sample from an underground mine (Shipton Mine) was published by Lord et al (1913) (Table 2). Only 5.5 ft of a 7.5 ft thick coal seam were included in the analysis.

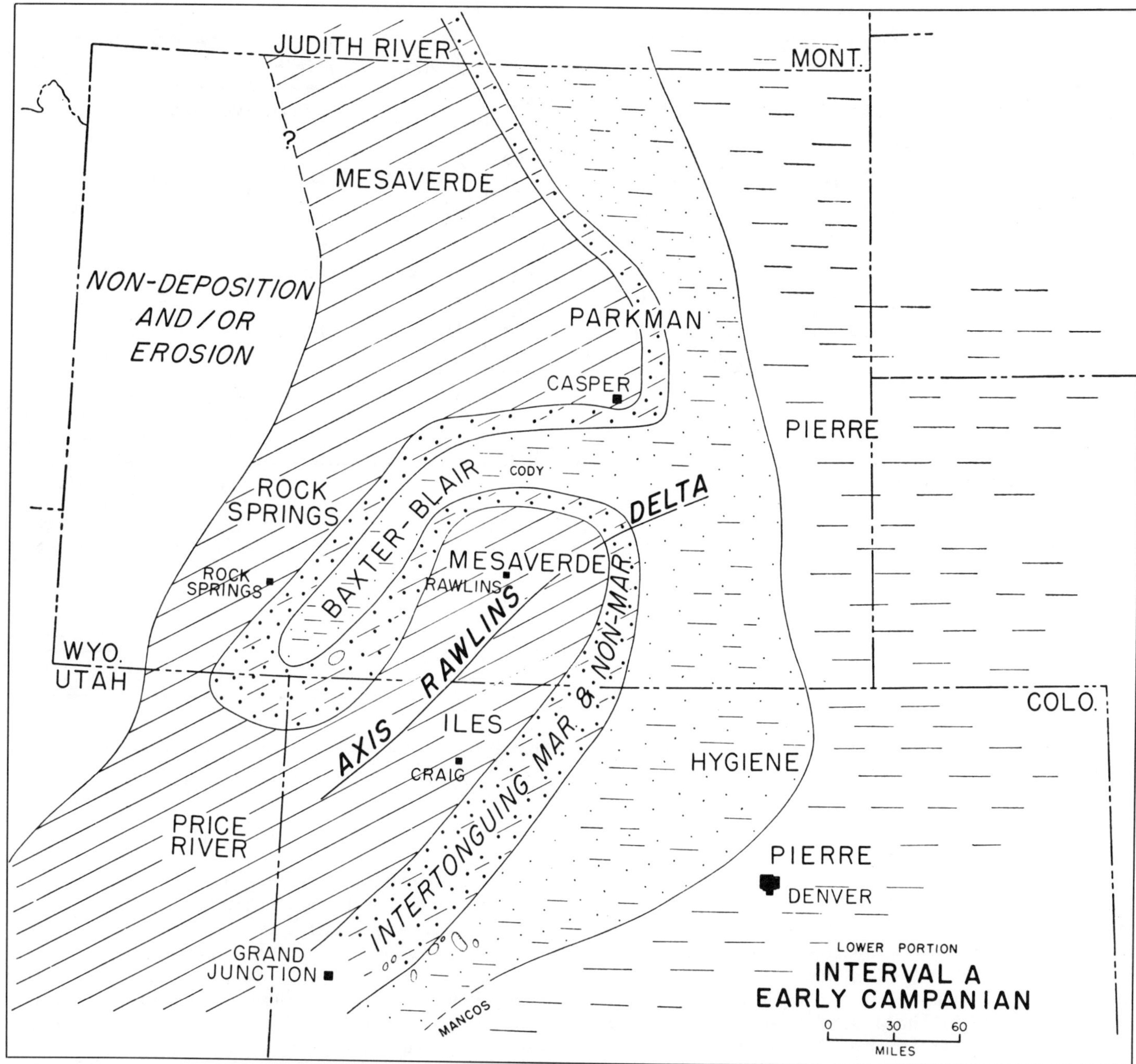

Figure 14—Rawlins delta and lithofacies map. Weimer, 1961. Used with the permission of the Wyoming Geological Association.

Meeteetse Formation

The Upper Cretaceous Meeteetse Formation is found only in the Muddy Creek coal field in the northwestern corner of the basin. Meeteetse coals common to that area are almost as thick as the Mesaverde coals of the Alkali Butte field. Meeteetse coals are reported to occur as much as 16 ft thick, although they average less than 4 ft (Woodruff and Winchester, 1912). The thickest Meeteetse coals occur in persistent shaly carbonaceous zones where numerous thin coals coalesce into interbedded shale and coal units (Keefer and Troyer, 1964). Because these carbonaceous units contain several thin coalbeds separated by shale partings or one or two relatively clean coalbeds of greater thickness, it is difficult to calculate and classify the coal reserves in them without a large number of measurements along each coal zone.

Meeteetse Depositional Environment—The Meeteetse Formation was deposited in widespread swamps, broad flood plains, and lagoons that lay along the west edge of the Late Cretaceous sea (Keefer, 1965). The thin tongues of marine shale and sandstone of the Lewis Shale represent minor readvances of the sea into the eastern part of the Wind River Basin. Little evidence exists of local tectonic activity during deposition of the Meeteetse and Lewis, except possibly some subsidence of the major trough area along the present north margin of the basin. The Meeteetse Formation is easily distinguished from the Fort Union Formation by its lack of conglomerate beds.

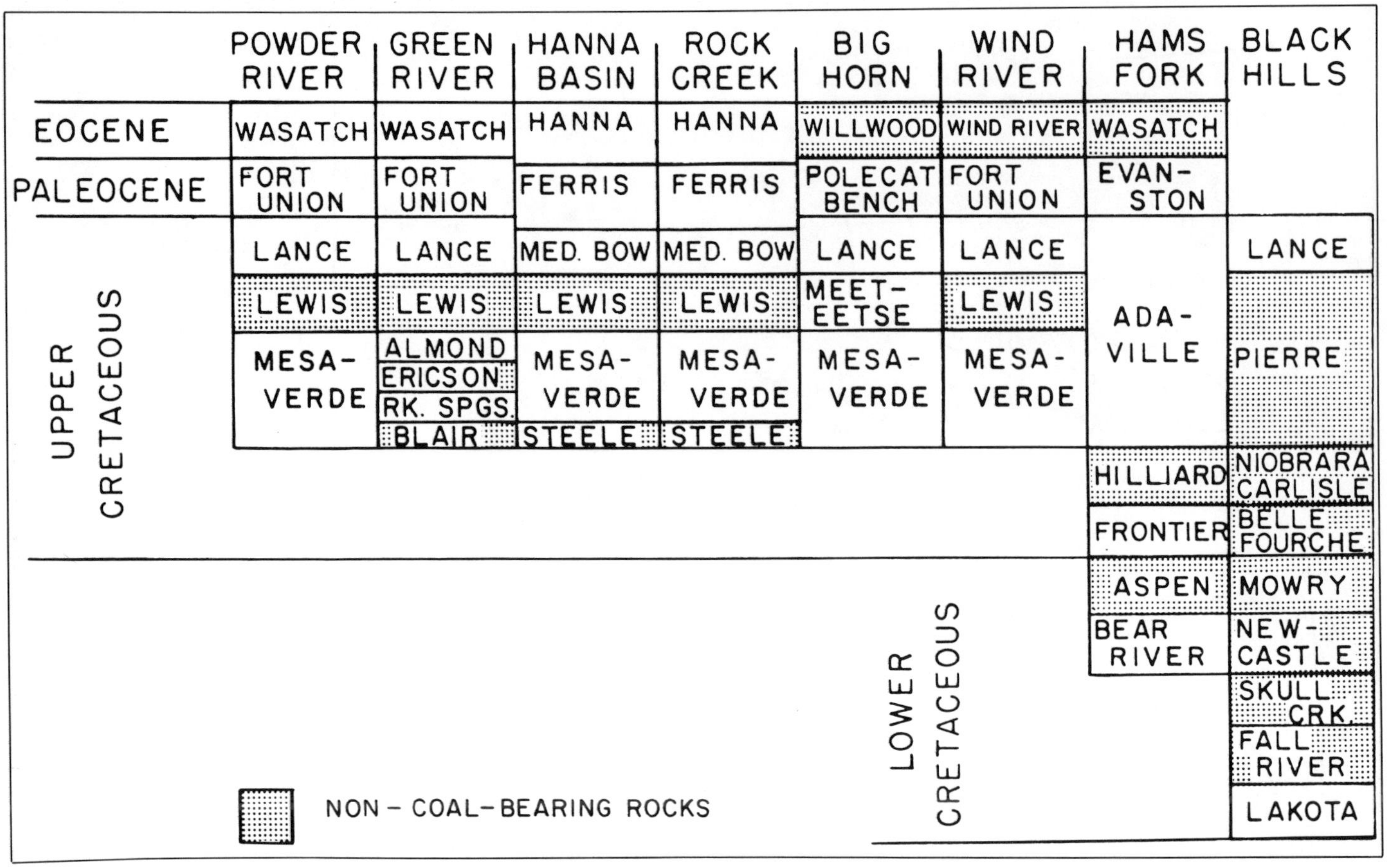

		POWDER RIVER	GREEN RIVER	HANNA BASIN	ROCK CREEK	BIG HORN	WIND RIVER	HAMS FORK	BLACK HILLS
EOCENE		WASATCH	WASATCH	HANNA	HANNA	WILLWOOD	WIND RIVER	WASATCH	
PALEOCENE		FORT UNION	FORT UNION	FERRIS	FERRIS	POLECAT BENCH	FORT UNION	EVAN-STON	
UPPER CRETACEOUS		LANCE	LANCE	MED. BOW	MED. BOW	LANCE	LANCE	ADA-VILLE	LANCE
		LEWIS	LEWIS	LEWIS	LEWIS	MEET-EETSE	LEWIS		PIERRE
		MESA-VERDE	ALMOND	MESA-VERDE	MESA-VERDE	MESA-VERDE	MESA-VERDE		
			ERICSON						
			RK. SPGS.						
			BLAIR	STEELE	STEELE				
								HILLIARD	NIOBRARA CARLISLE
								FRONTIER	BELLE FOURCHE
LOWER CRETACEOUS								ASPEN	MOWRY
								BEAR RIVER	NEW-CASTLE
									SKULL CRK.
									FALL RIVER
									LAKOTA

(Stippled pattern:) NON – COAL-BEARING ROCKS

Figure 15—Known major coal-bearing formations of the Wind River Basin and correlation to other areas in Wyoming and the Black Hills. Modified from Glass, 1978. Used with the permission of the Geological Survey of Wyoming.

Meeteetse Coalbed Stratigraphy—The Meeteetse coalbeds are lenticular and in a few places in the Muddy Creek field extend at minable thickness for more than a mile (Berryhill et al, 1950). Because the boundaries of the Meeteetse Formation cannot be reliably traced into the subsurface, the Meeteetse and the overlying Lance Formation are commonly mapped together (Fig. 19) (Paape, 1968). The Meeteetse Formation is overturned in the northern portion of the basin (Windolph, 1981, personal communication).

Welton Coalbed—The Welton coalbed is the thickest Meeteetse coal in the Wind River coal basin. Ranging up to 16.5 ft thick, it contains a thin shale parting separating the basal 2 ft of the bed from the upper part (Keefer and Troyer, 1964). The coal crops out for more than 2 mi where it dips eastward at 19 to 28°. It apparently splits and thins northward and has been deep mined in the past (Glass, 1978).

In 1913, an analysis was made of 3.8 ft of a 12.2 ft thick single channel sample of the Welton coal (Lord et al, 1913) (Table 2).

Lance Formation

The Lance Formation is a sequence of gray, medium-grained sandstones and chocolate-brown, carbonaceous shales and coals that weathers to a conspicuous blue. It is the uppermost Cretaceous formation in the basin and although sparsely fossiliferous, it yields a diagnostic plant, *Sequoia reichenbachi*, that distinguishes the formation from overlying Paleocene or younger strata (Thompson and White, 1952).

The Lance Formation overlaps the Mesaverde in a 5° angular unconformity from the Alkali Butte anticline eastward to Conant Creek. On the west side of the Alkali Butte anticline, the Lance Formation is overlapped and cut out entirely by the Fort Union Formation, which rests on older strata with up to 30° angular discordance. Paape (1968) stated that the Lance Formation ranges in thickness from 4,500 ft along the northern flank of the basin, thinning to a wedge along the southwestern flank caused in part by erosion, but mostly by depositional processes. In areas where the Lance Formation is thickest, sandstone predominates in the lower part and shale and claystone in the upper part.

Lance Depositional Environment—Keefer (1965) proposed that the depositional pattern of the Lance Formation was significantly influenced by local tectonic movements representing the start of basin growth, in contrast to the history of the earlier Cretaceous.

The most prominent feature was an extensive downwarp along the present north margin of the Wind River Basin, as up to 6,000 ft of fine-grained clastic strata accumulated in this area

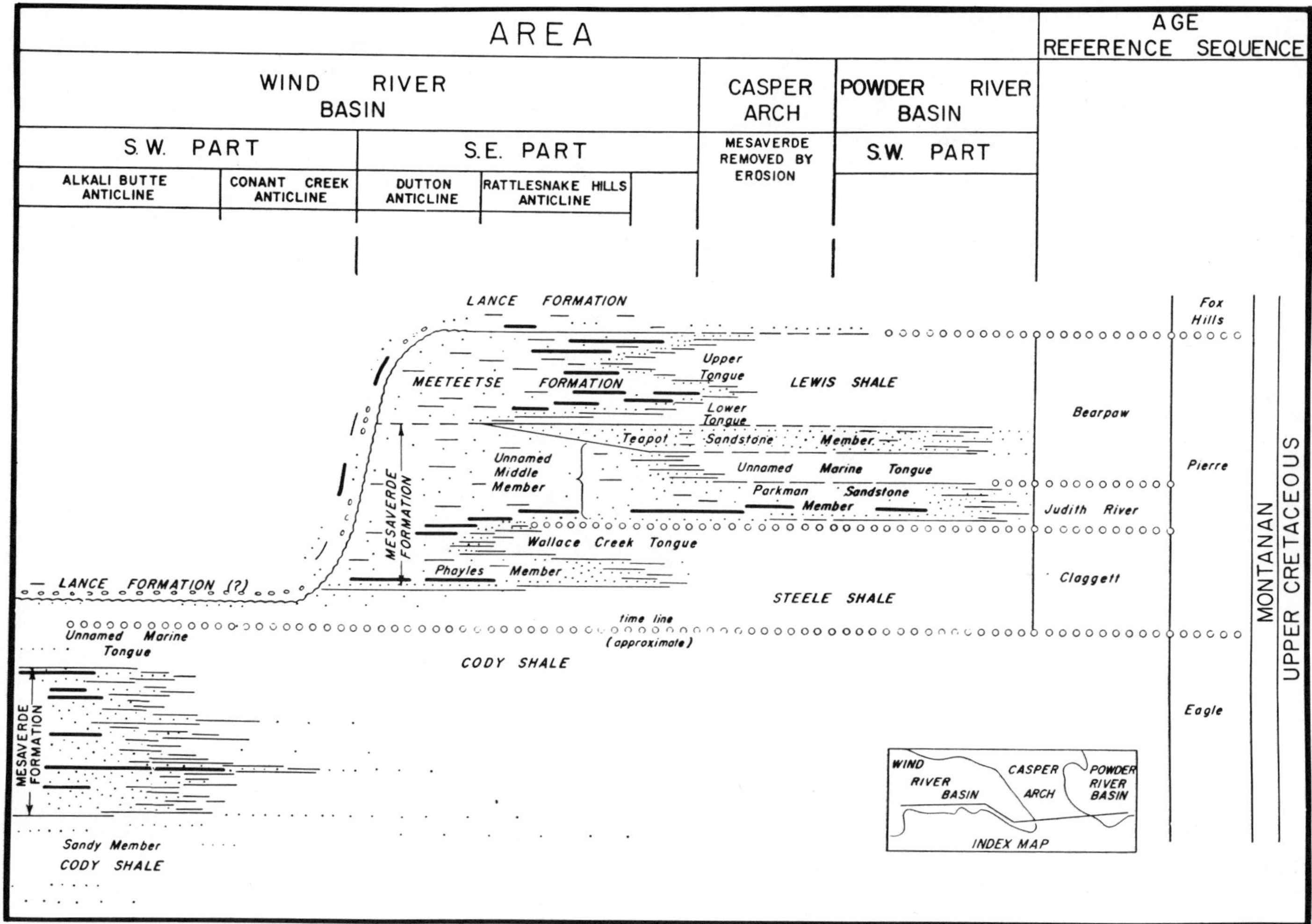

Figure 16—Diagrammatic cross section showing the coalbed continuity and stratigraphic relationship in the Mesaverde Formation. Barwin, 1961. Used with the permission of the Wyoming Geological Association.

during Lance time (Fig. 20). Although most of these strata suggest broad river flood-plain and swamp environments, some of the thick, noncarbonaceous gray shale and claystone beds in the upper part of the formation may have originated in large lakes.

Lance Coalbed Stratigraphy—Coals are documented in the Upper Cretaceous Lance Formation only in the Shotgun Butte area and the Alkali Butte coal field. The most persistent coal in the Alkali Butte coal field lies 62 ft above the unconformable contact between the Mesaverde and Lance Formations. Most of these coals are very thin and shaley. Glass (1978) stated that because no recent detailed mapping has been carried out in the Arminto coal field of the easternmost Wind River Basin, some coals identified as Mesaverde in that field could be Lance coals, but this has not been confirmed.

Only one potentially minable Lance coal has been identified in the Alkali Butte coal field. Thompson and White (1952) noted that a Lance coal, 3 to 6 ft thick, extended for more than 2 mi along its outcrop.

Fort Union Formation

The Paleocene-age Fort Union Formation is a minor coal-bearing formation in the Wind River coal basin. Fort Union coals are documented only in the south-central and eastern portions of the basin. Even in those areas, Fort Union coals are uncommon and thin, seldom reaching 3 ft in thickness. The Fort Union is divisible into "three" parts (Fig. 21). This formation ranges in thickness from a thin edge to possibly 8,000 ft in the subsurface along the north flank of the basin (Fig. 22).

Marked angular unconformities above and below the Fort Union Formation facilitate surface field identification on the basis of structural relationships.

Fort Union Depositional Environment—Hyne et al (1979) suggest that the Paleocene Waltman Lake and Fort Union Formation are an ancient analog of Lake Maracaibo and the Catatumbo River delta. The Catatumbo River delta displays a coarsening-upward sequence; however, it lacks the slurry

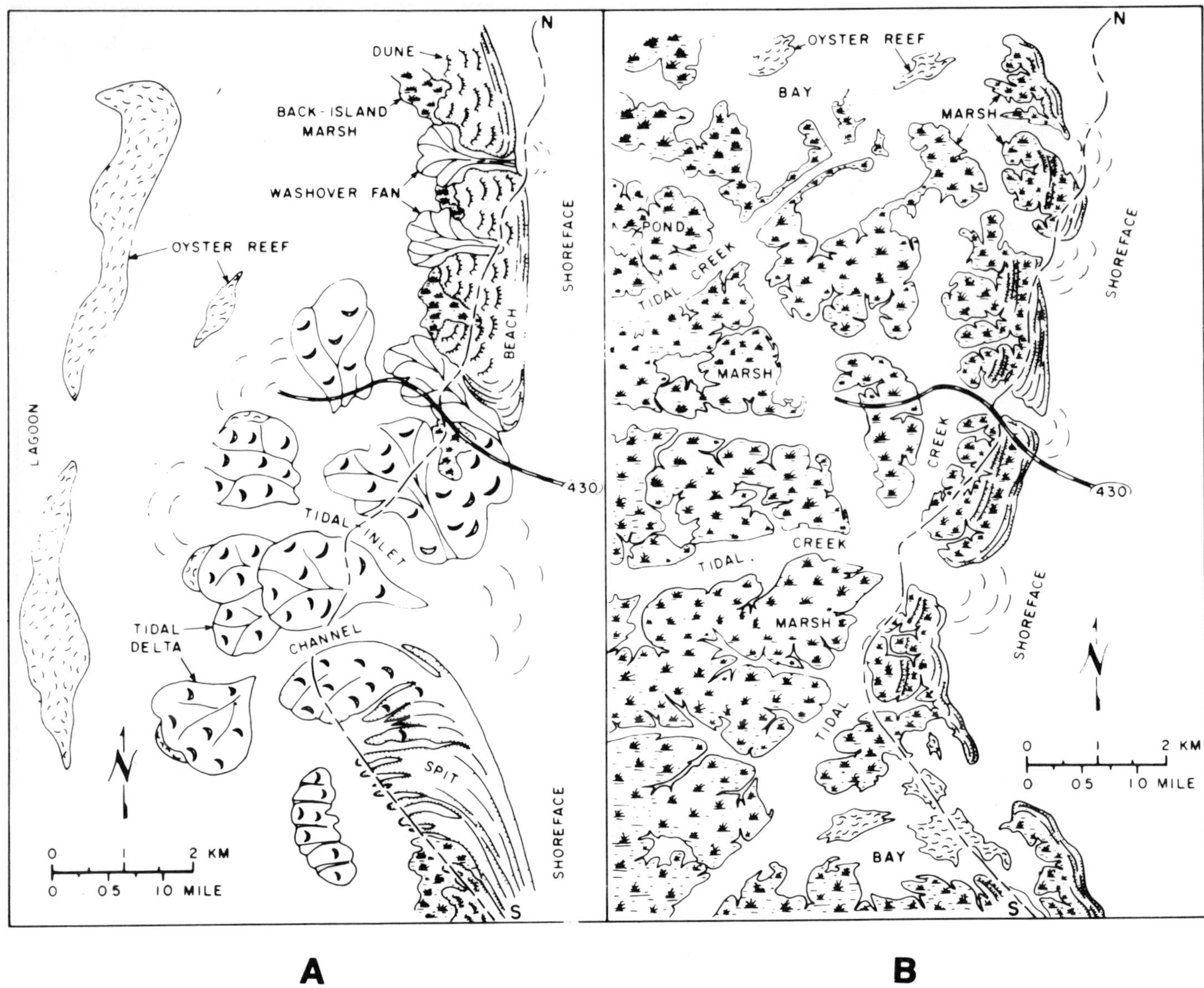

Figure 17—Diagrammatic paleographic maps of Mesaverde coalbed depositional environments.

deposits near the base of the column and the tidal deposits near the top that are common in oceanic deltas. The three recognizable members of the Fort Union Formation correspond to a lower fluvial environment (lower unnamed member), lacustrine (Waltman Shale), and the fluvial deltaic environment (Shotgun Member).

The tectonic movements that began in latest Cretaceous time in the Wind River Basin continued to influence the pattern of sedimentation through the Paleocene. The Fort Union Formation was deposited during the uplift of the surrounding mountains and subsidence of the Wind River Basin, resulting in the deposition of thick fluvial and extensive lacustrine deposits. As the late Paleocene Waltman Lake spread westward and southward, the quiet-water Waltman Shale was deposited. The Shotgun and Waltman Shale Members intertongue extensively, and the Shotgun Member is, in part, deltaic in origin (Hyne et al, 1979). Deltas were prograding into a fresh- to brackish-water lake that drained into an eastward-flowing river.

Topography and sediment infill determined the configuration of the lake (Paape, 1961). Sediments filling the basin were transported to the east from sources in the north, west, and south (Owl Creek, Washakie, Wind River, and Granite Mountains). The shoreline of Waltman Lake migrated east to west across the Wind River Basin and apparently extended eastward (and perhaps northeastward) beyond the present limits of the basin. Sedimentation occurred largely under quiet-water conditions in an environment that prevailed for a long time. By the end of the Paleocene, the lake had largely disappeared,

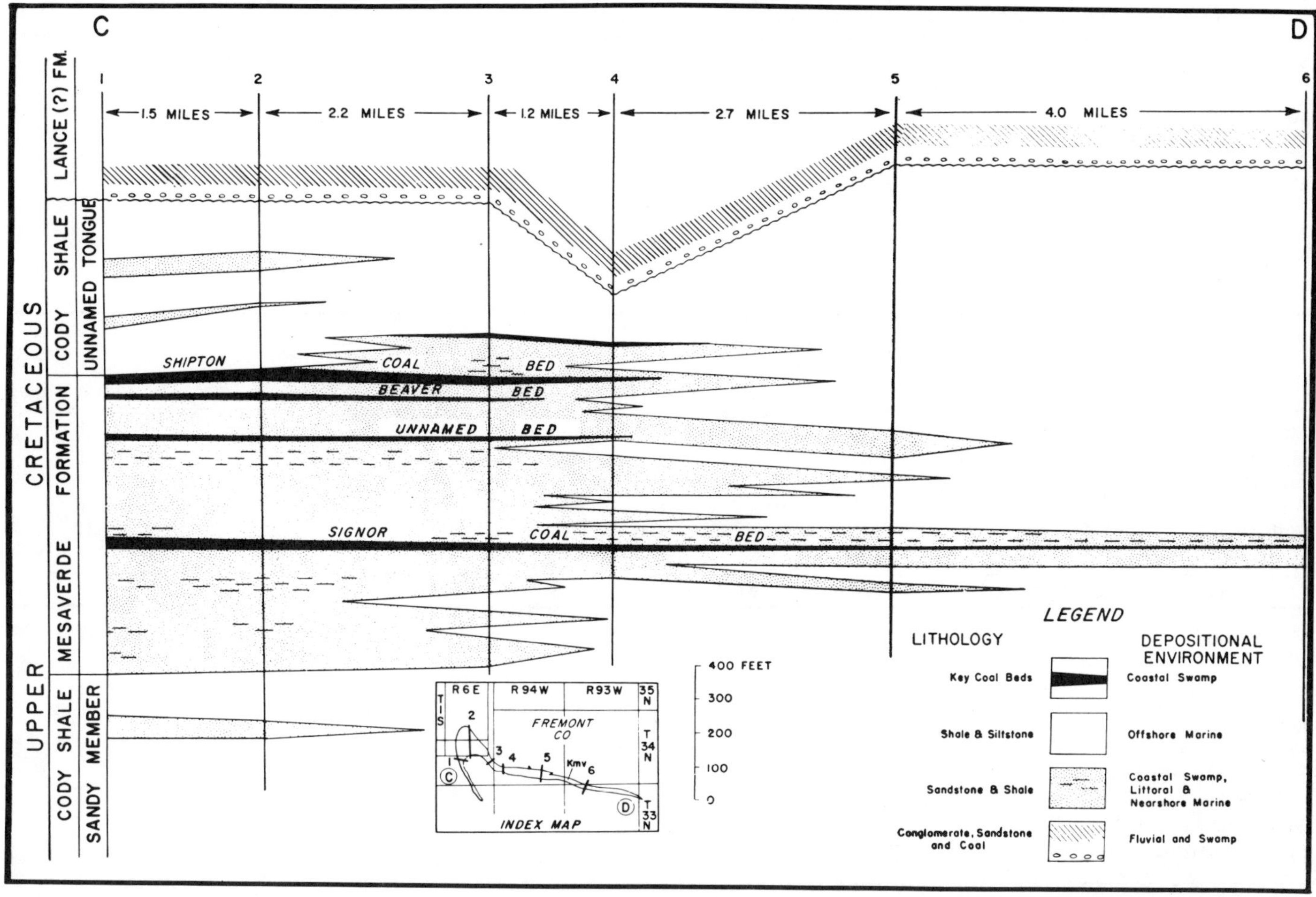

Figure 18—Diagrammatic cross section showing the continuity of the key coals in the south-southwestern part of the Wind River Basin. Barwin, 1961. Used with the permission of the Wyoming Geological Association.

although vestiges probably remained in the central part of the basin.

Concomitant downwarping along the present northeast margin of the basin resulted in the accumulation of a thick sequence of fine-grained fluviatile sediments in the major basin trough area during early Paleocene time and of several thousand feet of lacustrine or marine deposits in the middle and late Paleocene (Keefer, 1965).

Fort Union Coalbed Stratigraphy—Several coalbeds in the lowest part of the formation are present in the subsurface sections but are either incompletely exposed on the surface or are represented by a brown carbonaceous shale. Although coal is not present in exposed rocks of the Fort Union Formation at Big Sand Draw Alkali Butte, it is present in the subsurface in the Beaver Creek coal field, in some Atlantic wells on Riverton Dome, and in a Stanolind well on North Alkali Butte (Fig. 5). In the southernmost part of the Beaver Creek Field, drill holes show as many as three coals in the Fort Union Formation, but they are only 2 to 3 ft thick (Thompson and White, 1952). One prospect in the Arminto field exposed 4.5 ft of Paleocene coal (Woodruff and Winchester, 1912). About 25 mi east of Alkali Butte, coal has been mined in the Castle Gardens area in Fort Union Formation rocks.

Wind River Formation

The Wind River Formation is a variable Eocene sequence of continental beds exposed at the surface throughout most of the Wind River Basin. Within the mapped area, the formation is nearly always present in the synclines. In some places, it covers older rocks on the anticlines but commonly has been eroded from the top and crops out only around the flanks.

The Eocene Wind River Formation is also a minor coal-bearing formation in the Wind River Basin (Glass, 1978). Drilling in the southernmost part of the Beaver Creek coal field,

Table 2—Summary of Kinnear and Shipton coal analyses from Pilot Butte, Alkali Butte, and Shotgun Butte coal fields, respectively, in Fremont County, Wyoming. Modified from Glass, 1978. Used with the permission of the Geological Survey of Wyoming.

Kinnear Coal

As-Received Basis	Single Analysis
Moisture (%)	14.8
Volatile Matter (%)	34.0
Fixed Carbon (%)	42.6
Ash (%)	8.6
Sulfur (%)	0.9
Heat Value (Btu/lb)	10,193

Shipton Coal

	As-Received Basis[1]	Moisture-Free Basis
Moisture (%)	34.1	46.2
Volatile Matter (%)	30.4	46.2
Fixed Carbon (%)	29.3	44.4
Ash (%)	6.2	9.4
Sulfur (%)	0.6	0.9
Heat Value (Btu/lb)	6,080	9,230

[1]The sample was probably weathered as it was only 45 ft from the outcrop.

Welton Coal

As-Received Basis	Single Analysis
Moisture (%)	15.7
Volatile Matter (%)	28.5
Fixed Carbon (%)	47.7
Ash (%)	8.1
Sulfur (%)	0.35
Heat Value (Btu/lb)	9,920

however, documents a few 2 to 3 ft thick coals (Thompson and White, 1952). Most Wind River coals are only a few inches thick and of little economic value.

Coalbed methane resources associated with these deeply buried coalbeds potentially could be very large. Methane characterization data, however, are limited to only certain widely scattered locations where shallow subbituminous Mesaverde coals have been sampled. Although these coals are of low rank, indicating small amounts of adsorbed methane per unit volume of coal, the beds are continuous and thicken locally. These conditions are commonly favorable for large volumes of methane to be contained in a unit area of coalbed. In some areas of the Wind River Basin, numerous thick coalbeds are associated with sandstones and carbonaceous shales and can be intercepted by a single well. This small, limited sample is only a first step in the development of an exploration rationale required to characterize overall distribution of the coalbed methane in the Wind River Basin.

POTENTIAL METHANE RESOURCE

Previous Studies/Analyses

No previous studies have determined the presence of coalbed methane gas in the Wind River Basin. Results presented are from cooperative arrangement between the USGS coal resource investigations group and the MRCP.

Gas emission data are nonexistent for the abandoned small mines along the rim of the coal basin. Minor amounts of methane gas have been reported from the Wyoming Central Mine in the Lander (Signor) coalbed south of Hudson, Wyoming (Fig. 23).

Wind River MRCP Well Test Data

Methane Recovery from Coalbeds Project data on the methane content of Wind River Basin coals presently are limited to coring and well testing on the Wind River Indian Reservation, Fremont County, Wyoming (Fig. 24). Seven Mesaverde coal samples have been desorbed by TRW from the three USGS wells:

- WR-11 (U); 17, 5N., 1W.: 1 core (184–190 ft)
- WR-18 (DD); 22, 33N., 95W.: 2 cores (158–164 ft) (188–192 ft)
- WR-27 (A-1); 29, 1S., 6E.: 4 cores (572–578 ft) (644–648 ft) (830–831 ft) (965–969 ft)

The desorption test results are inconclusive. No gas was obtained from the cores, but proximate analyses and rank designations will be done.

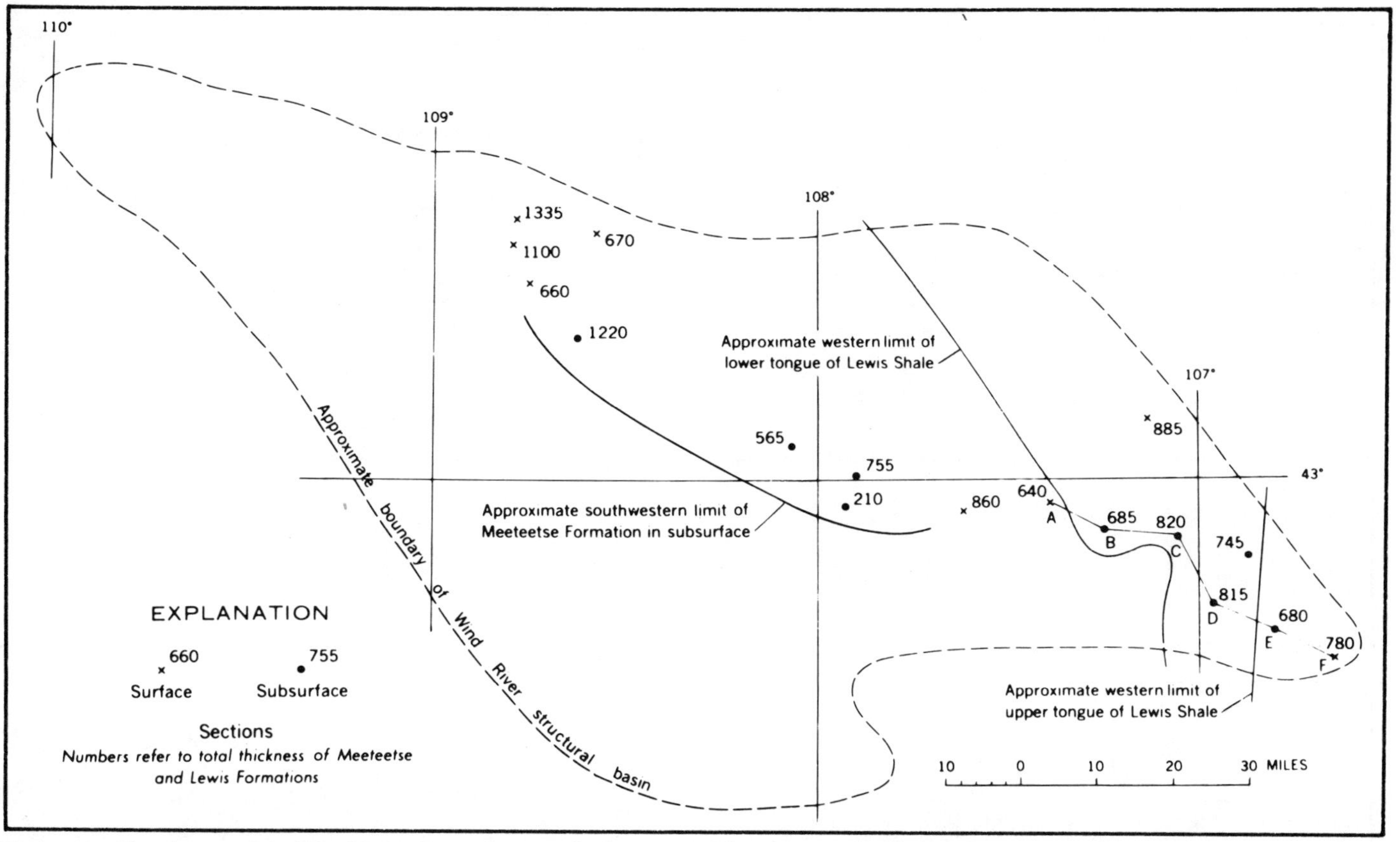

Figure 19—Sketch map of the Wind River Basin showing thicknesses and distribution of Meeteetse Formation and Lewis Shale. Keefer, 1965.

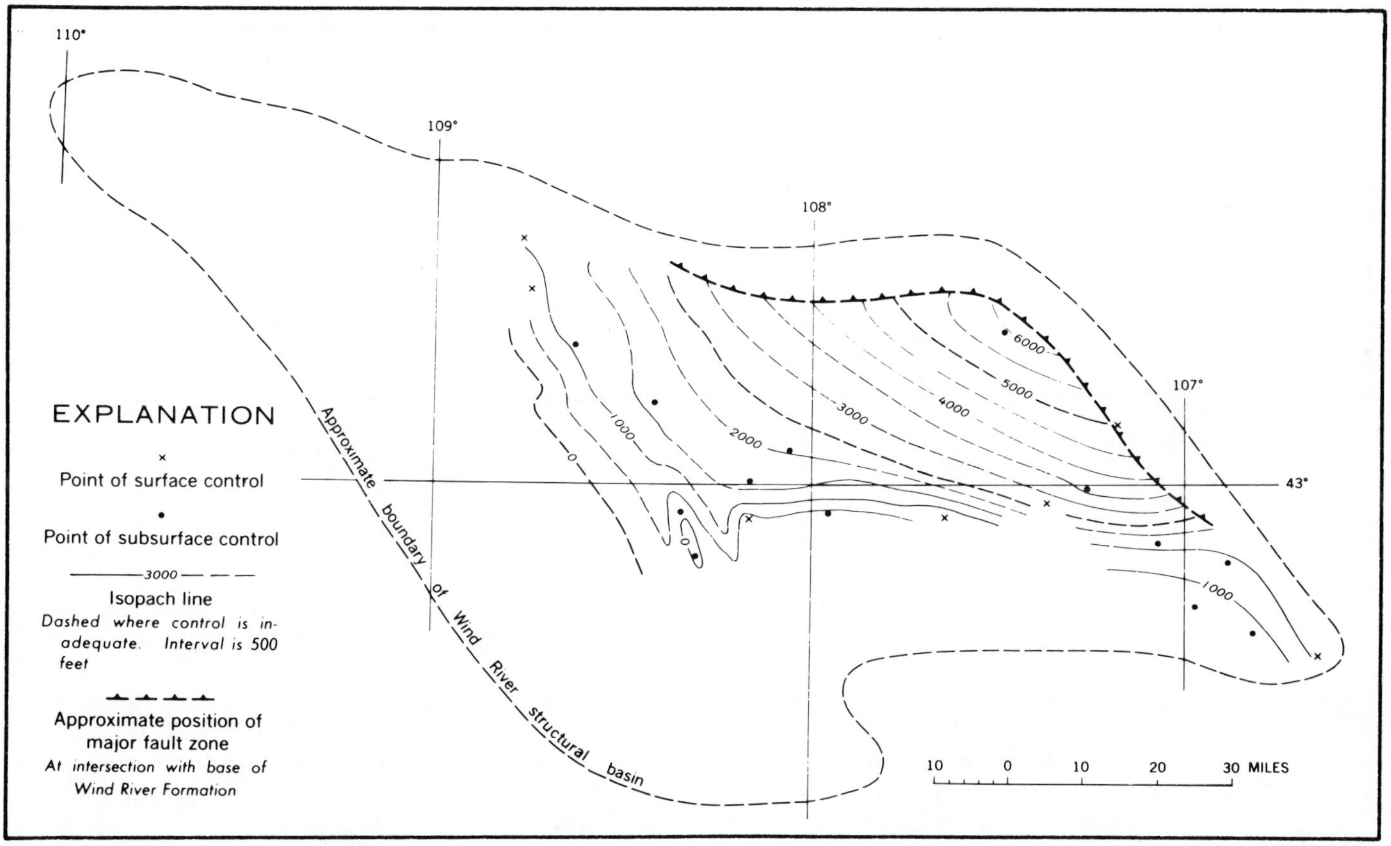

Figure 20—Isopach map of the Lance Formation. Keefer, 1965.

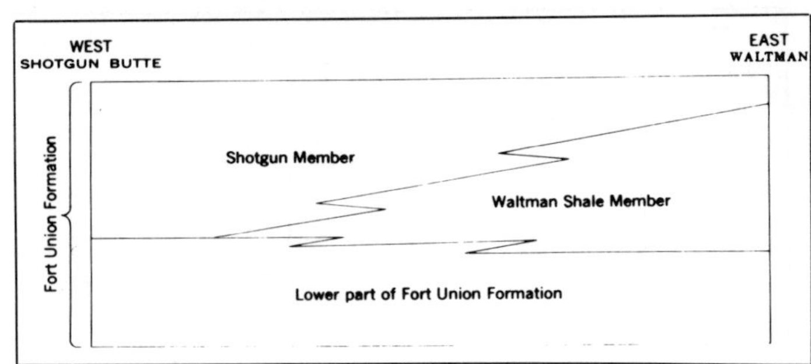

Figure 21—Sketch showing general relations of various members of the Fort Union Formation from west to east along north margin of the Wind River Basin. Keefer and Troyer, 1964.

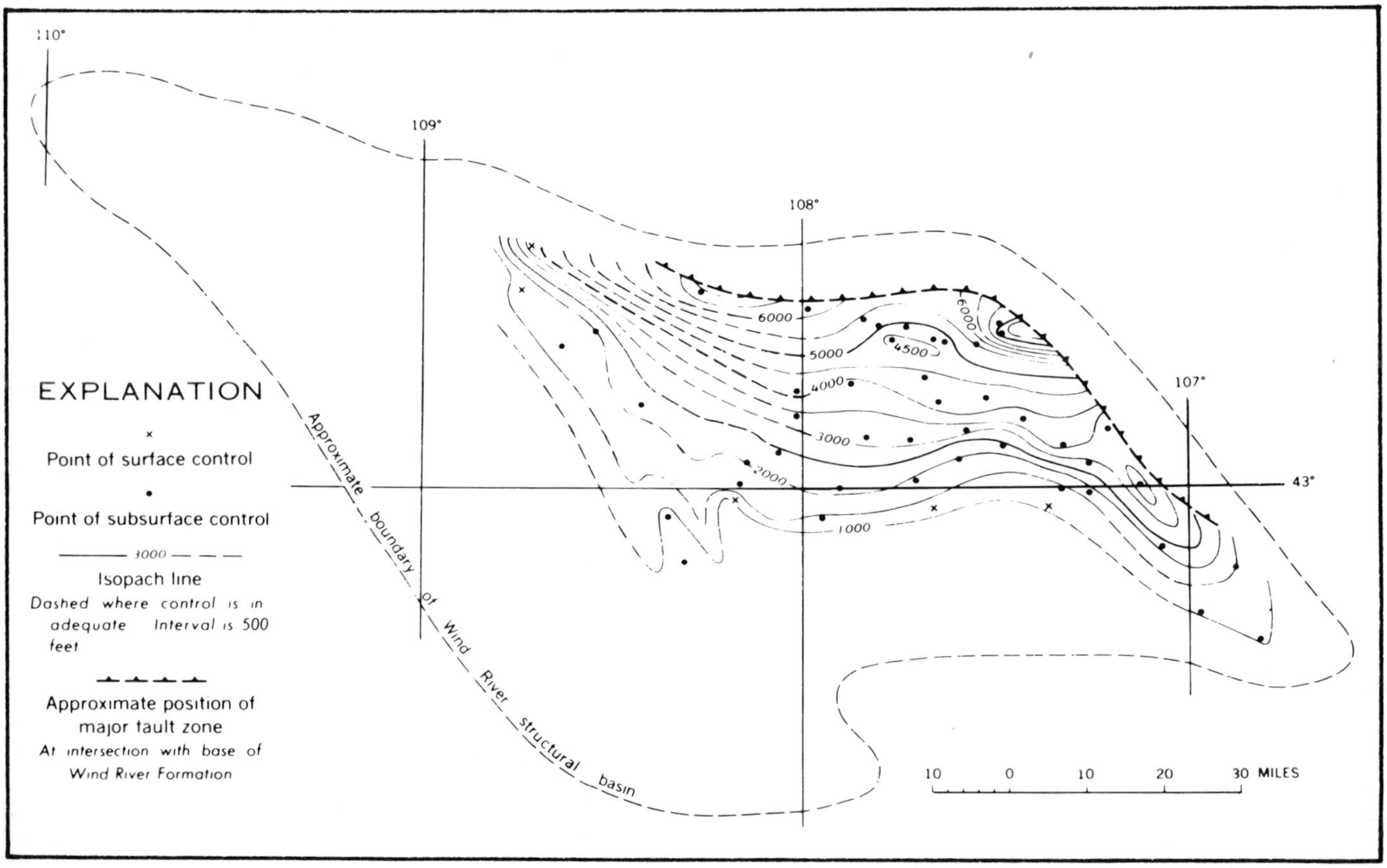

Figure 22—Isopach map of the Fort Union Formation. Keefer, 1965.

Estimate Resource Volume

Absence of methane gas in the tested shallow coalbeds makes it difficult to construct a resource estimate for the Wind River Basin. USGS drill hole T (Fig. 24) had a documented gas show in the Frontier Formation. The gas blew out the drill mud and cement in the well (Windolph, 1981, personal communication), but it is unknown whether the gas was associated with the Frontier coals.

Wind River Coal Resource Assessment

Early USGS Estimates—In 1917, Campbell established the first resource estimates of the coal present in the Wind River Basin (Table 3). His estimate of 17 billion short tons included the Big Horn Basin, Wyoming, and the Red Lodge coal field in Montana.

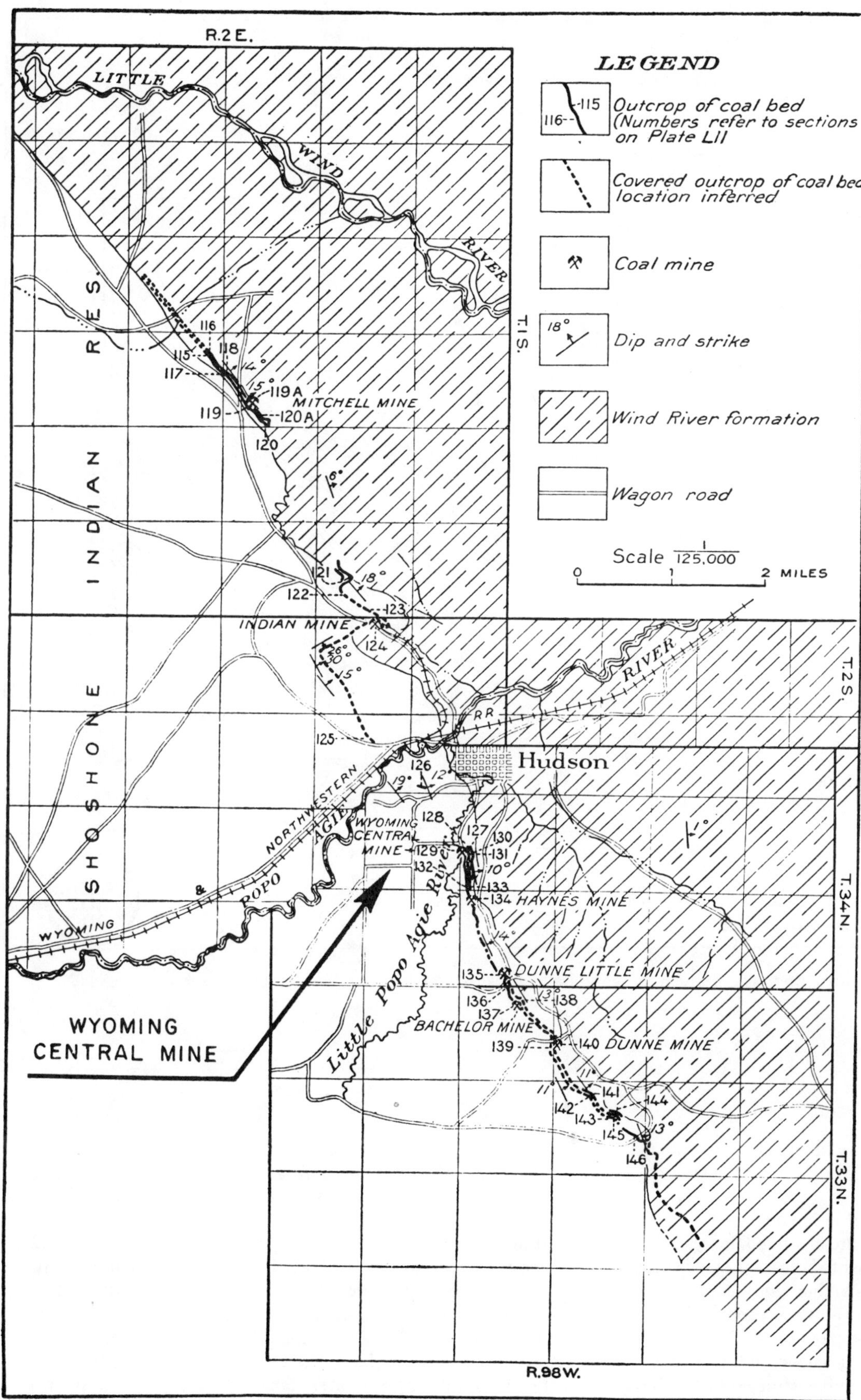

Figure 23—Map of the Hudson coal field, Fremont County, Wyoming. Woodruff and Winchester, 1912.

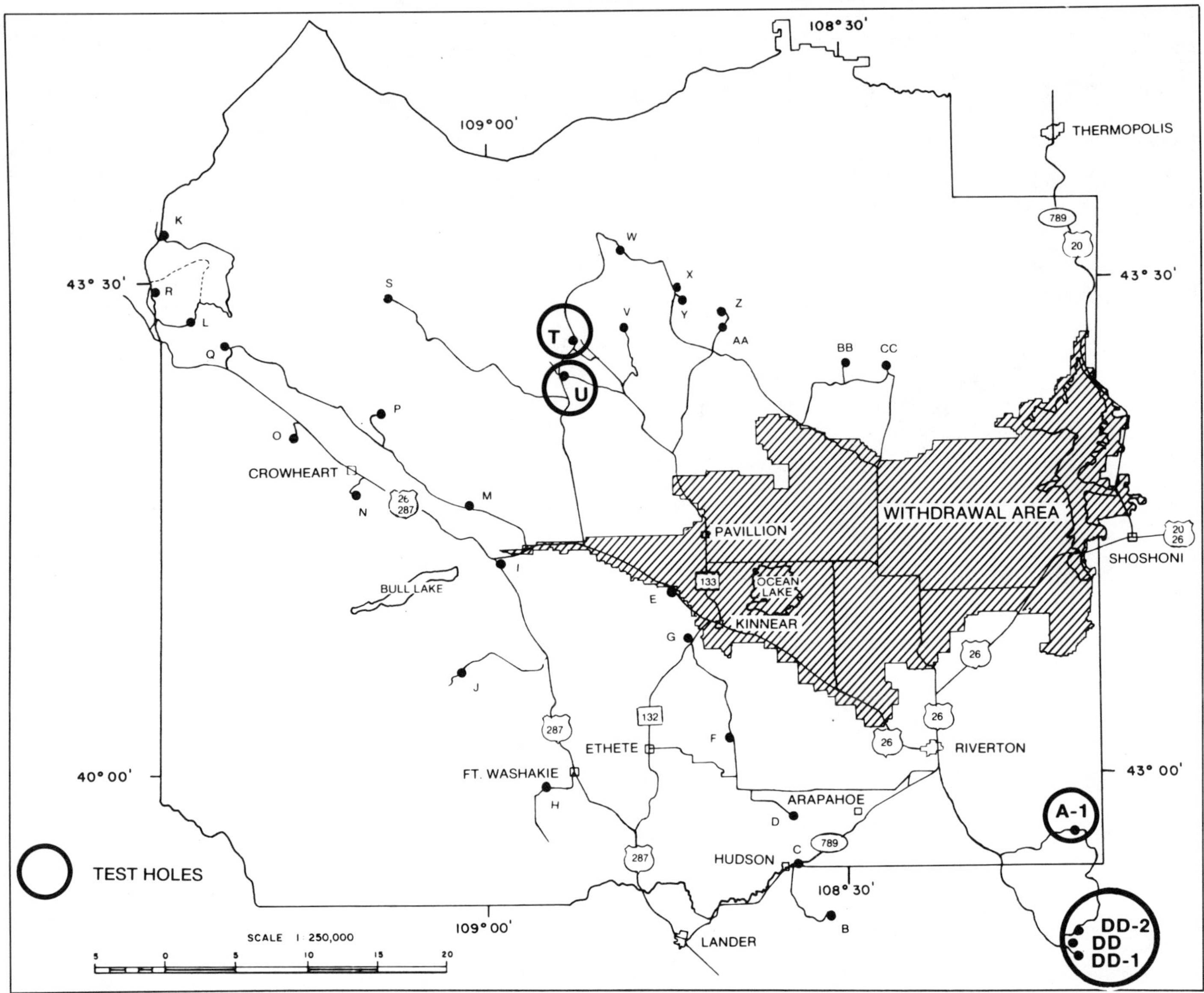

Figure 24—Location map of the USGS test holes from which coal cores were recovered and desorbed. Well designations A-1, DD, and U are USGS field designations.

Missouri River Basin Study—Berryhill et al (1950) estimated the coal resources in the Wind River Basin on the basis of township and range locations, classes of reserves, and the reliability of available data. The three coal reserve categories identifying the reliability of data are measured, indicated, and inferred. The reserve classes presented are subdivided according to coal rank, bed thickness, and overburden thickness.

Berryhill et al (1950) reported that certain counties in Wyoming contain large areas that probably contain minable coal at depths less than 3,000 ft but for which little or no information is available. Although Berryhill et al could have made rough calculations on a regional basis of the coal in such areas, they omitted them from the accompanying estimates, which thus are confined to areas where extent and thickness of the coalbeds were available. The minimum thickness of coal included in the estimates of reserves is 14 inches for bituminous and 2.5 ft for subbituminous coal and lignite. Coal thicker than the minimum for each rank is divided into three thickness groups. For bituminous coal, the ranges are 14 to 28, 28 to 42, and more than 42 inches. For subbituminous coal and lignite, the thickness ranges are 2.5 to 5, 5 to 10, and over 10 ft. The estimated reserves are subdivided into three depth ranges: less than 1,000, 1,000 to 2,000, and 2,000 to 3,000 ft. No coal under more than 3,000 ft of cover is included in the Missouri River Basin Study estimates. Table 4 provides this information for Fremont and Natrona counties, Wyoming. Tables 5, 6, and 7 show the results of the Berryhill study reported on a township and range basis and by coal field. The total for Fremont County

Table 3—Original and present quantity of coal in the Wind River Basin including the Big Horn Basin and Red Lodge field. Campbell, 1917.

Anthracite and Semianthracite	0	Coal below surface from 3,000 to 6,000 ft	12,010,000,000
Semibituminous Coal	0	Production in 1913	1,604,524
Bituminous Coal	608,800,000	Production in 1914	1,612,941
Subbituminous Coal	4,706,700,000	Total production to end of 1914	15,633,450
Lignite	0		
Total coal of all ranks	5,315,500,000	Estimated Supply within 3,000 ft of surface	5,291,000,000

Table 4—Areas included in and omitted from reserve estimates, by counties. Berryhill et al, 1950.

County	Total Coal-bearing Area in Square Miles	Coal-bearing Area Included in Estimates in Square Miles	Coal-bearing Area Omitted from Estimates in Square Miles	Percentage of Coal-bearing Area Omitted from Estimates
Fremont	2,467	140	2,327	94.32
Natrona	1,297	655	642	49.50
Total	3,764	795	2,969	
Average	—	—	—	72

is 733,760,000 tons and for Natrona County, 167,160,000 tons. The Wind River Basin has a total of 900,920,000 tons of coal. Berryhill et al (1950) estimated 875,660,000 tons of coal in established coal fields within the coal basin (Table 8).

Distinction Between Original, Remaining, and Recoverable Reserves—The estimates given by Berryhill et al (1950) are of original coal reserves in the ground before mining. They include the narrow weathered zone at the outcrops and coal under railways and roads but exclude all known areas of burned coal. Calculation on the basis of original reserves, rather than reserves remaining as of a certain date, is the only practicable procedure in Wyoming, where many coal-bearing areas have not been surveyed and where total production is small compared with total reserves in the Wind River Basin. Estimates of remaining and recoverable reserves at any given time can be ascertained more readily, therefore, if records of original reserves and amount of coal mined and lost in mining are kept separately.

Alkali Butte and Big Sand Draw Coal Reserve Study

Thompson and White (1952) estimated original coal reserves for the Alkali Butte and Big Sand Draw coal fields. Their study also included data from Berryhill et al (1950) for the Beaver Creek coal field. Coal reserves were assigned to the Signor, Shipton, Lower Lance, and Downey coalbeds (Tables 9, 10).

Reserve Calculations—Because most data on coal in the Alkali Butte and the Big Sand Draw fields were obtained by surface mapping of the coal outcrops, and because many beds are lenticular, Thompson and White (1952) found it necessary to make some assumptions in calculating coal reserves. Measured coal was defined as coal lying in a block 1/4 to 1/2 mi downdip behind a block of measured coal. Inferred coal was defined as coal in a block 1/2 to 1 mi downdip behind a block of measured and indicated coal or behind a block of indicated coal. In general, no measured coal is more than 1/4 mi from the outcrop; no indicated coal is more than 1/2 mi from the outcrop; and no inferred coal is more than 1 mi from the outcrop.

A somewhat more conservative set of definitions was formulated for coal in the Lance bed, which is poorly exposed, contains more ash, and is probably more lenticular than other beds in the two fields (Thompson and White, 1952). A further consideration is the fact that the Lance Formation containing the Lance coalbed was deposited unconformably on underlying Cretaceous rocks, and the Lance bed may have been locally eroded during Lance time. Coal in the Lance bed was thus classed only as indicated or inferred; indicated coal was projected only 1/4 mi from the outcrop, and inferred coal was projected only 1/2 mi from the outcrop.

Thompson and White (1952) reported total measured, indicated, and inferred reserves for the following coalbeds:

Table 5—Indicated original reserves of subbituminous coal in the Wind River Basin, in millions of short tons. Berryhill et al, 1950.

	0 to 1,000 ft Overburden				1,000 to 2,000 ft Overburden				2,000 to 3,000 ft Overburden				Total in all Overburden Categories			
	Thickness of Bed				Thickness of Bed				Thickness of Bed				Thickness of Bed			
Township	2.5–5 ft	5–10 ft	10+ ft	Total	2.5–5 ft	5–10 ft	10+ ft	Total	2.5–5 ft	5–10 ft	10+ ft	Total	2.5–5 ft	5–10 ft	10+ ft	Township Total
Freemont County																
T34N, R89W	1.25	—	—	1.25	—	—	—	—	—	—	—	—	1.25	—	—	—
T34N, R90W	2.45	.19	—	2.64	.20	—	—	.20	—	—	—	—	2.65	.19	—	2.84
T35N,R90W	2.69	—	—	2.69	2.94	—	—	2.94	—	—	—	—	5.63	—	—	5.63
T34N, R91W	9.98	5.63	—	15.61	12.77	1.13	—	13.90	11.10	—	—	11.10	33.85	6.76	—	40.61
T34N, R95W	4.62	1.79	5.38	11.79	1.98	1.93	2.31	6.22	2.01	.26	—	2.27	8.61	3.08	7.69	20.28
T33N, R96W	—	—	—	—	32.45	27.11	45.96	105.52	20.91	12.24	69.44	102.59	53.36	39.35	115.40	208.11
T33N, R98W	6.01	.99	—	7.00	5.26	1.67	—	6.93	2.48	—	—	2.48	13.75	2.66	—	16.41
T34N, R98W	3.03	.56	—	3.59	3.21	2.79	—	6.00	5.70	.56	—	6.26	11.94	3.91	—	15.85
T3N, R1W	.36	—	—	.36	—	—	—	—	—	—	—	—	.36	—	—	.36
T5N, R1W	—	.92	—	.92	—	.46	—	.46	—	—	—	—	—	1.38	—	1.38
T6N, R2W	4.02	4.09	—	8.10	2.73	3.81	—	6.54	1.40	1.73	—	3.13	8.15	9.62	—	17.77
T6N, R1E	3.20	12.25	—	15.45	—	—	—	—	—	—	—	—	3.20	12.25	—	15.45
T6N, R2E	3.99	—	—	3.99	—	—	1.17	—	—	—	—	5.16	—	—	3.16	
T1S, R2E	1.37	—	—	1.37	—	—	—	—	—	—	—	—	1.37	—	—	1.36
T2S, R2E	3.63	—	—	3.63	—	—	—	—	—	—	—	—	3.63	—	—	3.63
T6N, R3E	2.74	—	—	2.74	—	—	—	—	—	—	—	—	2.74	—	—	2.74
T1S, R6E	—	—	—	—	.95	—	—	.95	4.83	—	—	4.83	5.78	—	—	5.78
T2S, R6E	3.47	6.18	—	9.60	6.20	.37	—	6.66	2.75	—	—	2.75	12.51	6.55	—	19.05
County Total	52.81	32.59	5.38	90.78	69.95	39.27	48.27	157.49	51.18	14.79	69.44	135.41	173.94	86.65	123.09	383.68
Natrona County																
T31N, N82W	2.32	.19	—	2.51	.34	—	—	.34	—	—	—	—	2.66	.19	—	2.85
T32N, R82W	.27	.39	—	.66	.24	—	—	.24	—	—	—	—	.51	.39	—	.90
T33N, R83W	.03	—	—	.03	—	—	—	—	—	—	—	—	.03	—	—	.03
T35N, R84W	2.72	3.89	—	6.61	2.03	3.67	—	5.70	1.00	3.53	—	4.53	5.75	11.08	—	16.84
T35N, R84W	1.83	—	—	1.83	1.39	—	—	1.39	.68	—	—	.68	3.90	—	—	3.90
T35N, R85W	14.15	2.65	—	16.80	11.09	.25	—	11.34	7.59	—	—	7.59	32.83	2.90	—	35.73
T36N, R85W	3.78	3.91	—	7.69	2.76	2.47	—	5.23	1.83	1.14	—	2.97	8.37	7.52	—	15.89
T33N, R86W	4.39	.12	—	4.51	.51	—	—	.51	—	—	—	—	4.90	.12	—	5.02
T36N, R86W	9.18	2.94	—	12.12	8.26	2.17	—	10.43	4.98	.99	—	5.97	22.42	6.10	—	28.32
T37N, R86W	—	8.33	—	8.33	—	7.14	—	7.14	—	3.57	—	3.57	—	19.04	—	19.04
T33N, R87W	5.51	2.66	—	8.17	5.84	—	—	5.84	3.18	—	—	3.16	14.51	2.66	—	17.17
T34N, R87W	.02	—	—	.02	.74	—	—	.74	1.66	—	—	1.66	2.42	—	—	2.42
T37N, R87W	2.74	—	—	2.74	2.35	—	—	2.35	1.17	—	—	1.17	6.26	—	—	6.26
T34N, R88W	4.90	1.70	—	6.60	3.52	—	—	3.52	2.45	—	—	2.45	10.87	1.70	—	12.67
County Total	51.86	26.78	—	78.62	39.07	15.70	—	54.777	24.52	9.23	—	33.75	115.42	51.71	—	176.14
Basin Total																550.82

Table 6—Inferred original reserves of subbituminous coal in the Wind River Basin, in millions of short tons. Berryhill et al, 1950.

Township	0 to 1,000 ft Overburden				1,000 to 2,000 ft Overburden				2,000 to 3,000 ft Overburden				Total in all Overburden Categories			Township Total
	Thickness of Bed				Thickness of Bed				Thickness of Bed				Thickness of Bed			
	2.5–5 ft	5–10 ft	10+ ft	Total	2.5–5 ft	5–10 ft	10+ ft	Total	2.5–5 ft	5–10 ft	10+ ft	Total	2.5–5 ft	5–10 ft	10+ ft	
Fremont County																
T27N, R90W	—	.67	—	.67	—	—	—	—	—	—	—	—	—	.67	—	.67
T28N, R90W	1.56	—	—	1.56	—	—	—	—	—	—	—	—	1.56	—	—	1.56
T27N, R93W	—	—	—	—	3.32	2.21	—	5.53	7.52	5.75	2.88	16.15	10.84	7.96	2.88	21.68
T27N, R94W	1.33	—	—	1.33	—	—	—	—	—	—	—	—	1.33	—	—	1.33
T34N, R94W	—	—	—	—	—	—	—	—	3.36	—	—	3.36	3.36	—	—	3.36
T33N, R96W	—	—	—	—	—	—	—	—	48.36	33.48	122.63	204.47	48.36	33.48	122.63	204.47
T34N, R96W	—	—	—	—	—	—	—	—	19.58	10.54	64.77	94.89	19.58	10.54	64.77	94.89
T1S, R6E	—	—	—	—	—	—	—	—	.41	—	—	.41	.41	—	—	.41
County Total	2.89	.67	—	3.56	3.32	2.21	—	5.53	79.23	49.77	190.28	319.28	85.44	52.65	190.28	328.37
Basin Total																328.47

Table 7—Original reserves, in all categories of subbituminous coal in the Wind River Basin, in millions of short tons. Berryhill et al, 1950.

Township	0 to 1,000 ft Overburden				1,000 to 2,000 ft Overburden				2,000 to 3,000 ft Overburden				Total in all Overburden Categories			Township Total
	Thickness of Bed				Thickness of Bed				Thickness of Bed				Thickness of Bed			
	2.5–5 ft	5–10 ft	10+ ft	Total	2.5–5 ft	5–10 ft	10+ ft	Total	2.5–5 ft	5–10 ft	10+ ft	Total	2.5–5 ft	5–10 ft	10+ ft	
Fremont County																
T34N, R89W	1.25	—	—	1.25	—	—	—	—	—	—	—	—	1.25	—	—	1.25
T27N, R90W	—	0.67	—	0.67	—	—	—	—	—	—	—	—	—	0.67	—	0.67
T28N, R90W	1.56	—	—	1.56	—	—	—	—	—	—	—	—	1.56	—	—	1.56
T34N, R90W	2.45	0.19	—	2.64	0.20	—	—	0.20	—	—	—	—	2.65	0.19	—	2.84
T35N, R90W	2.69	—	—	2.69	2.94	—	—	2.94	—	—	—	—	5.63	—	—	5.63
T27N, R93W	—	—	—	—	3.32	2.21	—	5.63	7.52	5.75	2.88	16.15	10.84	7.96	2.88	21.68
T27N, R94W	1.33	—	—	1.33	—	—	—	—	—	—	—	—	1.33	—	—	1.33
T34N, R94W	9.98	5.63	—	15.61	12.77	1.13	—	13.00	14.46	—	—	14.46	37.21	6.76	—	43.97
T34N, R95W	4.62	1.79	5.38	11.79	1.98	1.93	2.31	6.22	2.01	0.26	—	8.61	3.98	7.69	20.28	
T33N, R96W	—	—	—	—	32.45	27.11	45.96	105.52	69.27	45.72	192.07	307.06	101.72	72.83	238.03	412.58
T34W, R96W	—	—	—	—	—	—	—	—	19.58	10.54	64.77	94.89	19.58	10.54	64.77	94.89
T33N, R98W	7.03	2.38	9.29	18.70	5.26	3.06	0.89	9.21	2.48	—	—	2.48	14.77	5.44	10.18	30.39
T34N, R96W	3.91	1.95	1.78	7.64	3.21	5.58	0.89	9.68	5.70	0.56	—	6.26	12.82	8.09	2.67	23.58
T3N, R1W	0.36	—	—	0.36	—	—	—	—	—	—	—	—	0.36	—	—	0.36
T3N, R1W	—	0.92	—	0.92	—	0.46	—	0.45	—	—	—	—	—	1.38	—	1.38
T6N, R2W	4.02	4.08	—	8.10	2.73	3.81	—	6.54	1.40	1.73	—	3.13	8.15	9.62	—	17.77
T6N, R1E	3.20	12.25	—	16.45	—	—	—	—	—	—	—	—	3.20	12.25	—	15.45
T6N, R2E	3.99	—	—	3.99	1.17	—	—	1.17	—	—	—	—	5.16	—	—	5.16
T1S, R2E	1.37	—	—	1.37	—	—	—	—	—	—	—	—	1.37	—	—	1.37
T2S, R2E	3.63	—	—	3.63	—	—	—	—	—	—	—	—	3.63	—	—	3.63
T6W, R3E	2.74	—	—	2.74	—	—	—	—	—	—	—	—	2.74	—	—	2.74
T1S, R6E	—	—	—	—	0.95	—	—	0.95	5.24	—	—	5.24	6.19	—	—	6.19
T2S, R6E	3.47	6.18	—	9.65	6.29	0.37	—	6.66	2.75	—	—	2.75	12.51	6.55	—	19.06
County Total	57.60	36.04	16.45	110.09	73.27	45.66	50.05	168.98	130.41	64.56	259.72	454.69	261.28	146.26	326.22	733.76

continued

Table 7— (Continued)

Township	0 to 1,000 ft Overburden				1,000 to 2,000 ft Overburden				2,000 to 3,000 ft Overburden				Total in all Overburden Categories			Township Total
	Thickness of Bed				Thickness of Bed				Thickness of Bed				Thickness of Bed			
	2.5–5 ft	5–10 ft	10+ ft	Total	2.5–5 ft	5–10 ft	10+ ft	Total	2.5–5 ft	5–10 ft	10+ ft	Total	2.5–5 ft	5–10 ft	10+ ft	
Natrona County																
T31N, R82W	2.32	0.19	—	2.51	0.34	—	—	—	—	2.66	0.19	—	2.86			
T32N, R82W	0.27	0.39	—	0.66	0.24	—	—	0.24	—	—	—	—	0.51	0.39	—	0.90
T33N, R82W	0.03	—	—	0.03	—	—	—	—	—	—	—	—	0.03	—	—	0.03
T33N, R83W	2.72	3.89	—	6.61	2.03	3.67	—	5.70	1.00	3.53	—	4.53	5.75	11.09	—	16.84
T35N, R84W	1.83	—	—	1.83	1.39	—	—	1.39	0.68	—	—	0.68	3.90	—	—	3.90
T35N, R84W	14.15	2.65	—	16.80	11.09	0.25	—	11.34	7.59	—	—	7.59	32.83	2.90	—	35.73
T36N, R85W	3.78	3.91	—	7.69	2.76	2.47	—	5.23	1.83	1.14	—	2.97	8.37	7.52	—	15.89
T33N, R86W	4.39	0.12	—	4.51	0.51	—	—	0.51	—	—	—	—	4.90	0.12	—	5.02
T36N, R86W	9.18	2.94	—	12.12	8.26	2.17	—	10.43	4.98	0.99	—	5.97	22.42	6.10	—	28.52
T37N, R86W	—	8.33	—	8.33	—	7.14	—	7.14	—	3.57	—	3.57	—	19.04	—	19.04
T33N, R87W	5.51	2.66	—	8.17	5.84	—	—	5.84	3.16	—	—	3.16	14.51	2.66	—	17.17
T34N, R87W	0.02	—	—	0.02	0.74	—	—	0.74	1.66	—	—	1.66	2.42	—	—	2.42
T37N, R87W	2.74	—	—	2.74	2.35	—	—	2.35	1.17	—	—	1.17	6.26	—	—	6.26
T34N, R88W	4.90	1.70	—	6.60	3.52	—	—	3.52	2.45	—	—	2.45	10.87	1.70	—	12.57
County Total	51.84	26.78	—	78.62	39.07	15.70	—	54.77	24.52	9.23	—	33.75	115.43	51.71	—	167.16
Basin Total																900.92

Table 8—Estimated original coal reserves in the Wind River Basin coal basin, by coal field. Berryhill et al, 1950.

Field	Bituminous				Subbituminous				Total in all Ranks and Categories
	Measured	Indicated	Inferred	Total	Measured	Indicated	Inferred	Total	
Muddy Creek	—	—	—	—	—	42.50	—	42.50	42.50
Pilot Butte	—	—	—	—	—	.36	—	.36	.36
Hudson	—	—	—	—	21.71	37.26	—	58.97	58.97
Alkali Butte	—	—	—	—	—	85.73	3.77	89.50	89.50
T33-34N, R96W	—	—	—	—	—	208.11	299.36	507.47	507.47
Arminto (Powder River)	—	—	—	—	—	176.86	—	176.86	176.86
Total	—	—	—	—	21.71	550.82	303.13	875.66	875.66

- Downey—71,830,000
- Signor—58,780,000
- Beaver—3,470,000
- Shipton—40,520,000
- Lance Coal—5,780,000

Recent Estimates

Glass and Roberts (1978) combined coal resource estimates from Berryhill et al (1950), Thompson and White (1952), and Keefer and Troyer (1964) to arrive at a total original coal resource (> 2.5 ft thick) of about one billion tons (Table 11). This revised estimate is 150 million tons more than the original estimate made by Berryhill et al (1950) and includes coal lying between 0 and 3,000 ft of the surface.

Of this original resource, 100.7 million remaining tons is 1) in the measured or indicated categories of reliability; 2) less than 1,000 ft below the surface; and 3) 5 or more ft thick. These three parameters delimit today's potentially recoverable portion of the resource: the reserve base. Allowing for future mining losses equal to production, this resource reduces to a recoverable reserve base of about 50.4 million tons.

The only known strippable coal resources in the Wind River Coal Basin are in the Signor bed of the Alkali Butte coal field and in the Wilton coalbed in the Muddy Creek coal field. Combined, they total an estimated one million tons of strippable coal in these two coalbeds. Additional strippable coal reserves exist, but documentation of these reserves awaits more detailed geologic and/or mining reports. Because of the relatively steep dips associated with the outcropping coals in the Wind River Basin, however, strippable reserves probably are not large (Glass and Roberts, 1978).

Wind River Coalbed Methane Resource Base

No quantitative data are available on the specific amount of methane present in coalbeds of the Wind River Basin. The in-place methane is probably much greater than that indicated by desorption analysis of the USGS coal samples.

Coalbeds at very shallow depths (< 1,000 ft) in the Wind River Basin are low rank and exhibit none of the maturation characteristics needed for methane gas generation. The shallow coals in the Wind River do not contain methane gas like those in the Arkoma Basin of Oklahoma and Arkansas, which was discovered during the MRCP/USGS cooperative well tests completed in late 1980. However, in the deeper part of the basin below the over-pressured zone, methane gas has probably been generated in sufficient quantities.

Miller and VerPloeg (1980) conducted an inventory of undeveloped natural gas still in Wyoming's tight gas sands. They estimated that the Wind River Basin has about 15 billion cubic feet (Tcf) of gas-in-place. The stratigraphic intervals containing substantial amounts of natural gas in unconventional reservoirs in the Wind River Basin have been identified and include much of the Fort Union, most of the sandstone facies in the Mesaverde Group, and individual sandstone units in the Lower Cretaceous rocks (Miller and VerPloeg, 1980). These are the same intervals that contain coalbeds, although Miller and VerPloeg (1980) did not mention the methane associated with these coalbeds.

To develop a rationale for establishing the coalbed resource base for the Wind River Basin, the MRCP results from the neighboring Green River Basin must be examined. Coalbed methane values in the Green River Basin range from 2.9 cubic feet/ton (cf/ton) (minimum value) to 539 cf/ton (maximum value) (McCord, 1980). Using these data, a triangular distribution plot can be made to estimate the probable distribution of gas in that basin. Application of these values to the Wind River Basin is reasonable to provide a test case for speculative coalbed methane resource values. Figure 25 was developed from these data using the principle of triangular distributions by assigning probabilities to the amount of methane contained in coalbeds, as discussed by Megill (1977). Figure 25 shows the cumulative percent frequency of occurrence for gas contents in Green River coalbeds. The cumulative percent frequency values are considered probability values that will allow an approximation of a real distribution from only three points: minimum, most likely, and maximum (Megill, 1977).

The maximum amount of coal present in Wind River Basin is uncertain. However, using the reasonable estimates of the total coal resource base, two scenarios were generated based on the range in values in the Green River Basin data. The conservative approach addresses the shallow coal resources that lie entirely above the overpressured zones in the basin. Estimates of 5 to 100 cf/ton yield an estimated in-place gas resource of 5.2 to 103 billion cubic feet (Bcf) of gas. The second optimistic scenario deals with deep-lying coals, most of which lie directly above and below the deep over-pressured zones. These estimates of the gas content range from 5 to 539 cf/ton and give an in-place gas resource of 20 to 2,220 Bcf of gas. These estimates are presented in Table 12 and are reasonable in light of Miller and VerPloeg's methane resource estimations for the Wind River tight sands.

Table 9—Estimated original coal reserves in the Mesaverde Formation in the Alkali Butte field, Fremont County, Wyoming, in millions of short tons. (All coal is of subbituminous rank.) Thompson and White, 1952.

Bed	Area (acres)	Wgted. Average Thickness in ft	Overburden	Measured Reserves: Thickness of Bed				Indicated Reserves: Thickness of Bed				Inferred Reserves: Thickness of Bed				Total Measured Indicated and Inferred Reserves
				2.5–5 ft	5–10 ft	10+ ft	Total	2.5–5 ft	5–10 ft	10+ ft	Total	2.5–5 ft	5–10 ft	10+ ft	Total	
							T2S, R6E (Partial Township)									
Signor	78	8	0–1,000	—	1.10	—	1.10	—	—	—	—	—	—	—	—	1.10
Signor	656	6.5	0–1,000	—	—	—	—	—	7.53	—	—	—	—	7.53		
Signor	104	6	1,000–2,000	—	—	—	—	—	1.10	—	—	—	—	1.10		
Signor	6	7	0–1,000	—	—	—	—	—	—	—	—	—	0.74	—	0.74	.74
Signor	922	7	1,000–2,000	—	—	—	—	—	—	—	—	—	11.42	—	11.42	11.42
Signor	625	7	1,000–2,000	—	—	—	—	—	—	—	—	—	4.03	—	4.03	4.03
Beaver	557	3	0–1,000	—	—	—	—	2.96	—	—	2.96	—	—	—	—	2.96
Beaver	96	3	1,000–2,000	—	—	—	—	.51	—	—	.51	—	—	—	—	.51
Shipton	219	9	0–1,000	—	3.49	—	3.49	—	—	—	—	—	—	—	—	3.49
Shipton	421	9	0–1,000	—	—	—	—	—	6.71	—	6.71	—	—	—	—	6.71
Shipton	253	9	1,000–2,000	—	—	—	—	—	4.03	—	4.03	—	—	—	—	4.03
Shipton	230	9	0–1,000	—	—	—	—	—	—	—	—	—	3.66	—	3.66	3.66
Shipton	653	9	1,000–2,000	—	—	—	—	—	—	—	—	—	10.40	—	10.40	10.40
Shipton	226	9	2,000–3,000	—	—	—	—	—	—	—	—	—	3.60	—	3.60	3.60
Total					4.59		4.59	3.47	19.37		22.84		33.85		33.85	61.28
							T1S, R6E (Partial Township)									
Signor	29	7	2,000–3,000	—	—	—	—	—	—	—	—	—	0.36	—	0.36	0.36
Shipton	208	5	1,000–2,000	—	—	—	—	—	—	—	—	1.84	—	—	1.84	1.84
Shipton	38	8	2,000–3,000	—	—	—	—	—	—	—	—	—	.54	—	.54	1.84
Total												1.84	0.90		2.74	2.74

continued

Table 9—(Continued)

Bed	Area (acres)	Wgted. Average Thickness in ft	Overburden	Measured Reserves: Thickness of Bed 2.5–5 ft	Measured: 5–10 ft	Measured: 10+ ft	Measured: Total	Indicated Reserves: Thickness of Bed 2.5–5 ft	Indicated: 5–10 ft	Indicated: 10+ ft	Indicated: Total	Inferred Reserves: Thickness of Bed 2.5–5 ft	Inferred: 5–10 ft	Inferred: 10+ ft	Inferred: Total	Total Measured Indicated and Inferred Reserves
								T34N, R95W (Partial Township)								
Signor	50	13	0–1,000	—	—	1.15	1.15	—	—	—	—	—	—	—	—	1.15
Signor	13	13	0–1,000	—	—	—	—	—	—	0.29	0.29	—	—	—	—	0.29
Signor	42	13	1,000–2,000	—	—	—	—	—	—	.97	.97	—	—	—	—	0.97
Signor	32	13	1,000–2,000	—	—	—	—	—	—	—	—	—	—	0.74	0.74	0.74
Signor	83	15	0–1,000	—	—	2.20	2.20	—	—	—	—	—	—	—	—	2.20
Signor	107	3.5	0–1,000	0.66	—	—	.66	—	—	—	—	—	—	—	—	.66
Signor	14.4	15	0–1,000	—	—	—	—	—	—	.38	.38	—	—	—	—	.38
Signor	115	11	0–1,000	—	—	—	—	—	—	2.24	2.24	—	—	—	—	2.24
Signor	80	3.5	0–1,000	—	—	—	—	.50	—	—	.50	—	—	—	—	.50
Signor	51	9	1,000–2,000	—	—	—	—	—	0.81	—	0.81	—	—	—	—	.81
Signor	51	9	1,000–2,000	—	—	—	—	—	0.81	—	0.81	—	—	—	—	.81
Signor	53	9	0–1,000	—	—	—	—	—	—	—	—	—	0.84	—	.84	.84
Signor	86	9	1,000–2,000	—	—	—	—	—	—	—	—	—	1.37	—	1.37	1.37
Signor	160	3.5	1,000–2,000	—	—	—	—	—	—	—	—	0.99	—	—	.99	.99
Signor	130	9	2,000–3,000	—	—	—	—	—	—	—	—	—	2.07	—	2.07	2.07
Shipton	70	9.5	0–1,000	—	1.18	—	1.18	—	—	—	—	—	—	—	—	1.18
Shipton	118	9.5	0–1,000	—	—	—	—	—	1.98	—	1.98	—	—	—	—	1.98
Shipton	6	9.5	1,000–2,000	—	—	—	—	—	.10	—	.10	—	—	—	—	.10
Shipton	16	9.5	0–1,000	—	—	—	—	—	—	—	—	—	.27	—	.27	.27
Shipton	149	9.5	1,000–2,000	—	—	—	—	—	—	—	—	—	2.50	—	2.50	2.50
Shipton	13	9.5	2,000–3,000	—	—	—	—	—	—	—	—	—	.22	—	.22	.22
Total				0.66	1.18	3.35	5.19	0.50	2.89	3.88	7.27	0.99	7.27	0.74	9.00	21.46
								T34N, R94W (Southwest Corner)								
Signor	88	13	1,000–2,000	—	—	—	—	—	—	—	—	—	—	2.04	—	2.04
Signor	173	9	0–1,000	—	2.76	—	2.76	—	—	—	—	—	—	—	—	2.76
Signor	221	9	0–1,000	—	—	—	—	—	3.52	—	3.52	—	—	—	—	3.52
Signor	43	9	1,000–2,000	—	—	—	—	—	.68	—	.68	—	—	—	—	.68
Signor	349	6	0–1,000	—	—	—	—	—	—	—	—	—	3.71	—	3.71	3.71
Signor	362	6	1,000–2,000	—	—	—	—	—	—	—	—	—	3.64	—	3.84	3.84
Signor	70	6	2,000–3,000	—	—	—	—	—	—	—	—	—	.74	—	.74	.74
Total					2.76		2.76		4.20		4.20		8.29	2.04	10.33	17.29
Grand Total				0.66	8.53	3.35	12.54	3.97	26.46	3.88	34.31	2.83	50.31	2.78	55.92	102.77

Table 10—Estimated original coal reserves in the Lance Formation in the Alkali Butte field and the Downey Bed in the Big Sand Draw field, Fremont County, Wyoming, in millions of short tons. Thompson and White, 1952.

Bed	Area (acres)	Wtd. Avg. Thickness in ft.	Overburden	Measured Reserves				Indicated Reserves				Total Measured and Indicated Reserves
				Thickness of Bed				Thickness of Bed				
				2.5–5 ft	5–10 ft	10+ ft	Total	2.5–5 ft	5–10 ft	10+ ft	Total	
T2S, R6E (Partial Township)												
Lance Coal	42	4	0–1,000	0.29	—	—	0.29	—	—	—	—	0.29
Lance Coal	11	4	0–1,000	—	—	—	—	0.08	—	—	0.08	0.08
Total			0.29			0.29	0.08			0.08	0.37	
T1S, R6E (Partial Township)												
Lance Coal	205	4	0–1,000	1.45	—	—	1.45	—	—	—	—	1.45
Lance Coal	566	4	0–1,000	—	—	—	—	3.96	—	—	3.96	3.96
Total				1.45			1.45	3.96			3.96	5.41
Grand Total				1.74			1.74	4.04			4.04	5.78

Bed	Area (acres)	Wtd. Avg.[1] Thickness in ft.	Overburden	Measured Reserves		Indicated Reserves		Inferred Reserves		Total Measured Indicated and Inferred Reserves
				Thickness of Bed		Thickness of Bed		Thickness of Bed		
				10+ ft	Total	10+ ft	Total	10+ ft	Total	
T33N, R6E										
Downey	179	28	0–2,000	8.87[2]	8.87	—	—	—	—	8.87
	379	28	0–2,000	—	—	18.78	18.78	—	—	18.78
	1,248	20	0–3,000	—	—	—	—	44.18	44.18	44.18
Total				8.87	8.87	18.78	18.78	44.18	44.18	71.83

[1]Thickness figures supplied by George Downey, February 1951, and not checked by writers because mine was caved in.
[2]Estimate of 10,000 short tons taken from Downey Mine. Computed February 1951.

Abnormal Fluid Pressure Zones

The development of overpressuring in sedimentary rocks may involve several processes, one of which is the active generation of large amounts of wet gas. Deep drilling in the Wind River Basin has encountered high-pressured shales, high temperatures, and high-pressure gradients. Over-pressured zones have been identified in the Lower Fort Union, Meeteetse, Mesaverde, Cody, and Frontier (Bilyeu, 1978; Miller and VerPloeg, 1980).

Law et al (1980) evaluated the generation of gas in the Green River Basin overpressured zones and found that high reservoir pressure in gas-bearing sandstones greatly increases the potential for large in-place gas resources. Coalbeds in such an environment would be natural candidates for generating large amounts of gaseous hydrocarbons.

CONCLUSIONS

Exploration Criteria

Gas content of a coalbed is related to gas formation during coalification and the post-deposition geologic history of the coalbed. The degree of fracturing, distance to outcrop, depositional environment, depth of burial, bed thickness, rate of burial, and permeability of adjacent strata can also be important. Based on this understanding, the following criteria were developed to use in designating areas with the greatest potential methane resource:

- Thick continuous coalbeds
- High cumulative coal thickness
- High coal rank
- Deep coals
- High geothermal gradient
- Abnormally high fluid pressure environments.

Target Areas

An initial target area of approximately 8,100 sq mi of the Wind River structural basin was selected because it contained ample opportunities for testing coals with the greatest potential for commercialization of the coalbed methane resource. The primary target areas now believed to meet these criteria best are shown in Figure 26. The reasons for their selection are:

Table 11—Coal resources and reserves base in the Wind River coal basin of Wyoming. Glass and Roberts, 1978. Used with the permission of the Wyoming Geological Association.

Original Subbituminous Coal Resources (millions of tons)[1]				
Coal Thickness (ft)	0–1,000 ft of Overburden[2]	1,000–2,000 ft of Overburden[2]	2,000–3,000 ft of Overburden[2]	Total[2]
2.5–5	112.32	109.91	147.41	396.64
5–10	86.04	91.29	79.35	256.68
10	41.98	76.71	280.78	399.47
Total	240.34	277.91	507.54	1,025.79
Reserve base (coal beds 5 ft or greater in thickness, under less than 1,000 ft of cover, only measured and indicated categories of reliability				108.67
Production 1970–present[3]				3.98
Mining losses (equal to production)				3.98
Remaining reserve base				100.71
Remaining recoverable reserve base (assume future mining losses equal future production)				50.36

[1]Includes measured, indicated, and inferred categories of reliability.
[2]Berryhill et al (1950) modified by Thompson and White (1952) and Keefer and Troyer (1964).
[3]Assumes most mining occurred on beds 5 ft or greater in thickness.

Resource Base (millions of tons)[1]		
	Coal Resource[1]	Remaining Reserve Base
Alkali Butte Coal Field	166.74	34.53
Arminto Coal Field	176.86	26.97
Beaver Creek Coal Field	507.47	0
Big Sandy Draw Coal Field	71.83	13.82
Hudson Coal Field	58.97	7.46
Muddy Creek Coal Field	43.56	17.93
Pilot Butte Coal Field	.36	0
	1,025.79	100.71

[1]Includes measured, indicated, and inferred resources for coals greater than 2.5 ft thick between 0–3,000 ft of cover.

- Existence of overpressured zones
- Reports of thick coalbeds lying between 3,000 and 14,000 ft
- Presence of gas and oil in great quantities
- Continuity of major coal-bearing zones.

ACKNOWLEDGMENTS

TRW gratefully acknowledges the assistance of those organizations and persons who provided information, data, and advice during preparation of this report. Among these, special thanks are due to John F. Windolph, Jr. (U.S. Geological Survey, Reston, Virginia); Gary B. Glass and Alan J. VerPloeg (Geological Survey of Wyoming); and J. D. Love (U.S. Geological Survey, Laramie, Wyoming).

In addition, the total cooperation of the U.S. Geological Survey and personnel on three test wells located on the Wind River Indian Reservation is gratefully appreciated.

REFERENCES CITED

Averitt, P., 1975, Coal resources of the United States, January 1, 1974: U.S. Geological Survey Bulletin 1412, 131 p.

Barwin, J. R., 1961, Stratigraphy of the Mesaverde Formation in the southern part of the Wind River Basin, Wyoming, *in* Symposium on Late Cretaceous rocks of Wyoming: Wyoming Geological Association Guidebook, 16th Annual Field Conference, p. 171–179.

Berg, R. R., 1961, Laramide tectonics of the Wind River Mountains, *in* Symposium on Late Cretaceous rocks of Wyoming: Wyoming Geological Association Guidebook, 16th Annual Field Conference, p. 70–80.

Berryhill, Jr., H. L., D. M. Brown, A. Brown, and D. A. Taylor, 1950, Coal resources of Wyoming: U.S. Geological Survey Circular 71, 78 p.

Bilyeu, B. D., 1978, Deep drilling practices—Wind River Basin of Wyoming: Wyoming Geological Association Guidebook, 30th Annual Field Conference, p. 13–24.

Blackstone, Jr., D. L., 1948, The structural pattern of the Wind River Basin, Wyoming, *in* Wind River Basin: Wyoming Geological Association Guidebook, 3rd Annual Field Conference, p. 69–78.

Brenner, R. W., 1972, Petroleum and natural gas: The Wind River Basin, *in* Geological atlas of the Rocky Mountain Region: Denver, Colorado, Rocky Mountain Association of Geologists, p. 273–274.

Campbell, M. R., 1917, The coal fields of the United States: U.S. Geological Survey Professional Paper 100-A, 33 p.

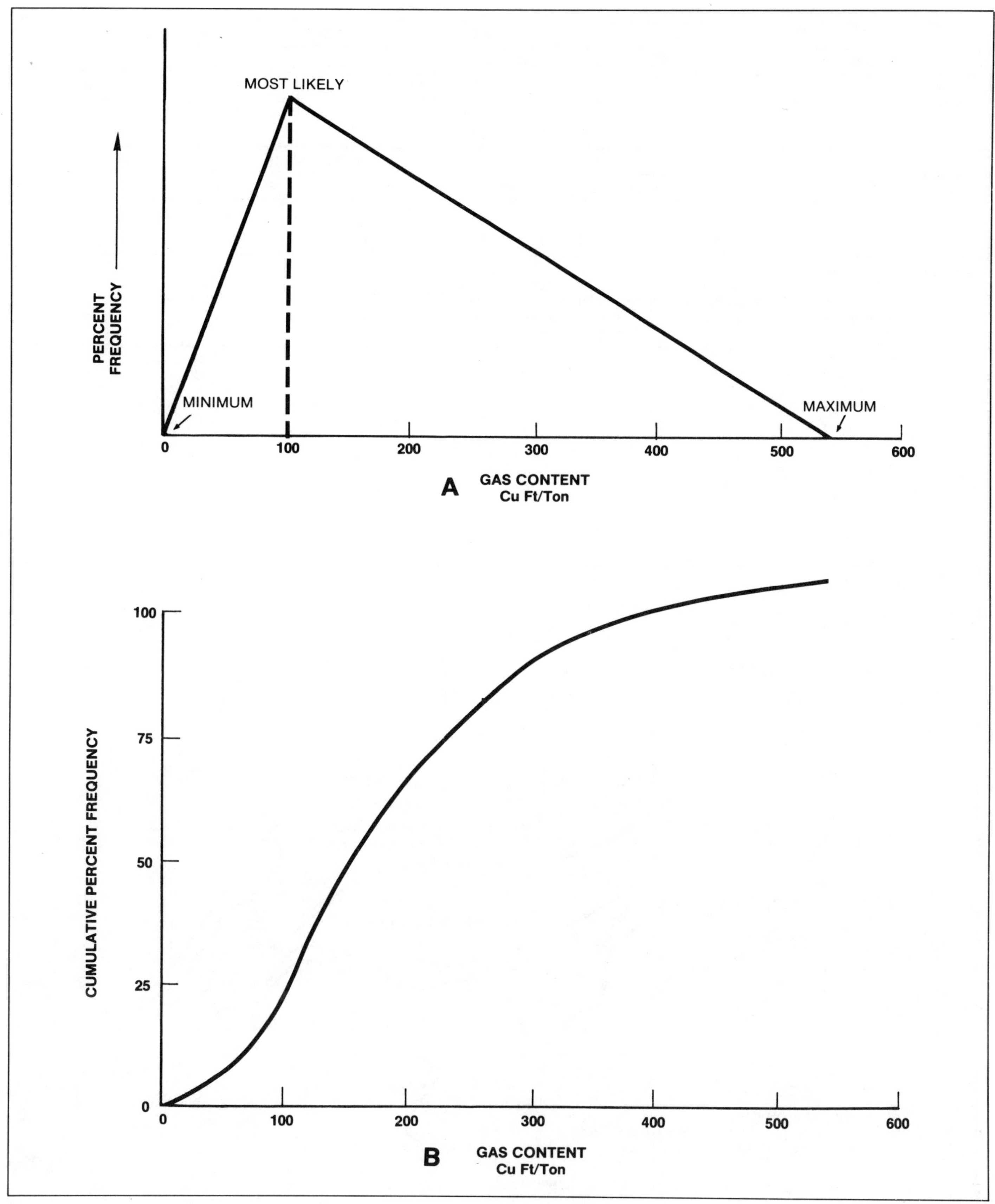

Figure 25—Basis for a statistical speculation of methane gas resources in the Wind River Basin.

Table 12—Estimated coalbed methane resource in the Wind River Coal Basin—two scenarios.

Conservative Scenario (Original Shallow (≤3,000 ft) Coal Resources—See Table 11)					
Coalbeds >2.5 ft	Gas Resource in Cubic Feet				
Short Tons	5 cf/ton	10 cf/ton	20 cf/ton	50 cf/ton	100 cf/ton
1.03×10^9 2.06×10^9	5.15×10^9 10.30×10^9	10.03×10^9 20.60×10^9	20.06×10^9 40.12×10^9	51.50×10^9 103×10^9	103×10^9 206×10^9
Optimistic Scenario {Original Deep (≥ 3,000 ft) Coal Resources}					
Coalbeds >2.5 ft	Gas Resource in Cubic Feet				
Short Tons	5 cf/ton	100 cf/ton	121 cf/ton	273 cf/ton	539 cf/ton
4.23×10^9	0.02×10^{12}	0.41×10^{12}	0.50×10^{12}	1.12×10^{12}	2.22×10^{12}

Assumptions: Shallow coalbeds maximum original coal resources is twice report for the whole basin.
Maximum desorbed gas value is the maximum value assigned to subbituminous coals in the Green River Basin.
Deep coal beds are assumed to be 2 times that of the shallow beds, based on reported very deep lying thick coals.
Maximum desorbed gas value corresponds to that for deep lying Green River coals.
Mostly value is 273 cf/ton derived from Green River data.

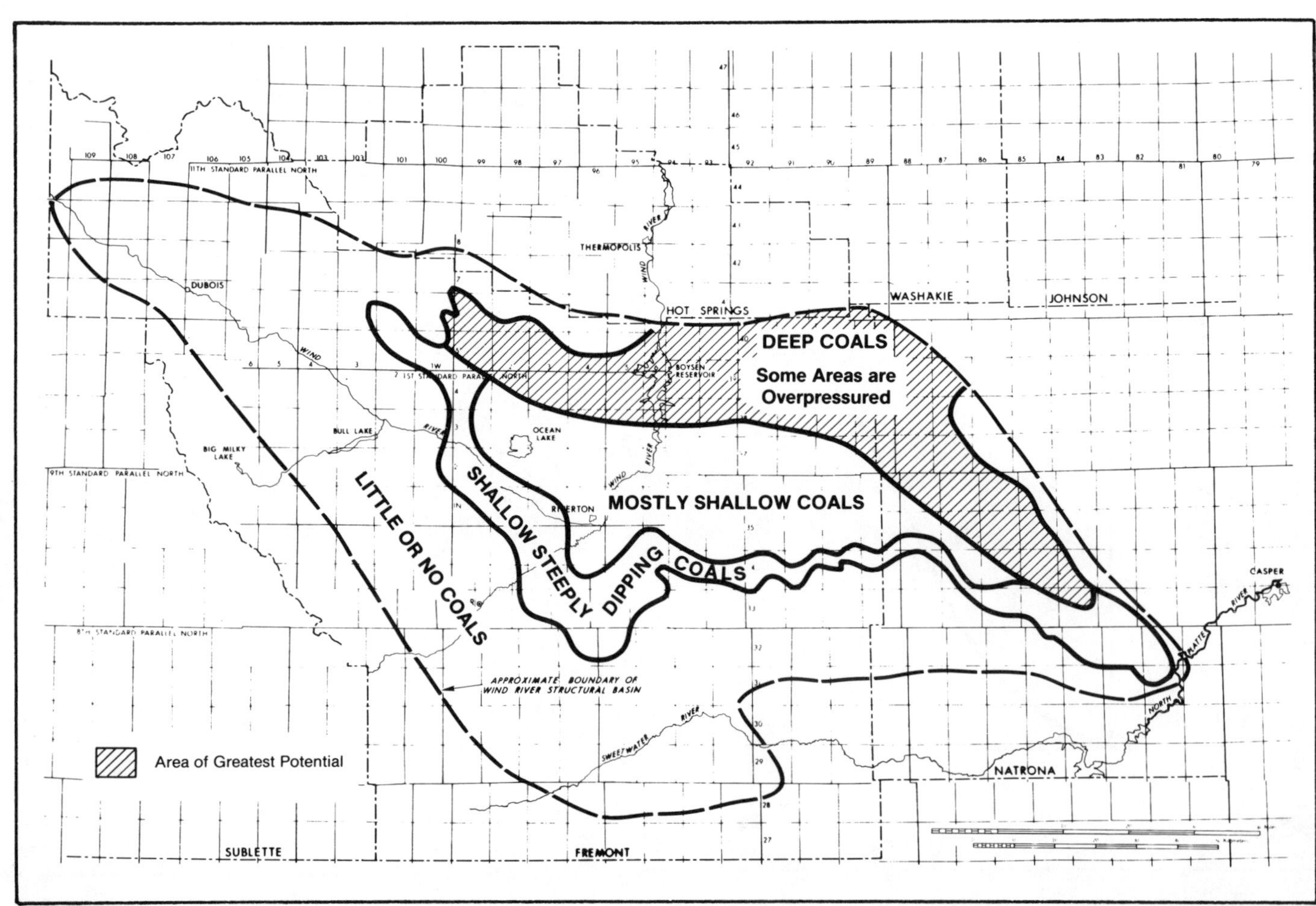

Figure 26—Primary target areas in the Wind River Basin.

Glass, G. B., 1975, Analyses and measured sections of 54 Wyoming coal samples (collected in 1974): Geological Survey of Wyoming Report of Investigations No. 11.
——— , 1978, Wyoming coal fields: Geological Survey of Wyoming Public Information Circular No. 9.
——— , and J. T. Roberts, 1978, Update on the Wind River Basin, *in* Resources of the Wind River: Wyoming Geological Association Guidebook, 30th Annual Field Conference, 414 p.
Hyne, N. J., W. A. Cooper, and P. A. Dickey, 1979, Stratigraphy of intermontane, lacustrine delta, Catatumbo River, Lake Maracaibo, Venezuela: Bulletin of the American Association of Petroleum Geologists, v. 63, n. 11, p. 2042–2057.
Keefer, W. R., 1957, Geology of the Du Noir area, Fremont County, Wyoming: U.S. Geological Survey Professional Paper 294-E, p. 155–221.
——— , 1965, Geologic history of the Wind River Basin, central Wyoming: Bulletin of the American Association of Petroleum Geologists, v. 49, n. 11, p. 1878–1892.
——— , 1970, Structural geology of the Wind River Basin, Wyoming: U.S. Geological Survey Professional Paper 495-D, 35 p.
——— , 1972, Frontier, Cody, and Mesaverde Formations in the Wind River and southern Bighorn Basins, Wyoming: U.S. Geological Survey Professional Paper 495-E, 23 p.
——— , and E.I. Rich, 1957, Stratigraphy of the Cody shale and younger Cretaceous and Paleocene rocks in the western and southern parts of the Wind River Basin, Wyoming, *in* Southwest Wind River Basin: Wyoming Geological Association Guidebook, 12th Annual Field Conference, p. 71–78.
——— , and M. L. Troyer, 1964, Geology of the Shotgun Butte area Fremont County, Wyoming: U.S. Geological Survey Bulletin 1157, 123 p.
——— , and J. A. Van Lieu, 1966, Paleozoic Formations in the Wind River Basin, Wyoming: U.S. Geological Survey Professional Paper 495-B, 60 p.
Lane, D. W., 1973, The Phosphoria and Goose Egg Formations in Wyoming: Geological Survey of Wyoming Preliminary Report No. 12, 7 p.
Law, B. E., C. W. Spencer, and N. H. Bostick, 1980, Evaluation of organic matter, subsurface temperature, and pressure with regard to gas generation in low-permeability Upper Cretaceous and Lower Tertiary sandstones in Pacific Creek area, Sublette and Sweetwater counties, Wyoming: The Mountain Geologist, v. 17, n. 2, p. 23–35.
Lord, N. W., J. A. Holmes, F. M. Stanton, A. C. Fieldner, and S. Sanford, 1913, Analyses of coals in the United States: U.S. Bureau of Mines Bulletin 22, 321 p.
Love, J. D., 1939, Geology along the southern margin of the Absaroka Range, Wyoming: Geological Society of America Special Paper 20, 134 p.
McCord, J. P., 1980, Geologic overview, coal and coalbed methane resources of the greater Green River coal region, Wyoming and Colorado: TRW Inc.
McGreevy, L. H., W. G. Hodson, and S. J. Rucker, IV, 1969, Ground-water resources of the Wind River Indian Reservation, Wyoming: U.S. Geological Survey Water-Supply Paper 1576-I, 145 p.
McKelvey, V. E., 1959, The Phosphoria, Park City, and Shedhorn Formations in the western phosphate field: U.S. Geological Survey Professional Paper 313-A, p. 1–47.
Megill, R. E., 1977, An introduction to risk analysis: Tulsa, Oklahoma, Petroleum Publishing Co., 199 p.
Miller, D. N., and A. J. VerPloeg, 1980, Tight gas sand inventory of Wyoming: Geological Survey of Wyoming Open-File Report WGS81-1, 19 p.
Paape, D. W., 1961, Tectonics and its influence on Late Cretaceous and Tertiary sedimentation, Wind River Basin, Wyoming, *in* Symposium on Late Cretaceous rocks of Wyoming: Wyoming Geological Association Guidebook, 16th Annual Field Conference, p. 187–194.
——— , 1968, Geology of Wind River Basin of Wyoming and its relationship to natural gas accumulation: American Association of Petroleum Geologists Memoir 9, p. 760–779.
Picard, M. D., 1978, Stratigraphy of Triassic rocks in west-central Wyoming, *in* Resources of the Wind River: Wyoming Geological Association Guidebook, 30th Annual Field Conference, 414 p.
Rich, E. J., 1958, Stratigraphic relation of Late Cretaceous rocks in parts of Powder River, Wind River and Big Horn basins, Wyoming: Bulletin of the American Association of Petroleum Geologists, v. 42, n. 10, p. 2424–2444.
Schilt, S., J. Oliver, L. Brown, S. Kaufman, D. Albaugh, J. Brewer, F. Cook, L. Jensen, P. Krumhansl, G. Long, and D. Steiner, 1979, The heterogeneity of the continental crust: Results from deep crustal seismic reflection profiling using the Vibrosis technique: Review of Geophysics and Space Physics, v. 17, n. 2, p. 354–368.
Shapurji, S. S., 1978, Depositional environments and correlation of the Mesaverde Formation, Wind River Basin, *in* Resources of the Wind River: Wyoming Geological Association Guidebook, 30th Annual Field Conference, 414 p.
Shelton, J. W., and J. W. Ross, 1971, Graphic analysis of basins and uplifts: The Mountain Geologist, v. 8, n. 4, p. 183–187.
Smithson, S. B., and R. J. Ebens, 1971, Interpretation of data from a 3.05 KM borehole in Precambrian crystalline rocks, Wind River Mountains, Wyoming: Journal of Geophysical Research, v. 76, p. 7079–7087.
Thomas, H. D., 1948, Summary of Paleozoic stratigraphy of the Wind River Basin, Wyoming, *in* Wind River Basin: Wyoming Geological Association Guidebook, 3rd Annual Field Conference, p. 79–95.
Thomas, G. E., 1971, Continental plate tectonics southwest Wyoming: Wyoming Geological Association Guidebook, 23rd Annual Field Conference, p. 103–123.
Thompson, R. M., and V. L. White, 1952, Coal deposits of Alkali Butte, Big Sand Draw, and Beaver Creek fields, Fremont County, Wyoming: U.S. Geological Survey Oil and Gas Investigations Chart 36.
——— , J. D. Love, and H. A. Tourtelot, 1949, Stratigraphic sections of pre-Cody Cretaceous rocks in central Wyoming: U.S. Geological Survey Circular 152, 24 p.
Unfer, Jr., L., 1951, Study of factors influencing the rank of Wyoming coals: Unpublished Master's Thesis, University of Wyoming, Laramie, 54 p.
Weimer, R. J., 1961, Uppermost Cretaceous rocks in central and southern Wyoming, and northwest Colorado, *in* Symposium on Late Cretaceous rocks of Wyoming: Wyoming Geological Association Guidebook, 16th Annual Field Conference, p. 17–28.
Windolph, Jr., J. R., N. L. Hickling, and R. C. Warlow, 1980,

Report of 1978-1979 field investigations of coal resources in the Wind River Indian Reservation, Fremont County, Wyoming: U.S. Geological Survey Open-File Report 80-674, 12 p.

Woodruff, E. G., and D. E. Winchester, 1912, Coal fields of the Wind River region, Fremont and Natrona counties, Wyoming: U.S. Geological Survey Bulletin 471-G, p. 516–564.

Geologic Overview, Coal Deposits, and Potential for Methane Recovery from Coalbeds—Powder River Basin

R. Choate
C. A. Johnson
J. P. McCord

The Powder River Basin contains the most important coal deposits in the United States. Though the unit gas concentration of Powder River Basin coals is uniformly low, the sheer magnitude of the coal resource ensures a large potential gas resource.

INTRODUCTION

Powder River Basin is an intermontane structural basin lying between the Black Hills on the east and the Bighorn Mountains on the west (Fig. 1). Basin area is approximately 25,800 sq mi. Bounding the basin on the north, in Montana, is the Yellowstone River, and on the south, in Wyoming, is the North Platte River.

Major structural features of the basin were formed during the Laramide orogeny of Late Cretaceous and Early Tertiary time. The deeply synclinal structure is not reflected in the topography. Most topographic features are a result of differences in erosional characteristics of the Tertiary rocks, stream erosion, and subsidence because of the burning of coal seams. The smaller stream channels are parallel to a N 35° W fracture system resulting in a northwest-trending grain in the topography (Mapel, 1959). Many small streams have carved the soft shale and sandstone into local badlands. The natural burning of coal seams has formed many resistant baked rock and red clinker layers that cap ridges and divides. Average surface elevation in Powder River Basin is 5,000 ft and ranges from 3,700 ft in the valley of the Powder River to 5,700 ft in the foothills of the Bighorn Mountains. The summit of Cloud Peak in the Bighorn Mountains is at 13,175 ft.

Powder River Basin lies within a semiarid region of pronounced seasonal variation in temperature and precipitation. Vegetation is sparse (in places, barren) and consists of grass, sagebrush, and prickly pear, with willows and cottonwoods along the valley bottoms. The annual rate of evapotranspiration exceeds the area's rainfall, resulting in drought conditions and a short growing season. Irrigation is therefore necessary to revegetate strip-mined areas (Toy and Munson, 1978).

GEOLOGY

Basin Structure

Powder River Basin is an elongate northerly trending structural basin lying along the eastern front of the Middle Rocky Mountains (Fig. 2). When considering the entire sequence of sedimentary rocks deeply buried in the basin, it is probably most accurately visualized as a broad, flat-bottomed, but slightly tilted, structural trough downwarped between the Bighorn Uplift on the west and the Black Hills Uplift on the east (Fig. 3).

Powder River Basin is only one of a line of basins that lie along the entire front of the Rocky Mountains, extending from northern New Mexico to central Montana. In general, these basins—especially Denver Basin to the south, which is separated from Powder River Basin by the Hartville Uplift—have similar asymmetries, in that their structural axes lie very close to the Rocky Mountain front. Of these basins, Powder River is the deepest, containing in excess of 18,000 ft of sedimentary rocks. At Buffalo, Wyoming, structural relief of the Precambrian surface across the mountain-basin interface is in excess of 25,000 ft. This relief occurs in a horizontal distance of only 25 mi, extending from a height of 13,175 ft above sea level at Cloud Peak in the Bighorns to a depth greater than 12,000 ft below sea level at the basin axis.

Basin Stratigraphy

Sedimentary rocks within Powder River Basin include a thick sequence of Paleozoic and Mesozoic rocks mostly of marine origin and a thinner sequence of Late Cretaceous and Cenozoic rocks of continental origin. Figure 4 is a representative stratigraphic column for Powder River Basin. Only the Late

Figure 1—Index maps showing location of Powder River Basin and showing Powder River Basin in relation to geographic and cultural features.

Cretaceous and Cenozoic units that include the primary economic coal-bearing strata will be discussed below.

Coal-Bearing Strata

The initial continental deposits of Late Cretaceous age are called the Lance Formation in Wyoming and the Hell Creek Formation in southeast Montana. These two formations consist of alternating, massively bedded sandstone and dark-colored claystone and shale. The formations thicken southward from 500 to 600 ft in Big Horn County, Montana, to 2,500 ft in Converse County, Wyoming. Although evidence of Laramide deformation is present in rocks of this age in other areas, there is no evidence in Powder River Basin of the Laramide Orogeny during Lance time. The basal part of the Lance Formation contains small, local beds of impure coal in the southwestern part of the Powder River Basin. Some of the coal seams are reported to average 3 to 6 ft in thickness (Mapel, 1958; Curry, 1971; Glass, 1976).

Early Tertiary time began with a large influx of sediment into flood plains, estuaries, and swamps of the newly formed Powder River Basin. The open sea was to the east and northeast. The stratigraphic position of the Cretaceous-Tertiary boundary has been disputed over the years. It is now generally accepted, however, that Tertiary rocks begin with the "lowest persistent bed of lignite" found directly above the latest dinosaur remains (Calvert, 1912; Brown, 1958; Matson and Pinchock, 1976). Tertiary rocks covering most of Powder River Basin are the Fort Union Formation of Paleocene age and the Wasatch Formation of Eocene age. Fort Union and Wasatch time was characterized by cyclic deposition in a near-shore environment that was periodically undergoing uplift and subsidence. During periods of stability, extensive coal-forming swamps developed.

The Fort Union crops out in a band around the margin of the basin (Fig. 5), averaging 60 mi wide on the north, 25 mi on the east, and 5 mi on the south and west sides. Structural deformation and its effects on Eocene deposition are responsible for limited exposures on the west side of the basin along the Bighorn Mountains (Obernyer, 1979; Mapel, 1958). The Fort Union, about 2,100 ft thick in the east, has its maximum thickness of 5,215 ft at the center of the basin (Curry, 1971; Grazis, 1977). Dips are 10 to 25° near the Bighorns and otherwise relatively flat, averaging 2 to 3° (Mapel, 1958).

The Fort Union is subdivided into three members. In ascending order, they are the Tullock, Lebo Shale, and Tongue River. The Tullock Member consists of friable, light-colored sandstone; dark- to light-gray shale; and thin beds of coal. It is as much as 1,100 ft thick in the southern part of the basin and thins northward to about 200 ft to 250 ft in Montana. The

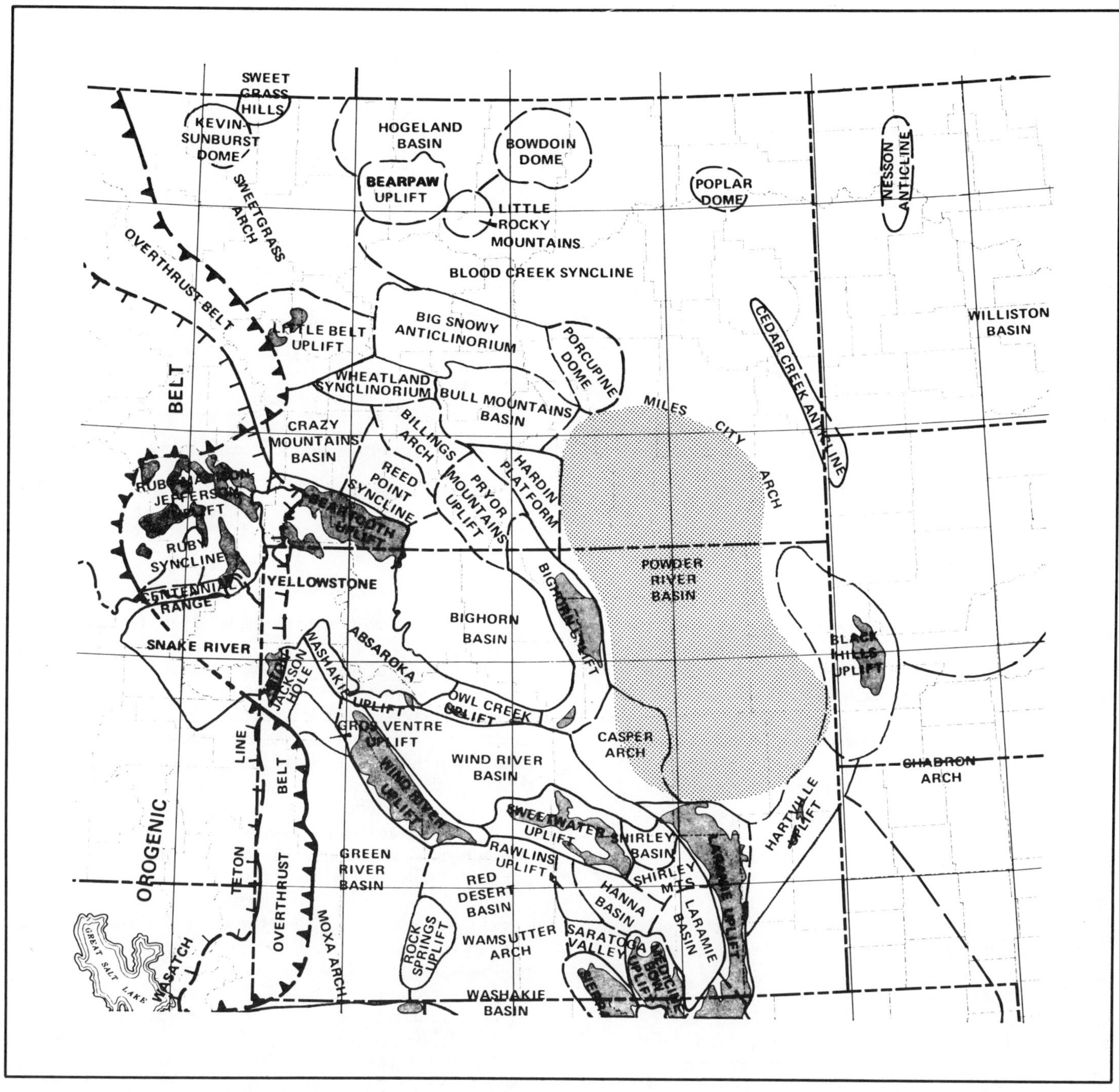

Figure 2—Major tectonic elements associated with development of Powder River Basin. Modified from Grose. Copyright © 1972. The Rocky Mountain Association of Geologists. Used with permission.

overlying Lebo Shale Member is 150 to 400 ft thick and consists mostly of soft clay-like shale and thin beds of sandstone and siltstone that locally contain ferruginous concretions and thin beds of coal. The Tongue River Member consists of 2,000 ft of yellow-weathering sandstone and coal in Montana that thin southward to about 800 ft at Sheridan and 600 ft near Gillette (Mapel, 1958). The most persistent coalbeds in the basin occur in the Tongue River Member of the Fort Union. The largest Fort Union coal seam, the Wyodak-Anderson, commonly ranges from 50 to 100 ft thick (Glass, 1976).

The contact of the Fort Union Formation of Paleocene age and the overlying Wasatch Formation of Eocene age is still controversial. In some areas, the contact is a recognizable unconformity, but in other areas the contact is conformable and

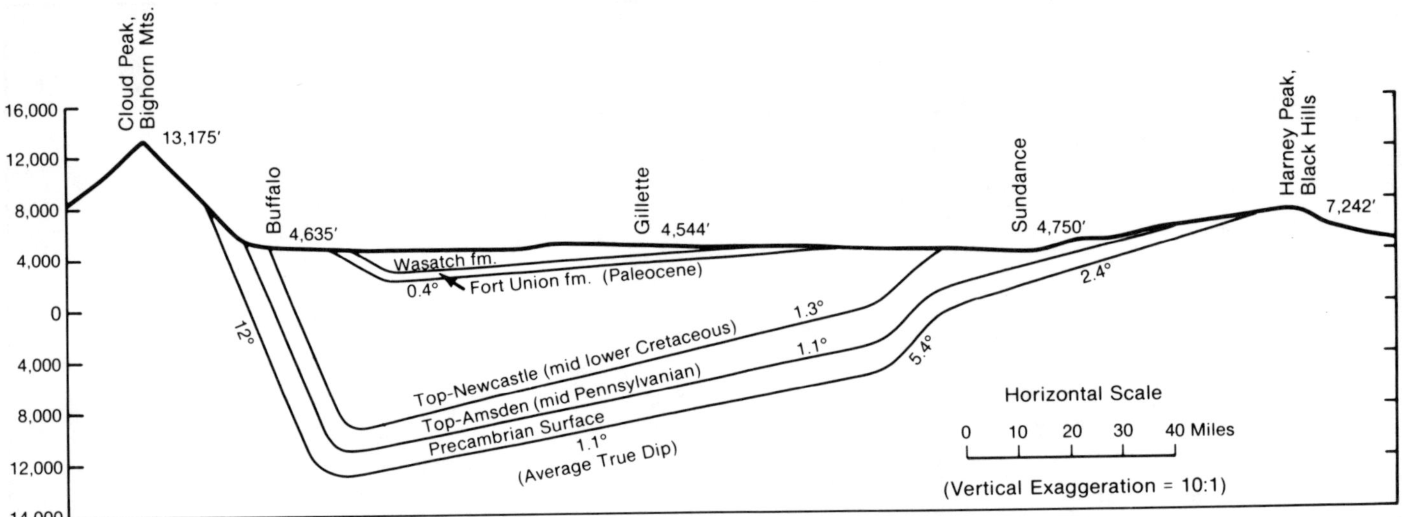

Figure 3—East-west section across Powder River Basin showing slightly tilted, trough-like structure in older sedimentary rocks.

not easily defined even after detailed study. The Wasatch is 1,000 to 2,000 ft thick and is lithologically similar to the Tongue River Member of the Fort Union. It is composed of thin lenticular sandstone, gray claystone, carbonaceous shale, and coal, and is drab-yellow to light-gray in outcrop. The Wasatch is "another prolific coal-bearing formation" (Glass, 1976). It is reported to contain the thickest coalbed in the nation. The Lake De Smet coalbed in the Lake De Smet area is reported to be as much as 220 ft thick, locally.

COAL RESOURCES

Regional Setting

Powder River Basin contains the most important coal deposits in the United States. Because of their size and accessibility, these deposits could make up the entire near- and intermediate-term energy shortfall of the United States. The most recent coal resource estimate for the basin, utilizing available subsurface data, is 1.3 trillion tons. Most of this resource is in thick beds that are relatively near-surface, with most coalbeds at a depth of 2,500 ft or less, even in the basin center (Schell, USGS, 1976, oral communication).

Most of the earlier coal-resource calculations were restricted to outcrop areas along strike of the coalbeds and for relatively short distances downdip. Until recently, little was known about the extent of the coalbeds downdip to the center of Powder River Basin. For that reason, early calculations of coal-resources grossly understated the coal that almost certainly exists (Averitt, 1975).

Subsurface logs indicate that strata in the basin contain five major coal zones, which generally are concentrated in a vertical section of 1,000 to 1,200 ft. In addition, these studies have indicated that probably 50% or more of the total coal in the basin would be in beds 15 to 50 ft thick, and 30 to 40% in beds greater than 50 ft thick. As yet, no detailed evaluation has been made by the USGS on an individual bed basis (Rupert et al, 1975).

Coal in Powder River Basin ranges in rank from lignite A through subbituminous A. As-received heat values from coals in the Montana portion of the basin range from 6,500 Btu per pound in the Knowlton deposit, just outside the northeast boundary of the basin, to 9,850 Btu in the Decker deposit, on the west side of the basin. In the Wyoming portion of the basin, coals range from 7,550 Btu per pound in Converse County to 9,300 Btu per pound in the Sheridan area. The average of the Wyoming coals is relatively low—8,300 Btu per pound as-received (Glass, 1978; Matson and Blumer, 1973).

Generally, Powder River Basin coals are low in sulfur and have low to moderate ash contents, although the ash content can vary considerably. The coals usually have an as-received moisture content of 20 to 30%, and volatile matter and fixed-carbon contents of 30 to 40%. The coal loses moisture, slacks, and can ignite spontaneously when exposed to air.

Most of the thick coalbeds occur in the upper member—the Tongue River Member—of the Fort Union Formation and the overlying Wasatch Formation. These two sequences reach a maximum thickness of 3,970 ft near the Buffalo-Lake De Smet area (Curry, 1971); however, most of the thick coalbeds range from surface exposures to less than 2,500 ft in depth. Coalbeds do occur in the lower members of the Fort Union Formation, as well as the Lance and Mesaverde Formations, but they are generally thinner and less continuous. Some of these older coalbeds crop out in the southwestern portion of Powder River Basin, near Glenrock and Douglas.

The total number of coalbeds in the basin is difficult to determine because the beds split, coalesce, and are commonly discontinuous, with beds pinching out and new beds appearing. In the Sheridan area of Wyoming, as many as 11 persistent coalbeds occur in the Wasatch Formation. In other areas, as many as 12 to 18 coalbeds occur, most of them within the Fort

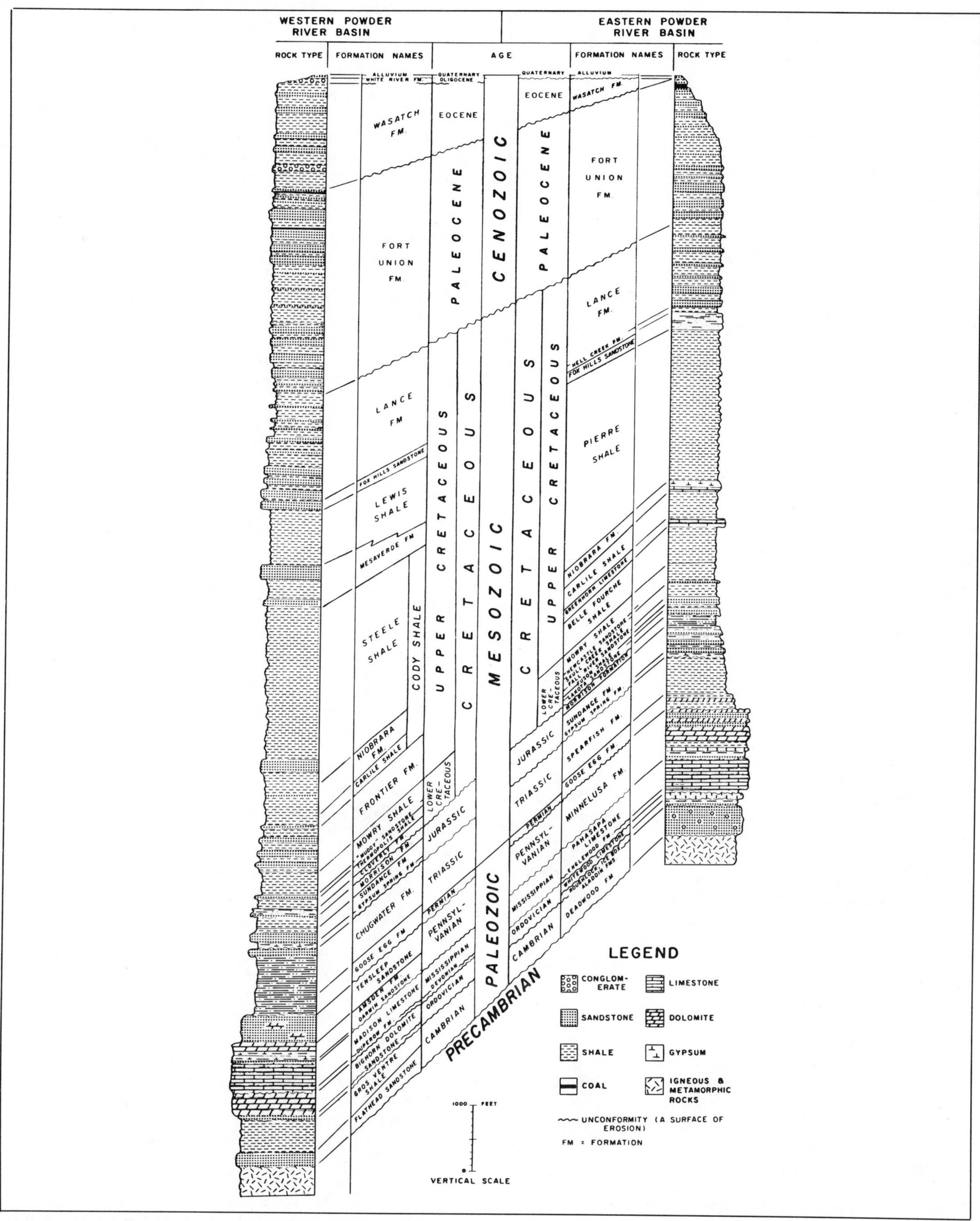

Figure 4—Representative stratigraphic columns for the Powder River Basin, Wyoming. Breckenridge et al, 1974. Used with the permission of the Geological Survey of Wyoming.

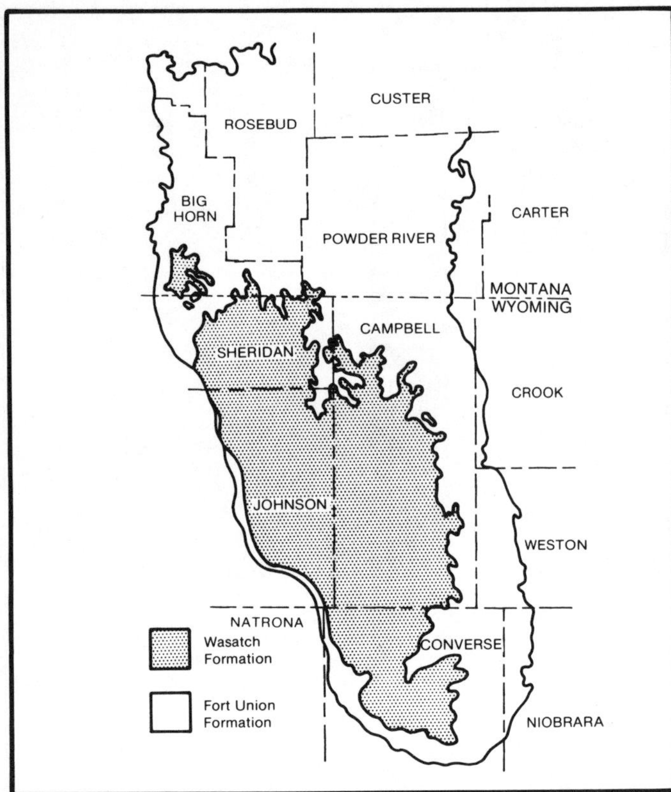

Figure 5—Map showing extent of the Wasatch Formation of Eocene age and the Fort Union Formation of Paleocene age. Modified from Mapel. Copyright © 1958 Wyoming Geological Association. Used with permission.

Union Formation. Most of the Wasatch beds occur under less than 200 ft of overburden.

Two of the largest coalbeds in the basin are the Wyodak-Anderson and the Lake De Smet bed. The Wyodak-Anderson coalbed crops out over a north-south distance of 120 mi in the Gillette Coal Field. It, and the beds correlative to it, persist downdip to the deepest part of the basin. The Wyodak-Anderson bed is locally up to 150 ft thick but averages 50 to 100 ft. Based on these figures, the bed contains at least 100 billion tons of coal to a depth of 2,000 ft. This is the largest tonnage in a single continuous coalbed anywhere in the United States (Averitt, 1975).

The Lake De Smet bed, which occurs in the Buffalo Coal Field, is thought to be the thickest coalbed in the United States and second thickest in the world. It is 15 mi in length, 70 to 220 ft thick, and 0.5 to 2 mi wide (Obernyer, 1978).

Coal Resources by Area

Powder River Basin contains 21 coal fields (Fig. 6). In Wyoming, coal field boundaries have been clearly defined. In Montana, however, coal fields and their boundaries are much less clearly defined. The coalbeds in each field are typically given different names, and, until recently, very little correlation between these fields had been attempted.

There are 14 operating coal mines (as of December 1, 1979) in Powder River Basin. There are 20 more in various stages of permit and construction. Most of these mines are in the Gillette and Sheridan areas of Wyoming. There are 19 proposed and existing mines along the north-south trend of the Wyodak-Anderson coalbed in Campbell County.

Eastern Coal Fields

Coal fields grouped into the eastern fields include the Gillette, Pumpkin Buttes, Powder River, Little Powder River, and Spotted Horse (Fig. 6). These fields, mostly in Campbell County, are grouped together because they all lie along the central portion of the very gently dipping eastern limb of the asymmetrical syncline containing the nonmarine, coal-bearing strata of the Fort Union and Wasatch Formations. Regional dip of these beds is only 0.4° to the west, as measured along a 25-mi, east-west section of the Wyodak coalbed (contact between the Fort Union and Wasatch Formations in recent USGS mapping). Because of small folds within Fort Union and Wasatch strata, dips generally in the range of 2 to 4° occur locally.

Within the five coal fields comprising the eastern fields of Powder River Basin, the principal coalbeds include:

- Spotted Horse field (Olive, 1957)
  Eleven named beds: Ulm No. 1, Ulm No. 2, Scott, Felix, Arvada, Roland, Smith, Anderson, Dietz No. 1, Canyon, and Wall
- Pumpkin Buttes field (Wegemann et al, 1928)
  Eight named beds: A through H.
- Gillette field (Dobbin and Barnett, 1927)
  Nineteen named beds, as composited from exposures in the northern, central, and southern parts of the field: A through S.
- Little Powder River field (Davis, 1912)
  Ten named beds: A through J.
- Powder River field (Stone and Lupton, 1910)
  Five named beds: Lower Ulm, Felix, Arvada, Roland, and Smith.

Stratigraphic columns for coalbeds occurring in Campbell County coal fields, recently compiled by the Wyoming and U. S. Geological Surveys, are shown in Figure 7. A fence diagram correlating coalbeds between the eastern and western coal fields is shown in Figure 8. Coal rank generally ranges from lignite to subbituminous C.

Western Coal Fields

The western Powder River Basin region includes the Sheridan, Buffalo, Barber, and Sussex Coal fields (Fig. 6). In this region, as many as 11 persistent coalbeds occur in the Wasatch Formation of Eocene age and as many as 12 coalbeds occur in the upper half of the Fort Union Formation of Paleocene age. Stratigraphic columns for coalbeds in Johnson and Sheridan counties are shown in Figures 9 and 10. Coals also occur in the lower half of the Fort Union but are thinner and less persistent. Very thin and discontinuous coal lenses occur in the Lance Formation of Cretaceous age.

There are three existing coal mines in western Powder River Basin. The Black Mountain Mine is a three-person, locally owned and operated mine near Sheridan. The Ash Creek Mine is a strip mine located 15 mi north of Sheridan. The Bighorn Strip Mine is 8 mi south of Sheridan and produces 3 million tons a year. The Fort Union Monarch and Dietz No. 3 beds

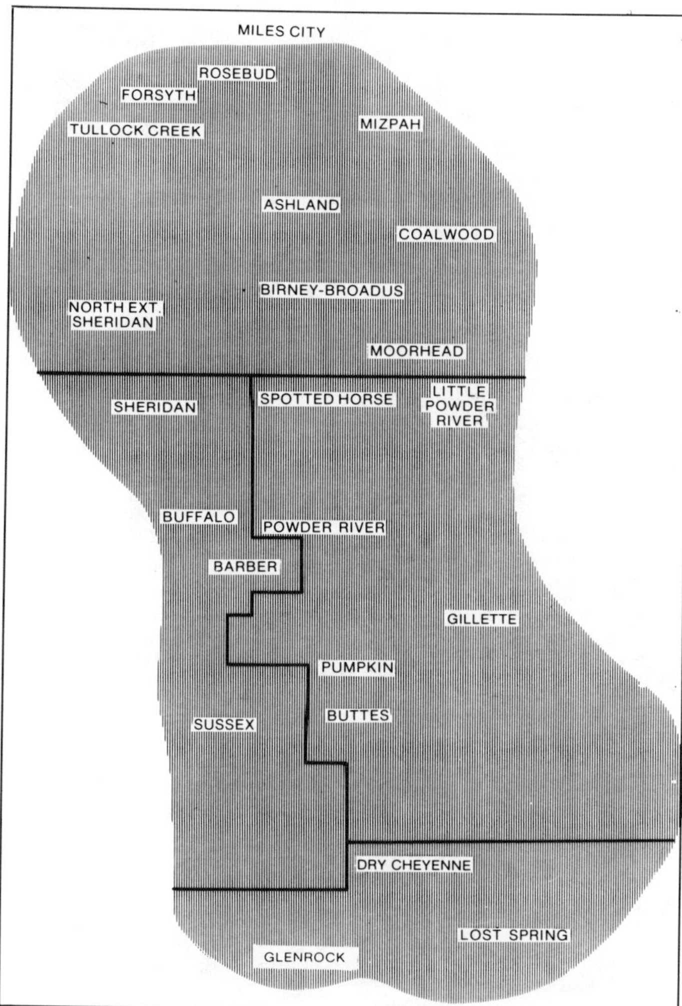

Figure 6—Location of Powder River Basin coal fields.

coalesce at the Big Horn Mine to form a bed 44 ft thick that is being mined together with the overlying 12 ft thick Dietz No. 2 bed. At least three other mines have been proposed for the Sheridan coal field; additionally, Texaco, Inc., has tentative plans for a coal gasification project on the Lake De Smet bed in the Buffalo Field.

The total number of coalbeds in the Wasatch Formation is difficult to determine, because the beds split, coalesce, and are sometimes discontinuous, with beds pinching out and new beds appearing. The coals are generally subbituminous C or in the upper lignite range. Much of the Wasatch coal is under less than 200 ft of overburden, although some may be appreciably deeper. Maximum thickness of the Wasatch Formation is 1,500 ft in the Buffalo-Lake De Smet area. The principal coalbeds of the Wasatch Formation include the Monument Peak, Walters, Healy, Cameron, Murray, Ucross, and Lake De Smet beds.

In addition, four small coal zones occur near the base of the Wasatch Formation in the Sheridan coal field and have been described by Culbertson and Mapel (1976). They are the Bar N, Burgess, Heppner, and Wyarno.

The coalbeds of the Fort Union Formation are deeply buried in the Buffalo and Barber Coal Fields and have not been individually identified in these fields. A few small beds near the top of the Fort Union are identified in the Sussex Field to the south, where they are designated by the letters "A" through "H." These are thought to occur in the stratigraphic interval between the Roland and Anderson coalbeds, although they have not been correlated positively. In the northern part of the region, in the Sheridan Field, several coalbeds are described by Taff (1909) in the uppermost 2,000 ft of the Fort Union Formation. Some of these Fort Union coals are identified in recent maps, coal sections, and electric log correlations. Fort Union coal is generally subbituminous C in rank.

Northern Coal Fields

The northern Powder River region includes all of the Montana of Powder River Basin and contains nine coal fields (Fig. 6). They are the Rosebud, Forsyth, Tullock Creek, Mizpah, Ashland, Coalwood, Birney Broadus, Moorhead, and Sheridan Northern Extension. Coal is present in all three members of the Fort Union Formation, as well as the overlying Wasatch Formation; however, all of the important coalbeds occur in the

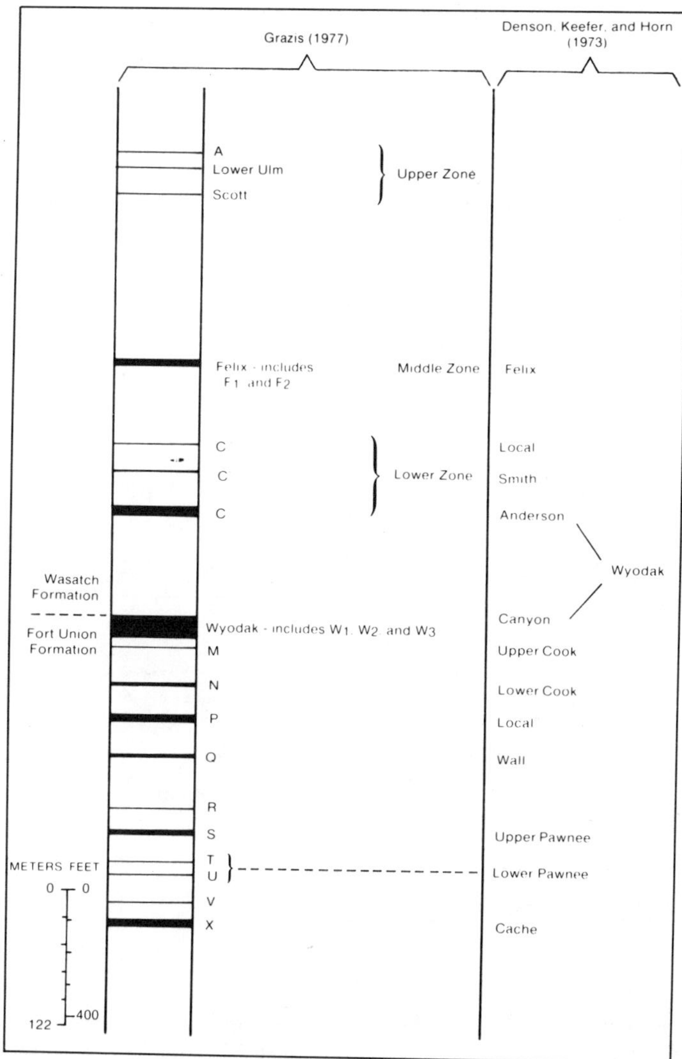

Figure 7—Stratigraphic columns of Fort Union and Wasatch coalbeds in Campbell County as compiled from drill hole geophysical logs by the U.S. Geological Survey. Difficulties in correlation of beds and placement of the Fort Union-Wasatch contact are demonstrated.

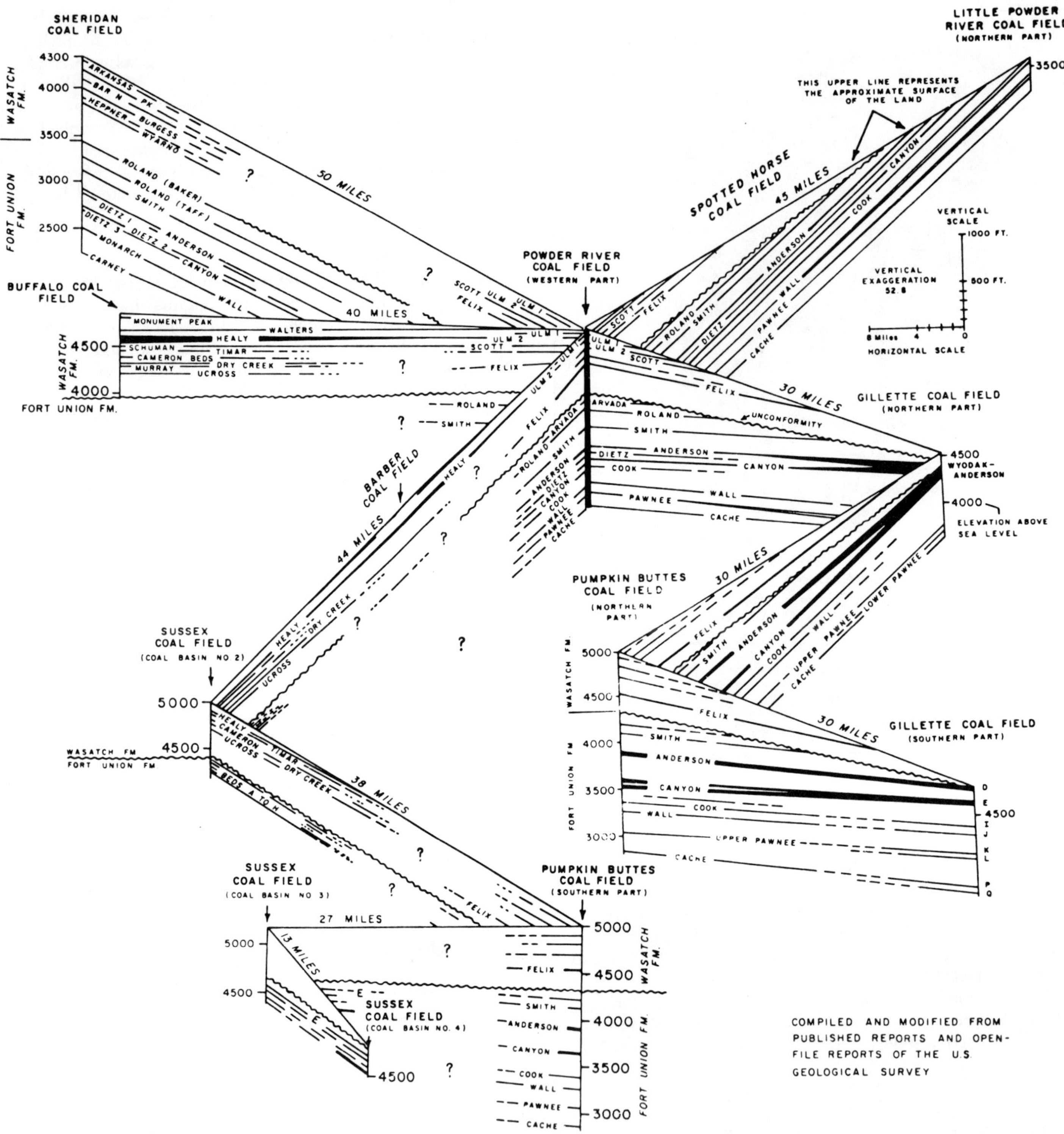

Figure 8—Fence diagram showing stratigraphic correlations of coalbeds between major eastern and western Powder River Basin coal fields. From Glass. Copyright © 1976 Wyoming Geological Association. Used with permission.

Tongue River Member of the Fort Union Formation. Figure 11 is a stratigraphic diagram showing the distribution and thickness of the important coalbeds. There are six coalbeds considered most important in respect to the quality and magnitude of the coal reserves they contain: the Anderson, Dietz, Canyon, Wall, Knobloch, and Rosebud. Numerous smaller coalbeds are present locally. The thickest deposits and best-quality coal are found in the Decker area.

Coal in the northern Powder River Region increases in rank from lignite A in the northeastern part of the region to subbituminous southwestward. The coal is noncoking and nonagglomerating, with low ash and low sulfur contents.

Southern Coal Fields

Three coal fields comprise the southern portion of Powder River Basin: the Lost Spring, Dry Cheyenne, and Glenrock fields (Fig. 6). The Dave Johnston Mine is 14 mi north of Glenrock, in the Glenrock field. Coal-bearing rocks of this area include the Fort Union and Wasatch Formations of Tertiary age and the marine Lance and Mesaverde Formations of Cretaceous age. The coal-bearing rocks dip basinward, 15 to 25° around the periphery of the basin, and flatten to 1 to 5° within a few miles of the basin margin. Most of the coal zones have not been correlated positively with those of other fields to the north, because there has been very little recent geologic work performed in the area. All the coal is generally subbituminous in rank. However, Mesaverde Formation coal is described as blacker, cleaner, more brittle, and more lustrous than the overlying Lance and Tertiary coal.

POTENTIAL METHANE RESOURCE

Methane Encountered in Shallow Drill Holes

Fifteen shallow drill holes or wells (Fig. 12) have been identified to date (1979) in Campbell County that contain anomalously large amounts of methane in product fluids. These holes were drilled either as water wells for local use or as shallow investigations of near-surface coalbeds (<500′ depth); none was drilled for the much deeper lying oil and gas deposits of the basin. Fourteen of these wells are in the Recluse area of northeastern Campbell County and define a northwest-southeast–trending elliptical area approximately 33 mi long and 16 mi wide. The fifteenth well is separated from the above group by a distance of 50 mi; it lies 25.5 mi south of Gillette. Following is a brief description of these 15 holes.

Holes 1–7 were all air-drilled by the USGS in July and August of 1975 (Hobbs, 1978) to depths ranging between 400 and 500 ft. All holes were drilled to penetrate the Anderson and Canyon seams (splits of the Wyodak) and the Cook seam that underlies the Canyon. During drilling, methane shows at the surface were monitored with a hand-held methanometer. Methane was encountered in all seven holes, starting in sandstone beds just above the top coalbed and continuing through all coalbeds and interlying sandstone-shale beds, to TD. Methane in the return flow at the surface ranged from as little as 0.1 to 0.2% in shale interbeds, to massive blowouts where gas flow rates were in excess of 1,000,000 cubic feet per day (cf/day).

Hole 6 probably encountered the greatest concentrations of methane of any of the holes. Methane surges occurred when drilling through the Canyon seam, and finally below the seam, at a depth of 325 ft, the hole blew out during core retrieval. Gas flow rate was estimated at 1,000,000 cf/day, with the gas coming from a soft, poorly cemented sandstone. This occurred where methane concentration in the return flow at the surface just prior to the blowout had been measured at 1 to 2%. The blowout continued for about one-half to one hour before water inflow to the hole stopped the gas flow. The hole was capped, and during a drawdown test several months later, all water pumped out was highly charged with methane.

Hole 5 blew out at two different depths: at about 190 ft, below the Anderson seam; and at about 390 ft within the Canyon seam, near its base. Water inflow from the Anderson and Canyon seams, respectively, stopped both blowouts. Gas flow rates were estimated to be between 800,000 and 900,000 cf/day. Especially important in this hole is the fact that the second blowout occurred within a major coal seam (the Canyon).

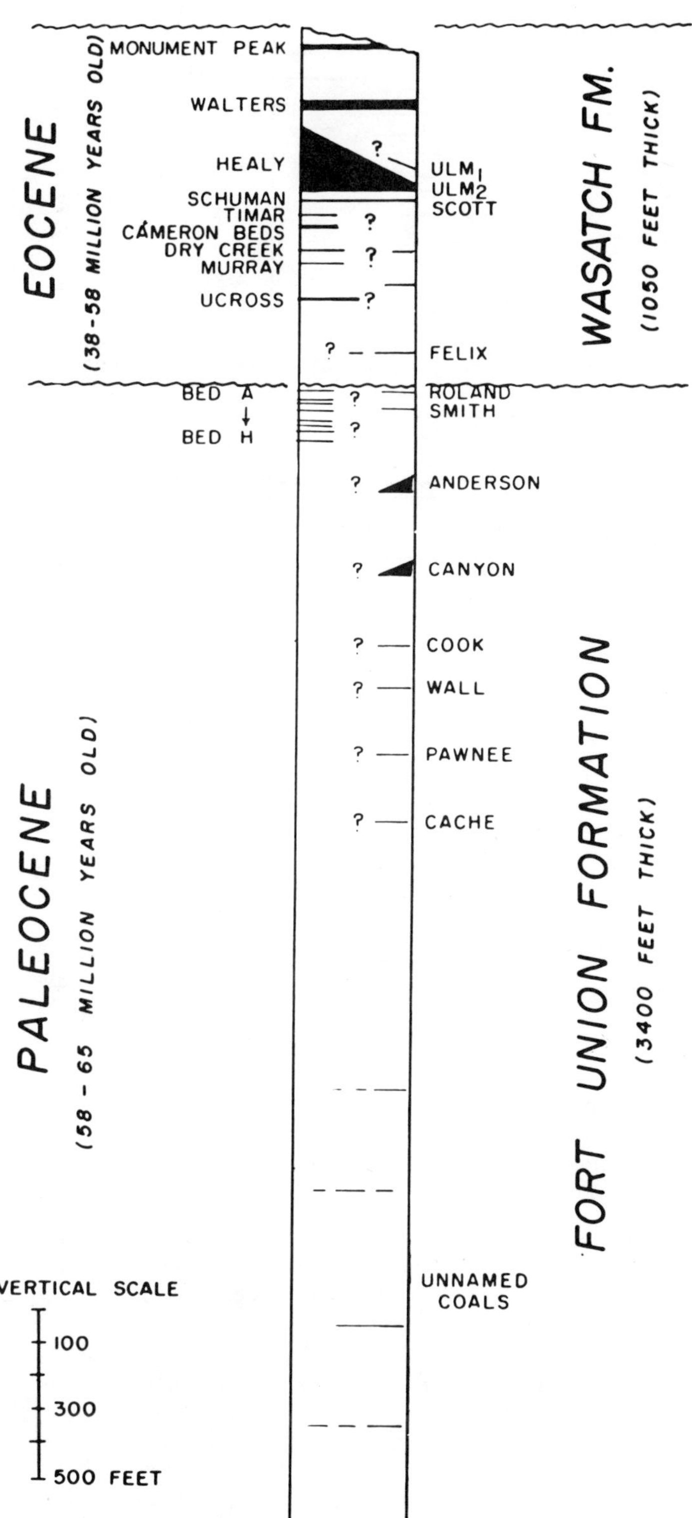

Figure 9—Stratigraphic column showing coalbeds in Johnson County, Wyoming. Wendel et al, 1976. Used with the permission of the Geological Survey of Wyoming.

In Hole 2, methane concentrations up to 1% of the return air

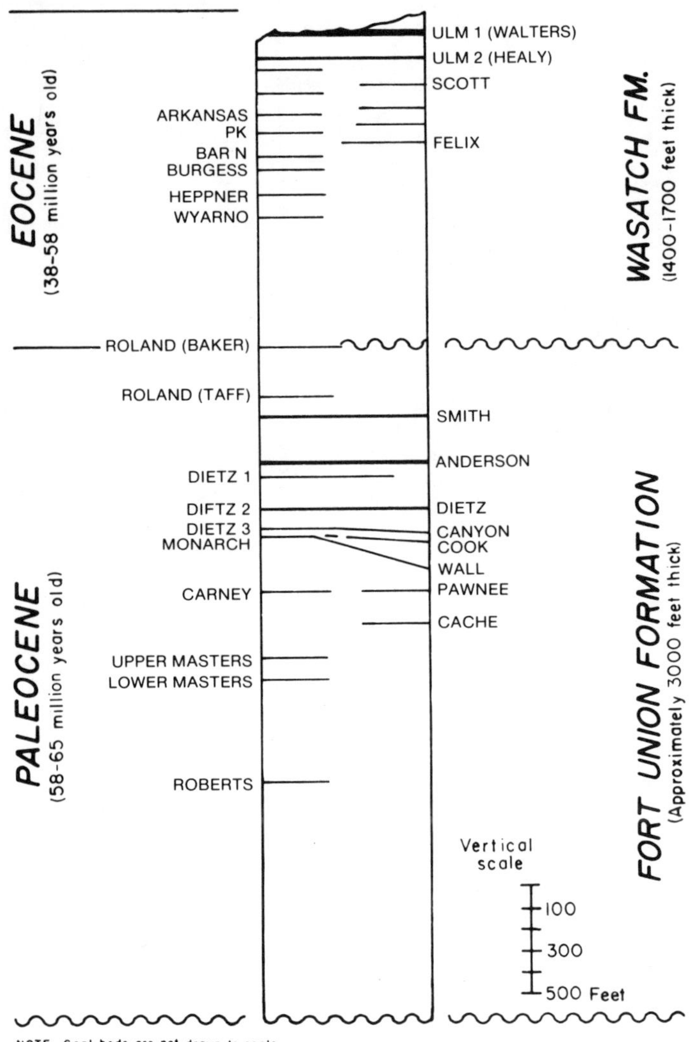

Figure 10—Stratigraphic column showing coalbeds in Sheridan County, Wyoming. Lageson et al, 1978. Used with the permission of the Geological Survey of Wyoming.

were measured in the lower 200 ft of hole. In Hole 1, during geophysical logging, gas flow out of the hole was very low, but methane content of the gas flowing registered the 60% maximum methanometer reading.

Holes 8 through 10 are listed in the report by Olive (1957) on the Spotted Horse coal field as shallow wells in which substantial gas flows were encountered. Though not specifically stated by Olive, it is presumed that these wells were water wells because of their shallow depths—total depths given for two of the wells being 245 and 415 ft. Olive states that ranchers in the area use sealed tanks for capture of the natural gas produced in their water wells; the gas is used for domestic and ranch heating purposes.

Initial gas flow rate of Holes 8, 9, and 10 was 1,000,000, 1,000,000, and 500,000 cf/day from depths of 225, 400, and 265 ft, respectively. Gas associated with other shallow water wells drilled into near-surface coal zones is also reported by Olive to occur in areas adjacent to Powder River and Clear Creek.

Hole 11 is a water well mentioned in Olive's report on the Spotted Horse coal field. This well, in the SE 1/4, Sec. 30, T58N, R75W, was drilled for the Dobrenz ranch in 1916 and since then has been supplying by-product gas to the ranch. Olive reported that gas pressure from the well was sufficient to lift a weight of 600 pounds (there is no mention, however, of what pressure to which this force relates). The Dobrenz ranch house's location is shown on Olive's geologic map of the Spotted Horse field, which places the well in the center of a local structural low.

Hole 12 is a well drilled to provide water for the school at Recluse. The school building is shown on the 7.5-ft Recluse topographic quadrangle map as being 0.5 mi west of the hamlet of Recluse. We have not been able to determine any quantitative data on gas flow from the well, only that substantial gas flow was encountered. It was drilled by a local water-well driller, and no geotechnical people were on site at the time.

Hole 13 is reported to have hit gas in or adjacent to the Canyon bed. This hole, drilled by local drillers with no geotechnical personnel in attendance, is about 7 mi northwest of Hole 6. The hole blew out, forming a surface crater 6 to 8 ft in diameter. Gas flow was so great that the only way it could be stopped was by pumping cement into the hole.

Hole 14 is a water well in Section 7, T55N, R73W, where a large gas flow was encountered at 90 to 100 ft depth in a sandstone bed between the Smith and Anderson coalbeds.

Hole 15 is the USGS Hole No. US-754 (drilled in 1975). Matson (Montana Geological Survey, 1980, personal communication) states that this hole was drilled with the intent of reaching the lower split of the Wyodak seam (the Canyon bed) but that they hit gas at 280 ft and were only able to drill another 10 ft and had to stop because of the large gas flow that blew the hole out. The hole had intersected coal at 100 ft (5 ft of coal) and at 180 ft (32 ft of coal—probably the upper split of the Wyodak, i.e., the Anderson seam). Hole location was the W 1/2, NW 1/4, Sec. 24, T46N, R71W (Neil Butte 7.5-ft quadrangle).

Methane in Artesian Wells

A large number of flowing artesian wells occur within the Powder River Basin. A number of these wells are known to contain substantial amounts of natural gas. Origin of this gas is the coalbeds and carbonaceous zones in the Fort Union Formation (Lowry and Cummings, 1966; Whitcomb et al, 1966).

It was shown by Meinzer (1942) that gas in an aquifer increases the height to which water rises in a well and can create a flowing artesian well by raising the water to the surface. This lifting action is created by expansion of the gas as pressure is reduced in the water flowing from the aquifer. It is stated by Lowry and Cummings (1966) and Whitcomb et al (1966) that, because of this lifting action, some portion of the artesian wells in the basin flow that would not otherwise do so.

Assuming that some significant—but as yet unknown—portion of the flowing artesian wells are flowing because they are tapping coalbeds with anomalously high methane content, such artesian wells might be a useful exploration aid. Hydrologic maps covering the Powder River Basin show locations of at least 198 flowing artesian wells within the boundary of the Fort Union Formation. Distribution of these 198 wells is shown on the map of Figure 13. This map indicates a nonrandom distribution of the wells, with 68 in the eastern half of Sheridan County, 71 in northeastern Johnson County, 29 in the northern half of Campbell County, and 16 located mostly in two separate areas in Converse County. Of the flowing artesian wells in Sheridan and Johnson counties, at least five are known to contain methane. The area of northern Campbell County with anomalously large numbers of flowing artesian wells also coincides with the elliptical area around Recluse that contains the 14 shallow drill holes in which anomalously large quantities of methane were encountered.

Of the flowing artesian wells in Sheridan, Johnson, and northern Campbell counties, most are located along the major drainages of Tongue River, Powder River, and two principal tributaries of the Powder: Little Powder River and Clear Creek. A secondary zone of occurrences comprises a narrow belt along the western margin of the Fort Union Formation, extending from north of Tongue River to south of Clear Creek. Reason for apparent concentration of flowing wells along the main drainage channels is probably threefold:

- Concentration of rural development (and wells) along river alluvial flats
- Lower wellhead elevations along river channels than on the intervening ridges
- Probable concentration of free methane in coalbed open fractures along the structurally controlled major drainage channels.

Location of four of the flowing artesian wells known to contain methane, along the Powder River south from Clear Creek, and one well near the Tongue River provides tentative evidence that reason three above is real; i.e., that free methane is more abundant in fractured coal-bearing strata adjacent to structurally controlled drainage channels than it is in land areas between such channels.

Possible Structural and Stratigraphic Controls of Methane in Coalbeds

All data collected to date have indicated that Powder River Basin should provide several favorable target areas for recovery of methane gas from coalbeds by use of shallow drill holes. The basin is one of the largest coal-bearing basins in the west and has the largest reserves of coal of any basin or region of the United States—1.3 trillion tons. Coal in the basin is mostly in thick beds, all within potentially economic drilling depths, and the deepest principal coalbeds are at 2,500 ft in the basin center. Tertiary structure of coal-bearing formations in the basin is relatively simple, comprising a generally northwest-southeast–trending gentle asymmetrical syncline lying between the Black Hills Uplift on the east and the Bighorn Mountains on the west. Because the synclinal axis lies close to the western margin of the shallow basin, dips on the eastern limb are quite low, averaging less than 1. Superimposed on this simple basin structure is a fracture system that generally comprises two fracture sets that control local surface drainage and topography. The dominant fracture set parallels the northwest-southeast trend of the basin axis, the secondary fracture set being approximately perpendicular to the basin axis. Roughly concordant with the northwest-southeast and northeast-southwest trends of the two fracture sets is a system of very local small synclinal-anticlinal folds that seem to be confined stratigraphically to the coal-bearing zones within the Fort Union and Wasatch Formations. These local folds are probably syngenetically related to the original deposition of the sediments. That is, they may in part have been "draped" over topographic undulations on the depositional surface and in part caused by slumping of saturated sediments on shallow basin slopes prior to

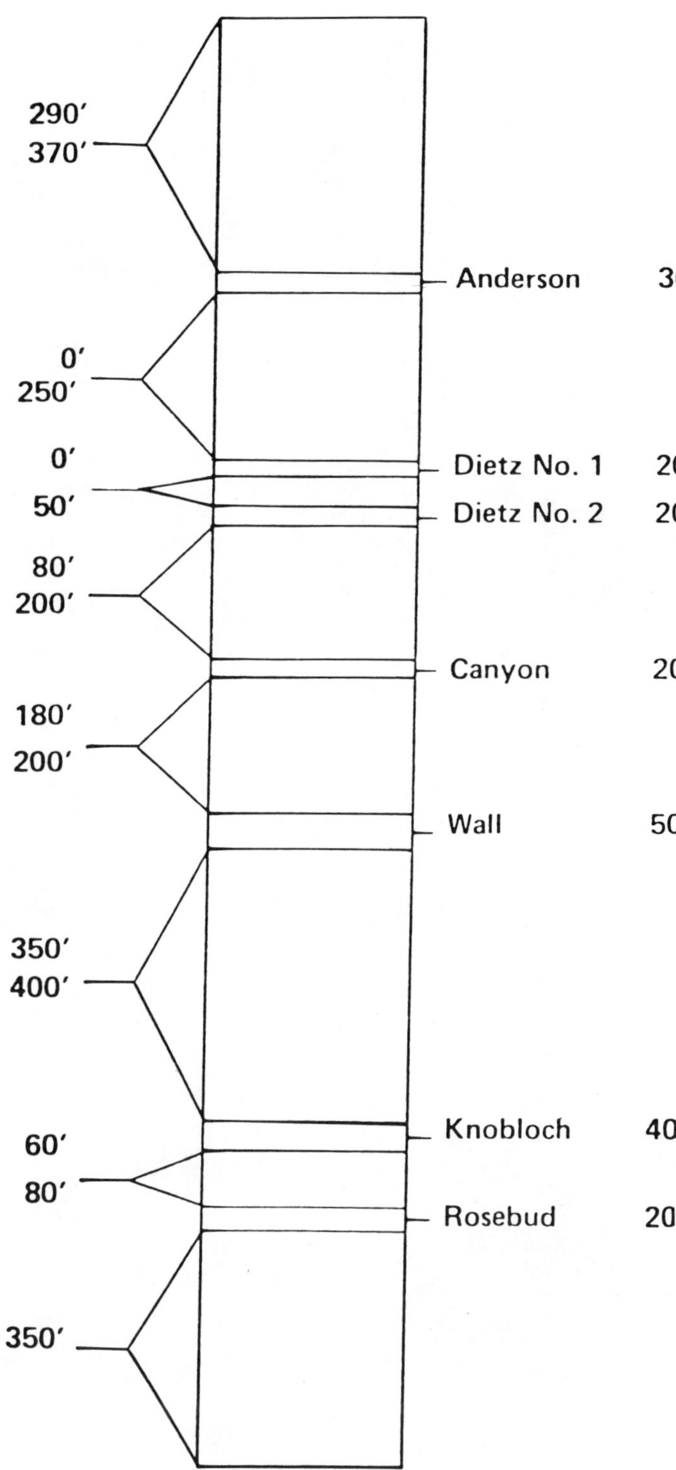

Figure 11—Stratigraphic diagram showing distribution and thickness of the important coalbeds in the Tongue River Member, northern Powder River Basin, Montana. Matson and Pinchock, 1976. Used with the permission of the Colorado Geological Survey.

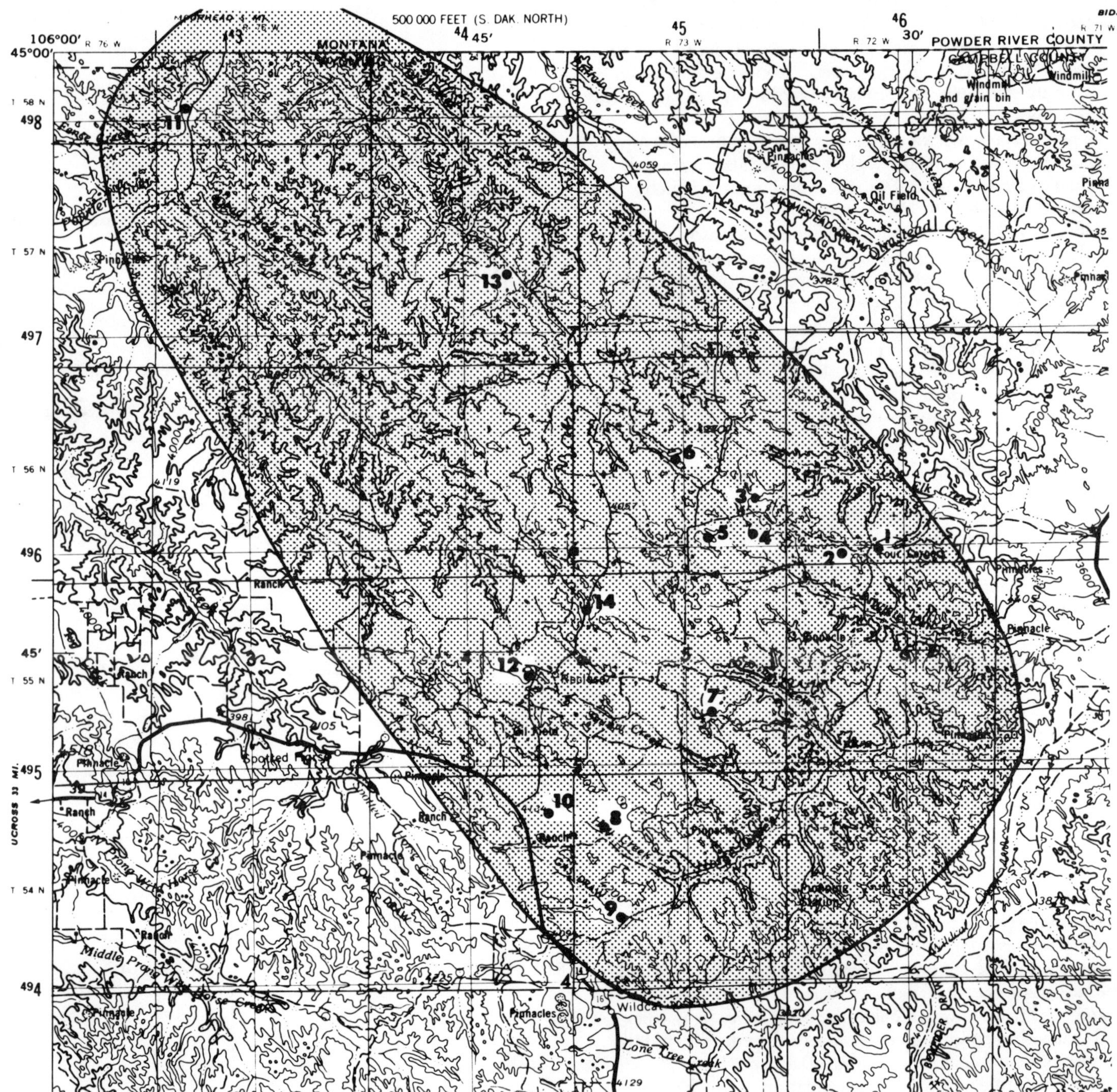

Figure 12—Shallow wells in the Recluse area, Campbell County, Wyoming, with anomalously large concentrations of methane gas. Base, NE Corner of USGS 1:250,000 Gillette Topographic Map.

lithification. It is the resulting small but numerous anticlinal highs that seem such promising specific targets for collecting any methane adsorbed within Fort Union and Wasatch coalbeds, as in the Spotted Horse and Sheridan fields, Wyoming, and northern extension of the Sheridan field in Montana.

Coal in the Paleocene Fort Union Formation and the immediately overlying Eocene Wasatch Formation occurs in at least 18 stratigraphically separable beds. Because the basin is so large and the study of subsurface coal relationships so limited, correlation of coalbeds from place to place in the basin is still uncertain. What is known, though, is that the basin contains the greatest concentration of thick coalbeds in the nation. In at least three centers of deposition within the basin, individual coalbeds are coalesced into massive "superbeds." These three centers are:

- A 20-sq-mi area just west of the Decker Mine, Big Horn County, Montana, where the Anderson, Dietz No. 1, and Dietz No. 2 coal seams are coalesced into a bed 80 ft thick.
- A north-south–trending belt approximately 15 mi long and 1 to 2 mi wide at Lake De Smet, Sheridan County, Wyoming, where five coal seams are coalesced into a bed whose thickness ranges from 70 to more than 220 ft. This combined bed—now called the Lake De Smet Bed—is the thickest known coalbed in the United States and thought to be the second thickest in the world.
- The area surrounding Gillette, Campbell County, Wyoming, where the Anderson, Canyon "A," and Canyon "B" seams coalesce into a single bed (the "Wyodak" or "Anderson Wyodak") ranging in thickness from 70 to more than 125 ft.

The methane recovery target area at Recluse defined by drill holes 1 through 14 occurs at the margin of one of these depositional centers, where a superbed splits into two or more component seams; that is, at the northern margin of the Wyodak superbed centered on Gillette. Hole 15 lies just to the south of the Gillette depositional center. At such locations, it may be that methane gas release from the coal seams, and collection within interbedded coal seam and porous sandstone sequences, may be optimal. Identification of such natural gas collection traps should be useful for indicating areas within the Powder River Basin where underlying coalbeds could have anomalously large amounts of adsorbed methane.

In the Recluse area (Hobbs, 1978), the apparent preferential sequence of individual coal and sandstone beds for production of methane from the shallow wells drilled is:

- The sandstone beds between the Anderson and Canyon bed(s)
- The Canyon coalbeds
- The sandstone bed below the Canyon
- The sandstone sequence between the Canyon coalbeds
- The Anderson coalbed.

Shallow drill hole data in the Recluse area indicate that coalbeds and porous sandstone sequences interbedded with and closely overlying the coalbeds all act as reservoirs for methane desorbed from the coal as well as reservoirs for ground water. Though poorly cemented, highly porous sandstone beds seem to be the most efficient reservoirs in the Recluse area coal-zone sequence. The coalbeds themselves are also good-to-excellent reservoirs.

Methane Recovery from Coalbeds Project Data

At the time of the writing of the original Powder River Basin Report (December 1979), no data had been obtained from the basin in the Methane Recovery from Coalbeds Project. The only two drill core coal samples collected to that date for determination of adsorbed methane were from the Recluse area. Desorption tests on these samples indicated only a small methane content of the samples, about 8 to 9 cubic feet per ton (cf/ton) of coal (Hobbs, 1978). The samples were from the Canyon and Cook coal seams collected at depths of approximately 375 ft. Coals of the Spotted Horse Field range in rank mostly between lignite and subbituminous C. Because of their low rank and never having been buried much, if any, deeper than their present shallow depths, the coals are very porous. Thus, any coal samples probably will lose much of their adsorbed methane very quickly; i.e., during drilling and retrieval, prior to placing of the samples in sealed canisters at the surface.

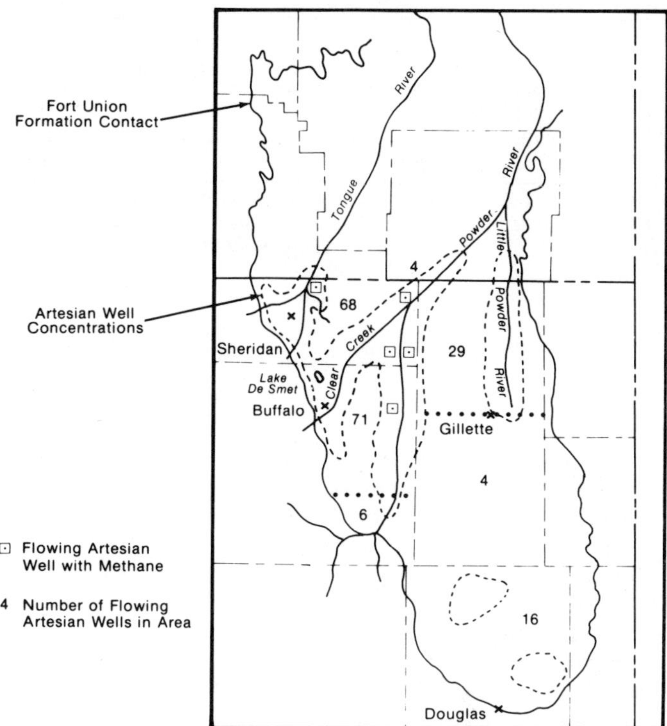

Figure 13—Map showing distribution of flowing artesian wells within the Powder River Basin area underlain by the Fort Union Formation.

Estimated Resource Volume

At the time of preparation of the original report, there was almost no quantitative data available on the specific amount of methane present in coalbeds of the Powder River Basin. As stated previously, the in-place methane was thought to be higher than that indicated in the three desorption analyses available. This was thought, in part, to be due to the low rank, shallow depth, and relative high porosity of coals of Powder River Basin.

Using estimates of the total coal resource for various beds and areas in the Powder River Basin, however, it was possible to present reasonable estimates of expected in-place methane resource.

Schell (1976, personal communication), based on an extensive USGS program for interpretation of oil and gas well geophysical logs, estimated the total coal resource of Powder River Basin to be 1.3 trillion tons. Glass (1976) and Mapel and Swanson (1977) gave more conservative estimates for the Wyoming and Montana portions of the basin, totalling 788 billion tons. Using these figures as high and low estimates, respectively, for the total coal resource of the basin, we assumed that approximately half of the coal would not be amenable to methane drainage, possibly because of shallow depth or the discontinuous nature of the coalbeds. That indicated that between 394 billion and 650 billion tons of coal might be available in Powder River Basin for methane recovery.

Table 1—Estimated in-place methane resource, Powder River Basin.

Coal Bed or Area	Assumed Coal Resource (tons)	Gas Resource in Cubic Feet @15 cf/ton	@25 cf/ton	@50 cf/ton	@100 cf/ton
Powder River Basin[1]	6.50×10^{11}	9.75×10^{12}	1.63×10^{13}	3.25×10^{13}	6.50×10^{13}
Powder River Basin[2]	3.94×10^{11}	5.91×10^{12}	9.85×10^{12}	1.97×10^{13}	3.94×10^{13}
Wyodak-Anderson coal bed[3]	1.00×10^{11}	1.50×10^{12}	2.50×10^{12}	5.00×10^{12}	1.00×10^{13}

[1]Half of Elmer Schell's (USGS) coal resource estimate of 1.3 trillion tons.
[2]Half of combined coal resource estimate of Glass (1976) and Mapel and Swanson (1977) of 788 billion tons.
[3]Total estimate of coal resource of Wyodak-Anderson coal bed (Averitt, 1975).

A reasonable estimate was used for the minimum average in-place methane content of the coalbeds of the basin at about 15 cf/ton of coal. This was twice the amount indicated in measured desorbed gas in three near-surface coal cores. It was thought that prime methane target areas—as outlined previously—and deeper coals would contain greater concentrations of methane. Table 1 presents the estimates made in the original report of the probably developable methane resource of the Powder River Basin, as well as the methane resource of the Wyodak coalbed, the largest single coalbed in the United States.

Subsequent to preparation of the original report in December 1979, TRW collected samples for desorption measurement from ten wells in Powder River Basin: eight wells under the DOE-sponsored Coalbed Methane Recovery Project and two wells on the Crow Indian Reservation for the Council of Energy Resource Tribes. Results of these measurements are summarized in Table 2. Data in the table include well location, formation name, coalbed name (where known), depth to the top of each coalbed and bed thickness from which samples were collected, the number of samples from each well, and the average gas content for all samples from each bed. The gas content listed is comprised only of the measured desorbed gas and the calculated lost gas. Because of the low gas content and porous nature of the samples, any residual gas in the samples had bled out prior to their being measured after shipment to the USBM lab in Pittsburgh, PA.

The data in Table 2 tend to confirm the statements made in the original basin report (1979) that gas concentrations in Powder River coalbeds are probably low. Of the seven wells (IAA–IAG) in Table 2 for which depth and gas concentration data are listed, there are 26 samples for depths less than 500 ft, 18 samples from between 500 and 1,000 ft depths, and 5 samples from deeper than 1,000 ft. For the samples taken between 500 and 1,000 ft, average gas concentration is 19 cf/ton and for samples from greater than 1,000 ft, average gas concentration is 32 cf/ton. Because the maximum depth for most of the thick coalbeds at basin center is about 2,500 ft, a reasonable value for the average gas concentration for all coal in the basin is probably about 25 cf/ton. Using Schell's estimate of 1.3 trillion tons of coal in the basin, based on interpretation of geophysical logs, then the total methane content in the coalbeds is approximately 30 trillion cubic feet of gas (cfg). Assuming a 50% recovery factor, the Powder River Basin coalbed methane resource that is possibly recoverable can then be estimated to be approximately 15 trillion cfg.

CONCLUSIONS

Powder River Basin of Wyoming and Montana presents an unusually promising opportunity for collecting large amounts of methane gas from relatively shallow coalbeds. The basin is very large and contains the greatest concentration of thick coalbeds in the nation. Though the coals are of low rank, which normally indicates small amounts of adsorbed methane per unit volume of coal, the great thickness of many of the coal seams indicates large volumes of methane contained per unit area of land surface. In addition to great thickness for individual seams, the basin is noted for the large number of thick seams intercepted by a single drill hole at shallow depths. For example, a single drill hole in Campbell County, Wyoming, intercepts 16 coalbeds greater than 5 ft thick, of which 10 beds are greater than 20 ft thick and one is greater than 60 ft thick (375 ft of coal within 2,500 ft of the surface; Apache-1 Geer wildcat well; Sec. 31, T47N, R73E). Because of the high porosity of these low-rank coals and the great total thickness of coal encountered at shallow depths, large amounts of methane should be recoverable per unit length of drill hole.

It is important to determine if methane gas can be practicably collected from shallow Powder River coalbeds as quickly as possible. These coalbeds are the principal target of the great strip mines being put into operation along the northwestern and entire eastern rim of the basin. As these pits are opened, any methane contained within and adjacent to stripping areas will be lost to the atmosphere. Thus, if such methane gas is to be salvaged, it must be done before new pits are opened and the entire rim of the Powder River Basin coalbeds is exposed to updip migration of methane contained in the deeper, more central portion of the basin.

The area that we consider the prime methane exploration target within the Powder River Basin is outlined in Figure 14. This area encompasses portions of Campbell, Sheridan, and Johnson counties in Wyoming and Big Horn and Powder River counties in Montana. Factors used in delimiting the boundaries of this target area include:

- Restricting the area to lands underlain by the Tongue River Member of the Fort Union Formation.
- Areas in which shallow drill holes are known to flow anomalously large amounts of methane.
- Areas in which flowing artesian wells are concentrated.
- Areas in which three or more individually thick coalbeds are coalesced together into a single superbed.

Table 2—Summary of gas desorption results for core and drilling chip samples collected from 10 wells, subsequent to publication of the original Powder River Basin Report.

Well Number	Location: Sec, T & R	Location: County	Formation	Coal Bed	Depth to Top of Bed	Bed Thickness (ft)	Type of Sample	Number of Samples	Gas Content (cf/ton) Average
IAA	22, 8S, 47E	Powder River, MT	Ft Union	Anderson	243	52.6	Core	3	2
			Ft Union	Canyon A&B	337	24.5		3	2
IAB	5, 7S, 40E	Big Horn, MT	Ft Union		121	24.5	Core	3	4
			Ft Union		347	5.5		1	1
			Ft Union		390	12.5		2	1
			Ft Union		620	53.5		3	7
IAC	7, 9S, 40E	Big Horn, MT	Ft Union		155	19.8	Core	2	2
			Ft Union		424	79.4		5	3
			Ft Union		539	64.2		1	7
			Ft Union		743	27.0		2	11
IAD	20, 55N, 75W	Campbell, WY	Ft Union	Smith	295	34.5	Core	2	<1
			Ft Union	Anderson/ Canyon	681	74.4		4	29
IAE	5, 54N, 76W	Sheridan, WY	Ft Union	Arvada	63	10.0	Core	1	1
			Ft Union	Smith, Upper	265	10.8		1	6
			Ft Union	Smith, Lower	300	18.1		1	8
			Ft Union	Anderson	592	45.0		3	38
IAF	3, 54N, 77W	Sheridan, WY	Ft Union	Smith Upper	202	12.0	Core	1	13
			Ft Union	Smith, Lower	297	14.0		1	13
			Ft Union	Anderson	588	25.0		1	23
			Ft Union	Canyon	673	16.0		1	3
IAG[1]	7, 52N, 77W	Johnson, WY	Wasatch	1	557	9.0	Drilling[1] Chips	1	not measured
			Wasatch	2	693	34.0		1	14
			Wasatch	3	831	25.0		1	19
			Wasatch	4	924	4.0		1	7
			Wasatch	5	1,058	23.0		1	8
			Ft Union	1	1,235	48.0		1	71
			Ft Union	2	1,622	26.0		1	23
			Ft Union	3	1,796	58.0		1	37
			Ft Union	4	2,145	17.0		1	23
IAH[1]	20, 51N, 81W	Johnson, WY	Ft Union		800–2,700		Drilling[1] Chips	7	<1
CERT 1	23, 53N, 37E	Big Horn, MT	Ft Union		220–605		Core	7	<1
CERT 2	7, 8S, 38W	Big Horn, MT	Ft Union		330–754		Core	7	<1

[1]Gas measured from drilling - chip samples commonly is only 20 to 80% of the gas measured from core samples, because of rapid outgassing and because of sample contamination.

Within the regional target delimited on the map of Figure 14, smaller initial target areas can be defined based on probable structural and stratigraphic controls, including the following:

- Major drainage channels that parallel any of the known fracture sets, especially those channels along which flowing artesian wells are concentrated. Included principally are Tongue River, Little Powder River, and especially Powder River and Clear Creek. These drainages probably are fault- or, at least, fracture-controlled. Coal-bearing strata, along and adjacent to these channel ways, are probably fractured to a greater degree and should contain more free methane than rocks away from the channel ways.
- Secondary channel ways with anomalously linear development parallel to the major, northwest-southeast or northeast-southwest fracture-set directions and along which are located flowing artesian wells or shallow-drill holes known to emit methane. An example is Hay Creek, southeast of Recluse, which follows an extremely straight northwest-southeast–trending path and along which is located a 245 ft deep hole that had an initial gas emission rate of a million cf/day (Hole 8).
- Zones along anticlinal fold axes, where rocks under tensional stress could be more highly fractured than rocks on fold limbs. Anticlinal crests present highly favorable zones where fractured rocks can act as hosts for methane migrating updip from coalbeds along fold limbs.
- Zones along synclinal fold axes where methane desorbed from coal-particle surfaces can collect in-place in open fractures along the stressed rocks.

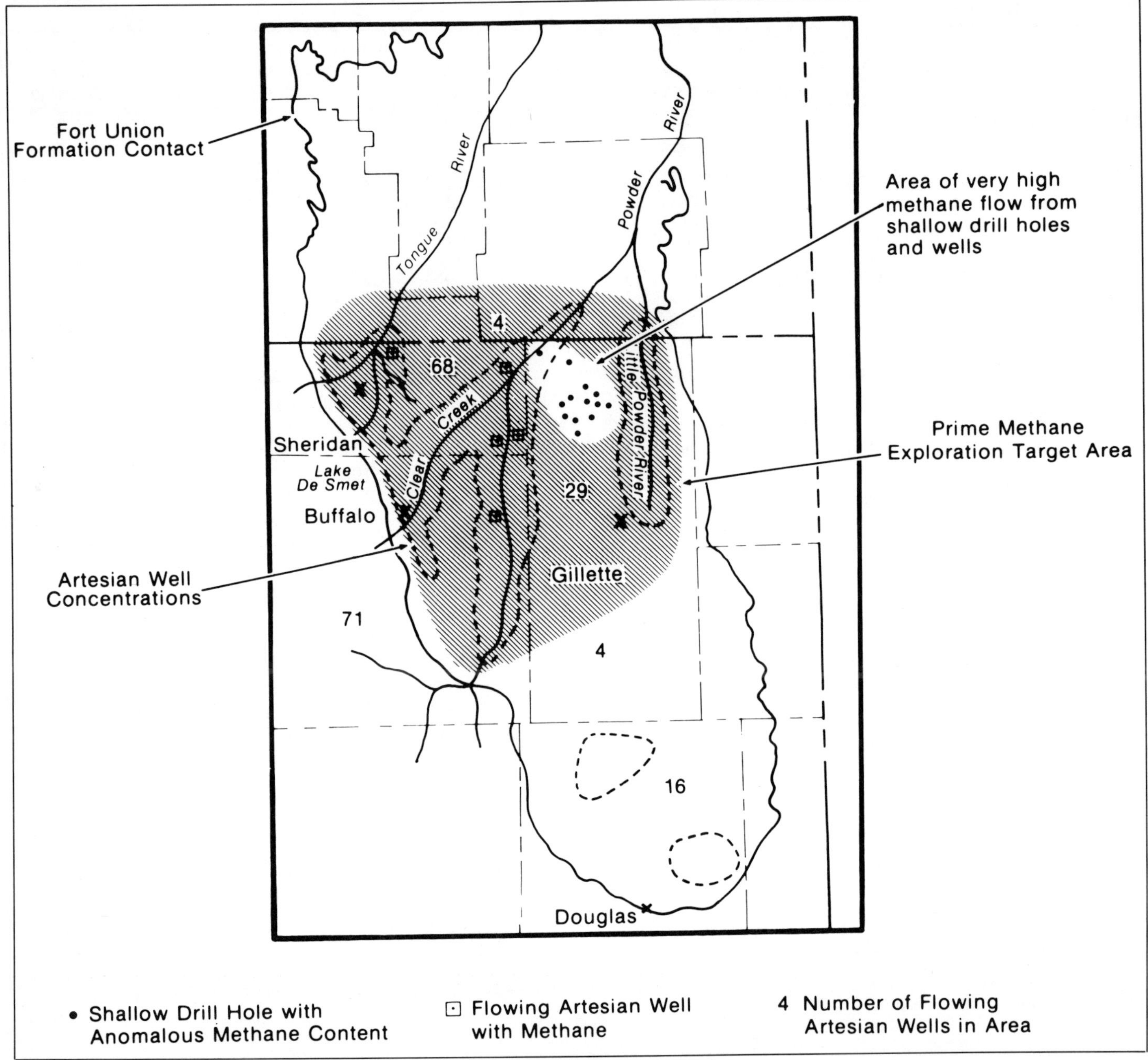

Figure 14—Methane exploration target area in Tongue River Member of the Fort Union Formation, central Powder River Basin.

- Stratigraphic traps created where superbeds split into two or more individual coalbeds. An example is the area northwest of the Decker Mine, Big Horn County, Montana, where the Anderson splits from the Dietz No. 1 and No. 2 coalbeds.
- Coalbeds in the deeper portion of the basin, paralleling the basin synclinal axis. Adsorbed methane concentration should be greatest in such beds because of the greater hydrostatic pressure and the probable higher coal rank in the more deeply buried coals.
- The northwest-trending, 33-mile-long, elliptical area in Campbell County in which shallow drill holes flowing methane gas are concentrated. Geologic controls for this concentration probably are partly structural and partly stratigraphic.

Initial individual drill sites for collecting coalbed methane data could include:

- Sites beside holes known to flow methane or beside flowing artesian wells containing methane.
- Sites on the crests of small structural closures along fold axes.
- Sites at the intersections of two or more favored linear structural features; e.g., the intersection of a fold axis and a linear stream channel or the intersection of two linear stream channels such as the conflux of Powder River and Clear Creek.

ACKNOWLEDGMENTS

We wish to thank the U.S. Geological Survey, the Montana Bureau of Mines and Geology, Anschutz Corp., and Bass Enterprises Production for the opportunity to collect core and drilling-chip samples during their drilling programs.

REFERENCES CITED

Averitt, P., 1975, Coal resources of the United States, January 1, 1974: U.S. Geological Survey Bulletin 1412, 131 p.

Breckenridge, R. M., G. B. Glass, F. K. Root, and W. G. Wendell, 1974, Campbell County, Wyoming: Wyoming Geological Survey County Resource Series CRS-3, 9 color plates.

Brown, R. W., 1958, Fort Union Formation in the Powder River Basin, Wyoming, *in* Powder River Basin, 1958: Wyoming Geological Association Guidebook, 13th Annual Field Conference, p. 111–113.

Calvert, W. R., 1912, Geology of certain lignite fields in eastern Montana: U.S. Geological Survey Bulletin 471, p. 187–201.

Culbertson, W. C., and W. J. Mapel, 1976, Coal in the Wasatch Formation, northwest part of the Powder River Basin near Sheridan, Sheridan County, Wyoming: Wyoming Geological Association Guidebook, 28th Annual Field Conference, p. 193–201.

Curry, III, W. H., 1971, Laramie structural history of the Powder River Basin, Wyoming: Wyoming Geological Association Guidebook, Wyoming Tectonics Symposium, p. 49–60.

Davis, J. A., 1912, Little Powder River Coal Field: U.S. Geological Survey Bulletin 471, p. 423–440.

Dobbin, C. E., and V. H. Barnett, 1927, The Gillette Coal Field, northeastern Wyoming, with a chapter on the Minturn district and the northwestern part of the Gillette Field by W. T. Thom, Jr.: U.S. Geological Survey Bulletin 796-A.

Glass, G. B., 1976, Update on the Powder River Coal Basin: Wyoming Geological Association Guidebook, 28th Annual Field Conference, p. 209–220.

——— , 1978, Wyoming coal fields, 1978: Geological Survey of Wyoming Public Information Circular 9, 91 p.

Grazis, S. L., 1977, Geologic maps and coal resources of four quadrangles, Campbell County, Wyoming: U.S. Geological Survey Coal Investigations Maps C-76, C-77, C-78, C-79. Scale 1:24,000.

Grose, L. T., 1972, Tectonics, *in* Geologic Atlas of the Rocky Mountain Region United States of America: Rocky Mountain Association of Geologists, p. 35–44.

Hobbs, R. G., 1978, Methane occurrences, hazards, and potential resources: Recluse geologic analysis area, northern Campbell County, Wyoming: U.S. Geological Survey Open-File Report 78-401, 20 p.

Lageson, D. R., A. J. Ver Ploeg, G. B. Glass, W. D. Hausel, R. M. Breckenridge, D. R. Gaylord, R. W. Mars, J. K. King, and A. L. Madina, 1978, Sheridan County, Wyoming: Geological Survey of Wyoming County Resource Series CRS-5, 9 plates.

Lowry, M. E., and T. R. Cummings, 1966, Ground water resources of Sheridan County, Wyoming: U.S. Geological Survey Water-Supply Paper 1807, 77 p.

Mapel, W. J., 1958, Coal in the Powder River Basin: Wyoming Geological Association Guidebook, 13th Annual Field Conference, p. 218–224.

——— , 1959, Geology and coal resources of the Buffalo-Lake De Smet area, Johnson and Sheridan counties, Wyoming: U.S. Geological Survey Bulletin 1078, 148 p.

——— , and V. E. Swanson, 1977, Summary of the geology, mineral resources, environmental geochemistry, and engineering geology characteristics of the northern Powder River coal region, Montana: U.S. Geological Survey Open File Report 77-292, 124 p., 2 plates.

Matson, R. E., and J. W. Blumer, 1973, Quality and reserves of strippable coal, selected deposits, southeastern Montana: Montana Bureau of Mines and Geology Bulletin 91, 135 p.

——— , and J. M. Pinchock, 1976, Geology of the Tongue River Member, Fort Union Formation of Eastern Montana, *in* Symposium on the geology of Rocky Mountain coal: Colorado Geological Survey Resource Series RS-1, p. 91–114.

Meinzer, 0. E., 1942, Occurrence, origin, and discharge of ground water, *in* Hydrology: New York, Dover Publications, p. 385–443.

Obernyer, S. L., 1978, Basin-margin depositional environments of the Wasatch Formation in the Buffalo-Lake De Smet area, Johnson County, Wyoming, *in* Proceedings of the Second Symposium on the Geology of Rocky Mountain Coal—1977: Colorado Geological Survey Resource Series 4, p. 49–65.

——— , 1979, Basin-margin depositional environments of the Fort Union and Wasatch Formations (Tertiary) in the Buffalo-Lake De Smet area, Johnson County, Wyoming: U.S. Geological Survey Open-File Report 79-712, 132 p., 2 plates.

Olive, W. W., 1957, The Spotted Horse Coal Field, Sheridan and Campbell counties, Wyoming: U.S. Geological Survey Bulletin 1050, 83 p.

Rupert, R. C., R. Choate, S. Cohen, A. A. Lee, J. Lent, and J. R. Spraul, 1975, Energy extraction from coal in situ—a five year plan, v. III, Resources: U.S. Energy Research and Development Agency Report No. E(49-18)-2041, NTIS No. T1D-27023-P3 P4-45-4-57.

Stone, R. W., and C. T. Lupton, 1910, The Powder River Coal Field, Wyoming, adjacent to the Burlington Railroad: U.S. Geological Survey Bulletin 381-B, p. 115–136.

Taff, J. A., 1909, The Sheridan Coal Field, Wyoming: U.S. Geological Survey Bulletin 341, p. 123–150.

Toy, T. J., and B. F. Munson, 1978, Climate appraisal maps of the rehabilitation potential of strippable coal lands in the Powder River Basin, Wyoming, and Montana: U.S. Geological Survey Miscellaneous Field Studies Map MF-932. Scale 1:1,000,000.

Wegemann, C. H., R. W. Howell, and C. E. Dobbin, 1928, The Pumpkin Buttes Coal Field, Wyoming: U.S. Geological Survey Bulletin 806-A, p. 1–14.

Wendell, W. G., G. B. Glass, R. M. Breckenridge, F. K. Root, and D. Lageson, 1976, Johnson County, Wyoming: Geological Survey of Wyoming County Resource Series CRS-4, 9 plates.

Whitcomb, H. A., T. R. Cummings, and R. A. McCullough, 1966, Ground water resources and geology of northern and central Johnson County, Wyoming: U.S. Geological Survey Water-Supply Paper 1806, 99 p.

Geologic Overview, Coal, and Coalbed Methane Resources of the Western Washington Coal Region

R. Choate
C. A. Johnson
J. P. McCord

The Western Washington coalbed methane province defined in this study comprises three broad regional target areas totaling approximately 6,500 sq mi and encloses four targets with a total area of approximately 1,800 sq mi. The methane resource contained in this province is estimated to range at a minimum between 0.3 and 3.0 trillion cubic feet and at a maximum between 2.6 and 24 trillion cubic feet.

INTRODUCTION

The state of Washington, as well as the entire Northwest—including Oregon, Idaho, western Montana, and northern California and northern Nevada—is deficient in energy resources. Within the overall area, there are few major occurrences of coal, petroleum, natural gas, uranium, oil shale, or tar sands. The region's dominant energy resource has been the hydroelectric potential of its major rivers. In fact, 80% of all electricity to residents of the Pacific Northwest is from hydroelectric plants; and more than 90% of all new homes being built in the Pacific Northwest will use electricity for heating purposes (Blundell, 1980). However, much of this hydroelectric potential already has been developed, and there are conflicting claims on many of the remaining potentially developable river areas; e.g., White Water Rivers Preservation, Wild Rivers Act, Wilderness and Primitive Area set-asides, National Park and National Recreation Area inclusions, and planned expansions. Moreover, with delays affecting 10 of the 13 coal- and nuclear-powered plants proposed for the area, energy shortages are predicted to average 9% over the next 7 years (Blundell, 1980).

Besides its hydroelectric potential, the major energy resource that can be readily developed is the north-south–trending belt of coal deposits stretching along the western foothills of Washington's Cascade Mountains, from the Canadian border on the north to the Oregon border on the south. This belt comprises a wide range of coal types, including the anthracite to bituminous coals of the Bellingham-Cokedale-Hamilton coal fields in the north; the bituminous to subbituminous coals of the Green River-Wilkeson-Carbonado-Ashford fields in the central portion of the belt; and the subbituminous to lignitic coals of the Centralia-Chehalis-Kelso-Castle Rock fields in the south. Of economic importance is the proximity of this coal belt to the major population corridor bordering the belt on the west that includes the major cities of Seattle, Tacoma, Olympia, and Portland.

REGIONAL SETTING

The western Washington coal region consists of a series of discontinuous coal fields extending north-south along the western foothills of the Cascade Range, with the exception of the Roslyn and Taneum-Manastash fields on the eastern flank of the Cascades. Methane-recovery target areas discussed in this report comprise part of the coal fields defined by Beikman et al (1961) (Fig. 1). Coal-bearing rocks, however, almost certainly cover a much larger portion of Washington State than that within the boundaries of known fields defined by Beikman et al (1961).

Two physiographic provinces represented in the western Washington coal-bearing region are the Puget Lowlands and the Cascades. The Puget Lowlands are characterized by gently rolling hills with relief, typically less than 400 ft above sea level. South of Bellingham, mountains rise steeply and reach an elevation of 2,385 ft less than 2 mi from Bellingham Bay at Chuckanut Mountain. Elevation of the Glacier coal field ranges from 1,000 ft along the Nooksack River to over 5,000 ft at the southern end of the field halfway up the slopes of Mount Baker. Topography in the Cascade Range is rugged. Mountain ridges are deeply dissected with steep-walled, glaciated valleys, many of which head in glacial cirques. Elevations of the Cascades crests generally range from 6,000 ft to 8,500 ft.

Glaciation has strongly affected the Puget Lowlands landscape, resulting in the general north-south trend of large lakes, valleys, and ridges. The many waterways of the Puget Sound were scoured out by glacial ice to depths of 1,000 ft and more. Numerous permanent and intermittent streams form a dendritic drainage pattern, but major rivers flow northwest or southeast, following a structural grain established during the Oligocene.

The Puget Lowland has the highest population density in the state. Major cities include Seattle (1,855,000, including suburban areas), Olympia (26,490), and Tacoma (157,800). Smaller communities within and near the coal fields include

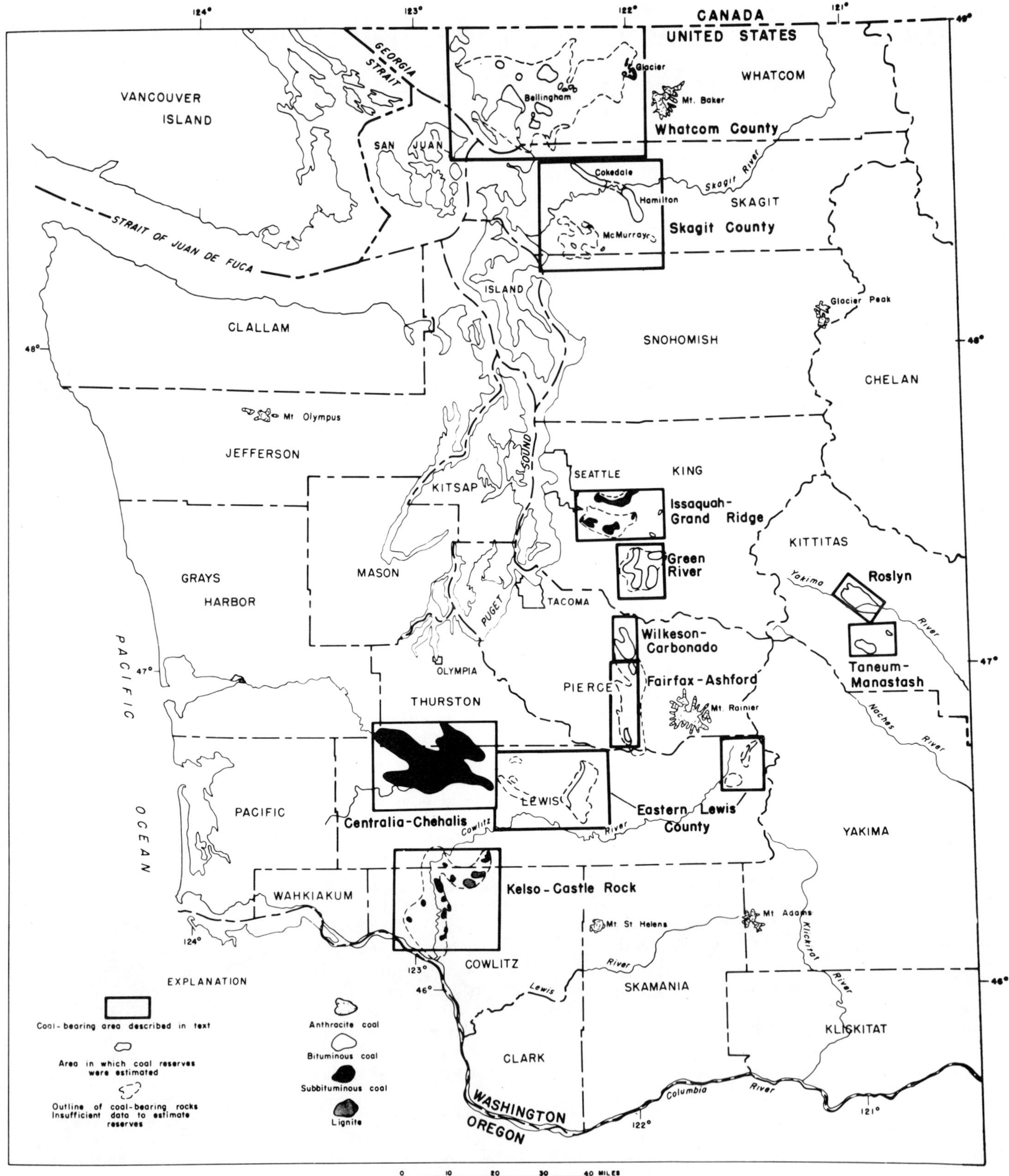

Figure 1—Index map of western Washington showing major coal-bearing areas. Beikman et al, 1961. Used with the permission of the State of Washington Department of Conservation.

Bellingham (43,400), Sedro Wooley (5,285), Mount Vernon (10,082), Enum Clau (5,145), Roslyn-Cle Elum (3,500), and Centralia (10,900).

REGIONAL GEOLOGY

Structural History of the Western Washington Region

Coal-bearing rocks of western Washington were deposited on the landward side along a north-south–trending coastline. The sequence developed as a fluvial-deltaic system in the Paleocene and the early Eocene. Deltaic sedimentation was interrupted by local volcanism in the late Eocene and ended during extensive volcanism in the Oligocene (Buckovic, 1979). This vast swampy coastal plain—the "Weaver Plain" (Mackin and Cary, 1965)—extended from the present position of the Puget Sound eastward across what is now the Cascade Range.

Coalbeds are thickest in the Centralia and Puget Lowlands areas and are thought to represent peat accumulation in poorly drained swamps of the lower delta plain. Thinner coalbeds in the foothills of the Cascades probably formed in a better drained paludal environment of the upper portion of the delta plain (Buckovic, 1979).

The extensive volcanism, tectonism, and shoaling of the coastal area during the late Eocene are attributed to the accretion of the western coastline area of Oregon and Washington onto the North American plate and the apparent westward shift of the Eocene subduction zone (Buckovic, 1979).

During the Oligocene, volcanic activity increased until the deltaic sediments were buried in pyroclastic flows. Coal swamps existed in only a few areas. Compression caused folding of the continental deposits and volcanic rocks and uplifting of the Calkins Range along northwest-southeast trends. This structural trend is still evident in the coal-bearing regions of Washington, as well as in the modern-day topography of the region.

Extensive volcanism during the Miocene is represented by the Columbia River Basalts that poured out over central Washington, burying the coalbeds of the region's southwestern portion. Small intrusives found in the coal fields of Washington are thought to have been emplaced at this time.

The Coast Range and Cascade Mountains were formed during the late Pliocene, when the Cascades were uplifted along a north-south trend. Associated folding and faulting is evident in the coal fields. Five large volcanic cones—Mt. Rainier, Mt. Adams, Mt. St. Helens, Mt. Baker, and Glacier Peak—all were formed during the Pleistocene. All except Mt. St. Helens are considered dormant. Small phreatic eruptions and seismic activity are still monitored at Mt. St. Helens, since a major eruption in mid-1980.

The Puget Lowlands continue to be a tectonically active region. During 1976, 307 earthquakes were recorded in western Washington (Crosson and Noson, 1978). The greatest concentration of these was in the lowlands between latitudes 47° and 48°. Faults, interpreted from geophysical data, are thought to be still active (Danes et al, 1965) and may be serving as conduits for methane migrating from underlying coalbeds.

Regional Stratigraphy

All important coalbeds of Washington occur in sediments of the Eocene Puget Group and equivalent formations (Table 1). Coalbeds represent a relatively small proportion of the Eocene continental arkosic rocks that are as much as 6,500 ft thick (Beikman et al, 1961). Locally, volcanic activity resulted in deposition of thick flows and volcanic sediment. One example is the Tukwila Formation, a large wedge-shaped deposit up to 7,000 ft thick in King County (Vine, 1969).

Marine sediments commonly interfinger with the near-shore facies. In the Skookumchuck Formation of the Centralia area, marine rocks form a wedge 1,000 ft thick, separating two groups of continental coal-bearing rocks. Similarly, the marine McIntosh Formation in the Centralia area can be traced eastward into Lewis County where its near-shore facies contains the coal seams of the Morton area (Snavely et al, 1951).

Most coal-bearing rocks are exposed in northwest-trending anticlinal and synclinal structures. Coalbeds, which generally dip 24 to 50°, may be vertical or overturned in places. Overlying the coal deposits are unconsolidated Miocene-Pliocene sediments, Pleistocene glacial deposits, and recent alluvium up to 1,000 ft thick in the central Puget Lowlands area.

COAL RESOURCES

Most principal coal deposits in Washington occur in a discontinuous zone along the western edge of the Cascade Range. They are primarily of Eocene age, interbedded with shale, siltstones, arkoses, and conglomerates. These coal deposits formed in swamps along the eastern shoreline of a north-trending basin. They also interfinger with, and grade into, barren rocks to the west. All coal-bearing areas have been structurally deformed by a series of folds and faults.

Although coal occurs in many locations throughout the state, known coal fields are restricted to the western half (Fig. 1). Of these fields, all but two lie in a north-south belt along the western side of the Cascade Range. These two are the Roslyn and Taneum-Manastash fields of Kittitas County and are located in the eastern flank of the Cascades.

Nearly 150 million tons of coal were produced in Washington from 1860 to 1960. Most came from coal fields in Kittitas, King, and Pierce counties during the early 1900s. Two mines were operating in the state in 1980: the Palmer Coking Company at Ravensdale (25,000 tons/year), and the Washington Irrigation and Development Company (WIDCO) open pit at Centralia (4.8 million tons/year). Mining activities at Centralia comprise a major strip mine operated in conjunction with a mine-mouth electric-generating power plant. The mine is operated by WIDCO, a subsidiary of Washington Water Power Company; the power plant is operated by Pacific Power and Light. Mining operations are centered about 5 mi northeast of Centralia, where approximately 9,400 acres will be stripped during the planned 35-year life of the operation. The associated power plant comprises two 700-MW generating units.

Coal resources in the state of Washington were estimated to be 6.2 billion tons (Beikman et al, 1961). Of this, 68% is subbituminous, 30% is bituminous, 2% is lignite, and less than 0.1% is anthracite. This estimate represents only coal at depths less than 3,000 ft, in beds thicker than 2.5 ft for lignite and subbituminous, and 14 inches thick for bituminous and anthracite. Additional calculations have raised the total reserves to 6.4 billion tons (Table 2).

Table 1—Stratigraphic correlation of coal-bearing rocks of the central and southern portion of the Puget Sound Lowlands. From Buckovic. Copyright © 1979 Society of Economic Paleontologists and Mineralogists. Used with permission.

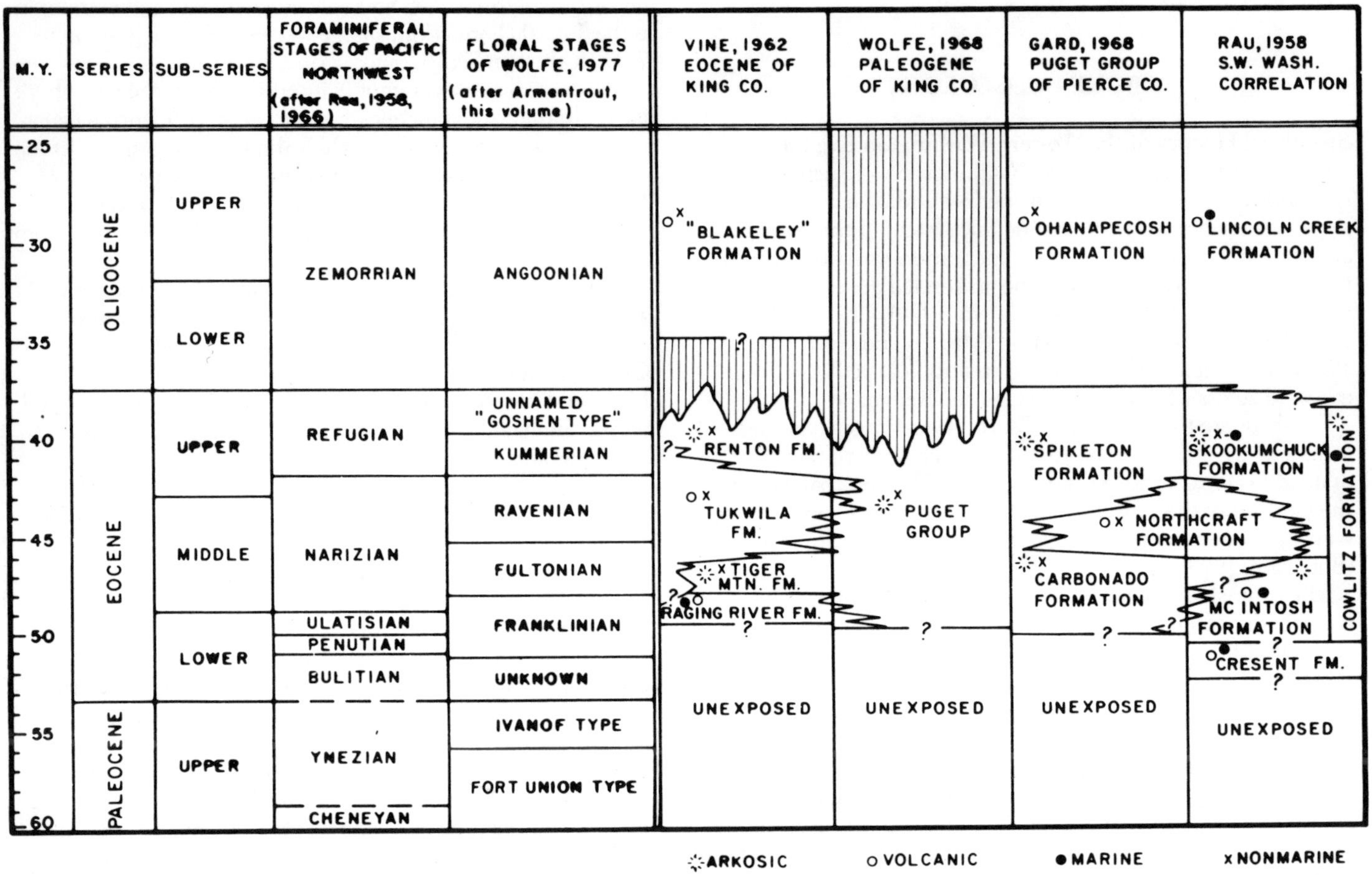

M.Y.	SERIES	SUB-SERIES	FORAMINIFERAL STAGES OF PACIFIC NORTHWEST (after Rau, 1958, 1966)	FLORAL STAGES OF WOLFE, 1977 (after Armentrout, this volume)	VINE, 1962 EOCENE OF KING CO.	WOLFE, 1968 PALEOGENE OF KING CO.	GARD, 1968 PUGET GROUP OF PIERCE CO.	RAU, 1958 S.W. WASH. CORRELATION
25–30	OLIGOCENE	UPPER	ZEMORRIAN	ANGOONIAN	o x "BLAKELEY" FORMATION		o x OHANAPECOSH FORMATION	o • LINCOLN CREEK FORMATION
35	OLIGOCENE	LOWER	ZEMORRIAN	ANGOONIAN	?			?
40	EOCENE	UPPER	REFUGIAN	UNNAMED "GOSHEN TYPE"; KUMMERIAN	? ✲ x RENTON FM.		✲ x SPIKETON FORMATION	✲ x • SKOOKUMCHUCK FORMATION; ✲ • COWLITZ FORMATION
45	EOCENE	MIDDLE	NARIZIAN	RAVENIAN; FULTONIAN	o x TUKWILA FM.; ✲ x TIGER MTN. FM.	✲ x PUGET GROUP	o x NORTHCRAFT FORMATION; ✲ x CARBONADO FORMATION	NORTHCRAFT FORMATION; ? o • MC INTOSH FORMATION; COWLITZ FORMATION
50	EOCENE	LOWER	ULATISIAN; PENUTIAN; BULITIAN	FRANKLINIAN; UNKNOWN	? • o RAGING RIVER FM. ?	?	?	? o • CRESENT FM. ?
55–60	PALEOCENE	UPPER	YNEZIAN; CHENEYAN	IVANOF TYPE; FORT UNION TYPE	UNEXPOSED	UNEXPOSED	UNEXPOSED	UNEXPOSED

✲ ARKOSIC o VOLCANIC • MARINE x NONMARINE

Very little is known about some of the coal fields, and Beikman et al (1961) calculated the reserves by use of statistics. Thus, 82% of total reserves for Skagit County are "inferred," as are 90% of the reserves of the Melmont area in Pierce County; 91% of the reserves of the Tiger Mountain area in King County; and all of the reserves from the Ashford area of King County. Beikman also estimates that because of the scarcity of drill-hole data, the measured reserves probably comprise less than 10% of the total reserves that may be present. Beikman et al suggest that the coal-bearing rocks are more extensive than generally recognized and that undiscovered coalbeds may exist.

This 6.2-billion-ton estimate is probably only a small part of Washington's coal. Vonheeder (1975) suggests that probably about 10% of the coal actually present in the state has been discovered and described. For example, of the 6.2-billion-ton reserves estimate of Beikman, calculations were limited to only those areas in solid shading in Figure 1. Areas known to be underlain by coal, but for which there are no quantitative data, are shown by dashed lines on this map. Based on these factors, the total western Washington coal resources are estimated to approach 60 billion tons.

To simplify description of the coal fields, they have been divided into six regions:

- Whatcom and Skagit Counties
- King County
- Pierce County
- Kittitas County
- Lewis and Thurston Counties
- Others.

The general setting and geology of each region are unique and complex. The structural geology of the western Washington coal region is so complex that coalbeds frequently cannot be correlated from one mine to another, and no coalbeds have been correlated among the six coal-bearing regions.

Whatcom and Skagit Counties

Whatcom and Skagit counties in the northwest corner of Washington are bounded on the north by British Columbia and on the west by Puget Sound. The two counties are in the northern portion of the Puget Lowlands and Cascade Mountain physiographic provinces. Much of the coal-bearing region lies in the lowlands, where glaciated hills seldom rise more than 500 ft above sea level.

Coal-bearing rocks of Whatcom and Skagit counties are found only on the west side of the Cascade Range and underlie areas of 500 sq mi in Whatcom County and 700 sq mi in Skagit County. The exact number and extent of the coalbeds are unknown. As many as 15 coalbeds may be present in Whatcom County (Jenkins, 1924), and 8 are known in the Cokedale area of Skagit County. Coalbeds range from seams of less than 1 to over 15 ft in thickness. Coal rank ranges from lignite to anthracite but is generally high volatile C bituminous.

Table 2—Summary of reserves in western Washington coal region (millions of tons).

Coal Field Region or District	County	Reserves Calculated by Beikman, et al, 1961	Additional Reserves Calculated	Total
Bellingham and Glacier coal fields	Whatcom	324.87	35.75 (Glaeser, 1962; Moen, 1969) 20.44 (Vonheeder, 1977)	381.06
Cokedale; Cumberland-Day Creek, Mount Vernon-Big Lake-McMurray, and Rock Creek areas	Skagit	506.96	0	506.96
Issaquah-Grand Ridge and Green River Districts	King	826.00	161.38 (Vine, 1969)	987.38
Wilkeson-Carbonado coal field, Spiketon, Melmont, Fairfax-Montezuma, and Ashford areas	Pierce	361.95	0	361.95
Roslyn coal field and Taneum and Manastash areas	Kittitas	281.80	0	281.80
Centralia-Chehalis District	Lewis-Thurston	3,693.78	0	3,694.00
Alpha-Cinnabar, Morton-Mineral Lake, Packwood, and Carlton Pass areas	Lewis	47.32	0	47.32
Kelso-Castle Rock area*	Cowlitz and Lewis	149.19	0	149.19
Grand Total		6,191.87	217.57	6,409.44

*Area not discussed in this report because coal is low rank, generally lignite or subbituminous C.

Skagit and Whatcom counties contain six coal fields:

- Bellingham
- Glacier
- Cokedale
- Cumberland-Day Creek
- Mount Vernon-Big Lake McMurray
- Rick Creek.

Two main coal zones are recognized in these counties. Coal at the Blue Canyon Mine and in the Glacier and Cokedale-Hamilton areas occurs near the base of the Chuckanut Formation, known as the Blue Canyon Coal Zone (Beikman et al, 1961). Coal at Bellingham and along the Van Wyck Syncline occurs within the overlying Huntingdon Formation and is known as the Bellingham Coal Zone (Vonheeder, 1975). These two zones are stratigraphically about 10,000 ft apart. All coal mines in the Chuckanut Formation are reputed to be very gassy.

In 1960, Beikman et al (1961) estimated the reserves of Whatcom County to be 324.87 million tons, 36% of which is measured and indicated and 64% of which is inferred. Based on drilling by Puget Sound Power and Light (Glaeser, 1962; Moen, 1969), an additional 35.75 million tons were added to reserves; and drilling by Washington Department of Natural Resources on state lands has defined an additional 20.44 million tons (Vonheeder, 1977). These reserves total 381.06 million tons, of which 37% is measured and indicated and 63% is inferred.

Beikman et al (1961) also estimated the reserves for Skagit County to be 506.96 million tons, of which 18% is measured or indicated and 82% is inferred. No additional reserves have been calculated for Skagit County.

King County

Coal deposits of King County are grouped into two main areas: the Issaquah-Grand Ridge, and the Green River Districts (Fig. 1). The Isaquah-Grand Ridge District is further subdivided into six subareas: Newcastle-Grand Ridge, Cedar Mountain, Renton, Tiger Mountain, Niblock, and Taylor. The Green River District is subdivided into nine areas: Black Diamond-Green River, Dale, Ravensdale-Georgetown, Kummer, Elk, Alta, Palmer-Bayne, Occidental, and Cumberland. In most parts of the Renton, Newcastle-Grand Ridge, and Cedar Mountain areas, deformation has been moderate and most coalbeds dip less than 35°. Elsewhere—especially in the Green River area—deformation has been more intense, and dips up to 50° are common. Figure 2 shows coal-interval stratigraphic sections for the various areas in King County.

Coalbeds of the Green River District occur within two zones of the Puget Group: the Kummer and the Franklin coal zones. The Kummer coal zone comprises about 1,200 ft of the upper part of the sequence, and the Franklin coal zone comprises 3,000 ft of the middle of the sequence, although few thick coalbeds occur below the Franklin coal zone. Nearly all the coalbeds of the Kummer coal zone in the Green River area are subbituminous. Coalbeds of the Franklin zone generally are high-volatile A or B bituminous in rank and have been the most important economically in King County. Most of the coalbeds in the Green River district are quite gassy. During the years 1901 to 1914, 27 mine explosions were recorded in this area.

Coalbeds in the Issaquah-Grand Ridge District are found in the Renton Formation and, in a general way, correlate with the Kummer coal zone and the upper part of the Franklin coal zone of the Black Diamond-Green River area. Coal quality ranges from subbituminous A to high-volatile C bituminous.

Total coal reserves for King County are about 826 million tons, and total reserves remaining in the Green River District of southern King County are estimated to be 357 million tons (Beikman et al, 1961). Estimates by Vine (1969) for the three-quadrangle area of Cumberland, Hobart, and Maple Valley total 613 million tons. This area includes all of the coal-bearing areas of the Green River District plus Cedar Mountain, Taylor, and Tiger Mountain and is an increase of 161 million tons over that of Beikman et al (1961) for the same area. Vine speculates that the potential coal resource for the three-quadrangle area to a depth of 6,000 ft is about 5 billion tons—nearly as much as Beikman had estimated for the whole state. Calculated reserves for the county, including the Beikman et al estimate of 826 millions tons—and the additional 161 million tons calculated by Vine—total 987 million tons.

Pierce County

Over 11,000 ft of Eocene to Miocene sedimentary and volcanic rocks occur in central Pierce County. They are, from oldest to youngest, the Carbonado, Northcraft, and Spiketon Formations of the Puget Group and the overlying Ohanapecosh Formation. All of the important coal seams occur in the Carbonado Formation, although some seams also have been mined from the Spiketon Formation.

The Carbonado Formation consists of more than 5,000 ft of interbedded sandstone, siltstone, mudstone, and shale, with lesser amounts of carbonaceous shale and coal. Lithologies generally occur as lenses, and changes in thickness and lithology occur over short distances along strike. Coalbeds within the formation are 1 to 5 ft thick, with a maximum thickness of 15 ft. The Carbonado is partly equivalent to the upper part of the McIntosh Formation of Centralia and to the Tiger Mountain and Raging River Formations of King County.

The Spiketon, named by Gard (1968), consists of over 3,000 ft of alternating beds of arkosic sandstone, siltstone, mudstone, and shale with some carbonaceous shale and coal. Coalbeds of the Spiketon are generally lower in rank, coke poorly, and have heating values of 9,000 to 12,000 Btu per pound. The Spiketon may be correlative with the Skookumchuck Formation of Centralia and possibly to the Renton Formation of King County (Gard, 1968).

Coal deposits of Pierce County occur in five areas. They are the Wilkeson-Carbonado Field; Spiketon; Melmont; Fairfax-Montezuma; and Ashford areas (Fig. 1). Throughout the coal-bearing region, extensive deformation has produced important coking coal deposits, commonly associated with beds dipping 60° or more on steep fold flanks.

Essentially all Pierce County coal is bituminous in rank; however, the composition varies widely within a single coalbed as well as between beds. Some anthracite has been found, but it represents local metamorphism by igneous intrusives. Several coalbeds have good coking qualities, and they constitute the largest known U.S. reserves of coking coal on the Pacific Coast. Coals range in rank from low-volatile bituminous to high-volatile A bituminous and have relatively high ash contents, averaging 11 to 15%. All Pierce County coals have

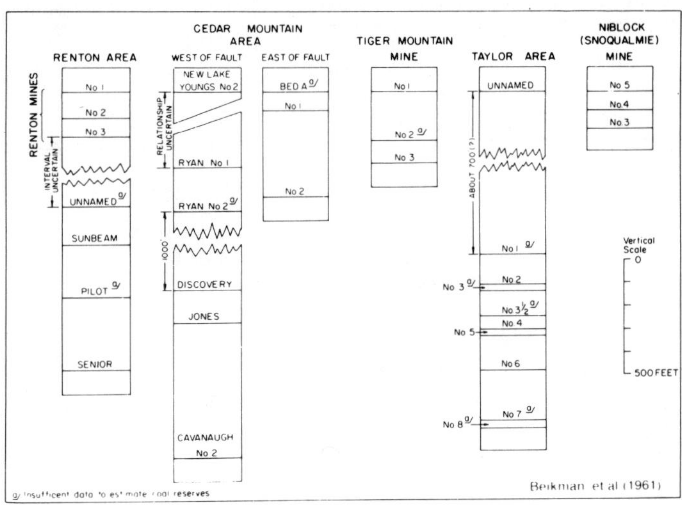

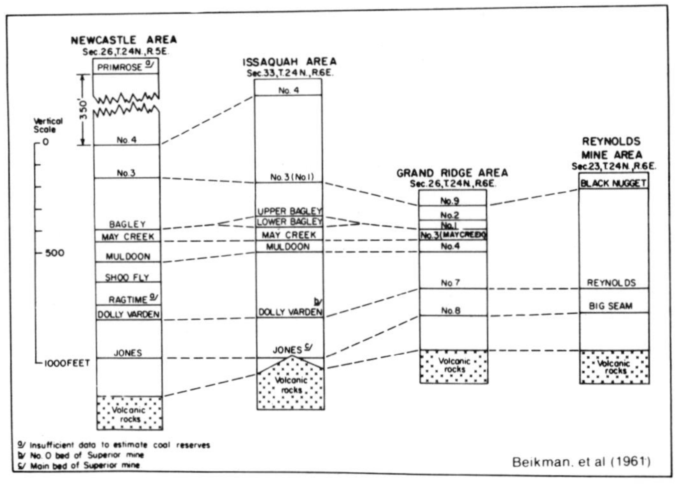

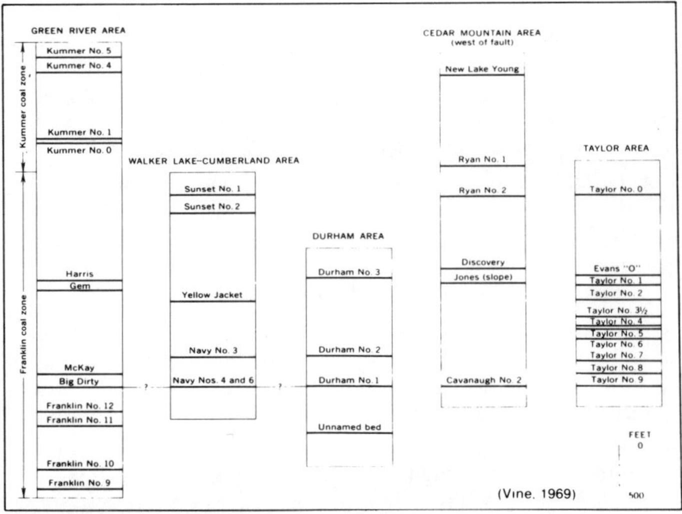

Figure 2—Stratigraphic sections showing correlation of coalbeds for the various areas in King County.

been mined underground because of the steep dips and complex structure. Figure 3 shows coal-interval stratigraphic sections for the various areas in Pierce County.

Total coal reserves for Pierce County, estimated by Beikman et al (1961), are 362 million tons, about 37% classified as measured and indicated, and 63% inferred. Much Pierce County coal is of good coking quality, although no coal is currently produced from there.

Kittitas County

The coal fields areas of Kittitas County are the only important coal deposits on the east side of the Cascade Range. The Roslyn coal field encompasses an area of 30 sq mi in central Kittitas County. Although small, the field was the state's major coal-producing area until production ceased in 1963. Other coal-bearing areas of Kittitas County are the Taneum and Manastash areas, located about 8 mi south of the Roslyn coal field. Very little coal was produced from these latter two areas.

The coal in the Roslyn coal field occurs in eight beds within the upper part of the Roslyn Formation. Stratigraphic relationships of these beds are shown in Figure 4. Five of the beds are of potential economic importance: the Nos. 1, 5, 6, 7, and 8 beds. Significant mining has occurred in the No. 1 or "Big Dirty" bed and the No. 5 or "Roslyn" bed, almost entirely by underground methods.

Ninety percent of all recorded production has been from the "Roslyn" bed. The Roslyn or No. 5 coalbed contains an average of 51 inches of coal, with one or two thin partings. Normally, the coal occurs in an upper bench about 29 inches thick and a lower bench 20 inches thick with a 4- to 6-inch parting. The coal changes considerably in character and quality from the eastern to western ends of the field. At Cle Elum, the coal is banded, the ash content is higher, and Btu content is lower. It is high-volatile B bituminous in rank and does not coke well. The rank, however, gradually increases toward the west to high-volatile A bituminous, and the coal becomes denser. Coals at the western end of the field have good coking characteristics.

The Taneum and Manastash areas lie about 8 to 10 mi south of the Roslyn coal field. Three coal-bearing areas are exposed in the Taneum area where Taneum Creek has eroded windows through the overlying volcanic rocks. The coal-bearing formation is thought to be the Manastash Formation (Saunders, 1914), although Beikman et al (1961) suggest that this area might be continuous with the Roslyn coal field.

Coal-bearing rocks of the Manastash area are exposed along Manastash Creek, where the creek has eroded the overlying volcanics parallel to the axis of an anticlinal fold. Coal is exposed here at elevations of 4,500 to 5,000 ft over an area of 8 sq mi. In this area, the Manastash Formation is at least 1,900 ft thick, with coal occurring in the lower part. The rocks are folded into a large anticlinal structure, which Saunders (1914) speculates may be part of an even larger syncline that would also include the Taneum area. The Manastash area may be continuous with the Roslyn field or may constitute a separate basin.

Throughout the length of the field, a large diabase dike trending northwest-southeast cuts the coal-bearing rocks. It has displaced and crushed the coals and is a serious problem to mining. The coal is high-volatile A bituminous in rank, with a relatively high ash content. Reserves in this area were estimated for one unnamed bed 1 to 2 ft in thickness to total 39.4 million tons. Thickness and extent of other beds are not known.

Coal reserves for Kittitas County are estimated by Beikman et al (1961) to total 507 million tons, 18% of which is measured and indicated and 82% of which is inferred. In a drilling program in the Roslyn coal field during 1966, Tuck and Boyd (1966) discovered that many of the coalbeds decrease in thickness and increase in ash content to the east. Although their estimate of reserves for the Roslyn field was reduced to less than half of that given by Beikman et al (1961), the area of coal-bearing rocks may actually extend well beyond the previously defined boundaries, and reserves may in fact greatly exceed those of Beikman et al.

Lewis and Thurston Counties

The coal-bearing region of the Centralia-Chehalis Field straddles the Lewis and Thurston County line in the Puget Lowland of southwestern Washington. It includes the old mining districts of Centralia and Chehalis in Lewis County, and Tono, Tenino, Mendota, and Bucoda in Thurston County.

In these counties, the coal-bearing strata of the Skookumchuck Formation crop out in a series of en echelon folds whose axes trend northwest-southeast. Within this en echelon fold complex, anticlines are relatively tightly folded compared with the broad, open synclines. The entire fold complex is cut by a series of high-angle reverse faults roughly paralleling fold axes; the faults tend to occur on steeper fold limbs close to anticlinal axes. Fault dip is predominantly northeast, with the southwest block downthrown. Stratigraphic displacements along these principal faults range between 100 and 1,500 ft, although most are 200 to 500 ft. Along several of the faults, especially the Kopiah and Doty, coalbeds have steep dips, ranging between 30 and 90°.

The late Eocene Skookumchuck Formation consists of 3,500 ft of marine and continental sediment that overlies the Northcraft Formation through an apparent local angular unconformity. The Skookumchuck Formation generally consists of a lower and upper sandstone sequence separated by a wedge of siltstone that thickens to the west. It is characterized by alternating and interfingering marine and continental deposits. Coalbeds range from a few inches to over 40 ft in thickness and span a stratigraphic interval of 2,500 ft. Within this interval, coalbeds occur in upper and lower groups, separated by 1,000 ft of non-coal-bearing rocks (Snavely et al, 1958; Geer, 1963).

At least 13 separate coalbeds are known to occur within the district. Bed thicknesses average 6 to 8 ft; individual beds may be as much as 40 ft thick, with partings and dips ranging from horizontal to vertical. Names, stratigraphic positions, and relative thickness for most of these beds are given in Figure 5.

Coal rank within the district ranges from lignite to subbituminous B; the bulk is subbituminous C. Average composition of the three thickest and principal coal-producing beds (Tono No. 1, Upper Thompson, and Big Dirty) is: moisture, 26.1%; volatile matter, 32.1%; fixed carbon, 32.1%; ash, 9.6%; sulfur, 0.9%; and heating value 8,108 Btu per pound.

The Centralia-Chehalis district of Lewis and Thurston counties contains more than half the calculated reserves of the state. Beikman et al (1961) estimate the total reserves of the district to be nearly 3.7 billion tons, 55% of which is measured and indicated and 45% of which is inferred. Almost all of the coal is subbituminous; minor amounts are lignite.

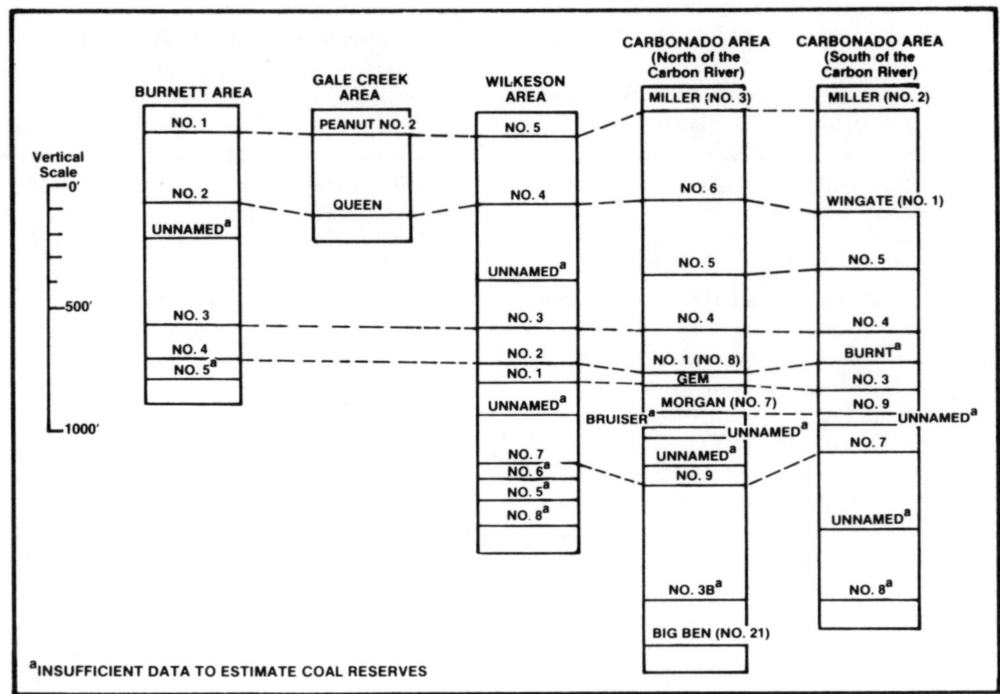

a) Wilkeson-Carbonado Coal Field

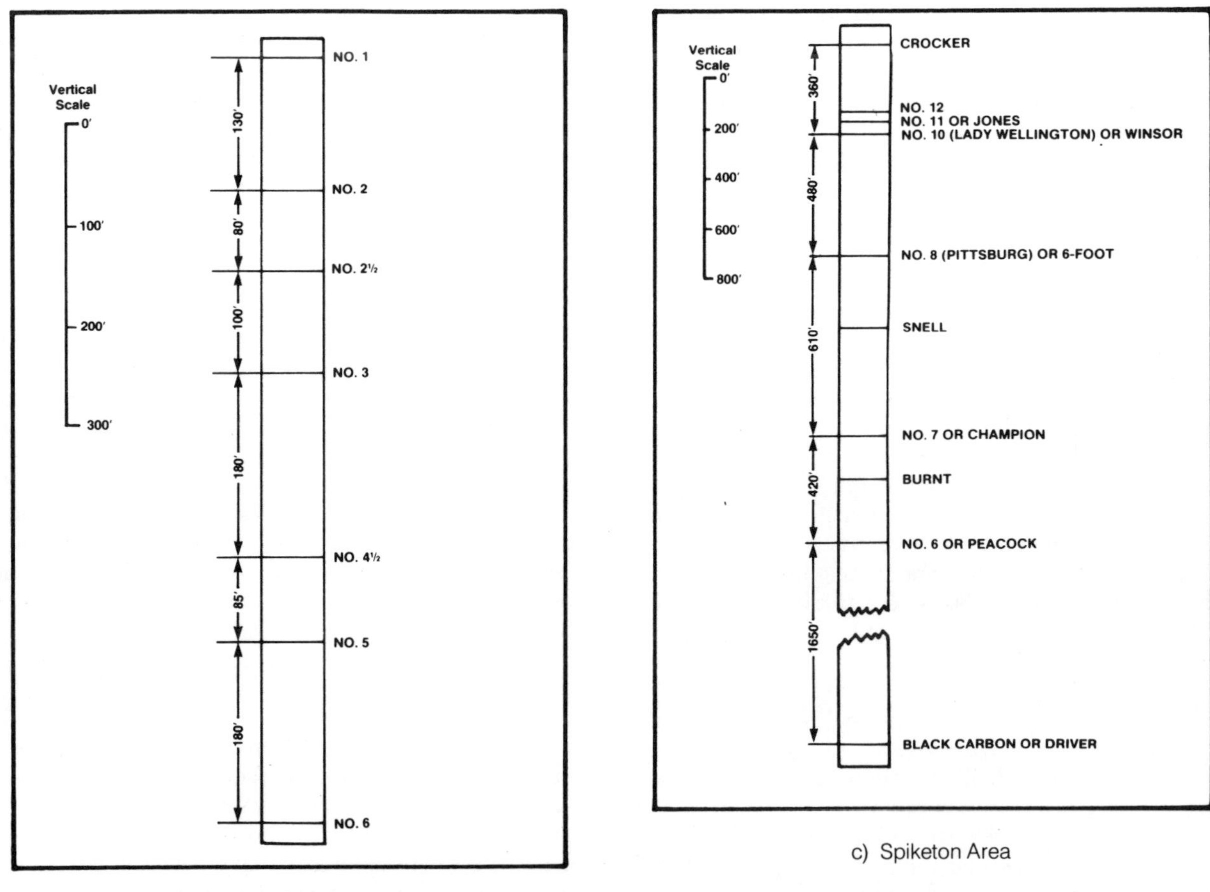

b) Melmont Area

c) Spiketon Area

Figure 3—Stratigraphic sections showing coalbeds of Pierce County, Washington. From Vonheeder, 1979; after Beikman et al, 1961. Used with the permission of the State of Washington Department of Conservation and Department of Natural Resources.

OTHER COAL-BEARING AREAS

Central Lewis County

Coal-bearing regions of central Lewis County include the Alpha-Cinnabar and the Morton-Mineral Lake areas that lie immediately east of the Centralia-Chehalis Coal Field. The coalbeds of this region have not been described in any recent literature; however, the coal-bearing formation is thought to be an eastward near-shore facies of the Eocene McIntosh Formation (Snavely et al, 1951).

In the Cinnabar-Alpha area, ten coalbeds are reported to occur. Details on their extent and thickness are lacking, and therefore no estimates of reserves were calculated for the area by Beikman et al (1961). Culver (1919) describes the coal-bearing sequence as folded and refolded in a relatively intricate pattern, further complicated by faulting and igneous intrusion. Several of the coalbeds are reported to be 6 ft or more in thickness. Vein No. 5 along Sherman Creek is described by Culver (1919) to be 9 ft thick, and all of the coalbeds exposed in tunnels in the area are somewhat gassy. No analyses are available, although the coal is thought to be bituminous in rank.

The Morton-Mineral Lake area is dominated by a major anticlinal fold with several episodes of minor parallel folds. At least three coalbeds have been mined, but the total number is not known, because they have not been correlated from one prospect to another. Coalbeds are intruded by igneous rocks and described as being near-anthracite in rank (Culver, 1919). Coalbeds up to 9 ft thick are reported, and analyses indicate that most of the coal is high-volatile bituminous. Ash content ranges from 10.4 to 45.7% and averages 20%. Reserves, calculated by Beikman et al (1961), total about 47 million tons, 85% of which is inferred; and 15% of which is measured and indicated.

Eastern Lewis County

The Packwood and Carlton Pass (or Cowlitz Pass) fields are two small coal fields at the extreme eastern end of Lewis County described by Smith (1911), Culver (1919), and Beikman et al (1961).

Coal at the Packwood coal field is exposed along the banks of the upper Cowlitz River. Coalbeds are generally flat-lying to gently dipping and thin-bedded. The coal-bearing rocks are overlain by igneous rocks, which in nearby areas show evidence of intrusive origin. The coal occurs in thin layers containing considerable amounts of carbonaceous shale and bone. Most seams are only a few inches thick. The clean coal is generally anthracite in rank but because of the numerous partings, the ash content is high. Analysis from a 3-ft bed of bony coal reported by Smith (1911) showed an ash content of 36% and a heating value of 8,200 Btu per pound on an as-received basis.

The Carlton Pass (or Cowlitz) coal field is located north and east of the Packwood field. The coalbeds are exposed along Summit, Carlton, Coal, and Clear Creeks and form the western limb of a north-trending anticline 10 mi in length. The area is one of considerable relief and extends across the summit of the Cascades. Stream valleys are 1,500 to 2,000 ft below the mountain peaks.

A section exposed along Summit Creek (Culver, 1919) contains 19 coalbeds, ranging from a few inches to 39 ft in thickness, although most contain considerable shale and bone. Seams of clean coal are generally very thin. The coal is semianthracitic to anthracite in rank, but most analyses reflect numerous partings. Ash contents range from 17.1 to 35.8% and average 26% (Beikman et al, 1961). Gas is reported to occur in the Primrose Bed along Summit Creek and is reported by Smith (1911) to be bubbling from some of the coalbeds in the bottom of the creek.

The total coal reserves in all the areas discussed above are summarized in Table 2.

POTENTIAL METHANE RESOURCE

A major methane-in-coalbed province exists in western Washington, and this methane could be present in sufficient quantities to partly fill the critical needs of the Seattle-to-Portland metropolitan area for natural gas in the future.

In work preceding this study, one author (Choate) investigated the possibilities for developing the energy potential of western Washington coal deposits by in situ techniques (Choate and Lent, 1975, 1977). These studies principally concerned the potential for in situ underground gasification of coalbeds in the western United States. Washington coal deposits were of particular interest because a large percentage of the state's total resources occur as steeply dipping beds. Conventional recovery of such coals by underground mining is dangerous and expensive to the point of being uneconomical. Thus, an energy recovery technique that avoids the problems of underground mining and is potentially economical merits serious consideration. Two in situ energy recovery methods considered for these resources are underground gasification and coalbed methane recovery. Of the two, coalbed methane recovery now seems the simplest, involving the least environmental impact and requiring no new technology development. Nevertheless, evaluation of locations and areal extent and determination of the quantities of methane present are essential tasks. Material covered in this section comprises:

1. A search of the literature for data involving methane accumulation, including underground mine accident reports, old reports of gassy underground conditions, methane in water wells, and methane encountered in oil and gas exploration wells in coal-bearing sedimentary rocks. A more detailed compilation of the data summarized in this section can be found in Choate and Johnson (1980).
2. Data collected for the Coalbed Methane Recovery Project (MRCP) in the state of Washington.
3. Geologic elements controlling the development of western Washington coalbeds and the methane they contain, along with the definition of regional and local methane target areas.
4. An estimate of the total methane resource contained in western Washington coalbeds.

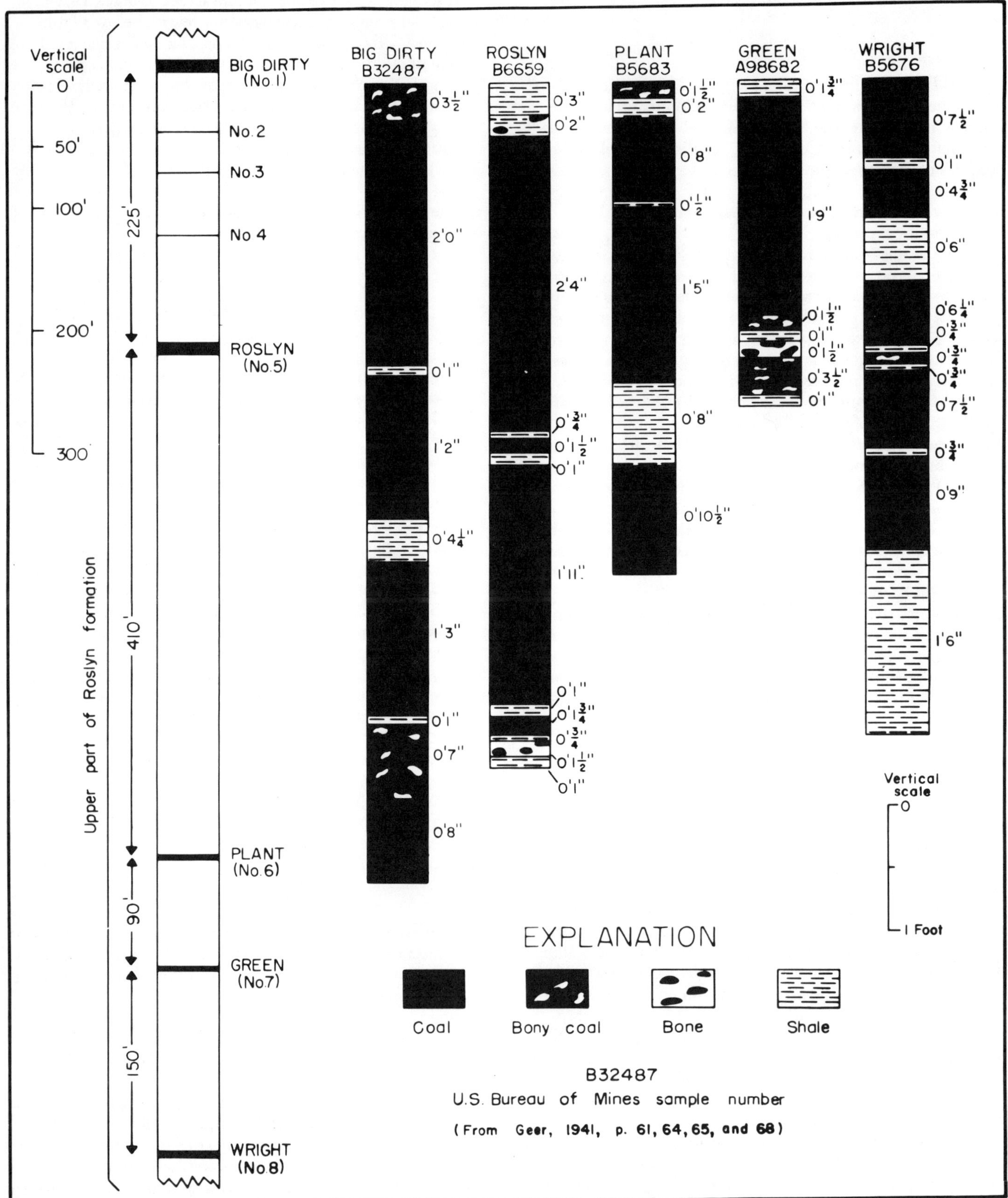

Figure 4—Generalized section showing principal coalbeds in the Roslyn field and detailed sections of minable portions of individual coalbeds. Beikman et al, 1961. Used with the permission of the State of Washington Department of Conservation.

Methane Data in Existing Literature

Underground Mine Accidents

In locating areas with unusually gassy coalbeds, underground mines worked in the early 1900s were identified that were known to be unusually gassy. Identifying such mines was accomplished principally by searching the biennial and annual reports of the State Inspector of Coal Mines for the years 1906 to 1919 for data on underground mine accidents involving accumulations of methane gas (Botting 1909, 1911, 1913; Bagley, 1915, 1917, 1918, 1919, 1920).

A total of 115 accidents were identified, and detailed data for the years 1907 through 1912 and 1914 provided specific mine locations for both fatal and nonfatal accidents. For the years 1913 and 1915 through 1919, locations were mentioned only for fatal accidents. In these 115 accidents, 119 miners were killed and 153 were injured. For those years in which detailed data were available, methane-related accidents are highly localized in: 1) the mining area of north-central Pierce County centering on the Wilkeson-Carbonado district; 2) the Green River region of southern King County; 3) the Roslyn coal field of Kittitas County; and 4) two smaller mining areas—at Ashford in southern Pierce County and at Ladd in north-central Lewis County.

In the 10 mi long, northwest-trending mining belt centered on Wilkeson and Carbonado, 52 of the accidents involved the Burnett, Spiketon, Wilkeson, Carbonado, and Fairfax districts. In an area about 5 mi wide in the Green River region, 27 accidents occurred, mostly in the Ravensdale, Black Diamond, Palmer, and Franklin districts. At the Roslyn field—a northwest-trending belt about 8 mi long—18 accidents occurred. The principal mines involved were the Ravensdale No. 1 in King County; the Gale Creek, Wilkeson, Carbonado, Burnett, Spiketon, Wingate, and Fairfax mines in Pierce County; the Roslyn Nos. 3, 4, 5, and 7 mines in Kittitas County; and the adjoining Ladd-East Creek properties in Lewis County.

Although these areas of methane-related accidents appear restricted to locales of relatively small areal extent, coal mining in these years was similarly restricted to local areas of high-value coals, such as the bituminous steam coals of the Green River and Roslyn fields and the bituminous coking coals of the Wilkeson-Carbonado area.

While methane is known to be present in the old mines of Whatcom and Skagit counties, mining occurred here mainly in years not covered by these data. For example, underground mining in the Bellingham area was principally practiced between 1853 and 1878 and again between 1917 and 1955. The state's second worst mine explosion—in 1894 at the Blue Canyon Mine, south of the Bellingham area—indicates that the bituminous coals of Whatcom and Skagit counties may be as gassy as those to the south in Pierce, King, and Kittitas counties.

Identification of Gassy Mines in the Literature

The principal publication that notes the methane gas danger in specific underground mines is a report on the coal fields of King County (Evans, 1912), which contains 11 references to underground gassy conditions. These data correlate quite well with the underground mine accidents reported by the State Inspector of Mines. In examining Evans' remarks, however, the true dangers of methane gas accumulation are understated in King County underground workings. Nonetheless, these observations indicate that methane gas concentration is lowest in the subbituminous coals in the western portion of the county and increases toward the eastern portion, where regional metamorphism has converted the coals to bituminous rank. Evans' observations also indicate that methane gas concentrations also rise with increased depth of the underground workings.

Data from both Evans (1912) and the State Inspector of Mines indicate that coal districts closest to the base of the Cascade Range have the greatest methane concentrations; i.e., the Ravensdale, Black Diamond, Franklin, Durham, Palmer, and Cumberland Districts. The principal mines involved in both data sets are the Ravensdale, Black Diamond, Franklin, Durham, Occidental, and Pocahontas Mines. The principal coalbeds involved appear to have been the McKay and possibly the Gem (of lesser importance).

Water Wells and Springs Containing Methane Gas

A total of 39 water wells and one spring that emit methane gas have been identified in the Puget Lowlands Province. Many of these wells are in close geographic association to known coal fields. Because the methane in many, if not most, of them probably had its origin in underlying coalbeds, such wells can be considered possible indicators of methane-in-coalbed targets.

Of these wells, 14 are in Whatcom County, 10 in King, 8 in Lewis, 4 in Pierce, 2 in Skagit, and 1 in Thurston. In 36 wells where depths are known, total depth ranges from 28 to 775 ft. In 31 of these wells, methane was encountered at depths of 250 ft or less. In the remaining 5 wells, the total depth or depth at which methane was encountered ranged from 325 to 685 ft.

Of possible interest in the search for coalbed methane targets in western Washington is the large number of flowing artesian wells concentrated in specific localities within the Puget Lowlands Province. Choate and Johnson (1979) studied the methane-in-coalbed potential of the Powder River Basin of Wyoming-Montana and found a close correlation between concentrations of flowing artesian wells and areas where shallow exploration wells and water wells showed methane in intimate association with coalbeds. Meinzer (1942) showed that gas in an aquifer increases the height to which water rises in a well and can create a flowing artesian well by raising the water to the surface. This lifting action is created by expansion of the gas as pressure is reduced in the water flowing from the aquifer. Lowry and Cummings (1966) and Whitcomb et al (1966) stated that this lifting action is responsible for several flowing artesian wells in the Powder River Basin.

One area in western Washington containing a large number of flowing artesian wells is the lowland region lying between the Green River and Wilkeson-Carbonado coal fields on the east and Puget Sound on the west (Fig. 6). Locations of flowing artesian wells on this map are plotted from tabulated data in reports on ground-water occurrence in southwestern King County (Luzier, 1969) and central Pierce County (Walters and Kimmel, 1968). In the nine-township area within King County on Figure 6, 38 flowing artesian wells were identified, as well as 33 other wells with positive static heads of up to 100 ft above ground surface. Such wells would be flowing, if uncapped. Of these 71 flowing artesian wells, 15 were found to have a sulfurous odor, and 8 showed noticeable iron content. In central

Pierce County, an additional 111 flowing artesian wells are identified.

As shown in Figure 6, these 182 flowing artesian wells do not form a random pattern but instead are concentrated along two linear paths that follow structurally controlled river valleys. One of the linear paths in the valley that forms the head of Puyallup River has the same northwest-southeast direction as the dominant structural trend of most Washington coal fields. The second linear path has a northern trend and lies within the valley formed by the present—and historical—bed of the Duwamish River. All but one of the wells with a sulfurous odor, and all but one of the wells with noticeable iron content, lie along the linear path that follows Duwamish Valley; also, eight of the water wells known to contain methane lie along this path, and two of these methane-bearing wells are also flowing artesian wells. A trend line of N 7° E, drawn through this linear path of wells, parallels the axis of the Black Diamond anticline to the east, along which wildcat wells have encountered methane.

There are other, perhaps better, reasons for artesian flow in Puget Lowlands water wells: for example, positive hydrostatic head caused by local topographic relief. Moreover, these water wells are all tapping aquifers in unconsolidated sediments above bedrock, and depth-to-bedrock for much of this area is 1,000 ft or more. However, the close association of structural trends, methane occurrence in water wells, concentration of flowing artesian wells, and occurrence of hydrogen sulfide and iron as well as methane in many of these flowing artesian wells seem to require special conditions to explain their origin. Possibly the simplest explanation is that: 1) Coalbeds underlie the sediments; 2) these coalbeds are highly fractured along fault lines; 3) the river valleys developed along these fault lines; 4) methane from such coalbeds migrates upward into the overlying sediments; and 5) ground waters moving along these fault lines also act as carriers of sulfide and iron formed from organic sulfur and from inorganic iron- and sulfur-bearing minerals—particularly pyrite—in the coalbeds.

In the northern Bellingham area, similar unconsolidated sediments cover bedrock to depths ranging from 400 to 800 ft, and water wells and gas exploration wells have encountered methane trapped in porous sand lenses beneath impervious clay seams within glacial drift overlying coalbeds in bedrock. In the Bellingham area, coalbeds are the only logical source of the methane, because all sedimentary rocks above the metamorphic basement are nonmarine.

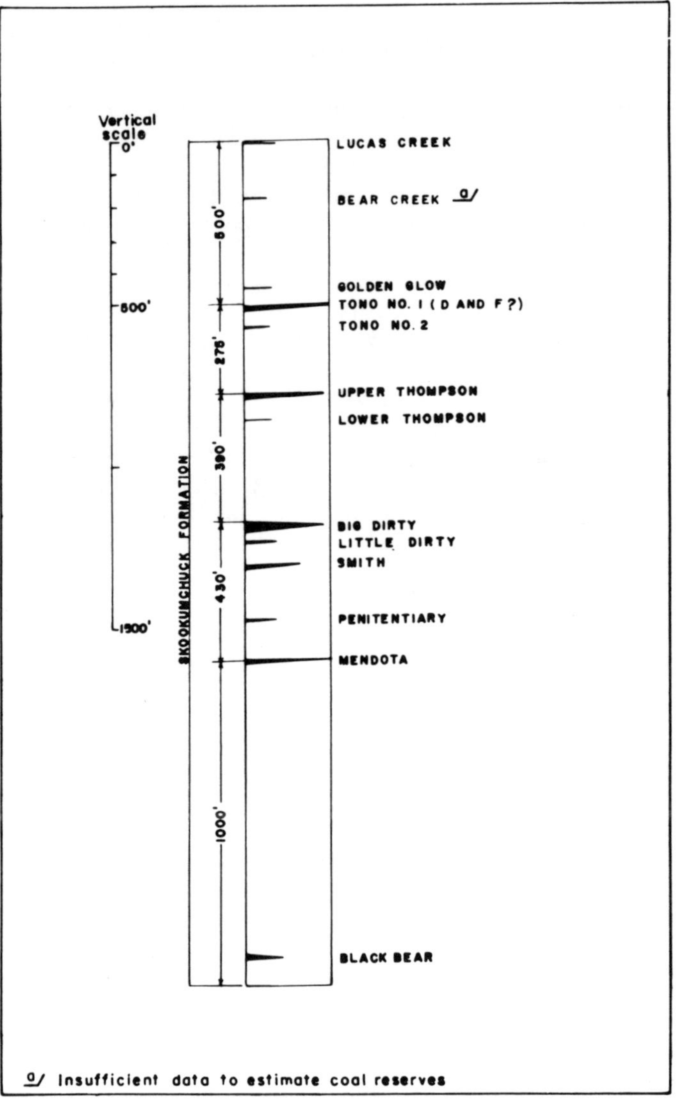

Figure 5—Generalized section showing positions of principal coalbeds in the Skookumchuck Formation. Beikman et al, 1961 (used with the permission of the State of Washington Department of Conservation); modified from Snaveley et al, 1958.

Oil and Gas Wells

Oil and gas exploration wells within the Puget Lowlands Province are significant because methane found in many of the wells almost certainly originated from coalbeds in nonmarine sedimentary rocks and not from marine oil- and gas-bearing deposits.

Methane has been encountered in 67 oil and gas exploration wells drilled in the Puget Lowlands Province. Of these, 43 are in Whatcom County, 13 in King, 5 in Lewis, 2 in Thurston, and 1 in Pierce. From this group of wells, methane was encountered at depths less than 500 ft in 25 wells; less than 1,000 ft in 38 wells; and less than 2,000 ft in 50 wells.

In the area of western Whatcom County, methane has been produced from very shallow wells. Methane has been found here not only in shallow, nonmarine, coal-bearing rocks of the Chuckanut Formation, but also in unconsolidated glacial drift overlying the bedrock. Methane degassing from the 7 to 9 ft thick Bellingham, and from other coalbeds, has migrated upward into the glacial overburden to be trapped in porous sand lenses capped by impervious clay seams.

East of Ferndale (Secs. 27 and 28, T39N, R2E), gas has been produced commercially from such unconsolidated deposits at depths ranging from 166 to 193 ft and at well flow rates ranging from 750,000 to 5,000,000 cubic feet per day (cf/day). Two other areas of known shallow methane concentrations in western Whatcom County are: 1) the area bordering Birch Bay (Sec. 32, T40N, R1E), and 2) the area south of Lawrence (Secs. 28, 32 and 33, T39N, R4E). At Birch Bay, methane occurs in oil and gas exploration wells at depths ranging from 130 to 430 ft with pressures up to 55 psi. In the Lawrence area, methane has been discovered in shallow water wells ranging in depth from 50 to 150 ft. In these and other parts of western Whatcom County, methane has been used domestically from wells tapping deposits

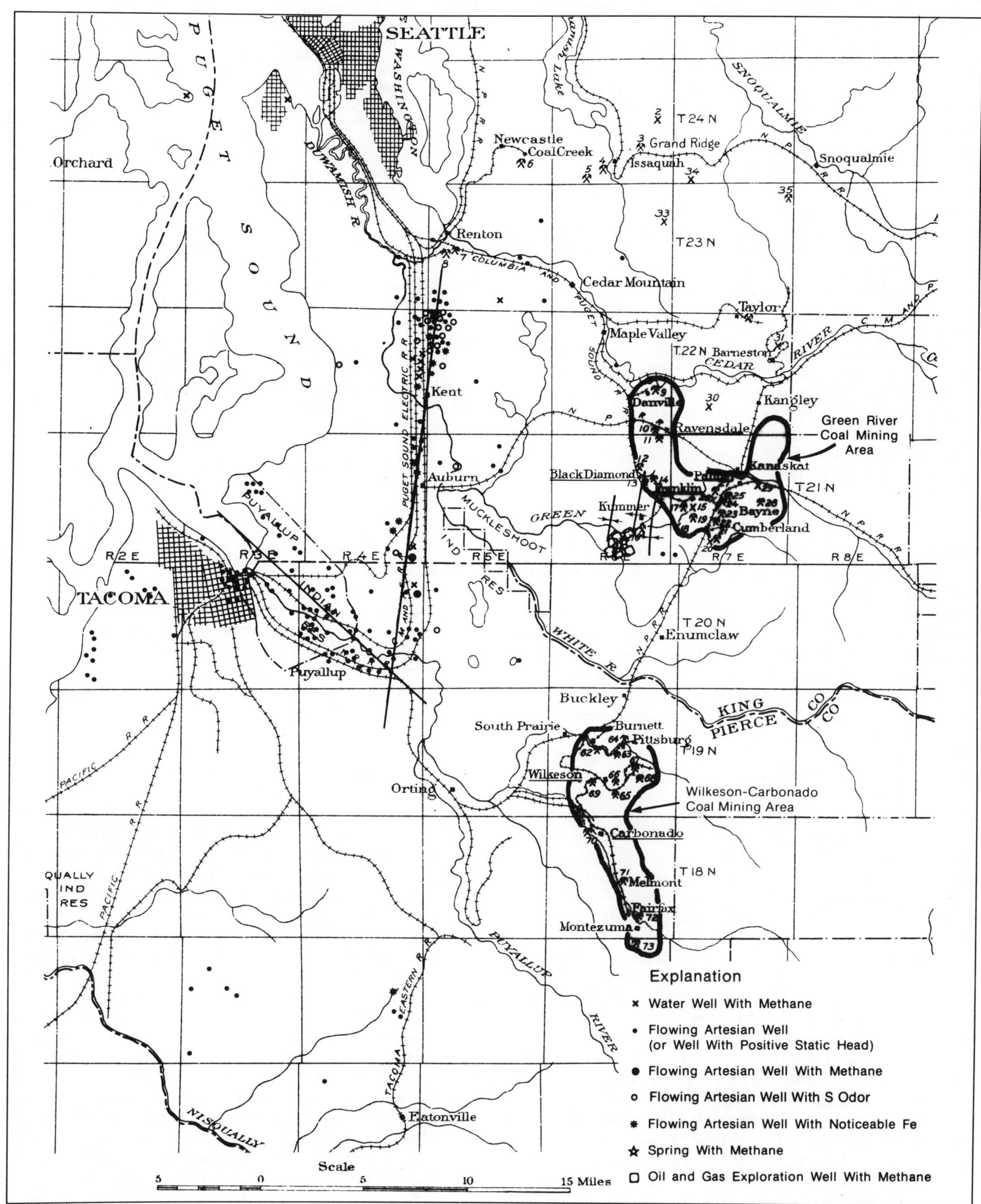

Figure 6—Flowing artesian well locations in portions of King and Pierce counties. Note alignment of wells and association of CH_4, S, and Fe.

in both glacial overburden and in bedrock, at depths ranging from 250 to 2,000 ft.

In western Whatcom County, drill-hole cores and logs indicate that the nonmarine coal-bearing formations directly overlie basement metamorphics, and thus, coalbeds must be the source for the shallow methane. Additional evidence is provided by the extremely large nitrogen component in the gases. In 13 analyses of 9 different wells, nitrogen concentration ranges from 2.5 to 70% and averages 33.2%. Such high nitrogen concentrations are normally associated only with gas produced from coalbeds as the nitrogen forms from the breakdown of ammonia. In the geochemical processes involved in the conversion of plant materials to coal, the principal gases produced are methane (80% of total gases) and ammonia (15%). Highly soluble in water, ammonia may migrate away from the methane, thus accounting for the different methane-nitrogen concentration ratios found.

While oil has been discovered and produced in small amounts in the marine sedimentary rocks west of the Puget Lowlands Province, it has not been discovered anywhere in the predominantly nonmarine rocks of the Puget Lowlands. Of the 67 wells within the province where methane is reported, none have oil except as showings, trace amounts, or films, and most of these are thought to have originated from the drilling process itself.

Methane Recovery from Coalbeds Project Data

During this project, coalbed methane content measurements have been made at only one location in Washington. These measurements of four core samples were collected in November 1979, from one hole drilled through the Big Dirty bed at Tono Basin in the Centralia-Chehalis Field (DOE Hole No. 7; NW/4, NW/4, Sec. 21, T15N, R1W).

Desorption measurements of these four core samples indicate an unusually large methane content for a coalbed of such low rank. Heating values for Big Dirty samples collected from four locations in the coal field (Snavely et al, 1958) ranged from 8,320 to 9,070 Btu and averaged 8,620 Btu, indicating that the Big Dirty bed is in the lower third of the subbituminous C classification of coal rank, just above lignite. However, as shown in Table 3, methane content measured in the bed at Tono Basin ranged from 75 cubic feet per ton (cf/ton) near the coalbed roof to 20 cf/ton near the floor. For the top 8 ft of coal in the bed, methane content averaged 73 cf/ton and in the bottom 5 ft averaged 21 cf/ton. The above data do not include values for gas lost during core recovery or bound as residual gas.

As indicated in the corehole schematic of Figure 7, the Big Dirty at DOE Hole No. 7 is 21.7 ft thick, lying 601 ft below the surface at the roof. The bed contains several stringers of siltstone and thin partings of altered tuff. Roof rock comprises 53 ft of relatively uniform fine-grained siltstone, which is overlain by at least 146 ft of uniform, fine- to medium-grained sandstone. A possible reason for the high methane content is the ground-water table within a few feet of the ground surface. Thus, hydrostatic head above the coalbed is probably about 600 ft. Because roof rocks overlying the Big Dirty are relatively soft and porous, the hydrostatic head should convert directly to a pressure of approximately 260 psi; i.e., the pressure related to methane adsorption to the coal surfaces.

The methane desorption core samples at Tono Basin were collected in conjunction with a field investigation of the

Table 3—Summary of methane desorption results for samples obtained from drillhole DOE No. 7, Tono Basin, Thurston County, Washington (NWNW Section 21 T15N R1W).

Sample Depth Interval (ft below ground surface)	Sample Weight (gm)	Desorbed Gas Content* (cf/ton)
602.8–603.8	629	74.7
610.0–611.0	741	72.3
617.9–618.9	842	22.5
623.2–623.9	639	19.8

*Does not include the estimated gas lost during core retrieval nor residual gas.

underground gasification potential of Centralia-Chehalis district coalbeds. A total of seven geologic holes and three hydrologic observation wells were drilled in Tono Basin during 1979 by Sandia Laboratories, assisted by Lawrence Livermore Laboratories and Laramie Energy Technology Center. The methane desorption samples were collected from the last hole to be cored during the field season, after methane was observed bubbling in the hydrologic observation wells.

Geologic Elements in Washington Coalbed Methane Development

The following is an interpretation of major elements in the geologic development, present distribution, and physical conditions affecting western Washington coalbeds containing methane:

1. *Initial development of a broad coastal depositional plain.* In the Paleocene and Eocene, Washington's coal-bearing sediments were deposited upon a broad coastal area of low relief, now called the Weaver Plain (Mackin and Cary, 1965). This plain stretched from beyond the present boundaries of Canada on the north into Oregon on the south. The plain, ranging from 100 to 150 mi wide, extended from a coastline nominally coincident with the present eastern shoreline of Puget Sound, well east of today's Cascade Mountains.
2. *Development of extensive swamps on the plain.* Coal-forming forests and swamps developed on the plain and extended across the present site of the Cascade Mountains to areas east of the present sites of Wenatchee and Yakima. Dynamic equilibrium between deposition and subsidence permitted accumulation of thousands of feet of sediment in coastal-plain swamps, extending from Paleocene through Eocene to early Oligocene times. Although the shoreline migrated eastward during this time, sediments deposited east of the present site of Seattle were dominantly nonmarine; those to the west were marine deposits typical of major geosynclines. Coal-forming sediments included all of western Washington's major coal fields: Bellingham, Cokedale-Hamilton, the Green River area, Wilkeson-Carbonado, Roslyn, and Centralia-Chehalis.
3. *Rise of major pre-Cascade mountain range.* During late stages of the Weaver Plain's existence, northeast-southwest compressional stresses caused gradual folding of the thick sheet of buried sediments previously deposited on the plain. These folded sediments were uplifted into a major mountain chain now called the Calkins Range (Mackin and Cary, 1965). The northwest axial trend in the Calkins

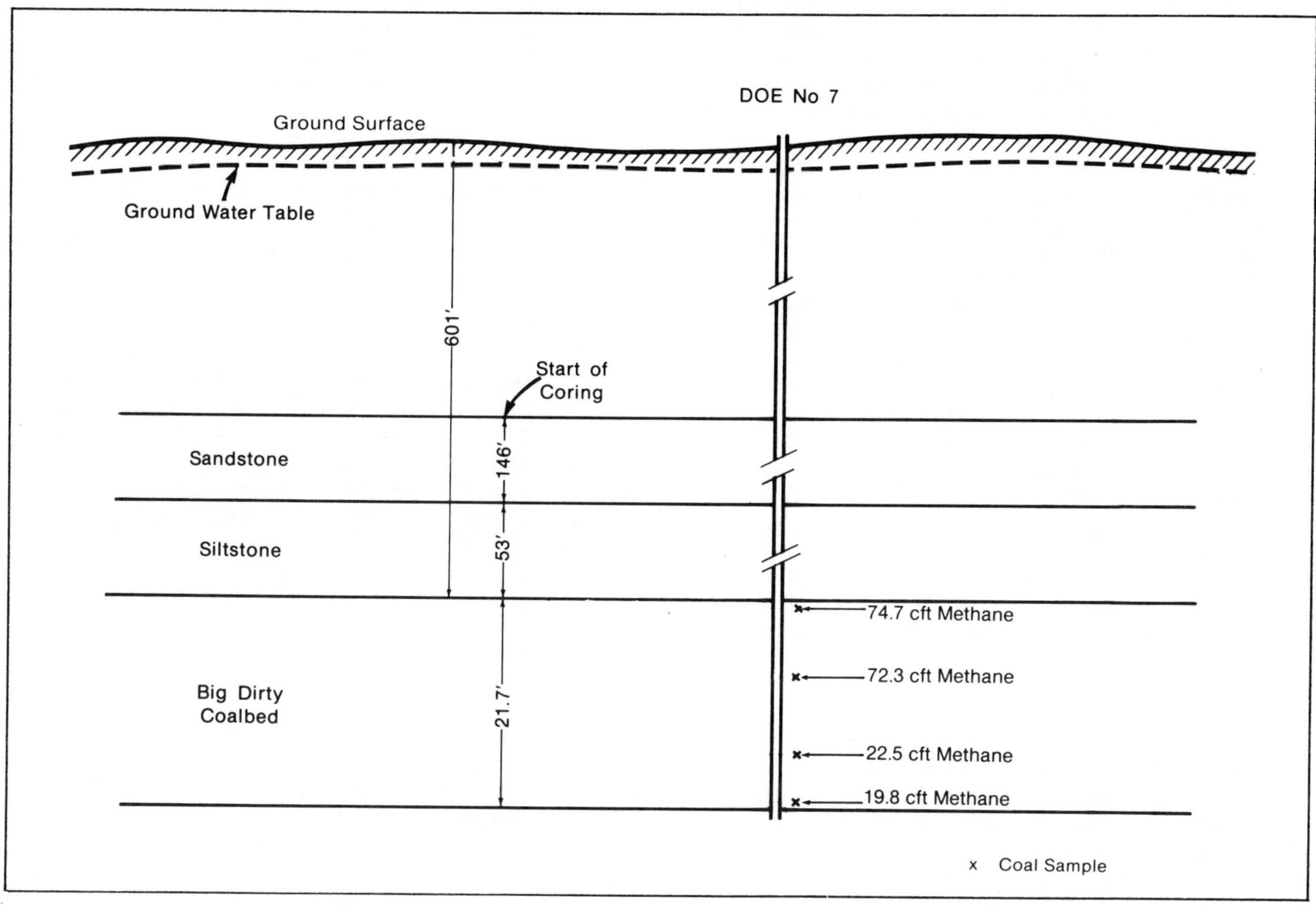

Figure 7—Schematic stratigraphic relationships and methane sample points, DOE Hole No. 7, Tono Basin, Thurston County, Washington.

Range controlled drainage development during erosion of these ancient mountains and still remains in today's pattern of major rivers. This northwest trend is also the dominant structural control in most of the state's present coal fields.

4. *Development of ancestral Cascade stratovolcanoes*. Coincident with uplift of the Calkins Range, a series of large volcanoes erupted along a line parallel to the continental margin and coincident with today's Cascade Range. These volcanoes covered much of the coal-bearing sedimentary rocks of the southern Calkins Range with andesitic and basaltic flows.
5. *Intrusion of large igneous plutons*. In conjunction with the development of the ancestral stratovolcanoes, a series of large igneous plutons was intruded. The heat created and released by these stratovolcanoes and the cooling of the associated igneous plutons are the dominant factors involved in the major metamorphism of most Washington higher rank coals. Deposits such as those found at Centralia-Chehalis far from the line of volcanism and igneous intrusion were little affected. Deposits as are found at Wilkeson-Carbonado and Bellingham, close to the line of igneous activity, were converted to coals of bituminous and anthracitic rank.
6. *Injection of small igneous intrusive bodies*. In late Oligocene times, a large number of small igneous bodies in the form of stocks, sills, and dikes were intruded along the north-south line occupied by the plutons described above and into the coal-bearing strata in the coal fields at Wilkeson-Carbonado; in the Green River area; in the deposits north of Green River at Tiger Mountain, Issaquah, and Renton; and even as far west as Centralia-Chehalis. Local heating of host rocks accompanying such intrusions may account in part for the anomalously large methane content found in these coalbeds, as at Tono Basin in the Centralia-Chehalis district.
7. *Massive extrusion of basalt flows*. Following long erosion of the Calkins Range, extrusion of the great Columbia River basalt sheets occurred in the Miocene. Because of the low viscosity of these lavas, there was little heating of existing near-surface rocks during extrusion of these flows. The major effect of these flows on Washington coal deposits was the burial of any such coalbeds in the area now occupied by the southern and southeastern flanks of the present Cascade Range.
8. *Uplift of the Cascade Range*. Beginning in the early to mid-Pliocene and continuing for approximately 6 million

years, well into the Pleistocene, regional uplift occurred along the previous line of volcanism and igneous pluton emplacement. This uplift set the stage for later alpine glaciation and the erosional sculpturing of the present Cascade Range with exposure of coalbeds at the very crest of the Cascades.

9. *Development of today's Cascade Range stratovolcanoes*. The five major stratovolcanoes of Washington's central Cascades had their beginnings in the Pliocene and continued to the present. Their high relief was caused by the viscous nature of their andesitic lavas. The growth of these great piles of volcanic rocks must have been accompanied by substantial regional heating in near-surface rocks. This buildup of heat may have contributed to the conversion of coal deposits in nearby rocks to higher ranks. This may be especially true for the existence of bituminous coals of coking quality in the Carbonado-Wilkeson field adjacent to Mount Rainier and the anthracites at the Glacier Field adjacent to Mount Baker.
10. *Continental glaciation*. During the Pleistocene, continental glaciers covered the lowlands on both sides of the Cascade Range. The major effect on coal-bearing rocks in the Puget Lowlands and along the Cascade foothills was to cover them uniformly with a great sheet of rock debris, under which much of the state's coal resource remains hidden.
11. *Alpine glaciation*. Concurrent with advance of the continental ice sheet, alpine glaciers carved the rugged topography of the high Cascades. Along many of the highest ridges, this glacial erosion has exposed coalbeds originally deposited on the old Weaver plain.
12. *Formation of thick sediment cover in the Puget Lowlands*. In recent times, a thick blanket of unconsolidated sediments has been deposited covering all coal-bearing rocks lying along the south and east sides of Puget Sound. These sediments have maximum depths of 2,000 ft south of Tacoma, 3,600 ft in Seattle, and 1,200 ft northeast of Bellingham.
13. *Development of thick forest and dense underbrush cover*. Over most of the Puget Lowlands and the western Cascade foothills, extremely thick forest and heavy underbrush have developed, covering all outcrops of coal-bearing rocks.
14. *Development of shallow water tables*. Throughout most of the Puget Lowlands and the western Cascade foothills, abundant precipitation maintains water-table levels near the ground surface. This near-surface water-table level thus maintains maximum hydrostatic pressure heads on buried coalbeds, which promotes maximum retention of methane adsorbed to coal surfaces.

Definition of Western Washington Methane Province

In this section, all the individual geologic elements previously discussed are consolidated in the accumulation of coalbed methane in the state of Washington. Assessment of these elements led to the delineation of three regional methane target areas.

These three broad targets are shown on the map of Figure 8 and include:

1. *The southern Puget Lowland area*. An elliptical area, trending N 20 E, that extends from the Columbia River at Kelso to Snohomish, a small town 35 mi northeast of Seattle. This 5,100-sq-mi area has a length of 183 mi with a maximum width of 55 mi and an average width of approximately 35 mi.
2. *The northern Puget Lowlands area*. An irregular hourglass-shaped area of approximately 1,050 sq mi with a length of 55 mi and a maximum width of 35 mi.
3. *The Roslyn area*. An irregular semirectangular area of about 400 sq mi with maximum dimensions of 26 mi by 20 mi.

The southern Puget Lowlands area includes all of the bituminous coal fields along the western Cascade foothills in King, Pierce, and Lewis counties; the subbituminous coals of the Centralia-Chehalis district; and the subbituminous and lignite coals of the Toledo-Castle Rock District. In addition, however, a very large area is included containing few coal outcrops, but where methane has been encountered in shallow water wells and at shallow depths in oil- and gas-exploration wells. The most probable origin for this methane is in coalbeds hidden beneath the thick blanket of unconsolidated sediments east and south of Puget Sound or beneath soil and forest cover or volcanic flows, further to the south.

The northern Puget Lowlands area includes the bituminous coal deposits surrounding Bellingham, the anthracite field at Glacier, and the bituminous coals of Skagit County at Hamilton and Cokedale. Additionally, boundaries of the area were drawn to include locales containing water wells and oil and gas wells known to contain methane at shallow depths.

The Roslyn area includes the coal fields at Ronald-Roslyn-Cle Elum and at Taneum-Manastash, and also areas both northeast and southwest of Roslyn known to contain rocks of the coal-bearing Roslyn Formation.

Following definition of the regional methane target areas, boundaries were then drawn around those limited target areas considered most favorable for coalbed methane exploration. Four such limited targets are shown on Figure 8. Listed in possible order of exploration interest, they are:

1. The Bellingham-Glacier
2. The Green River-Wilkeson-Carbonado central foothills
3. The Roslyn
4. The Centralia-Chehalis.

Approximate areas of these four targets are 500, 750, 40, and 500 sq mi, respectively.

In summary, the western Washington methane province as defined in this report comprises three broad regional target areas with a total size of approximately 6,500 sq mi, within which are four more favorable targets of limited size with a total area of approximately 1,800 sq mi.

ESTIMATED RESOURCE VOLUME

Coal reserves in the state of Washington were estimated by Beikman et al (1961) to be 6.2 billion tons (Table 2); additional estimates of reserves have increased this value to 6.4 billion tons. Of this reserve, 68% is subbituminous, 30% is bituminous, 2% is lignite, and less than 0.1% is anthracite. This reserve

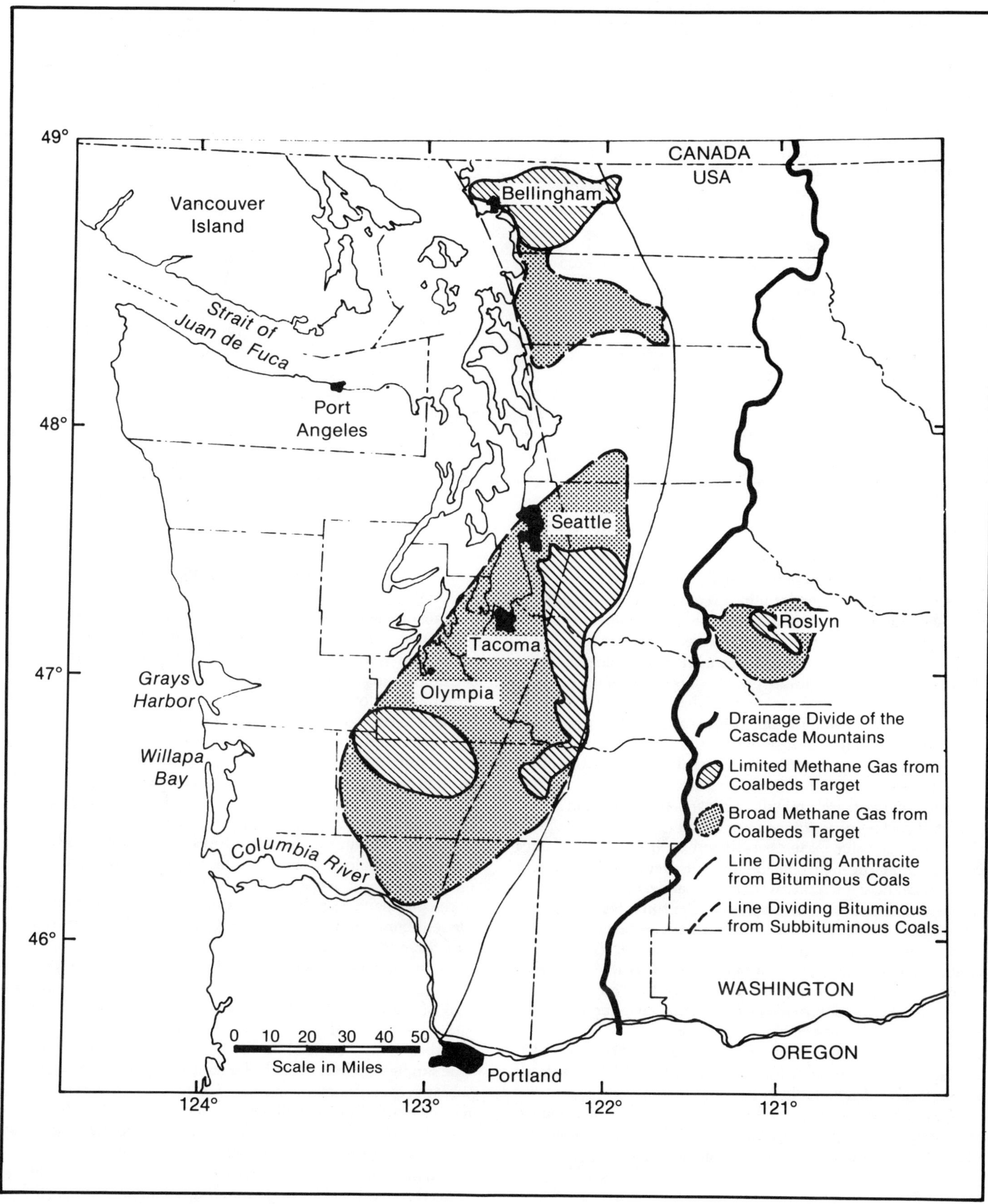

Figure 8—Coalbed methane target areas in western Washington.

Table 4. Estimated western Washington coalbed methane resource.

Coal Field Region or District	County	Coal Reserve Estimate (millions of tons)	Total Coal Resource (billions of tons)	Minimum Methane Resource @ 50 cf/ton	Maximum Methane Resource @ 400 cf/ton
Bellingham and Glacier coal fields	Whatcom	381	—	19×10^9 cf	152×10^9 cf
Cokedale; Cumberland-Day Creek, Mount Vernon-Big Lake-McMurray, and Rock Creek Areas	Skagit	507	—	25×10^9 cf	203×10^9 cf
Issaquah-Grand Ridge and Green River Districts	King	987	—	49×10^9 cf	395×10^9 cf
Wilkeson-Carbonado coal field, Spiketon, Melmont, Fairfax-Montezuma, and Ashford Areas	Pierce	362	—	18×10^9 cf	145×10^9 cf
Roslyn coal field and Taneum and Manastash Areas	Kittitas	282	—	14×10^9 cf	113×10^9 cf
Centralia-Chehalis District	Lewis-Thurston	3,694	—	185×10^9 cf	$1{,}478 \times 10^9$ cf
Alpha-Cinnabar, Morton-Mineral Lake, Packwood, and Carlton Pass Areas	Lewis	47	—	2×10^9 cf	19×10^9 cf
Kelso-Castle Rock Area*	Cowlitz and Lewis	149	—	7×10^9 cf	60×10^9 cf
Western Washington (see Table 4-0)	Total	6,409	—	0.3×10^{12} cf	2.6×10^{12} cf
Western Washington (Choate and Lent, 1977)		—	60	3.0×10^{12} cf	24×10^{12} cf

1) 1×10^9 cf = 1 billion cubic feet of gas
2) 1×10^{12} cf = 1 trillion cubic feet of gas
*Area not discussed in this report because coal is low rank, generally lignite or subbituminous C.

represents only coal at depths less than 3,000 ft and in beds thicker than 2.5 ft for subbituminous coal and 14 inches for bituminous and anthracite. Additionally, these values are for coal deposits contained within the limited areas of well-defined coal districts, and percentage values are non-representatively skewed toward subbituminous class because of the large reserves of near-surface coals in the Centralia-Chehalis district.

This 6.4-billion-ton estimate probably represents only a small part of the coal in the state, about 10% of the coal actually present. In Table 4, therefore, the value of 6.4 billion tons of coal has been used as a minimum coal reserve and ten times that figure (or approximately 60 billion tons) as the maximum potential coal resource for western Washington.

The average value for methane desorbed* from four coal cores collected from the Centralia-Chehalis district is 47.2 cf/ton of coal. Since these four cores were from coal near the low end of the subbituminous C rank, a value of 50 cf/ton is believed to be a reasonable minimum for in-place methane resource for subbituminous coal. Higher rank bituminous coals should contain more methane, and also prime target areas would have even higher methane concentrations. Therefore, on the basis of methane content data from other areas, 400 cu ft of methane per ton of coal is a reasonable estimate for the maximum in-place methane resource. Using these figures, the estimated in-place methane resource for the western Washington coal region is tabulated in Table 4.

The minimum estimated value for coalbed methane resource in western Washington ranges from 0.3 to 3.0 trillion cubic feet (Tcf); the maximum value ranges from 2.6 to 24 Tcf.

*Value is only for methane actually measured during desorption and does not include gas lost during core retrieval or residual gas.

CONCLUSIONS

A major energy resource that can be developed in the Northwest is that belt of coal deposits lying along the western foothills of Washington's Cascade Range. Known reserves in the limited areas where coalbeds crop out are estimated at 6.4 billion tons. However, these reserves are only a small portion of the total coal resources thought to be present in the state, hidden beneath thick strata of unconsolidated alluvial deposits, glacial drift, and volcanic flows—much of which, in turn, is covered by dense forests and heavy underbrush. Total coal resources of the state have been estimated at approximately 60 billion tons.

Except for limited areas in southwestern Washington where some of the coalbeds near the surface are amenable to strip mining, most of the coalbeds are highly folded and faulted and in the past have been developed only by underground mining. Underground mining in most Washington coal deposits is so expensive and dangerous that it is considered both economically impractical and environmentally unacceptable.

The most promising means for energy development from the bulk of the state's coal deposits would be in situ techniques: either underground gasification or recovery of methane adsorbed in coalbeds, or possibly some combination of the two. The simplest technique with the least impact on surface and subsurface environments would be drilling and recovering methane from buried coalbeds.

In this study, the authors defined a possible coalbed methane province in western Washington comprising three regional targets, with a total area of approximately 6,500 sq mi. Within these three regional areas are four limited targets totaling approximately 1,800 sq mi. The coalbed methane resource contained in western Washington is estimated to range from a minimum of 0.3 to 3.0 Tcf to a maximum of 2.6 to 24 Tcf.

REFERENCES CITED

Bagley, J., 1915, Report of the State Inspector of Coal Mines, biennial period ending December 31, 1914: State of Washington Sixteenth Biennial Publication.

——, 1917, Report of the State Inspector of Coal Mines, biennial period ending December 31, 1916: State of Washington Seventeenth Biennial Publication.

——, 1918, Annual report of coal mines for the year ending December 31, 1917: State of Washington, Report of the State Mine Inspector.

——, 1919, Annual report of coal mines for the year ending December 31, 1918: State of Washington, Report of the State Mine Inspector.

——, 1920, Annual report of coal mines for the year ending December 31, 1919: State of Washington, Report of the State Mine Inspector.

Beikman, H. M., H. D. Gower, and T. A. M. Dana, 1961, Coal Reserves of Washington: Washington Division of Mines and Geology Bulletin 47, 115 p.

Blundell, W. E., 1980, Electricity hogs: The Wall Street Journal, January 16, 1980, p. 1, 13.

Botting, D. C., 1909, Report of the State Inspector of Coal Mines from September 30, 1906 to December 31, 1908: State of Washington Thirteenth Biennial Publication.

——, 1911, Report of the State Inspector of Coal Mines for the biennial period ending December 31, 1910: State of Washington Fourteenth Biennial Publication.

——, 1913, Report of the State Inspector of Coal Mines for the biennial period ending December 31, 1912: State of Washington Fifteenth Biennial Publication.

Buckovic, W. A., 1979, The Eocene deltaic system of west-central Washington in Cenozoic paleogeography of the western United States: Pacific Coast Paleogeography Symposium 3, p. 147–163.

Choate, R., and J. Lent, 1975, Coal resources of the United States suitable for underground gasification, v. III, *in* Energy extraction from coal in situ—A national five-year plan; Redondo Beach, California, TRW Systems and Energy, for Energy Resources Development Agency, 158 p.

—— and ——, 1977, Steeply dipping coalbeds suitable for in-situ gasification; a resource definition: Proceedings of the Third Annual Symposium on Underground Coal Degasification, Energy Resources Development Agency (presented at Fallen Leaf Lake California, June 6-9).

Choate, R., and C. A. Johnson, 1979, Geologic overview, coalbed description and potential for methane recovery from coalbeds, Powder River Basin, Wyoming-Montana: McLean, Virginia, TRW Energy Systems Group, for U.S. Department of Energy.

—— and ——, 1980, Western Washington coal region report: McLean, Virginia, TRW Energy Systems Group, for U.S. Department of Energy.

Crosson, R. S., and L. J. Noson, 1978, Compilation of earthquake hypocenters in western Washington—1976: Washington Department of Natural Resources Information Circular 65, 13 p.

Culver, H. E., 1919, The coal fields of southwestern Washington: Washington Geological Survey Bulletin 19, 155 p.

Danes, Z. F., M. M. Bonno, E. Brau, W. D. Gilham, T. F. Hoffman, D. Johansen, M. H. Jones, B. Malfait, J. Masten, and G. O. Teague, 1965, Geophysical investigation of the southern Puget Sound area, Washington: Journal of Geophysical Research, v. 70, n. 22, p. 5573–5580.

Evans, G. W., 1912, The coal fields of King County: Washington Geological Survey Bulletin 3, 247 p.

Gard, Jr., L. M. 1968, Bedrock geology of the Lake Tapps quadrangle, Pierce County, Washington: U.S. Geological Survey Professional Paper 388-B, 33 p.

Geer, M. R., 1963, State-by-state reports on coal west of the Mississippi including Canada–Washington: Coal Age, v. 78, n. 5, p. 177–179, 182–183, 186.

Glaeser, O. A., 1962, Second progress report on Washington coal investigations, north Bellingham area, Whatcom County: Report prepared for Puget Sound Power and Light Company by United States Smelting Refining and Mining Company, USSRAM Division, 9 p.

Jenkins, O. P., 1924, Geological investigation of the coal fields of Skagit County, Washington: Washington Division Geology Bulletin 29, 63 p.

Lowry, M. E., and T. R. Cummings, 1966, Ground water resources of Sheridan County, Wyoming: U.S. Geological Survey Water Supply Paper 1807, 77 p.

Luzier, J. E., 1969, Geology and ground-water resources of southwestern King County, Washington: Washington Department of Water Resources Water Supply Bulletin 28, 260 p.

Mackin, J. H., and A. S. Cary, 1965, Origin of Cascade landscapes: Washington Department of Conservation Information Circular No. 41, 35 p.

Meinzer, O. E., 1942, Occurrence, origin and discharge of ground water, *in* Hydrology: New York, Dover Publications, p. 385–443.

Moen, W. S., 1969, Mines and mineral deposits of Whatcom County, Washington: Washington Department of Natural Resources Bulletin no. 57, 134 p.

Saunders, E. J., 1914, The coal fields of Kittitas County, Washington: Washington Geological Survey Bulletin 9, 204 p.

Smith, E. E., 1911, Coals of the State of Washington: U.S. Geological Survey Bulletin 474, 206 p.

Snavely, Jr., P. D., W. W. Rau, L. Hoover, Jr., and A. E. Roberts, 1951, McIntosh Formation, Centralia-Chehalis coal district, Washington: Bulletin of the American Association of Petroleum Geologists; v. 35, n. 5, p. 1052–1061.

——, R. D. Brown, Jr., A. E. Roberts, and W. W. Rau, 1958, Geology and coal resources of the Centralia-Chehalis district, Lewis and Thurston counties, Washington: U.S. Geological Survey Bulletin 1053, 159 p.

Tuck, R., and G. A. Boyd, 1966, Report on 1966 drilling in the Roslyn-Cle Elum coal field, Kittitas County, Washington: Washington Department of Water Resources Open-File Report, 21 p.

Vine, J. D., 1969, Geology and coal resources of the Cumberland, Hobart, and Maple Valley quadrangles, King County, Washington: U.S. Geological Survey Professional Paper 624, 67 p.

Vonheeder, E. R., 1975, Coal reserves of Whatcom County, Washington: Washington Department of Natural Resources Open-File Report OF 75-9, 78 p.

——, 1977, Whatcom County coal reserves: Washington

Department of Natural Resources Open-File Report OF 77-3, 3 sheets.

——, 1979, Pierce County, Washington, Coal Reserves: Washington Department of Natural Resources, Open File Map 79-4, 5 sheets.

Walters, K. L., and G. E. Kimmel, 1968, Ground-water occurrence and stratigraphy of unconsolidated deposits of central Pierce County Washington, Washington Department of Ecology Water Supply Bulletin 22, 428 p.

Whitcomb, H. A., T. R. Cummings, and R. A. McCullough, 1966, Ground-water resources and geology of northern and central Johnson County, Wyoming: U.S. Geological Survey Water Supply Paper 1806, 99 p.

Conclusions on the Development of Coalbed Methane

G. E. Eddy

As the preceding chapters have shown, coalbed methane production is possible in many areas within the United States. Commercial production is well under way by several producers in the Warrior, San Juan, and Piceance Basins and on an experimental level in the the Northern Appalachian, Central Appalachian, and Uinta Basins. Production testing is under way in the Raton Basin.

The coal basins in the United States can be divided by different criteria into two major groups: eastern coal basins and western coal basins. Coalbeds in the eastern basins (Northern Appalachian, Central Appalachian, Warrior, Arkoma, and Illinois) are rarely thicker than 8 to 10 ft or deeper than 2,500 ft. In the western basins, coalbeds 25 ft thick are common and are found at depths below 10,000 ft in some of the basins.

Most of the coal in the eastern United States was formed during the Pennsylvanian period but some coal was deposited during the Triassic period in the Triassic basins. The bulk of the coal in the western United States was formed during the Cretaceous and Tertiary periods. The rank of the coalbeds in the east is fairly consistent and is due to their uniform burial history. All of the coal is bituminous, except in the part of Pennsylvania where the pressure of intense folding formed the famous Anthracite beds of central Pennsylvania. In western basins, the rank of the coal ranges from lignite through anthracite and can vary greatly in short distances, sometimes within the same coalbed, because of the greater variation in depth of burial and complex thermal history. An example of this was shown by Choate and Rightmire (1982), who related the distribution of coal rank and gas content of the Piceance, San Juan, and Raton basins to the intrusion of the San Juan Mountain igneous complex. Although the dominant factors controlling gas content of the coal in those basins are depth of burial and coal rank, as in all basins, the coal rank has been modified by the thermal activity. Figure 1 is a map summarizing the results of this study. The San Juan Mountain intrusive complex and the Rio Grande Rift Valley have increased the coal rank and therefore the gas content and coalbed methane production potential of the southeastern part of the Piceance Basin, the western part of the Raton Basin, and the northern part of the San Juan Basin. The relatively young, Tertiary coalbeds of high rank, up to anthracite, in the western Washington coal region are another example of the upgrading of coal rank by igneous thermal activity. Studies like that of Choate and Rightmire relating the coal rank and gas content to the regional geologic setting may be needed to further delineate the target areas in other basins.

In spite of the great potential in many coal basins, more coalbed wells have been drilled in the Warrior basin of Alabama than in any other coal basin. For example, the two largest fields, the Brookwood and the Oak Grove degasification fields, had a total of 79 wells as of May 1983, producing 268,899,000 cu ft of gas with a cumulative production of 2.7 billion cubic feet (Bcf). This is an average of 110 thousand cubic feet (Mcf)/day/well during May 1983. These wells were all less than 2,500 deep. These production figures illustrate only one of the positive economic features of coalbed wells. Another is demonstrated by the wells that have been drilled by coal companies to degas coalbeds prior to mining. Degasification in this manner increases productivity by making the coal safer to mine and saves on ventilation costs. Some coal companies have found degasification to be economically beneficial even without selling the gas.

One of the greatest advantages of coalbed wells is that very few are dry holes. Coalbeds usually are more continuous and their thickness and gas content are more uniform or predictable than conventional reservoirs. This makes exploration for coalbed gas easier in some respects than for conventional gas and results in few dry holes. This uniformity of coal thickness and gas content in most basins makes coalbed methane prospects fairly easy to evaluate. One or two core holes are usually adequate to test the potential of a large area.

In spite of these positive aspects of coalbed wells, most conventional oil and gas operators have been slow to develop the coalbed methane resource. One reason is that in some states, especially in the northeast, the legal ownership of coalbed gas is not clearly defined. This problem should be resolved in the near future but has been a deterrent to the development of this resource. The most important reasons are that the economics and production of coalbed methane wells differ somewhat from conventional wells. Coalbed wells are usually smaller producers than conventional wells although some produce over 1 million cubic feet (MMcf)/day. The average production in Alabama is just over 100 Mcfd.

The initial production of a conventional gas well is usually its peak production and it decreases from then on. Most coalbed wells produce water along with the gas, and water in the coalbed retards gas production. The coalbed must be dewatered before the gas production can increase to a positive economic level. It may take several months to pump off enough water to allow the gas production to increase to an economic level. This process differs from conventional wells, which are usually judged by their initial production. Not only does the production of a coalbed well increase, often dramatically, after the initial pumping off of water but it often continues to increase for many years. The negative decline demonstrated by most coalbed wells is thought to be caused by shrinkage of the coal causing an increase in the number of fractures and lengthening of existing ones. The shrinkage is caused by dewatering and/or degassing of the coals. Some wells in the San Juan Basin have had negative decline curves over periods of more than 20 years. Therefore, initial production figures for coalbed wells are relatively meaningless.

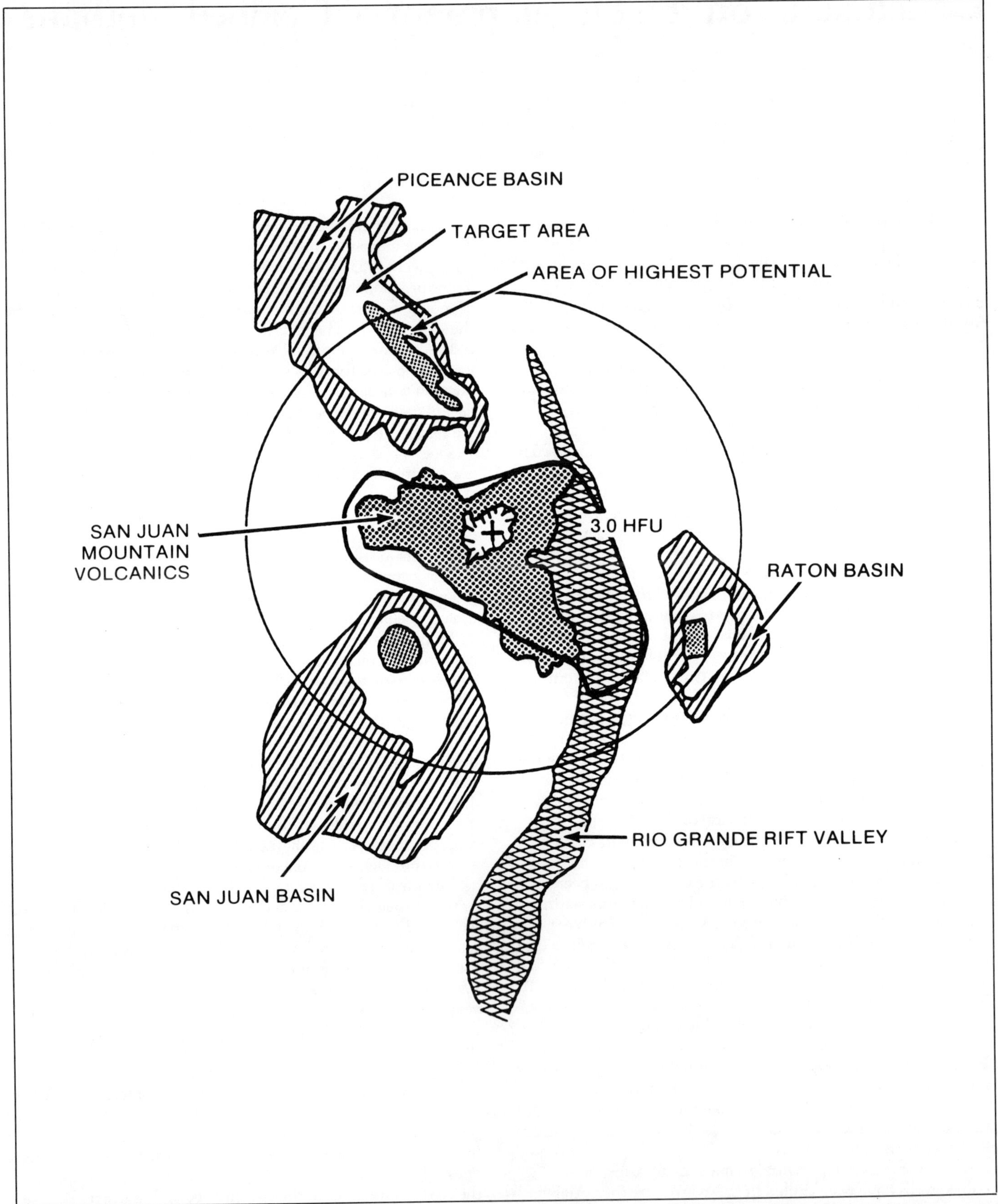

Figure 1—Map of coalbed methane target areas relative to center of high heat flow represented by the San Juan Mountains and the Rio Grande Rift Valley. The 125-mi radius is centered in the La Garita Caldera. Choate and Rightmire, 1982.

Before most coalbed wells can be economically productive, they must be stimulated. In most areas of the eastern United States, the coalbeds are less than 10 ft thick and several are present. This makes multiple zone stimulations necessary to maximize gas recovery. A positive is that the stimulations are of smaller volume than those performed in a conventional well of equivalent depth.

At the present time, economically successful coalbed wells are limited to depths less than 3,500 ft. Wells have been completed at greater depths, but they have been plagued by engineering problems. The most common problem has been plugging of the wells by coal fines flowing or caving into the well bore. Apparently coal becomes plastic at depths below 3,500 ft (Willis, 1978). The Gas Research Institute is funding a research project to solve some of the problems associated with the drilling and completion of coalbed wells below 3,500 ft. When these problems are solved, coal in several basins, especially in the western United States, will be opened to production that is now too deep to be considered.

In summary, there is no technical reason why gas production from coalbeds cannot be done economically in several of the coal basins in the United States and in nearly all of them within the next 4 years. This is a valuable resource that will add an appreciable amount of gas to the gas reserves of this country.

REFERENCES CITED

Choate, R., and C. T. Rightmire, 1982, Influence of the San Juan Mountain geothermal anomaly and other Tertiary igneous events on the coalbed methane potential in the Piceance, San Juan, and Raton Basins, Colorado and New Mexico; Proceedings of the Unconventional Gas Recovery Symposium, May 16–18, 1982, Pittsburgh, Pennsylvania: Society of Petroleum Engineers/U.S. Department of Energy, 10805, p. 151–164.

Willis, C., 1978, Rocky Mountain coal seams call for special drilling techniques: The Oil and Gas Journal, July 10, 1978, p. 143–152.

Index

A reference is indexed according to its important or "key," words.

Three columns are to the left of a keyword entry. The first column, a letter entry, represents the AAPG book series from with the reference originated. In this case, ST stands for Studies in Geology Series. Every five years, AAPG will merge all its indexes together, and the letter ST will differentiate this reference from those of the AAPG Memoir Series (M) or the AAPG Bulletin (B).

The following number is the series number. In this case 0017 represents a reference from AAPG Studies in Geology Series. The third column lists the page number of this volume on which the reference can be found.
